Topics in Hyperplane Arrangements

Mathematical
Surveys
and
Monographs

Volume 226

Topics in Hyperplane Arrangements

Marcelo Aguiar
Swapneel Mahajan

American Mathematical Society
Providence, Rhode Island

2010 *Mathematics Subject Classification.* Primary 05E10, 06A07, 06A11, 06C10, 16G10, 17B01, 20F55, 20M10, 20M25, 52C35;
Secondary 05A18, 05A30, 05C25, 05E05, 05E18, 05E45, 18D50, 18G35, 20B30.

For additional information and updates on this book, visit
www.ams.org/bookpages/surv-226

Library of Congress Cataloging-in-Publication Data

Names: Aguiar, Marcelo, 1968- | Mahajan, Swapneel Arvind, 1974-
Title: Topics in hyperplane arrangements / Marcelo Aguiar, Swapneel Mahajan.
Description: Providence, Rhode Island : American Mathematical Society, [2017] | Series: Mathematical surveys and monographs ; volume 226 | Includes bibliographical references and index.
Identifiers: LCCN 2017019308 | ISBN 9781470437114 (alk. paper)
Subjects: LCSH: Incidence algebras. | Algebraic spaces. | Hyperspace. | Geometry, Plane. | AMS: Combinatorics – Algebraic combinatorics – Combinatorial aspects of representation theory. msc | Order, lattices, ordered algebraic structures – Ordered sets – Combinatorics of partially ordered sets. msc | Order, lattices, ordered algebraic structures – Ordered sets – Algebraic aspects of posets. msc | Order, lattices, ordered algebraic structures – Modular lattices, complemented lattices – Semimodular lattices, geometric lattices. msc | Associative rings and algebras – Representation theory of rings and algebras – Representations of Artinian rings. msc | Nonassociative rings and algebras – Lie algebras and Lie superalgebras – Identities, free Lie (super)algebras. msc | Group theory and generalizations – Special aspects of infinite or finite groups – Reflection and Coxeter groups. msc | Group theory and generalizations – Semigroups – General structure theory. msc | Group theory and generalizations – Semigroups – Semigroup rings, multiplicative semigroups of rings. msc | Convex and discrete geometry – Discrete geometry – Arrangements of points, flats, hyperplanes. msc | Combinatorics – Enumerative combinatorics – Partitions of sets. msc | Combinatorics – Enumerative combinatorics – q-calculus and related topics. msc | Combinatorics – Graph theory – Graphs and abstract algebra (groups, rings, fields, etc.). msc | Combinatorics – Algebraic combinatorics – Symmetric functions and generalizations. msc | Combinatorics – Algebraic combinatorics – Group actions on combinatorial structures. msc | Combinatorics – Algebraic combinatorics – Combinatorial aspects of simplicial complexes. msc | Category theory; homological algebra – Categories with structure – Operads. msc | Category theory; homological algebra – Homological algebra – Chain complexes. msc | Group theory and generalizations – Permutation groups – Symmetric groups. msc
Classification: LCC QA251.3 .A356 2017 | DDC 516/.11–dc23
LC record available at https://lccn.loc.gov/2017019308

∞ The paper used in this book is acid-free and falls within the guidelines
established to ensure permanence and durability.
Visit the AMS home page at `http://www.ams.org/`

10 9 8 7 6 5 4 3 2 1 22 21 20 19 18 17

Contents

Preface

Synopsis

The goal of this monograph is to study the interplay between various algebraic, geometric and combinatorial aspects of real hyperplane arrangements. The text contains many new ideas and results. It also gathers and organizes material from various sources in the literature, sometimes highlighting previously unnoticed connections. We briefly outline the contents below. They are explained in more detail in the main introduction.

We provide a detailed discussion on faces, flats, chambers, cones, gallery intervals, lunes, the support map, the case and base maps, and other geometric notions associated to real hyperplane arrangements. We show that any cone can be optimally decomposed into lunes. We introduce the category of lunes. This beautiful structure is intimately related to the substitution product of chambers a generalization of the classical associative operad). The classical case is obtained by specializing to the braid arrangement. We give several generalizations of the classical identity of Witt from Coxeter theory under the broad umbrella of descent and lune identities. The topological invariant involved here is the Euler characteristic of a relative pair of cell complexes. We generalize a well-known factorization theorem of Varchenko to cones, and also initiate an abstract approach to distance functions on chambers.

The main algebraic objects are the Birkhoff monoid and the Tits monoid, and their linearized algebras. The former is commutative and its elements are the flats of the arrangement, while the latter is not commutative and its elements are the faces. A module whose elements are chambers also plays a central role. Both monoids carry natural partial orders. The Birkhoff monoid is a lattice and its product is the join operation in the lattice. One may think of the Tits monoid as a noncommutative lattice. Its abelianization is the Birkhoff monoid, via the map that sends a face to the flat which supports it. We introduce the Janus monoid which is built out of the Tits and Birkhoff monoids.

We initiate a noncommutative Möbius theory of the Tits monoid and relate it to the representation theory of its linerization which is the Tits algebra. The central object is the lune-incidence algebra, which is a certain reduced incidence algebra of the poset of faces. It contains noncommutative zeta functions characterized by lune-additivity, and noncommutative Möbius functions characterized by the noncommutative Weisner formula. This theory lifts the usual Möbius theory for lattices, where the central object is the incidence algebra of the lattice of flats.

We introduce Lie and Zie elements. The latter belong to the Tits algebra, and the former to the module of chambers. The space of Zie elements is a right ideal of the Tits algebra. Any special Zie element defines an idempotent operator on chambers whose image is the space of Lie elements. To any generic half-space, we associate a special Zie element called the Dynkin element. Its action on chambers generalizes the left bracketing operator in classical Lie theory. We define a substitution product and establish a presentation of Lie. This generalizes the familiar presentation of the classical Lie operad. Antisymmetry is encoded in the notion of orientation of the rank-one arrangement and the Jacobi identity in the form of a linear relation among chambers obtained by "unbracketing" lines of the rank-two arrangements. This is same as saying that the space of Lie elements is isomorphic, up to orientation, to the top cohomology of the lattice of flats. This generalizes a celebrated theorem due to the combined work of Joyal, Klyachko and Stanley. We introduce the Lie-incidence algebra and show that it is isomorphic to the Tits algebra. This is intimately connected to the two-sided Peirce decomposition of the Tits algebra. The latter can be understood in terms of left and right Peirce decompositions of chambers and Zie elements respectively.

The Birkhoff algebra is split-semisimple. For the Tits algebra, complete systems of primitive orthogonal idempotents are in correspondence with algebra sections of the support map. We obtain many interesting characterizations of such sections. This aspect of the theory generalizes the classical theory of Eulerian idempotents. Noncommutative zeta and Möbius functions, and special Zie families are among the various concepts in correspondence. For reflection arrangements, there is a similar theory for the subalgebra of the Tits algebra invariant under the action of the Coxeter group. (The opposite of this algebra is the Solomon descent algebra.)

Precedents

This work benefits from and builds on some important recent developments. For the representation theory of the Tits algebra, we mention work of Brown, Diaconis and Saliola propelled by a landmark paper of Bidigare, Hanlon and Rockmore. (Older work of Solomon on the descent algebra has also been influential.) Some of these results are given in the generality of left regular bands and even bands. Further generalizations of this kind appear in work of Steinberg. For Lie theory, we mention work of Barcelo, Bergeron, Björner, Garsia, Patras, Reutenauer and Wachs. Saliola's work also implicitly contains elements of Lie theory. Explicit references to Lie are made only for the braid arrangement and the reflection arrangement of type B. The work of Joyal, Klyachko and Stanley relating Lie to order homology is for the braid arrangement. On the other hand, related results on order homology in the literature are usually given in the generality of arrangements or beyond. There have been several other contributors; most of them are mentioned in the main introduction. Two new entrants are the mathematicians Janus and Zie.

Organization

The text is organized in two parts. In Part I, the emphasis is on set-theoretic objects associated to hyperplane arrangements such as posets, monoids and the action of monoids on sets. In Part II, the emphasis is on linear objects such as algebras and their modules. There is a Notes section at the end of each chapter where detailed references to the literature, including discussions on alternative terminology and notation, are provided. Background information on topics such as Möbius functions, incidence algebras, representation theory of algebras and bands is provided in Appendices at the end of the main text. A notation index and a subject index are provided at the end of the book. Pictures and diagrams form an important component of our exposition which has a distinct geometric flavor. Numerous exercises are interspersed throughout the book.

The text is not meant to be read linearly from start to finish. We encourage readers to take up a particular chapter or section of their interest and backtrack as necessary. As an aid, the diagram of interdependence of chapters and appendices is displayed below.

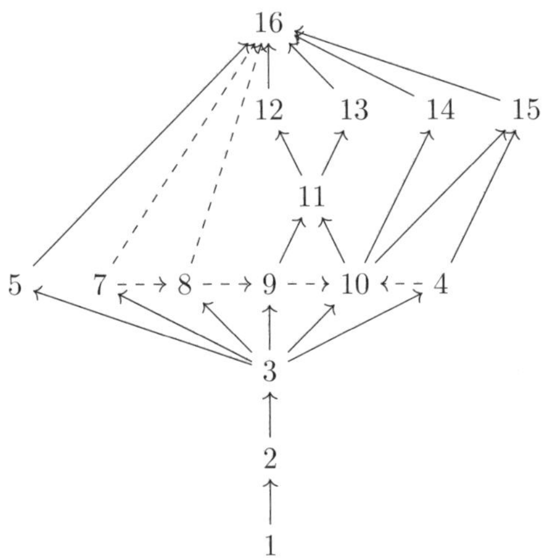

A $\longrightarrow$ 1,2,7

B $\longrightarrow$ 1,2,3,4

C $\longrightarrow$ 1,15,16

D $\longrightarrow$ 9,11,12,13,16

E $\longrightarrow$ 1,8

A directed path from i to j indicates that some basic familiarity with Chapter i is necessary before proceeding to Chapter j. A dashed arrow from i to j means that the dependence of Chapter j on Chapter i is minimal, that is, restricted to some section or example.

Chapter 6 is not shown in the above diagram. It discusses the braid arrangement, the reflection arrangement of type B and other examples. They are employed frequently in later chapters for illustration.

Readership

We have strived to keep the text self-contained and with minimum prerequisites with the objective of making it accessible to advanced undergraduate and beginning graduate students. We hope it also serves as a useful reference on hyperplane arrangements to experts. The book touches upon several fields of mathematics such as representation theory of monoids and associative algebras, posets and their incidence algebras, lattice theory, random walks, invariant theory, discrete geometry, algebraic and geometric combinatorics, and algebraic Lie theory.

Scope

The theory of hyperplane arrangements has grown enormously in several different directions in the past two decades. The text is not meant to be a comprehensive survey of the entire theory. For instance, topics such as singularities, integral systems, hypergeometric functions and resonance varieties find no mention in the book. For these, one may look at [**15, 121, 138, 157, 177, 329, 413**] and references therein.

Future directions

Our constructions are all based on the choice of a real hyperplane arrangement. It is apparent, moreover, that a central role is played by the Tits monoid of faces of the arrangement. It is tempting to try to extend the theory to more general classes of monoids, particularly bands and left regular bands. We have kept our focus on arrangements, although such generalizations offer a promising line of research. We also mention the Janus monoid, the category of lunes and noncommutative Möbius functions as important objects worthy of further study. Our choice of topics has mainly been guided by applications to the theory of species, operads and Hopf algebras which we plan to develop in future work.

Acknowledgements

We warmly thank our mentors Ken Brown and Steve Chase for their advice and encouragement over the years. Our gratitude goes also to Lou Billera and Persi Diaconis whose support we greatly value. We thank Frédéric Patras, Vic Reiner, Richard Stanley, and many others for sharing insights or providing useful comments and references. The second author thanks all students at IIT Mumbai who attended his lectures on the subject for their valuable feedback.

Thanks to the American Mathematical Society for publishing our work in this series, and to Ina Mette, Marcia Almeida and Isabel Stafford for their support during the production of this book. We also thank the reviewers for their helpful and critical comments on a preliminary draft of the text.

Aguiar was supported by NSF grants DMS-1001935 and DMS-140111.

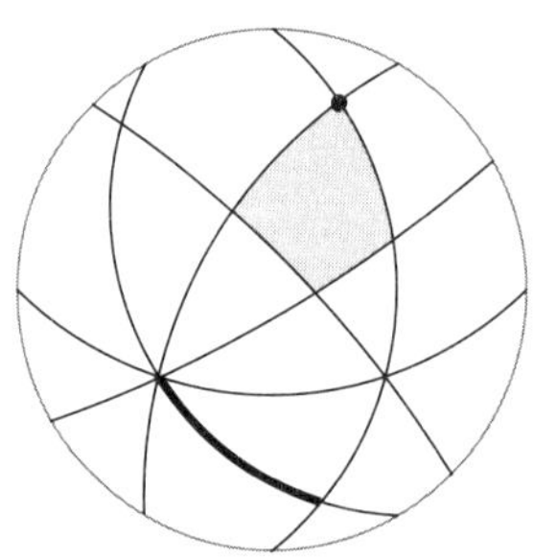

Introduction

Part I

Arrangements. (Chapter 1.) A hyperplane arrangement $\mathcal{A}$ is a set of hyperplanes (codimension-one subspaces) in a fixed real vector space. We assume that the number of hyperplanes is finite and all of them pass through the origin. The intersection of all hyperplanes is the central face. The rank of an arrangement is the dimension of the ambient vector space minus the dimension of the central face. An arrangement has rank 0 if it has no hyperplanes, rank 1 if it has one hyperplane, and rank 2 if it has at least two hyperplanes and all of them pass through a codimension-two subspace.

Flats and faces. (Chapter 1.) Subspaces obtained by intersecting hyperplanes are called the *flats* of the arrangement. We let $\Pi[\mathcal{A}]$ denote the set of flats. It is a graded lattice with partial order given by inclusion. The minimum element is the central face and the maximum element is the ambient space. The codimension-one flats are the hyperplanes. Each hyperplane divides the ambient space into two half-spaces. Their intersection is the given hyperplane. Subsets obtained by intersecting half-spaces, with at least one half-space chosen for each hyperplane, are called the *faces* of the arrangement. We let $\Sigma[\mathcal{A}]$ denote the set of faces. It is a graded poset under inclusion. The central face is the minimum element. However, there is no unique maximum face, so $\Sigma[\mathcal{A}]$ is *not* a lattice. A maximal face is called a *chamber.* We let $\Gamma[\mathcal{A}]$ denote the set of chambers. The linear span of any face is a flat. This defines a surjective map

$$\mathrm{s} : \Sigma[\mathcal{A}] \twoheadrightarrow \Pi[\mathcal{A}].$$

We call this the *support map*. It is order-preserving.

Birkhoff monoid and Tits monoid. (Chapter 1.) We view the lattice of flats $\Pi[\mathcal{A}]$ as a (commutative) monoid with product given by the join operation. We call this the *Birkhoff monoid.* For flats X and Y, their Birkhoff product is $\mathrm{X} \vee \mathrm{Y}$. The poset of faces $\Sigma[\mathcal{A}]$ is not a lattice. Nonetheless, it carries a (noncommutative) monoid structure. We call this the *Tits monoid.* It is an example of a left regular band (since it satisfies the axiom $xyx = xy$). For faces F and G, we denote their Tits product by FG. The set of chambers $\Gamma[\mathcal{A}]$ is a left $\Sigma[\mathcal{A}]$-set, that is, for F a face and C a chamber, FC is a chamber. The support map is a monoid homomorphism.

Janus monoid. (Chapter 1.) A *bi-face* is a pair (F, F') of faces such that F and F' have the same support. Let $\mathrm{J}[\mathcal{A}]$ denote the set of bi-faces. The operation

$$(F, F')(G, G') := (FG, G'F')$$

turns $J[\mathcal{A}]$ into a monoid. We call it the *Janus monoid.* [1] It is the fiber product of the Tits monoid $\Sigma[\mathcal{A}]$ and its opposite $\Sigma[\mathcal{A}]^{\mathrm{op}}$ over the Birkhoff monoid $\Pi[\mathcal{A}]$. This can be pictured as follows.

Janus monoid

↙ ↘

Tits monoid   opposite Tits monoid

↘ ↙

Birkhoff monoid

The Janus monoid is a band (since every element is idempotent) which is neither left regular nor right regular in general.

Arrangements under and over a flat. (Chapter 1.) From a flat X of an arrangement $\mathcal{A}$, one may construct two new arrangements: $\mathcal{A}^{\mathrm{X}}$, the arrangement *under* X, and $\mathcal{A}_{\mathrm{X}}$, the arrangement *over* X. The former is the arrangement obtained by intersecting the hyperplanes in $\mathcal{A}$ with X, while the latter is the subarrangement consisting of those hyperplanes which contain X. For flats $\mathrm{X} \leq \mathrm{Y}$, the arrangement under Y in $\mathcal{A}_{\mathrm{X}}$ is the same as the arrangement over X in $\mathcal{A}^{\mathrm{Y}}$. We denote this arrangement by $\mathcal{A}^{\mathrm{Y}}_{\mathrm{X}}$.

Cones. (Chapter 2.) Subsets obtained by intersecting half-spaces (with no restriction) are called the *cones* of the arrangement. In particular, faces and flats are cones. (A hyperplane is the intersection of the two half-spaces it bounds.) Let $\Omega[\mathcal{A}]$ denote the set of all cones. It is a lattice under inclusion. The support map extends to an order-preserving map

$$\mathrm{c} : \Omega[\mathcal{A}] \to \Pi[\mathcal{A}].$$

We call this the *case map.* It sends a cone to the smallest flat containing that cone. The case map is the left adjoint of the inclusion map $\Pi[\mathcal{A}] \to \Omega[\mathcal{A}]$. There is another order-preserving map

$$\mathrm{b} : \Omega[\mathcal{A}] \to \Pi[\mathcal{A}]$$

which we call the *base map.* It sends a cone to the largest flat which is contained in that cone. The base map is the right adjoint of the inclusion map. Note that the base and case of a flat is the flat itself.

Cones whose case is the maximum flat are called top-cones. The poset of top-cones is a join-semilattice which is join-distributive, and in particular, graded and upper semimodular (Theorems 2.56, 2.58 and 2.60).

Lunes. (Chapters 3 and 4.) A cone is a *lune* if it has the property that for any hyperplane containing its base, the entire cone lies on one side of that hyperplane. Faces and flats are lunes. In general, any cone can be optimally cut up into lunes by using hyperplanes containing the base of the cone (Theorem 3.27). Finer decompositions can be obtained by using hyperplanes containing a fixed flat lying inside the base (Proposition 3.22). For instance, it is possible to cut a lune itself into smaller lunes. The optimal decomposition of a flat X is X itself (since it is a lune). An instance of a finer decomposition is to write X as a union of faces having support X.

[1] Janus Bifrons is a Roman god with two faces.

Lunes which are top-cones are called top-lunes. The poset of top-lunes under inclusion is graded (Theorem 4.9). We consider two partial orders on lunes. The first partial order is the inclusion of lune closures (and is the restriction of the partial order on cones), while the second is the inclusion of lune interiors. Both extend the partial order on top-lunes and are graded (Theorems 4.12 and 4.26).

Lunes can be composed when the case of the first lune equals the base of the second lune. This yields a category whose objects are flats and morphisms are lunes. We call it the *category of lunes*. It is internal to posets under the second partial order on lunes (Proposition 4.31). It also admits a nice presentation (Proposition 4.42). A lune with base X and case Y is the same as a chamber in the arrangement $\mathcal{A}_{\mathrm{X}}^{\mathrm{Y}}$. Using this, composition of lunes can be recast as follows. For any flat X, there is a map

$$\Gamma[\mathcal{A}^{\mathrm{X}}] \otimes \Gamma[\mathcal{A}_{\mathrm{X}}] \to \Gamma[\mathcal{A}].$$

We call this the substitution product of chambers, see (4.18).

Braid arrangement. (Chapters 5 and 6.) The braid arrangement is the motivating example for many of our considerations. The key observation is that for this arrangement, geometric notions of faces, flats, top-cones, and so on can be encoded by combinatorial notions of set compositions, set partitions, partial orders and so on. This correspondence between geometry and combinatorics is summarized in Table 6.2. The braid arrangement is an example of a reflection arrangement whose associated Coxeter group is the group of permutations. In the Coxeter case, one can define face-types and flat-types. Face-types are orbits of the set of faces under the Coxeter group action. Similarly, flat-types are orbits of the set of flats. For the braid arrangement, face-types and flat-types correspond to integer compositions and integer partitions.

Descent equation and lune equation. (Chapter 7.) Fix chambers C and D. The *descent equation* is $HC = D$. In other words, we need to solve for faces H such that the Tits product of H and C equals D. (This is related to descents of permutations in the case of the braid arrangement which motivates our terminology.) More generally, we can fix faces F and G, and consider the equation $HF = G$. In fact, one can do the following. For any left $\Sigma[\mathcal{A}]$-set h, the descent equation is $H \cdot x = y$, where x and y are fixed elements of h, the variable is H, and $\cdot$ denotes the action of $\Sigma[\mathcal{A}]$ on h. Apart from finding the solutions, there is also interest in computing the sum $\sum(-1)^{\mathrm{rk}(H)}$ as H ranges over the solution set, with $\mathrm{rk}(H)$ denoting the rank of H. For this, we attach to the solution set a relative pair (X, A) of cell complexes whose Euler characteristic is the given sum, see (7.32). By construction X is either a ball or sphere, but the topology of A is complicated in general. In our starting examples h is either $\Gamma[\mathcal{A}]$ or $\Sigma[\mathcal{A}]$. In these cases, A also has the topology of a ball or sphere. This leads to explicit identities, see (7.10) and (7.11a).

Fix a face H and a chamber D. The *lune equation* is $HC = D$. The difference is that now we need to solve for C. For a solution to exist H must be smaller than D. Assuming this condition, the solution set is precisely the set of chambers contained in some top-lune (which explains our terminology). More generally, an arbitrary lune can be obtained as the solution set of the equation $HF = G$ for some fixed H and G. Since lunes have the topology of a ball or sphere, we can again compute $\sum(-1)^{\mathrm{rk}(F)}$ explicitly, see (7.12a). An analysis with relative pairs, similar to the one for the descent equation, can be carried out for right $\Sigma[\mathcal{A}]$-sets h, see (7.41). The lune equation in this case is $x \cdot F = y$, with $x, y \in \mathrm{h}$.

Distance function and Varchenko matrix. (Chapter 8.) A hyperplane separates two chambers if they lie on its opposite sides. The distance between two chambers is defined to be the number of hyperplanes which separate them. Fix a scalar q, and define a bilinear form on the set of chambers $\Gamma[\mathcal{A}]$ by

$$\langle C, D\rangle := q^{\mathrm{dist}(C,D)}.$$

Here C and D are chambers and $\mathrm{dist}(C, D)$ denotes the distance between them. The determinant of this matrix factorizes with factors of the form $1-q^i$, see (8.41). In particular, the bilinear form is nondegenerate if q is not a root of unity.

More generally, assign a weight to each half-space, and define $\langle C, D\rangle$ to be the product of the weights of all half-spaces which contain C but do not contain D. Setting each weight to be q recovers the previous case. A factorization of the determinant of this matrix was obtained by Varchenko (Theorem 8.11). (He worked in the special case when the two opposite half-spaces bound by each hyperplane carry the same weight.) Lunes play a key role in the proof. The Varchenko matrix can be formally inverted using non-stuttering paths, see (8.30).

It is fruitful to consider a more general situation where we start with an arbitrary top-cone, and restrict the Varchenko matrix to chambers of this top-cone. The determinant of this matrix also factorizes. This more general result is given in Theorem 8.12. Specializing the top-cone to the ambient space recovers the previous situation. The special case of weights on hyperplanes is given in Theorem 8.22. This latter result has been obtained recently by Gente independent of our work.

Part II

Birkhoff algebra and Tits algebra. (Chapter 9.) The linearization of a monoid over a field $\Bbbk$ yields an algebra. Let $\mathsf{\Pi}[\mathcal{A}]$ denote the linearization of $\Pi[\mathcal{A}]$, and $\mathsf{\Sigma}[\mathcal{A}]$ denote the linearization of $\Sigma[\mathcal{A}]$ over $\Bbbk$. We call these the *Birkhoff algebra* and the *Tits algebra*, respectively. These are finite-dimensional $\Bbbk$-algebras (since the original monoids are finite). The linearization of $\Gamma[\mathcal{A}]$, denoted $\mathsf{\Gamma}[\mathcal{A}]$, is a left module over $\mathsf{\Sigma}[\mathcal{A}]$. One can linearize the support map as well to obtain an algebra homomorphism $\mathrm{s} : \mathsf{\Sigma}[\mathcal{A}] \twoheadrightarrow \mathsf{\Pi}[\mathcal{A}]$.

The Birkhoff algebra $\mathsf{\Pi}[\mathcal{A}]$ is isomorphic to $\Bbbk^n$, where n is the number of flats. In other words, $\mathsf{\Pi}[\mathcal{A}]$ is a split-semisimple commutative algebra (Theorem 9.2). (By a result of Solomon, this holds for any algebra obtained by linearizing a lattice.) The coordinate vectors of $\Bbbk^n$ yield a unique complete system of primitive orthogonal idempotents of $\mathsf{\Pi}[\mathcal{A}]$. We denote them by Q_{X}, as X varies over flats. The simple modules over $\mathsf{\Pi}[\mathcal{A}]$ are all one-dimensional, and given by $\mathsf{Q}_{\mathrm{X}} \cdot \mathsf{\Pi}[\mathcal{A}]$. Further, any module h is a direct sum of simple modules. More precisely, we have the Peirce decomposition [2]

$$\mathsf{h} = \bigoplus_{\mathrm{X}} \mathsf{Q}_{\mathrm{X}} \cdot \mathsf{h},$$

[2] A decomposition of a module using an orthogonal family of idempotents is called a Peirce decomposition.

and the simple module $\mathtt{Q}_{\mathrm{X}} \cdot \mathsf{\Pi}[\mathcal{A}]$ occurs in the summand $\mathtt{Q}_{\mathrm{X}} \cdot \mathsf{h}$ with multiplicity equal to its dimension (Theorems 9.7 and 9.8). As a consequence, the action of any element of $\mathsf{\Pi}[\mathcal{A}]$ on any module h is diagonalizable (Theorem 9.9).

The largest nilpotent ideal of an algebra A is called its radical, denoted $\mathrm{rad}(A)$. The Birkhoff algebra has no nonzero nilpotent elements, so $\mathrm{rad}(\mathsf{\Pi}[\mathcal{A}]) = 0$. In contrast, the Tits algebra has many nilpotent elements. In fact, $\mathrm{rad}(\Sigma[\mathcal{A}])$ is precisely the kernel of the (linearized) support map s, hence

$$\Sigma[\mathcal{A}]/\,\mathrm{rad}(\Sigma[\mathcal{A}]) \cong \mathsf{\Pi}[\mathcal{A}].$$

This was proved by Bidigare. We say that $\Sigma[\mathcal{A}]$ is an elementary algebra since the quotient by its radical is a split-semisimple commutative algebra. The simple modules over $\Sigma[\mathcal{A}]$ coincide with those over $\mathsf{\Pi}[\mathcal{A}]$ (since $\mathrm{rad}(\Sigma[\mathcal{A}])$ is forced to act by zero on such modules). However, a module of $\Sigma[\mathcal{A}]$ does not split as a direct sum of simple modules in general. (An example is provided by the module of chambers $\mathsf{\Gamma}[\mathcal{A}]$.) Similarly, the action of an element of $\Sigma[\mathcal{A}]$ on a module h is not diagonalizable in general. Nonetheless, by taking a filtration of h, one can gain detailed information about the eigenvalues and multiplicities of the action (Theorem 9.44). This result for $\mathsf{h} := \mathsf{\Gamma}[\mathcal{A}]$ was first obtained by Bidigare, Hanlon and Rockmore (Theorem 9.46); their motivation for considering this problem came from random walks. The above line of argument was given by Brown.

Any left module h over the Tits algebra has a *primitive part* which we denote by $\mathcal{P}(\mathsf{h})$. It consists of those elements of h which are annihilated by all faces except the central face (which acts by the identity). Dually, any right module h has a *decomposable part* which we denote by $\mathcal{D}(\mathsf{h})$. The duality is made precise in Proposition 9.61.

Janus algebra. (Chapter 9.) Let $\mathsf{J}[\mathcal{A}]$ denote the linearization of $\mathrm{J}[\mathcal{A}]$. We call this the *Janus algebra*. Just like the Tits algebra, the Janus algebra is elementary, and its split-semisimple quotient is the Birkhoff algebra. Interestingly, the Janus algebra admits a deformation by a scalar q. When q is not a root of unity, the q-Janus algebra is in fact split-semisimple, that is, isomorphic to a product of matrix algebras over $\mathbb{k}$. There is one matrix algebra for each flat X, with the size of the matrix being the number of faces with support X (Theorem 9.75). As a consequence, the q-Janus algebra, for q not a root of unity, is Morita equivalent to the Birkhoff algebra (Theorem 9.76). This is completely different from what happens for $q = 1$.

Eulerian idempotents. (Chapter 11.) Let us go back to the Tits algebra $\Sigma[\mathcal{A}]$. An *Eulerian family* $\mathtt{E}$ is a complete system of primitive orthogonal idempotents of $\Sigma[\mathcal{A}]$. Eulerian families are in correspondence with algebra sections $\mathsf{\Pi}[\mathcal{A}] \hookrightarrow \Sigma[\mathcal{A}]$ of the support map s. The construction of such sections is the idempotent lifting problem in ring theory. For elementary algebras, lifts always exist and any two lifts are conjugate by an invertible element in the algebra. For each X, we let $\mathtt{E}_{\mathrm{X}}$ denote the image of $\mathtt{Q}_{\mathrm{X}}$ under an algebra section, thus, $\mathrm{s}(\mathtt{E}_{\mathrm{X}}) = \mathtt{Q}_{\mathrm{X}}$. The $\mathtt{E}_{\mathrm{X}}$ are called *Eulerian idempotents* and constitute the Eulerian family $\mathtt{E}$. Apart from being elementary, the Tits algebra is also the linearization of a left regular band. This allows for many interesting characterizations of Eulerian families (Theorems 11.20, 11.40 and 15.44). A highlight here is a construction of Saliola which produces an Eulerian family starting with a homogeneous section of the support map. (A homogeneous section is equivalent to an assignment of a scalar $\mathtt{u}^F$ to each face F such that for any flat X, the sum of $\mathtt{u}^F$ over all F with support X is 1.) This construction employs the *Saliola lemma* (Lemma 11.12), which is an important property of any Eulerian family. For a good reflection arrangement, we give cancelation-free formulas for the Eulerian idempotents arising from the uniform homogeneous section (Theorem 11.53).

Diagonalizability. (Chapter 12.) An element of an algebra is diagonalizable if it can be expressed as a linear combination of orthogonal idempotents. All elements of the Birkhoff algebra are diagonalizable. However, that is not true for the Tits algebra. For instance, no nonzero element of the radical of $\Sigma[\mathcal{A}]$ is diagonalizable. Following another method of Saliola, one can characterize diagonalizable elements using existence of eigensections (Corollary 12.15). Examples include nonnegative elements (Theorem 12.20) and separating elements (Theorem 12.17). The separating condition was introduced by Brown. For separating elements, there is a formula for the eigensection (arising from the Brown-Diaconis stationary distribution formula (12.6)), and a formula for the Eulerian idempotents due to Brown, see (12.12) and (12.13). Apart from these families, we also consider diagonalizability of specific elements such as the Takeuchi element (12.23) and the Fulman elements (12.38). For the braid arrangement, these include the Adams elements; their diagonalization is given in (12.49).

Lie elements and JKS. (Chapters 10 and 14.) Recall that the Tits algebra $\Sigma[\mathcal{A}]$ acts on the space of chambers $\Gamma[\mathcal{A}]$. We put

$$\mathsf{Lie}[\mathcal{A}] := \mathcal{P}(\Gamma[\mathcal{A}]),$$

the primitive part of $\Gamma[\mathcal{A}]$. This is the space of *Lie elements*. We refer to this description of $\mathsf{Lie}[\mathcal{A}]$ as the Friedrichs criterion. There are other characterizations of $\mathsf{Lie}[\mathcal{A}]$ such as the top-lune criterion and the descent criterion. In the case of the braid arrangement, $\mathsf{Lie}[\mathcal{A}]$ is the space of classical Lie elements (the multilinear part of the free Lie algebra). The top-lune criterion extends a classical result of Ree for the free Lie algebra, while the descent criterion extends a result of Garsia. The top-lune criterion says the following: A Lie element is an assignment of a scalar x^C to each chamber C such that the sum of these scalars in any top-lune (containing more than one chamber) is zero. In fact, by cutting a top-lune into smaller top-lunes, it suffices to restrict to top-lunes whose base is of rank 1. The dimension of $\mathsf{Lie}[\mathcal{A}]$ equals the absolute value of the Möbius number of $\mathcal{A}$. There are many ways to deduce this, see for instance (10.24) or (11.63). There are also many interesting bases for $\mathsf{Lie}[\mathcal{A}]$. We discuss the Dynkin basis (which depends on a generic half-space) and the Lyndon basis (which depends on a choice function).

For any flat X, there is a map

$$\mathsf{Lie}[\mathcal{A}^{\mathrm{X}}] \otimes \mathsf{Lie}[\mathcal{A}_{\mathrm{X}}] \to \mathsf{Lie}[\mathcal{A}].$$

We call this the substitution product of Lie, see (10.28). It is obtained by restricting the substitution product of chambers. All Lie elements of $\mathcal{A}$ can be generated by repeated substitutions starting with Lie elements of rank-one arrangements (which incorporate antisymmetry), subject to the Jacobi identities in rank-two arrangements (Theorem 14.41). Antisymmetry can be visualized as follows.

$$(\overset{1}{\bullet} \qquad \overset{\bar{1}}{\bullet}) + (\overset{\bar{1}}{\bullet} \qquad \overset{1}{\bullet}) = 0.$$

(By convention, $\overline{1}$ denotes -1.) The two vertices are the two chambers of a rank-one arrangement. The Jacobi identity for the hexagon and octagon (which are the spherical models of rank-two arrangements of 3 and 4 lines, respectively) are shown below. The figures show the coefficients of each chamber in a Lie element.

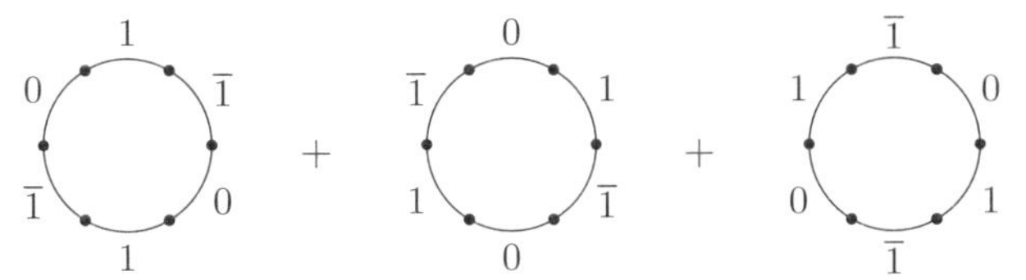

= 0.

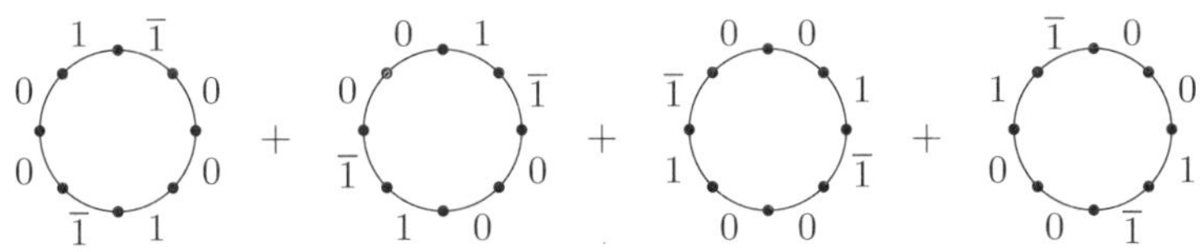

= 0.

The Ree criterion says that the sum of the scalars in any semicircle is 0.

A closely related object to $\mathsf{Lie}[\mathcal{A}]$ is the order homology of the lattice of flats $\Pi[\mathcal{A}]$. The latter is a well-studied object. The order homology is nonzero only in top rank and has dimension equal to the absolute value of the Möbius number of $\mathcal{A}$. Again, there are many bases for this space. We discuss the Björner-Wachs basis and the Björner basis. One of our main results, the *Joyal-Klyachko-Stanley theorem*, or JKS for short, states that up to the one-dimensional orientation space of $\mathcal{A}$, there is a natural isomorphism between $\mathsf{Lie}[\mathcal{A}]$ and the top-cohomology of $\Pi[\mathcal{A}]$ (Theorem 14.38). We write this as

$$\mathcal{H}^{\mathrm{top}}(\Pi[\mathcal{A}]) \otimes \mathsf{E}^{\mathrm{o}}[\mathcal{A}] \cong \mathsf{Lie}[\mathcal{A}].$$

The special case of the braid arrangement is a classical result due to separate work by Joyal, Klyachko and Stanley. Under the JKS isomorphism and the duality between homology and cohomology, the Dynkin basis corresponds to the Björner-Wachs basis (Corollary 14.39) while the Lyndon basis corresponds to the Björner basis (Propositions 14.51 and 14.52). The latter correspondence was used by Barcelo to give the first combinatorial proof of the classical JKS.

Zie elements. (Chapter 10, 11 and 14.) Consider the left action of the Tits algebra $\Sigma[\mathcal{A}]$ on itself, and put

$$\mathsf{Zie}[\mathcal{A}] := \mathcal{P}(\Sigma[\mathcal{A}]),$$

the primitive part of $\Sigma[\mathcal{A}]$. This is the space of *Zie elements* (defined using the Friedrichs criterion). In analogy with $\mathsf{Lie}[\mathcal{A}]$, we also have other criteria such as the lune and descent criteria. A Zie element is a particular element of the Tits algebra. It is called special if its coefficient of the central face is 1. The space $\mathsf{Zie}[\mathcal{A}]$ is a right ideal of $\Sigma[\mathcal{A}]$ generated by any special Zie element (Lemma 10.21). Any special Zie element is an idempotent. In fact, an element of the Tits algebra is a special Zie element iff it is an idempotent whose support is $\mathtt{Q}_\perp$ (Lemma 10.24). The first Eulerian idempotent $\mathtt{E}_\perp$ of any Eulerian family is a special Zie element, and conversely every special Zie element arises in this manner (Lemma 11.42). More generally, the higher Eulerian idempotent $\mathtt{E}_{\mathrm{X}}$ (of any Eulerian family) yields a special Zie element of the arrangement $\mathcal{A}_{\mathrm{X}}$ over X. This leads to a characterization of Eulerian families in terms of families of special Zie elements (Theorem 11.40).

For any left $\Sigma[\mathcal{A}]$-module h, a special Zie element projects h onto its primitive part $\mathcal{P}(\mathsf{h})$ (Proposition 10.35).

Given any generic half-space of $\mathcal{A}$, the alternating sum of faces contained in that half-space yields a special Zie element. We call this the *Dynkin element* (Proposition 14.1). It projects the module of chambers $\Gamma[\mathcal{A}]$ onto its primitive part which is $\mathsf{Lie}[\mathcal{A}]$. This generalizes the classical Dynkin operator (left nested bracketing) in the case of the braid arrangement. Under this projection, the images of chambers in the half-space opposite to the given generic half-space yields a basis of $\mathsf{Lie}[\mathcal{A}]$. This is precisely the Dynkin basis mentioned earlier (Proposition 14.16).

Loewy series and Peirce decompositions. (Chapter 13.) The *primitive series* of a left $\Sigma[\mathcal{A}]$-module h is a specific filtration of h with the primitive part $\mathcal{P}(\mathsf{h})$ as the first nontrivial term from the bottom. Dually, the *decomposable series* of a right $\Sigma[\mathcal{A}]$-module h is a specific filtration of h with the decomposable part $\mathcal{D}(\mathsf{h})$ as the first nontrivial term from the top. The primitive series and decomposable series are both examples of Loewy series (Propositions 13.4 and 13.6). By general theory, they are trapped between the radical and socle series; see Lemmas 13.8 and 13.18. The left module of chambers is rigid, that is, its radical, primitive and socle series coincide (Theorem 13.68). The right module of Zie elements is also rigid (Theorem 13.83).

For any left $\Sigma[\mathcal{A}]$-module h, we have the left Peirce decomposition

$$\mathsf{h} = \bigoplus_{\mathrm{X}} \mathtt{E}_{\mathrm{X}} \cdot \mathsf{h}.$$

This depends on the choice of the Eulerian family $\mathtt{E}$. However, the summand indexed by the minimum flat $\perp$ is independent of this choice. More precisely,

$$\mathtt{E}_{\perp} \cdot \mathsf{h} = \mathcal{P}(\mathsf{h}),$$

see Proposition 13.21. This is consistent with the earlier statement that a special Zie element projects h onto $\mathcal{P}(\mathsf{h})$. In general, the components $\mathtt{E}_{\mathrm{X}} \cdot \mathsf{h}$ are related to the primitive series of h (Proposition 13.22). Similarly, one can relate the components of the right Peirce decomposition of a right $\Sigma[\mathcal{A}]$-module to its decomposable series (Proposition 13.24).

The components of the left Peirce decompositions of $\Gamma[\mathcal{A}]$ and $\Sigma[\mathcal{A}]$ relate to Lie and Zie elements as follows.

$$\mathtt{E}_{\mathrm{X}} \cdot \Gamma[\mathcal{A}] \cong \mathsf{Lie}[\mathcal{A}_{\mathrm{X}}] \quad \text{and} \quad \mathtt{E}_{\mathrm{X}} \cdot \Sigma[\mathcal{A}] \cong \mathsf{Zie}[\mathcal{A}_{\mathrm{X}}].$$

See Lemmas 13.26 and 13.30. The former yields an algebraic form of the Zaslavsky formula, see (13.8). Similarly, the components of the right Peirce decompositions of $\mathsf{Zie}[\mathcal{A}]$ and $\Sigma[\mathcal{A}]$ relate to Lie and chamber elements as follows.

$$\mathsf{Zie}[\mathcal{A}] \cdot \mathtt{E}_{\mathrm{Y}} \cong \mathsf{Lie}[\mathcal{A}^{\mathrm{Y}}] \quad \text{and} \quad \Sigma[\mathcal{A}] \cdot \mathtt{E}_{\mathrm{Y}} \cong \Gamma[\mathcal{A}^{\mathrm{Y}}].$$

See Lemmas 13.43 and 13.40. The latter is present in work of Saliola. Combining these decompositions yields a vector space isomorphism

$$\mathtt{E}_{\mathrm{X}} \cdot \Sigma[\mathcal{A}] \cdot \mathtt{E}_{\mathrm{Y}} \cong \mathsf{Lie}[\mathcal{A}_{\mathrm{X}}^{\mathrm{Y}}].$$

See Proposition 13.50 and Table 13.1. These are components of the two-sided Peirce decomposition of $\Sigma[\mathcal{A}]$. By taking direct sum over all $\mathrm{X} \leq \mathrm{Y}$, we obtain an algebra isomorphism

$$\Sigma[\mathcal{A}] \cong \bigoplus_{X \le Y} \mathsf{Lie}[\mathcal{A}_X^Y].$$

In the rhs, elements in the (X, Y)-summand are multiplied with elements in the (Y, Z)-summand by substitution; the remaining products are all zero. This isomorphism is given in Theorem 13.54. As an application, we obtain the quiver of the Tits algebra (Theorem 13.73). This is a result of Saliola, who proved it by linking the Tits algebra to the top-cohomology of the lattice of flats.

Lune-incidence algebra and noncommutative zeta and Möbius functions. (Chapter 15.) A *nested flat* is a pair of flats (X, Y) with $Y \geq X$. Let $I_{\mathsf{flat}}[\mathcal{A}]$ denote the incidence algebra of the poset of flats $\Pi[\mathcal{A}]$. We call it the *flat-incidence algebra*. It consists of functions f on nested flats, with the product of f and g given by

$$(fg)(X, Z) = \sum_{Y:\, X \le Y \le Z} f(X, Y) g(Y, Z).$$

The zeta function $\zeta \in I_{\mathsf{flat}}[\mathcal{A}]$ is defined to be identically 1. It is invertible and its inverse is the Möbius function $\mu \in I_{\mathsf{flat}}[\mathcal{A}]$. The Möbius function satisfies the Weisner formula, and in fact is completely characterized by it. A standard way to prove this formula is to use the split-semisimplicity of the Birkhoff algebra.

We propose a noncommutative version of this theory with $\Pi[\mathcal{A}]$ replaced by $\Sigma[\mathcal{A}]$. A *nested face* is a pair of faces (A, F) with $F \geq A$. Let $I_{\mathsf{face}}[\mathcal{A}]$ denote the incidence algebra of $\Sigma[\mathcal{A}]$. We call it the *face-incidence algebra*. It consists of functions f on nested faces, with the product of f and g given by

$$(fg)(F, H) = \sum_{G:\, F \le G \le H} f(F, G) g(G, H).$$

We say two nested faces (A, F) and (B, G) are equivalent if $AB = A$, $BA = B$, $AG = F$ and $BF = G$. Equivalence classes are indexed by lunes (Proposition 3.13). Let $I_{\mathsf{lune}}[\mathcal{A}]$ denote the subalgebra of $I_{\mathsf{face}}[\mathcal{A}]$ consisting of those f which take the same value on equivalent nested faces. In particular, $I_{\mathsf{lune}}[\mathcal{A}]$ has a basis indexed by lunes. It is an example of a reduced incidence algebra. We call it the *lune-incidence algebra*. It can also be interpreted as the incidence algebra of the category of lunes (Proposition 15.7).

We define noncommutative zeta functions $\boldsymbol{\zeta}$ and noncommutative Möbius functions $\boldsymbol{\mu}$ as particular elements of the lune-incidence algebra. They are no longer unique; zeta functions are characterized by lune-additivity (15.23) and Möbius functions by the noncommutative Weisner formula (15.27). They correspond to each other under taking inverses in the lune-incidence algebra (Theorem 15.28). We relate this result to the representation theory of the Tits algebra. This circle of ideas is summarized in the important Theorem 15.44, which states in particular that noncommutative zeta and Möbius functions are in bijection with Eulerian families. Also see Table 15.1.

The flat-incidence algebra and lune-incidence algebra are both elementary and their quivers are acylic with vertices indexed by flats (Proposition 15.1 and Theorem 15.2, and Proposition 15.10 and Theorem 15.14).

Lie-incidence algebra. (Chapter 15.) We introduce the Lie-incidence algebra $\mathrm{I}_{\mathrm{Lie}}[\mathcal{A}]$. It is a subalgebra of the lune-incidence algebra (Proposition 15.51). It is isomorphic to the Tits algebra (Theorem 15.56). We also introduce additive and Weisner functions on lunes. These are linear subspaces of the lune-incidence algebra which respectively contain noncommutative zeta and Möbius functions as affine subspaces. Further, they are right and left modules respectively over $\mathrm{I}_{\mathrm{Lie}}[\mathcal{A}]$ with action induced from the product of $\mathrm{I}_{\mathrm{lune}}[\mathcal{A}]$ (Propositions 15.62 and 15.66). Moreover, they are isomorphic to the right and left regular representations of $\mathrm{I}_{\mathrm{Lie}}[\mathcal{A}]$ (Propositions 15.63 and 15.67).

Invariant objects. (Chapter 16.) In the discussion so far, the arrangement $\mathcal{A}$ and the field $\Bbbk$ have been arbitrary. Suppose now that $\mathcal{A}$ is a reflection arrangement with associated Coxeter group W, and the characteristic of $\Bbbk$ does not divide the order of W. Here W acts on both $\Sigma[\mathcal{A}]$ and $\Pi[\mathcal{A}]$ giving rise to the invariant subalgebras $\Sigma[\mathcal{A}]^W$ and $\Pi[\mathcal{A}]^W$. We call these the *invariant Tits algebra* and *invariant Birkhoff algebra*, respectively. The former is elementary, and the latter is its split-semisimple quotient. They have a basis indexed by face-types and flat-types. Complete systems of primitive orthogonal idempotents of $\Sigma[\mathcal{A}]^W$ (also called invariant Eulerian families) can be characterized in a manner similar to $\Sigma[\mathcal{A}]$ (Theorem 16.48). (The hypothesis on the characteristic of $\Bbbk$ is clarified by Lemma 16.42.) The Garsia-Reutenauer idempotents are the Eulerian idempotents which arise by specializing to the braid arrangement and taking the invariant homogeneous section to be uniform. By linking the invariant Tits algebra to invariant Lie elements, one can obtain information on the quiver of the invariant Tits algebra. The related result given in Proposition 16.55 is due to Saliola.

The Coxeter group acts on the lune-incidence algebra giving rise to the *invariant lune-incidence algebra*. This algebra can also be viewed as a reduced incidence algebra of the poset of face-types. It has a basis indexed by lune-types. For those noncommutative zeta and Möbius functions which belong to this algebra, lune-additivity and the noncommutative Weisner formula can be reformulated using face-types, see (16.41) and (16.42). The structure constants of the invariant Tits algebra intervene in this description.

There is an injective map from the Tits algebra to the space indexed by pairs of chambers. Taking invariants induces an injective map from $\Sigma[\mathcal{A}]^W$ to W (the group algebra of W). The image of this map is a subalgebra of W which is known as the *Solomon descent algebra*. This induces an isomorphism between the invariant Tits algebra and the opposite of the Solomon descent algebra (Theorem 16.8). This was proved by Bidigare. Invariant Eulerian families of the Solomon descent algebra appeared in work of Bergeron, Bergeron, Howlett and Taylor (Theorem 16.43).

Projective objects. Every arrangement carries a symmetry of order 2 given by the opposition map (which sends a point to its negative). The *projective Tits algebra* is the subalgebra of the Tits algebra which is invariant under the opposition map. It is elementary with the Birkhoff algebra as its split-semisimple quotient (Proposition 9.27). Its quiver is given in Theorem 13.75. A complete system of the projective Tits algebra is a projective Eulerian family. Its characterizations in terms of projective analogues of noncommutative zeta and Möbius functions, homogeneous sections and so on are summarized in Theorem 15.47. These results assume that the field characteristic is not 2.

Part I.

CHAPTER 1

Hyperplane arrangements

Some basic geometric objects associated to a hyperplane arrangement are summarized below.

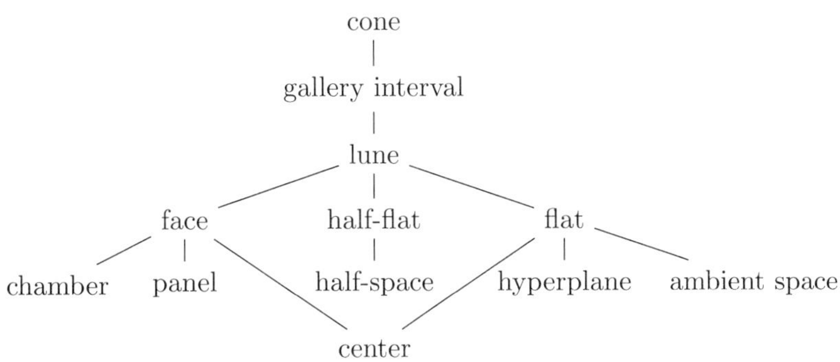

The diagram is to be read as follows. The most general object is a cone. Every gallery interval is a cone, and every lune is a gallery interval. Faces, half-flats and flats are lunes. Chambers and panels are faces, a half-space is a half-flat, hyperplanes and the ambient space are flats, and the center is both a face and a flat. Some other important objects which are not seen in the diagram are bi-faces, nested faces, charts, dicharts and partial-flats.

In this chapter, we discuss

- faces, chambers, flats and bi-faces,
- Tits monoid, Birkhoff monoid and Janus monoid,
- minimal galleries and gate property,
- arrangements under and over a flat of an arrangement,
- other operations on arrangements such as adjoint and cartesian product,
- enumerative aspects such as Möbius functions and characteristic polynomial.

Cones, gallery intervals, lunes, charts, dicharts and partial-flats are discussed in Chapters 2 and 3.

Many of the sets that we consider, such as the set of faces, carry the structure of a poset. All our posets are finite. For posets, we will employ the terminology given in Section B.1. Note in particular that 'smaller than' means $\leq$ and 'greater than' means $\geq$. Graded posets are reviewed in Section B.2.

Convention 1.1. An arrangement is usually denoted by $\mathcal{A}$. To show the dependence of an object on $\mathcal{A}$, we use $[\mathcal{A}]$. For instance, $\Sigma[\mathcal{A}]$ denotes the set of faces of $\mathcal{A}$ and $\Pi[\mathcal{A}]$ denotes the set of flats of $\mathcal{A}$. When $\mathcal{A}$ is understood from the context, we may simply write Σ, Π, and so on.

1.1. Faces

We briefly review hyperplane arrangements, and discuss the poset of faces.

1.1.1. Hyperplanes and half-spaces. Let V be a finite-dimensional vector space over $\mathbb{R}$. A codimension-one affine subspace of V is called a *hyperplane*. A *half-space* is a subset of V which lies on one side of some hyperplane. The bounding hyperplane is the *boundary* of the half-space. By convention, a half-space is closed, that is, it contains its boundary. The *interior* of the half-space is the half-space minus its boundary. Each hyperplane has two associated half-spaces which lie on its two sides.

The picture on the left shows a hyperplane in $\mathbb{R}^2$ which is the same as a line, while the one on the right shows a half-space.

1.1.2. Hyperplane arrangements. A *hyperplane arrangement* $\mathcal{A}$ is a finite set of hyperplanes in a finite-dimensional real vector space V. The latter is called the *ambient space* of $\mathcal{A}$. The arrangement is *central* if all its hyperplanes pass through the origin. Unless stated otherwise, all our arrangements are assumed to be central. The *center* of $\mathcal{A}$ is the subspace obtained by intersecting all hyperplanes of $\mathcal{A}$. The arrangement is *essential* if its center is the zero subspace.

The *rank* of $\mathcal{A}$, denoted $\operatorname{rk}(\mathcal{A})$, is the difference between the dimensions of the ambient space and the center. In particular, the rank of an essential arrangement equals the dimension of the ambient space.

The arrangement on the left consists of three lines in $\mathbb{R}^2$, while the one on the right consists of four lines in $\mathbb{R}^2$ (passing through the origin). Both are essential and have rank 2.

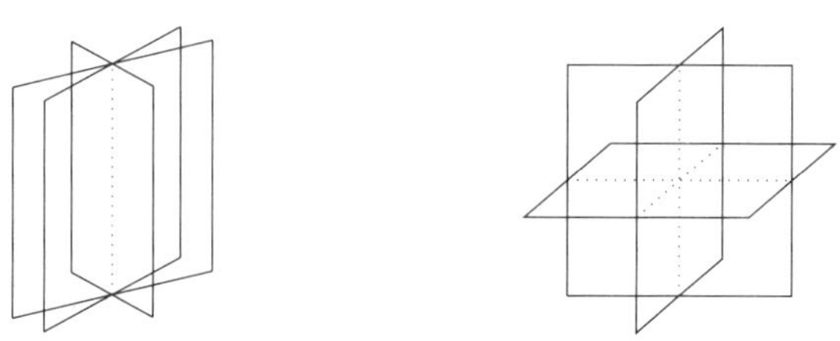

Both the above arrangements consist of three planes in $\mathbb{R}^2$. The one on the left is not essential and has rank 2, while the one on the right is essential and has rank 3.

1.1.3. Faces. A half-space of an arrangement $\mathcal{A}$ is a half-space of the ambient space V whose bounding hyperplane belongs to $\mathcal{A}$. A *face* of $\mathcal{A}$ is a subset of V obtained by intersecting half-spaces of $\mathcal{A}$, with at least one associated half-space chosen for each hyperplane. The *interior* of a face F is the subset of F obtained by intersecting F with the interiors of those half-spaces used to define F whose boundary does not contain F. Every point in the ambient space belongs to the interior of a unique face. In other words, the interiors of all faces partition the ambient space. The center of $\mathcal{A}$ is a face. We call it the *central face* and denote it by O. The interior of the central face is the central face itself.

Let $\Sigma[\mathcal{A}]$ denote the set of all faces. It is a graded poset under inclusion, with the central face O as its minimum element. Each face F has a dimension, and the rank of F is the dimension of F minus the dimension of O. We write this as

$$\operatorname{rk}(F) = \dim(F) - \dim(O).$$

The rank of the poset $\Sigma[\mathcal{A}]$ equals the rank of $\mathcal{A}$. A maximal face of $\Sigma[\mathcal{A}]$ is called a *chamber*. We denote the set of chambers by $\Gamma[\mathcal{A}]$. A rank-one face is called a *vertex*, a rank-two face is called an *edge*, while a corank-one face is called a *panel*.

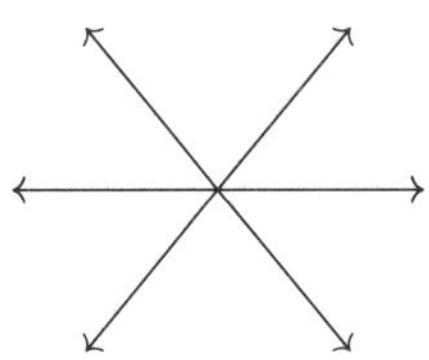

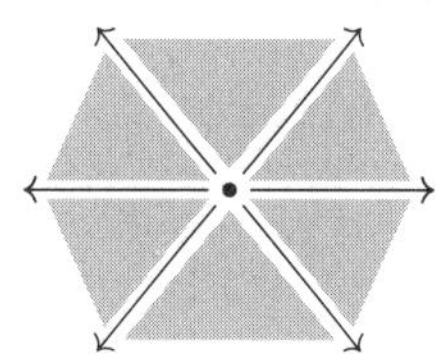

Let us return to the arrangement of three lines (hyperplanes) in the plane passing through the origin. The partition of the plane by face interiors is illustrated on the right. There are 13 faces: six edges or chambers (sectors), six vertices or panels (rays) and the central face (the origin). The Hasse diagram of the poset of faces is shown below. It has rank 2. The minimum element is the origin, the rank-one elements are the rays, and the rank-two elements are the sectors.

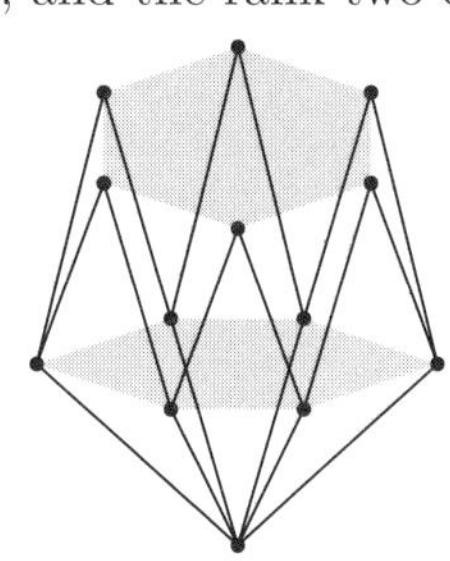

Faces of an arrangement will generally be denoted by the letters F, G, H and K, chambers by the letters C, D and E, and vertices by the letters P and Q. We will also employ the letter E to denote edges.

The intersection of two faces is a face, so meets exist in $\Sigma[\mathcal{A}]$. We denote the meet of F and G by $F \wedge G$. In contrast, joins may not exist. We denote the join of F and G by $F \vee G$ (whenever it exists). It exists precisely when F and G have a common upper bound. In particular, the join of two distinct chambers cannot exist. In summary, $\Sigma[\mathcal{A}]$ is a graded meet-semilattice; it is not a lattice unless the rank of $\mathcal{A}$ is 0.

We say F is a face of G if $F \leq G$. A vertex of G is a rank-one face of G. Similarly, F is a panel of G if F is a corank-one face of G, that is, $F \lessdot G$.

A face is the meet of all chambers greater than that face. Similarly, a face is the join of all its vertices. This follows from the fact that a face is the convex hull of its vertices.

1.1.4. Opposition map. Every face F has an *opposite face*, denoted $\overline{F}$, which is given by

$$\overline{F} := \{-x \mid x \in F\}.$$

The *opposition map on faces*

$$\Sigma[\mathcal{A}] \to \Sigma[\mathcal{A}], \qquad F \mapsto \overline{F} \tag{1.1}$$

sends every face to its opposite. It is an order-preserving involution. That is,

$$\overline{\overline{F}} = F \qquad \text{and} \qquad F \leq G \iff \overline{F} \leq \overline{G}.$$

Since chambers are maximal faces, the opposition map restricts to an involution on the set of chambers $\Gamma[\mathcal{A}]$. Thus, every chamber C has an *opposite chamber* $\overline{C}$.

Every half-space h also has an *opposite half-space*, denoted $\overline{\mathrm{h}}$, which is given by

$$\overline{\mathrm{h}} := \{-x \mid x \in \mathrm{h}\}.$$

Note that the bounding hyperplanes of h and $\overline{\mathrm{h}}$ coincide.

Exercise 1.2. For any face F, check that $F = \overline{F}$ iff $F = O$. Deduce that the number of faces in any arrangement is odd.

1.1.5. Projective faces. A *projective face* is an unordered pair consisting of a face and its opposite. We denote a projective face by $\{F, \overline{F}\}$. The number of projective faces equals half the number of noncentral faces plus 1.

Similarly, a *projective chamber* is an unordered pair consisting of a chamber and its opposite. It is denoted by $\{C, \overline{C}\}$. The number of projective chambers equals half the number of chambers (assuming $\mathcal{A}$ has rank at least one).

1.1.6. Isomorphism of arrangements. We consider two notions of isomorphism of arrangements, called gisomorphism and cisomorphism. The former is stronger than the latter.

We say two arrangements $\mathcal{A}$ and $\mathcal{A}'$ are geometrically isomorphic, or *gisomorphic* for short, if there is a linear isomorphism between their ambient spaces which induces a bijection between the two sets of hyperplanes. We refer to any such isomorphism as a *gisomorphism*.

We say two arrangements $\mathcal{A}$ and $\mathcal{A}'$ are combinatorially isomorphic, or *cisomorphic* for short, if the poset of faces $\Sigma[\mathcal{A}]$ and $\Sigma[\mathcal{A}']$ are isomorphic. We refer to any such isomorphism as a *cisomorphism*.

It is clear that gisomorphic implies cisomorphic. But the converse is not true.

1.1.7. Essentialization. If an arrangement $\mathcal{A}$ is not essential, then we can make it essential by taking quotient of the ambient space by its center. This is the *essentialization* of $\mathcal{A}$. This construction loses information about the dimension of the center, however it does not affect the poset of faces. Similar remark applies to the poset of flats, and other posets that we will encounter later.

If an arrangement is not essential, then its essentialization is cisomorphic but not gisomorphic to the original arrangement (since the two ambient spaces have different dimensions).

Consider the arrangement of three planes in $\mathbb{R}^3$ passing through the z-axis discussed in Section 1.1.2. Its essentialization is gisomorphic to an arrangement of three lines in $\mathbb{R}^2$.

1.1.8. Cell complex of an arrangement. Regular cell complexes are reviewed in Section A.1. The poset of faces $\Sigma[\mathcal{A}]$ has the structure of a regular cell complex. Further, this cell complex is pure, of the same rank as $\Sigma[\mathcal{A}]$, and is homeomorphic to the sphere of dimension one lower than the rank. The construction goes as follows.

We assume that $\mathcal{A}$ is essential. (If not, we repeat the following on the essentialization of $\mathcal{A}$.) Put a norm on the ambient space, cut the arrangement by the unit sphere, and identify faces of the arrangement with cells on the sphere to obtain the cell complex. This is illustrated below on the arrangement of 3 lines in the plane.

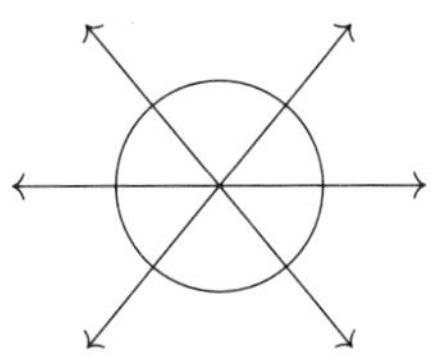

so we see

The central face $O = \{0\}$ is not visible in the spherical model; it corresponds to the unique face of rank 0 of the cell complex.

Another illustration for an arrangement of four planes in $\mathbb{R}^3$ is given below. The spherical model is shown on the right.

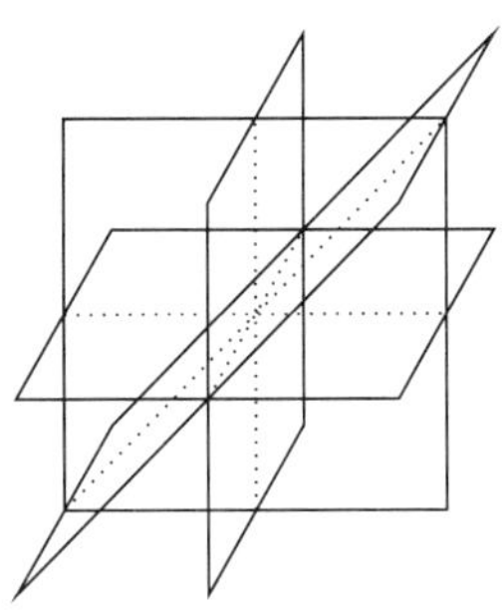

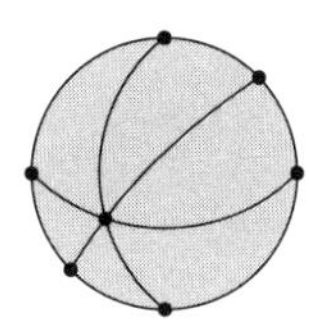

This construction explains the motivation for calling rank-one faces as vertices; they are indeed vertices of the associated cell complex, though they are rays in the original arrangement. Similar comment applies to edges.

As far as notation goes, from now on, we will use $\Sigma[\mathcal{A}]$ to denote faces of the arrangement (linear model) as well as faces of the associated cell complex (spherical model).

1.1.9. Simplicial arrangements. An essential arrangement is *simplicial* if each chamber is a simplicial cone, that is, a cone over a simplex (of the same rank as $\mathcal{A}$) with the origin as the cone-point. In general, an arrangement is said to be simplicial if its essentialization is simplicial. Equivalently, an arrangement is simplicial iff its associated cell complex is a pure simplicial complex.

We refer to faces of a simplicial arrangement as simplices. Any interval in the poset of faces of a simplicial arrangement is Boolean.

1.2. Arrangements of small rank

We look at arrangements of small rank, starting from rank zero and going up to rank three. Arrangements up to rank 2 are easy to classify up to cisomorphism. They are all simplicial.

1.2.1. Rank 0. An arrangement has rank 0 iff it has no hyperplanes. All these arrangements are clearly cisomorphic. There is only one arrangement of rank 0 which is essential, namely, the arrangement whose ambient space is zero-dimensional.

1.2.2. Rank 1. An arrangement has rank 1 iff it has exactly one hyperplane. All these arrangements are cisomorphic. An arrangement of rank 1 is essential iff its ambient space is one-dimensional (with the origin as the unique hyperplane). This is illustrated below.

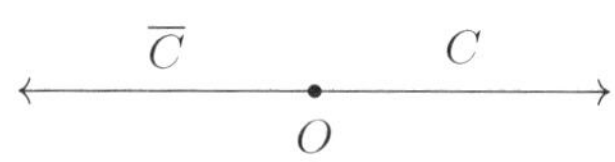

There are two chambers (rays) which we will usually denote by C and $\overline{C}$.

1.2.3. Rank 2. An arrangement has rank 2 iff it has at least two hyperplanes and they all pass through a codimension-two subspace of the ambient space. An essential arrangement of rank 2 consists of n lines through the origin in a two-dimensional space, with $n \geq 2$. The arrangement of three lines and four lines in the plane discussed in Section 1.1.2 are examples with $n = 3$ and $n = 4$. Any two rank-two arrangements with the same number of lines are cisomorphic. When the lines are equally spaced, the arrangement is called *dihedral*. Any two dihedral arrangements with the same number of lines are gisomorphic.

Exercise 1.3. Show that all arrangements of 3 lines in the plane (passing through the origin) are gisomorphic.

1.2.4. Rank 3. The figure below shows the spherical model of an arrangement of rank 3 consisting of five hyperplanes. A hyperplane in this case is the same as a great circle. Only one half of the arrangement is visible in the picture, the other half being on the backside.

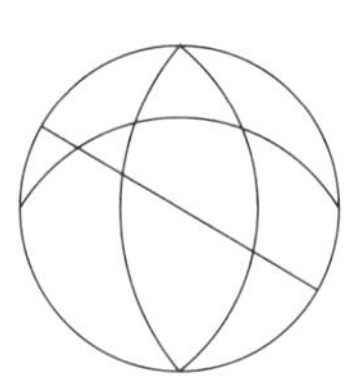

The chambers are either triangles or quadrilaterals, so the arrangement is not simplicial. Three of the hyperplanes pass through the north and south poles. Note that the region between two adjacent longitudes contains either 3 or 4 chambers. The fact that these numbers can differ is of importance.

Rank-three arrangements can be visualized in this manner, and are very useful to develop a geometric feel for notions that we discuss.

1.2.5. Smallest nonsimplicial arrangement. Let us go back to the arrangement of five hyperplanes discussed above. Removing one of the hyperplanes (passing through the north and south poles) gives an arrangement with four hyperplanes. This is also a nonsimplicial arrangement, and it is the smallest such in terms of number of hyperplanes. It has 14 chambers of which 8 are triangles and 6 are quadrilaterals. A standard way to picture this arrangement is shown below.

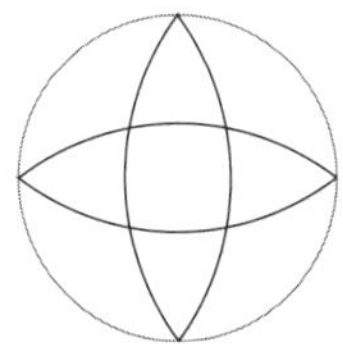

The outer circle is not a part of the arrangement. The four hyperplanes are those that pass through the four sides of the central quadrilateral.

1.3. Flats

We discuss flats, the support map from faces to flats, and a combinatorial approach to flats.

1.3.1. Flats. A *flat* of an arrangement $\mathcal{A}$ is a subspace of the ambient space obtained by intersecting some subset of hyperplanes of $\mathcal{A}$. In particular, a flat has a dimension. Let $\Pi[\mathcal{A}]$ denote the set of flats. It is a graded poset under inclusion, with the center as the minimum element, and the ambient space as the maximum element. (The center is the only subset which is both a face and a flat.) The rank of a flat X is the dimension of X minus the dimension of the center. Intersection of two flats is a flat, so meets exist in the poset of flats. Further, since there is a maximum element, joins exist as well. Thus $\Pi[\mathcal{A}]$ is a lattice.

We will use the letters X, Y, Z and W to denote flats. The minimum and maximum flats will be denoted $\bot$ and $\top$, respectively. We denote the meet of X and Y by $\mathrm{X} \wedge \mathrm{Y}$ and the join by $\mathrm{X} \vee \mathrm{Y}$. We write $[\mathrm{X}, \mathrm{Z}]$ for the interval consisting of all flats which lie between X and Z. Observe that $\mathrm{X} \wedge \mathrm{Y}$ is the intersection of the hyperplanes which contain X and the hyperplanes which contain Y, while $\mathrm{X} \vee \mathrm{Y}$ is the intersection of the hyperplanes which contain both X and Y.

The Hasse diagram of the poset of flats for the arrangement of 3 lines in the plane is shown below. It has rank 2. It consists of the minimum flat $\bot$ (center), the three lines (hyperplanes), and the maximum flat $\top$ (ambient space).

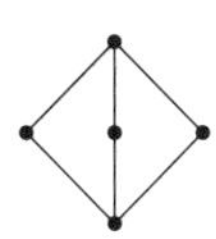

The intersection of a face and a flat is a face. For a face F and flat X, we write $F \wedge \mathrm{X}$ for their intersection. This can be interpreted as a meet in a larger poset which includes both faces and flats. This is the poset of cones, which is discussed in Chapter 2.

1.3.2. Support map. The *support* of a face F is the smallest flat which contains F. It is the intersection of all flats which contain F, or equivalently, the intersection of all hyperplanes which contain F, or equivalently, the linear span of F. The *support map*

$$\mathrm{s} : \Sigma[\mathcal{A}] \twoheadrightarrow \Pi[\mathcal{A}] \tag{1.2}$$

sends a face to its support. It is surjective and order-preserving.

We say a flat X supports a face F if $\mathrm{s}(F) = \mathrm{X}$, that is, the support of F is X. The minimum flat $\perp$ supports exactly one face, namely, the central face O. Any rank-one flat supports two vertices, which are opposite to each other. The maximum flat $\top$ supports chambers.

Exercise 1.4. Meets exist in both $\Sigma[\mathcal{A}]$ and $\Pi[\mathcal{A}]$. Give an example to show that the support map does not preserve meets in general.

Exercise 1.5. Show that an arrangement of rank n has at least n hyperplanes, at least n rank-one flats, and at least $2n$ vertices.

1.3.3. Combinatorial flats. A face F is a *top-dimensional face* of a flat X if X supports F. Note that a flat is the union of its top-dimensional faces. (For instance, a line through the origin is the union of its two opposite rays starting at the origin.) This suggests the following alternative approach to flats.

A *combinatorial flat* is a subset of $\Sigma[\mathcal{A}]$ consisting of all faces with the same support. In other words, a combinatorial flat precisely consists of the top-dimensional faces of some flat.

For a combinatorial flat X, define its *closure* to be

$$\mathrm{Cl}(\mathrm{X}) = \{F \in \Sigma[\mathcal{A}] \mid F \leq G \text{ for some } G \in \mathrm{X}\}.$$

This is the set of all faces contained in X viewed as a geometric flat (subset of the ambient space). Equivalently, it is the set of faces whose support is smaller than X. It follows that

$$\mathrm{X} \leq \mathrm{Y} \iff \mathrm{Cl}(\mathrm{X}) \subseteq \mathrm{Cl}(\mathrm{Y})$$

and

$$\mathrm{Cl}(\mathrm{X} \wedge \mathrm{Y}) = \mathrm{Cl}(\mathrm{X}) \cap \mathrm{Cl}(\mathrm{Y}).$$

Since a flat and a combinatorial flat are equivalent notions, we will usually just say a "flat" with the context determining which notion is being used.

1.4. Tits monoid and Birkhoff monoid

The lattice of flats carries a commutative monoid structure given by the join operation. We call this the Birkhoff monoid. The poset of faces also carries a monoid structure whose product can be viewed as a "noncommutative join". (Recall that joins of faces may or may not exist.) We call this the Tits monoid. Further, the support map from faces to flats is a morphism of monoids.

1.4.1. Sign sequences. For a hyperplane H, let us denote its two associated half-spaces by H^+ and H^-. The choice of $+$ and $-$ is arbitrary but fixed. It is convenient to let $\mathrm{H}^0 := \mathrm{H}$. Observe that $\mathrm{H}^0 = \mathrm{H}^+ \cap \mathrm{H}^-$. In this notation, a face of $\mathcal{A} = \{\mathrm{H}_i\}_{i \in I}$ is a subset of the ambient space of the form

$$F = \bigcap_{i \in I} \mathrm{H}_i^{\epsilon_i},$$

where $\epsilon_i \in \{+, 0, -\}$. Different choices of ϵ_i can yield the same face. However, there is a canonical way to write F in this form, namely,

$$F = \bigcap_{i \in I} \mathrm{H}_i^{\epsilon_i(F)},$$

where $\epsilon_i(F)$ is 0 if F lies in H_i, it is $+$ if the interior of F lies in the interior of H_i^+, and it is $-$ if the interior of F lies in the interior of H_i^-. We refer to

$$(\epsilon_i(F))_{i \in I}$$

as the *sign sequence* of F.

A possible selection of sign sequences for the arrangement of three lines is shown below. Note very carefully that not all sign sequences occur. For instance, no chamber has sign sequence $+-+$.

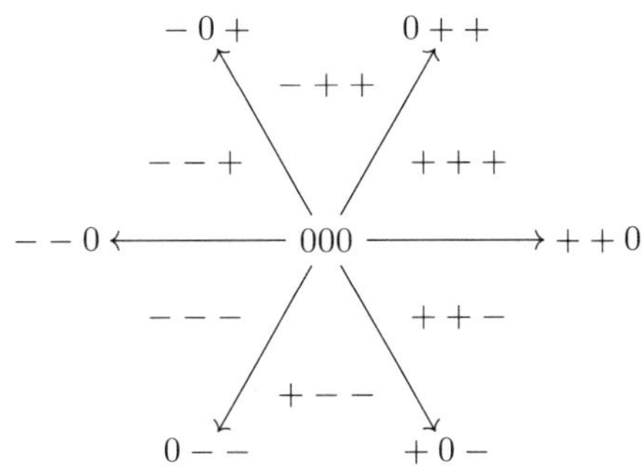

Let us return to the general case. The central face is the unique face F for which $\epsilon_i(F) = 0$ for each i, while a chamber is a face F for which $\epsilon_i(F) \neq 0$ for each i. The sign sequence of $\overline{F}$ is obtained by reversing each sign in the sign sequence of F:

$$\epsilon_i(\overline{F}) = -\epsilon_i(F). \tag{1.3}$$

Observe that

$$F \leq G \iff \epsilon_i(F) = \epsilon_i(G) \text{ whenever } \epsilon_i(F) \neq 0. \tag{1.4}$$

In other words, $F \leq G$ iff the sign sequence of F is obtained from that of G by replacing some $+$ and $-$ by 0.

1.4.2. Tits monoid. For faces F and G, define the face FG by

$$\epsilon_i(FG) := \begin{cases} \epsilon_i(F) & \text{if } \epsilon_i(F) \neq 0, \\ \epsilon_i(G) & \text{if } \epsilon_i(F) = 0. \end{cases} \tag{1.5}$$

We refer to FG as the *Tits product* of F and G. The product has a geometric meaning: if we move from an interior point of F to an interior point of G along a straight line then FG is the face that we are in after moving a small positive distance. Hence, we also say that FG is the *Tits projection* of G on F.

An example in the arrangement of three lines is shown below: F is a vertex (ray), C is an edge (sector) and FC is another edge (sector) which has F as a vertex.

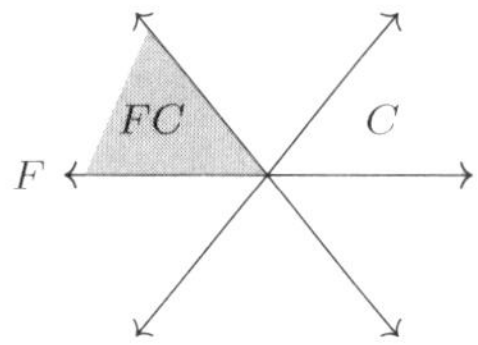

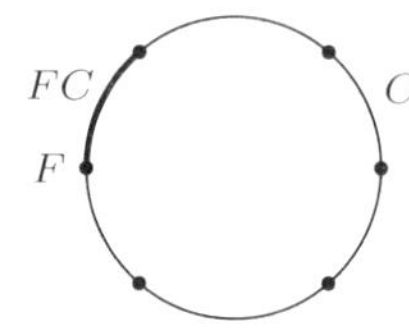

More illustrations of the Tits product in the spherical model of a rank-three arrangement are shown below.

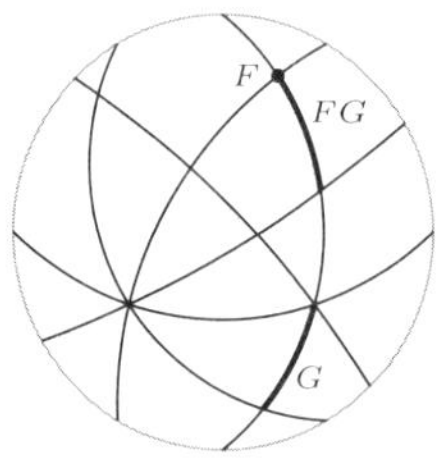

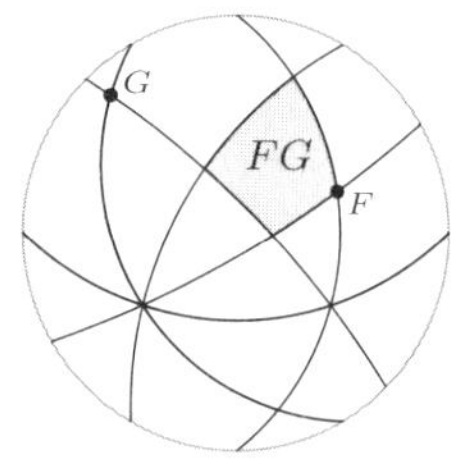

The Tits product is associative, that is,

$$F(GH) = (FG)H$$

for any faces F, G and H. Further, the central face is the identity element for the product. This follows from (1.5). Thus, the set of faces $\Sigma[\mathcal{A}]$ is a monoid. We call this the *Tits monoid*.

For any faces F and G, we have

$$FF = F \quad \text{and} \quad FGF = FG. \tag{1.6}$$

This follows from (1.5). The first identity is a special case of the second obtained by setting $G = O$.

For any faces F and G,

$$\overline{FG} = \overline{F}\,\overline{G}. \tag{1.7}$$

This follows from (1.3) and (1.5). Thus, the opposition map (1.1) is an automorphism of the Tits monoid.

For any faces F and G, a hyperplane contains FG iff it contains both F and G. In other words,

$$\epsilon_i(FG) = 0 \iff \epsilon_i(F) = 0 \text{ and } \epsilon_i(G) = 0 \tag{1.8}$$

for all $i \in I$. This follows from (1.5).

1.4.3. Ideal of chambers. Let C be a chamber. Then for any face F, the faces FC and CF are both chambers, and in fact $CF = C$. Thus, the set of chambers $\Gamma[\mathcal{A}]$ is a two-sided ideal of $\Sigma[\mathcal{A}]$. An illustration follows.

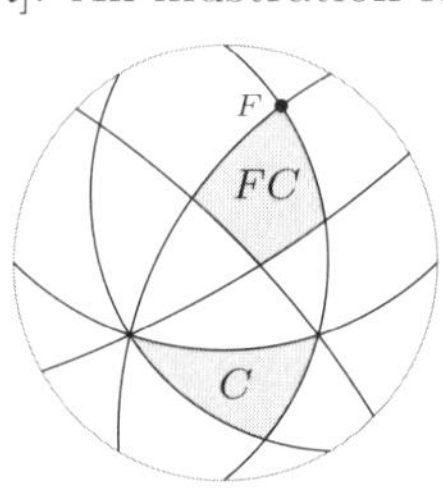

1.4.4. Tits monoid and poset of faces. The partial order on faces is completely determined by the Tits product: For any faces H and F,

$$HF = F \iff H \leq F, \tag{1.9a}$$

$$H\overline{F} = F \iff H = F. \tag{1.9b}$$

This follows from (1.3), (1.4) and (1.5). In particular, for any face F,

$$F\overline{F} = F. \tag{1.10}$$

Some further interactions between the product and the partial order are summarized below.

Lemma 1.6. *The following properties hold for any faces F, G, H and K.*
(1) *If $G \leq H$, then $FG \leq FH$. In particular, $F \leq FG$.*
(2) *If $FG = K$ and $F \leq H \leq K$, then $HG = K$.*
(3) *$FG \wedge F\overline{G} = F$. In particular, $G \wedge \overline{G} = O$.*
(4) *$F \wedge G = F\overline{G} \wedge G = F\overline{G} \wedge G\overline{F}$.*
(5) *$F \leq H$ and $G \leq H$ imply $FG \leq H$.*

PROOF. For (1), we can employ (1.9a). We are given $G \leq H$, that is, $GH = H$. Then $(FG)(FH) = FGH = FH$, that is, $FG \leq FH$. We used associativity and (1.6). Alternatively, one can also directly use (1.4) and (1.5).

The remaining assertions can be verified in a similar manner. □

Exercise 1.7. Use Lemma 1.6, item (3) to deduce the result of Exercise 1.2.

Exercise 1.8. Show that $FG = \bigwedge FC$, with the meet taken over over all chambers C which are greater than G. What happens when $F = G$?

Exercise 1.9. Show that in any arrangement of rank at least one, the assertion $G \leq H \implies GF \leq HF$ is false in general.

1.4.5. Birkhoff monoid. Recall the lattice of flats $\Pi[\mathcal{A}]$. It has the structure of a monoid: the product of X and Y is defined to be $\mathrm{X} \vee \mathrm{Y}$. The unit element is the minimum flat $\perp$. We call this the *Birkhoff monoid*. It is commutative.

The motivation for using the join (as opposed to the meet) for defining the product is explained below.

1.4.6. Support map. Recall the support map from faces to flats defined in (1.2). For any faces F and G,

$$\mathrm{s}(FG) = \mathrm{s}(F) \vee \mathrm{s}(G). \tag{1.11}$$

This follows from (1.8). Further, $\mathrm{s}(O) = \perp$. Thus, the support map is a homomorphism from the Tits monoid to the Birkhoff monoid.

Observe that FG and GF always have the same support. In particular, if $GF = G$, then FG and G have the same support. Similar useful observations are given below.

$$GF = G \iff \mathrm{s}(F) \leq \mathrm{s}(G). \tag{1.12}$$

Either of these conditions is equivalent to the condition

$$\epsilon_i(G) = 0 \text{ implies } \epsilon_i(F) = 0 \text{ for all } i \in I.$$

It follows that

$$FG = F \text{ and } GF = G \iff \mathrm{s}(F) = \mathrm{s}(G). \tag{1.13}$$

Either of these conditions is equivalent to the condition

$$\epsilon_i(F) = 0 \text{ iff } \epsilon_i(G) = 0 \text{ for all } i \in I.$$

To summarize: The relation

$$F \sim G \iff FG = F \text{ and } GF = G \tag{1.14}$$

is an equivalence relation on the set of faces whose equivalence classes correspond to flats. In fact, note that the equivalence classes are precisely combinatorial flats (Section 1.3.3).

Lemma 1.10. *The Birkhoff monoid is the abelianization of the Tits monoid with the support map being the abelianization map.*

Proof. To construct a commutative quotient of the Tits monoid, at the very least, we need to identify FG and GF. When F and G have the same support, by (1.13), $FG = F$ and $GF = G$, so we must identify F and G. Since these forced identifications yield the Birkhoff monoid, it is the largest commutative quotient. □

Exercise 1.11. Show that $FGF' = FG$ whenever $\mathrm{s}(F) = \mathrm{s}(F')$.

Exercise 1.12. Show that

$$HF = G \text{ and } FH = F \iff HF = G \text{ and } \mathrm{s}(F) = \mathrm{s}(G).$$

Exercise 1.13. Suppose F and G have the same support. Show that: If $F \leq K$, then GK is a face greater than G of the same support as K. In contrast, if $K \leq F$, then there may *not* exist a face $H \leq G$ of the same support as K.

Exercise 1.14. A hyperplane contains F iff it contains $\overline{F}$. Thus, F and $\overline{F}$ have the same support. Deduce this fact as a formal consequence of (1.13) and (1.10). Further deduce that FG and $F\overline{G}$ have the same support for any F and G.

Exercise 1.15. Suppose $\mathrm{X} \leq \mathrm{Y}$ and F is a face with support X. Show that there exists a face G with support Y such that $F \leq G$.

1.4.7. Bands. Bands are reviewed in Section E.1. By the first identity in (1.6), every element of the Tits monoid is idempotent, so it is a band. Further, the second identity in (1.6) says that it is a left regular band. Also, we see from (1.9a) that the partial order (E.2) coincides with the partial order on faces.

Every band has a support lattice. The support lattice of the Tits monoid is precisely the lattice of flats. Compare (1.13) with (E.4). Also note that the Birkhoff monoid arises from the lattice of flats via the construction of Example E.2.

1.4.8. Closures of combinatorial flats. Closures of combinatorial flats can be characterized using the Tits product as follows.

Proposition 1.16. *Let* A *be any set of faces of* $\mathcal{A}$. *Then* A *is the closure of a combinatorial flat iff the following properties hold.*

(1) $O \in \mathrm{A}$.
(2) *If* $G \in \mathrm{A}$ *and* $GF = G$, *then* $F \in \mathrm{A}$.
(3) *If* $F \in \mathrm{A}$ *and* $G \in \mathrm{A}$, *then* $FG \in \mathrm{A}$.

PROOF. Forward implication: Suppose A = Cl(X) for a combinatorial flat X. It consists of all faces with support smaller than X. Property (1) holds since $s(O) = \bot$, the minimum flat. Property (2) follows from (1.12) and (3) from (1.11).

Backward implication: By property (1), A is nonempty. By property (3), the largest faces in A are obtained by multiplying all the faces in A in different orders. By (1.11), they are all of the same support, say X. We claim that A = Cl(X). It is clear that A $\subseteq$ Cl(X). For the reverse inclusion, if F has support smaller than X, then pick a $G \in$ A with support X and use property (2) to conclude that $F \in$ A. $\square$

1.4.9. Join of faces. We say that two faces F and G are *joinable* if their join exists in $\Sigma[\mathcal{A}]$, or equivalently, if there is a face greater than both F and G.

Proposition 1.17. *Distinct subfaces of a face have distinct supports.*

PROOF. Suppose F and G are distinct subfaces of a face K. Then by (1.4), $\epsilon_i(F) = \epsilon_i(K)$ whenever $\epsilon_i(F) \neq 0$, and $\epsilon_i(G) = \epsilon_i(K)$ whenever $\epsilon_i(G) \neq 0$. Thus, the set of hyperplanes H_i for which $\epsilon_i(F) = 0$ differs from the one for which $\epsilon_i(G) = 0$. So by the sign sequence condition given after (1.13), the supports of F and G are distinct. $\square$

Proposition 1.18. *Two faces F and G are joinable iff $FG = GF$. In this situation,*

$$F \vee G = FG = GF.$$

In particular, faces with the same support are joinable iff they are equal.

PROOF. Any face which contains F and G must contain FG and GF by Lemma 1.6, item (5). Since FG and GF have the same support, the only way a face can be greater than both of them is if $FG = GF$. All claims follow. $\square$

In the figure, the vertices F and G are joinable, the edge connecting them is their join. On the other hand, the vertices F and H are not joinable since the edges FH and HF are distinct.

Proposition 1.18 suggests that one may view $\Sigma[\mathcal{A}]$ as a "noncommutative lattice" in which the role of the join is played by the Tits product (which is noncommutative in general). This suggestion is further substantiated by (1.11).

Proposition 1.19. *For any faces F and G, the faces FG and $F\overline{G}$ are joinable iff $FG = F\overline{G} = F$. In particular, G and $\overline{G}$ are joinable iff $G = \overline{G} = O$.*

PROOF. By Exercise 1.14, FG and $F\overline{G}$ have the same support. Hence, by Proposition 1.18, they are joinable iff they are equal. In this case, by Lemma 1.6, item (3), both must also equal F. $\square$

Exercise 1.20. Show that: Any face is the Tits product of all its vertices (with the product taken in any order).

Exercise 1.21. Show that: Suppose $G \leq K$. Then F and G are joinable with $F \vee G = K$ iff $FG = K$.

Exercise 1.22. Show that: If $H_1F = G$ and $H_2F = G$, then H_1 and H_2 are joinable and $(H_1 \vee H_2)F = G$.

Exercise 1.23. Show that: $FG = \overline{GF} \iff F = \overline{G}$.

1.4.10. Subarrangements. Let $\mathcal{A}'$ be a subarrangement of $\mathcal{A}$, that is, the set of hyperplanes in $\mathcal{A}'$ is a subset of the set of hyperplanes in $\mathcal{A}$. There is a morphism of monoids

$$\Sigma[\mathcal{A}] \to \Sigma[\mathcal{A}'] \tag{1.15}$$

which sends a face F of $\mathcal{A}$ to the unique face of $\mathcal{A}'$ whose interior contains the interior of F. In terms of sign sequences, this map simply restricts the sign sequence of F to hyperplanes in $\mathcal{A}'$.

The map (1.15) has the property that if F and F' have the same support, then so do their images. Hence, it factors through the support maps to yield a morphism of monoids

$$\Pi[\mathcal{A}] \to \Pi[\mathcal{A}']. \tag{1.16}$$

This map sends a flat X of $\mathcal{A}$ to the smallest flat of $\mathcal{A}'$ which contains X. The latter is the intersection of all hyperplanes in $\mathcal{A}'$ which contain X.

1.5. Bi-faces and Janus monoid

We introduce the Janus monoid indexed by bi-faces. It is the fiber product of the Tits monoid and its opposite over the Birkhoff monoid. The Tits monoid is a left regular band and hence its opposite is a right regular band. However, since the Janus monoid involves both of them, it is a band which is neither left regular nor right regular in general. Its support lattice is the lattice of flats.

1.5.1. Janus monoid. A *bi-face* is a pair (F, F') of faces such that F and F' have the same support. Let $\mathrm{J}[\mathcal{A}]$ denote the set of bi-faces. The binary operation

$$(F, F')(G, G') := (FG, G'F') \tag{1.17}$$

turns $\mathrm{J}[\mathcal{A}]$ into a monoid. The unit element is (O, O). We call it the *Janus monoid*. Since each element is an idempotent, it is a band. However, it is neither left regular, nor right regular.

Observe that the Janus monoid $\mathrm{J}[\mathcal{A}]$ is the fiber product of the Tits monoid $\Sigma[\mathcal{A}]$ and its opposite $\Sigma[\mathcal{A}]^{\mathrm{op}}$ over the Birkhoff monoid $\Pi[\mathcal{A}]$. We express this by

$$\mathrm{J}[\mathcal{A}] = \Sigma[\mathcal{A}] \times_{\Pi[\mathcal{A}]} \Sigma[\mathcal{A}]^{\mathrm{op}}.$$

In particular, we have a commutative diagram of monoids

$$\begin{array}{ccc} \mathrm{J}[\mathcal{A}] & \longrightarrow & \Sigma[\mathcal{A}]^{\mathrm{op}} \\ \downarrow & & \downarrow{\scriptstyle \mathrm{s}} \\ \Sigma[\mathcal{A}] & \xrightarrow[\mathrm{s}]{} & \Pi[\mathcal{A}] \end{array} \tag{1.18}$$

with s being the support map, and the maps from $\mathrm{J}[\mathcal{A}]$ being the projections on the two coordinates respectively.

We also deduce that the Janus monoid is canonically isomorphic to its opposite monoid via

$$\mathrm{J}[\mathcal{A}] \to \mathrm{J}[\mathcal{A}]^{\mathrm{op}}, \qquad (F, F') \mapsto (F', F).$$

This can be taken as one motivation for reversing the order of the product in the second coordinate in definition (1.17).

Exercise 1.24. Check that $\Sigma[\mathcal{A}]$ and $\Sigma[\mathcal{A}]^{\mathrm{op}}$ cannot be isomorphic if $\mathcal{A}$ has rank at least one.

Exercise 1.25. Check that the abelianization of the Janus monoid is the Birkhoff monoid.

Exercise 1.26. Check that the map (1.15) induces a map

$$\mathrm{J}[\mathcal{A}] \to \mathrm{J}[\mathcal{A}'] \tag{1.19}$$

which is a morphism of monoids.

1.5.2. Support lattice. Recall that every band has a partial order as well as a support lattice (Section E.1). For the Janus monoid, this works as follows.

Lemma 1.27. *For the Janus monoid, the partial order* (E.1) *is given by*

$$(F, F') \le (G, G') \iff F \le G \text{ and } F' \le G'.$$

Its support lattice is the lattice of flats, with the support map being the composite map in diagram (1.18).

PROOF. Employing (E.1) both for the Janus monoid and the Tits monoid, we obtain:

$$\begin{aligned}(F, F') \le (G, G') &\iff (FG, G'F') = (G, G') = (GF, F'G') \\ &\iff FG = G = GF \text{ and } G'F' = G' = F'G' \\ &\iff F \le G \text{ and } F' \le G'.\end{aligned}$$

Similarly, employing (E.4) both for the Janus monoid and the Tits monoid, we can deduce that two bi-faces (F, F') and (G, G') have the same support iff F, F', G and G' all have the same support. So the support lattice of the Janus monoid is the lattice of flats. □

1.5.3. Presentation.

Lemma 1.28. *The Janus monoid has a presentation given by generators e_F, one for each face F, subject to the relations*

$$e_O = \mathrm{id}, \quad e_F e_F = e_F, \quad e_G e_F = e_{GF} e_{FG},$$

for all F and G, and

$$e_H e_G e_F = e_H e_F$$

whenever F, G and H have the same support.

PROOF. Let us work with the presentation. For any faces F, G and H,

$$e_H e_G e_F = e_H e_{GF} e_{FG} = e_{HGF} e_{GFH} e_{FG} = e_{HGF} e_{GFH} e_{FGH} = e_{HGF} e_{FGH}.$$

We repeatedly used the third relation till all faces subscripting e had the same support, and then used the last relation to remove the intermediate e. More generally, by the same method, for any faces $F_1, \dots, F_k$,

$$e_{F_1} \dots e_{F_k} = e_{F_1 \dots F_k} e_{F_k \dots F_1}.$$

(The special case $k = 1$ is covered by the second relation.) Thus, any word in the generators, say w, can be written in the form $e_F e_{F'}$, with F and F' of the same support. Further, F and F' are unique, namely, F is the product of the faces in w in the order in which they appear, while F' is the product of the faces in w in reverse order. (This is because in any relation $u = v$, the two products for u respectively equal the two products for v. For instance, in the relation $e_G e_F = e_{GF} e_{FG}$, the products on both sides are GF and FG respectively. Hence, whenever we use a relation to replace a word by another, the products remain the same.) Thus, we can view $e_F e_{F'}$ as a normal form for w. It remains to see how words in normal form multiply. Observe that

$$(e_F e_{F'})(e_G e_{G'}) = e_{FG} e_{G'F'}.$$

Now identify $e_F e_{F'}$ with the bi-face (F, F'). □

Exercise 1.29. Deduce the relations $e_G e_F = e_G$ when $F \leq G$, and $e_F e_G e_F = e_{FG}$ for any F and G.

1.5.4. Rank one. Consider the set $\{\mathrm{id}, e, f, ef, fe\}$ with following monoid structure. All elements are idempotent, and they are multiplied using the relations $efe = e$ and $fef = f$. This is precisely the Janus monoid of the rank-one arrangement (with chambers C and $\overline{C}$). The identification is done as follows.

$$\mathrm{id} \leftrightarrow (O, O),\ e \leftrightarrow (C, C),\ f \leftrightarrow (\overline{C}, \overline{C}),\ ef \leftrightarrow (C, \overline{C}),\ fe \leftrightarrow (\overline{C}, C).$$

In this case, the Tits monoid is given by the set $\{\mathrm{id}, e, f\}$ of idempotents with $ef = e$ and $fe = f$. The opposite Tits monoid is the same set but with $ef = f$ and $fe = e$. The Birkhoff monoid consists of two idempotent elements, namely, id and $e = f$.

1.6. Order-theoretic properties of faces and flats

We now discuss some order-theoretic properties of the poset of faces and the lattice of flats. Semimodular and geometric lattices are reviewed in Section B.3. Strongly connected posets are reviewed in Section B.4.

1.6.1. Semimodularity. For flats Y and Y', let $\mathrm{Y} + \mathrm{Y}'$ denote their linear span. Note very carefully that $\mathrm{Y} + \mathrm{Y}'$ may not be a flat. For example, Y and Y' could be two lines in a three-dimensional essential arrangement, but the plane determined by Y and Y' may not be a hyperplane of the arrangement. Thus, in general, $\mathrm{Y} + \mathrm{Y}'$ is only a subspace of $\mathrm{Y} \vee \mathrm{Y}'$. For a concrete example, take Y and Y' to be the one-dimensional subspaces, respectively, passing through two non-adjacent vertices of a square in the nonsimplicial arrangement of Section 1.2.4. It follows that:

Lemma 1.30. *The lattice of flats $\Pi[\mathcal{A}]$ is lower semimodular, that is, for any flats Y and Y',*

$$\mathrm{rk}(\mathrm{Y}') + \mathrm{rk}(\mathrm{Y}) \leq \mathrm{rk}(\mathrm{Y}' \wedge \mathrm{Y}) + \mathrm{rk}(\mathrm{Y}' \vee \mathrm{Y}). \tag{1.20}$$

Equality holds iff $\mathrm{Y} + \mathrm{Y}' = \mathrm{Y} \vee \mathrm{Y}'$.

In addition, the poset opposite to $\Pi[\mathcal{A}]$ is atomic (that is, every flat can be expressed as the meet of corank-one flats (hyperplanes)), and hence is a geometric lattice.

Note also that $\Pi[\mathcal{A}]$ contains no 3-element intervals consistent with Proposition B.3.

1.6.2. Modular complements. We say that flats Y and Y′ are *modular complements* if

$$\mathrm{Y}' \wedge \mathrm{Y} = \bot, \quad \mathrm{Y}' \vee \mathrm{Y} = \top \quad \text{and} \quad \mathrm{rk}(\mathrm{Y}') + \mathrm{rk}(\mathrm{Y}) = \mathrm{rk}(\mathcal{A}).$$

Equivalently, the last two conditions can be replaced by the condition $\mathrm{Y}' + \mathrm{Y} = \top$. In this situation, we say that Y′ is a modular complement of Y. More generally, we say that Y and Y′ are *modular complements* in the interval [X, Z] if

$$\mathrm{Y}' \wedge \mathrm{Y} = \mathrm{X}, \quad \mathrm{Y}' \vee \mathrm{Y} = \mathrm{Z} \quad \text{and} \quad \mathrm{rk}(\mathrm{Y}') + \mathrm{rk}(\mathrm{Y}) = \mathrm{rk}(\mathrm{X}) + \mathrm{rk}(\mathrm{Z}).$$

Lemma 1.31. *Every flat has a modular complement. More precisely:*

- *The minimum flat $\bot$ has a modular complement, namely, the maximum flat $\top$.*
- *If* X *is a flat and* H *a hyperplane not containing* X, *then* X *has a modular complement contained in* H.

PROOF. The first case is clear. For the second case, we induct on the rank of $\mathcal{A}$. Suppose that X is a flat and H a hyperplane not containing X. Using (1.20), we obtain $\mathrm{rk}(\mathrm{X}) \leq \mathrm{rk}(\mathrm{X} \wedge \mathrm{H}) + 1$. In fact, equality must hold since $\mathrm{X} \wedge \mathrm{H} < \mathrm{X}$. Thus, $\mathrm{X} \wedge \mathrm{H}$ is of codimension one in X. By induction hypothesis, $\mathrm{X} \wedge \mathrm{H}$ has a modular complement, say Y, in the arrangement $\mathcal{A}^{\mathrm{H}}$ under H. (Arrangements under flats are defined in Section 1.7.1.) Then Y is a modular complement of X in $\mathcal{A}$ which is contained in H. □

Proposition 1.32. *Suppose* X *and* Y *are modular complements of each other. Then the following are equivalent.*

1. *Every hyperplane contains either* X *or* Y.
2. Y *is the unique modular complement of* X.
3. X *is the unique modular complement of* Y.

PROOF. It suffices to show that (1) and (2) are equivalent.

(1) $\implies$ (2). Suppose Z is a modular complement of X. Then Z is obtained by intersecting a set of hyperplanes none of which contain X. By hypothesis, all these hyerplanes contain Y implying $\mathrm{Z} \geq \mathrm{Y}$. Hence $\mathrm{Z} = \mathrm{Y}$ (since the two flats have the same rank).

(2) $\implies$ (1). Suppose H is a hyperplane which does not contain X. Then by the second part of Lemma 1.31, X has a modular complement contained in H, which by uniqueness must be Y. Hence H contains Y. □

1.6.3. Strong connectivity. We now show that the poset of faces and the lattice of flats are strongly connected.

Lemma 1.33. *For any arrangement $\mathcal{A}$, the poset of faces $\Sigma[\mathcal{A}]$ with a top element adjoined is the face lattice of a convex polytope.*

PROOF. See [**82**, Example 4.1.7], [**448**, Corollary 7.18] or [**105**, Appendix]. □

Lemma 1.34. *Let $\mathcal{A}$ be any arrangement. The poset of faces $\Sigma[\mathcal{A}]$ with a top element adjoined is strongly connected. In particular, $\Sigma[\mathcal{A}]$ is also strongly connected.*

PROOF. This follows from Lemma 1.33 and Lemma B.7. □

When $\mathcal{A}$ is simplicial, any interval in the face poset of $\mathcal{A}$ is Boolean, which is both lower semimodular and upper semimodular. So in this case, one may also use Lemma B.8 to deduce that $\Sigma[\mathcal{A}]$ is strongly connected.

Lemma 1.35. *For any arrangement $\mathcal{A}$, the lattice of flats $\Pi[\mathcal{A}]$ is strongly connected.*

PROOF. This follows from Lemma 1.30 and Lemma B.8. □

1.7. Arrangements under and over a flat

A flat of an arrangement gives rise to two smaller arrangements. The new arrangements single out the portions of the old arrangement under and over the given flat. We discuss these constructions and their combination, which focuses on the portion comprised between two flats. We also discuss the notion of star and top-star of a face. Faces of the arrangement over a flat X correspond to the star of any face of support X. Faces of the arrangement under X correspond to faces contained in X.

1.7.1. Under a flat. Let X be any fixed flat of $\mathcal{A}$. The *arrangement under* X is

$$\mathcal{A}^{\mathrm{X}} = \{\mathrm{H} \cap \mathrm{X} \mid \mathrm{H} \in \mathcal{A},\ \mathrm{X} \not\subseteq \mathrm{H}\}.$$

It is a hyperplane arrangement with ambient space X. Its hyperplanes are obtained by intersecting X with hyperplanes in $\mathcal{A}$ not containing it. Its center is the same as that of $\mathcal{A}$. An example follows.

The arrangement under a rank-two flat in a rank-three arrangement is illustrated below. It has eight vertices, four of which are visible in the picture.

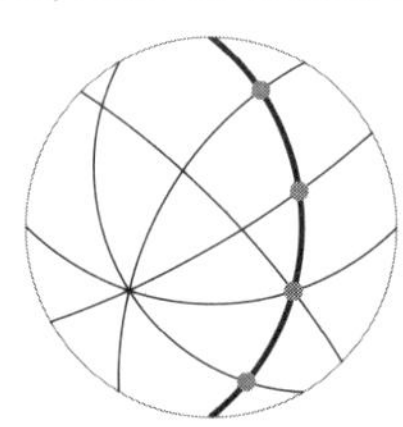

The arrangement $\mathcal{A}^{\mathrm{X}}$ singles out the portion of $\mathcal{A}$ below X. Faces, chambers and flats of $\mathcal{A}^{\mathrm{X}}$ are as follows.

$$\Sigma[\mathcal{A}^{\mathrm{X}}] = \{F \in \Sigma[\mathcal{A}] \mid \mathrm{s}(F) \leq \mathrm{X}\},$$
$$\Gamma[\mathcal{A}^{\mathrm{X}}] = \{F \in \Sigma[\mathcal{A}] \mid \mathrm{s}(F) = \mathrm{X}\},$$
$$\Pi[\mathcal{A}^{\mathrm{X}}] = \{\mathrm{Y} \in \Pi[\mathcal{A}] \mid \mathrm{Y} \leq \mathrm{X}\}.$$

For any face K, let $\mathcal{A}^K := \mathcal{A}^{\mathrm{s}(K)}$.

1.7.2. Over a flat. Let X be any fixed flat of $\mathcal{A}$. The *arrangement over* X is

$$\mathcal{A}_{\mathrm{X}} = \{\mathrm{H} \in \mathcal{A} \mid \mathrm{X} \subseteq \mathrm{H}\}.$$

It is a hyperplane arrangement with the same ambient space as $\mathcal{A}$. It consists of the hyperplanes of $\mathcal{A}$ which contain X. The center is X. An example follows.

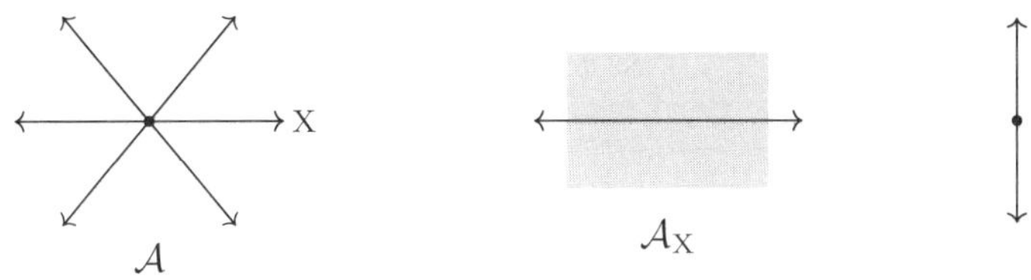

The essentialization of $\mathcal{A}_X$ (up to gisomorphism) is shown on the far right.

The arrangement over a rank-one flat of a rank-three arrangement is illustrated below. It consists of the three thick lines.

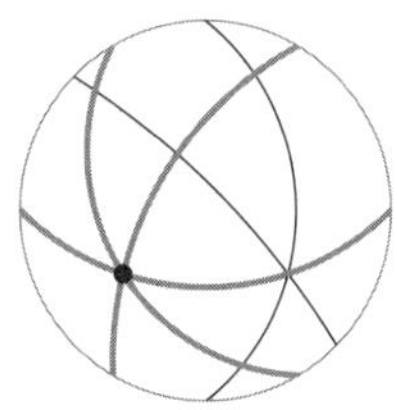

Roughly, the arrangement $\mathcal{A}_X$ singles out the portion of $\mathcal{A}$ above X. Flats of $\mathcal{A}_X$ are flats of $\mathcal{A}$ which contain X. Faces and chambers are in canonical correspondence with faces and chambers of $\mathcal{A}$ that contain any fixed face F of support X.

$$\Sigma[\mathcal{A}_X] \cong \{G \in \Sigma[\mathcal{A}] \mid F \leq G\},$$
$$\Gamma[\mathcal{A}_X] \cong \{C \in \Gamma[\mathcal{A}] \mid F \leq C\},$$
$$\Pi[\mathcal{A}_X] = \{Y \in \Pi[\mathcal{A}] \mid X \leq Y\}.$$

We elaborate on this below.

1.7.3. Stars and top-stars. For a face F, let $\Sigma[\mathcal{A}]_F$ denote the set of faces of $\mathcal{A}$ which are greater than F. This is the *star* of F. For clarity, we denote elements of $\Sigma[\mathcal{A}]_F$ by K/F, where K is a face greater than F. The star of a chamber is a singleton consisting of the chamber itself, while the star of the central face is the set of all faces. Let $\Gamma[\mathcal{A}]_F$ denote the set of chambers of $\mathcal{A}$ which are greater than F. This is the *top-star* of F.

The star of F is illustrated above in rank three. In the picture on the left, F is an edge and its top-star consists of two chambers, while in the picture on the right, F is a vertex and its top-star consists of six chambers.

Lemma 1.36. *The star $\Sigma[\mathcal{A}]_F$ is the right ideal of $\Sigma[\mathcal{A}]$ generated by F. It is also a monoid in its own right with unit element F. Further, it is a left regular band. The associated partial order* (E.2) *is the restriction of the partial order on faces.*

Proof. The right ideal generated by F consists of elements of the form FG. Using (1.9a) and Lemma 1.6, item (1), we see that these are precisely faces in the star of F. To see that F is the unit, we need $F(FG) = FG$ and $(FG)F = FG$. This holds by (1.6). Since the product is obtained by restricting the Tits product of $\Sigma[\mathcal{A}]$, the rest follows. $\square$

Lemma 1.37. *When F and G have the same support, we have an isomorphism*

$$\Sigma[\mathcal{A}]_F \xrightarrow{\cong} \Sigma[\mathcal{A}]_G, \qquad K/F \mapsto GK/G$$

of monoids, and hence of posets. The inverse is given by

$$\Sigma[\mathcal{A}]_G \xrightarrow{\cong} \Sigma[\mathcal{A}]_F, \qquad H/G \mapsto FH/F$$

Further, it restricts to a bijection

$$\Gamma[\mathcal{A}]_F \xrightarrow{\cong} \Gamma[\mathcal{A}]_G, \qquad C/F \mapsto GC/G.$$

PROOF. Let us denote the first map by f and the second by g. For faces K and K' both greater than F,

$$f(K)f(K') = (GK)(GK') = GKK' = f(KK').$$

The second step used associativity and (1.6). (In the above calculation, we did not write K/F, K'/F and so on to avoid cumbersome notation.) Thus, f is a monoid homomorphism. By symmetry, g is also a monoid homomorphism. For K greater than F,

$$(gf)(K) = g(GK) = FGK = FK = K.$$

We used associativity, (1.9a) and (1.13). Thus, $gf = \mathrm{id}$, and by symmetry, $fg = \mathrm{id}$. Thus, f and g are inverses. It is also clear that they restrict to inverse bijections between the top-stars of F and G. $\square$

Lemma 1.38. *Let* X *be a flat. Then for any face F with support* X*, there is an isomorphism*

$$\Sigma[\mathcal{A}]_F \xrightarrow{\cong} \Sigma[\mathcal{A}_{\mathrm{X}}]$$

of monoids and hence of posets. Further, when F and G both have support X*, the diagram*

$$\begin{array}{ccc} \Sigma[\mathcal{A}]_F & \longrightarrow & \Sigma[\mathcal{A}]_G \\ & \searrow \quad \swarrow & \\ & \Sigma[\mathcal{A}_{\mathrm{X}}] & \end{array}$$

commutes.

PROOF. We make use of some ideas from Chapter 3. The element K/F is the same as a nested face (F, K). It gives rise to the combinatorial lune $\mathrm{s}(F, K)$ with base X. The latter is the same as a face of $\mathcal{A}_{\mathrm{X}}$ (Lemma 3.2). This explains the map from the star of F to the set of faces of $\mathcal{A}_{\mathrm{X}}$. It is a bijection by (3.15). It is straightforward to check that it is also a monoid homomorphism. The commutativity of the above diagram says $\mathrm{s}(F, K) = \mathrm{s}(G, GK)$. This is contained in Proposition 3.13 (since $(F, K) \sim (G, GK)$). $\square$

For any face F, let $\mathcal{A}_F := \mathcal{A}_{\mathrm{s}(F)}$. Thus, there is no distinction between $\mathcal{A}_F$ and $\mathcal{A}_G$ when F and G have the same support. However, for book-keeping purposes, we would like to keep them apart. Hence, we identify faces of $\mathcal{A}_F$ with the star of F, and chambers with the top-star of F. Thus, K/F and C/F denote a face and chamber of $\mathcal{A}_F$. In this notation, the opposite of a face K/F of $\mathcal{A}_F$ is the face $F\overline{K}/F$. Also note that $\mathrm{rk}(K/F) = \mathrm{rk}(K) - \mathrm{rk}(F)$.

Exercise 1.39. Check that: For any faces F and G,

$$F \leq G \iff \Gamma[\mathcal{A}]_G \subseteq \Gamma[\mathcal{A}]_F.$$

More generally,

$$\Gamma[\mathcal{A}]_F \cap \Gamma[\mathcal{A}]_G = \begin{cases} \Gamma[\mathcal{A}]_{FG} = \Gamma[\mathcal{A}]_{GF} & \text{if } F \text{ and } G \text{ are joinable,} \\ \emptyset & \text{otherwise.} \end{cases}$$

In particular, distinct faces with the same support have disjoint top-stars (since they are not joinable by Proposition 1.18).

All of the above results hold with $\Gamma[\mathcal{A}]$ replaced by $\Sigma[\mathcal{A}]$.

Exercise 1.40. For any face F, describe the left ideal generated by F in the Tits monoid $\Sigma[\mathcal{A}]$. Show that it is two-sided, and in particular, contains the star of F. Say explicitly what happens when F is the central face and when F is a chamber. Check that faces F and G generate the same left ideal iff $\mathrm{s}(F) = \mathrm{s}(G)$.

Exercise 1.41. Describe the left, right and two-sided ideals generated by a bi-face in the Janus monoid. Check that: For bi-faces (F, F') and (G, G'), the right ideals they generate coincide iff $F = G$, the left ideals they generate coincide iff $F' = G'$, and the two-sided ideals they generate coincide iff they have the same support.

1.7.4. Between flats. The over and under constructions can be combined. Let X be a flat contained in another flat Y, that is, $\mathrm{X} \leq \mathrm{Y}$. Then X corresponds to a flat of $\mathcal{A}^{\mathrm{Y}}$, and Y to a flat of $\mathcal{A}_{\mathrm{X}}$. Thus, one may first consider the arrangement under Y and within it the arrangement over X, or the other way around. The resulting arrangements $(\mathcal{A}^{\mathrm{Y}})_{\mathrm{X}}$ and $(\mathcal{A}_{\mathrm{X}})^{\mathrm{Y}}$ are the same. We use $\mathcal{A}_{\mathrm{X}}^{\mathrm{Y}}$ to denote this arrangement. Flats of $\mathcal{A}_{\mathrm{X}}^{\mathrm{Y}}$ correspond to flats of $\mathcal{A}$ which lie between X and Y. In other words,

$$\Pi[\mathcal{A}_{\mathrm{X}}^{\mathrm{Y}}] = [\mathrm{X}, \mathrm{Y}].$$

We will also use the notations $\mathcal{A}_F^{\mathrm{X}}$ (when support of F is smaller than X) and $\mathcal{A}_F^G$ (when $F \leq G$). We identify their faces with faces of $\mathcal{A}$ which are greater than F and of support smaller than X.

Exercise 1.42. Put $r = \mathrm{rk}(\mathcal{A})$. Let $\bot = \mathrm{X}_0 \lessdot \mathrm{X}_1 \lessdot \cdots \lessdot \mathrm{X}_r = \top$ be a maximal chain of flats. Show that: There exists a maximal chain of faces $O = F_0 \lessdot F_1 \lessdot \cdots \lessdot F_r$ such that $\mathrm{s}(F_i) = \mathrm{X}_i$ for each i. (See Exercise 1.15.) Further, there are exactly 2^r maximal chains of faces with this property.

1.8. Cartesian product of arrangements

We review cartesian product of arrangements. An arrangement is prime if it cannot be expressed as a cartesian product of arrangements of strictly smaller rank. Every arrangement can be uniquely expressed as a cartesian product of prime arrangements. Modular complements play an important role in this discussion.

1.8.1. Cartesian product. Given two arrangements $\mathcal{A}$ and $\mathcal{A}'$, one can form their *cartesian product* $\mathcal{A} \times \mathcal{A}'$. Its ambient space is $V \oplus V'$, where V and V' are the ambient spaces of $\mathcal{A}$ and $\mathcal{A}'$. Its hyperplanes are codimension-one subspaces of the form $\mathrm{H} \oplus V'$ and $V \oplus \mathrm{H}'$, where H and H' are hyperplanes of $\mathcal{A}$ and $\mathcal{A}'$. Observe that

$$\mathrm{rk}(\mathcal{A} \times \mathcal{A}') = \mathrm{rk}(\mathcal{A}) + \mathrm{rk}(\mathcal{A}').$$

The cartesian product of the rank-one arrangement and the rank-two arrangement of 4 lines is shown below.

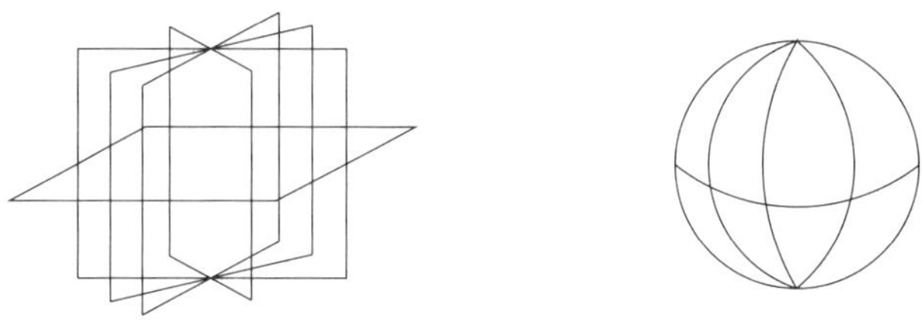

The operation of taking cartesian product is associative and commutative (up to gisomorphism). The essential arrangement of rank 0 serves as the unit. In general, taking cartesian product with a rank-zero arrangement has the effect of fattening up the center.

1.8.2. Faces and flats. A face of $\mathcal{A} \times \mathcal{A}'$ is the same as a pair (F, F'), where F is a face of $\mathcal{A}$ and F' is a face of $\mathcal{A}'$. In other words,

$$\Sigma[\mathcal{A} \times \mathcal{A}'] = \Sigma[\mathcal{A}] \times \Sigma[\mathcal{A}'].$$

This identification is an isomorphism of monoids, that is,

$$(F, F')(G, G') = (FG, F'G'). \tag{1.21}$$

One way to see this is to note that the sign sequence of (F, F') can be identified with the sign sequence of F followed by the sign sequence of F'. Either directly or as a formal consequence of (1.9a),

$$(F, F') \leq (G, G') \iff F \leq G \text{ and } F' \leq G'.$$

A chamber of $\mathcal{A} \times \mathcal{A}'$ is the same as a pair (C, C'), where C is a chamber of $\mathcal{A}$ and C' is a chamber of $\mathcal{A}'$. Thus,

$$\Gamma[\mathcal{A} \times \mathcal{A}'] = \Gamma[\mathcal{A}] \times \Gamma[\mathcal{A}'].$$

A flat of $\mathcal{A} \times \mathcal{A}'$ is the same as a pair $(\mathrm{X}, \mathrm{X}')$, where X is a flat of $\mathcal{A}$ and X' is a flat of $\mathcal{A}'$. Thus,

$$\Pi[\mathcal{A} \times \mathcal{A}'] = \Pi[\mathcal{A}] \times \Pi[\mathcal{A}'].$$

This identification is an isomorphism of posets, that is,

$$(\mathrm{X}, \mathrm{X}') \leq (\mathrm{Y}, \mathrm{Y}') \iff \mathrm{X} \leq \mathrm{Y} \text{ and } \mathrm{X}' \leq \mathrm{Y}'. \tag{1.22}$$

Let $\bot'$ and $\top'$ denote the minimum and maximum flats of $\mathcal{A}'$. Under the above identification, a hyperplane of $\mathcal{A} \times \mathcal{A}'$ is either $(\mathrm{H}, \top')$ with H an hyperplane of $\mathcal{A}$ or $(\top, \mathrm{H}')$ with H' an hyperplane of $\mathcal{A}'$. Also note that the flats $(\bot, \top')$ and $(\top, \bot')$ are modular complements.

1.8.3. Under and over a flat of a product. For a flat $(\mathrm{X}, \mathrm{X}')$ of $\mathcal{A} \times \mathcal{A}'$,

$$(\mathcal{A} \times \mathcal{A}')^{(\mathrm{X}, \mathrm{X}')} = \mathcal{A}^{\mathrm{X}} \times (\mathcal{A}')^{\mathrm{X}'} \quad \text{and} \quad (\mathcal{A} \times \mathcal{A}')_{(\mathrm{X}, \mathrm{X}')} = \mathcal{A}_{\mathrm{X}} \times \mathcal{A}'_{\mathrm{X}'}.$$

Specializing to the flat $(\top, \bot')$ of $\mathcal{A} \times \mathcal{A}'$, we get

$$(\mathcal{A} \times \mathcal{A}')^{(\top, \bot')} \cong \mathcal{A} \quad \text{and} \quad (\mathcal{A} \times \mathcal{A}')_{(\top, \bot')} \cong \mathcal{A}'.$$

A similar remark applies to the flat $(\bot, \top')$. Thus, $\mathcal{A}$ and $\mathcal{A}'$ may both be seen as arrangements under and over a flat of $\mathcal{A} \times \mathcal{A}'$, up to cisomorphism.

1.8.4. Factors. Roughly, a factor of an arrangement $\mathcal{A}$ is a flat X for which $\mathcal{A}$ can be expressed as a cartesian product of $\mathcal{A}^{\mathrm{X}}$ and $\mathcal{A}_{\mathrm{X}}$. This notion can be formalized using modular complements, as follows.

A flat is a *factor* of an arrangement if it has a unique modular complement. Suppose X is a factor. By Proposition 1.32, its unique modular complement, say Y, is also a factor. We say that Y is the complementary factor of X. Since X is also the complementary factor of Y, we say that X and Y are *complementary factors* of each other. Note that $\perp$ and $\top$ are complementary factors. We call these the *trivial factors*.

Proposition 1.43. *Let* X *and* Y *be complementary factors of an essential arrangement* $\mathcal{A}$. *Then there is a canonical gisomorphism*

$$\mathcal{A} = \mathcal{A}^{\mathrm{X}} \times \mathcal{A}^{\mathrm{Y}}. \tag{1.23}$$

Similarly, there is a canonical gisomorphism

$$\mathcal{A}_{\mathrm{Y}} = \mathcal{A}^{\mathrm{X}} \times \mathrm{Y}^0, \tag{1.24}$$

where Y^0 *denotes the rank-zero arrangement with ambient space* Y. *In particular,* $\mathcal{A}^{\mathrm{X}}$ *is the essentialization of* $\mathcal{A}_{\mathrm{Y}}$.

PROOF. This can be deduced from Proposition 1.32, condition (1). □

The isomorphism (1.24) is illustrated below in a rank-three arrangement with 4 hyperplanes.

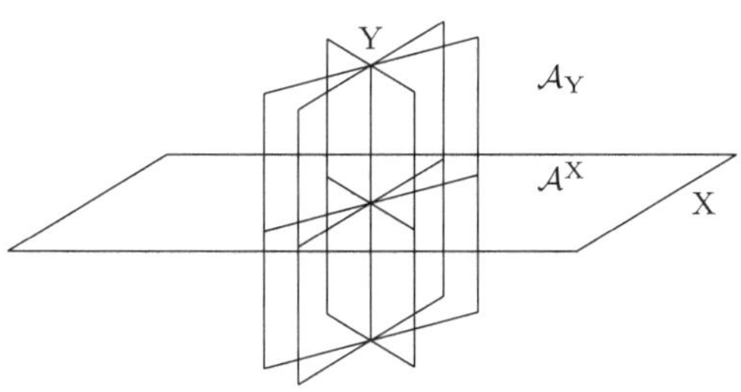

The flat Y is the vertical line, and $\mathcal{A}_{\mathrm{Y}}$ consists of three planes passing through Y. The flat X is the horizontal plane, and $\mathcal{A}^{\mathrm{X}}$ consists of three lines passing through the origin.

Proposition 1.44. *For complementary factors* X *and* Y *in any arrangement* $\mathcal{A}$ *(not necessarily essential), there are cisomorphisms*

$$\mathcal{A} \cong \mathcal{A}^{\mathrm{X}} \times \mathcal{A}^{\mathrm{Y}} \quad \textit{and} \quad \mathcal{A}_{\mathrm{Y}} \cong \mathcal{A}^{\mathrm{X}}. \tag{1.25}$$

In particular, for any factor X, *there is a cisomorphism*

$$\mathcal{A} \xrightarrow{\cong} \mathcal{A}^{\mathrm{X}} \times \mathcal{A}_{\mathrm{X}}. \tag{1.26}$$

Under this identification, X *maps to* (X, X).

The difference with the previous result is that here we not assuming the arrangement to be essential. So we only get cisomorphisms instead of gisomorphisms.

Exercise 1.45. Let H be a hyperplane in $\mathcal{A}$. Show that H is a factor of $\mathcal{A}$ iff H does not contain the intersection of the remaining hyperplanes. In this situation, there are cisomorphisms

$$\mathcal{A} \setminus \{\mathrm{H}\} \cong \mathcal{A}^{\mathrm{H}} \quad \text{and} \quad \mathcal{A} \cong \mathcal{A}^{\mathrm{H}} \times \mathcal{A}_1,$$

where $\mathcal{A}_1$ is the rank-one arrangement.

Lemma 1.46. *Let $\mathcal{A}$ and $\mathcal{A}'$ be two arrangements. Then $(\mathrm{X}, \mathrm{X}')$ is a factor of $\mathcal{A} \times \mathcal{A}'$ iff X is a factor of $\mathcal{A}$, and X′ is a factor of $\mathcal{A}'$. More precisely, $(\mathrm{X}, \mathrm{X}')$ and $(\mathrm{Y}, \mathrm{Y}')$ are complementary factors of $\mathcal{A} \times \mathcal{A}'$ iff X and Y are complementary factors of $\mathcal{A}$, and X′ and Y′ are complementary factors of $\mathcal{A}'$.*

PROOF. This is a straightforward observation. □

As an easy consequence: The flats $(\top, \bot')$ and $(\bot, \top')$ are complementary factors of $\mathcal{A} \times \mathcal{A}'$. Equivalently, they are the unique modular complements of each other.

Lemma 1.47. *If X and Y are factors, then so are $\mathrm{X} \vee \mathrm{Y}$ and $\mathrm{X} \wedge \mathrm{Y}$.*

PROOF. Let X and Y be two factors of $\mathcal{A}$. Let us denote the map (1.26) by φ_{X}. Put $(\mathrm{Y}_1, \mathrm{Y}_2) := \varphi_{\mathrm{X}}(\mathrm{Y})$. Then by Lemma 1.46, Y_1 is a factor of $\mathcal{A}^{\mathrm{X}}$, and Y_2 is a factor of $\mathcal{A}_{\mathrm{X}}$. Observe that

$$\varphi_{\mathrm{X}}(\mathrm{X} \vee \mathrm{Y}) = (\mathrm{X}, \mathrm{X}) \vee (\mathrm{Y}_1, \mathrm{Y}_2) = (\mathrm{X}, \mathrm{Y}_2).$$

Since X is a (trivial) factor of $\mathcal{A}^{\mathrm{X}}$, and Y_2 is a factor of $\mathcal{A}_{\mathrm{X}}$, again by Lemma 1.46, it follows that $\mathrm{X} \vee \mathrm{Y}$ is a factor of $\mathcal{A}$. Similarly,

$$\varphi_{\mathrm{X}}(\mathrm{X} \wedge \mathrm{Y}) = (\mathrm{X}, \mathrm{X}) \wedge (\mathrm{Y}_1, \mathrm{Y}_2) = (\mathrm{Y}_1, \mathrm{X}).$$

By the same reasoning, $\mathrm{X} \wedge \mathrm{Y}$ is a factor of $\mathcal{A}$. □

Lemma 1.48. *Suppose X is a factor of $\mathcal{A}$. The factors of $\mathcal{A}^{\mathrm{X}}$ correspond to factors of $\mathcal{A}$ which are contained in X. Similarly, the factors of $\mathcal{A}_{\mathrm{X}}$ correspond to factors of $\mathcal{A}$ which contain X.*

Suppose X and Z are factors with $\mathrm{X} \leq \mathrm{Z}$. The factors of $\mathcal{A}^{\mathrm{Z}}_{\mathrm{X}}$ correspond to factors of $\mathcal{A}$ which lie between X and Z.

PROOF. The second part follows by combining the two statements in the first part. For the first part, we proceed as in the previous proof. If Y is a flat contained in X, then $\varphi_{\mathrm{X}}(\mathrm{Y}) = (\mathrm{Y}, \mathrm{X})$. Since X is a trivial factor of $\mathcal{A}_{\mathrm{X}}$, by Lemma 1.46, Y is a factor of $\mathcal{A}$ iff Y is a factor of $\mathcal{A}^{\mathrm{X}}$. This proves the first statement. Similarly, for the second statement, we use that if Y is a flat containing X, then $\varphi_{\mathrm{X}}(\mathrm{Y}) = (\mathrm{X}, \mathrm{Y})$. □

1.8.5. Prime arrangements. A factor is *prime* if it is not the minimum flat and it cannot be written as a join of two distinct nontrivial factors. The convention that the minimum flat is not prime is analogous to the convention that 1 is not a prime number.

Proposition 1.49. *For an arrangement $\mathcal{A}$, the following conditions are equivalent.*

- *The maximum flat is prime.*
- *$\mathcal{A}$ has rank at least one and no nontrivial factors.*
- *$\mathcal{A}$ has rank at least one and is not cisomorphic to a cartesian product of two arrangements both of nonzero rank.*

PROOF. The equivalence of the first two conditions is clear. For the equivalence of the last two conditions, use that: Complementary factors yield a factorization (1.23), and conversely, a cartesian product $\mathcal{A}\times\mathcal{A}'$ has complementary factors $(\top, \bot')$ and $(\bot, \top')$. □

An arrangement is *prime* if any of the above equivalent conditions hold. By convention, an arrangement of rank 0 is *not* prime.

Observe that the notion of factors, prime factors, and hence primeness of an arrangement only depends on its lattice of flats. In particular, an arrangement $\mathcal{A}$ is prime iff the essentialization of $\mathcal{A}$ is prime. Note that any rank-one arrangement is prime.

Lemma 1.50. *Suppose* X *is a factor of* $\mathcal{A}$. *The prime factors of* $\mathcal{A}^{\mathrm{X}}$ *correspond to prime factors of* $\mathcal{A}$ *which are contained in* X.

Suppose X *and* Y *are complementary factors of* $\mathcal{A}$. *The prime factors of* $\mathcal{A}$ *correspond to disjoint unions of prime factors of* $\mathcal{A}^{\mathrm{X}}$ *and prime factors of* $\mathcal{A}^{\mathrm{Y}}$.

PROOF. The first claim follows from the first statement in Lemma 1.48. The second claim can be deduced from the first using (1.23). □

Proposition 1.51. *Let* $\mathrm{X}_1, \dots, \mathrm{X}_k$ *denote the prime factors of an essential arrangement* $\mathcal{A}$. *Then there is a gisomorphism*

$$\mathcal{A} = \bigtimes_{i=1}^{k} \mathcal{A}^{\mathrm{X}_i}. \tag{1.27}$$

For an arbitrary arrangement $\mathcal{A}$ *(not necessarily essential), there is a cisomorphism as in* (1.27).

PROOF. We induct on the rank of $\mathcal{A}$. The result is clear if $\mathcal{A}$ is prime. So suppose X and Y are nontrivial complementary factors of $\mathcal{A}$. Now start with the factorization (1.23) and apply the induction hypothesis to both $\mathcal{A}^{\mathrm{X}}$ and $\mathcal{A}^{\mathrm{Y}}$ to get a factorization of $\mathcal{A}$ into primes. Further, by Lemma 1.50, all prime factors of $\mathcal{A}$ appear in this factorization. Thus, we obtain (1.27). □

We refer to (1.27) as the *prime decomposition* of $\mathcal{A}$. Note that the arrangements $\mathcal{A}^{\mathrm{X}_i}$ that appear in the decomposition are prime.

Corollary 1.52. *The set of factors of* $\mathcal{A}$ *under inclusion is a Boolean poset on the set of prime factors of* $\mathcal{A}$. *Further, it is a sublattice of the lattice of flats.*

PROOF. Consider the prime decomposition (1.27). By Lemma 1.46 (extended to k-fold cartesian products), a factor of $\mathcal{A}$ corresponds to a k-tuple consisting of factors of each $\mathcal{A}^{\mathrm{X}_i}$. But $\mathcal{A}^{\mathrm{X}_i}$ being prime has only the two trivial factors. The first claim follows. The second claim is a restatement of Lemma 1.47. □

Any decomposition of an essential arrangement as a cartesian product necessarily arises from the prime decomposition (1.27) by partitioning the set of prime factors into two parts.

1.8.6. Irreducible arrangements. An arrangement is *irreducible* if it is not gisomorphic to a cartesian product of two arrangements both with nonzero ambient space.

An irreducible arrangement is necessarily essential. Also, an arrangement $\mathcal{A}$ of rank at least one is prime iff the essentialization of $\mathcal{A}$ is irreducible.

The essential rank-zero and rank-one arrangements are irreducible. For $n \geq 2$, the essential rank-two arrangement of n lines is irreducible iff $n > 2$. For $n = 2$, this arrangement is reducible being the cartesian product of two rank-one arrangements.

A rank-three arrangement is irreducible iff it is not gisomorphic to the cartesian product of the rank-one arrangement and the rank-two arrangement of n lines for some $n \geq 2$.

Exercise 1.53. Give an example of an arrangement $\mathcal{A}$ and a flat X where $\mathcal{A}$ is prime but $\mathcal{A}^{\mathrm{X}}$ is not prime. Similarly, give an example where $\mathcal{A}$ is prime but $\mathcal{A}_{\mathrm{X}}$ is not prime.

1.9. Generic hyperplanes and adjoints of arrangements

Every arrangement $\mathcal{A}$ has an adjoint arrangement $\widehat{\mathcal{A}}$. Chambers of $\widehat{\mathcal{A}}$ correspond to hyperplanes which are generic wrt $\mathcal{A}$ (under an appropriate notion of equivalence).

1.9.1. Generic hyperplane. Let $\mathcal{A}$ be any arrangement of rank at least 1 with ambient space V. A *generic hyperplane* wrt $\mathcal{A}$ is a codimension-one subspace of V which contains the central face O but does not contain any vertex of $\mathcal{A}$. A *generic half-space* wrt $\mathcal{A}$ is a half-space of V whose bounding hyperplane is generic wrt $\mathcal{A}$.

Adding a generic hyperplane, say H, to $\mathcal{A}$ yields a new arrangement $\mathcal{A}'$. Let us compare the set of faces of the two arrangements. We say H cuts a face F of $\mathcal{A}$ if there are points of F which lie strictly on both sides of H. A face F of $\mathcal{A}$ which is cut by H splits into three distinct faces of $\mathcal{A}'$: one face consists of those points of F which lie on H, while the remaining two consist of those points of F which lie on either side of H. In contrast, a face of $\mathcal{A}$ which is not cut by H remains a face of $\mathcal{A}'$.

An illustration is given below.

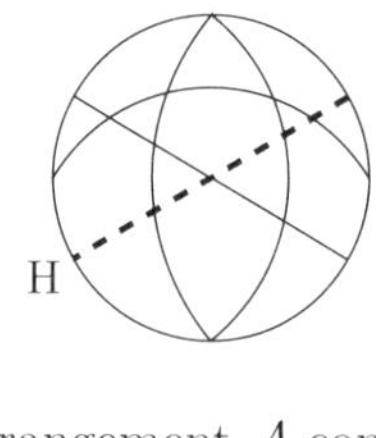

The figure shows a rank-three arrangement $\mathcal{A}$ consisting of five hyperplanes (great circles in this case). Only one half of the arrangement is visible in the picture, the other half being on the backside. The hyperplane H, shown as a dotted line, does not contain any vertex of $\mathcal{A}$, so it is generic. Observe that an edge which is cut by the dotted line gives rise to two edges and one vertex in the new arrangement $\mathcal{A}'$. Similarly, a chamber which is cut by the dotted line gives rise to two chambers and one edge. Edges and chambers not cut by the dotted line remain unchanged.

We say two generic half-spaces are equivalent if they contain the same set of vertices. Recall that a face is the convex hull of its vertices. Hence, a face is contained in a generic half-space h iff all its vertices are contained in h. As a consequence, two generic half-spaces are equivalent iff they contain the same set of faces.

1.9.2. Adjoint of an arrangement. For each subspace X of a vector space V, let

$$\mathrm{X}^{\perp} = \{f \in V^* \mid f(x) = 0 \text{ for all } x \in \mathrm{X}\}$$

be the orthogonal space. It is a subspace of the dual space V^*.

Let $\mathcal{A}$ be an arrangement with ambient space V and center O. For a rank-one flat X of $\mathcal{A}$, $\mathrm{X}^{\perp}$ is a hyperplane in $O^{\perp}$. Letting X run over all rank-one flats of $\mathcal{A}$, we obtain a hyperplane arrangement $\widehat{\mathcal{A}}$ with ambient space $O^{\perp}$. This is called the *adjoint* of $\mathcal{A}$.

The arrangement $\widehat{\mathcal{A}}$ is always essential. Indeed, if $\mathcal{A}$ has rank 0, then $O^{\perp}$ is the zero space. Otherwise, choose a chamber C of $\mathcal{A}$ and consider its vertices. The flats X that support them are of rank 1 and span V (the support of C), hence the corresponding orthogonal spaces $\mathrm{X}^{\perp}$ intersect trivially.

A non-generic hyperplane H wrt $\mathcal{A}$ corresponds to a line $\mathrm{H}^{\perp}$ in V^* contained in one of the hyperplanes $\mathrm{X}^{\perp}$. A generic hyperplane wrt $\mathcal{A}$ therefore corresponds to a line in the complement of the arrangement in V^*. Similarly, a generic half-space h wrt $\mathcal{A}$ corresponds to a ray $\mathrm{h}^{\perp}$ in the complement of the arrangement in V^*. This is defined by

$$\mathrm{h}^{\perp} = \{f \in \mathrm{H}^{\perp} \mid f(v) > 0\},$$

where H is the boundary of h and v is any vector in h.

Observe that:

Lemma 1.54. *Two generic half-spaces* h_1 *and* h_2 *wrt* $\mathcal{A}$ *are equivalent iff the rays* $\mathrm{h}_1^{\perp}$ *and* $\mathrm{h}_2^{\perp}$ *are contained in the same chamber of* $\widehat{\mathcal{A}}$. *Thus, equivalence classes of generic half-spaces wrt* $\mathcal{A}$ *correspond to chambers of* $\widehat{\mathcal{A}}$.

1.10. Separating hyperplanes, minimal galleries and gate property

Recall that an arrangement has an associated cell complex, so concepts related to cell complexes can be applied to arrangements. We now focus on minimal galleries and the gate property (Section A.2). This is intimately connected to the notion of separating hyperplanes.

1.10.1. Separating hyperplanes. We say that a hyperplane *separates* two faces if they lie on opposite sides of that hyperplane. In terms of sign sequences, the hyperplane H_i separates faces F and G if $\epsilon_i(F)$ and $\epsilon_i(G)$ have opposite signs, that is, one is $+$ and the other is $-$. Note:

- A hyperplane separates F and $\overline{F}$ iff that hyperplane does not contain F. This follows from (1.3). In particular, every hyperplane separates C and $\overline{C}$.
- If $H \leq F$ and $H \leq G$, then a hyperplane separating F and G necessarily contains H. This follows from (1.4).

A schematic illustration is shown in the picture below on the left; the hyperplane H separates chambers C and D, while H' does not separate C and D. The same is shown in the picture on the right in a concrete rank-three arrangement.

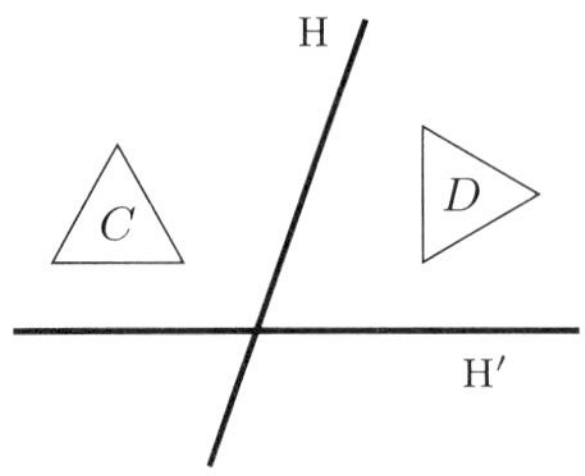

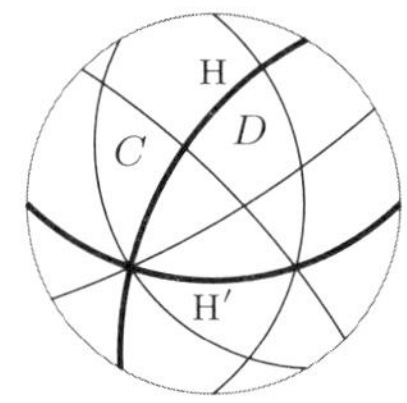

Lemma 1.55. *Suppose $H \leq D$ and C is a chamber. The following are equivalent.*

- $HC = D$.
- *If a hyperplane separates C and D, then it does not contain H.*
- *If a hyperplane contains H, then it does not separate C and D.*

PROOF. This can be deduced from (1.5). The third statement is the contrapositive of the second. □

Exercise 1.56. Check that: A hyperplane H separates chambers C and D iff $FC \neq FD$ for any panel F with support H.

1.10.2. Minimal galleries. The cell complex of an arrangement $\mathcal{A}$ is gallery connected. This can be deduced, for instance, from Lemma 1.34 and Lemma B.6. (Use the coatom connectedness property.) So we can talk of minimal galleries. A minimal gallery from C to D is shown in the picture below. In each step, we move from a chamber to an adjacent chamber along a panel.

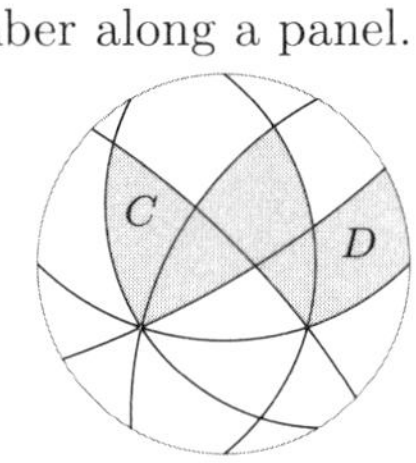

A key observation is that minimal galleries can be handled using separating hyperplanes:

(1.28) $\quad C \text{--} E \text{--} D \iff$ If a hyperplane separates C and E, then it also separates C and D.

Recall that the lhs means that there is a minimal gallery from C to D which passes through E.

Consider the following pictures.

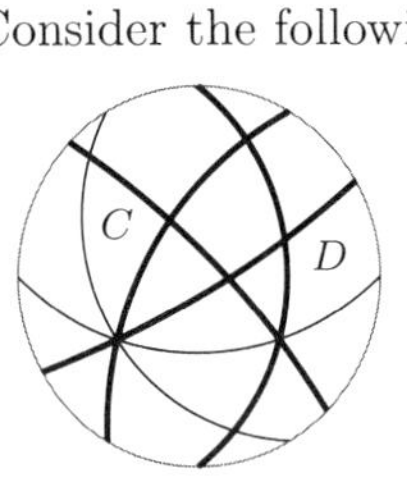

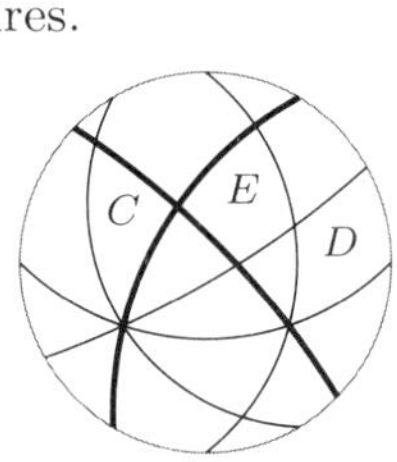

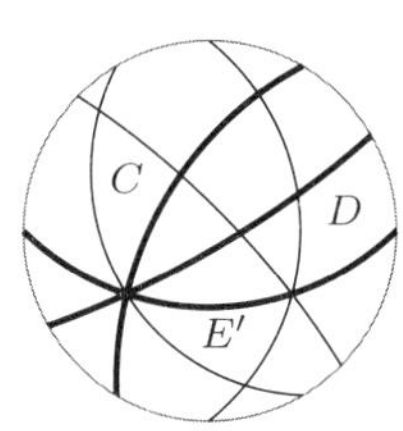

Four hyperplanes separate C and D. They are highlighted in the first picture. Exactly two of these separate C and E and hence $C \text{--} E \text{--} D$. This is shown in the second picture. Three hyperplanes separate C and E', but one of them does not separate C and D. So $C \text{--} E' \text{--} D$ fails. This is shown in the third picture.

The characterization (1.28) gives an algorithm to produce minimal galleries: Suppose we are given chambers C and D. Starting with C, at each step move to an adjacent chamber along a panel whose support separates C and D and this hyperplane has not been crossed before. By this procedure, we will eventually reach D via a minimal gallery. Further, any minimal gallery from C to D arises by this procedure.

Proposition 1.57. *For any face H and chamber C, there exists a minimal gallery $HC \text{ -- } C \text{ -- } \overline{H}C$. In particular, for any chambers C and D, we have $C \text{ -- } D \text{ -- } \overline{C}$. In other words, any minimal gallery starting at a chamber C can to extended to end at its opposite $\overline{C}$.*

PROOF. We employ (1.28). Suppose H_i separates C and HC. This means that $\epsilon_i(C)$ and $\epsilon_i(H)$ have opposite signs. Thus, $\epsilon_i(H)$ and $\epsilon_i(\overline{H})$ have opposite signs, and H_i separates HC and $\overline{H}C$. This proves the first statement. To get the second statement, take H to be a chamber. □

The opposition map preserves minimal galleries. That is,

$$C \text{ -- } D \text{ -- } E \iff \overline{E} \text{ -- } \overline{D} \text{ -- } \overline{C}.$$

This is because a hyperplane separates two chambers iff that hyperplane separates their opposites.

Lemma 1.58. *The Tits projection preserves minimal galleries. That is, for any face K,*

$$C \text{ -- } D \text{ -- } E \implies KC \text{ -- } KD \text{ -- } KE. \tag{1.29}$$

PROOF. We employ (1.28). Suppose H_i separates KC and KD. This means that $\epsilon_i(K) = 0$, and $\epsilon_i(C)$ and $\epsilon_i(D)$ have opposite signs. Since $C \text{ -- } D \text{ -- } E$, we deduce that $\epsilon_i(C)$ and $\epsilon_i(E)$ also have opposite signs, and hence so do $\epsilon_i(KC)$ and $\epsilon_i(KE)$. So H_i separates KC and KE. □

Corollary 1.59. *Suppose $C \text{ -- } D \text{ -- } E$, and K is a face of E such that $KC = E$. Then $KD = E$.*

The picture below on the left illustrates Lemma 1.58, while the one on the right illustrates Corollary 1.59.

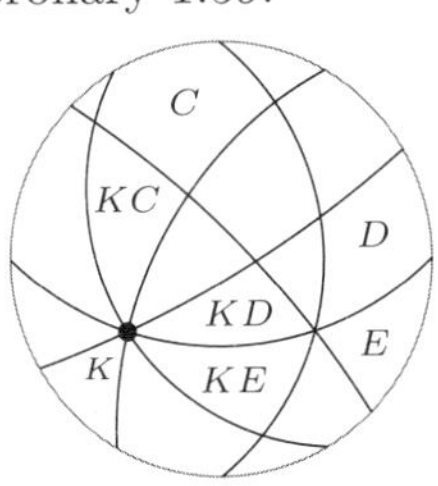

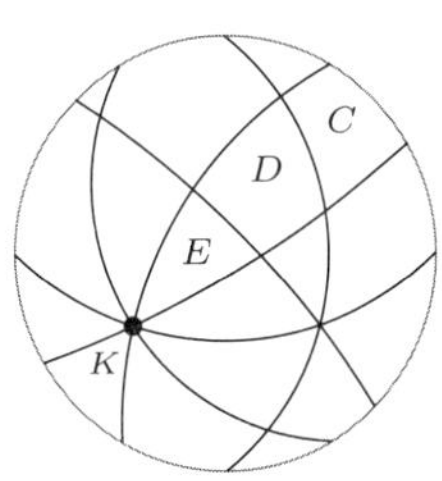

1.10.3. Gallery distance.

The gallery distance $\mathrm{dist}(C, D)$ is the minimum length of a gallery connecting C and D. It is equal to the number of hyperplanes which separate C and D. It verifies the familiar properties of a metric:

$$\mathrm{dist}(C, D) \geq 0 \text{ with equality iff } C = D, \tag{1.30a}$$

$$\mathrm{dist}(C, D) = \mathrm{dist}(D, C), \tag{1.30b}$$

$$\mathrm{dist}(C, E) \leq \mathrm{dist}(C, D) + \mathrm{dist}(D, E) \text{ with equality iff } C \text{ -- } D \text{ -- } E. \tag{1.30c}$$

The maximum gallery distance is $\text{dist}(C, \overline{C})$. It is independent of C and equal to the number of hyperplanes in the arrangement.

Two chambers C and D are adjacent iff $\text{dist}(C, D) = 1$ iff there is a unique hyperplane which separates C and D.

More generally: For any faces F and G, define $\text{dist}(F, G)$ to be the number of hyperplanes which separate F and G. Some simple observations are listed below.

$$\text{dist}(F, G) = \text{dist}(FG, GF). \tag{1.31}$$

$$\text{dist}(F, G) = 0 \iff FG = GF. \tag{1.32}$$

If F and G have the same support, and $F \leq C$, then

$$\text{dist}(F, G) = \text{dist}(C, GC). \tag{1.33}$$

For a fixed flat X, $\text{dist}(F, G)$, as F and G vary over faces with support X, is maximum when $G = \overline{F}$. This maximum value is the number of hyperplanes which do not contain X.

In the picture below on the left, $\text{dist}(F, G) = 3$, while in the picture on the right, $\text{dist}(F, G) = 2$.

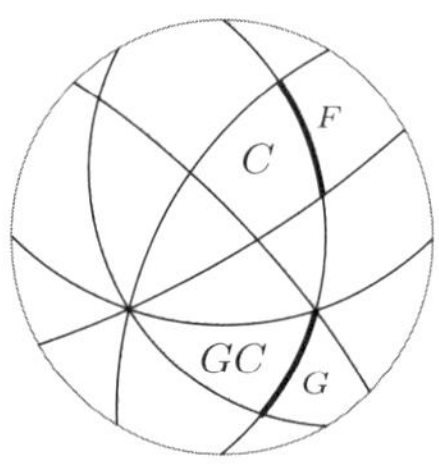

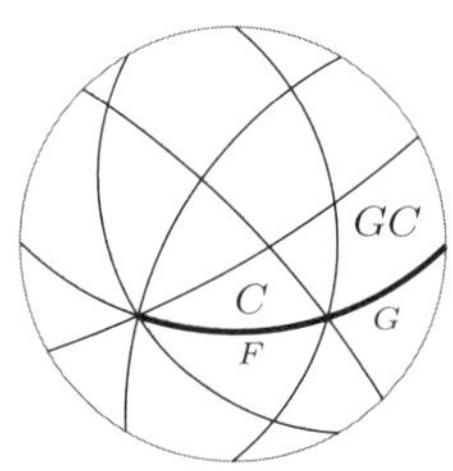

Warning. Faces F and G with the same support, say X, correspond to chambers of the arrangement $\mathcal{A}^{\mathrm{X}}$ under X. However, $\text{dist}(F, G)$ is in general *not* the gallery distance between F and G in $\mathcal{A}^{\mathrm{X}}$ but larger than it. Intuitively, this is because there is more room to move in $\mathcal{A}$ than in $\mathcal{A}^{\mathrm{X}}$.

1.10.4. Chamber graph. The *chamber graph* of an arrangement $\mathcal{A}$ is the graph which has chambers for vertices, and edges between adjacent chambers.

The chamber graph of $\mathcal{A}$ is connected. This is a reformulation of the fact that $\mathcal{A}$ is gallery connected. In addition, the chamber graph is bipartite. To see this, fix a chamber C, and put all chambers at an even distance from C in one part, and all chambers at an odd distance in the other. This works since two chambers both at an even (or odd) distance from C cannot be adjacent. The bipartition can also be characterized as follows: D and D' belong to the same part iff $\text{dist}(D, D')$ is even.

A bipartite graph is *balanced* if the two parts have the same cardinality.

Exercise 1.60. Show that the chamber graph of an arrangement with an odd number of hyperplanes is a balanced bipartite graph. (Use the opposition map.) The result is false in general for arrangements with an even number of hyperplanes. For instance, for the smallest nonsimplicial arrangement, the eight triangles are in one part, while the six quadrilaterals are in the other part.

A related result is given in Exercise 5.7 with further considerations in Section 8.2.2.

1.10.5. Gate property. The following fact is of fundamental importance:

Proposition 1.61. *For chambers C and D, and H any face of D, there exists a minimal gallery C -- HC -- D.*

PROOF. We employ (1.28). Suppose H_i separates C and HC. This means that $\epsilon_i(C)$ and $\epsilon_i(H)$ have opposite signs. Since H is a face of D, $\epsilon_i(H)$ and $\epsilon_i(D)$ have the same sign. Thus, H_i separates C and D. □

A schematic illustration is shown below.

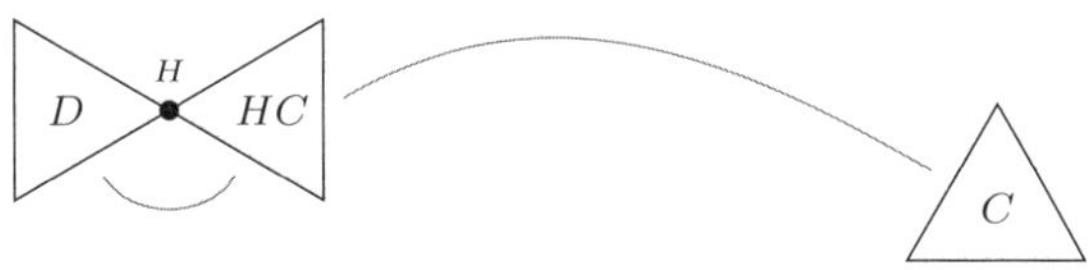

A concrete illustration in a rank-three arrangement is shown below.

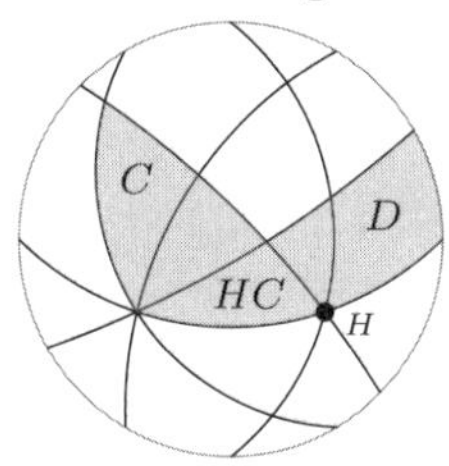

As a consequence: The cell complex of $\mathcal{A}$ satisfies the gate property (Definition A.4). The gate of the top-star of H wrt C is the chamber HC. In other words, HC is the chamber closest to C in the gallery metric having H as a face, and more precisely

$$\mathrm{dist}(C, D) = \mathrm{dist}(C, HC) + \mathrm{dist}(HC, D) \tag{1.34}$$

for any chamber D greater than H. This is another way to understand the geometric meaning of the Tits product.

Exercise 1.62. For $H \leq D$, there exists a minimal gallery from C to D passing through HC. However, a minimal gallery from C to D does not necessarily have to pass through HC. Give an example.

Exercise 1.63. By Proposition A.2, top-stars in an arrangement are also gated wrt each other. Describe all gate pairs.

1.10.6. Minimal galleries for faces. Minimal galleries also make sense for faces with the same support (by working in the arrangement under that support). Suppose F, G and H are faces with the same support, say X. Then F -- G -- H denotes a minimal gallery in $\mathcal{A}^{\mathrm{X}}$.

Proposition 1.64. *There exists a minimal gallery HG -- GH -- $G\overline{H}$ -- $\overline{H}G$ for any faces G and H. In particular, when G and H have the same support, we have H -- G -- $\overline{H}$.*

PROOF. First note that as required all faces involved in the gallery have the same support. So we may assume that they are all chambers. Apply Proposition 1.57 with $C = GH$, to get HG -- GH -- $\overline{H}G$. By the gate property, we have GH -- $G\overline{H}$ -- $\overline{H}G$. Now refine the first using the second. □

For faces F, G and H with the same support, and for any face K,

$$F \text{--} G \text{--} H \iff FK \text{--} GK \text{--} HK. \tag{1.35}$$

One can prove this using (1.28). The key observation is that a hyperplane separates F and G iff it separates FK and GK.

Exercise 1.65. Show: For any face K, $F \text{--} G \text{--} H$ implies $KF \text{--} KG \text{--} KH$.

1.10.7. Properties of separating hyperplanes. For chambers C and D, let $g(C, D)$ denote the set of hyperplanes which separate C and D. Note that $g(C, D) = g(D, C)$. For any chambers C, D and E and hyperplane H,

$$\mathrm{H} \in g(C,E) \iff \text{Either } \mathrm{H} \in g(C,D) \text{ and } \mathrm{H} \notin g(D,E), \quad \text{or } \mathrm{H} \notin g(C,D) \text{ and } \mathrm{H} \in g(D,E). \tag{1.36}$$

Also,

$$g(C,D) \subseteq g(C,E) \iff C \text{--} D \text{--} E \iff g(C,D) \sqcup g(D,E) = g(C,E). \tag{1.37}$$

The first equivalence is a reformulation of (1.28). The second can be deduced from the first by using (1.36).

Similarly, for chambers C and D, let $r(C, D)$ denote the set of half-spaces which contain C but do not contain D. In other words, $\mathrm{h} \in r(C, D)$ iff C lies in h while D lies in $\overline{\mathrm{h}}$. It follows that

$$\mathrm{h} \in r(C,D) \iff \overline{\mathrm{h}} \in r(D,C).$$

In contrast to the previous situation, the order in which C and D are written is crucial now. For any chambers C, D and E and half-space h,

$$\mathrm{h} \in r(C,E) \iff \text{Either } \mathrm{h} \in r(C,D) \text{ and } \overline{\mathrm{h}} \notin r(D,E), \quad \text{or } \overline{\mathrm{h}} \notin r(C,D) \text{ and } \mathrm{h} \in r(D,E). \tag{1.38}$$

Also,

$$r(C,D) \subseteq r(C,E) \iff C \text{--} D \text{--} E \iff r(C,D) \sqcup r(D,E) = r(C,E). \tag{1.39}$$

Note that $g(C, D)$ consists of hyperplanes bounding the half-spaces in $r(C, D)$. Similarly, $r(C, D)$ consists of half-spaces containing C whose bounding hyperplanes are in $g(C, D)$.

Basic properties of the sets $r(C, D)$ are listed below.

Proposition 1.66. *For any chamber C,*

$$r(C,C) = \emptyset. \tag{1.40a}$$

For any chambers C and D, and faces F and G with the same support,

$$r(FC, GC) = r(FD, GD). \tag{1.40b}$$

For any minimal gallery $C \text{--} D \text{--} E$,

$$r(C,E) = r(C,D) \sqcup r(D,E). \tag{1.40c}$$

For any C, and G a face of D,

$$r(C,D) = r(C,GC) \sqcup r(GC,D). \tag{1.40d}$$

For any D, and F a face of C,

$$r(C,D) = r(C,FD) \sqcup r(FD,D). \tag{1.40e}$$

For any chambers C and D, and faces F and G with the same support,

$$r(FC, FD) = r(GC, GD). \tag{1.40f}$$

For any chambers C and D,

$$r(C, D) = r(\overline{D}, \overline{C}). \tag{1.40g}$$

The situation in (1.40b) and (1.40f) is illustrated below.

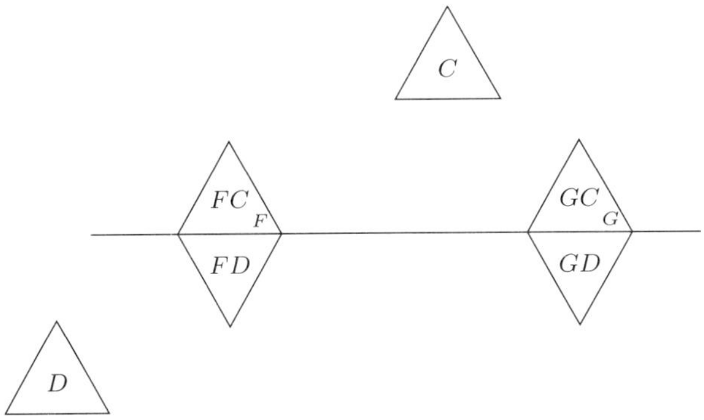

PROOF. Identity (1.40a) is clear. For (1.40b), note that

$$\mathrm{h} \in r(FC, GC) \iff \text{The bounding hyperplane of h separates } F \text{ and } G,$$
$$\text{with h containing } F \text{ and } \overline{\mathrm{h}} \text{ containing } G \iff \mathrm{h} \in r(FD, GD).$$

(1.40c) is contained in (1.39). For (1.40d) and (1.40e): By the gate property, for any face G of D, there exists a minimal gallery C -- GC -- D, and for any face F of C, there exists a minimal gallery C -- FD -- D. Now apply (1.40c). For (1.40f), note that for h to belong to either side, its bounding hyperplane must contain both F and G. Further, in this situation, FC lies in h iff GC lies in h, and similarly, FD lies in $\overline{\mathrm{h}}$ iff GD lies in $\overline{\mathrm{h}}$. For (1.40g), note that h contains C iff $\overline{\mathrm{h}}$ contains $\overline{C}$, and similarly, $\overline{\mathrm{h}}$ contains D iff h contains $\overline{D}$. □

Proposition 1.66 is also true with 'r' replaced by 'g' everywhere.

1.11. Combinatorially isomorphic arrangements

Recall that two arrangements are combinatorially isomorphic, or cisomorphic, if their posets of faces are isomorphic. Let us look at this notion in more detail.

Proposition 1.67. *Let $\mathcal{A}$ and $\mathcal{A}'$ be arrangements, and $\varphi : \Sigma[\mathcal{A}] \to \Sigma[\mathcal{A}']$ be any bijection between their sets of faces. Then φ is an isomorphism of posets iff φ is an isomorphism of monoids.*

PROOF. Let ψ denote the inverse of φ. Backward implication. Suppose $F \le G$. Then $G = FG$. Hence $\varphi(G) = \varphi(FG) = \varphi(F)\varphi(G)$. So $\varphi(F) \le \varphi(G)$. Thus, φ is order-preserving. By symmetry, ψ is also order-preserving.

Forward implication. First observe that φ and ψ send adjacent chambers to adjacent chambers, so they preserve galleries. This implies that they are both non-increasing wrt gallery distance. So in fact, they preserve gallery distances. Now let F be a face and C a chamber. The chambers greater than F are in bijection with chambers greater than $\varphi(F)$. Hence, from the gate property (1.34), we deduce that $\varphi(FC) = \varphi(F)\varphi(C)$. Thus, φ preserves Tits projection of chambers. For the general case, we employ Exercise 1.8 and the fact that φ preserves meets:

$$\varphi(FG) = \varphi(\bigwedge FC) = \bigwedge \varphi(FC) = \bigwedge \varphi(F)\varphi(C) = \varphi(F)\varphi(G).$$

Also φ preserves the central face since it is the minimum element. So φ is a morphism of monoids, and by symmetry, so is ψ. □

Corollary 1.68. *Two arrangements are cisomorphic iff their Tits monoids are isomorphic.*

Observe that a morphism of monoids $\varphi : \Sigma[\mathcal{A}] \to \Sigma[\mathcal{A}']$ induces a commutative diagram of monoids

$$\begin{array}{ccc} \Sigma[\mathcal{A}] & \xrightarrow{\varphi} & \Sigma[\mathcal{A}'] \\ {\scriptstyle \mathrm{s}}\big\downarrow & & \big\downarrow{\scriptstyle \mathrm{s}} \\ \Pi[\mathcal{A}] & \xrightarrow[\overline{\varphi}]{} & \Pi[\mathcal{A}']. \end{array}$$

Further, if φ is an isomorphism, then so is $\overline{\varphi}$. In conjunction with Proposition 1.67, we obtain:

Corollary 1.69. *A cisomorphism of arrangements induces an isomorphism between their posets of flats (and hence a bijection between the two sets of hyperplanes).*

Corollary 1.70. *A cisomorphism preserves gallery distances, and more generally, distances between faces. It also preserves opposite faces.*

PROOF. Let φ be a cisomorphism. In the proof of Proposition 1.67, we saw that φ preserves gallery distances. To see that φ preserves distances between faces: Say F and G have same support and $F \leq C$. Then $\varphi(F) \leq \varphi(C)$, and $\varphi(F)$ and $\varphi(G)$ have the same support. Employing (1.33),

$$\begin{aligned} \mathrm{dist}(\varphi(F), \varphi(G)) = \mathrm{dist}(\varphi(C), \varphi(G)\varphi(C)) &= \mathrm{dist}(\varphi(C), \varphi(GC)) \\ &= \mathrm{dist}(C, GC) = \mathrm{dist}(F, G). \end{aligned}$$

For arbitrary faces F and G, we employ (1.31) and apply the above result to FG and GF (which have the same support). The last claim also follows since opposite faces are the farthest apart within their support. □

1.12. Partial order on pairs of faces

Fix an arrangement $\mathcal{A}$. Recall the set of faces $\Sigma[\mathcal{A}]$, the set of chambers $\Gamma[\mathcal{A}]$, and the set of flats $\Pi[\mathcal{A}]$. We now define partial orders on pairs of chambers, pairs of faces, and on pairs of flats.

1.12.1. Pairs of chambers. We begin with the partial order on pairs of chambers. We say that $(C_1, D_1) \leq (C_2, D_2)$ in $\Gamma[\mathcal{A}] \times \Gamma[\mathcal{A}]$ if

(i) $D_1 = D_2 = D$ (say),
(ii) $C_2 \,\text{--}\, C_1 \,\text{--}\, D$.

This is illustrated below.

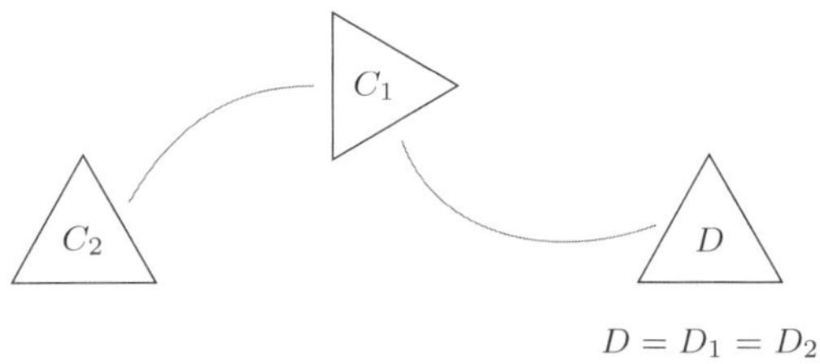

Lemma 1.71. *For a face G, and chambers E, E' and D with $G \le D$ and $\overline{G} \le E'$,*

$$(E', D) \ge (E, D) \iff (GE', D) \ge (GE, D).$$

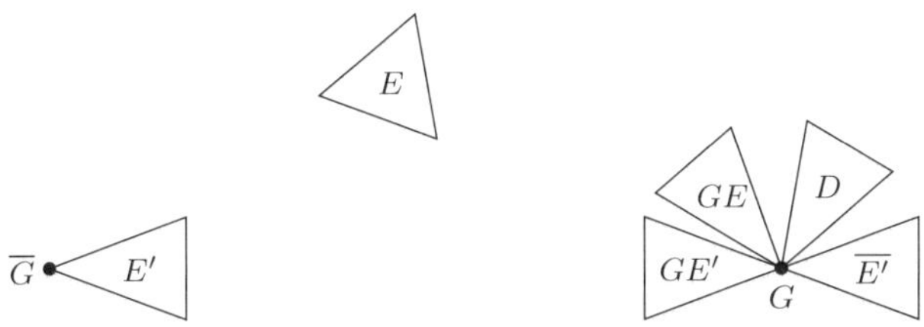

PROOF. The forward implication follows from (1.29) by projecting $E' \text{ -- } E \text{ -- } D$ on G. We now prove the backward implication. We are given $GE' \text{ -- } GE \text{ -- } D$. Since GE' and $\overline{E'}$ are opposite chambers in the star of G, by Proposition 1.57, the minimal gallery can be extended to $GE' \text{ -- } GE \text{ -- } D \text{ -- } \overline{E'}$. By restricting, we have $GE \text{ -- } D \text{ -- } \overline{E'}$. Now since E' and $\overline{E'}$ are opposite chambers, we have $E' \text{ -- } E \text{ -- } \overline{E'}$. Since G is a face of $\overline{E'}$, by the gate property, this refines to $E' \text{ -- } E \text{ -- } GE \text{ -- } \overline{E'}$. Combining this with $GE \text{ -- } D \text{ -- } \overline{E'}$, we obtain $E' \text{ -- } E \text{ -- } D$ as required. □

1.12.2. Pairs of faces. We now define a partial order on pairs of faces which extends the partial order on chambers. We say that $(H_1, K_1) \le (H_2, K_2)$ in $\Sigma[\mathcal{A}] \times \Sigma[\mathcal{A}]$ if

(i) $K_1 = K_2 = K$ (say),
(ii) $H_2H_1 = H_2$ and $KH_1 = KH_2$,
(iii) $H_2K \text{ -- } H_1K \text{ -- } KH_2$.

It is clear from (ii) above that H_2K, H_1K and KH_2 all have the same support; so condition (iii) makes sense.

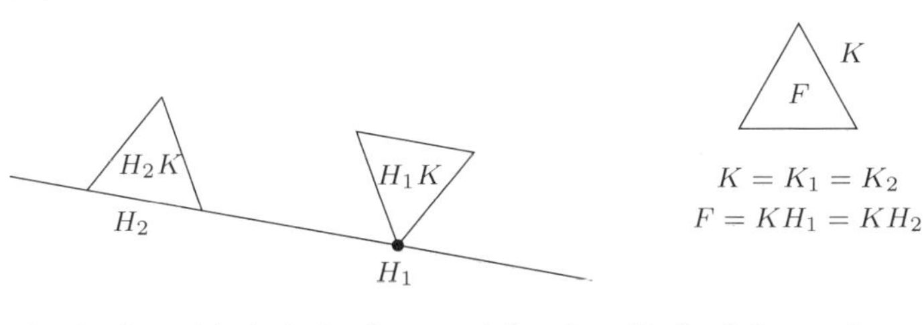

Let us check that this is indeed a partial order. Reflexivity and transitivity are clear. To check antisymmetry, suppose $(H_1, K) \le (H_2, K)$ and $(H_2, K) \le (H_1, K)$. Then H_1 and H_2 have the same support, and $H_1K = H_2K$. So by Proposition 1.17, $H_1 = H_2$.

Lemma 1.72. *For any faces G, H, H' and K with $G \leq K$ and $\overline{G} \leq H'$,*

$$(H', K) \geq (H, K) \iff (GH', K) \geq (GH, K).$$

PROOF. Let us first look at condition (ii) for the lhs and for the rhs. Since $\overline{G} \leq H'$, we deduce that H' and GH' have the same support. It contains the support of H iff it contains the support of GH. Next, since $G \leq K$, we have $KH = KGH$ and $KH' = KGH'$. Hence $KH = KH'$ iff $KGH = KGH'$.

To check condition (iii), we may assume that H, H' and K are chambers. In this case, the claim reduces to Lemma 1.71. □

Lemma 1.73. *Let G, H, F and K be any faces with $G \leq H$. Then*

$$(F, K) \geq (H, K) \iff \begin{array}{c} (GF, GK) \geq (H, GK), \ FG = F, \text{ and} \\ FK \text{ -- } GFK \text{ -- } GKF \text{ -- } KF. \end{array}$$

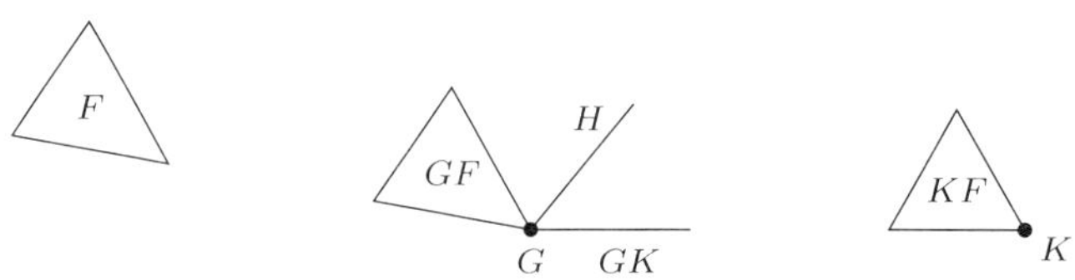

PROOF. The conditions in the lhs can be explicitly written as

(a) $$KF = KH, \quad \mathrm{s}(H) \leq \mathrm{s}(F), \quad FK \text{ -- } HK \text{ -- } KF.$$

Similarly, the conditions in the rhs can be explicitly written as

$$GKH = GKF, \quad \mathrm{s}(H) \leq \mathrm{s}(GF) = \mathrm{s}(F),$$
$$GFK \text{ -- } HK \text{ -- } GKF, \quad FK \text{ -- } GFK \text{ -- } GKF \text{ -- } KF.$$

The two minimal galleries can be combined as $FK \text{ -- } GFK \text{ -- } HK \text{ -- } GKF \text{ -- } KF$, which by the gate property is equivalent to $FK \text{ -- } HK \text{ -- } KF$. Also the support condition can be simplified to $\mathrm{s}(H) \leq \mathrm{s}(F)$. So conditions in the rhs can be rewritten as

(b) $$GKH = GKF, \quad \mathrm{s}(H) \leq \mathrm{s}(F), \quad FK \text{ -- } HK \text{ -- } KF.$$

It is clear that (a) implies (b). For the reverse implication, apply Corollary 1.59 to $FK \text{ -- } HK \text{ -- } KF$ with $C = FK$, $D = HK$ and $E = KF$ to deduce $KF = KH$. □

For any face A, we use $\leq_A$ to denote the restriction of the above partial order to faces greater than A. Using convexity of stars, one can see that this agrees with the partial order on pairs of faces in $\mathcal{A}_A$. Lemma 1.72 may be reformulated as follows.

Lemma 1.74. *For any faces G, H and K (all greater than A) with $G \leq K$, there is a bijection between the sets*

$$\{H' \mid (H', K) \geq_A (H, K), \ A\overline{G} \leq H'\} \longrightarrow \{H'' \mid (H'', K) \geq_G (GH, K)\}.$$

It sends H' to GH', and the inverse sends H'' to $A\overline{G}H''$.

1.12.3. Faces and chambers. There are two more interesting posets that lie between the poset on pairs of chambers and the poset on pairs of faces. They are obtained by taking faces in one coordinate and chambers in the other coordinate. Explicitly, the partial orders are as follows.

We say that $(H_1, D_1) \leq (H_2, D_2)$ in $\Sigma[\mathcal{A}] \times \Gamma[\mathcal{A}]$ if

(i) $D_1 = D_2 = D$ (say),
(ii) $H_2 H_1 = H_2$,
(iii) $H_2D \text{ -- } H_1D \text{ -- } D$.

We say that $(C_1, K_1) \leq (C_2, K_2)$ in $\Gamma[\mathcal{A}] \times \Sigma[\mathcal{A}]$ if

(i) $K_1 = K_2 = K$ (say),
(ii) $KC_1 = KC_2 = D$ (say),
(iii) $C_2 \text{ -- } C_1 \text{ -- } D$.

The inclusion maps in the commutative diagram

$$\begin{array}{ccc} \Gamma[\mathcal{A}] \times \Gamma[\mathcal{A}] & \longrightarrow & \Gamma[\mathcal{A}] \times \Sigma[\mathcal{A}] \\ \downarrow & & \downarrow \\ \Sigma[\mathcal{A}] \times \Gamma[\mathcal{A}] & \longrightarrow & \Sigma[\mathcal{A}] \times \Sigma[\mathcal{A}] \end{array}$$

are all order-preserving.

1.12.4. Pairs of flats. We now define a partial order on pairs of flats. We say that $(\mathrm{X}_1, \mathrm{Y}_1) \leq (\mathrm{X}_2, \mathrm{Y}_2)$ in $\Pi[\mathcal{A}] \times \Pi[\mathcal{A}]$ if

(i) $\mathrm{Y}_1 = \mathrm{Y}_2 = \mathrm{Y}$ (say),
(ii) $\mathrm{X}_1 \leq \mathrm{X}_2$, and $\mathrm{Y} \vee \mathrm{X}_1 = \mathrm{Y} \vee \mathrm{X}_2$,
(ii') $\mathrm{X}_1 \leq \mathrm{X}_2 \leq \mathrm{Y} \vee \mathrm{X}_1$.

Conditions (ii) and (ii') are equivalent.

For any flat Z, we use $\leq_{\mathrm{Z}}$ to denote the restriction of the above partial order to flats greater than Z.

Exercise 1.75. Check that the map $\mathrm{s} \times \mathrm{s} : \Sigma[\mathcal{A}] \times \Sigma[\mathcal{A}] \twoheadrightarrow \Pi[\mathcal{A}] \times \Pi[\mathcal{A}]$ is order-preserving.

1.13. Characteristic polynomial and Zaslavsky formula

We now look at some enumerative aspects of arrangements. A key role is played by the Möbius function. This is conveniently encoded in a polynomial called the characteristic polynomial. The value of this polynomial at -1 is (up to sign) the number of chambers in the arrangement. This is known as the Zaslavsky formula. We also discuss the Whitney numbers of the first kind which are obtained as coefficients of the characteristic polynomial.

The Möbius function of a poset is reviewed in Section C.1.

Notation 1.76. For any arrangement $\mathcal{A}$, let $c(\mathcal{A})$ denote the number of chambers in $\mathcal{A}$. Recall that there are a number of arrangements associated to $\mathcal{A}$ such as $\mathcal{A}_F, \mathcal{A}^{\mathrm{X}}$, and so on. If $\mathcal{A}$ is understood from the context, then we will allow ourselves to write c_F instead of $c(\mathcal{A}_F)$, c^{X} instead of $c(\mathcal{A}^{\mathrm{X}})$, and so on.

1.13.1. Euler characteristic. Recall that faces of an arrangement $\mathcal{A}$ are cells in a regular cellular decomposition of a sphere of dimension $\mathrm{rk}(\mathcal{A})-1$. Taking reduced Euler characteristics (A.1), we obtain

$$\sum_{F\in\Sigma[\mathcal{A}]} (-1)^{\mathrm{rk}(F)} = (-1)^{\mathrm{rk}(\mathcal{A})}. \tag{1.41}$$

For any flat X,

$$\sum_{\mathrm{Y}:\,\mathrm{Y}\leq\mathrm{X}} (-1)^{\mathrm{rk}(\mathrm{Y})} c^{\mathrm{Y}} = (-1)^{\mathrm{rk}(\mathrm{X})}, \tag{1.42}$$

where c^{Y} is the number of faces of support Y.

The two identities are equivalent. Applying the first to $\mathcal{A}^{\mathrm{X}}$ yields the second, while applying the second to $\mathrm{X} = \top$ yields the first.

Recall the notion of Eulerian poset from Section C.1.6. The poset of faces $\Sigma[\mathcal{A}]$ of any arrangement $\mathcal{A}$ is Eulerian, that is,

$$\mu(H,G) = (-1)^{\mathrm{rk}(G)-\mathrm{rk}(H)} \tag{1.43}$$

for $H \leq G$. This is clear for a simplicial arrangement since each interval in the poset of faces is Boolean. For the general case, we can use Lemma 1.33.

Proposition 1.77. *In any arrangement, for faces $O < F \leq G$,*

$$\sum_{H:\,HF=G} (-1)^{\mathrm{rk}(H)} = 0.$$

PROOF. Any interval in the poset of faces is a lattice. Apply the Weisner formula (C.7a) to the lattice $[O,G]$ with $z := G$ to obtain

$$\sum_{H:\,H\vee F=G} \mu(O,H) = 0.$$

(The lattice $[O,G]$ consists of all faces smaller than G.) Now use Exercise 1.21 and (1.43). □

As a special case: For any face G,

$$\sum_{F:\,F\leq G} (-1)^{\mathrm{rk}(F)} = \begin{cases} 1 & \text{if } G = O, \\ 0 & \text{otherwise.} \end{cases} \tag{1.44}$$

Proposition 1.78. *In any arrangement of rank at least* 1,

$$\sum_{\mathrm{Y}\in\Pi[\mathcal{A}]} (-1)^{\mathrm{rk}(\mathrm{Y})} c^{\mathrm{Y}} c_{\mathrm{Y}} = 0 \qquad \textit{or equivalently} \qquad \sum_{F} (-1)^{\mathrm{rk}(F)} c_F = 0. \tag{1.45}$$

PROOF. Using the definition of c_F and (1.44),

$$\sum_F (-1)^{\mathrm{rk}(F)} c_F = \sum_C \sum_{F:\,F\leq C} (-1)^{\mathrm{rk}(F)} = 0.$$

(The assumption on the rank ensures that $C \neq O$.) The fact that the inside sum is zero can also be seen from the Eulerian property (1.43) and (C.5b). □

1.13.2. Möbius number. For any arrangement $\mathcal{A}$, define

$$\mu(\mathcal{A}) := \mu(\bot, \top).$$

We refer to this as the *Möbius number* of $\mathcal{A}$. It is the value of the Möbius function on the largest interval in the lattice of flats $\Pi[\mathcal{A}]$.

Proposition 1.79. *Suppose* $\mathrm{Y} > \bot$. *Then for any flat* Z,

$$\sum_{\mathrm{X}:\, \mathrm{Y} \vee \mathrm{X} = \mathrm{Z}} \mu(\mathcal{A}^{\mathrm{X}}) = 0. \tag{1.46a}$$

Suppose $\mathrm{Y} < \top$. *Then for any flat* Z,

$$\sum_{\mathrm{X}:\, \mathrm{Y} \wedge \mathrm{X} = \mathrm{Z}} \mu(\mathcal{A}_{\mathrm{X}}) = 0. \tag{1.46b}$$

PROOF. This is the Weisner formula (Proposition C.4) specialized to the lattice of flats. □

Proposition 1.80. *For any arrangement* $\mathcal{A}$,

$$(-1)^{\mathrm{rk}(\mathcal{A})} \mu(\mathcal{A}) = |\mu(\mathcal{A})| \neq 0. \tag{1.47}$$

In other words, the Möbius number of an arrangement is nonzero and its sign is the same as the parity of its rank.

PROOF. In view of Lemma 1.30, this is a special case of Proposition C.6. □

1.13.3. Zaslavsky formula. The *Zaslavsky formula* counts the number of chambers in an arrangement in terms of the absolute values of the Möbius function of the lattice of flats. It is given as follows.

Theorem 1.81. *For any arrangement* $\mathcal{A}$,

$$\sum_{\mathrm{X} \in \Pi[\mathcal{A}]} |\mu(\mathrm{X}, \top)| = \sum_{\mathrm{X} \in \Pi[\mathcal{A}]} |\mu(\mathcal{A}_{\mathrm{X}})| = c(\mathcal{A}), \tag{1.48}$$

where $c(\mathcal{A})$ *is the number of chambers of* $\mathcal{A}$.

PROOF. For each flat X, put

$$f(\mathrm{X}) := (-1)^{\mathrm{rk}(\mathrm{X})} c^{\mathrm{X}},$$

where c^{X} is the number of faces of support X. Then, by (1.42),

$$g(\mathrm{Y}) := \sum_{\mathrm{X}:\, \mathrm{X} \leq \mathrm{Y}} f(\mathrm{X}) = \sum_{\mathrm{X}:\, \mathrm{X} \leq \mathrm{Y}} (-1)^{\mathrm{rk}(\mathrm{X})} c^{\mathrm{X}} = (-1)^{\mathrm{rk}(\mathrm{Y})}.$$

Now, by Möbius inversion (C.12),

$$f(\top) = \sum_{\mathrm{X}} g(\mathrm{X}) \mu(\mathrm{X}, \top).$$

The result now follows by applying Proposition 1.80 to each $\mathcal{A}_{\mathrm{X}}$. □

There is a similar formula for face enumeration which is given below.

Corollary 1.82. *For any arrangement* $\mathcal{A}$,

$$\sum_{\mathrm{X} \leq \mathrm{Y}} |\mu(\mathrm{X}, \mathrm{Y})| = \sum_{\mathrm{X} \leq \mathrm{Y}} |\mu(\mathcal{A}_{\mathrm{X}}^{\mathrm{Y}})| = f(\mathcal{A}), \tag{1.49}$$

where $f(\mathcal{A})$ *is the number of faces of* $\mathcal{A}$. *(The sum is over both* X *and* Y.*)*

PROOF. Each face is a chamber of the arrangement under its support. So the result follows by applying the Zaslavsky formula (1.48) to $\mathcal{A}^{\mathrm{Y}}$ for each flat Y. □

Exercise 1.83. Show that

$$\sum_{\mathrm{rk(Y)}=i, \mathrm{X} \leq \mathrm{Y}} |\mu(\mathrm{X}, \mathrm{Y})| \tag{1.50}$$

equals the number of faces of rank i. The sum is over both X and Y with i fixed.

Lemma 1.84. *We have*

$$\mu(\mathcal{A}) = \sum_{T} (-1)^{|T|}, \tag{1.51}$$

where the sum is over all subsets T of the set of hyperplanes such that the intersection of the elements in T is the minimum flat.

PROOF. The formula is clear if $\mathcal{A}$ has no hyperplanes. So assume that $\mathcal{A}$ has rank at least one. The alternating sum as in the rhs above but taken over all subsets is zero. Now apply induction to $\mathcal{A}_{\mathrm{X}}$ for each flat $\mathrm{X} > \perp$, and use (C.5b). □

For instance, for the smallest nonsimplicial arrangement $\mathcal{A}$ of four hyperplanes, the intersection of any three or all four hyperplanes is the minimum flat, hence $\mu(\mathcal{A}) = (-1)^4 + 4(-1)^3 = -3$.

Theorem 1.85. *For any arrangement $\mathcal{A}$,*

$$c(\mathcal{A}) = \sum_{T} (-1)^{d(T)}, \tag{1.52}$$

where the sum is over all subsets T of the set of hyperplanes, and $d(T)$ is the cardinality of T minus the codimension of the flat obtained by intersecting elements in T. (If T is the empty subset, then $d(T) = 0$.)

PROOF. From (1.47) and (1.51), we see that (1.52) is equivalent to the Zaslavsky formula (1.48). □

We refer to (1.52) as the *Winder formula*.

1.13.4. Characteristic polynomial.

For any arrangement $\mathcal{A}$, define a polynomial with integer coefficients in the variable t by

$$\chi(\mathcal{A}, t) := \sum_{\mathrm{Y}} \mu(\mathrm{Y}, \top)\, t^{\mathrm{rk(Y)}}. \tag{1.53}$$

This is the *characteristic polynomial* of $\mathcal{A}$. Its degree equals the rank of $\mathcal{A}$.

If $\mathcal{A}$ has rank 0, then $\chi(\mathcal{A}, t) = 1$, independent of t.

Let us now consider the values $t = 0, 1, -1$. For $t = 0$, only the summand for $\mathrm{Y} = \perp$ contributes to the rhs of (1.53). Thus,

$$\chi(\mathcal{A}, 0) = \mu(\mathcal{A}). \tag{1.54a}$$

For $t = 1$, using (C.5a) and (C.5b),

$$\chi(\mathcal{A}, 1) = \begin{cases} 1 & \text{if } \mathcal{A} \text{ has rank } 0, \\ 0 & \text{otherwise.} \end{cases} \tag{1.54b}$$

The case $t = -1$ is nontrivial. Using (1.47) and the Zaslavsky formula (1.48),

$$\chi(\mathcal{A}, -1) = (-1)^{\mathrm{rk}(\mathcal{A})} c(\mathcal{A}), \tag{1.54c}$$

where $c(\mathcal{A})$ is the number of chambers in $\mathcal{A}$.

Lemma 1.86. *For any flat* Z *of an arrangement* $\mathcal{A}$,

$$t^{\mathrm{rk}(\mathrm{Z})} \chi(\mathcal{A}_{\mathrm{Z}}, t) = \sum_{\mathrm{X}:\, \mathrm{X} \vee \mathrm{Z} = \top} \chi(\mathcal{A}^{\mathrm{X}}, t). \tag{1.55}$$

PROOF. The rhs can be manipulated as follows.

$$\begin{aligned} \sum_{\mathrm{X}:\, \mathrm{X} \vee \mathrm{Z} = \top} \chi(\mathcal{A}^{\mathrm{X}}, t) &= \sum_{\mathrm{X}:\, \mathrm{X} \vee \mathrm{Z} = \top} \; \sum_{\mathrm{Y}:\, \mathrm{Y} \leq \mathrm{X}} \mu(\mathrm{Y}, \mathrm{X})\, t^{\mathrm{rk}(\mathrm{Y})} \\ &= \sum_{\mathrm{Y}} t^{\mathrm{rk}(\mathrm{Y})} \sum_{\mathrm{X}:\, \mathrm{X} \vee \mathrm{Z} = \top,\, \mathrm{X} \geq \mathrm{Y}} \mu(\mathrm{Y}, \mathrm{X}). \end{aligned}$$

Now split this sum into two, depending on whether $\mathrm{Y} \geq \mathrm{Z}$ or not. The second sum is zero by the Weisner formula (1.46a): Use $\top$ for Z, Y for $\perp$, and $\mathrm{Y} \vee \mathrm{Z}$ for Y. In the first sum, since $\mathrm{Y} \geq \mathrm{Z}$, X is forced to be $\top$, and so the sum becomes

$$\sum_{\mathrm{Y}:\mathrm{Y} \geq \mathrm{Z}} \mu(\mathrm{Y}, \top)\, t^{\mathrm{rk}(\mathrm{Y})} = t^{\mathrm{rk}(\mathrm{Z})} \chi(\mathcal{A}_{\mathrm{Z}}, t)$$

as required. □

We refer to (1.55) as the *Crapo identity* for the characteristic polynomial.

1.13.5. Whitney numbers of the first kind. For any arrangement $\mathcal{A}$ and any integer $0 \leq k \leq \mathrm{rk}(\mathcal{A})$, define

$$\mathrm{wy}(\mathcal{A}, k) := \sum_{\mathrm{X}:\, \mathrm{rk}(\mathrm{X}) = k} \mu(\mathrm{X}, \top) = \sum_{\mathrm{X}:\, \mathrm{rk}(\mathrm{X}) = k} \mu(\mathcal{A}_{\mathrm{X}}). \tag{1.56}$$

These are the *Whitney numbers of the first kind*.

Lemma 1.87. *The number* $\mathrm{wy}(\mathcal{A}, k)$ *is the coefficient of* t^k *in the characteristic polynomial* $\chi(\mathcal{A}, t)$.

This follows from the definitions. In the sum (1.56), observe from (1.47) that all summands have the same sign, namely, $(-1)^{\mathrm{rk}(\mathcal{A}) - k}$. The Whitney numbers are positive or negative depending on this sign. Also note that

$$\mathrm{wy}(\mathcal{A}, 0) = \mu(\mathcal{A}), \tag{1.57a}$$

$$\mathrm{wy}(\mathcal{A}, \mathrm{rk}(\mathcal{A})) = 1, \tag{1.57b}$$

$$\sum_k \mathrm{wy}(\mathcal{A}, k) = 0 \quad \text{for } \mathrm{rk}(\mathcal{A}) > 0, \tag{1.57c}$$

$$\sum_k (-1)^k \mathrm{wy}(\mathcal{A}, k) = (-1)^{\mathrm{rk}(\mathcal{A})} c(\mathcal{A}). \tag{1.57d}$$

The last identity follows from Lemma 1.87 and (1.54c).

1.13.6. Examples. The Möbius number and characteristic polynomial of an arrangement only depend on its lattice of flats. Hence, cisomorphic arrangements have the same Möbius number and characteristic polynomial.

For the rank-one arrangement, we have

$$c(\mathcal{A}) = 2, \quad f(\mathcal{A}) = 3, \quad \mu(\mathcal{A}) = -1, \quad \chi(\mathcal{A}, t) = t - 1. \tag{1.58}$$

For the rank-two arrangement of n lines, with $n \geq 2$, we have

$$\begin{aligned} c(\mathcal{A}) &= 2n, \\ f(\mathcal{A}) &= 4n + 1, \\ \mu(\mathcal{A}) &= n - 1, \\ \chi(\mathcal{A}, t) &= t^2 - nt + n - 1. \end{aligned} \tag{1.59}$$

Exercise 1.88. For rank-two arrangements, verify formulas (1.45), (1.52) and (1.55) directly.

Notes

Arrangements. Hyperplane arrangements are treated by Abramenko and Brown [**2**, Chapter 1], Orlik [**324, 325**], Orlik and Terao [**328**] and Stanley [**400**]. Among earlier references, we mention [**214**], [**440**], [**111**] and [**243**]. Short introductions can be found in [**8**, Sections 1.1 and 1.2], [**19**, Section 6], [**103**, Appendix A], [**215**, Chapter 18] and [**401**, Section 3.11]. Oriented matroids and convex polytopes are two notions closely related to hyperplane arrangements. The standard reference for oriented matroids is [**82**]. For convex polytopes, see [**215, 311, 448**].

An arrangement of hyperplanes has an associated matroid. The arrangements under and over a flat correspond to the matroid operations of contraction and restriction, respectively. There is no uniform terminology or notation in the literature for these operations at the level of arrangements. The term localization is sometimes used for $\mathcal{A}_X$ [**121**]. Several authors use restriction for $\mathcal{A}^X$, including Orlik and Terao [**328**, Definition 1.13] (even though this conflicts with the usage in matroid theory).

The lattice of flats is often called the intersection lattice in the literature. Many authors order flats by reverse inclusion, contrary to our convention. More precisely, we view flats as subsets of the ambient space and order them by inclusion. Many authors view them instead as closed subsets of the ground set of the associated matroid (the set of hyperplanes in the arrangement), and order them by inclusion as such. The two choices lead to opposite partial orders, with the latter resulting in a geometric lattice. Chambers are often called regions. They are also called topes in the oriented matroid literature.

The connection between arrangements and polytopes in Lemma 1.33 was brought forth by Ziegler [**447**]. For each hyperplane in the arrangement, pick a nonzero normal segment. Their Minkowski sum is the *zonotope* of the arrangement. The poset of nonempty faces of this polytope is opposite to the poset of faces of the arrangement (when both are ordered by inclusion).

Tits monoid and left regular bands. The Tits product appeared in the work of Tits on Coxeter complexes and buildings [**415**, Section 2.30]. He used the notation $\mathrm{proj}_F G$ instead of FG, since he viewed this operation as a geometric tool rather than as a product. The associativity and Lemma 1.6, item (1) is given in [**416**, Proposition 1]. Bland considered this product in the context of oriented matroids [**87**, Section 5, page 62]. Also see [**328**, Definition 2.21 and Proposition 2.22] and [**2**, Section 1.4.6].

The fact that the Tits monoid is a LRB was first observed by Brown [**103**]. The fact $\Sigma_F = \Sigma_G$ implies $F = G$ (contained in Exercise 1.39) says that the Tits monoid is $\mathcal{R}$-trivial, that is, the Green's relation $\mathcal{R}$ is simply equality. This is true for any LRB.

The result of Exercise 1.40 shows that the Green's relations $\mathcal{L}$ and $\mathcal{J}$ are not trivial in general. Similarly, the result of Exercise 1.41 shows that the Janus monoid is not $\mathcal{R}$-trivial, $\mathcal{L}$-trivial or $\mathcal{J}$-trivial in general. The original source for Green's relations is [**206**]. For information related to $\mathcal{R}$-trivial monoids, see [**405**, Chapter 2] and references therein. More references on LRBs can be found in the notes to Appendix E.

Simplicial arrangements. Information related to classification of simplicial rank-three arrangements can be found in the survey article [**216**] by Grünbaum, with advances in recent work of Cuntz [**132**]. An interesting class of simplicial arrangements with close connections to reflection arrangements has arisen in recent years in work of Cuntz, Heckenberger and Welker [**131, 133, 226**].

Subarrangements. The monoid morphism (1.15) is considered in [**296**, Section 2.6 and Appendix A.5], along with examples. A similar idea occurs in [**29**, proof of Proposition 3.1].

Chamber graph, separating hyperplanes and weak order. The fact that the chamber graph is bipartite goes back to Eberhard [**168**], for affine arrangements of lines in the plane. See also [**383, 213**]. For more information on separating hyperplanes, see [**400**, Chapter 6]. For each chamber D, one can define a partial order on the set of chambers: $C_1 \leq C_2$ if $C_2 \,\text{--}\, C_1 \,\text{--}\, D$. This is often called the weak order in the literature. Phrased in terms of separating hyperplanes, it appears in work of Edelman [**170**]. In the more general context of oriented matroids, it appears in Mandel's thesis [**300**]. For reflection arrangements, the weak order on chambers corresponds to the weak order on the Coxeter group. For the latter, see (5.8) and for more details [**80**, Chapter 3]. For simplicial arrangements, Björner, Edelman and Ziegler [**81**] showed that this poset is always a lattice; also see [**82**, Proposition 4.4.5]. For further results, see [**352, 353, 309**]. The weak order is a connected component of the poset of pairs of chambers in Section 1.12.1. The latter is considered in [**8**, Definition 5.2.2]. Lemma 1.71 is [**8**, Fact 7.3.1]. The partial order on pairs of faces dates back to the same time.

Characteristic polynomial. Information on the characteristic polynomial can be found in [**400**] and [**19**, Section 6.2]. This includes the *deletion-restriction* or *deletion-contraction theorem*. We present this result in Section 12.4.7, as an application of some of the theory to be developed. An important technique discussed in those references but not here is the *finite field method*. For more on this, see [**26**] and [**18**].

Kung provides a wealth of information on the characteristic polynomial of geometric lattices in [**267**, Section 5]. The identity (1.55) is a special case of a result of Crapo [**128**, Corollary 5]. The version in [**267**, Section 5.2] involves an arbitrary subarrangement instead of one of the form $\mathcal{A}_{\mathrm{Z}}$.

Information on the Whitney numbers of the first and second kind of a graded poset is given in [**13**, Pages 155-156]. It follows from Proposition 1.80 that the Whitney numbers of the first kind of an arrangement alternate in sign. An expression for these quantities in terms of volumes is given in [**257**, Theorem 5]. The absolute values of the Whitney numbers of the first kind constitute a *log-concave sequence*. This long-standing conjecture has been settled in recent years by Huh [**235**] and extended to matroids in [**236**] and [**3**].

Zaslavsky formula. Enumeration of faces in arrangements has a long history going back at least to Steiner [**406**]. See [**214**, page 46] for a more complete list of references.

Theorem 1.81 (and Corollary 1.82) as well as its proof is due to Zaslavsky [**440**, Theorem A]. The proof has its origins in work of Buck [**108**]. The result was discovered independently by Las Vergnas in the more general context of oriented matroids [**271**, Proposition 8.1] or [**274**, Theorem 3.1]. The Zaslavsky formula is also discussed in [**82**, Theorem 4.6.1], [**328**, Theorem 2.68] or [**400**, Theorem 2.5]. Zaslavsky also considered very general topological dissections. His fundamental theorem of dissection theory is given in [**439**, Theorem 1] or [**441**, Theorem 1.2]. For related recent work, see [**175**, Theorems 3.6 and 3.11] and [**147**, Theorems 4.6 and 4.9]. The latter paper also presents a nice survey. For generalizations to left regular bands, see [**304**, Section 9] and references therein.

Formula (1.52) is due to Winder [**434**]. Formula (1.51) is given in [**328**, Lemma 2.35]. It is a special case of the Crosscut Theorem [**401**, Corollary 3.9.4], also see [**401**, Proposition 3.11.3].

The Zaslavsky formula expresses the number of chambers, or more generally the number of faces of a given rank, in terms of the lattice of flats (see Exercise 1.83). More generally, the *flag f-vector* of the face poset is determined by the lattice of flats, see [**48**, Theorem 3.4, Corollary 4.6] and [**64**, Theorem 3.1]. For a related result, see [**82**, Proposition 4.6.2]. More references related to the Zaslavsky formula are given in the notes to Chapter 6 in connection with graphic arrangements.

CHAPTER 2

Cones

Cones form a nice general class of objects which encompass many other objects associated to hyperplane arrangements. For instance, faces, flats and half-spaces are examples of cones. Cones are related to the geometric notion of convexity. Each cone has a convexity dimension. Cones of convexity dimension either 1 or 2 are called gallery intervals. The support map on faces extends to cones. We call it the case map. Top-cones are cones whose case is the maximum flat. There is another map from cones to flats which we call the base map. The poset of top-cones is join-distributive, and in particular, upper semimodular and graded. We also discuss charts and dicharts and relate them to flats and cones, respectively. Finally, we introduce the notion of a partial-flat as an interpolating object between faces and flats. Partial-flats are also cones.

Adjunctions between posets play an important role in this chapter. Our examples include adjunctions between faces and top-cones, between cones and flats, between flats and charts, between charts and dicharts and between cones and dicharts. Background information on adjunctions is given in Section B.5.

2.1. Cones and convexity

We begin with cones. Cones bear to half-spaces the same relation that flats bear to hyperplanes. They can be characterized by using a combinatorial notion of convexity involving minimal galleries.

2.1.1. Cones. A *cone* of an arrangement $\mathcal{A}$ is a subset of the ambient space which can be obtained by intersecting some subset of half-spaces in the arrangement.

Let $\Omega[\mathcal{A}]$ denote the set of all cones. It is a poset under inclusion. The center of $\mathcal{A}$ is the minimum element. Since the intersection of two cones is a cone, meets exist in this poset. Further, it has a maximum element, namely, the ambient space, so joins exist as well, and $\Omega[\mathcal{A}]$ is a lattice. Explicitly, the join of two cones is the intersection of those half-spaces which contain both of them. We will usually denote cones by V and W; we will denote their meet by $\mathrm{V} \wedge \mathrm{W}$ and join by $\mathrm{V} \vee \mathrm{W}$.

Recall the poset of faces $\Sigma[\mathcal{A}]$. Every face is a cone. Further, if a cone V is smaller than a face G, that is, if $\mathrm{V} \leq G$, then V is necessarily a face. It follows that $\Sigma[\mathcal{A}]$ is a convex subposet of $\Omega[\mathcal{A}]$. Hence the inclusion $\Sigma[\mathcal{A}] \hookrightarrow \Omega[\mathcal{A}]$ always preserves meets. It preserves joins whenever they exist in $\Sigma[\mathcal{A}]$. This makes the notations $F \vee G$ and $F \wedge G$ unambiguous. Note very carefully that $F \vee G$ is in general only a cone (and not a face).

Recall the lattice of flats $\Pi[\mathcal{A}]$. Every flat is a cone. This yields a map $\Pi[\mathcal{A}] \hookrightarrow \Omega[\mathcal{A}]$. This is a lattice homomorphism. This makes the notations $\mathrm{X} \vee \mathrm{Y}$ and $\mathrm{X} \wedge \mathrm{Y}$ unambiguous. We point out that $\Pi[\mathcal{A}]$ is *not* a convex subposet of $\Omega[\mathcal{A}]$ in general: the ambient space and the center are both flats, but clearly not everything in-between is a flat.

2.1.2. Combinatorial cones. Recall that every point in the ambient space lies in the interior of a unique face. Suppose V is a cone. If a point lies in V, then it follows that the corresponding face also lies in V. Thus every cone is a union of faces. Also observe that if faces F and G lie in V, then FG also lies in V. By multiplying all the faces lying in V in different orders we obtain the "largest" faces lying in V. They are all of the same support. We call these the top-dimensional faces of V. Since faces are closed by convention, we obtain:

Proposition 2.1. *A cone is the union of its top-dimensional faces.*

This allows us to take a combinatorial approach to cones, generalizing what we did for flats.

A *combinatorial cone* is a subset of the set of faces consisting of precisely the top-dimensional faces of some cone. A combinatorial cone and a cone are equivalent notions but they are different kinds of objects. The former is a set of faces (all of the same dimension) while the latter is a subset of the ambient space.

2.1.3. Top-cones. A cone has a dimension, namely, the dimension of any of its top-dimensional faces. A cone of maximum dimension is called a *top-cone*.

Proposition 2.2. *For a cone* V*, the following are equivalent.*

- V *is a top-cone.*
- V *contains at least one chamber.*
- *The top-dimensional faces of* V *are chambers.*
- *No hyperplane contains* V*.*

A *combinatorial top-cone* is a combinatorial cone which consists of chambers. This notion is equivalent to that of a top-cone.

The figure on the left shows a top-cone with four chambers, while the figure on the right shows a top-cone with eight chambers (three of which are only partly visible).

Let $\widehat{\Omega}[\mathcal{A}]$ denote the set of all top-cones. It is a poset under inclusion. It is an upper set in the poset of all cones, and hence a join-semilattice. The ambient space is the maximum element, while each chamber is a minimal element. Note that the meet of two distinct chambers does not exist in this poset.

2.1.4. Convexity. In any pure regular cell complex which is gallery connected, one has the notion of convexity (Section A.2.2).

Proposition 2.3. *Let* A *be a nonempty set of chambers. Then the following are equivalent.*

(1) A *is a combinatorial top-cone.*
(2) A *is convex.*
(3) *For any* $C, D \in \mathrm{A}$ *and* $F \leq C$*, we have* $FD \in \mathrm{A}$.
(4) *For any* $C, D \in \mathrm{A}$ *and* F *a panel of* C*, we have* $FD \in \mathrm{A}$.

PROOF. (1) implies (2). This is a consequence of (1.28). We could also do this argument in two steps. First use (1.28) to deduce that a combinatorial half-space is convex. Next use that intersection of convex sets is convex.

(2) implies (3). This follows from the gate property (Proposition 1.61).

(3) implies (4). Clear.

(4) implies (1). Let C and E be adjacent chambers such that C belongs to A but E does not belong to A, and let h be the half-space which contains C but does not contain E. Let V be the cone obtained by intersecting all half-spaces h which can be obtained in this manner. We claim that A is the set of chambers contained in V.

For the purpose of argument, let B denote the set of chambers contained in V. We claim that any chamber in A is contained in any of the chosen half-spaces h: Let h, C and E be as above, and let F be the common panel of C and E. Suppose $D \in \mathrm{A}$. Since $E \notin \mathrm{A}$, we have $FD \neq E$. Thus $FD = C$, and D belongs to h. Hence, A is a subset of B. For the reverse containment, observe that if C belongs to A, and D is adjacent to C and belongs to B, then in fact D belongs to A. Since B is a combinatorial cone, it is convex and hence gallery connected. It follows that A equals B, as required. □

By passage to arrangements under flats, Proposition 2.3 yields a characterization of all combinatorial cones:

Proposition 2.4. *Let* A *be a set of faces all with the same support, say* X*. Then the following are equivalent.*

(1) A *is a combinatorial cone.*
(2) A *is a convex set of chambers in* $\mathcal{A}^{\mathrm{X}}$.
(3) *For any* $G, H \in \mathrm{A}$ *and* $F \leq G$*, we have* $FH \in \mathrm{A}$.

Exercise 2.5. A subset of a real vector space is *convex* if it is closed under nonnegative linear combinations of vectors. Let A be any set of chambers in an arrangement, and X be the corresponding subset of the ambient space. Show that A is convex iff X is convex.

Exercise 2.6. Show that the convex closure of a set of chambers is their join in the poset of top-cones.

2.1.5. Closure of combinatorial cones.

For a combinatorial cone V, define its *closure* to be

$$\mathrm{Cl}(\mathrm{V}) = \{F \in \Sigma[\mathcal{A}] \mid F \leq G \text{ for some } G \in \mathrm{V}\}. \tag{2.1}$$

This is the same as the set of faces contained in the corresponding (geometric) cone. It follows that

$$\mathrm{V} \leq \mathrm{W} \iff \mathrm{Cl}(\mathrm{V}) \subseteq \mathrm{Cl}(\mathrm{W}) \tag{2.2}$$

and

$$\mathrm{Cl}(\mathrm{V} \wedge \mathrm{W}) = \mathrm{Cl}(\mathrm{V}) \cap \mathrm{Cl}(\mathrm{W}). \tag{2.3}$$

Compare the above with the discussion on combinatorial flats in Section 1.3.3. The following characterization parallels Proposition 1.16.

Proposition 2.7. *Let* A *be any set of faces of* $\mathcal{A}$. *Then* A *is the closure of a combinatorial cone iff the following properties hold.*

(1) $O \in \mathrm{A}$.
(2) *If* $G \in \mathrm{A}$ *and* $F \leq G$, *then* $F \in \mathrm{A}$.
(3) *If* $F \in \mathrm{A}$ *and* $G \in \mathrm{A}$, *then* $FG \in \mathrm{A}$.

Proof. The necessity of the properties is clear. For sufficiency: By property (1), A is nonempty. By property (3), the largest faces in A are obtained by multiplying all the faces in A in different orders. They are all of the same support. It remains to show that this set is a combinatorial cone. This follows from the characterization given in Proposition 2.4. □

Exercise 2.8. Let F be a face and V be a cone. Then F belongs to the closure of V iff all vertices of F belong to the closure of V.

Exercise 2.9. If V and W are combinatorial top-cones with nonempty intersection, that is, there is a chamber common to both, then $\mathrm{V} \wedge \mathrm{W}$ is a top-cone and $\mathrm{V} \wedge \mathrm{W} = \mathrm{V} \cap \mathrm{W}$.

2.1.6. Faces and top-stars. Recall from Section 1.7.3 the top-star $\Gamma[\mathcal{A}]_F$ of a face F. It consists of chambers greater than F. Applying Proposition A.5, we deduce that it is convex. Combining with Proposition 2.3, we obtain: Any top-star is a combinatorial top-cone.

Consider the maps

$$\varphi : \Sigma[\mathcal{A}] \to \widehat{\Omega}[\mathcal{A}], \qquad F \mapsto \Gamma[\mathcal{A}]_F \tag{2.4}$$

and

$$\psi : \widehat{\Omega}[\mathcal{A}] \to \Sigma[\mathcal{A}], \qquad \mathrm{V} \mapsto \bigwedge_{C \in \mathrm{V}} C. \tag{2.5}$$

Both maps are order-reversing. Further, for any face F and top-cone V,

$$\mathrm{V} \leq \varphi(F) \iff F \leq \psi(\mathrm{V}).$$

Both sides are equivalent to the statement that all chambers of V are greater than F. Also $\psi\varphi = \mathrm{id}$. In particular, φ is injective, while ψ is surjective.

By reversing the partial order on either $\Sigma[\mathcal{A}]$ or $\widehat{\Omega}[\mathcal{A}]$, both φ and ψ become order-preserving and define an adjunction (Section B.5). Since left (right) adjoints preserve joins (meets) whenever they exist, we deduce the following. For joinable faces F and G, and any top-cones V and W,

$$\varphi(F \vee G) = \varphi(F) \wedge \varphi(G) \quad \text{and} \quad \psi(\mathrm{V} \vee \mathrm{W}) = \psi(\mathrm{V}) \wedge \psi(\mathrm{W}).$$

The first identity is the same as the second identity in Exercise 1.39. Facts relevant to the second identity: The poset of top-cones is a join-semilattice, while the poset of faces is a meet-semilattice.

The maps φ and ψ induce inverse bijections between the set of faces and the set of top-stars. In fact, the poset of faces is isomorphic to the dual of the poset of top-stars, viewed as a subposet of top-cones. This is a specialization of (B.4). The above fact is also given in the first identity in Exercise 1.39.

2.1.7. Walls of a cone. Let V be a combinatorial top-cone. A *wall* of V is a hyperplane H for which there exist adjacent chambers C and D whose common panel has support H, and such that C belongs to V but D does not belong to V. This notion extends to any cone: use this definition in the arrangement under the flat which supports the top-dimensional faces of the given cone.

Proposition 2.10. *A cone is determined by its walls. More precisely, for a cone* V*,*

$$\mathrm{V} = \bigcap \mathrm{H}^{\epsilon},$$

where the intersection is over all walls H *of* V*, and* H^{ϵ} *denotes the half-space containing* V *whose bounding hyperplane is* H*. Further, this expression of* V *is minimal. That is, if* V *is the intersection of some half-spaces, then the bounding hyperplanes of these half-spaces must include the walls of* V*. (The half-spaces and hyperplanes are in the arrangement under a flat.)*

PROOF. This follows from the proof of Proposition 2.3. □

In the picture below, the top-cone on the left has three walls (indicated by the thick lines), while the one on the right has two.

We note some other simple examples.

- Any chamber C is a top-cone. The walls of C are precisely the supports of the panels of C.
- Let F be a face. A wall of the top-star $\Gamma[\mathcal{A}]_F$ is precisely a wall of some chamber E greater than F which does not contain F.
- The ambient space has no walls. More generally, a flat has no walls. In fact, a cone has no walls iff that cone is a flat.

2.1.8. Opposition map on cones. Recall that every half-space h has an opposite half-space $\overline{\mathrm{h}}$ (Section 1.1.4). More generally, every cone V has an *opposite cone*, denoted $\overline{\mathrm{V}}$, which is given by

$$\overline{\mathrm{V}} := \{-x \mid x \in \mathrm{V}\}.$$

This is the cone obtained by intersecting the half-spaces opposite to those that define V. If V is a face, then $\overline{\mathrm{V}}$ is precisely its opposite face. In other words, the opposition map (1.1) extends to

$$\Omega[\mathcal{A}] \to \Omega[\mathcal{A}], \qquad \mathrm{V} \mapsto \overline{\mathrm{V}}. \tag{2.6}$$

It continues to be an order-preserving involution. In particular,

$$\overline{\mathrm{V} \wedge \mathrm{W}} = \overline{\mathrm{V}} \wedge \overline{\mathrm{W}} \quad \text{and} \quad \overline{\mathrm{V} \vee \mathrm{W}} = \overline{\mathrm{V}} \vee \overline{\mathrm{W}}.$$

Proposition 2.11. *Suppose* V *is a cone. Then* V *is a flat iff* $\overline{\mathrm{V}} = \mathrm{V}$*. In particular, for any cone* V*, the cones* $\mathrm{V} \wedge \overline{\mathrm{V}}$ *and* $\mathrm{V} \vee \overline{\mathrm{V}}$ *are flats.*

PROOF. The forward implication is clear, since a flat is a subspace of the ambient space. For the backward implication: V is an intersection of some subset of half-spaces. For any such half-space h, V must be contained in h as well as in $\overline{\mathrm{h}}$, so it is contained in the boundary of h. Thus V is an intersection of some subset of hyperplanes and hence a flat. Alternatively: Suppose F is any top-dimensional face of V. Then so is $\overline{F}$, by hypothesis. By Proposition 1.64, $F \,\text{--}\, G \,\text{--}\, \overline{F}$ for any face G with the same support as F, hence by Proposition 2.4, all such G are top-dimensional faces of V. So V is a flat. □

Exercise 2.12. Show that: Cones V and $\overline{\mathrm{V}}$ are either equal, or do not have any top-dimensional faces in common.

Exercise 2.13. Show that: Cones V and $\overline{\mathrm{V}}$ have the same set of walls. In particular, chambers C and $\overline{C}$ have the same set of walls.

A *projective cone* is an unordered pair consisting of a cone and its opposite. We denote a projective cone by $\{\mathrm{V}, \overline{\mathrm{V}}\}$. The number of projective cones equals the number of flats plus half the number of cones which are not flats.

2.2. Case and base maps

The support map from faces to flats extends to a map from cones to flats. We call it the case map. It is the left adjoint to the inclusion from flats into cones. The inclusion map also has a right adjoint which we call the base map. The case of a cone is the join of that cone with its opposite, while the base of a cone is the meet of that cone with its opposite.

2.2.1. Case map.

We extend the support map on faces to cones. The *support* or *case* of a cone V is the smallest flat which contains V. It is the meet of all flats which contain V, or equivalently, the intersection of all hyperplanes which contain V. The *case map*

$$\mathrm{c} : \Omega[\mathcal{A}] \twoheadrightarrow \Pi[\mathcal{A}] \tag{2.7}$$

sends a cone to its case. It is order-preserving.

Lemma 2.14. *For any cone* V,

$$\mathrm{c}(\mathrm{V}) = \mathrm{V} \vee \overline{\mathrm{V}}. \tag{2.8}$$

In particular, $\mathrm{c}(\mathrm{V}) = \mathrm{c}(\overline{\mathrm{V}})$.

PROOF. This follows from two facts. A flat which contains V necessarily contains $\overline{\mathrm{V}}$, and $\mathrm{V} \vee \overline{\mathrm{V}}$ is a flat (Proposition 2.11). □

By Proposition 2.1, the case of V is the support of any of its top-dimensional faces.

The case of a face is its support. In particular, the case of any chamber is the maximum flat, and the case of a panel is a hyperplane. The case of any flat is the flat itself. Note that a cone is a top-cone iff its case is the maximum flat.

By construction, for any cone V and flat X,

$$\mathrm{c}(\mathrm{V}) \leq_{\Pi[\mathcal{A}]} \mathrm{X} \iff \mathrm{V} \leq_{\Omega[\mathcal{A}]} \mathrm{X}.$$

Thus, the case map and the inclusion map define an adjunction, with the case map being the left adjoint of the inclusion map, see (B.2). As a formal consequence, the case map preserves joins:

$$\mathrm{c}(\mathrm{V} \vee \mathrm{W}) = \mathrm{c}(\mathrm{V}) \vee \mathrm{c}(\mathrm{W}).$$

It does not preserve meets in general. For instance, consider two cones with the same case $\mathrm{X} \neq \perp$ whose meet is $\perp$. Or, consider two top-cones whose meet is not a top-cone.

Exercise 2.15. Show that: The case of V is the smallest subspace of the ambient space which contains V.

2.2.2. Base map. The *base* of a cone V is the largest flat contained in V. It is the join of all flats contained in V. The *base map*

$$\mathrm{b} : \Omega[\mathcal{A}] \twoheadrightarrow \Pi[\mathcal{A}] \tag{2.9}$$

sends a cone to its base. It is order-preserving.

Lemma 2.16. *For any cone* V*,*

$$\mathrm{b}(\mathrm{V}) = \mathrm{V} \wedge \overline{\mathrm{V}}. \tag{2.10}$$

In particular, $\mathrm{b}(\mathrm{V}) = \mathrm{b}(\overline{\mathrm{V}})$.

Proof. This follows from two facts. A flat which is contained in V is necessarily contained in $\overline{\mathrm{V}}$, and $\mathrm{V} \wedge \overline{\mathrm{V}}$ is a flat (Proposition 2.11). □

The base of any face is the minimum flat, the base of any flat is the flat itself, and the base of a half-space is its bounding hyperplane. We deduce that a flat is contained in a half-space iff it is contained in the base of that half-space. (For the forward implication, apply the base map.)

By construction, for any cone V and flat X,

$$\mathrm{X} \leq_{\Omega[\mathcal{A}]} \mathrm{V} \iff \mathrm{X} \leq_{\Pi[\mathcal{A}]} \mathrm{b}(\mathrm{V}).$$

Thus, the base map and the inclusion map define an adjunction, with the base map being the right adjoint of the inclusion map, see (B.2). As a formal consequence, the base map preserves meets:

$$\mathrm{b}(\mathrm{V} \wedge \mathrm{W}) = \mathrm{b}(\mathrm{V}) \wedge \mathrm{b}(\mathrm{W}).$$

It does not preserve joins in general. For instance, consider a noncentral face and its opposite.

Proposition 2.17. *The base of a cone* V *is the intersection of the bases of any set of half-spaces which define* V*. In particular, the base of a cone is the intersection of its walls.*

Proof. Let A be any set of half-spaces whose intersection is V. Let X denote the intersection of the bases of the half-spaces in A. Then X is a flat and it is contained in V, hence it is contained in b(V). Conversely: Since V is contained in the half-spaces in A, b(V) is contained in their bases, and hence in X. Therefore $\mathrm{X} = \mathrm{b}(\mathrm{V})$.

For the second part: The base of a cone remains unchanged if we pass to the arrangement under the case of the cone. Now apply Proposition 2.10. □

Exercise 2.18. Let V be a cone. Show that b(V) is the minimum flat iff Cl(V) does not contain a pair of opposite vertices. Give an example of a top-cone which is not a chamber but whose base is the minimum flat.

Exercise 2.19. Show that if the base of a cone is a hyperplane then it is either a half-space or a hyperplane.

Exercise 2.20. Show that: The base of V is the largest subspace of the ambient space which is contained in V.

2.2.3. Rank one. The poset of cones for the rank-one arrangement has four elements, namely, the minimum and maximum flats and the two chambers. The case and base maps are illustrated below.

The case map takes both chambers to the maximum flat, while the base map takes them to the minimum flat. Both maps preserve the minimum and maximum flats.

2.2.4. Half-flats. We introduce the notion of a half-flat which generalizes the notion of a half-space.

A *half-flat* is a cone V such that b(V) is a codimension-one subspace of c(V), that is, $\mathrm{b(V)} \lessdot \mathrm{c(V)}$ in the poset of flats. Equivalently, a cone is a half-flat iff it can be written in the form $\mathrm{X} \wedge \mathrm{h}$, where X is a flat and h is a half-space whose base does not contain X. For any flat X, half-flats with case X correspond to half-spaces in the arrangement $\mathcal{A}^{\mathrm{X}}$.

Vertices of $\mathcal{A}$ are half-flats: their base has rank 0 while their case has rank 1.

2.2.5. Subarrangements. Let $\mathcal{A}'$ be a subarrangement of $\mathcal{A}$, that is, the set of hyperplanes in $\mathcal{A}'$ is a subset of the set of hyperplanes in $\mathcal{A}$. It follows that the set of flats (cones) of $\mathcal{A}'$ is contained in the set of flats (cones) of $\mathcal{A}$. Further, these containments are compatible with the inclusion, base and case maps. That is, the following diagrams commute.

$$\begin{array}{ccc} \Omega[\mathcal{A}'] & \longrightarrow & \Omega[\mathcal{A}] \\ {\scriptstyle \mathrm{i}}\uparrow & & \uparrow{\scriptstyle \mathrm{i}} \\ \Pi[\mathcal{A}'] & \longrightarrow & \Pi[\mathcal{A}] \end{array} \qquad \begin{array}{ccc} \Omega[\mathcal{A}'] & \longrightarrow & \Omega[\mathcal{A}] \\ {\scriptstyle \mathrm{c}}\downarrow & & \downarrow{\scriptstyle \mathrm{c}} \\ \Pi[\mathcal{A}'] & \longrightarrow & \Pi[\mathcal{A}] \end{array} \qquad \begin{array}{ccc} \Omega[\mathcal{A}'] & \longrightarrow & \Omega[\mathcal{A}] \\ {\scriptstyle \mathrm{b}}\downarrow & & \downarrow{\scriptstyle \mathrm{b}} \\ \Pi[\mathcal{A}'] & \longrightarrow & \Pi[\mathcal{A}] \end{array}$$

To see this, view all sets as subsets of the power set of the ambient space. The first diagram is clear, while the remaining two follow from Exercises 2.15 and 2.20.

Exercise 2.21. Let $\mathcal{A}'$ be a subarrangement of $\mathcal{A}$. Show that: A top-cone in $\mathcal{A}'$ is also a top-cone in $\mathcal{A}$ with the same set of walls. Conversely, a top-cone in $\mathcal{A}$ is a top-cone in $\mathcal{A}'$ iff all its walls belong to $\mathcal{A}'$.

Exercise 2.22. Show that: Given any cone V of $\mathcal{A}$, there exists a subarrangement $\mathcal{A}'$ which has V as a face. There will be many choices for $\mathcal{A}'$ in general.

2.3. Topology of a cone

In the spherical model, a cone has the topology of either a ball or a sphere. We also discuss related notions of boundary and interior of a cone.

2.3.1. Boundary and interior of a cone. Let V be a combinatorial cone. A face F in the closure of V is in the *boundary* of V if it is contained in some wall of V, else it is in the *interior* of V. We write

$$\mathrm{Cl}(\mathrm{V}) = \mathrm{V}^o \sqcup \mathrm{V}^b, \tag{2.11}$$

where V^o is the set of faces in the interior, and V^b is the set of faces in the boundary.

Note that the top-dimensional faces of V belong to V^o. Recall that flats are precisely those cones which have no walls. Hence, for a cone V,

$$\mathrm{V} \text{ is a flat} \iff O \in \mathrm{V}^o \iff \mathrm{Cl}(\mathrm{V}) = \mathrm{V}^o.$$

Proposition 2.23. *Let* V *be a combinatorial cone and H be a face. Then:*

- *If* $\mathrm{s}(H) \leq \mathrm{b}(\mathrm{V})$ *and* $G \in \mathrm{V}^b$, *then* $HG \in \mathrm{V}^b$.
- *If* $H \in \mathrm{Cl}(\mathrm{V})$ *and* $G \in \mathrm{V}^o$, *then* $HG \in \mathrm{V}^o$.

PROOF. Suppose $\mathrm{s}(H) \leq \mathrm{b}(\mathrm{V})$ and $G \in \mathrm{V}^b$. So G is contained in some wall, say H, of V. Since H is contained in the base, it is contained in every wall, and in particular in H. Therefore HG is contained in H, and hence it belongs to the boundary.

Suppose $H \in \mathrm{Cl}(\mathrm{V})$ and $G \in \mathrm{V}^o$. If HG belonged to the boundary, then it would be contained in some wall of V, which would then force G to belong to this wall, and hence belong to the boundary. Thus HG belongs to the interior. □

Exercise 2.24. Let V be a combinatorial top-cone and H be a face. Show that the following are equivalent.

- $H \in \mathrm{V}^o$.
- All chambers greater than H belong to V.
- There exists a chamber D greater than H such that D and $H\overline{D}$ both belong to V.
- For some flat X containing H, all faces greater than H and with support X belong to V^o.

For example: Any chamber of V lies in the interior of V. A face F always lies in the interior of its top-star Γ_F.

Exercise 2.25. Show that: For cones V and W,

$$\mathrm{Cl}(\mathrm{V}) \subseteq \mathrm{Cl}(\mathrm{W}) \iff \text{Either } \mathrm{V}^o \subseteq \mathrm{W}^o \text{ or } \mathrm{Cl}(\mathrm{V}) \subseteq \mathrm{W}^b.$$

Note that the two conditions in the rhs are mutually exclusive. In particular, if V and W have the same support, then $\mathrm{Cl}(\mathrm{V}) \subseteq \mathrm{Cl}(\mathrm{W})$ iff $\mathrm{V}^o \subseteq \mathrm{W}^o$.

2.3.2. Topology of a cone. Put a norm on the ambient space of an arrangement. If the arrangement has at least one hyperplane, then the intersection of any chamber C with the unit sphere is a topological ball: This is well-known in the essential case. In the general case, the intersection is the join of a sphere and a ball, which is a ball. (The sphere comes from the center, and the ball from the essential part of the arrangement.)

Proposition 2.26. *Put a norm on the ambient space of an arrangement. The intersection of any cone* V *of the arrangement with the unit sphere is either a topological ball or sphere. The latter happens iff* V *is a flat.*

We adopt the convention that the empty set is the sphere of dimension -1.

PROOF. We may assume that V is a top-cone. If V equals the ambient space, then the intersection is clearly a topological sphere. Assume that this is not the case. Consider the subarrangement whose hyperplanes are the bases of the half-spaces which define V. Its center is b(V), while V is a chamber of this subarrangement (which has rank at least one). So the intersection is a topological ball. □

Thus:

Proposition 2.27. *In the spherical model, a cone* V *of an arrangement is either a topological ball or sphere. The latter happens iff* V *is a flat.*

The dimension of the sphere or ball is $\dim(\mathrm{V}) - \dim(O) - 1$. If the arrangement is essential, then this is simply $\dim(\mathrm{V}) - 1$.

For any cone V, there is a cell complex whose cells correspond to faces in Cl(V). By taking negative of its reduced Euler characteristic (A.1), we obtain: For any cone V,

$$\sum_{F \in \mathrm{Cl}(\mathrm{V})} (-1)^{\mathrm{rk}(F)} = \begin{cases} (-1)^{\dim(\mathrm{V})-\dim(O)} & \text{if V is a flat,} \\ 0 & \text{otherwise.} \end{cases} \tag{2.12}$$

We have encountered special cases of this identity before when V is either a flat or a face, see (1.41), (1.42) and (1.44).

Exercise 2.28. Show that: For any cone V,

$$\sum_{F \in \mathrm{V}^o} (-1)^{\mathrm{rk}(F)} = (-1)^{\dim(\mathrm{V})-\dim(O)}. \tag{2.13}$$

2.4. Cutting and separating hyperplanes and gated sets

We discussed separating hyperplanes for chambers in Section 1.10. We now extend these ideas to top-cones.

2.4.1. Cutting and separating hyperplanes. Recall that a hyperplane separates two chambers if they lie on opposite sides of that hyperplane. We say that a hyperplane *cuts* a top-cone if it separates some two chambers contained in that top-cone. The two extremes cases are noted below. For a top-cone V,

$$\text{V is a chamber} \iff \text{no hyperplane cuts V,} \tag{2.14a}$$

$$\text{V is the ambient space} \iff \text{all hyperplanes cut V.} \tag{2.14b}$$

We say that a hyperplane *separates* two top-cones if it separates every chamber in one top-cone from every chamber in the other top-cone. Equivalently, H separates V and W if V lies on one side of H and W on the other side.

Lemma 2.29. *Let* V *be a top-cone and* C *a chamber not contained in* V*. Then there exists a hyperplane which separates* V *and* C*.*

PROOF. Consider all half-spaces which contain V. Since their intersection is V, at least one of them, say h, does not contain C. The bounding hyperplane of h then separates V and C. □

Exercise 2.30. Prove the following generalization of Lemma 2.29. Let V be a top-cone, and let W be a top-cone made up of two (adjacent) chambers C and D neither of which is contained in V. Then there exists a hyperplane which separates V and W.

Exercise 2.31. It is clear that top-cones which can be separated by a hyperplane cannot have any common chamber. Show that the converse is false. In other words, give an example of two top-cones which do not have any common chamber and which cannot be separated by a hyperplane.

Lemma 2.32. *Let* V *be a top-cone. Then:*

- *A hyperplane cuts* V *iff it cuts* $\overline{\mathrm{V}}$.
- *A hyperplane does not cut* V *iff it separates* V *and* $\overline{\mathrm{V}}$.

PROOF. The first claim holds because a hyperplane separates C and D iff it separates $\overline{C}$ and $\overline{D}$. For the second claim: A hyperplane H does not cut V means that V lies on one side of H. In this case, $\overline{\mathrm{V}}$ lies on the other side of H which means that H separates V and $\overline{\mathrm{V}}$. □

More generally, we say that: A hyperplane *cuts* a cone if it separates some two top-dimensional faces contained in that cone. A hyperplane *separates* two cones if it separates every top-dimensional face in one cone from every top-dimensional face in the other cone.

2.4.2. Gated sets. Recall the notion of gated sets from Section A.2.4.

Lemma 2.33. *Let C be a chamber in a top-cone* V, *and D a chamber in a top-cone* W. *Then the following are equivalent.*

(1) V *and* W *are gated wrt each other with* (C, D) *as a gate pair.*
(2) *A hyperplane separates* V *and* W *iff it separates C and D.*

PROOF. We employ (1.28).

(1) implies (2). Suppose H separates C and D. By hypothesis, D is the gate of W wrt C, so $C \text{--} D \text{--} E$ for any $E \in \mathrm{W}$. Thus H separates C and W. By symmetry, H separates D and V. Combining the two, H separates V and W.

(2) implies (1). Suppose H separates C and D. Then by hypothesis, H separates V and W. In particular, H separates C and E for any $E \in \mathrm{W}$. Thus, $C \text{--} D \text{--} E$, and D is the gate of W wrt C. By symmetry, C is the gate of V wrt D. □

Exercise 2.34. By Proposition A.1, a gated set of chambers is convex and hence a combinatorial top-cone. Show that the converse is false. In other words, give an example of a top-cone V and a chamber C such that V is not gated wrt C.

2.4.3. Case and base maps. We now record an interesting result involving separating hyperplanes that connects the case and base maps.

Proposition 2.35. *Let $\mathcal{A}$ be a simplicial arrangement. Then for any chambers C and D, the support of $C \wedge D$ equals the intersection of all hyperplanes which separate C and D. In other words, by Proposition* 2.17,

$$\mathrm{s}(C \wedge D) = \mathrm{b}(\overline{C} \vee D).$$

The join $\overline{C} \vee D$ is in the poset of top-cones.

PROOF. Let $\mathrm{X}(C, D)$ denote the intersection of all hyperplanes which separate C and D. We do an induction on the distance of D from C. The base step is $C = D$. In this case, no hyperplane separates C and D, so $\mathrm{X}(C, D)$ is equal to the maximum flat, which is the same as the support of $C \wedge D = C$.

For the induction step, suppose $C \text{--} D \text{--} E$, with D strictly between C and E. Put $F := C \wedge D$ and $K := D \wedge E$. By hypothesis, $\mathrm{s}(F) = \mathrm{X}(C, D)$ and

$s(K) = X(D, E)$. By convexity of stars, if a face is common to C and E, then it will also be common to C and D. Hence, $C \wedge E \leq C \wedge D$. Similarly, $C \wedge E \leq D \wedge E$, and we deduce that $C \wedge E = F \wedge K$. Since $X(C, E) = X(C, D) \wedge X(D, E)$, we are reduced to showing that $s(F \wedge K) = s(F) \wedge s(K)$. This holds because F and K are faces of D, and D is a simplex. This completes the induction step. □

More generally:

Proposition 2.36. *Let $\mathcal{A}$ be a simplicial arrangement. Suppose* V *and* W *are top-cones which are gated wrt each other. Then*

$$c(V \wedge W) = b(\overline{V} \vee W). \tag{2.15}$$

The rhs is the same as the intersection of all hyperplanes which separate V *and* W.

PROOF. Let F be any face contained in $V \wedge W$. Then any hyperplane which separates V and W must contain F. This shows that the lhs is contained in the rhs. This is true in general, and does not require the gated assumption. For the reverse inclusion: Let (C, D) be a gate pair. By Lemma 2.33, a hyperplane separates C and D iff it separates V and W. Hence the rhs is the intersection of all hyperplanes which separate C and D. By Proposition 2.35, this is $s(C \wedge D)$, which is contained in the lhs. □

Exercise 2.37. Show that Proposition 2.35 fails in general for nonsimplicial arrangements.

Exercise 2.38. Give an example of top-cones V and W in a simplicial arrangement for which (2.15) fails.

Exercise 2.39. Show that (2.15) holds for any cones V and W which are opposite to each other. This hypothesis is different from the gated hypothesis in Proposition 2.36.

2.5. Gallery intervals

In a pure regular cell complex which is gallery connected, one can talk of gallery intervals (Section A.2.3). We now discuss gallery intervals for hyperplane arrangements. They are examples of cones.

2.5.1. Gallery intervals. A set of chambers is a gallery interval if it can be expressed in the form

$$[C\!:\!D] := \{E \mid C \text{ -- } E \text{ -- } D\}.$$

Gallery intervals also make sense for faces with the same support. If F and G have the same support, then

$$[F\!:\!G] := \{H \mid F \text{ -- } H \text{ -- } G\}.$$

Note that $[C : \overline{C}]$ is the set of all chambers. More generally, $[F : \overline{F}]$ is the set of all faces whose support equals the support of F. In other words, it is a combinatorial flat. We write

$$s(F) = [F\!:\!\overline{F}].$$

We now show that all gallery intervals are combinatorial cones.

Proposition 2.40. *For any chambers C and D, the gallery interval $[C : D]$ is convex and hence a combinatorial top-cone. In other words, $[C:D]$ is the convex closure of $\{C, D\}$. More generally: For faces F and G with the same support, $[F:G]$ is a combinatorial cone.*

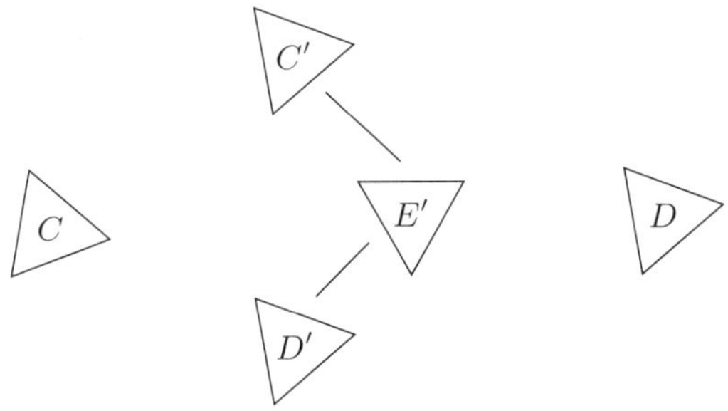

PROOF. We prove the first part. Let C' and D' be any chambers in $[C:D]$. Let E' be a chamber such that $C' \text{--} E' \text{--} D'$. We want to show that E' lies in $[C:D]$. We employ (1.36) and (1.37). Suppose $\mathrm{H} \in g(C, E')$. It is not possible that both $\mathrm{H} \in g(C', E')$ and $\mathrm{H} \in g(D', E')$. Hence, either $\mathrm{H} \in g(C, C')$ or $\mathrm{H} \in g(C, D')$. Since $C \text{--} C' \text{--} D$ and $C \text{--} D' \text{--} D$, in either case we deduce $\mathrm{H} \in g(C, D)$. Hence $C \text{--} E' \text{--} D$ as required.

The second part follows by applying the first part to an arrangement under a flat. ☐

Proposition 2.41. *Let V be a top-cone, and C and D be chambers in V. Then the following are equivalent.*

(1) *V is a gallery interval with $\mathrm{V} = [C:D]$.*
(2) *A hyperplane cuts V iff it separates C and D.*
(3) *A hyperplane separates V and $\overline{\mathrm{V}}$ iff it separates C and $\overline{D}$.*
(4) *V and $\overline{\mathrm{V}}$ are gated wrt each other with $(C, \overline{D})$ as a gate pair.*

PROOF. There are different ways in which one can proceed with the implications.

(1) implies (2). Suppose H separates C and D. Then clearly it cuts V. Conversely, suppose H cuts V. Let C' and D' be chambers in V separated by H. In other words, $\mathrm{H} \in g(C', D')$. By (1.36), either $\mathrm{H} \in g(C, C')$ or $\mathrm{H} \in g(C, D')$. Since $C \text{--} C' \text{--} D$ and $C \text{--} D' \text{--} D$, in either case, by (1.37), $\mathrm{H} \in g(C, D)$. Thus, H separates C and D.

(2) iff (3). Clear from Lemma 2.32.

(3) iff (4). This is a special case of Lemma 2.33.

(4) implies (1). Suppose E is a chamber in V. Then $\overline{D} \text{--} C \text{--} E$, since C is a gate wrt $\overline{D}$. This implies $C \text{--} E \text{--} D$ showing that $\mathrm{V} = [C:D]$. ☐

An illustration is given below.

The top-cone on the left is a gallery interval. The two hyperplanes cutting it separate C and D. The top-cone on the right is *not* a gallery interval. There are three hyperplanes which cut it but at most two of them separate C and D, no matter what D is. This is a top-cone whose convexity dimension is 3 (Section 2.5.3).

The result below is contained in Proposition 2.41 but worth stating separately.

Proposition 2.42. *Let* V *be a top-cone. Then* V *is a gallery interval iff* V *and* $\overline{\mathrm{V}}$ *are gated wrt each other.*

Exercise 2.43. Check that in a rank-two arrangement all top-cones are gallery intervals. The illustration above on the right shows that this fails in rank 3.

Exercise 2.44. Verify Proposition 2.41 directly when V is the ambient space.

Exercise 2.45. Show that for any face F, its top-star Γ_F is a gallery interval. In fact, for any chamber C greater than F,

$$\Gamma_F = [C : F\overline{C}].$$

Thus, a top-star is a nice example of a top-cone which can be realized as a gallery interval in multiple ways.

2.5.2. Join of faces in the poset of cones. The join of two distinct chambers does not exist in the poset of faces $\Sigma[\mathcal{A}]$. But it exists in the poset of top-cones $\widehat{\Omega}[\mathcal{A}]$ (since the latter is a join-semilattice). Explicitly, by Proposition 2.40 and Exercise 2.6,

$$C \vee D = [C : D]. \tag{2.16}$$

We emphasize that this is an identity of combinatorial top-cones. It follows that a top-cone V is a gallery interval iff V is the join of two minimal elements of $\widehat{\Omega}[\mathcal{A}]$.

If a cone contains F and G in its closure, then it also contains FG and GF. It follows that

$$F \vee G = FG \vee GF = [FG : GF], \tag{2.17}$$

with the join taken in the poset of cones $\Omega[\mathcal{A}]$. If the join of F and G exists in $\Sigma[\mathcal{A}]$, then

$$F \vee G = FG = GF$$

as already noted in Proposition 1.18.

The following result is contained in (2.16) and (2.17).

Lemma 2.46. *If a cone* V *contains* F *and* G *in its closure, then* $[FG : GF]$ *is smaller than* V*. In particular, if a top-cone* V *contains* C *and* D*, then* $[C : D]$ *is smaller than* V*.*

Exercise 2.47. Use Proposition 2.35 to check that: In a simplicial arrangement, for any chambers C and D,

$$[C \wedge D : \overline{D} \wedge \overline{C}] = [C : \overline{D}] \wedge [D : \overline{C}].$$

The meets are taken in the poset of cones.

2.5.3. Convexity dimension. The *convexity dimension* of a top-cone V is the least number k such that there exists a set of k chambers whose convex closure is V (or equivalently, whose join is V).

Observe that a top-cone V is a gallery interval iff the convexity dimension of V is either 1 or 2. Those of convexity dimension 1 are precisely chambers.

2.6. Charts and dicharts

We define charts and dicharts for an arrangement. The set of charts is a Boolean poset, and so is the set of dicharts. Further, they relate to each other and to the lattice of flats and the lattice of cones through a commutative square of order-preserving maps.

2.6.1. Charts. A *chart* in $\mathcal{A}$ is a subset of the set of hyperplanes in $\mathcal{A}$. This is the same as a subarrangement of $\mathcal{A}$. Let $\mathrm{G}[\mathcal{A}]$ denote the set of charts in $\mathcal{A}$. We partially order this set by reverse inclusion: $g \leq h$ if h is a subset of g. This is a Boolean poset. The minimum element is the chart consisting of all hyperplanes, while the maximum element is the chart with no hyperplanes.

The *center* of a chart g is the flat obtained by intersecting all hyperplanes in g. A chart is *connected* if its center is the minimum flat. Let $\mathrm{cG}[\mathcal{A}]$ denote the set of connected charts in $\mathcal{A}$. A chart is *coordinate* if it is connected and has size r, where $r := \mathrm{rk}(\mathcal{A})$.

For any chart g and flat X, let g_{X} denote the chart consisting of those hyperplanes in g which contain X, and let g^{X} denote the chart in $\mathcal{A}^{\mathrm{X}}$ obtained by intersecting the hyperplanes in g with X.

2.6.2. Dicharts. A *dichart* in $\mathcal{A}$ is a subset of the set of half-spaces in $\mathcal{A}$. Let $\vec{\mathrm{G}}[\mathcal{A}]$ denote the set of dicharts in $\mathcal{A}$. We partially order this set by reverse inclusion: $r \leq s$ if s is a subset of r. This is a Boolean poset. The minimum element is the dichart consisting of all half-spaces, while the maximum element is the dichart with no half-spaces.

2.6.3. Adjunctions. Adjunctions between posets are reviewed in Section B.5. Recall the lattice of flats $\Pi[\mathcal{A}]$ and the lattice of cones $\Omega[\mathcal{A}]$. There are two commutative diagrams of order-preserving maps, namely,

$$
\begin{array}{ccc}
\mathrm{G}[\mathcal{A}] & \xrightarrow{\lambda'} & \vec{\mathrm{G}}[\mathcal{A}] \\
{\scriptstyle\lambda}\uparrow & & \uparrow{\scriptstyle\vec{\lambda}} \\
\Pi[\mathcal{A}] & \xrightarrow[\mathrm{i}]{} & \Omega[\mathcal{A}]
\end{array}
\qquad\qquad
\begin{array}{ccc}
\mathrm{G}[\mathcal{A}] & \xleftarrow{\rho'} & \vec{\mathrm{G}}[\mathcal{A}] \\
{\scriptstyle\rho}\downarrow & & \downarrow{\scriptstyle\vec{\rho}} \\
\Pi[\mathcal{A}] & \xleftarrow[\mathrm{b}]{} & \Omega[\mathcal{A}]
\end{array}
\tag{2.18}
$$

Maps in the first diagram are injective while those in the second diagram are surjective. Any injective map is a section of the corresponding surjective map. Further, any corresponding pair of injective and surjective maps is an adjunction, with the injective map as the left adjoint and the surjective map as the right adjoint. In particular, maps in the first diagram preserve joins, while those in the second diagram preserve meets.

We have already seen the adjunction (i, b) between flats and cones in Section 2.2.2. The map i is the inclusion map while b is the base map. Let us now understand the rest of the picture.

2.6.4. Flats and charts. Flats and charts are related by order-preserving maps

$$\lambda : \Pi[\mathcal{A}] \to \mathrm{G}[\mathcal{A}], \qquad \mathrm{X} \mapsto \{\mathrm{H} \mid \mathrm{H} \geq \mathrm{X}\}, \tag{2.19}$$

and

$$\rho : \mathrm{G}[\mathcal{A}] \to \Pi[\mathcal{A}], \qquad g \mapsto \bigcap_{\mathrm{H} \in g} \mathrm{H}. \tag{2.20}$$

In other words, $\rho(g)$ is the center of g. Observe that for any flat X and chart g,

$$\lambda(\mathrm{X}) \leq g \iff \mathrm{X} \leq \rho(g).$$

Thus, (λ, ρ) is an adjunction between the posets $\Pi[\mathcal{A}]$ and $\mathrm{G}[\mathcal{A}]$, with λ as the left adjoint and ρ as the right adjoint. It follows that λ preserves joins, and ρ preserves meets:

$$\lambda(\mathrm{X} \vee \mathrm{Y}) = \lambda(\mathrm{X}) \cap \lambda(\mathrm{Y}) \quad \text{and} \quad \rho(g \cup h) = \rho(g) \wedge \rho(h).$$

(Due to reverse inclusion, the join in $\mathrm{G}[\mathcal{A}]$ is intersection, while meet is union.)

2.6.5. Charts and dicharts. We now relate charts and dicharts. Define the map

$$\lambda' : \mathrm{G}[\mathcal{A}] \to \vec{\mathrm{G}}[\mathcal{A}] \tag{2.21}$$

which sends g to r, where r is the set of half-spaces whose base is in g. Define the map

$$\rho' : \vec{\mathrm{G}}[\mathcal{A}] \to \mathrm{G}[\mathcal{A}] \tag{2.22}$$

which sends r to g, where g is the set of hyperplanes which are bases of the half-spaces in r. Both maps are order-preserving, and for any chart g and dichart r,

$$\lambda'(g) \leq r \iff g \leq \rho'(r).$$

Thus, (λ', ρ') is an adjunction between the posets $\mathrm{G}[\mathcal{A}]$ and $\vec{\mathrm{G}}[\mathcal{A}]$.

The map λ' also has a left adjoint given by the order-preserving map

$$\rho'' : \vec{\mathrm{G}}[\mathcal{A}] \to \mathrm{G}[\mathcal{A}]$$

which sends r to g, where g is the set of hyperplanes both of whose associated half-spaces are in r. Thus, (ρ'', λ') is an adjunction.

Observe that λ' preserves both meets and joins, ρ' only preserves meets (unions by our convention), and ρ'' only preserves joins (intersections). This is consistent with general properties of adjunctions.

By composing adjunctions, we know that $(\lambda'\lambda, \rho\rho')$ is an adjunction between the posets $\Pi[\mathcal{A}]$ and $\vec{\mathrm{G}}[\mathcal{A}]$. The composite maps are

$$\lambda'\lambda : \Pi[\mathcal{A}] \to \vec{\mathrm{G}}[\mathcal{A}], \qquad \mathrm{X} \mapsto \{\mathrm{h} \mid \mathrm{b}(\mathrm{h}) \geq \mathrm{X}\},$$

and

$$\rho\rho' : \vec{\mathrm{G}}[\mathcal{A}] \to \Pi[\mathcal{A}], \qquad r \mapsto \bigcap_{\mathrm{h} \in r} \mathrm{b}(\mathrm{h}).$$

(Recall that $\mathrm{b}(\mathrm{h})$ denotes the base or bounding hyperplane of the half-space h.) One may also directly check that $(\lambda'\lambda, \rho\rho')$ is an adjunction.

Exercise 2.48. Show that λ is tight but not supertight is general, while λ' is supertight (Definition B.13).

2.6.6. Cones and dicharts. Finally, we relate cones and dicharts. The order-preserving maps are

$$\vec{\lambda} : \Omega[\mathcal{A}] \to \vec{\mathrm{G}}[\mathcal{A}], \qquad \mathrm{V} \mapsto \{\mathrm{h} \mid \mathrm{h} \geq \mathrm{V}\}, \tag{2.23}$$

and

$$\vec{\rho} : \vec{\mathrm{G}}[\mathcal{A}] \to \Omega[\mathcal{A}], \qquad r \mapsto \bigcap_{\mathrm{h} \in r} \mathrm{h}. \tag{2.24}$$

For any cone V and dichart r,

$$\vec{\lambda}(\mathrm{V}) \leq r \iff \mathrm{V} \leq \vec{\rho}(r).$$

Thus, $(\vec{\lambda}, \vec{\rho})$ is an adjunction between the posets $\Omega[\mathcal{A}]$ and $\vec{\mathrm{G}}[\mathcal{A}]$, with $\vec{\lambda}$ as the left adjoint and $\vec{\rho}$ as the right adjoint. It follows that $\vec{\lambda}$ preserves joins, and $\vec{\rho}$ preserves meets:

$$\vec{\lambda}(\mathrm{V} \vee \mathrm{W}) = \vec{\lambda}(\mathrm{V}) \cap \vec{\lambda}(\mathrm{W}) \quad \text{and} \quad \vec{\rho}(r \cup s) = \vec{\rho}(r) \wedge \vec{\rho}(s). \tag{2.25}$$

(Due to reverse inclusion, the join in $\vec{\mathrm{G}}[\mathcal{A}]$ is intersection, while meet is union.)

2.6.7. Summary. The different maps are summarized below.

$$\begin{array}{l|l}
\lambda(\mathrm{X}) = \{\mathrm{H} \mid \mathrm{H} \geq \mathrm{X}\} & \rho(g) = \bigcap_{\mathrm{H} \in g} \mathrm{H} \\
\lambda'(g) = \{\mathrm{h} \mid \mathrm{b(h)} \in g\} & \rho'(r) = \{\mathrm{H} \mid \mathrm{H} = \mathrm{b(h)} \text{ for some } \mathrm{h} \in r\} \\
\vec{\lambda}(\mathrm{V}) = \{\mathrm{h} \mid \mathrm{h} \geq \mathrm{V}\} & \vec{\rho}(r) = \bigcap_{\mathrm{h} \in r} \mathrm{h}
\end{array}$$

With these definitions, it is clear that the first diagram in (2.18) commutes. By uniqueness of adjoints, the second diagram also commutes. Explicitly, it says that for any dichart r,

$$\bigcap_{\mathrm{h} \in r} \mathrm{b(h)} = \mathrm{b}\Big(\bigcap_{\mathrm{h} \in r} \mathrm{h}\Big). \tag{2.26}$$

This is same as the first statement in Proposition 2.17. For a direct argument, see the proof given there.

Exercise 2.49. Describe the (co)closure operators associated to the adjunctions discussed in this section.

Exercise 2.50. Check that the following diagrams commute.

$$\begin{array}{ccc}
\mathrm{G}[\mathcal{A}] & \xrightarrow{\lambda'} & \vec{\mathrm{G}}[\mathcal{A}] \\
{\scriptstyle \rho}\downarrow & & \downarrow{\scriptstyle \vec{\rho}} \\
\Pi[\mathcal{A}] & \xrightarrow[\mathrm{i}]{} & \Omega[\mathcal{A}]
\end{array}
\qquad\qquad
\begin{array}{ccc}
\mathrm{G}[\mathcal{A}] & \xleftarrow{\rho''} & \vec{\mathrm{G}}[\mathcal{A}] \\
{\scriptstyle \lambda}\uparrow & & \uparrow{\scriptstyle \vec{\lambda}} \\
\Pi[\mathcal{A}] & \xleftarrow[\mathrm{c}]{} & \Omega[\mathcal{A}]
\end{array}$$

They are adjoints of each other, so commutativity of one implies that of the other. (Here c is the case map of Section 2.2.1.)

2.7. Poset of top-cones

Recall the poset of top-cones $\widehat{\Omega}[\mathcal{A}]$. The partial order is inclusion, that is, $\mathrm{V} \leq \mathrm{W}$ iff W contains V. We now show that this poset is join-distributive. In particular, it is upper semimodular and hence graded. The key idea is to compare it with the Boolean poset of dicharts.

Graded posets, semimodularity and join-distributivity are reviewed in Sections B.2 and B.3.

2.7.1. Rank function. Consider the injective order-preserving map

$$\widehat{\Omega}[\mathcal{A}] \to \overrightarrow{\mathrm{G}}[\mathcal{A}], \qquad \mathrm{V} \mapsto r, \tag{2.27}$$

where r consists of those half-spaces that contain V. This is the restriction of the map $\vec{\lambda}$ defined in (2.23) to top-cones. We say that r is the dichart associated to V. Note that $\mathrm{V} = \vec{\rho}(r)$, where $\vec{\rho}$ is the map defined in (2.24). Explicitly, V is the intersection of the half-spaces in r. Hence, for any hyperplane, at most one (but not both) of its associated half-spaces can belong to r. Since $\vec{\lambda}$ is injective, it is also strictly order-preserving.

By composing (2.27) with (2.22), we obtain the order-preserving map

$$\widehat{\Omega}[\mathcal{A}] \to \mathrm{G}[\mathcal{A}], \qquad \mathrm{V} \mapsto g, \tag{2.28}$$

where g consists of those hyperplanes which do not cut V. In particular, the walls of V belong to g. We say that g is the chart associated to V. Note that (2.28) preserves minimum and maximum elements: A chamber maps to the chart consisting of all hyperplanes, while the ambient space maps to the chart with no hyperplanes. This follows from (2.14).

Consider the rank-two arrangement of 4 lines. Let V be the top-cone shown above consisting of two adjacent edges. Its associated dichart r consists of three half-spaces which are shown on the left, while its associated chart g consists of three lines, shown on the right. Note that these lines include the two walls of V.

Proposition 2.51. *Let* V *be a top-cone with associated dichart* r. *Let* s *be obtained by deleting from* r *a half-space whose base is a wall of* V. *Then* $\mathrm{W} := \vec{\rho}(s)$ *is the unique top-cone covering* V *whose associated dichart is* s.

PROOF. Uniqueness of W is clear since the map $\vec{\lambda}$ is injective. By Proposition 2.10, $\mathrm{V} < \mathrm{W}$. Applying $\vec{\lambda}$, we obtain $r < \vec{\lambda}\vec{\rho}(s)$. But $\vec{\lambda}\vec{\rho}(s) \leq s$. Since s covers r, we obtain $\vec{\lambda}(\mathrm{W}) = s$, and we also deduce that $\mathrm{V} \lessdot \mathrm{W}$. □

Proposition 2.52. *Let* V *and* W *be top-cones with associated dicharts* r *and* s, *and associated charts* g *and* h, *respectively. Then*

$$\mathrm{V} \lessdot \mathrm{W} \iff r \lessdot s.$$

In this case: $g \lessdot h$. *More precisely,* h *is obtained by deleting from* g *the unique wall of* V *which cuts* W.

PROOF. Forward implication. Suppose $\mathrm{V} \lessdot \mathrm{W}$. By Proposition 2.10, there exists a wall, say H, of V which cuts W. Let h be the half-space in r with base H, and r' be the dichart obtained by deleting h from r. We have $r \lessdot r' \leq s$. Now we employ Proposition 2.51. First applying $\vec{\rho}$ yields $\mathrm{V} \lessdot \vec{\rho}(r') \leq \mathrm{W}$. This implies $\vec{\rho}(r') = \mathrm{W}$. Next applying $\vec{\lambda}$ yields $r' = s$, and hence $r \lessdot s$.

Backward implication. Suppose $r \lessdot s$. Applying $\vec{\rho}$, we have $\mathrm{V} < \mathrm{W}$. Since $\vec{\lambda}$ is strictly order-preserving, we further deduce that $\mathrm{V} \lessdot \mathrm{W}$.

It is clear that $r \lessdot s$ implies $g \lessdot h$. The last claim also follows. □

The following is an illustration of the cover relation in top-cones.

Both pictures show a top-cone W (with 5 triangles) containing a top-cone V (shown in dark shade and containing 3 and 2 triangles, respectively). In the first picture, $\mathrm{V} \lessdot \mathrm{W}$, but this is false in the second picture.

Propositions 2.51 and 2.52 yield the following.

Corollary 2.53. *Given a top-cone* V*, the set of top-cones which cover* V *are indexed by walls of* V*.*

This is illustrated below in the rank-two arrangement of 4 lines. The top-cone V consists of two adjacent edges. It is covered by two top-cones each consisting of three edges. They correspond to the two walls of V.

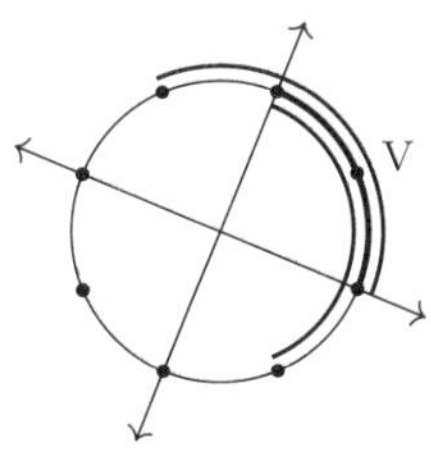

Two other cases to note are: If V is a half-space consisting of four edges, then it has only one cover, namely, the ambient space and only one wall, namely, its base. If V is the ambient space, then it has no covers and no walls.

Exercise 2.54. Let V and W be top-cones with associated charts g and h, respectively. Show that: $g \lessdot h$ implies $\mathrm{V} \lessdot \mathrm{W}$ is false in general.

Exercise 2.55. Recall the notion of convex closure from Section A.2.2. Let V and W be top-cones. Show that: $\mathrm{V} \lessdot \mathrm{W}$ iff there exists a chamber D which is not in V but adjacent to a chamber in V such that W is the convex closure of $\mathrm{V} \cup \{D\}$.

By Proposition 2.52, the map (2.28) preserves cover relations. Hence composing it with the rank function of $\mathrm{G}[\mathcal{A}]$ yields a rank function for $\widehat{\Omega}[\mathcal{A}]$. Thus:

Theorem 2.56. *The poset of top-cones* $\widehat{\Omega}[\mathcal{A}]$ *is graded, with the rank of a top-cone being the number of hyperplanes which cut that top-cone. In particular, the rank of* $\widehat{\Omega}[\mathcal{A}]$ *is the number of hyperplanes in* $\mathcal{A}$*. The map* (2.28) *is rank-preserving.*

The Hasse diagram of the poset of top-cones for the rank-two arrangement of 3 lines is shown below on the left.

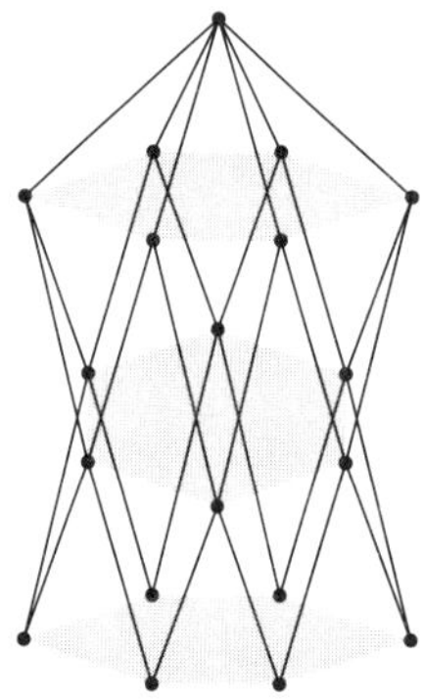

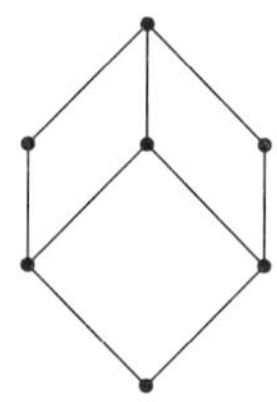

There are 6 elements of rank 0 (chambers), 6 elements of rank 1 (pairs of adjacent chambers), 6 elements of rank 2 (half-spaces), and 1 element of rank 3 (ambient space). Observe that every element is the join of two rank-zero elements, in agreement with Exercise 2.43. The figure on the right shows the Hasse diagram of the subposet of all top-cones which are greater than a fixed chamber.

More generally, for the rank-two arrangement of n lines, there are $2n$ elements of each rank from 0 to $n-1$, and one element of rank n.

Exercise 2.57. Check that the poset of *all* cones for the rank-two arrangement of n lines is *not* graded for $n \geq 3$. (First do the case $n = 3$ and then generalize).

2.7.2. Semimodularity.

Theorem 2.58. *Any interval in the poset of top-cones $\widehat{\Omega}[\mathcal{A}]$ is upper semimodular. Equivalently, for any top-cones* V *and* W *which have a chamber in common,*

$$\mathrm{rk}(\mathrm{V}) + \mathrm{rk}(\mathrm{W}) \geq \mathrm{rk}(\mathrm{V} \wedge \mathrm{W}) + \mathrm{rk}(\mathrm{V} \vee \mathrm{W}).$$

Proof. Let us prove the reverse inequality for the corank function of top-cones. The corank of a top-cone V equals the cardinality of $\vec{\lambda}(\mathrm{V})$, which is the number of half-spaces which contain V. This follows from Theorem 2.56 which says that the corank of V is the number of hyperplanes which do not cut V. Note that

$$\vec{\lambda}(\mathrm{V} \wedge \mathrm{W}) \supseteq \vec{\lambda}(\mathrm{V}) \cup \vec{\lambda}(\mathrm{W}) \quad \text{and} \quad \vec{\lambda}(\mathrm{V} \vee \mathrm{W}) = \vec{\lambda}(\mathrm{V}) \cap \vec{\lambda}(\mathrm{W}),$$

The first holds since $\vec{\lambda}$ is order-preserving, and the second since it preserves joins, see (2.25). The result now follows from the modularity of the Boolean poset. $\square$

If every interval in a finite poset with a top element is upper semimodular, then the poset is graded. This follows from Proposition B.2. Thus, Theorem 2.58 contains the result that the poset of top-cones is graded.

2.7.3. Join-distributivity.

Proposition 2.59. *Let* $\mathrm{V} \leq \mathrm{W}$. *The following are equivalent.*

(1) *The interval* $[\mathrm{V}, \mathrm{W}]$ *is a Boolean poset of rank* k.
(2) *There are exactly* k *top-cones in* $[\mathrm{V}, \mathrm{W}]$ *which cover* V *and their join is* W.
(3) $\vec{\lambda}(\mathrm{W})$ *is obtained from* $\vec{\lambda}(\mathrm{V})$ *by deleting* k *distinct half-spaces whose bases are walls of* V.

Proof. Put $\vec{\lambda}(\mathrm{V}) = r$ and $\vec{\lambda}(\mathrm{W}) = s$.

(1) implies (2). Clear.

(2) implies (3). Let $V_1, \ldots, V_k$ be the k top-cones in question. Put $\vec{\lambda}(V_i) = r_i$. By Proposition 2.52, each r_i is obtained by deleting a distinct half-space from r whose base is a wall of V. Since $\vec{\lambda}$ preserves joins and $W = V_1 \vee \ldots \vee V_k$, we have $s = r_1 \vee \ldots \vee r_k$.

(3) implies (1). Let $r_1, \ldots, r_k$ be the dicharts obtained by deleting from r the k half-spaces in question, one at a time. The interval $[r, s]$ is a Boolean poset of rank k with the r_i as the join-irreducibles. By Proposition 2.51, there exist unique top-cones V_i with $\vec{\lambda}(V_i) = r_i$. The map $[V, W] \to [r, s]$ induced by $\vec{\lambda}$ is order-preserving and injective (since $\vec{\lambda}$ is injective). It is also surjective since it is join-preserving and its image contains the join-irreducibles. Hence, it is an isomorphism. □

As a consequence:

Theorem 2.60. *The join-semilattice of top-cones $\widehat{\Omega}[\mathcal{A}]$ is join-distributive. Equivalently, for any chamber C, the interval $[C, \top]$ is a join-distributive lattice.*

Any interval in a join-distributive join-semilattice is upper semimodular (Proposition B.5). Thus, Theorem 2.58 may also be deduced from Theorem 2.60.

Exercise 2.61. Show that: Given a top-cone V, there exists a top-cone W such that the interval $[V, W]$ is a Boolean poset of rank equal to the number of walls of V. This is the largest interval in $\widehat{\Omega}[\mathcal{A}]$ with bottom element V which is a Boolean poset.

Exercise 2.62. Consider the smallest nonsimplicial arrangement in rank three (Section 1.2.5). Deduce the following using Proposition 2.59. For any quadrilateral C, the interval $[C, \top]$ in the poset of top-cones is a Boolean poset of rank 4. (First locate the four top-cones which cover C.) For any triangle D and V a half-space whose base is not a wall of D, the interval $[D, V]$ is a Boolean poset of rank 3. Also verify these facts directly.

2.7.4. Convex geometries.

Join-distributive lattices are intimately connected to convex geometries (Section B.5.5). This gives an alternative way to think about and prove Theorem 2.60 as sketched below.

Fix a top-cone V. Consider the Boolean poset on $\vec{\lambda}(V)$, the set of all half-spaces containing V. The map $r \mapsto \vec{\lambda}\vec{\rho}(r)$ defines a closure operator on this poset. Its closed sets are in correspondence with top-cones which are greater than V. In addition, the result of Exercise 2.63 below precisely says that this closure operator is a convex geometry. By Proposition B.17, the poset of convex sets is meet-distributive. The opposite of this poset is the interval $[V, \top]$, which is then join-distributive as required. Also observe that the extreme points of a top-cone W (viewed as a convex set in this convex geometry) are the walls of W. Using this fact, Proposition 2.59 follows from Proposition B.18.

Exercise 2.63. Fix top-cones $V \leq W$. Suppose h and h′ are two distinct half-spaces which contain V and whose base cuts W. Then $h \in \vec{\lambda}(W \wedge h')$ implies $h' \notin \vec{\lambda}(W \wedge h)$. This is illustrated below.

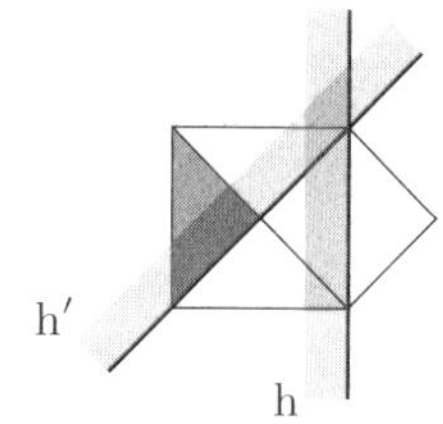

The top-cone W consists of 5 triangles, while V is the triangle which is shaded in dark. The top-cone $\mathrm{W} \wedge \mathrm{h}'$ (which has 2 triangles) is contained in h, while $\mathrm{W} \wedge \mathrm{h}$ (which has 4 triangles) is not contained in h′.

2.7.5. Möbius function. Join-distributivity which we showed in Theorem 2.60 can be used to gain information about the Möbius function of the poset of top-cones. Specializing Lemma C.7, we obtain:

Corollary 2.64. *Let* $\mathrm{V} \leq \mathrm{W}$. *Then*

$$\mu(\mathrm{V},\mathrm{W}) = \begin{cases} (-1)^{\mathrm{rk}(\mathrm{W})-\mathrm{rk}(\mathrm{V})} & \text{if } [\mathrm{V},\mathrm{W}] \text{ is a Boolean poset,} \\ 0 & \text{otherwise.} \end{cases}$$

Equivalent formulations of the first alternative are given in Proposition 2.59.

Exercise 2.65. Work out the Möbius function of the poset of top-cones for the rank-two arrangement of n lines.

2.8. Partial-flats

A partial-support relation is an equivalence relation on the set of faces such that the equivalence classes are combinatorial cones (contained in combinatorial flats) which inherit the Tits product on faces. We refer to these classes as partial-flats. They interpolate faces and flats.

A subarrangement gives rise to a partial-support relation. Partial-support relations which arise from subarrangements can be characterized by imposing an additional geometric condition. Further, in this case, the relation is completely determined by its restriction to the set of chambers, thus simplifying the theory. We make this the starting point of our discussion.

Warning. Subarrangements have been called charts in Section 2.6.

2.8.1. Partial-support relations on chambers. Recall the set of chambers $\Gamma[\mathcal{A}]$. A *partial-support relation on chambers* is an equivalence relation on $\Gamma[\mathcal{A}]$, denoted $\sim$, such that

$$C \sim D \implies FC \sim FD. \tag{2.29}$$

To develop some intuition for this condition, consider the rank-two arrangement of four lines shown below.

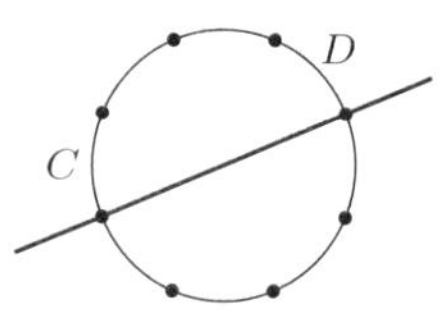

Suppose $C \sim D$, with C and D as shown in the figure. Then by projecting C and D on different vertices, we can deduce that the four edges on any one side of the line belong to the same equivalence class.

Proposition 2.66. *There is a bijection between subarrangements of $\mathcal{A}$ and partial-support relations on chambers of $\mathcal{A}$. In addition, equivalence classes under $\sim$ are in bijection with chambers of the corresponding subarrangement $\mathcal{A}'$.*

PROOF. For a subarrangement $\mathcal{A}'$ of $\mathcal{A}$, define a partial-support relation on chambers of $\mathcal{A}$ as follows:

$$C \sim D \iff C \text{ and } D \text{ are on the same side of every hyperplane in } \mathcal{A}'. \tag{2.30}$$

The rhs equivalently says that C and D belong to the same chamber of $\mathcal{A}'$. If C and D are on the same side of every hyperplane in $\mathcal{A}'$, then so are FC and FD, for any face F. This can be readily checked using (1.5). Thus $\sim$ is indeed a partial-support relation. It is clear that distinct subarrangements give rise to distinct partial-support relations.

Conversely, suppose $\sim$ is a partial-support relation. Define a subarrangement $\mathcal{A}'$ as follows. $\mathrm{H} \in \mathcal{A}'$ if H is the common wall of some pair of adjacent chambers C and D which belong to different equivalence classes. For this to be the inverse map, we need to check (2.30). Forward implication: Suppose $C \sim D$, and H is a hyperplane separating C and D. Let C' and D' be any adjacent chambers with common panel F whose support is H. Then $FC = C'$ and $FD = D'$ (or vice-versa), and hence $C' \sim D'$. Then by construction, $\mathrm{H} \notin \mathcal{A}'$. Backward implication: Suppose C and D are on the same side of every hyperplane in $\mathcal{A}'$. Pick any minimal gallery joining C and D. Then all chambers in this gallery lie on the same side of every hyperplane in $\mathcal{A}'$. We claim that any two adjacent chambers in this gallery are equivalent: If not, then the support of their common panel would belong to $\mathcal{A}'$. Therefore, by transitivity, $C \sim D$. □

In conjunction with the discussion in Section 2.2.5, we obtain:

Corollary 2.67. *Equivalence classes wrt a partial-support relation on chambers are combinatorial top-cones. They are closed under taking opposites, that is, if $C \sim D$, then $\overline{C} \sim \overline{D}$. They all have the same base (equal to the center of the corresponding subarrangement).*

The first part can also be deduced directly from Proposition 2.3.

Corollary 2.68. *Let $\sim$ be a partial-support relation on chambers. Let $A \leq C$, $A \leq D$, $\mathrm{s}(A) = \mathrm{s}(B)$ and $C \sim BC$. Then $D \sim BD$.*

PROOF. By (1.5), for any hyperplane H, the chambers C and BC are on the same side of H iff D and BD are on the same side of H. The result now follows from Proposition 2.66. □

In view of Corollary 2.67, one may use the term $\sim$-top-cones to refer to equivalence classes wrt $\sim$. Let $\Gamma_\sim[\mathcal{A}]$ denote the set of $\sim$-top-cones. As a shorthand, we will also write $\Gamma_\sim$ with $\mathcal{A}$ understood. This interpolates between two extreme cases:

$$\Gamma[\mathcal{A}] \twoheadrightarrow \Gamma_\sim[\mathcal{A}] \twoheadrightarrow \mathrm{E}[\mathcal{A}],$$

where $\mathrm{E}[\mathcal{A}]$ denotes a singleton set. It corresponds to the subarrangement with no hyperplanes.

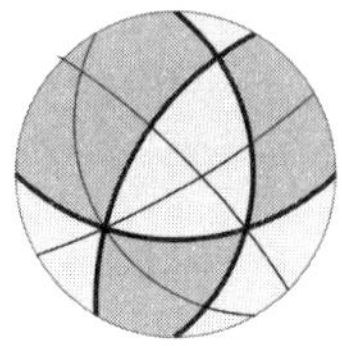

The figure shows a subarrangement of three hyperplanes. Each chamber of this arrangement is a $\sim$-top-cone, where $\sim$ is the corresponding partial-support relation.

2.8.2. Partial-support relations on faces. Recall the set of faces $\Sigma[\mathcal{A}]$. A *partial-support relation on faces* is an equivalence relation on $\Sigma[\mathcal{A}]$, denoted $\sim$, which satisfies the following properties.

$$F \sim G \implies \mathrm{s}(F) = \mathrm{s}(G) \tag{2.31a}$$

$$F \sim G \implies FH \sim GH \tag{2.31b}$$

$$G \sim G' \implies FG \sim FG' \tag{2.31c}$$

A partial-support relation is *geometric* if

$$\mathrm{s}(F) = \mathrm{s}(G) \text{ and } FH \sim GH \text{ for some } H \implies F \sim G. \tag{2.31d}$$

We refer to an equivalence class of a partial-support relation $\sim$ as a *partial-flat*. We denote it by letters x, y, z and w. If $F \in$ x, then we say that x is the *partial-support* of F. By axiom (2.31a), each partial-flat has a well-defined support. We say a partial-flat is maximal if its support is the maximum flat.

Let $\Sigma_\sim[\mathcal{A}]$ denote the set of partial-flats. As a shorthand, we will also write $\Sigma_\sim$ with $\mathcal{A}$ understood.

Example 2.69. The two canonical examples of (geometric) partial-support relations are as follows.

- $F \sim G$ iff $F = G$. That is, partial-flats are faces. Thus $\Sigma_\sim[\mathcal{A}] = \Sigma[\mathcal{A}]$. (We point out that checking axiom (2.31d) requires Proposition 1.17.)
- $F \sim G$ iff $\mathrm{s}(F) = \mathrm{s}(G)$. That is, partial-flats are flats. Thus $\Sigma_\sim[\mathcal{A}] = \Pi[\mathcal{A}]$.

In general, a partial-support relation lies somewhere between these two extreme cases with maps

$$\Sigma[\mathcal{A}] \twoheadrightarrow \Sigma_\sim[\mathcal{A}] \twoheadrightarrow \Pi[\mathcal{A}]. \tag{2.32}$$

For instance,

- $F \sim G$ iff either $F = G$, or F and G are both chambers,

is a (non-geometric) partial-support relation. In this case, a partial-flat is either the maximum flat, or a proper face.

Some basic properties of partial-support relations are listed below.

Proposition 2.70. *Let $\sim$ be a partial-support relation. Then, for any faces F, G, F', G' and H,*

$$F \leq G \text{ and } G \sim G' \implies G \sim FG' \sim G', \tag{2.33}$$

$$G \leq H \text{ and } G \sim G' \implies H \sim G'H, \tag{2.34}$$

$$F \sim G \text{ and } F' \sim G' \implies FF' \sim GG'. \tag{2.35}$$

PROOF. Property (2.33) follows from (2.31c), and (2.34) from (2.31b). For (2.35): Suppose $F \sim G$ and $F' \sim G'$. Then, by (2.31b), $FF' \sim GF'$, and by (2.31c), $GF' \sim GG'$. Thus, by transitivity, $FF' \sim GG'$. $\square$

Exercise 2.71. Show that (2.31a) and (2.34) imply (2.31b).

Exercise 2.72. Show that a partial-support relation $\sim$ is geometric iff

$$\mathrm{s}(F) = \mathrm{s}(G) \text{ and } FC \sim GC \text{ for some chamber } C \implies F \sim G,$$

or equivalently,

$$\mathrm{s}(F) = \mathrm{s}(G) \text{ and } H \sim K \text{ for some } H \geq F \text{ and } K \geq G \implies F \sim G.$$

2.8.3. Product and partial order on partial-flats. For a partial-support relation $\sim$, define a product on $\Sigma_\sim[\mathcal{A}]$ as follows. For equivalence classes x and y, let xy be the equivalence class of FG, with $F \in \mathrm{x}$ and $G \in \mathrm{y}$. By (2.35), the result does not depend on the particular choice of F and G. This turns $\Sigma_\sim[\mathcal{A}]$ into a monoid, with the unit element being the class of the central face. It follows from (1.6) that for any partial-flats x and y, we have $\mathrm{xx} = \mathrm{x}$ and $\mathrm{xyx} = \mathrm{xy}$, so $\Sigma_\sim[\mathcal{A}]$ is a left regular band. It interpolates the Tits monoid and the Birkhoff monoid; the maps (2.32) are morphisms of monoids.

Lemma 2.73. *For partial-flats* x *and* y*, the following are equivalent.*

(1) $\mathrm{xy} = \mathrm{y}$.
(2) *Every face in* x *is smaller than some face in* y.
(3) *Some face in* x *is smaller than some face in* y.

PROOF. (1) implies (2). Let $F \in \mathrm{x}$ and $G \in \mathrm{y}$. Then, by hypothesis, $FG \in \mathrm{y}$, and $F \leq FG$.

(2) implies (3). Clear.

(3) implies (1). Suppose $F \in \mathrm{x}$, $G \in \mathrm{y}$ and $F \leq G$. Then $FG = G$ which is in y. Hence $\mathrm{xy} = \mathrm{y}$. $\square$

For partial-flats x and y, we say $\mathrm{x} \leq \mathrm{y}$ if any of the equivalent conditions of Lemma 2.73 holds. This defines a partial order on $\Sigma_\sim[\mathcal{A}]$. It has a minimum element given by the class of the central face. It follows that the maps (2.32) are order-preserving.

Lemma 2.74. *Let* x *and* x' *be equivalence classes of a geometric partial-support relation. Then* $\mathrm{xx}' = \mathrm{x}'\mathrm{x}$ *iff* x *and* x' *have an upper bound iff* x *and* x' *have a join. In this situation,*

$$\mathrm{x} \vee \mathrm{x}' = \mathrm{xx}' = \mathrm{x}'\mathrm{x}.$$

PROOF. Note that $\mathrm{x} \leq \mathrm{xx}'$ and $\mathrm{x}' \leq \mathrm{x}'\mathrm{x}$. So if $\mathrm{xx}' = \mathrm{x}'\mathrm{x}$, then this element is an upper bound for x and x'. Conversely: Suppose $\mathrm{x} \leq \mathrm{y}$ and $\mathrm{x}' \leq \mathrm{y}$. Pick $F \in \mathrm{x}$, $F' \in \mathrm{x}'$, and $G, G' \in \mathrm{y}$ with $F \leq G$ and $F' \leq G'$. Then $FG', F'G \in \mathrm{y}$, so $FG' \sim F'G$. Now FF' and $F'F$ have the same support and $FF' \leq FG'$ and $F'F \leq F'G$. Hence, by (2.31d), $FF' \sim F'F$, which implies $\mathrm{xx}' = \mathrm{x}'\mathrm{x}$. It is also clear that this element is smaller than y, so it must be the join. $\square$

Lemma 2.75. *For a geometric partial-support relation* $\sim$*, the poset* $\Sigma_\sim[\mathcal{A}]$ *is a meet-semilattice.*

PROOF. Suppose x and y are partial-flats wrt $\sim$. Since $\Sigma_\sim[\mathcal{A}]$ has a minimum element, x and y have a lower bound. Let z be the product of all lower bounds of x and y (the order in which the product is taken does not matter in view of Lemma 2.74). Suppose w is a lower bound for x and y. Then pick an expression for z starting with w and use $\mathrm{ww} = \mathrm{w}$ to deduce that $\mathrm{wz} = \mathrm{z}$. Thus z is the meet of x and y. □

Exercise 2.76. What can one say about Lemma 2.74 if the relation is not geometric?

Exercise 2.77. For a geometric partial-support relation, show that $\mathrm{xy} = \mathrm{yx}$ and $\mathrm{z} \leq \mathrm{y}$ implies $\mathrm{xz} = \mathrm{zx}$.

2.8.4. Partial-flats as cones. It is clear that a partial-support relation on faces of $\mathcal{A}$ induces a partial-support relation on faces of $\mathcal{A}_\mathrm{X}$ and of $\mathcal{A}^\mathrm{X}$.

Proposition 2.78. *Partial-flats are combinatorial cones. They are closed under taking opposites, that is, if $F \sim G$, then $\overline{F} \sim \overline{G}$. Further, partial-flats with the same support have the same base.*

PROOF. A partial-flat with support X is a maximal partial-flat of the arrangement $\mathcal{A}^\mathrm{X}$. This is the same as an equivalence class wrt the restricted partial-support relation on the chambers of $\mathcal{A}^\mathrm{X}$. The result now follows from Corollary 2.67. The first part can also be deduced from (2.33) and Proposition 2.4. □

In view of this result, one may use the term $\sim$-cones to refer to partial-flats wrt $\sim$. For a $\sim$-cone x, we let $\overline{\mathrm{x}}$ denote its opposite $\sim$-cone.

Proposition 2.79. *Let $\sim$ be a geometric partial-support relation. If* x *and* y *are $\sim$-cones, then so is* $\mathrm{x} \wedge \mathrm{y}$*, with the meet being taken in the lattice of cones* $\Omega[\mathcal{A}]$.

PROOF. Put $\mathrm{V} := \mathrm{x} \wedge \mathrm{y}$. Fix $F \in \mathrm{V}$. Then F is smaller than some face in x and some face in y.

- Now if $F \sim G$, then by Lemma 2.73, G is also smaller than some face in x and some face in y, and hence $G \in \mathrm{V}$. So the $\sim$-cone of F is a subset of V.
- Conversely, suppose $G \in \mathrm{V}$. Pick $H \geq F$ and $K \geq G$ such that H and K are both in y. Then since y is a cone, by Proposition 2.4, GH also belongs to y. By property (2.31d), we conclude that $F \sim G$.

Thus V is the $\sim$-cone of F. □

Corollary 2.80. *For a geometric partial-support relation $\sim$, $\Sigma_\sim[\mathcal{A}]$ is a subposet of $\Omega[\mathcal{A}]$ closed under taking meets.*

PROOF. Lemma 2.73, item (2) implies that $\Sigma_\sim[\mathcal{A}]$ is a subposet of $\Omega[\mathcal{A}]$. This does not require geometric. The closure under meets follows from Proposition 2.79. □

Note that Lemma 2.75 is a consequence of Corollary 2.80.

Exercise 2.81. Show that the forward implication of (1.9b) does not hold for partial-flats in general. How can one fix this? Which properties stated in Lemma 1.6 hold?

Exercise 2.82. Generalize Corollary 2.68 to partial-support relations on faces.

Exercise 2.83. Show that Proposition 2.79 may fail if $\sim$ is not geometric.

Exercise 2.84. Suppose x and y are $\sim$-cones and $x \leq V \leq y$ for some cone V. Then is V necessarily a $\sim$-cone?

2.8.5. Geometric partial-flats and subarrangements. Note that a geometric partial-support relation on faces restricts to a partial-support relation on chambers. Conversely, a partial-support relation on chambers extends to a geometric partial-support relation on faces: For $s(F) = s(G)$,

$$F \sim G :\iff FC \sim GC \text{ for some chamber } C \iff FC \sim GC \text{ for all chambers } C. \tag{2.36}$$

(The second equivalence holds in view of Corollary 2.68.) The properties (2.31a)-(2.31d) are straightforward to check. Further, the two constructions are inverses of each other. So there is a bijection between partial-support relations on chambers, and geometric partial-support relations on faces. In view of Proposition 2.66, we obtain:

Proposition 2.85. *There is a bijection between subarrangements of $\mathcal{A}$ and geometric partial-support relations on faces of $\mathcal{A}$.*

Explicitly, from (2.30) and (2.36), the geometric partial-support relation associated to a subarrangement $\mathcal{A}'$ is given by:

$$F \sim G \iff F \text{ and } G \text{ have the same support and the same sign wrt every hyperplane in } \mathcal{A}'.$$

The rhs equivalently says that F and G have the same support and their interiors belong to the interior of the same face of $\mathcal{A}'$.

Exercise 2.86. Check directly using (1.5) that the above defines a geometric partial-support relation.

Fix a subarrangement $\mathcal{A}'$. Let $\sim$ denote the corresponding geometric partial-support relation. Each face of $\mathcal{A}'$ determines a unique cone of $\mathcal{A}$. For the moment, let us identify faces of $\mathcal{A}'$ with cones of $\mathcal{A}$. Define an operator

$$\Sigma_\sim[\mathcal{A}] \to \Sigma_\sim[\mathcal{A}] \tag{2.37}$$

which sends x to the smallest face of $\mathcal{A}'$ which contains x. This is a closure operator whose closed sets are precisely the faces of $\mathcal{A}'$.

The picture shows a subarrangement of three hyperplanes inside a rank-three arrangement. The thick segment consisting of two edges is a partial-flat and its closure is the shaded region.

Exercise 2.87. Check that: For any subarrangement $\mathcal{A}'$ of $\mathcal{A}$, the map (1.15) factors as

$$\begin{array}{ccc} \Sigma[\mathcal{A}] & \xrightarrow{\qquad} & \Sigma[\mathcal{A}'] \\ & \searrow \quad \nearrow & \\ & \Sigma_{\sim}[\mathcal{A}] & \end{array}$$

where $\sim$ is the associated geometric partial-support relation.

Exercise 2.88. Show that a partial-support relation on faces restricts to a partial-support relation on chambers. Conversely, show that a partial-support relation on chambers extends to a partial-support relation on faces by letting all each proper face be a partial-flat. Is this extension geometric?

Exercise 2.89. For a geometric partial-support relation, define a partial order on pairs of partial-flats which interpolates those on pairs of faces and on pairs of flats given in Section 1.12.

2.8.6. Janus monoid for partial-flats. The Janus monoid can be generalized as follows. Fix two partial-support relations, say $\sim$ and $\sim'$. A *partial-bi-flat* is a pair $(\mathrm{x}, \mathrm{x}')$ such that x is a partial-flat wrt $\sim$, x' is a partial-flat wrt $\sim'$, and x and x' have the same support. Let $\mathrm{J}_{\sim,\sim'}[\mathcal{A}]$ denote the set of partial-bi-flats. The operation

$$(\mathrm{x}, \mathrm{x}')(\mathrm{y}, \mathrm{y}') := (\mathrm{xy}, \mathrm{y}'\mathrm{x}')$$

turns $\mathrm{J}_{\sim,\sim'}[\mathcal{A}]$ into a monoid. It is the fiber product of $\Sigma_{\sim}[\mathcal{A}]$ and the opposite of $\Sigma_{\sim'}[\mathcal{A}]$ over the Birkhoff monoid $\Pi[\mathcal{A}]$. In particular, we have a commutative diagram of monoids

$$\begin{array}{ccc} \mathrm{J}_{\sim,\sim'}[\mathcal{A}] & \longrightarrow & \Sigma_{\sim'}[\mathcal{A}]^{\mathrm{op}} \\ \downarrow & & \downarrow \\ \Sigma_{\sim}[\mathcal{A}] & \longrightarrow & \Pi[\mathcal{A}] \end{array}$$

with the bottom horizontal map and right vertical map as in (2.32), and the other two maps being the projections on the two coordinates, respectively.

Observe that the Tits monoid, the Birkhoff monoid, and more generally the monoid $\Sigma_{\sim}[\mathcal{A}]$ are all special cases of this construction (by taking $\Sigma_{\sim'}[\mathcal{A}] = \Pi[\mathcal{A}]$). By construction, there is a surjective map of monoids

$$\mathrm{J}[\mathcal{A}] \twoheadrightarrow \mathrm{J}_{\sim,\sim'}[\mathcal{A}], \qquad (F, F') \to (\mathrm{x}, \mathrm{x}'),$$

where x is the partial-flat wrt $\sim$ which contains F, while x' is the partial-flat wrt $\sim'$ which contains F'.

Notes

Convexity. For Propositions 2.3, 2.7 and 2.40, see [**2**, Exercises 1.65, 1.68 and 1.66]. These ideas appeared in work of Tits on reflection arrangements [**415**, Theorem 2.19]. For the convex geometry described in Section 2.7.4, see [**172**, Section 3, Example I] and references therein. For an essential arrangement, a cone whose base is the minimum flat is often called *salient*.

CHAPTER 3

Lunes

We now study an important family of cones known as lunes. Faces, flats and half-flats are examples of lunes. Geometrically, lunes are cones which cannot be cut along their base. They can be identified with faces of arrangements over their base. A nested face is a pair of faces one contained in the other. There is a support map from nested faces to lunes which parallels the support map from faces to flats. This allows us to study lunes using the Tits product. Lunes are gallery intervals; in fact, they can be realized as gallery intervals in multiple ways. Lunes also serve as building blocks of cones. More precisely, any cone can be decomposed into lunes by cutting it along a fixed flat contained in its base. In particular, cutting along its base yields the optimal decomposition.

We also continue the discussion on cones. Cones offer the flexibility for interesting local operations, which we call restriction and extension. There is also a notion of conjugate top-cones; an important example of conjugate pairs is provided by top-stars and top-lunes. We also introduce top-star-lunes which are top-cones constructed inductively by using the extension operation. Both top-stars and top-lunes are examples of top-star-lunes. We discuss in detail the compatibility of cones, gallery intervals and lunes with the cartesian product of arrangements.

3.1. Lunes

We begin by defining lunes geometrically as cones which cannot be cut along their base, and then relate them to faces of arrangements over their base.

3.1.1. Lunes. Recall that a cone is a subset of the ambient space obtained by intersecting some half-spaces. The base of a cone V, denoted b(V), is the largest flat contained in that cone. For a hyperplane H, let H^+ and H^- denote its two associated half-spaces. A *lune* is a cone V with the following property.

(3.1) If a hyperplane H contains b(V),

then either H^+ contains V or H^- contains V.

In other words, for a cone to be a lune, a hyperplane containing the base of the cone is not allowed to cut the cone. In (3.1), it is possible that both H^+ and H^- contain V in which case $H = H^+ \cap H^-$ contains V.

Since lunes are cones, they have a *base* and a *case*. A *top-lune* is a lune whose top-dimensional faces are chambers, or equivalently, whose case is the maximum flat. In other words, a top-lune is a lune which is a top-cone.

Just as with cones, one can take a combinatorial approach to lunes. A *combinatorial lune* is a subset of the set of faces consisting of precisely the top-dimensional faces of some lune.

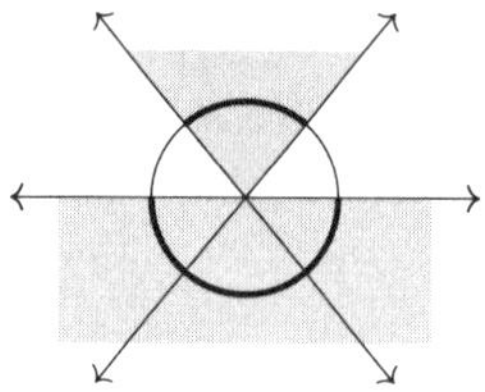 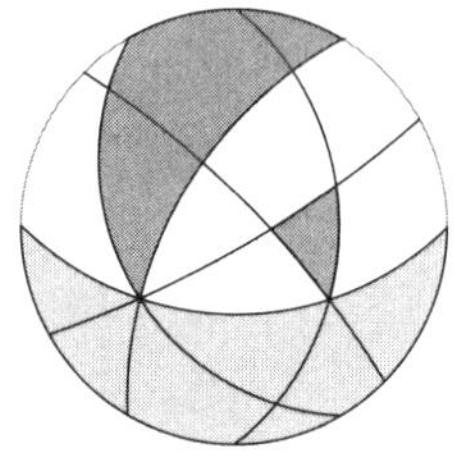

The figure on the left shows two top-lunes in rank 2 in both the spherical and the linear models. The figure on the right shows three top-lunes in rank 3 each with a different base. In rank 3, great circles, or great semicircles (such as the longitudes) are examples of lunes which are not top-lunes (since their case is a hyperplane).

The top-lunes (in the spherical model) in ranks 2 and 3 are described below.

Rank of the base	Top-lune in rank 2	Top-lune in rank 3
0	edge	chamber
1	semicircle	region between adjacent semicircles
2	circle	hemisphere
3	—	sphere

A *vertex-based lune* is a lune whose base is the support of a vertex. Similar terminology is employed for other faces. In rank 2, we have central-face-based, vertex-based and edge-based top-lunes. The latter two are the same as panel-based and chamber-based top-lunes. In rank 3, we have central-face-based, vertex-based, edge-based (or panel-based) and chamber-based top-lunes. In general: A central-face-based top-lune is a chamber (see exercise below), a panel-based top-lune is a half-space, while a chamber-based top-lune is the maximum flat.

Let $\Lambda[\mathcal{A}]$ denote the set of all lunes, and $\widehat{\Lambda}[\mathcal{A}]$ denote the set of all top-lunes. Lunes will usually be denoted by the letters L, M and N.

Exercise 3.1. Let L be a lune. Show that L is a face iff b(L) is the minimum flat.

3.1.2. Arrangements under and over a flat. Lunes of $\mathcal{A}$ correspond to faces and chambers of arrangements under and over various flats of $\mathcal{A}$. The precise relationship is stated below.

Lemma 3.2. *There are correspondences*

$$\text{Lunes of } \mathcal{A} \text{ with base } \mathrm{X} \longleftrightarrow \text{Faces of } \mathcal{A}_{\mathrm{X}},$$
$$\text{Top-lunes of } \mathcal{A} \text{ with base } \mathrm{X} \longleftrightarrow \text{Chambers of } \mathcal{A}_{\mathrm{X}},$$
$$\text{Lunes of } \mathcal{A} \text{ with base } \mathrm{X} \text{ and case } \mathrm{Y} \longleftrightarrow \text{Chambers of } \mathcal{A}_{\mathrm{X}}^{\mathrm{Y}}.$$

PROOF. We explain the first statement from which the next two follow. Recall that a face of $\mathcal{A}_{\mathrm{X}}$ is a subset of the ambient space obtained by intersecting half-spaces whose bounding hyperplanes contain X, with either H^+ or H^- or both chosen for each H containing X. Using the defining property (3.1), we see that these are precisely lunes with base X. □

3.1.3. Slack of a lune. Given a lune L, define its *slack* by

$$\mathrm{sk}(\mathrm{L}) = \mathrm{rk}(\mathrm{c}(\mathrm{L})) - \mathrm{rk}(\mathrm{b}(\mathrm{L})). \tag{3.2}$$

The rhs refers to rank in the lattice of flats. Lunes of slack 0 are precisely flats. Lunes of slack 1 are precisely half-flats. The slack of a face is its rank in the poset of faces. The slack of a general cone may be defined by the same formula as above.

3.2. Nested faces and lunes

The top-dimensional faces of a lune (as well as its closure, interior and boundary) can be described using the Tits product. In fact, lunes can be characterized using an equivalence relation on nested faces in the same manner that flats can be characterized using faces via (1.14). We first explain this for top-lunes and top-nested faces, and then deduce the general case by passing to an arrangement under a flat.

3.2.1. Top-nested faces and top-lunes. Let H be any face and D be a chamber greater than H. We refer to the pair (H, D) as a *top-nested face*. We define the *support* of such a top-nested face to be

$$\mathrm{s}(H, D) := \{C \mid HC = D\}. \tag{3.3}$$

By Lemma 1.58, this is a convex set of chambers, hence a combinatorial top-cone by Proposition 2.3.

Recall from (2.1) the notion of closure of a combinatorial cone.

Lemma 3.3. *The closure of* $\mathrm{s}(H, D)$ *is given by*

$$\mathrm{Cl}(\mathrm{s}(H, D)) = \{F \mid HF \leq D\}. \tag{3.4}$$

PROOF. Suppose F belongs to the closure. That is, for some C, we have $HC = D$, and $F \leq C$. Then $HF \leq HC$, so F belongs to the rhs above. Conversely, suppose F satisfies $HF \leq D$. Then $HFD = D$. So FD belongs to the top-cone and F belongs to its closure. □

The pictures below show the support of a top-nested face (H, D) in rank 3 in the cases when H is a vertex and H is an edge.

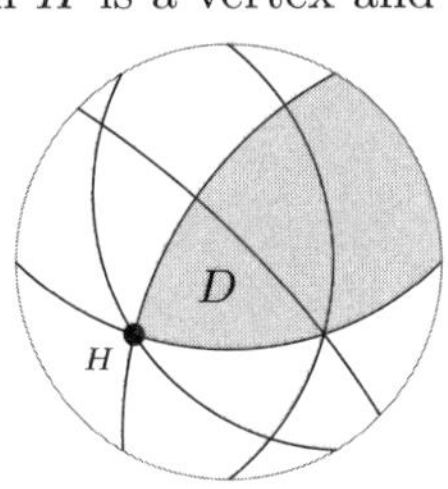

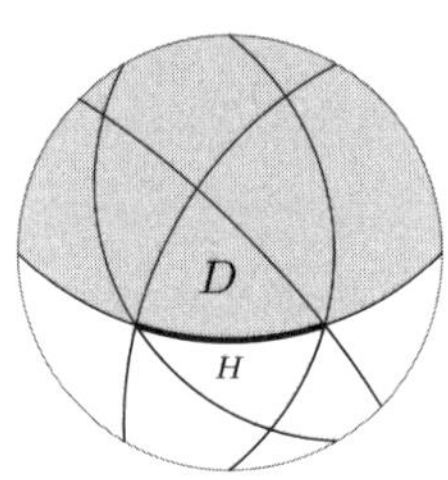

Note that both supports are in fact top-lunes. This is true in general, as we will see below.

Proposition 3.4. *Let* (H, D) *be a top-nested face. Then for any chamber* C,

$$\overline{H}D \text{ -- } C \text{ -- } D \iff HC = D.$$

Recall that the lhs means that there is a minimal gallery from $\overline{H}D$ to D passing through C.

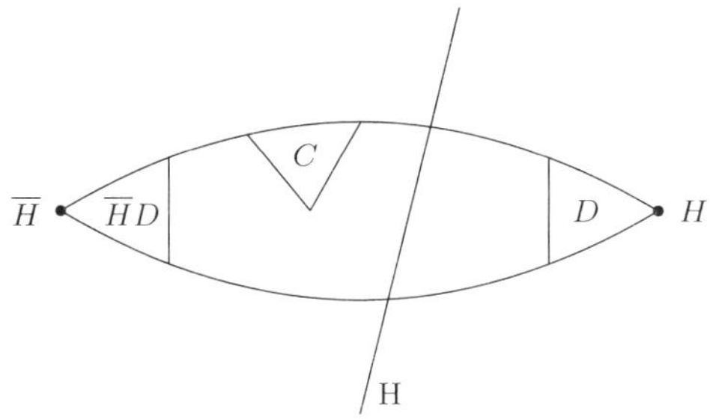

The figure illustrates the situation in rank 3 when H is a vertex.

PROOF. Since $H\overline{H}D = HD = D$, the forward implication follows from Corollary 1.59. For the backward implication: Since $HC = D$, we have $\overline{H}C = \overline{H}HC = \overline{H}D$. Now apply Proposition 1.57. □

This shows that the support of a top-nested face is a gallery interval. More precisely:

$$\mathrm{s}(H, D) = [D{:}\overline{H}D]. \tag{3.5}$$

Convexity of $\mathrm{s}(H, D)$ can also be deduced from Proposition 2.40.

Proposition 3.5. *For any top-nested face (H, D), the support $\mathrm{s}(H, D)$ is a combinatorial top-lune. The geometric lune corresponding to $\mathrm{s}(H, D)$ is the intersection of those half-spaces which contain D and whose base contains H. In particular, its base is $\mathrm{s}(H)$, the support of H.*

PROOF. Using Lemma 1.55, we deduce that a chamber belongs to $\mathrm{s}(H, D)$ iff it belongs to all half-spaces which contain D and whose base passes through H. The defining property of a lune (3.1) can now be readily verified. The last claim follows from Proposition 2.17. □

Corollary 3.6. *Let* H *be a hyperplane. Then* H *is a wall of* $\mathrm{s}(H, D)$ *iff* H *contains a panel of D which is greater than H.*

PROOF. Put $\mathrm{L} := \mathrm{s}(H, D)$. Suppose H is a wall of L. Then by Propositions 2.17 and 3.5, H contains the face H. Suppose C and C' are adjacent chambers with common panel K supported by H such that C belongs to L while C' does not belong to L. Then $HC = D$ and HK is a panel of D greater than H which is contained in H.

Conversely, suppose H contains a panel of D, say F, which is greater than H. Let E be the chamber adjacent to D with common panel F. Then $HE = E$, so E cannot belong to L. Hence H is a wall of L. □

Recall from Section 2.3 the notion of interior and boundary of a combinatorial cone.

Lemma 3.7. *The interior and the boundary of* $\mathrm{s}(H, D)$ *are given by*

$$\mathrm{s}(H, D)^o = \{F \mid HF = D\} \quad \textit{and} \quad \mathrm{s}(H, D)^b = \{F \mid HF < D\}. \tag{3.6}$$

The union of these two sets is the closure of $\mathrm{s}(H, D)$ *given in* (3.4).

PROOF. In view of (2.11), it suffices to prove the claim about the boundary. Suppose $HF < D$. Then HF is a face of some panel of D, so by Corollary 3.6, HF and hence F is contained in some wall of $\mathrm{s}(H, D)$. Thus F belongs to the boundary. Conversely, if F belongs to the boundary, then by reversing these steps, we see that $HF < D$. □

We now show that every top-lune is the support of a (not necessarily unique) top-nested face.

Proposition 3.8. *Let* V *be a combinatorial top-cone,* L *a combinatorial top-lune, and* H *a face.*

- *Suppose* $\mathrm{s}(H) \leq \mathrm{b}(\mathrm{V})$ *and* D *is a chamber in* V *which is greater than* H*. Then* $\mathrm{s}(H, D) \subseteq \mathrm{V}$.
- *Suppose* $\mathrm{s}(H) = \mathrm{b}(\mathrm{L})$*. Then there exists a unique chamber* D *in* L *which is greater than* H*, and further* $\mathrm{L} = \mathrm{s}(H, D)$.

PROOF. First part: Suppose C is a chamber such that $HC = D$. Let h be any half-space which contains V. Then it contains D and its base contains H. Since $HC = D$, h must also contain C. So C belongs to V, as required. Alternatively: By Proposition 2.7, item (3), the closure of a combinatorial cone is closed under taking products. Further, $\mathrm{s}(\overline{H})$ is also smaller than $\mathrm{b}(\mathrm{V})$, so $\overline{H}D$ belongs to V. By Proposition 2.3, V is convex, so it contains $[D\!:\!\overline{H}D]$ which is the same as $\mathrm{s}(H, D)$ by (3.5).

Second part: We first show that D exists. For that, take any chamber C' in L. Again by Proposition 2.7, item (3), $D := HC'$ belongs to L, and hence is a chamber in L which contains H. We now show that D is unique: Suppose E is another chamber in L which contains H. Then there exists a hyperplane containing H which separates D and E. This is not possible by the defining property of a lune (3.1). This further implies that $HC = D$ for any chamber C in L. Thus, $\mathrm{L} \subseteq \mathrm{s}(H, D)$. The reverse inclusion holds by the first part, so we have equality. □

Consider the following relation on the set of top-nested faces:

$$(H, D) \sim (G, C) \iff HG = H,\ GH = G,\ HC = D \text{ and } GD = C. \tag{3.7}$$

Using (1.13), note that $(H, D) \sim (G, C)$ iff H and G have the same support, and $HC = D$ and $GD = C$.

Proposition 3.9. *Equivalence classes for the relation* (3.7) *are in one-to-one correspondance with top-lunes. The class of* (H, D) *is the top-lune* $\mathrm{s}(H, D)$.

This statement makes two claims. The first claim is that every top-lune is the support of a top-nested face. This is contained in Proposition 3.8. The second claim is that: Two top-nested faces (H, D) and (G, C) have the same support iff $(H, D) \sim (G, C)$. This is proved below.

PROOF. Suppose $\mathrm{s}(H, D) = \mathrm{s}(G, C) = \mathrm{L}$ (say). Then $\mathrm{s}(H) = \mathrm{b}(\mathrm{L}) = \mathrm{s}(G)$, so H and G have the same support. Further, C and D both belong to L, and hence so do HC and GD. Proposition 3.8 then forces $HC = D$ and $GD = C$. Thus, $(H, D) \sim (G, C)$.

Conversely: Suppose $(H, D) \sim (G, C)$. We claim that for any chamber E, $HE = D$ iff $GE = C$. Assuming $HE = D$, we get $GE = GHE = GD = C$, and similarly assuming $GE = C$, we get $HE = HGE = HC = D$. Thus, $\mathrm{s}(H, D) = \mathrm{s}(G, C)$. □

The picture below shows two top-nested faces (H, D) and (G, C) in rank 3 with the same support. The support is a hemisphere.

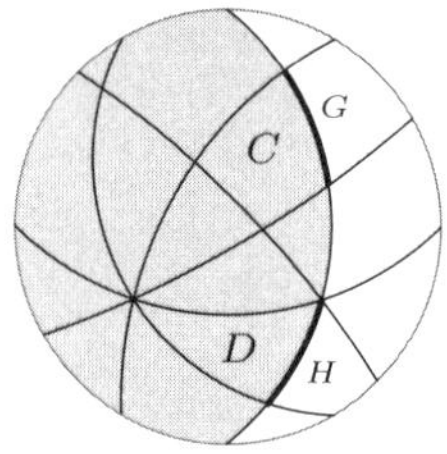

Try to visualize all top-nested faces with this support. They are all lined up along the base of the hemisphere which is a great circle.

Exercise 3.10. Let H be a face with support X, F a face with support Y, and $\mathrm{X} \vee \mathrm{Y} = \top$. Show that HF is a chamber and $\mathrm{s}(H, HF)$ is a top-lune with base X.

3.2.2. Nested faces and lunes. Recall that any cone can be viewed as a top-cone of an arrangement under a flat. The above study of top-lunes then directly yields the following results.

A *nested face* is a pair of faces (H, G) such that $H \leq G$. We define the *support* of such a nested face to be

$$\begin{aligned} \mathrm{s}(H, G) &:= \{F \mid HF = G \text{ and } \mathrm{s}(F) = \mathrm{s}(G)\} \\ &= \{F \mid HF = G \text{ and } FH = F\}. \end{aligned} \tag{3.8}$$

(The second equality holds by Exercise 1.12.) This is a gallery interval, namely,

$$\mathrm{s}(H, G) = [G : \overline{H}G]. \tag{3.9}$$

Further, it is a combinatorial lune with base $\mathrm{s}(H)$ and case $\mathrm{s}(G)$. In particular, it is a flat iff $H = G$. Its closure is

$$\mathrm{Cl}(\mathrm{s}(H, G)) = \{F \mid HF \leq G\}, \tag{3.10}$$

with its interior and boundary given by

$$\mathrm{s}(H, G)^o = \{F \mid HF = G\} \quad \text{and} \quad \mathrm{s}(H, G)^b = \{F \mid HF < G\}, \tag{3.11}$$

respectively. Also note that

$$\mathrm{sk}(\mathrm{s}(H, G)) = \mathrm{rk}(G) - \mathrm{rk}(H), \tag{3.12}$$

where sk denotes the slack of a lune (3.2).

Proposition 3.11. *The combinatorial lune* $\mathrm{s}(H, G)$ *is a topological sphere if* $H = G$*, and a topological ball if* $H < G$*. In the latter case, the boundary sphere is* $\mathrm{s}(H, G)^b$.

PROOF. Since $\mathrm{s}(H, G)$ is a flat iff $H = G$, the first part follows from Proposition 2.27. The second part is clear. □

Proposition 3.12. *Let* V *be a combinatorial cone,* L *a combinatorial lune, and* H *a face.*

- *Suppose* $\mathrm{s}(H) \leq \mathrm{b}(\mathrm{V})$ *and* K *is a top-dimensional face of* V *which is greater than* H*. Then* $\mathrm{s}(H, K) \subseteq \mathrm{V}$*, and in particular,* $\mathrm{s}(H, K)^o \subseteq \mathrm{Cl}(\mathrm{V})$.
- *Suppose* $\mathrm{s}(H) = \mathrm{b}(\mathrm{L})$*. Then there exists a unique top-dimensional face* K *of* L *which is greater than* H*, and further* $\mathrm{L} = \mathrm{s}(H, K)$.

Consider the following relation on the set of nested faces:

$$(H,G) \sim (K,F) \iff HK = H,\ KH = K,\ HF = G \text{ and } KG = F. \tag{3.13}$$

Note that $(H,G) \sim (K,F)$ iff H and K have the same support, $HF = G$ and $KG = F$.

Proposition 3.13. *Equivalence classes for the relation* (3.13) *are in one-to-one correspondance with lunes. The class of* (H,G) *is the lune* $\mathrm{s}(H,G)$.

Proposition 3.14. *Given faces A and A' with the same support, there exists a bijection*

$$\psi : \{F : A \leq F\} \to \{F' : A' \leq F'\}$$

such that $(A,F) \sim (A', \psi(F))$, *and in particular,* $\mathrm{s}(F) = \mathrm{s}(\psi(F))$ *for every* $F \geq A$.

PROOF. Define $\psi(F) = A'F$. This is the bijection in Lemma 1.37. □

Exercise 3.15. Given $A \leq F \leq G$ and $A' \leq F' \leq G'$,

$$\mathrm{s}(A,F) = \mathrm{s}(A',F') \text{ and } \mathrm{s}(F,G) = \mathrm{s}(F',G') \implies \mathrm{s}(A,G) = \mathrm{s}(A',G').$$

Conversely, given $\mathrm{s}(A,G) = \mathrm{s}(A',G')$ and F between A and G, there exists a unique F' between A' and G' such that

$$\mathrm{s}(A,F) = \mathrm{s}(A',F') \text{ and } \mathrm{s}(F,G) = \mathrm{s}(F',G').$$

In fact, either of the above two conditions suffices to force uniqueness of F'. In particular, given two distinct faces F and F' between A and G, we have $\mathrm{s}(A,F) \neq \mathrm{s}(A,F')$ and $\mathrm{s}(F,G) \neq \mathrm{s}(F',G)$.

3.2.3. Examples. Faces, flats and half-flats are lunes. The table below elaborates on how they arise as supports of nested faces.

Lune	Support of a nested face
Center	$\mathrm{s}(O,O)$
Face	$\mathrm{s}(O,F)$
Chamber	$\mathrm{s}(O,C)$
Flat	$\mathrm{s}(F,F)$
Half-space	$\mathrm{s}(H,D)$ with $H \lessdot D$
Half-flat	$\mathrm{s}(H,G)$ with $H \lessdot G$

Among these, chambers and half-spaces are top-lunes. Note that they are supports of top-nested faces.

3.2.4. Opposition map on lunes. The opposition map on cones (2.6) restricts to lunes. More precisely, the opposite of the lune $\mathrm{s}(H,G)$ is the lune $\mathrm{s}(H,H\overline{G})$. In particular, for the relation (3.13),

$$(H,G) \sim (K,F) \iff (H,H\overline{G}) \sim (K,K\overline{F}). \tag{3.14}$$

A lune equals its opposite iff it is a flat. This is a special case of Proposition 2.11.

The picture below shows two opposite vertex-based top-lunes in a rank-three arrangement. For proper visualization, fold the two lunes on a sphere so that the two vertices marked $\overline{H}$ coincide at the point antipodal to H.

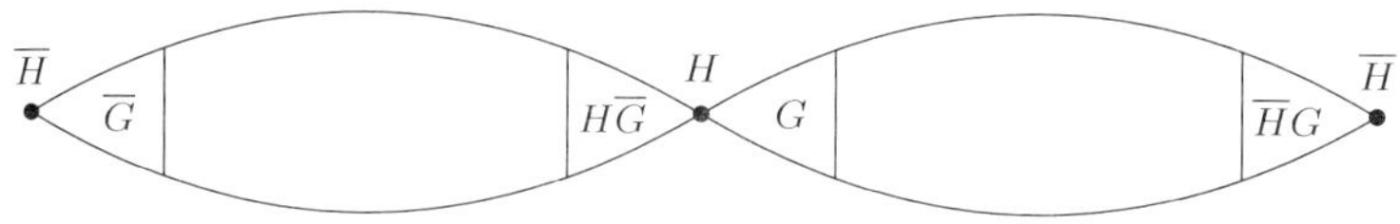

Let us go back to the general case. Specializing (2.10) and (2.8), we obtain

$$\mathrm{s}(H,G) \wedge \mathrm{s}(H,H\overline{G}) = \mathrm{s}(H) \quad \text{and} \quad \mathrm{s}(H,G) \vee \mathrm{s}(H,H\overline{G}) = \mathrm{s}(G),$$

with meet and join taken in the poset of cones.

A *projective lune* is an unordered pair consisting of a lune and its opposite. We denote a projective lune by $\{\mathrm{L},\overline{\mathrm{L}}\}$. The number of projective lunes equals the number of flats plus half the number of lunes which are not flats.

3.2.5. Lunes with a fixed base. Fix a face A. By Proposition 3.12, there is a bijection

$$\{F \mid F \geq A\} \longleftrightarrow \{\mathrm{L} \mid \mathrm{b}(\mathrm{L}) = \mathrm{s}(A)\}, \qquad F \mapsto \mathrm{s}(A,F). \tag{3.15}$$

The lhs consists of faces greater than A, while the rhs consists of lunes with base $\mathrm{s}(A)$.

Now view both sides as posets, the lhs as a subposet of the poset of faces, and the rhs as a subposet of the poset of cones.

Lemma 3.16. *The map* (3.15) *is a poset isomorphism. In other words: For nested faces* (A,F) *and* (A,G)*, we have*

$$\mathrm{s}(A,F) \leq \mathrm{s}(A,G) \iff F \leq G.$$

Equivalently: For lunes L *and* M *with the same base,* $\mathrm{L} \leq \mathrm{M}$ *iff* $\mathrm{L} = \mathrm{s}(A,F)$ *and* $\mathrm{M} = \mathrm{s}(A,G)$ *for some* $A \leq F \leq G$.

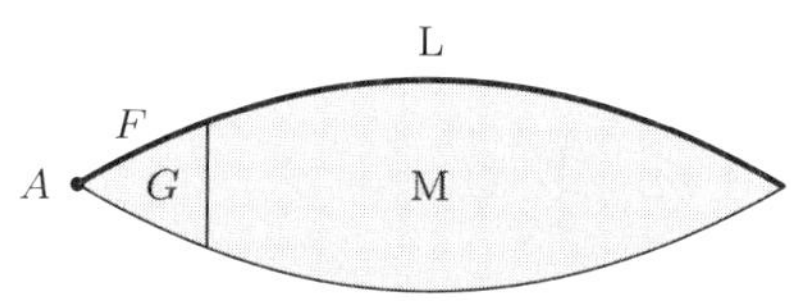

In the picture, L is the thick line, while M is the shaded region.

Proof. This follows from (2.2) and (3.10). □

Corollary 3.17. *Distinct lunes with the same base and same case are incomparable in the poset of cones. In other words, if* $\mathrm{L} \leq \mathrm{M}$*,* $\mathrm{b}(\mathrm{L}) = \mathrm{b}(\mathrm{M})$ *and* $\mathrm{c}(\mathrm{L}) = \mathrm{c}(\mathrm{M})$*, then* $\mathrm{L} = \mathrm{M}$.

3.2.6. Gallery intervals. Lunes are gallery intervals. This is elaborated below.

Proposition 3.18. *Lunes are gallery intervals. More precisely: Given a top-lune* L*, and a face* H *supported by the base of* L*, there exists a unique chamber* D *greater than* H *such that* $\mathrm{L} = [D\!:\!\overline{H}D]$*. More generally, given a lune* L*, and a face* H *supported by the base of* L*, there exists a unique face* G *greater than* H *such that* $\mathrm{L} = [G\!:\!\overline{H}G]$.

Proof. The first statement follows from (3.5) and Proposition 3.8. The second follows from the first by working in an arrangement under a flat. □

Thus, a lune is a nice example of a cone which can be realized as a gallery interval in multiple ways. In fact, there are lunes which can be realized as gallery intervals beyond the possibilities listed in Proposition 3.18. For instance, consider the top-lune shown below. It is based at the vertices H and $\overline{H}$. It consists of four chambers. Apart from $[D : \overline{H}D]$, it can also be realized as a gallery interval as $[C : C']$. In fact, this top-lune is also the top-star of the unique vertex in its interior.

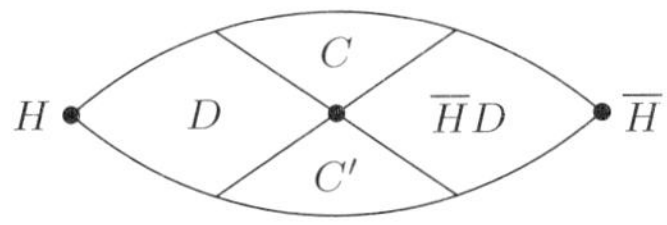

Such a top-lune occurs in the smallest nonsimplicial arrangement of four hyperplanes (Section 1.2.5). A slightly more complicated example is shown below.

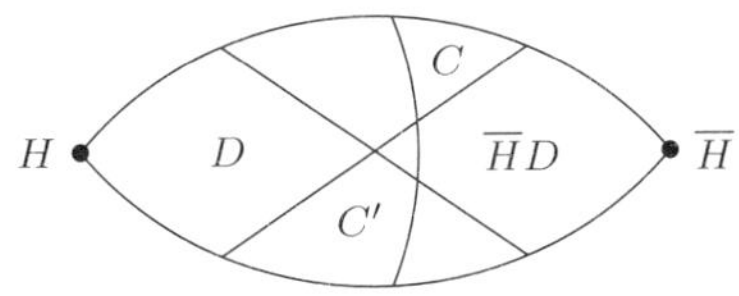

This top-lune is *not* a top-star. There are three hyperplanes which cut it, and these are precisely the hyperplanes which separate C and C', so it equals $[C : C']$.

In general, not every gallery interval is a lune. A pair of adjacent chambers whose base is the minimum flat is an example of a gallery interval which is *not* a top-lune. For instance, take any pair of adjacent chambers in the rank-two arrangement of n lines for $n \geq 3$.

Exercise 3.19. Check: A cisomorphism of arrangements preserves cones, gallery intervals, lunes, half-spaces and half-flats. (See the discussion in Section 1.11.)

3.3. Decomposition of a cone into lunes

We show that every cone can be optimally decomposed into lunes. In this sense, lunes serve as building blocks of cones. More generally, every cone has a lune decomposition over a flat contained in its base. The optimal decomposition arises when the flat equals the base. Special cases include decomposition of the ambient space into lunes, of a flat into lunes, of a lune into smaller lunes, and of a cone into vertex-based lunes.

3.3.1. Decomposition of the ambient space into lunes. The support of a top-nested face (H, D) is the (combinatorial) top-lune consisting of chambers whose projection on H is D. Hence by fixing H and varying D, we obtain a partition of the set of chambers by top-lunes:

$$\Gamma[\mathcal{A}] = \bigsqcup_{D:\, H \leq D} \mathrm{s}(H, D). \tag{3.16}$$

Geometrically, this is a decomposition of the ambient space into top-lunes.

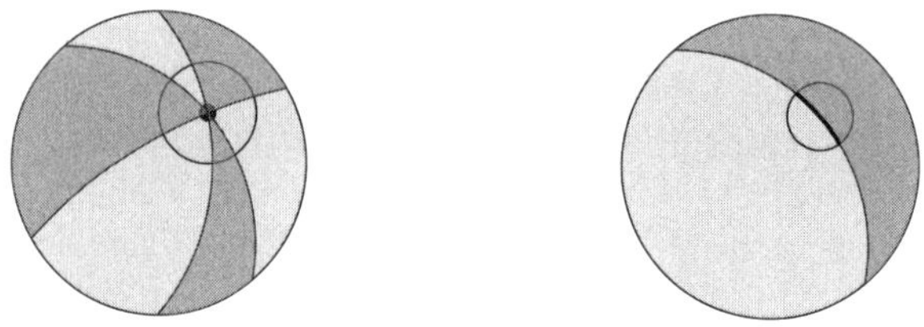

The figure on the left shows a decomposition of the sphere into six top-lunes. Only those hyperplanes and faces which are relevant to the decomposition are shown. The vertex H is shown in black and its top-star is the inner circle containing the six chambers which index the top-lunes. In the figure on the right, H is the thick black edge, its top-star has two chambers, so the resulting decomposition has two top-lunes, which are hemispheres.

One can also decompose the set of all faces using lune interiors as follows.

$$\Sigma[\mathcal{A}] = \bigsqcup_{K:\, H \leq K} \mathrm{s}(H,K)^o. \tag{3.17}$$

This follows from the definitions.

Using Proposition 3.8, the decomposition (3.16) can also be expressed as follows. For any flat X,

$$\Gamma[\mathcal{A}] = \bigsqcup_{\mathrm{L}:\, \mathrm{b}(\mathrm{L})=\mathrm{X}} \mathrm{L}. \tag{3.18}$$

The sum is over all (combinatorial) top-lunes L with base X. Thus, distinct (combinatorial) top-lunes with the same base are disjoint, and they partition the set of chambers.

3.3.2. Decomposition of a flat into lunes.

Let Y be a combinatorial flat. Fix a face H of support X smaller than Y. Then

$$\mathrm{Y} = \bigsqcup_{G:\, H \leq G,\, \mathrm{s}(G)=\mathrm{Y}} \mathrm{s}(H,G) = \bigsqcup_{\mathrm{L}:\, \mathrm{b}(\mathrm{L})=\mathrm{X},\, \mathrm{c}(\mathrm{L})=\mathrm{Y}} \mathrm{L}. \tag{3.19}$$

The first decomposition is clear. The second follows from Proposition 3.12. Setting Y to be the maximum flat recovers (3.16) and (3.18).

As a consequence:

Corollary 3.20. *Distinct (combinatorial) lunes with the same base and the same case are disjoint.*

This contains the result of Corollary 3.17.

3.3.3. Decomposition of a lune into smaller lunes.

For any $H \leq G \leq D$, we have

$$\mathrm{s}(G,D) = \bigsqcup_{C:\, H \leq C,\, GC=D} \mathrm{s}(H,C). \tag{3.20}$$

The top-lune in the lhs has base $\mathrm{s}(G)$ and it has been decomposed into top-lunes with a smaller base, namely, $\mathrm{s}(H)$. This identity is straightforward to check. The main point is that if a chamber E belongs to the lhs, then it belongs to the summand in the rhs indexed by $C := HE$. Note that the indexing set consists of chambers in the top-lune $\mathrm{s}(G/H, D/H)$. This leads to the following result.

Proposition 3.21. *For faces $H \leq G$, any G-based combinatorial top-lune can be written as a disjoint union of H-based combinatorial top-lunes indexed by chambers of a G/H-based combinatorial top-lune.*

More generally, for any $H \leq G \leq K$, we have

$$\mathrm{s}(G, K) = \bigsqcup_{\substack{F:\, H \leq F,\, GF = K \\ \mathrm{s}(F) = \mathrm{s}(K)}} \mathrm{s}(H, F). \tag{3.21}$$

The indexing set can be identified with $\mathrm{s}(G/H, K/H)$.

There is a similar decomposition of lune interiors. For any $H \leq G \leq K$, we have

$$\mathrm{s}(G, K)^o = \bigsqcup_{F:\, H \leq F,\, GF = K} \mathrm{s}(H, F)^o. \tag{3.22}$$

The indexing set can be identified with $\mathrm{s}(G/H, K/H)^o$.

3.3.4. Decomposition of a cone into lunes. We now turn to arbitrary cones, generalizing all that we have done so far.

Proposition 3.22. *Let* V *be a combinatorial cone and* X *a flat with* $\mathrm{X} \leq \mathrm{b}(\mathrm{V})$. *Then* V *can be written uniquely as a disjoint union of combinatorial lunes, each with base* X *and case* $\mathrm{c}(\mathrm{V})$. *Explicitly: Let H be any face with support* X. *Then*

$$\mathrm{V} = \bigsqcup_{K:\, K \in \mathrm{V}, K \geq H} \mathrm{s}(H, K). \tag{3.23}$$

In other words, the sum is over all top-dimensional faces K of V *which are greater than H.*

PROOF. By Proposition 3.12, $\mathrm{s}(H, K) \subseteq \mathrm{V}$. Conversely, for any $G \in \mathrm{V}$, HG must be one of the K and hence $G \in \mathrm{s}(H, K)$. This yields the decomposition (3.23). For uniqueness, we note that any lune contained in V with base X and case c(V) appears in the rhs of (3.23). This follows from Proposition 3.12. □

We call (3.23) the *lune decomposition* of the cone V over the flat X. Suppose $\mathrm{X} \leq \mathrm{Y} \leq \mathrm{b}(\mathrm{V})$. Then one may first decompose V over Y, and then decompose each of the resulting lunes (with base Y) over X. By uniqueness, this is the same as decomposing V directly over X.

An equivalent way of expressing the lune decomposition is given below.

Proposition 3.23. *Let* V *be a combinatorial cone and* X *a flat with* $\mathrm{X} \leq \mathrm{b}(\mathrm{V})$. *Then*

$$\mathrm{V} = \bigsqcup_{\substack{\mathrm{L}:\, \mathrm{b}(\mathrm{L}) = \mathrm{X},\, \mathrm{L} \leq \mathrm{V} \\ \mathrm{c}(\mathrm{L}) = \mathrm{c}(\mathrm{V})}} \mathrm{L}. \tag{3.24}$$

The condition $\mathrm{L} \leq \mathrm{V}$ *refers to the partial order on cones.*

Similar decompostion results for the interior, boundary and closure of a combinatorial cone are given below.

Proposition 3.24. *For* V *a combinatorial cone and H a face with* $\mathrm{s}(H) \leq \mathrm{b}(\mathrm{V})$,

$$\mathrm{V}^o = \bigsqcup_{K:\, K \in \mathrm{V}^o, K \geq H} \mathrm{s}(H, K)^o \quad \text{and} \quad \mathrm{V}^b = \bigsqcup_{K:\, K \in \mathrm{V}^b, K \geq H} \mathrm{s}(H, K)^o, \tag{3.25}$$

and

$$\mathrm{Cl}(\mathrm{V}) = \bigsqcup_{K:\, K \in \mathrm{Cl}(\mathrm{V}), K \geq H} \mathrm{s}(H,K)^o. \tag{3.26}$$

PROOF. By Proposition 2.23, faces in Cl(V) can be classified as boundary or interior faces by looking at their projection on H. In view of Proposition 3.12, we obtain both identities in (3.25) and their union which is (3.26). □

We now discuss some special cases.

- In Proposition 3.22, one can always take X to be the minimum flat. Recall that lunes whose base is the minimum flat are the same as faces, hence the decomposition (3.23) amounts to writing V as the union of its top-dimensional faces.
- Suppose V is the maximum flat. Then b(V) = V. So by Proposition 3.22, each flat X gives rise to a decomposition of V. In (3.23), H is any face with support X, and the K are chambers which contain H. This is precisely the top-lune decomposition given by (3.16). Note that (3.26) specializes to (3.17). More generally, an arbitrary flat V yields (3.19).
- Suppose V is a lune. Then (3.23) specializes to (3.21) while the first identity in (3.25) specializes to (3.22). When $\mathrm{s}(H) = \mathrm{b}(\mathrm{V})$, (3.23) is a tautology, but (3.26) is nontrivial and can be expressed as

$$\mathrm{Cl}(\mathrm{s}(H,G)) = \bigsqcup_{K: H \leq K \leq G} \mathrm{s}(H,K)^o. \tag{3.27}$$

  The summand indexed by $K = H$ accounts for the base of the lune. The summand indexed by $K = G$ accounts for the interior of the lune, while the remaining summands account for the boundary.

3.3.5. Decomposition of a cone into vertex-based lunes.

Corollary 3.25. *Let* V *be a combinatorial cone whose base is not the minimum flat. Then* V *can be written as a disjoint union of vertex-based combinatorial lunes. More precisely: Let $H \leq G$, where H is a vertex. Then any G-based combinatorial cone* V *is the disjoint union of some H-based combinatorial lunes.*

PROOF. By hypothesis, b(V) has rank at least 1. Apply Proposition 3.22 with $\mathrm{X} = \mathrm{s}(H)$, where H is any vertex contained in b(V). (Two vertices will yield the same decomposition iff they are opposites of each other.) In the second formulation, $\mathrm{b}(\mathrm{V}) = \mathrm{s}(G)$. Since $H \leq G$, $\mathrm{X} = \mathrm{s}(H)$ is smaller than b(V). □

A similar result which follows from (3.25) is given below.

Corollary 3.26. *Let* V *be a combinatorial cone whose base is not the minimum flat. Then* V^o *can be written as a disjoint union of the interiors of vertex-based combinatorial lunes. More precisely: Let $H \leq G$, where H is a vertex. Then the interior of any G-based combinatorial cone* V *is the disjoint union of the interiors of some H-based combinatorial lunes.*

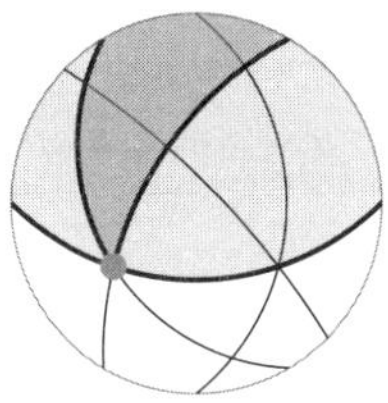

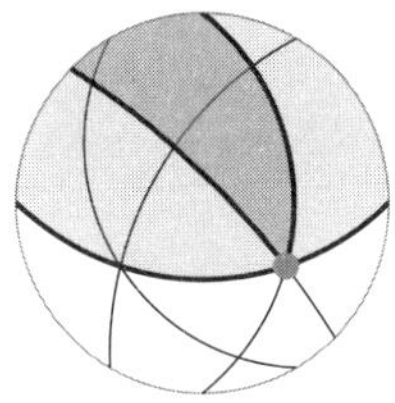

The pictures above show two distinct decompositions of the hemisphere by vertex-based lunes. The base of any of these lunes is the support of the vertex shown in gray.

3.3.6. Optimal decomposition of a cone. Applying Proposition 3.23 with $\mathrm{X} := \mathrm{b}(\mathrm{V})$, we obtain:

Theorem 3.27. *For any combinatorial cone* V,

$$\mathrm{V} = \bigsqcup_{\substack{\mathrm{L}:\, \mathrm{b}(\mathrm{L})=\mathrm{b}(\mathrm{V}),\, \mathrm{L}\leq \mathrm{V} \\ \mathrm{c}(\mathrm{L})=\mathrm{c}(\mathrm{V})}} \mathrm{L}. \tag{3.28}$$

We refer to (3.28) as the *optimal decomposition* of a cone (since any other decomposition of V involves decomposing the lunes in the optimal decomposition of V into smaller lunes). Similarly, one can optimally decompose the interior, boundary and closure of V by taking $\mathrm{s}(H) = \mathrm{b}(\mathrm{V})$ in Proposition 3.24.

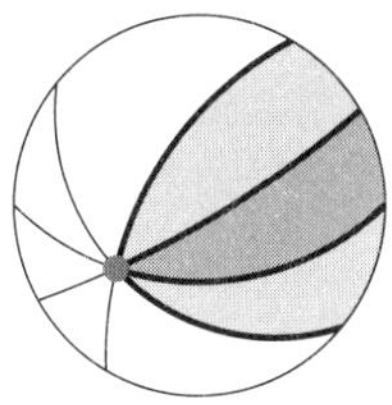

The figure shows the optimal decomposition of a vertex-based cone in rank 3.

Exercise 3.28. Let V be a cone. Then check that the following are equivalent.

(1) V is a lune.
(2) The optimal decomposition of V has only one lune.
(3) There is exactly one top-dimensional face of V which contains any given top-dimensional face of b(V).

Exercise 3.29. Any cone V can be written as the union of its top-dimensional faces. Show that this is the optimal decomposition iff b(V) is the minimum flat.

Exercise 3.30. Recall from Lemma 3.2 that a lune with base X and support Y corresponds to a chamber of the arrangement $\mathcal{A}^{\mathrm{Y}}_{\mathrm{X}}$. Show that: The optimal decomposition of a cone with base X and support Y into lunes corresponds to writing a top-cone of $\mathcal{A}^{\mathrm{Y}}_{\mathrm{X}}$ as a union of its chambers.

Exercise 3.31. Let L and M be two lunes with the same base and the same support. Fix a top-dimensional face F of L. Define $\Sigma_{\mathrm{L,M}}$ to be the set of all faces G such that GF is a top-dimensional face of M. Note that $\Sigma_{\mathrm{L,M}} \subseteq \mathrm{Cl}(\mathrm{M})$. Show that $\Sigma_{\mathrm{L,M}}$ is the disjoint union of certain summands of the optimal decomposition of Cl(M) given in (3.26). Deduce that $\Sigma_{\mathrm{L,M}}$ does not depend on the particular choice of F.

3.4. Restriction and extension of cones

A top-cone can be as small as a chamber and as large as the set of all chambers. This feature makes it possible to assign two meanings to what it means for a top-cone to be local to a face F. The first meaning is that it is contained in the top-star of F. The second meaning is that it contains F in its closure. Note that the first meaning is stricter than the second. These two kinds of local top-cones can be related to each other by the operations of restriction and extension. We also discuss the special case when the top-cones are gallery intervals and top-lunes. These considerations generalize to arbitrary cones.

3.4.1. Restriction. Fix a face F. Suppose W is a combinatorial top-cone whose closure contains F. Define

$$\mathrm{W}_F := \mathrm{W} \cap \Gamma_F,$$

where Γ_F is the top-star of F. Since F belongs to the closure of W, the intersection in the rhs is nonempty, so W_F is a top-cone. It is the meet of W and Γ_F in the poset of top-cones. (See Exercise 2.9.) We call it the *restriction* of W to F. It consists of those chambers of W which are greater than F.

3.4.2. Extension. Fix a face F. Suppose V is a combinatorial top-cone contained in the top-star of F. Define

$${}_F\mathrm{V} := \{C \mid FC \in \mathrm{V}\}.$$

This is a top-cone which contains V. (This can be checked using Proposition 2.3, items (2) or (3).) We call it the *extension* of V from F.

3.4.3. Adjunction. We now relate the two constructions. Let $\widehat{\Omega}_F$ denote the set of all top-cones contained in the top-star of F, and let ${}_F\widehat{\Omega}$ denote the set of all top-cones whose closure contains F. We view both of them as subposets of $\widehat{\Omega}$ (the poset of all top-cones). Note that $\widehat{\Omega}_F$ can be identified with the set of all top-cones of the arrangement $\mathcal{A}_F$.

Restriction and extension define order-preserving maps

$${}_F\widehat{\Omega} \to \widehat{\Omega}_F, \quad \mathrm{W} \mapsto \mathrm{W}_F \qquad \text{and} \qquad \widehat{\Omega}_F \to {}_F\widehat{\Omega}, \quad \mathrm{V} \mapsto {}_F\mathrm{V}.$$

Moreover, for any $\mathrm{V} \in \widehat{\Omega}_F$ and $\mathrm{W} \in {}_F\widehat{\Omega}$,

$$\mathrm{W}_F \leq \mathrm{V} \iff \mathrm{W} \leq {}_F\mathrm{V}. \tag{3.29}$$

In other words, restriction and extension form an adjunction, with restriction as the left adjoint and extension as the right adjoint, see (B.2).

PROOF. Forward implication. Suppose $\mathrm{W}_F \leq \mathrm{V}$. Let C be any chamber in W. Then by Proposition 2.7, FC is in W_F, and hence in V. Hence C belongs to ${}_F\mathrm{V}$. This shows that $\mathrm{W} \leq {}_F\mathrm{V}$.

Backward implication. Suppose $\mathrm{W} \leq {}_F\mathrm{V}$. Let D be any chamber in W_F. In particular, it belongs to W and hence to ${}_F\mathrm{V}$. So FD belongs to V, but $FD = D$, so D belongs to V. This shows that $\mathrm{W}_F \leq \mathrm{V}$. □

As a specialization of (B.3):

$$({}_F\mathrm{V})_F \leq \mathrm{V} \quad \text{and} \quad \mathrm{W} \leq {}_F(\mathrm{W}_F).$$

In fact, observe that the first inequality is always an equality, that is, $({}_F\mathrm{V})_F = \mathrm{V}$. Thus, ${}_F\mathrm{V}$ is the largest top-cone whose restriction to F is V. The second inequality can be strict. A necessary and sufficient condition for equality to hold is given below.

$$\mathrm{W} = {}_F(\mathrm{W}_F) \iff \mathrm{s}(F) \leq \mathrm{b}(\mathrm{W}). \tag{3.30}$$

Backward implication is equivalent to the lune decomposition of W over the flat $\mathrm{s}(F)$, see (3.23). Forward implication is implied by the following.

For $\mathrm{V} \in \widehat{\Omega}_F$, the support of F is contained in the base of ${}_F\mathrm{V}$.

To see this, fix a $D \in \mathrm{V}$. Let G be any face with the same support as F. Then $F(GD) = FD = D$, so $GD \in {}_F\mathrm{V}$. This shows that G lies in the closure of ${}_F\mathrm{V}$. Thus, the support of F is contained in the closure of ${}_F\mathrm{V}$ and hence lies in its base.

As a specialization of (B.4):

Proposition 3.32. *For any face F, there is a bijection between top-cones contained in the top-star of F, and top-cones whose base contains the support of F. The latter are the closed sets of the closure operator which sends* W *to* ${}_F(\mathrm{W}_F)$.

When F is the central face, the posets $\widehat{\Omega}_F$ and ${}_F\widehat{\Omega}$ coincide with $\widehat{\Omega}$, and both restriction and extension equal the identity map. The other extreme is when F is a chamber, say C. Then $\widehat{\Omega}_C$ is a singleton, its only element is the top-cone C. In contrast, ${}_C\widehat{\Omega}$ consists of all top-cones which contain C. The extension map sends C to the largest top-cone, namely, the set of all chambers. (The latter is the only top-cone whose base is the maximum flat.) The restriction map sends all top-cones containing C to C.

Exercise 3.33. Let F be a face and W a combinatorial top-cone with $\mathrm{s}(F) \leq \mathrm{b}(\mathrm{W})$. Show that: $C \in \mathrm{W}$ iff $FC \in \mathrm{W}$.

Exercise 3.34. Let F belong to the closure of W. Show that ${}_F(\mathrm{W}_F) = \Gamma$ iff F belongs to the interior of W.

Exercise 3.35. Give an example of a face K and a combinatorial top-cone V such that $\{KC \mid C \in \mathrm{V}\}$ is *not* a combinatorial top-cone.

3.4.4. Inclusion maps. Fix a face F. We can also consider the order-preserving inclusion map from $\widehat{\Omega}_F$ to ${}_F\widehat{\Omega}$. Observe that it is the left adjoint of the restriction map.

Similarly, there is an inclusion map from ${}_F\widehat{\Omega}$ to $\widehat{\Omega}$. It has a left adjoint which sends a top-cone V to the smallest top-cone greater than V which has F in its closure. Such a smallest top-cone exists and equals the meet of all top-cones with this property. This can be deduced from (2.3). Note that this adjunction can be composed with the adjunction given by restriction and extension.

3.4.5. Restriction and extension of arbitrary cones. The above considerations generalize to arbitrary cones.

Fix a face F. Let us first define Ω_F and ${}_F\Omega$. A cone V belongs to Ω_F iff all the top-dimensional faces of V are greater than F. A cone W belongs to ${}_F\Omega$ iff the closure of W contains F. For $\mathrm{V} \in \Omega_F$ and $\mathrm{W} \in {}_F\Omega$, define

$$\mathrm{W}_F = \{G \geq F \mid G \in \mathrm{W}\} \quad \text{and} \quad {}_F\mathrm{V} = \{G \mid FG \in \mathrm{V}, \mathrm{s}(G) = \mathrm{c}(\mathrm{V})\}.$$

(The same definitions as before are employed but in the arrangement under the case of the cone.) We check below that (3.29) continues to hold.

PROOF. Forward implication. Suppose $W_F \leq V$. Let $G \in W$. Then by Proposition 2.7, $FG \in W_F$. Hence, there exists $K \in V$ such that $FG \leq K$. Then GK satisfies $F(GK) = K$ and its support is the same as the case of V, so $GK \in {}_F V$. This shows that $W \leq {}_F V$.

Backward implication. Suppose $W \leq {}_F V$. Let $H \in W_F$. In particular, $H \in W$. Hence, there exists $K \in {}_F V$ such that $H \leq K$. So $FK \in V$, but $FK = K$, so $K \in V$. This shows that $W_F \leq V$. □

Thus, restriction and extension form an adjunction between Ω_F and ${}_F\Omega$.

Lemma 3.36. *Fix faces F and G. Let V be a cone whose top-dimensional faces are greater than F. Then G belongs to the closure of ${}_F V$ iff FG belongs to the closure of V.*

PROOF. This can be checked directly, or deduced from (3.26). □

3.4.6. Restriction and extension of gallery intervals. We now show that the operations of restriction and extension preserve gallery intervals.

Lemma 3.37. *Suppose V is a gallery interval, and F belongs to the closure of V. Then V_F, the restriction of V to F, is also a gallery interval. More precisely, if $V = [C\!:\!D]$, then $V_F = [FC\!:\!FD]$.*

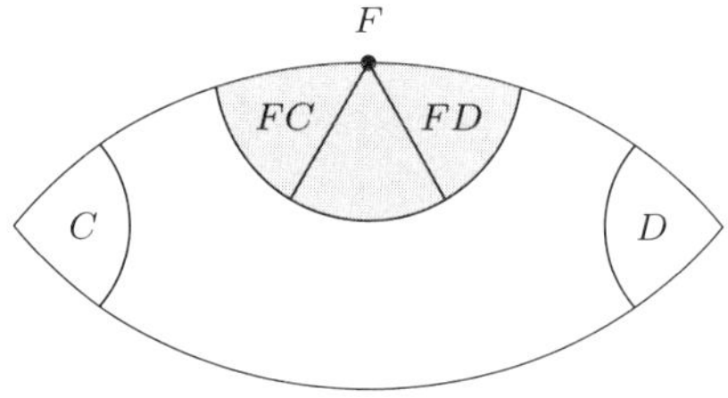

PROOF. Let $V = [C\!:\!D]$. Since F belongs to the closure of V, by Proposition 2.7, FC and FD belong to V, and hence to V_F. Now let E be any chamber in V_F. By hypothesis, $C \text{ -- } E \text{ -- } D$. By projecting this minimal gallery on F and using (1.29), we obtain $FC \text{ -- } E \text{ -- } FD$ since $FE = E$. Alternatively, this also directly follows from the gate property. Thus, $V_F = [FC\!:\!FD]$ as required. □

Lemma 3.38. *Suppose V is a gallery interval contained in a top-star of F for some face F. Then ${}_F V$, the extension of V from F, is also a gallery interval. More precisely, if $V = [C : D]$ for some chambers C and D greater than F, then ${}_F V = [C\!:\!\overline{F}D]$.*

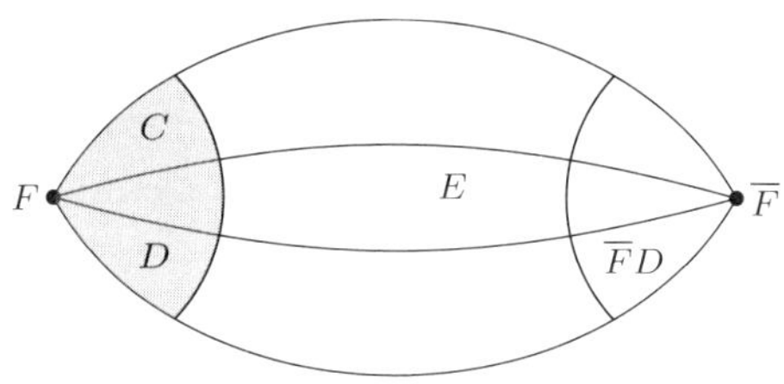

PROOF. Let $V = [C : D]$. Note that $F(\overline{F}D) = FD = D$, so $\overline{F}D \in {}_F V$. Now let E be any chamber in ${}_F V$. Then $FE \in V$ and hence by hypothesis, $C \text{ -- } FE \text{ -- } D$. Also, by the gate property, $C \text{ -- } D \text{ -- } \overline{F}D$. This yields $C \text{ -- } FE \text{ -- } \overline{F}D$. Again by the gate property, this can be refined to $C \text{ -- } FE \text{ -- } \overline{F}E \text{ -- } \overline{F}D$. By Proposition 1.57, this further refines to $C \text{--} FE \text{--} E \text{--} \overline{F}E \text{--} \overline{F}D$. Thus, $C \text{--} E \text{--} \overline{F}D$, and we obtain ${}_F\mathrm{V} = [C : \overline{F}D]$ as required. □

We mention that this fact can also be obtained more directly by employing Lemma 1.71.

3.4.7. Restriction and extension of top-lunes. Now let us consider the further subclass of top-lunes. This case is slightly different because there is a distinction between top-lunes contained in the top-star of F, and top-lunes in the arrangement $\mathcal{A}_F$. We need to work with the latter.

Suppose L is a combinatorial top-lune containing F in its closure. Then its restriction L_F may not be a top-lune, but it will be a top-lune when viewed as a top-cone in $\mathcal{A}_F$. This can be phrased as follows.

Lemma 3.39. *Let $HF \leq D$. Then*

$$\mathrm{s}(FH/F, FD/F) = \mathrm{s}(H, D)_F.$$

PROOF. We need to check that for any chamber E greater than F, we have $HE = D$ iff $FHE = FD$. For the forward implication, we premultiply by F. For the converse, we premultiply by H and use $HFH = HF$ and $HFD = D$. □

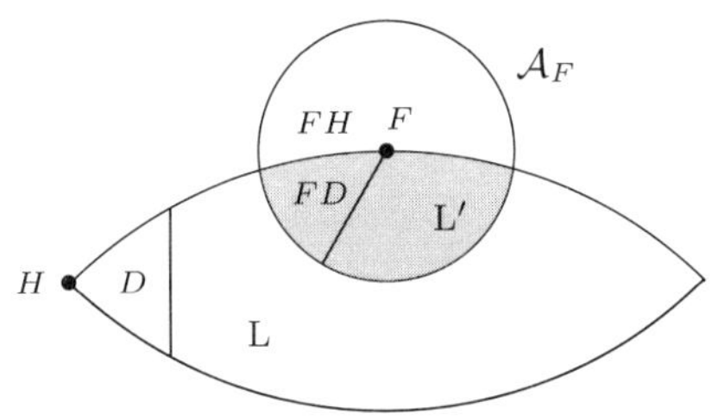

In the above picture, the top-lune L is the lunar region, while the top-star of F is the circular region. Their intersection is the shaded region marked L′. It is also a top-lune but in $\mathcal{A}_F$.

Suppose L is a top-lune in $\mathcal{A}_F$. View it as a top-cone contained in the top-star of F. Then its extension ${}_F\mathrm{L}$ is a top-lune. This can be phrased as follows.

Lemma 3.40. *For $F \leq H \leq D$,*

$${}_F(\mathrm{s}(H/F, D/F)) = \mathrm{s}(H, D).$$

In particular, $\mathrm{s}(H, D)$ *is the extension of D from H.*

PROOF. This is equivalent to saying $HC = D$ iff $H(FC) = D$, which is clear. □

This result can be viewed as a reformulation of (3.20) and the discussion leading to Proposition 3.21.

3.5. Top-star-lunes

We introduce top-star-lunes. Their relation to the other geometric objects that we have discussed is shown below.

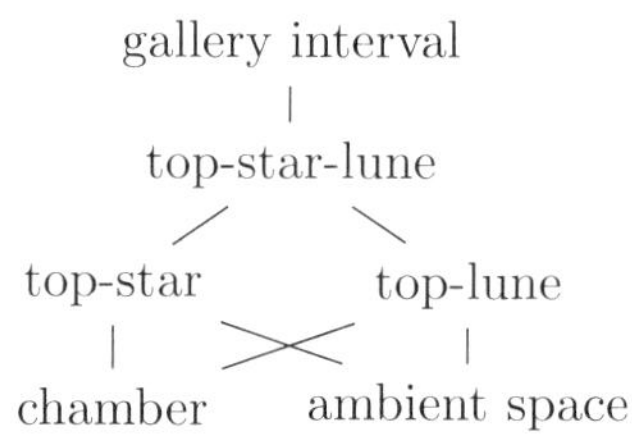

Thus, a top-star-lune is a gallery interval, while top-stars and top-lunes are top-star-lunes. Recall that a chamber or the ambient space is both a top-star and a top-lune, and in particular, these are also top-star-lunes.

Top-star-lunes of a given arrangement are defined inductively from top-star-lunes in arrangements of smaller rank as follows. The arrangement of rank zero has a unique top-cone. By definition, it is a top-star-lune. Inductively, a top-cone W in $\mathcal{A}$ is a *top-star-lune* in $\mathcal{A}$ if there exists a noncentral face F in $\mathcal{A}$ and a top-star-lune V in $\mathcal{A}_F$ such that

$$\text{either } \mathrm{W} = \mathrm{V} \text{ or } \mathrm{W} = {}_F\mathrm{V},$$

where recall that ${}_F\mathrm{V}$ is the extension of V from F. We refer to these two possibilities as the series alternative and parallel alternative, respectively. In the parallel alternative, note that the support of F must be smaller than the base of W.

Note very carefully that the F in the above definition is *not* required to be unique. Thus, it may happen that a given top-star-lune arises from a F by the series alternative, and at the same time from a F' by the parallel alternative.

Lemma 3.41. *Any top-star is a top-star-lune. Any top-lune is a top-star-lune.*

Proof. Any chamber C is a top-star-lune in the rank-zero arrangement $\mathcal{A}_C$, and hence a top-star-lune in $\mathcal{A}$ (by the series alternative). Hence, chambers are top-star-lunes.

For a top-nested face (H, D), the chamber D/H is a top-star-lune in $\mathcal{A}_H$ (by what we just saw). By Lemma 3.40, $\mathrm{s}(H, D)$ is the extension of D from H, and hence a top-star-lune in $\mathcal{A}$ (by the parallel alternative). Hence, top-lunes are top-star-lunes.

The top-star Γ_F of a face F is a top-lune in $\mathcal{A}_F$, hence a top-star-lune in $\mathcal{A}_F$ (by what we just saw), and hence a top-star-lune in $\mathcal{A}$ (by the series alternative). Hence, top-stars are top-star-lunes. $\square$

Lemma 3.42. *A top-star-lune is a gallery interval.*

Proof. We induct on the rank of the arrangement. The rank-zero case is clear. For the induction step, let W be a top-star-lune in $\mathcal{A}$. Let F be a noncentral face and V a top-star-lune in $\mathcal{A}_F$ which give rise to W. By induction hypothesis, V is a gallery interval in $\mathcal{A}_F$, and hence in $\mathcal{A}$. In the series alternative, W = V and hence W is a gallery interval. In the parallel alternative, we use Lemma 3.38 to reach the same conclusion. $\square$

In the arrangement of rank 1, all top-cones are top-star-lunes. In fact, the only top-cones are the two chambers and the ambient space.

In the rank-two arrangement of n lines, a top-cone is a top-star-lune iff the top-cone has either 1 edge, or 2 edges, or n edges or $2n$ edges. The smallest top-cone which is *not* a top-star-lune is shown above. It consists of three contiguous edges of the octagon.

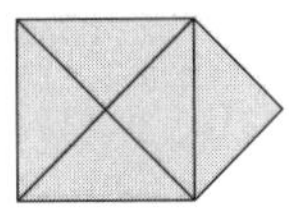

The above is a typical picture in rank three of a top-cone which is *not* a top-star-lune. It is neither localized enough to be in the top-star of a vertex, nor has it spread out enough to contain two opposite vertices (assuming the arrangement has enough hyperplanes).

3.6. Conjugate top-cones

Two top-cones are conjugate when the set of hyperplanes cutting them are complementary. Gallery intervals always have conjugates. Further, any top-lune has a conjugate which is a top-star, and vice versa. Similarly, any top-star-lune has a conjugate which is a top-star-lune. An arrangement satisfies the conjugate-meet property if any two conjugates share a common chamber. Stronger results can be proved under this assumption.

3.6.1. Conjugate top-cones. Two top-cones are *conjugate* if any hyperplane in the arrangement cuts exactly one of the two top-cones. In other words, V and W are conjugate if any hyperplane cuts either V or W but not both. In this situation, we say that W is a conjugate of V.

Recall from Lemma 2.32 that a hyperplane cuts V iff it cuts $\overline{\mathrm{V}}$. Hence, if V and W are conjugate, then so are V and $\overline{\mathrm{W}}$.

Any chamber has a unique conjugate, namely, the ambient space. Similarly, a top-cone is a conjugate of the ambient space iff that top-cone is a chamber. This follows from (2.14).

Any top-cone which is *not* a chamber has an even number of conjugates (possibly zero). This is because if V is a conjugate of the given top-cone, then so is $\overline{\mathrm{V}}$, and by Proposition 2.11, these are distinct since V is not the ambient space.

3.6.2. Gallery intervals. Any gallery interval has a conjugate which is also a gallery interval. More precisely, for any chambers C and D, the gallery intervals

$$\mathrm{V} = [C\!:\!D] \quad \text{and} \quad \mathrm{W} = [C\!:\!\overline{D}]$$

are conjugate. This is because any hyperplane separates exclusively either C and D, or C and $\overline{D}$. Observe that C and D can be recovered from V and W as

$$\mathrm{V} \wedge \mathrm{W} = C \quad \text{and} \quad \mathrm{V} \wedge \overline{\mathrm{W}} = D.$$

An example in rank 2 is shown below.

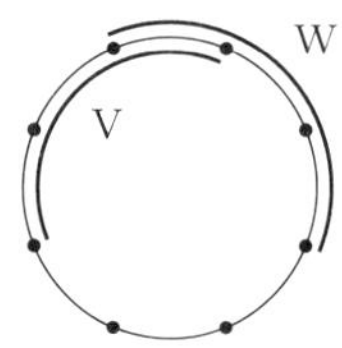

The edge common to both V and W is C, while the edges in V and W opposite to each other are D and $\overline{D}$.

3.6.3. Top-stars and top-lunes. Any top-lune has a conjugate which is a top-star, and conversely, any top-star has a conjugate which is a top-lune. More precisely, for any top-nested face (H, D), the top-star and top-lune, namely,

$$\Gamma_H = [D\!:\!H\overline{D}] \quad \text{and} \quad \mathrm{s}(H, D) = [D\!:\!\overline{H}D]$$

are conjugate. A hyperplane cuts the top-star Γ_H iff it contains H, and cuts the top-lune $\mathrm{s}(H, D)$ iff it does not contain H.

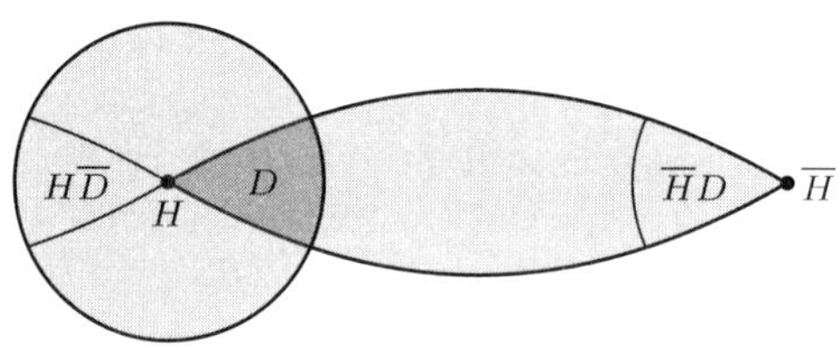

The situation in rank three when H is a vertex is illustrated above. The circular region is the top-star, while the lunar region is the top-lune. They have exactly one common chamber D. The chambers $H\overline{D}$ and $\overline{H}D$ are opposite.

When H is either O or D, we recover the fact that the chamber D and the ambient space are conjugate.

Lemma 3.43. *Every conjugate of a top-star is a top-lune. More precisely, the conjugates of the top-star of H are indexed by chambers D greater than H and given by the top-lunes* $\mathrm{s}(H, D)$.

Proof. Let V denote the top-star of H. We have seen that the $\mathrm{s}(H, D)$ are conjugates of V. It remains to show that these are all. For that, suppose W is a conjugate of V. Pick a chamber C in W and put $D := HC$. We claim that $\mathrm{W} = \mathrm{s}(H, D)$. First note that any chamber C' in W satisfies $HC' = D$. If not, then there exists a hyperplane passing through H which separates D and HC', and hence also D and C'. This hyperplane then cuts both V and W which is a contradiction. Thus, $\mathrm{W} \leq \mathrm{s}(H, D)$. Further, equality must hold. If not, then by Theorem 2.56, there exists a hyperplane which cuts $\mathrm{s}(H, D)$ but not W. This hyperplane then cuts neither V nor W which is a contradiction. $\square$

In contrast, not every conjugate of a top-lune is a top-star. An example is given below.

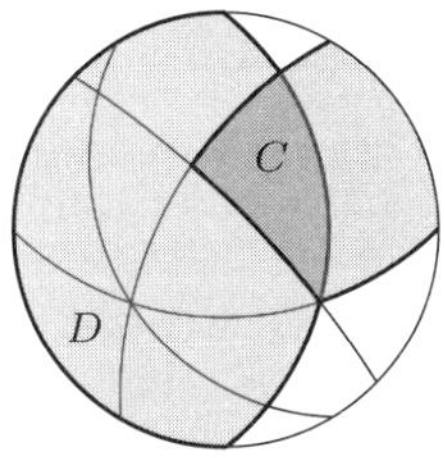

Consider the gallery interval $[C:D]$. It is a vertex-based top-lune with the north and south poles in its base. The conjugate $[C:\overline{D}]$ has three chambers, with the chamber $\overline{D}$ on the backside. It is not a top-star. In fact, it is not even a top-star-lune. This is relevant to the discussion below.

3.6.4. Top-star-lunes. Recall that top-star-lunes are defined inductively. The following observation about conjugates is of a similar nature. Recall that ${}_F\mathrm{V}$ denotes the extension of V from F.

Lemma 3.44. *Suppose* V *and* V′ *are contained in the top-star of* F. *Then* V *and* V′ *are conjugate in* $\mathcal{A}_F$ *iff* ${}_F\mathrm{V}$ *and* V′ *are conjugate in* $\mathcal{A}$.

PROOF. Let A denote the set of hyperplanes of $\mathcal{A}$. Write $\mathrm{A} = \mathrm{B} \sqcup \mathrm{C}$, where B consists of those hyperplanes which contain F, and C consists of those which do not contain F. Let g, h and g' denote the sets of hyperplanes which cut V, ${}_F\mathrm{V}$ and V′, respectively. Then $g, g' \subseteq \mathrm{B}$ and $h = g \sqcup \mathrm{C}$. It follows that V and V′ are conjugate in $\mathcal{A}_F$ iff $g \sqcup g' = \mathrm{B}$ iff $h \sqcup g' = \mathrm{A}$ iff ${}_F\mathrm{V}$ and V′ are conjugate in $\mathcal{A}$. □

Lemma 3.45. *Any top-star-lune has a conjugate which is also a top-star-lune.*

PROOF. We induct on the rank of the arrangement. The rank-zero case is clear. For the induction step, let W be a top-star-lune in $\mathcal{A}$. Let F be a noncentral face and V a top-star-lune in $\mathcal{A}_F$ which give rise to W. By induction hypothesis, V has a conjugate V′ in $\mathcal{A}_F$ which is a top-star-lune. We now employ Lemma 3.44. In the series alternative, W = V which is conjugate to ${}_F\mathrm{V}'$ and which is a top-star-lune by the parallel alternative. In the parallel alternative, W = ${}_F\mathrm{V}$ which is conjugate to V′ and which is a top-star-lune by the series alternative. □

In other words, we obtain a conjugate of V by repeating the inductive construction of V (which may not be unique) interchanging the series and parallel alternatives at each step.

Every conjugate of a top-star-lune is not necessarily a top-star-lune. The example given above of a top-lune with a conjugate which is not a top-star works.

3.6.5. Conjugate-meet property. An arrangement satisfies the *conjugate-meet property* if any of the following equivalent conditions hold.

- If V and W are conjugate, then $\mathrm{V} \wedge \mathrm{W} = C$, for some chamber C. That is, any two conjugate top-cones have a unique chamber in common.
- If V and W are conjugate, then $\mathrm{V} \wedge \mathrm{W}$ exists in the poset of top-cones. That is, any two conjugate top-cones have a chamber in common.

The first condition clearly implies the second. Conversely, if V and W had a common chamber apart from C, say E, then any hyperplane separating C and E would cut both V and W, which is not possible.

The conjugate-meet property fails precisely if there exist conjugate top-cones which do not have any chamber in common, or equivalently, whose meet (in the poset of cones) is *not* a top-cone.

Proposition 3.46. *Assume the conjugate-meet property. Let* V *and* W *be top-cones. Then* V *and* W *are conjugate iff there exist chambers* C *and* D *such that* $\mathrm{V} = [C : D]$ *and* $\mathrm{W} = [C : \overline{D}]$. *In particular, in this situation,* V *and* W *are both gallery intervals.*

PROOF. We have seen the backward implication. For the forward implication: Suppose V and W are conjugate. By the conjugate-meet property, $\mathrm{V} \wedge \mathrm{W} = C$, for some chamber C. Since V and $\overline{\mathrm{W}}$ are also conjugate, $\mathrm{V} \wedge \overline{\mathrm{W}} = D$, for some chamber D. Since cones are convex, by Proposition 2.40, V contains $[C:D]$ and W contains $[C:\overline{D}]$. In fact, equality must hold, that is, $\mathrm{V} = [C:D]$ and $\mathrm{W} = [C:\overline{D}]$. Suppose not. Say V is strictly greater than $[C:D]$. Then by Theorem 2.56, there exists a hyperplane which cuts V but not $[C : D]$. But then this hyperplane cuts $[C:\overline{D}]$ and hence W, contradicting the fact that V and W are conjugate. $\square$

Corollary 3.47. *Assume the conjugate-meet property. A top-cone* V *is a gallery interval iff there exists a top-cone which is conjugate to* V.

Arrangements of rank up to 2 have the conjugate-meet property. But it can fail in higher ranks. For instance, the conjugate-meet property fails for the smallest nonsimplicial arrangement in rank three (Section 1.2.5). In fact, in this arrangement, for any vertex-based top-lune with three chambers, there exists a conjugate vertex-based top-lune with three chambers such that the meet of the two top-lunes is the central face.

Lemma 3.48. *If* $\mathcal{A}$ *satisfies the conjugate-meet property, then so does* $\mathcal{A}_F$ *for any face* F.

PROOF. Suppose V and W are conjugate in $\mathcal{A}_F$. Then by Lemma 3.44, ${}_F\mathrm{V}$ and W are conjugate in $\mathcal{A}$. So by hypothesis, they have a common chamber, say C. Then $C \in \mathrm{W}$ implies that C is greater than F which then implies that $C \in \mathrm{V}$. Thus, V and W have a common chamber as required. $\square$

3.7. Cartesian product of cones, gallery intervals and lunes

Cartesian product of arrangements was discussed in Section 1.8. We will now see that cones, gallery intervals and lunes behave nicely under this operation. In this discussion, $\mathcal{A}$ and $\mathcal{A}'$ are arrangements, and $\mathcal{A} \times \mathcal{A}'$ is their cartesian product.

3.7.1. Cones. Recall that a hyperplane of $\mathcal{A} \times \mathcal{A}'$ is either $(\mathrm{H}, \top')$ with H an hyperplane of $\mathcal{A}$ or $(\top, \mathrm{H}')$ with H' an hyperplane of $\mathcal{A}'$. Thus, the set of hyperplanes of $\mathcal{A} \times \mathcal{A}'$ is the disjoint union of the set of hyperplanes of $\mathcal{A}$ and $\mathcal{A}'$. The same comment applies to half-spaces. Hence, by taking power sets and employing notations of Section 2.6, we get

$$\vec{\mathrm{G}}[\mathcal{A} \times \mathcal{A}'] = \vec{\mathrm{G}}[\mathcal{A}] \times \vec{\mathrm{G}}[\mathcal{A}'] \quad \text{and} \quad \mathrm{G}[\mathcal{A} \times \mathcal{A}'] = \mathrm{G}[\mathcal{A}] \times \mathrm{G}[\mathcal{A}'].$$

Since cones are obtained by intersecting half-spaces, and flats by intersecting hyperplanes, we deduce that

$$\Omega[\mathcal{A} \times \mathcal{A}'] = \Omega[\mathcal{A}] \times \Omega[\mathcal{A}'] \quad \text{and} \quad \Pi[\mathcal{A} \times \mathcal{A}'] = \Pi[\mathcal{A}] \times \Pi[\mathcal{A}']. \tag{3.31}$$

The second identification has been noted earlier. The first one says that a cone of $\mathcal{A} \times \mathcal{A}'$ is the same as a pair $(\mathrm{V}, \mathrm{V}')$, where V is a cone of $\mathcal{A}$ and V' is a cone of $\mathcal{A}'$. Also note that if V and V' are top-cones, then so is $(\mathrm{V}, \mathrm{V}')$, and the set of chambers in $(\mathrm{V}, \mathrm{V}')$ is the cartesian product of the set of chambers in V with the set of chambers in V'.

The maps in diagrams (2.18) are compatible with these identifications.

3.7.2. Minimal galleries. Let C and D be chambers of $\mathcal{A}$, and C' and D' be chambers of $\mathcal{A}'$. We make some elementary observations.

A hyperplane H in $\mathcal{A} \times \mathcal{A}'$ separates (C, C') and (D, D') iff either H corresponds to a hyperplane in $\mathcal{A}$ and separates C and D, or H corresponds to a hyperplane in $\mathcal{A}'$ and separates C' and D'.

The two alternatives are illustrated above. The chambers (C, C') and (D, D') are drawn shaded, while the separating hyperplanes are drawn thick.

Thus, the set of hyperplanes separating (C, C') and (D, D') is the union of the set of hyperplanes separating C and D, and the set of hyperplanes separating C' and D'. In particular,

$$\mathrm{dist}((C, C'), (D, D')) = \mathrm{dist}(C, D) + \mathrm{dist}(C', D').$$

Therefore: Chambers (C, C') and (D, D') are adjacent in $\mathcal{A} \times \mathcal{A}'$ iff either $C' = D'$ and C and D are adjacent in $\mathcal{A}$, or $C = D$ and C' and D' are adjacent in $\mathcal{A}'$.

By employing (1.28), we can now deduce all of the following.

For any chambers (C, C'), (D, D') and (E, E') of $\mathcal{A} \times \mathcal{A}'$,

$$(C, C') \mathrel{--} (E, E') \mathrel{--} (D, D') \iff C \mathrel{--} E \mathrel{--} D \text{ and } C' \mathrel{--} E' \mathrel{--} D'. \tag{3.32}$$

A minimal gallery from (C, C') to (D, D') yields a pair of galleries, one from C to D, and another from C' to D'. The gallery from C to D is obtained by projecting on the first coordinate and deleting repeated entries, while the gallery from C' to D' is obtained similarly by projecting on the second coordinate.

Conversely, given a minimal gallery from C to D, and another from C' to D', one can construct minimal galleries from (C, C') to (D, D') as follows. At each stage, we either change the first coordinate using the first gallery keeping the second coordinate fixed or vice versa.

Continuing with the previous picture, there are three minimal galleries from (C, C') to (D, D') and they are shown below.

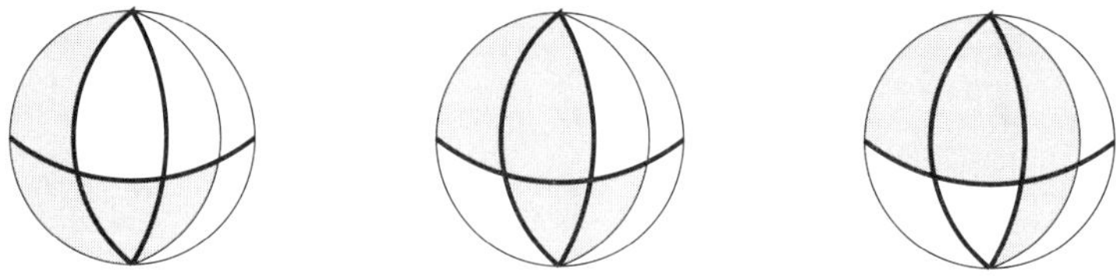

At a given stage in the minimal gallery, we either cross the horizontal hyperplane, or one of the two vertical hyperplanes drawn thick. The horizontal hyperplane can be interleaved with the vertical ones in any manner.

3.7.3. Gallery intervals. For chambers C and D of $\mathcal{A}$, and C' and D' of $\mathcal{A}'$,

$$[(C, C'):(D, D')] = ([C:D], [C':D']).$$

This is an identity of top-cones in $\mathcal{A} \times \mathcal{A}'$. It is a restatement of (3.32). It implies that a top-cone $(\mathrm{V}, \mathrm{V}')$ of $\mathcal{A} \times \mathcal{A}'$ is a gallery interval iff V is a gallery interval in $\mathcal{A}$ and V' is a gallery interval in $\mathcal{A}'$.

3.7.4. Top-stars and top-lunes. For a face F of $\mathcal{A}$, and F' of $\mathcal{A}'$,

$$\Gamma_{(F,F')} = (\Gamma_F, \Gamma_{F'}).$$

It implies that a top-cone $(\mathrm{V}, \mathrm{V}')$ of $\mathcal{A} \times \mathcal{A}'$ is a top-star iff V is a top-star in $\mathcal{A}$ and V' is a top-star in $\mathcal{A}'$.

For a top-nested face (H, D) of $\mathcal{A}$, and (H', D') of $\mathcal{A}'$,

$$\mathrm{s}((H, H'), (D, D')) = (\mathrm{s}(H, D), \mathrm{s}(H', D')).$$

This follows from (1.21). It implies that a top-cone $(\mathrm{V}, \mathrm{V}')$ of $\mathcal{A} \times \mathcal{A}'$ is a top-lune iff V is a top-lune in $\mathcal{A}$ and V' is a top-lune in $\mathcal{A}'$.

More generally, for a nested face (H, G) of $\mathcal{A}$, and (H', G') of $\mathcal{A}'$,

$$\mathrm{s}((H, H'), (G, G')) = (\mathrm{s}(H, G), \mathrm{s}(H', G')).$$

Thus, a cone $(\mathrm{V}, \mathrm{V}')$ of $\mathcal{A} \times \mathcal{A}'$ is a lune iff V is a lune in $\mathcal{A}$ and V' is a lune in $\mathcal{A}'$. In other words, the identification of cones in (3.31) restricts to

$$\Lambda[\mathcal{A} \times \mathcal{A}'] = \Lambda[\mathcal{A}] \times \Lambda[\mathcal{A}'].$$

3.7.5. Restriction and extension of cones. The operations of restriction and extension of cones are also well-behaved under cartesian product. Let (F, F') be a face and $(\mathrm{V}, \mathrm{V}')$ a top-cone of $\mathcal{A} \times \mathcal{A}'$. Then (F, F') belongs to the closure of $(\mathrm{V}, \mathrm{V}')$ iff F belongs to the closure of V and F' belongs to the closure of V'. In this situation,

$$(\mathrm{V}, \mathrm{V}')_{(F,F')} = (\mathrm{V}_F, \mathrm{V}'_{F'}).$$

Similarly, $(\mathrm{V}, \mathrm{V}')$ is contained in the top-star of (F, F') iff V is contained in the top-star of F and V' is contained in the top-star of F'. In this situation,

$$_{(F,F')}(\mathrm{V}, \mathrm{V}') = ({}_F\mathrm{V}, {}_{F'}\mathrm{V}').$$

3.7.6. Top-star-lunes. A top-cone $(\mathrm{V}, \mathrm{V}')$ of $\mathcal{A} \times \mathcal{A}'$ is a top-star-lune iff V is a top-star-lune in $\mathcal{A}$ and V' is a top-star-lune in $\mathcal{A}'$. This can be established by an induction on the sum of the ranks of $\mathcal{A}$ and $\mathcal{A}'$. The base case is when either $\mathcal{A}$ or $\mathcal{A}'$ has rank zero. For the induction step, we explain one case for the forward implication. Let $(\mathrm{W}, \mathrm{W}')$ be a top-star-lune of $\mathcal{A} \times \mathcal{A}'$. Suppose $(\mathrm{W}, \mathrm{W}')$ is constructed from (F, F') and a top-star-lune $(\mathrm{V}, \mathrm{V}')$ of $\mathcal{A}_{(F,F')}$ by the parallel alternative. (Recall that $\mathcal{A}_{(F,F')} = \mathcal{A}_F \times \mathcal{A}_{F'}$.) Hence, by induction hypothesis, V and V' are top-star-lunes in $\mathcal{A}_F$ and $\mathcal{A}_{F'}$, respectively. Then

$$(\mathrm{W}, \mathrm{W}') = {}_{(F,F')}(\mathrm{V}, \mathrm{V}') = ({}_F\mathrm{V}, {}_{F'}\mathrm{V}').$$

Thus, W is constructed from F and V, and W' from F' and V' by the parallel alternative. Hence, W and W' are both top-star-lunes. The other cases are similar.

3.7.7. Conjugate top-cones. A hyperplane H in $\mathcal{A} \times \mathcal{A}'$ cuts $(\mathrm{V}, \mathrm{V}')$ iff either H corresponds to a hyperplane in $\mathcal{A}$ and cuts V, or H corresponds to a hyperplane in $\mathcal{A}'$ and cuts V'. It follows that $(\mathrm{V}, \mathrm{V}')$ and $(\mathrm{W}, \mathrm{W}')$ are conjugate top-cones in $\mathcal{A} \times \mathcal{A}'$ iff V and W are conjugate top-cones in $\mathcal{A}$, and V' and W' are conjugate top-cones in $\mathcal{A}'$.

Similarly, one can deduce that $\mathcal{A} \times \mathcal{A}'$ has the conjugate-meet property iff both $\mathcal{A}$ and $\mathcal{A}'$ have the conjugate-meet property. The forward implication also follows from Lemma 3.48 (since $\mathcal{A}$ and $\mathcal{A}'$ can be viewed as arrangements over a flat of $\mathcal{A} \times \mathcal{A}'$).

Notes

Top-nested faces and top-lunes. Top-lunes for affine arrangements are considered by Varchenko [**420**, Section 7]. The dictionary between his terminology and ours is: cone $\leftrightarrow$ top-lune, edge $\leftrightarrow$ flat, vertex (of the cone) $\leftrightarrow$ base (of the top-lune), sharpness (of the cone) $\leftrightarrow$ slack (of the top-lune), marked cone $\leftrightarrow$ top-lune written in the form $\mathrm{s}(H, D)$. Top-lunes are used later by Bidigare, Hanlon and Rockmore [**61**]; our Proposition 3.5 is very closely related to their Lemma 4.5. The term "lune" was coined by Billera, Brown, and Diaconis [**63**, Section 6.1]. They work only in rank 3, but the general definition is clear from this case. For a labeled simplicial arrangement, the number of chambers in top-lunes is related to the flag f and flag h vector of the arrangement [**296**, Lemma 5 and Figure 1.9]. More information is given in the notes to Chapter 7.

Top-nested faces appear prominently in the description of the Salvetti complex by Arvola [**23**, Sections 5 and 7]; see also [**328**, page 175], [**179**, Chapter 5] and more recently [**148**, Sections 2.1 and 3.6].

The relevance of top-nested faces and top-lunes to combinatorial Hopf algebras was brought forth in [**8, 9**]. Top-nested faces are called directed faces, while top-lunes are called directed flats in [**9**]. Some basic theory of top-lunes (for any LRB) is developed in [**8**, Sections 2.3, 2.4 and 2.5]. Equations (3.4) and (3.7) are given in [**8**, (2.6) and (2.4)], and Propositions 3.4, 3.5 and 3.9 are given in [**8**, Fact 5.2.1, Lemmas 2.3.2 and 2.3.3]. The set in (3.4) is called a lunar region in this reference.

Reiner, Saliola and Welker consider an interesting family of bilinear forms on chambers [**358**, Definition II.1.1]. They can be phrased using top-lunes as follows. Define $\langle C, D\rangle$ to be the number of top-lunes which contain both C and D. More generally, for any subset A of the set of flats, define $\langle C, D\rangle_{\mathrm{A}}$ to be the number of top-lunes which contain both C and D, and whose base belongs to A. When A is the set of all hyperplanes, $\langle C, D\rangle_{\mathrm{A}}$ is the number of half-spaces which contain both C and D, which is the same as the number of hyperplanes in the arrangement minus $\mathrm{dist}(C, D)$.

CHAPTER 4

Category of lunes

We study two interesting partial orders on lunes. The first one is the restriction of the partial order on cones, and thus is defined by inclusion of lune closures. The second partial order is defined by inclusion of lune interiors. Both partial orders are graded. In the first case, the rank of a lune is the sum of the ranks of its base and its case, while in the second, it is just the rank of its base. The poset of top-lunes is an upper set in either partial order.

Lunes can be composed when the case of the first lune equals the base of the second lune. This yields the category of lunes whose objects are flats and morphisms are lunes. Further, this category is internal to posets under the second partial order on lunes. Also it has a nice presentation with generators being lunes of slack 1 (half-flats) subject to quadratic relations involving lunes of slack 2. In addition, the Birkhoff monoid acts on the category of lunes.

Recall that a lune is the same as a chamber in the arrangement over its base and under its case. Thus, composition of lunes is equivalent to an operation on chambers in arrangements over and under flats. We call this the substitution product of chambers. Using the same idea, one can also multiply chambers and faces, and top-lunes and chambers.

We consider the categories associated to the poset of faces and to the poset of flats. Since these posets are strongly connected, both categories have nice presentations. We also relate them to the category of lunes by functors which are internal to posets.

These ideas are further developed in Chapter 15.

4.1. Poset of top-lunes

Let us begin with the set of top-lunes $\widehat{\Lambda}[\mathcal{A}]$. It is a poset under inclusion, that is, $\mathrm{L} \leq \mathrm{M}$ iff M contains L (as subsets of the ambient space). The poset of top-lunes has a maximum element, namely, the ambient space. Each chamber is a minimal element. In combinatorial terms,

$$\mathrm{L} \leq \mathrm{M} \iff \mathrm{Cl}(\mathrm{L}) \subseteq \mathrm{Cl}(\mathrm{M}). \tag{4.1}$$

By definition, $\widehat{\Lambda}[\mathcal{A}]$ is a subposet of the poset of top-cones $\widehat{\Omega}[\mathcal{A}]$. Recall from Section 2.7 that the poset of top-cones is graded. We now proceed to show that the same is true for the poset of top-lunes. The strategy remains the same, namely, to find an order-preserving map to a graded poset which preserves cover relations. In the case at hand, this will be accomplished by the base map

$$\mathrm{b} : \widehat{\Lambda}[\mathcal{A}] \to \Pi[\mathcal{A}] \tag{4.2}$$

obtained by restricting (2.9).

The connection between top-nested faces and top-lunes was treated in Section 3.2.1. Lune decompositions were discussed in Section 3.3. These ideas will be used in the discussion below.

Proposition 4.1. *For top-lunes* L *and* M*, the following are equivalent.*

(1) $\mathrm{L} \leq \mathrm{M}$.
(2) $\mathrm{b}(\mathrm{L}) \leq \mathrm{b}(\mathrm{M})$*, and* L *appears in the lune decomposition of* M *over the flat* $\mathrm{b}(\mathrm{L})$.
(3) *There exist faces* $H \leq G$ *and* $H \leq C$ *such that* $\mathrm{L} = \mathrm{s}(H, C)$ *and* $\mathrm{M} = \mathrm{s}(G, GC)$.

Proof. (1) implies (2). Suppose $\mathrm{L} \leq \mathrm{M}$. Since the base map is order-preserving, we have $\mathrm{b}(\mathrm{L}) \leq \mathrm{b}(\mathrm{M})$. Further from (3.18), distinct top-lunes with the same base are disjoint, so L must appear in the lune decomposition of M over the flat $\mathrm{b}(\mathrm{L})$. Alternatively, we may apply (3.24) to $\mathrm{V} = \mathrm{M}$.

(2) implies (3). This follows from (3.20).

(3) implies (1). Suppose the chamber E belongs to L, that is, $HE = C$. Then $GE = GHE = GC$, so E belongs to M. Thus, $\mathrm{L} \leq \mathrm{M}$. □

Lemma 4.2. *For top-nested faces* (H, D) *and* (H', D'),

$$\begin{aligned} \mathrm{s}(H, D) \leq \mathrm{s}(H', D') \iff & HC = D \text{ implies } H'C = D' \\ \iff & H'H = H' \text{ and } H'D = D'. \end{aligned}$$

Proof. The first equivalence follows from the definition (3.3). We now prove the second equivalence. Forward implication. Taking $C = D$ yields $H'D = D'$. In particular, $H'H$ is a face of D'. Similarly, taking $C = \overline{H}D$, we get that $H'\overline{H}$ is a face of D'. Hence, $H'H = H'\overline{H} = H'$ by Proposition 1.19. Backward implication. Suppose $HC = D$. Then $H'C = H'HC = H'D = D'$. □

In order to state the next result fully, we make use of the action of the Birkhoff monoid on lunes discussed in Section 4.7.2.

Lemma 4.3. *For* $\mathrm{X} \leq \mathrm{Y}$*, and* L *a top-lune with base* X*, there exists a unique top-lune* M *with base* Y *such that* $\mathrm{L} \leq \mathrm{M}$*, and it is given by* $\mathrm{M} = \mathrm{Y} \cdot \mathrm{L}$*. In particular, for a chamber* D *and a flat* Y*, there exists a unique top-lune with base* Y *which contains* D.

Proof. We can use Lemma 4.2. Write $\mathrm{L} = \mathrm{s}(H, D)$ where $\mathrm{s}(H) = \mathrm{X}$. Pick any face H' with support Y. The desired lune is $\mathrm{M} = \mathrm{s}(H', H'D)$. By definition, this equals $\mathrm{Y} \cdot \mathrm{L}$.

Alternatively: Apply (3.18) to Y, and then further take the lune decomposition of each summand over X. This is the same as applying (3.18) directly to X. The unique summand whose lune decomposition contains L is the required M. □

Lemma 4.3 says that the base map from top-lunes to flats is a covering map in the sense of Section C.4.5 (provided the partial orders on both posets are reversed).

Lemma 4.4. *For* $\mathrm{X} \leq \mathrm{Y}$*, and* M *a top-lune with base* Y*, the top-lunes* L *with base* X *such that* $\mathrm{L} \leq \mathrm{M}$ *are precisely those that appear in the lune decomposition of* M *over the flat* X.

Proof. This follows from Proposition 4.1 or Proposition 3.23. □

Lemma 4.5. *For top-lunes* L, M *and* N,

$$\mathrm{L} \le \mathrm{M},\ \mathrm{L} \le \mathrm{N} \text{ and } \mathrm{b(M)} \le \mathrm{b(N)} \iff \mathrm{L} \le \mathrm{M} \le \mathrm{N}.$$

In particular: For top-lunes M *and* N,

$$\mathrm{M} \le \mathrm{N} \iff \mathrm{b(M)} \le \mathrm{b(N)}, \text{ and N contains a chamber of M.}$$

PROOF. Backward implication is clear. The forward implication is a formal consequence of Lemma 4.3. Let N′ be the unique top-lune with base b(N) such that M ≤ N′. Now N and N′ are both top-lunes with base b(N) which are greater than L, so N = N′, and hence M ≤ N. □

Lemma 4.6. *For top-lunes* L ≤ N, *and a flat* Y *such that* b(L) ≤ Y ≤ b(N), *there exists a unique top-lune* M *with base* Y *such that* L ≤ M ≤ N.

PROOF. By Lemma 4.3, there is a unique M with base Y such that L ≤ M. Now, by Lemma 4.5, L ≤ M ≤ N. □

An illustration is provided below.

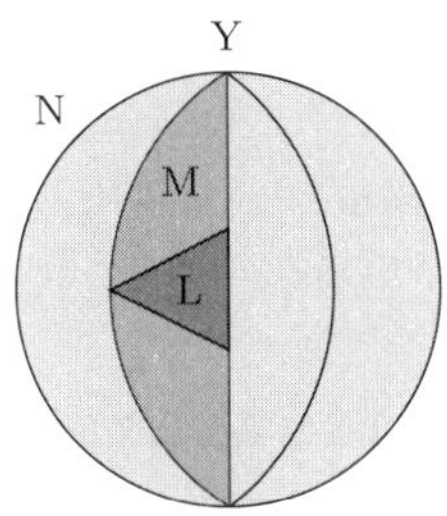

In the picture, N is a hemisphere, Y is the rank-one flat consisting of the north and south pole, and L is a chamber contained in N. (The base of L is the minimum flat.) The lune decomposition of N over Y has four top-lunes, and exactly one of them contains L. This is the desired top-lune M.

Exercise 4.7. First derive Lemma 4.6 as a formal consequence of Lemma 4.3, and then deduce Lemma 4.5 from it.

Lemma 4.8. *The base map* (4.2) *preserves cover relations. That is, for top-lunes* L *and* M,

$$\mathrm{L} \lessdot \mathrm{M} \implies \mathrm{b(L)} \lessdot \mathrm{b(M)}.$$

PROOF. By Corollary 3.17 or 3.20, distinct top-lunes with the same base are incomparable, so b(L) < b(M). Now use Lemma 4.6. □

As a consequence, composing (4.2) with the rank function of $\Pi[\mathcal{A}]$ yields a rank function for $\widehat{\Lambda}[\mathcal{A}]$. Thus:

Theorem 4.9. *The poset of top-lunes* $\widehat{\Lambda}[\mathcal{A}]$ *is graded with the rank of a top-lune being the rank of its base. In particular, the rank of* $\widehat{\Lambda}[\mathcal{A}]$ *equals the rank of* $\mathcal{A}$. *The map* (4.2) *is rank-preserving.*

Remark 4.10. One must be careful while talking of the rank of a top-lune L. It could refer to the rank of L either in the poset of top-lunes $\widehat{\Lambda}[\mathcal{A}]$ or in the poset of top-cones $\widehat{\Omega}[\mathcal{A}]$. The former is the rank of the flat b(L), while the latter is the number of hyperplanes which cut L. The latter is always greater than the former.

4.2. Two partial orders on lunes

Recall the set of lunes $\Lambda[\mathcal{A}]$. We consider two partial orders on it; both extend the partial order on the set of top-lunes. The first partial order is the restriction of the partial order on cones, that is, $\mathrm{L} \leq \mathrm{M}$ iff the closure of M contains the closure of L. The second partial order is as follows. $\mathrm{L} \preceq \mathrm{M}$ iff the interior of M contains the interior of L. The goal of this section is to show that both partial orders are graded.

4.2.1. First partial order on lunes. For lunes L and M, define

$$\mathrm{L} \leq \mathrm{M} \iff \mathrm{Cl}(\mathrm{L}) \subseteq \mathrm{Cl}(\mathrm{M}). \tag{4.3}$$

In other words, $\mathrm{L} \leq \mathrm{M}$ iff the closure of L is contained in the closure of M. This defines a partial order on lunes which extends the partial order on top-lunes (4.1). It is also the restriction of the partial order on cones in view of (2.2).

Consider the map

$$\Lambda[\mathcal{A}] \to \Pi[\mathcal{A}] \times \Pi[\mathcal{A}], \qquad \mathrm{L} \mapsto (\mathrm{b}(\mathrm{L}), \mathrm{c}(\mathrm{L})). \tag{4.4}$$

Since the base and case maps are order-preserving, this map is also order-preserving. We call this the *base-case map* and denote it by bc. In terms of nested faces, it can be expressed as

$$\mathrm{s}(H, G) \mapsto (\mathrm{s}(H), \mathrm{s}(G)). \tag{4.5}$$

Lemma 4.11. *The map* (4.4) *preserves cover relations. That is, for lunes* L *and* M,

$$\mathrm{L} \lessdot \mathrm{M} \implies \textit{Either } \mathrm{b}(\mathrm{L}) \lessdot \mathrm{b}(\mathrm{M}) \textit{ and } \mathrm{c}(\mathrm{L}) = \mathrm{c}(\mathrm{M}), \quad \textit{or } \mathrm{b}(\mathrm{L}) = \mathrm{b}(\mathrm{M}) \textit{ and } \mathrm{c}(\mathrm{L}) \lessdot \mathrm{c}(\mathrm{M}).$$

PROOF. Consider two cases.

$\underline{\mathrm{c}(\mathrm{L}) = \mathrm{c}(\mathrm{M})}$. Then, by passing to the arrangement under this flat and using Lemma 4.8, we see that $\mathrm{b}(\mathrm{L}) \lessdot \mathrm{b}(\mathrm{M})$.

$\underline{\mathrm{c}(\mathrm{L}) < \mathrm{c}(\mathrm{M})}$. We employ Lemma 3.16. Write $\mathrm{L} = \mathrm{s}(H, G)$, and let K be any top-dimensional face of M. Then, $\mathrm{s}(H, GK)$ is a lune strictly greater than L since $G < GK$ and less than M by Proposition 3.12. Hence, $\mathrm{M} = \mathrm{s}(H, GK)$. In particular, $\mathrm{b}(\mathrm{L}) = \mathrm{b}(\mathrm{M})$. Further, $G \lessdot GK$, so $\mathrm{c}(\mathrm{L}) \lessdot \mathrm{c}(\mathrm{M})$. □

As a consequence, composing (4.4) with the rank function of $\Pi[\mathcal{A}] \times \Pi[\mathcal{A}]$ yields a rank function for $\Lambda[\mathcal{A}]$. Thus:

Theorem 4.12. *The set of lunes $\Lambda[\mathcal{A}]$ under the partial order $\leq$ is graded with the rank of a lune being the sum of the ranks of its base and its case. In particular, the rank of $\Lambda[\mathcal{A}]$ is $2\,\mathrm{rk}(\mathcal{A})$. The map* (4.4) *is rank-preserving.*

Exercise 4.13. Show that: For any rank-two arrangement of n lines with $n \geq 3$, the set of lunes $\Lambda[\mathcal{A}]$ under $\leq$ is *not* a lattice. Give concrete examples of two lunes whose meet (join) does not exist. In particular, the meet (join) of two lunes in the lattice of cones may not be a lune.

The picture below shows two top-lunes in rank three whose meet (consisting of two adjacent triangles) is not a lune.

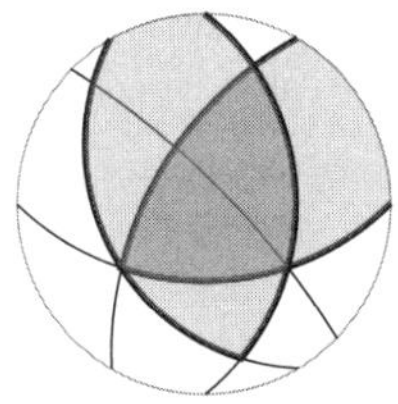

Locate other examples of this kind in the picture.

Exercise 4.14. Give an example of a lune and a flat whose meet in the lattice of cones is *not* a lune. In contrast, the join of a lune and a flat is a lune, see Proposition 4.47.

Lemma 4.15. *For nested faces* (H, G) *and* (H', G'),

$$\begin{aligned} \mathrm{s}(H,G) \leq \mathrm{s}(H',G') &\iff HK \leq G \text{ implies } H'K \leq G' \\ &\iff H'H = H' \text{ and } H'G \leq G'. \end{aligned}$$

This generalizes the result of Lemma 4.2. The proof is similar.

Exercise 4.16. Deduce that: For lunes L and L′,

$$\begin{aligned} \mathrm{L} \leq \mathrm{L}' \iff & \text{ There exist nested faces } (H,G) \text{ and } (H',G') \\ & \text{with supports L and L}', \text{ respectively, such that } H \leq H' \text{ and } H'G \leq G'. \end{aligned}$$

(The equivalence between items (1) and (3) in Proposition 4.1 is a special case.)

4.2.2. Second partial order on lunes. For lunes L and M, define

$$\mathrm{L} \preceq \mathrm{M} \quad \text{if} \quad \mathrm{L} \leq \mathrm{M} \text{ and } \mathrm{b}(\mathrm{M}) \vee \mathrm{c}(\mathrm{L}) = \mathrm{c}(\mathrm{M}). \tag{4.6}$$

The relation $\preceq$ defines another partial order on lunes. This follows from Lemma 4.18 below. It is also a special case of Lemma E.3.

Lemma 4.17. *For nested faces* (H, G) *and* (H', G'),

$$\mathrm{s}(H,G) \preceq \mathrm{s}(H',G') \iff H'H = H' \text{ and } H'G = G'.$$

Compare with Lemma 4.15.

Lemma 4.18. *For lunes* L *and* M,

$$\mathrm{L} \preceq \mathrm{M} \iff \mathrm{L}^o \subseteq \mathrm{M}^o. \tag{4.7}$$

In other words, L $\preceq$ M *iff the interior of* L *is contained in the interior of* M.

PROOF. Write $\mathrm{L} = \mathrm{s}(H, G)$ and $\mathrm{M} = \mathrm{s}(H', G')$. We employ (3.11) and Lemma 4.17.

Forward implication: Suppose $F \in \mathrm{L}^o$. Thus $HF = G$. Then $H'F = H'HF = H'G = G'$. So $F \in \mathrm{M}^o$. Backward implication: Since $G \in \mathrm{L}^o$, we have $G \in \mathrm{M}^o$ and hence $H'G = G'$. Similarly, since $\overline{H}G \in \mathrm{L}^o$, we have $H'\overline{H}G = G'$. So $H'H$ and $H'\overline{H}$ are both smaller than G', hence joinable. So by Proposition 1.19, $H'H = H'\overline{H} = H'$. ∎

Observe from (4.6) that:

$$\text{If } \mathrm{c(L)} = \mathrm{c(M)}, \text{ then } \mathrm{L} \preceq \mathrm{M} \iff \mathrm{L} \leq \mathrm{M}.$$

In particular, the partial orders $\leq$ and $\preceq$ are identical when restricted to the set of top-lunes.

$$\text{If } \mathrm{b(L)} = \mathrm{b(M)}, \text{ then } \mathrm{L} \preceq \mathrm{M} \iff \mathrm{L} = \mathrm{M}.$$

This follows from Corollary 3.17. In particular, for faces F and G,

$$F \preceq G \iff F = G.$$

Also note that for a lune L and flat Y,

$$\mathrm{L} \preceq \mathrm{Y} \iff \mathrm{L} \leq \mathrm{Y} \iff \mathrm{c(L)} \leq \mathrm{Y}. \tag{4.8}$$

In particular, for flats X and Y,

$$\mathrm{X} \preceq \mathrm{Y} \iff \mathrm{X} \leq \mathrm{Y}.$$

Further, the set of flats is an upper set under $\preceq$. That is, $\mathrm{X} \preceq \mathrm{L}$ implies L is a flat.

Exercise 4.19. Recall that faces are lunes. Show that: For a face F and lune L,

$$F \leq \mathrm{L} \iff F \in \mathrm{Cl(L)} \quad \text{and} \quad F \preceq \mathrm{L} \iff F \in \mathrm{L}^o.$$

(Use (3.10) and (3.11).)

Example 4.20. The set of lunes for the rank-one arrangement has four elements, namely, the minimum and maximum flats and the two chambers. The Hasse diagrams of the two partial orders are shown below.

The partial order on the left is $\leq$, while the one on the right is $\preceq$.

Exercise 4.21. Show that: For cones V and W,

$$\mathrm{Cl(V)} \subseteq \mathrm{Cl(W)} \text{ and } \mathrm{b(W)} \vee \mathrm{c(V)} = \mathrm{c(W)} \implies \mathrm{V}^o \subseteq \mathrm{W}^o.$$

(Use Exercise 2.25.) The converse is false in general.

The results of Section 4.1 generalize to arbitrary lunes for the partial order $\preceq$. (The partial order $\leq$ does not work so well.) This is briefly explained below. We begin with the generalization of Lemma 4.3. To state this fully, we make use of the action of the Birkhoff monoid on lunes discussed in Section 4.7.2.

Lemma 4.22. *For* $\mathrm{X} \leq \mathrm{Y}$, *and* L *a lune with base* X, *there exists a unique lune* M *with base* Y *such that* $\mathrm{L} \preceq \mathrm{M}$, *and it is given by* $\mathrm{M} = \mathrm{Y} \cdot \mathrm{L}$.

PROOF. We can use Lemma 4.17. Write $\mathrm{L} = \mathrm{s}(H, G)$ where $\mathrm{s}(H) = \mathrm{X}$. Pick any face H' with support Y. The desired lune is $\mathrm{M} = \mathrm{s}(H', H'G)$. By definition, it equals $\mathrm{Y} \cdot \mathrm{L}$. $\square$

Exercise 4.23. Show that: For lunes $\mathrm{L} \leq \mathrm{N}$, there exists a unique lune M such that $\mathrm{L} \preceq \mathrm{M} \leq \mathrm{N}$ and $\mathrm{b(M)} = \mathrm{b(N)}$. It is given by $\mathrm{M} = \mathrm{b(N)} \cdot \mathrm{L}$.

Lemma 4.22 says that the base map from lunes to flats is a covering map in the sense of Section C.4.5 (provided the partial orders on both posets are reversed). As formal consequences:

Lemma 4.24. *For lunes* L, M *and* N,

$$\mathrm{L} \preceq \mathrm{M},\ \mathrm{L} \preceq \mathrm{N} \text{ and } \mathrm{b(M)} \leq \mathrm{b(N)} \iff \mathrm{L} \preceq \mathrm{M} \preceq \mathrm{N}.$$

In particular: For lunes M *and* N,

$$\mathrm{M} \preceq \mathrm{N} \iff \mathrm{b(M)} \leq \mathrm{b(N)}, \text{ and } \mathrm{M} \text{ and } \mathrm{N} \text{ have a common interior face.}$$

Use Exercise 4.19 to see that the second fact follows from the first.

Lemma 4.25. *For lunes* $\mathrm{L} \preceq \mathrm{N}$, *and a flat* Y *such that* $\mathrm{b(L)} \leq \mathrm{Y} \leq \mathrm{b(N)}$, *there exists a unique lune* M *with base* Y *such that* $\mathrm{L} \preceq \mathrm{M} \preceq \mathrm{N}$.

The base map from the set of lunes under the partial order $\preceq$ to the set of flats preserves cover relations. Hence:

Theorem 4.26. *The set of lunes* $\Lambda[\mathcal{A}]$ *under the partial order* $\preceq$ *is graded with the rank of a lune being the rank of its base. In particular, the rank of this poset is* $\mathrm{rk}(\mathcal{A})$.

Observe that: The ambient space is the maximum element of this poset. Each face is a minimal element.

Exercise 4.27. Let L be a lune with case X, and $\mathrm{X} \leq \mathrm{Y}$. Consider the set $\{\mathrm{M} \mid \mathrm{L} \preceq \mathrm{M},\ \mathrm{c(M)} = \mathrm{Y}\}$. Show that: This set is nonempty, with maximum element Y. However, it may not have a unique minimum element.

4.3. Maps involving lunes

We now collect together different maps involving lunes and explain the interrelationships between them. More precisely, we consider adjunctions between nested faces and faces, between nested flats and flats, and relate them to the adjunctions between lunes and flats.

4.3.1. Lunes and flats. Consider the set of lunes $\Lambda[\mathcal{A}]$ under the first partial order $\leq$. Recall from Section 2.2 the inclusion, base and case maps relating cones and flats. By restricting them to lunes, we obtain

$$\mathrm{i} : \Pi[\mathcal{A}] \hookrightarrow \Lambda[\mathcal{A}], \quad \mathrm{b} : \Lambda[\mathcal{A}] \twoheadrightarrow \Pi[\mathcal{A}] \quad \text{and} \quad \mathrm{c} : \Lambda[\mathcal{A}] \twoheadrightarrow \Pi[\mathcal{A}],$$

and adjunctions (c, i) and (i, b). That is, for any lune L and flat X,

$$\mathrm{c(L)} \leq \mathrm{X} \iff \mathrm{L} \leq \mathrm{X} \quad \text{and} \quad \mathrm{X} \leq \mathrm{L} \iff \mathrm{X} \leq \mathrm{b(L)}.$$

4.3.2. Nested faces and faces. View the set of nested faces $\mathrm{Q}[\mathcal{A}]$ as a poset componentwise. Consider the order-preserving maps

$$\begin{aligned} \mathrm{i} : \Sigma[\mathcal{A}] &\to \mathrm{Q}[\mathcal{A}] & \mathrm{b} : \mathrm{Q}[\mathcal{A}] &\to \Sigma[\mathcal{A}] & \mathrm{c} : \mathrm{Q}[\mathcal{A}] &\to \Sigma[\mathcal{A}] \\ F &\mapsto (F, F) & (H, G) &\mapsto H & (H, G) &\mapsto G. \end{aligned}$$

Then (c, i) and (i, b) are adjunctions. That is, for any nested face (H, G) and face F,

$$G \leq F \iff (H, G) \leq (F, F) \quad \text{and} \quad (F, F) \leq (H, G) \iff F \leq H.$$

The following diagrams commute.

$$
\begin{array}{ccc}
\mathrm{Q}[\mathcal{A}] & \xrightarrow{\mathrm{s}} & \Lambda[\mathcal{A}] \\
{\scriptstyle \mathrm{i}}\uparrow & & \uparrow{\scriptstyle \mathrm{i}} \\
\Sigma[\mathcal{A}] & \xrightarrow[\mathrm{s}]{} & \Pi[\mathcal{A}]
\end{array}
\qquad
\begin{array}{ccc}
\mathrm{Q}[\mathcal{A}] & \xrightarrow{\mathrm{s}} & \Lambda[\mathcal{A}] \\
{\scriptstyle \mathrm{b}}\downarrow & & \downarrow{\scriptstyle \mathrm{b}} \\
\Sigma[\mathcal{A}] & \xrightarrow[\mathrm{s}]{} & \Pi[\mathcal{A}]
\end{array}
\qquad
\begin{array}{ccc}
\mathrm{Q}[\mathcal{A}] & \xrightarrow{\mathrm{s}} & \Lambda[\mathcal{A}] \\
{\scriptstyle \mathrm{c}}\downarrow & & \downarrow{\scriptstyle \mathrm{c}} \\
\Sigma[\mathcal{A}] & \xrightarrow[\mathrm{s}]{} & \Pi[\mathcal{A}]
\end{array}
\tag{4.9}
$$

These diagrams relate the support map from nested faces to lunes to the support map from faces to flats.

4.3.3. Nested flats and flats. There is a similar story with flats instead of faces. A *nested flat* is a pair of flats (X, Y) with $\mathrm{X} \leq \mathrm{Y}$. Let $\mathrm{P}[\mathcal{A}]$ denote the set of nested flats. We view it as a poset componentwise. Consider the order-preserving maps

$$
\begin{array}{ccc}
\mathrm{i} : \Pi[\mathcal{A}] \to \mathrm{P}[\mathcal{A}] & \qquad \mathrm{b} : \mathrm{P}[\mathcal{A}] \to \Pi[\mathcal{A}] & \qquad \mathrm{c} : \mathrm{P}[\mathcal{A}] \to \Pi[\mathcal{A}] \\
\mathrm{X} \mapsto (\mathrm{X}, \mathrm{X}) & \qquad (\mathrm{Y}, \mathrm{Z}) \mapsto \mathrm{Y} & \qquad (\mathrm{Y}, \mathrm{Z}) \mapsto \mathrm{Z}.
\end{array}
$$

Then (c, i) and (i, b) are adjunctions.

Recall the base-case map from (4.4). It restricts to a map from lunes to nested flats. The following diagrams commute.

$$
\begin{array}{ccc}
\Lambda[\mathcal{A}] & \xrightarrow{\mathrm{bc}} & \mathrm{P}[\mathcal{A}] \\
{\scriptstyle \mathrm{i}}\uparrow & & \uparrow{\scriptstyle \mathrm{i}} \\
\Pi[\mathcal{A}] & \xrightarrow[\mathrm{id}]{} & \Pi[\mathcal{A}]
\end{array}
\qquad
\begin{array}{ccc}
\Lambda[\mathcal{A}] & \xrightarrow{\mathrm{bc}} & \mathrm{P}[\mathcal{A}] \\
{\scriptstyle \mathrm{b}}\downarrow & & \downarrow{\scriptstyle \mathrm{b}} \\
\Pi[\mathcal{A}] & \xrightarrow[\mathrm{id}]{} & \Pi[\mathcal{A}]
\end{array}
\qquad
\begin{array}{ccc}
\Lambda[\mathcal{A}] & \xrightarrow{\mathrm{bc}} & \mathrm{P}[\mathcal{A}] \\
{\scriptstyle \mathrm{c}}\downarrow & & \downarrow{\scriptstyle \mathrm{c}} \\
\Pi[\mathcal{A}] & \xrightarrow[\mathrm{id}]{} & \Pi[\mathcal{A}]
\end{array}
\tag{4.10}
$$

4.4. Category of lunes

Lunes can be composed when the case of the first lune equals the base of the second lune. This yields the category of lunes whose objects are flats and morphisms are lunes. Further, it is internal to the category of posets under the second partial order on lunes.

4.4.1. Category of lunes. We proceed to define the *category of lunes*. Its objects are flats and morphisms are lunes. More precisely, a morphism from Y to X is a lune whose base is X and whose case is Y. In particular, a morphism from Y to X exists iff $\mathrm{X} \leq \mathrm{Y}$. Composition of lunes is defined as follows.

Suppose L and M are lunes such that $\mathrm{c}(\mathrm{L}) = \mathrm{b}(\mathrm{M})$. First write $\mathrm{L} = \mathrm{s}(A, F)$ for some nested face (A, F). Since $\mathrm{s}(F) = \mathrm{c}(\mathrm{L}) = \mathrm{b}(\mathrm{M})$, by Proposition 3.12, there is a unique G greater than F such that $\mathrm{M} = \mathrm{s}(F, G)$. Define $\mathrm{L} \circ \mathrm{M} = \mathrm{s}(A, G)$. In other words,

$$
\mathrm{s}(A, F) \circ \mathrm{s}(F, G) = \mathrm{s}(A, G). \tag{4.11}
$$

This is well-defined in view of the first part of Exercise 3.15. We call $\mathrm{L} \circ \mathrm{M}$ the composite of L and M.

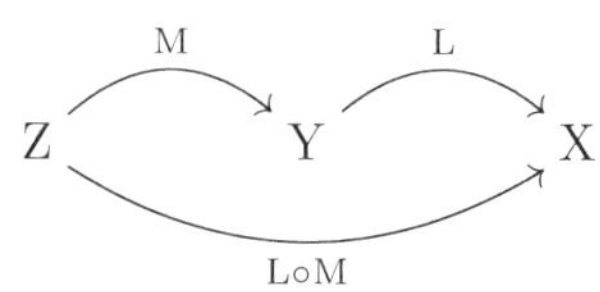

For any flat X, observe that $\mathrm{L} \circ \mathrm{X} = \mathrm{L}$ when $\mathrm{c}(\mathrm{L}) = \mathrm{X}$ and $\mathrm{X} \circ \mathrm{M} = \mathrm{M}$ when $\mathrm{X} = \mathrm{b}(\mathrm{M})$. Thus, X serves as the identity morphism for the object X. It is the only morphism from X to itself, since flats are the only lunes whose base and case are equal.

Example 4.28. The category of lunes for the rank-one arrangement is shown below. There are two flats, namely, $\bot$ and $\top$, while the chambers C and $\overline{C}$ are the only lunes which are not flats.

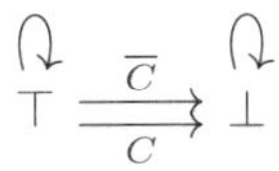

This is the category with two objects and two parallel (non-identity) arrows.

An illustration of lune composition in a rank-three arrangement is shown below.

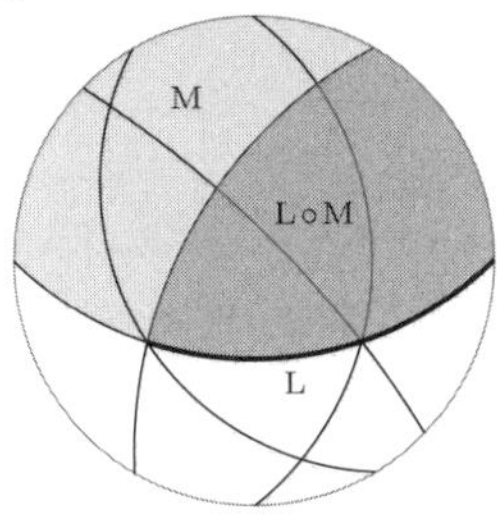

In the picture, L is a semicircle shown as a thick line, and M is the upper hemisphere. The composite $\mathrm{L} \circ \mathrm{M}$ is the vertex-based top-lune shown in darker shade.

Exercise 4.29. Since the category of lunes is finite, it is also locally finite in the sense that given N there are finitely-many pairs (L, M) such that $\mathrm{L} \circ \mathrm{M} = \mathrm{N}$. Show that the number of ways to express $\mathrm{N} = \mathrm{s}(A, G)$ as a composite equals the number of faces between A and G. (Use the converse in Exercise 3.15.) In particular, there is a unique way to decompose a flat X, namely, $\mathrm{X} = \mathrm{X} \circ \mathrm{X}$.

4.4.2. Interaction with partial orders. Recall that we have defined two partial orders on lunes, namely, $\leq$ and $\preceq$. We now study how the composition operation interacts with these partial orders.

Proposition 4.30. *Suppose* L *and* M *are lunes such that* $\mathrm{c(L)} = \mathrm{b(M)}$. *Then* $\mathrm{L} \circ \mathrm{M}$ *is the unique lune satisfying*

$$\mathrm{b(L)} = \mathrm{b(L \circ M)}, \quad \mathrm{L} \leq \mathrm{L} \circ \mathrm{M} \leq \mathrm{M}, \quad \mathrm{c(L \circ M)} = \mathrm{c(M)}. \tag{4.12}$$

PROOF. Write $\mathrm{L} = \mathrm{s}(A, F)$ and $\mathrm{M} = \mathrm{s}(F, G)$. Then it is easy to see that $\mathrm{L} \circ \mathrm{M} = \mathrm{s}(A, G)$ satisfies the conditions in (4.12). For uniqueness: Suppose N satisfies $\mathrm{b(L)} = \mathrm{b(N)}$, $\mathrm{L} \leq \mathrm{N} \leq \mathrm{M}$ and $\mathrm{c(N)} = \mathrm{c(M)}$. Since $\mathrm{s}(A) = \mathrm{b(L)} = \mathrm{b(N)}$, by Proposition 3.12, there is a unique G' such that $\mathrm{N} = \mathrm{s}(A, G')$. Further, since $\mathrm{L} \leq \mathrm{N}$, by Lemma 3.16, $F \leq G'$. Next, since $\mathrm{N} \leq \mathrm{M}$, G' belongs to the closure of M, so $FG' \leq G$. But $FG' = G'$, so $G' \leq G$. Finally, since $\mathrm{c(N)} = \mathrm{c(M)}$, G' and G have the same support which implies $G' = G$. Thus, $\mathrm{N} = \mathrm{L} \circ \mathrm{M}$ as required. □

A category is internal to the category of posets if the set of objects and the set of arrows are posets, and source, target, insertion of identities and composition are order-preserving.

Proposition 4.31. *The category of lunes is internal to posets under the partial order* $\preceq$. *In particular, for lunes* L, L′, M, M′,

$$\mathrm{L} \preceq \mathrm{L}' \text{ and } \mathrm{M} \preceq \mathrm{M}' \implies \mathrm{L} \circ \mathrm{M} \preceq \mathrm{L}' \circ \mathrm{M}'$$

whenever the composites are defined.

Proof. We employ Lemma 4.17. Write $\mathrm{L} = \mathrm{s}(A, F)$, $\mathrm{M} = \mathrm{s}(F, G)$, $\mathrm{L}' = \mathrm{s}(A', F')$ and $\mathrm{M}' = \mathrm{s}(F', G')$. We are given that $A'A = A'$, $A'F = F'$, $F'F = F'$ and $F'G = G'$. We want to show that $A'A = A'$ and $A'G = G'$. For the latter, $A'G = A'FG = F'G = G'$. □

Lemma 4.32. *Let* $\mathrm{L} \leq \mathrm{N}$ *with* $\mathrm{b}(\mathrm{L}) = \mathrm{b}(\mathrm{N})$. *Then there exists a unique lune* M *such that* $\mathrm{L} \circ \mathrm{M} = \mathrm{N}$.

Proof. We employ Lemma 3.16. Let $\mathrm{L} = \mathrm{s}(A, F)$ and $\mathrm{N} = \mathrm{s}(A, G)$ for some $A \leq F \leq G$. Now put $\mathrm{M} = \mathrm{s}(F, G)$ which yields $\mathrm{L} \circ \mathrm{M} = \mathrm{N}$ as required. Uniqueness is clear. □

Lemma 4.33. *Suppose* $\mathrm{L} \circ \mathrm{M} \preceq \mathrm{N}'$. *Then there exist unique lunes* L' *and* M' *such that* $\mathrm{L} \preceq \mathrm{L}'$, $\mathrm{M} \preceq \mathrm{M}'$ *and* $\mathrm{L}' \circ \mathrm{M}' = \mathrm{N}'$.

Proof. Write $\mathrm{L} = \mathrm{s}(A, F)$, $\mathrm{M} = \mathrm{s}(F, G)$ and $\mathrm{N}' = \mathrm{s}(A', G')$. By hypothesis, $A'A = A'$ and $A'G = G'$. To break N' as a composite, we need to look at faces between A' and G'. Put $F' := A'F$ and check that $\mathrm{L}' = \mathrm{s}(A', F')$ and $\mathrm{M}' = \mathrm{s}(F', G')$ works. Uniqueness of L' can be deduced from Lemma 4.22 from which uniqueness of M' follows by Lemma 4.32. □

Corollary 4.34. *For any lune* N', *there is a bijection*

$$\{(\mathrm{L}, \mathrm{M}) \mid \mathrm{L} \circ \mathrm{M} \preceq \mathrm{N}'\} \\ \longleftrightarrow \{(\mathrm{L}, \mathrm{M}, \mathrm{L}', \mathrm{M}') \mid \mathrm{L}' \circ \mathrm{M}' = \mathrm{N}',\ \mathrm{L} \preceq \mathrm{L}',\ \mathrm{M} \preceq \mathrm{M}',\ \mathrm{c}(\mathrm{L}) = \mathrm{b}(\mathrm{M})\}.$$

Further, for any $\mathrm{X} \leq \mathrm{Z}$, *the bijection restricts to the subsets defined by* $\mathrm{b}(\mathrm{L}) = \mathrm{X}$ *and/or* $\mathrm{c}(\mathrm{M}) = \mathrm{Z}$.

Exercise 4.35. Show that: For lunes L, L', M, M',

$$\mathrm{L} \leq \mathrm{L}' \text{ and } \mathrm{M} \leq \mathrm{M}' \implies \mathrm{L} \circ \mathrm{M} \leq \mathrm{L}' \circ \mathrm{M}'$$

whenever the composites are defined, and $\mathrm{b}(\mathrm{L}') \vee \mathrm{c}(\mathrm{L}) = \mathrm{c}(\mathrm{L}')$. (Use Lemma 4.15.) The implication may not hold if this last condition is dropped.

Exercise 4.36. Show that: For lunes L, L', M, M',

$$\mathrm{L} \leq \mathrm{L}' \iff \mathrm{L} \circ \mathrm{M} \leq \mathrm{L}' \circ \mathrm{M} \quad \text{and} \quad \mathrm{M} \leq \mathrm{M}' \iff \mathrm{L} \circ \mathrm{M} \leq \mathrm{L} \circ \mathrm{M}'$$

whenever the composites are defined. Deduce that in the category of lunes every arrow is both a monomorphism and an epimorphism.

Exercise 4.37. Show that: For lunes L, L', M, M',

$$\mathrm{L} \preceq \mathrm{L}' \text{ and } \mathrm{L} \circ \mathrm{M} \preceq \mathrm{L}' \circ \mathrm{M}' \implies \mathrm{M} \preceq \mathrm{M}'$$

whenever the composites are defined. (Use Lemma 4.17.)

Exercise 4.38. Show that: $\mathrm{L} \circ \mathrm{M} \preceq \mathrm{M}$ for composable lunes L and M. However, $\mathrm{L} \preceq \mathrm{L} \circ \mathrm{M}$ is false in general.

4.4.3. Opposition map on lunes. Recall from Section 3.2.4 that every lune L has an opposite lune $\overline{\mathrm{L}}$. Further, recall from Lemmas 2.14 and 2.16 that $\mathrm{b}(\mathrm{L}) = \mathrm{b}(\overline{\mathrm{L}})$ and $\mathrm{c}(\mathrm{L}) = \mathrm{c}(\overline{\mathrm{L}})$.

Lemma 4.39. *We have* $\overline{\mathrm{L} \circ \mathrm{M}} = \overline{\mathrm{L}} \circ \overline{\mathrm{M}}$ *for composable lunes* L *and* M.

PROOF. Write $\mathrm{L} = \mathrm{s}(A, F)$ and $\mathrm{M} = \mathrm{s}(F, G)$. Then

$$\overline{\mathrm{s}(A,F)} \circ \overline{\mathrm{s}(F,G)} = \mathrm{s}(A, A\overline{F}) \circ \mathrm{s}(F, F\overline{G}) = \mathrm{s}(A, A\overline{F}) \circ \mathrm{s}(A\overline{F}, A\overline{G})$$
$$= \mathrm{s}(A, A\overline{G}) = \overline{\mathrm{s}(A,G)}.$$

Note very carefully the second step which used $\mathrm{s}(F, F\overline{G}) = \mathrm{s}(A\overline{F}, A\overline{G})$. □

4.4.4. Cartesian product. Recall from Section 3.7.4 that the set of lunes of $\mathcal{A} \times \mathcal{A}'$ is the cartesian product of the set of lunes of $\mathcal{A}$ and of $\mathcal{A}'$:

$$\Lambda[\mathcal{A} \times \mathcal{A}'] = \Lambda[\mathcal{A}] \times \Lambda[\mathcal{A}'].$$

Further, this is an isomorphism of posets for both partial orders $\leq$ and $\preceq$. It is also compatible with lune composition (4.11). It follows that the category of lunes of $\mathcal{A} \times \mathcal{A}'$ is the cartesian product of the category of lunes of $\mathcal{A}$ and of $\mathcal{A}'$.

4.5. Categories associated to faces and flats

Recall that there is a category associated to any poset P (Section B.4.2). For the category associated to the poset of flats Π, objects are flats and there is a unique morphism from X to Y whenever $\mathrm{X} \leq \mathrm{Y}$. Similarly, we have the category associated to the poset of faces Σ. We now relate these two categories to the category of lunes.

The diagrams in (4.9) and (4.10) can be compactly written as follows.

$$\begin{array}{ccccc} \mathrm{Q}[\mathcal{A}] & \xrightarrow{\ \mathrm{s}\ } & \Lambda[\mathcal{A}] & \xrightarrow{\ \mathrm{bc}\ } & \mathrm{P}[\mathcal{A}] \\ {\scriptstyle \mathrm{b}}\downarrow\uparrow\downarrow{\scriptstyle \mathrm{c}} & & {\scriptstyle \mathrm{b}}\downarrow\uparrow\downarrow{\scriptstyle \mathrm{c}} & & {\scriptstyle \mathrm{b}}\downarrow\uparrow\downarrow{\scriptstyle \mathrm{c}} \\ \Sigma[\mathcal{A}] & \xrightarrow[\ \mathrm{s}\ ]{} & \Pi[\mathcal{A}] & \xrightarrow[\ \mathrm{id}\ ]{} & \Pi[\mathcal{A}] \end{array} \tag{4.13}$$

Each vertical triple can be viewed as a category: The set of objects is at the bottom, the set of morphisms is at the top, the two vertical arrows going down specify the source and target of a morphism, while the vertical arrow going up specifies the identity morphisms. The middle triple is the category of lunes. The first triple is the opposite of the category associated to the poset of faces, while the last triple is the opposite of the category associated to the poset of flats. By convention,

$$(A, F) \circ (F, G) = (A, G) \qquad \text{and} \qquad (\mathrm{X}, \mathrm{Y}) \circ (\mathrm{Y}, \mathrm{Z}) = (\mathrm{X}, \mathrm{Z}).$$

The pair of horizontal maps linking two triples can be viewed as a functor. The first pair given by the two support maps defines a functor from the opposite of the category associated to faces to the category of lunes. This follows from (4.11). Similarly, the second pair defines a functor from the category of lunes to the opposite of the category associated to flats which is identity on objects and the base-case map on morphisms.

For nested faces (H, G) and (H', G'), define

$$(H, G) \preceq (H', G') \quad \text{if} \quad (H, G) \leq (H', G') \text{ and } H'G = G'. \tag{4.14}$$

Similarly, for nested flats (X, Y) and $(\mathrm{X}', \mathrm{Y}')$, define

$$(\mathrm{X}, \mathrm{Y}) \preceq (\mathrm{X}', \mathrm{Y}') \quad \text{if} \quad (\mathrm{X}, \mathrm{Y}) \leq (\mathrm{X}', \mathrm{Y}') \text{ and } \mathrm{X}' \vee \mathrm{Y} = \mathrm{Y}'. \tag{4.15}$$

Proposition 4.31 can be extended as follows.

Proposition 4.40. *The categories associated to the posets of faces and flats are both internal to posets under the partial orders* $\preceq$. *Further, the functors relating these two categories with the category of lunes considered in* (4.13) *are internal to posets, that is, order-preserving on arrows as well as on objects.*

4.6. Presentation of categories

The categories associated to the poset of flats and the poset of faces and the lune category all have nice presentations. This is explained below.

Proposition 4.41. *The category associated to the poset of flats has a presentation given by generators* $\Delta : \mathrm{X} \to \mathrm{Y}$, *where* Y *covers* X, *and relations*

$$\begin{array}{ccc} \mathrm{X}' & \xrightarrow{\Delta} & \mathrm{Y} \\ {\scriptstyle\Delta}\uparrow & & \uparrow{\scriptstyle\Delta} \\ \mathrm{Z} & \xrightarrow[\Delta]{} & \mathrm{X} \end{array}$$

whenever Y *covers both* X *and* X', *and they in turn cover* Z.

Similar statement holds for the category associated to the poset of faces.

PROOF. The first claim follows from Lemma 1.35 and Proposition B.10 (applied to the poset of flats). The second claim follows similarly from Lemma 1.34 and Proposition B.10 (applied to the poset of faces). □

We now turn to the category of lunes. Observe that: If L and M are composable lunes, then

$$\mathrm{sk}(\mathrm{L} \circ \mathrm{M}) = \mathrm{sk}(\mathrm{L}) + \mathrm{sk}(\mathrm{M}), \tag{4.16}$$

where sk denotes the slack of a lune (3.2).

Proposition 4.42. *The category of lunes has the following presentation. It is generated by lunes of slack* 1 *(half-flats) subject to the quadratic relations*

$$\mathrm{L} \circ \mathrm{M} = \mathrm{L}' \circ \mathrm{M}',$$

where L, M, L' *and* M' *are lunes of slack* 1 *such that* $\mathrm{L} \circ \mathrm{M}$ *and* $\mathrm{L}' \circ \mathrm{M}'$ *both define the same lune of slack* 2.

PROOF. Let C denote the category of lunes, and D the category with the above presentation. Clearly, we have a functor $\mathsf{D} \to \mathsf{C}$. It is identity on objects.

Surjective on morphisms. Any lune L can be written as a composite $\mathrm{L}_1 \circ \cdots \circ \mathrm{L}_k$, with each L_i of slack 1. To see this, let $\mathrm{L} = \mathrm{s}(A, F)$. Pick a maximal chain $A = G_0 \lessdot G_1 \lessdot \cdots \lessdot G_k = F$ of faces starting at A and ending at F. Now set $\mathrm{L}_i = \mathrm{s}(G_{i-1}, G_i)$. This shows that the functor is surjective on morphisms.

Injective on morphisms. Suppose $\mathrm{L} = \mathrm{s}(A, F)$ equals both $\mathrm{L}_1 \circ \cdots \circ \mathrm{L}_k$ and $\mathrm{L}'_1 \circ \cdots \circ \mathrm{L}'_k$, where each L_i and L'_i has slack 1. These two composites correspond to maximal chains, say $A = G_0 \lessdot G_1 \lessdot \cdots \lessdot G_k = F$ and $A = G'_0 \lessdot G'_1 \lessdot \cdots \lessdot G'_k = F$. By Lemma 1.34, the two can be linked by a sequence of maximal chains in which two successive chains differ in exactly one position. Hence $\mathrm{L}_1 \circ \cdots \circ \mathrm{L}_k$ and

$L'_1 \circ \cdots \circ L'_k$ define the same morphism in D. This shows that the functor is injective on morphisms.

We conclude that C = D. □

Exercise 4.43. For any lune L, $\mathrm{sk}(L) \geq k$ iff there exist lunes $L_1, \ldots, L_k$ with $\mathrm{sk}(L_i) \geq 1$ for $1 \leq i \leq k$ such that $L = L_1 \circ \cdots \circ L_k$.

4.7. Action of the Birkhoff monoid on lunes

The Tits monoid acts on the set of nested faces componentwise. This descends to an action of the Birkhoff monoid on the set of lunes. We discuss how this action interacts with the two partial orders that we have defined on lunes.

4.7.1. Action on nested faces. The Tits monoid acts on the set of nested faces via

$$A \cdot (F, G) = (AF, AG). \tag{4.17}$$

Lemma 4.44. *The action* (4.17) *is compatible with the equivalence relations* (1.14) *and* (3.13)*: if* $A \sim A'$ *and* $(F, G) \sim (F', G')$*, then* $(AF, AG) \sim (A'F', A'G')$.

PROOF. The checks are straightforward. One of the required equalities may be derived as follows:

$$AFA'G' = AFG' = AG.$$

We employed the hypotheses and the property in Exercise 1.11. □

4.7.2. Action on lunes. As a consequence of Lemma 4.44: The Tits monoid acts on the set of lunes via

$$A \cdot \mathrm{s}(F, G) = \mathrm{s}(AF, AG).$$

Further, this induces an action of the Birkhoff monoid on the set of lunes. For a flat X and a lune L, we write $X \cdot L$ for the action of X on L.

An illustration in rank two when X has rank one is shown below.

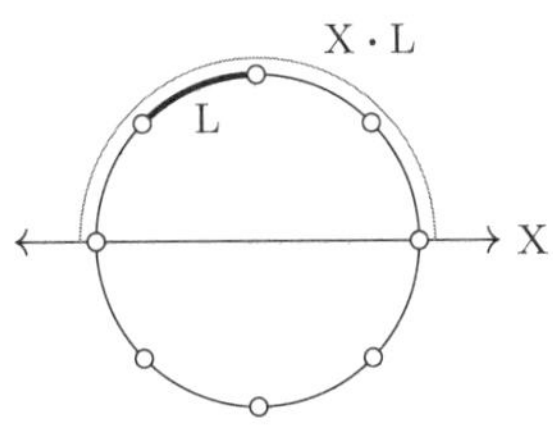

The lune L is the thick edge, and $X \cdot L$ is the semicircle.

Two illustrations in rank three are shown below. In both cases, X is the rank-one flat consisting of the north and south pole.

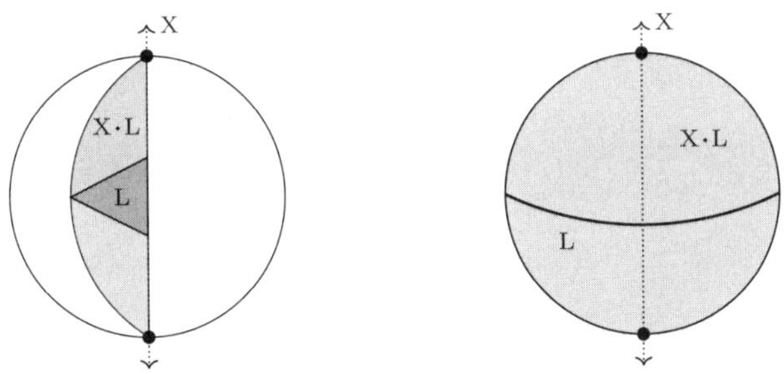

In the first picture, L is the shaded chamber, and $\mathrm{X} \cdot \mathrm{L}$ is the shaded vertex-based lune. In the second picture, L is a semicircle shown as a thick line, and $\mathrm{X} \cdot \mathrm{L}$ is the visible hemisphere.

4.7.3. Interaction with partial orders. Recall that we have defined two partial orders on lunes, namely, $\leq$ and $\preceq$. We now discuss how the action of the Birkhoff monoid interacts with these partial orders.

Proposition 4.45. *For any flat* X *and lune* L,

$$\mathrm{L} \leq \mathrm{X} \cdot \mathrm{L}, \quad \mathrm{b}(\mathrm{X} \cdot \mathrm{L}) = \mathrm{X} \vee \mathrm{b}(\mathrm{L}) \quad \textit{and} \quad \mathrm{c}(\mathrm{X} \cdot \mathrm{L}) = \mathrm{X} \vee \mathrm{c}(\mathrm{L}),$$

or equivalently,

$$\mathrm{L} \preceq \mathrm{X} \cdot \mathrm{L} \quad \textit{and} \quad \mathrm{b}(\mathrm{X} \cdot \mathrm{L}) = \mathrm{X} \vee \mathrm{b}(\mathrm{L}).$$

Further, $\mathrm{X} \cdot \mathrm{L}$ *is the unique lune with these properties.*

PROOF. Write $\mathrm{X} = \mathrm{s}(A)$, $\mathrm{L} = \mathrm{s}(F, G)$ and $\mathrm{X} \cdot \mathrm{L} = \mathrm{s}(AF, AG)$. Then

$$\mathrm{b}(\mathrm{X} \cdot \mathrm{L}) = \mathrm{s}(AF) = \mathrm{s}(A) \vee \mathrm{s}(F) = \mathrm{X} \vee \mathrm{b}(\mathrm{L}).$$

The assertion about $\mathrm{c}(\mathrm{X} \cdot \mathrm{L})$ can be derived similarly. The fact that $\mathrm{L} \leq \mathrm{X} \cdot \mathrm{L}$ can be deduced from the first equivalence in Lemma 4.15: $FK \leq G$ implies $AFK \leq AG$.

For uniqueness, suppose M is a lune satisfying

$$\mathrm{L} \leq \mathrm{M}, \quad \mathrm{b}(\mathrm{M}) = \mathrm{X} \vee \mathrm{b}(\mathrm{L}) \quad \text{and} \quad \mathrm{c}(\mathrm{M}) = \mathrm{X} \vee \mathrm{c}(\mathrm{L}).$$

Let K be any top-dimensional face of L. Since $\mathrm{L} \leq \mathrm{M}$, we deduce that AK is a top-dimensional face of M, where A is any face with support X. Now apply Corollary 3.20. □

SECOND PROOF. We employ Lemma 4.22. Let M be the unique lune such that $\mathrm{L} \preceq \mathrm{M}$ and $\mathrm{b}(\mathrm{M}) = \mathrm{X} \vee \mathrm{b}(\mathrm{L})$. Then $\mathrm{M} = (\mathrm{X} \vee \mathrm{b}(\mathrm{L})) \cdot \mathrm{L}$. Thus, $\mathrm{M} = \mathrm{X} \cdot (\mathrm{b}(\mathrm{L}) \cdot \mathrm{L}) = \mathrm{X} \cdot \mathrm{L}$. □

Exercise 4.46. Let F and G be two faces. Let $\mathrm{X} = \mathrm{s}(F)$. Show that F and $\mathrm{X} \cdot G$ are composable arrows in the category of lunes and that

$$F \circ (\mathrm{X} \cdot G) = FG.$$

This result links the product in the Tits monoid to composition in the category of lunes.

Observe that for any flats X and Y, we have $\mathrm{X} \cdot \mathrm{Y} = \mathrm{X} \vee \mathrm{Y}$. Thus, the action of flats on lunes extends the join operation on flats, which is the product in the Birkhoff monoid. More generally:

Proposition 4.47. *For any flat* X *and lune* L,

$$\mathrm{X} \cdot \mathrm{L} = \mathrm{X} \vee \mathrm{L},$$

with the latter being the join in the poset of cones.

PROOF. First note that $\mathrm{X} \cdot \mathrm{L}$ is greater than both X and L. Write $\mathrm{X} = \mathrm{s}(A)$, $\mathrm{L} = \mathrm{s}(F, G)$ and $\mathrm{X} \cdot \mathrm{L} = \mathrm{s}(AF, AG)$. By (3.9), $\mathrm{X} \cdot \mathrm{L} = [AG \colon \overline{AFG}]$. Now let V be any cone which is greater than both X and L. Then the closure of V contains A, F, $\overline{A}$, $\overline{F}$ and G (since X contains $\overline{A}$, and L contains $\overline{F}$). By Proposition 2.7, item (3), the closure of V also contains AG and $\overline{AFG}$, and finally by Lemma 2.46, it contains $[AG \colon \overline{AFG}]$. Therefore, V is greater than $\mathrm{X} \cdot \mathrm{L}$ as required. □

Recall that the inclusion map from flats to cones is join-preserving. Hence, for a flat X and cone V, the assignment

$$\mathrm{X} \cdot \mathrm{V} := \mathrm{X} \vee \mathrm{V}$$

specifies an action of the Birkhoff monoid on the set of cones. Proposition 4.47 says that the action of the Birkhoff monoid on the set of lunes is precisely the restriction of this action.

We note some more consequences of Proposition 4.47.

Corollary 4.48. *For any flat* X *and lune* M,

$$\mathrm{X} \cdot \mathrm{M} = \mathrm{M} \iff \mathrm{X} \cdot \mathrm{L} = \mathrm{M} \textit{ for some } \mathrm{L} \iff \mathrm{X} \leq \mathrm{b}(\mathrm{M}).$$

In particular: Given a flat X, *lunes of the form* X · L, *as* L *varies, are those whose base is greater than* X.

Corollary 4.49. *For any lunes* L *and* M *and flat* X,

$$\mathrm{L} \leq \mathrm{M} \implies \mathrm{X} \cdot \mathrm{L} \leq \mathrm{X} \cdot \mathrm{M}.$$

In other words, the Birkhoff monoid acts on the set of lunes under $\leq$ *(and also on the poset of top-lunes).*

Exercise 4.50. Prove Corollary 4.49 using the formulation of the partial order on lunes given in Lemma 4.15. Similarly, use Lemma 4.17 to check that Corollary 4.49 holds with $\leq$ replaced by $\preceq$.

The set of top-lunes is preserved under the action of the Birkhoff monoid. Further, the partial order on top-lunes can be captured using this action as follows.

Proposition 4.51. *For top-lunes* L *and* M, *the following are equivalent.*

(1) L $\leq$ M.
(2) b(M) · L = M.
(3) *There exists a flat* X *such that* X · L = M.

PROOF. (1) implies (2). Use Proposition 4.1, item (3). Alternatively, one can use the uniqueness assertion in Proposition 4.45.

(2) implies (3). Clear.

(3) implies (1). Follows from Proposition 4.45. □

Exercise 4.52. Check that: Proposition 4.51 generalizes to arbitrary lunes for the partial order $\preceq$. However, L $\leq$ M does not imply b(M) · L = M for arbitrary lunes.

Exercise 4.53. Given L $\preceq$ M and X, show that $\mathrm{X} \cdot \mathrm{L} = \mathrm{M} \iff \mathrm{X} \vee \mathrm{b}(\mathrm{L}) = \mathrm{b}(\mathrm{M})$.

4.7.4. Action on the category of lunes. A monoid acts on a category if it acts on the objects and on the morphisms such that source, target, composition and identities are preserved.

One may check that (4.17) yields an action of the Tits monoid on the category associated to the poset of faces. This action descends to an action of the Birkhoff monoid on the category of lunes and on the category associated to the poset of flats. In particular, for a flat X and composable lunes L and M,

$$\mathrm{X} \cdot (\mathrm{L} \circ \mathrm{M}) = (\mathrm{X} \cdot \mathrm{L}) \circ (\mathrm{X} \cdot \mathrm{M}).$$

Exercise 4.54. Use the above property to deduce the existence of L′ and M′ in Lemma 4.33.

4.8. Substitution product of chambers

We introduce the substitution product of chambers. It specifies a way to multiply chambers in arrangements under a flat with chambers in arrangements over a flat. This operation is equivalent to the composition operation on lunes. Its connection to the classical associative operad is given in Section 6.5.10. We also discuss related substitution products of chambers and faces, and of top-lunes and chambers.

4.8.1. Substitution product of chambers. Recall the set of chambers $\Gamma[\mathcal{A}]$. For any flat X, there is a map

$$\Gamma[\mathcal{A}^{\mathrm{X}}] \times \Gamma[\mathcal{A}_{\mathrm{X}}] \to \Gamma[\mathcal{A}]. \tag{4.18}$$

We call this the substitution product of chambers. To define this map, pick any face F with support X, consider the map

$$\Gamma[\mathcal{A}^{\mathrm{X}}] \times \Gamma[\mathcal{A}_F] \to \Gamma[\mathcal{A}], \qquad (H, C/F) \mapsto HC,$$

and identify $\Gamma[\mathcal{A}_F]$ with $\Gamma[\mathcal{A}_{\mathrm{X}}]$. The result does not depend on the particular choice of F. More directly,

$$\Gamma[\mathcal{A}^{\mathrm{X}}] \times \Gamma[\mathcal{A}_{\mathrm{X}}] \to \Gamma[\mathcal{A}], \qquad (H, \mathrm{L}) \mapsto D.$$

Here L is a top-lune with base X which is the same as a chamber in $\mathcal{A}_{\mathrm{X}}$ (Lemma 3.2), H is a face with support X, and D is the unique chamber in L which is greater than H (which exists by Proposition 3.8).

Let us consider the end cases when X is either the minimum flat or the maximum flat. The substitution products

$$\Gamma[\mathcal{A}^{\perp}] \times \Gamma[\mathcal{A}_{\perp}] \xrightarrow{\cong} \Gamma[\mathcal{A}] \quad \text{and} \quad \Gamma[\mathcal{A}^{\top}] \times \Gamma[\mathcal{A}_{\top}] \xrightarrow{\cong} \Gamma[\mathcal{A}]$$

are the canonical identifications. Since $\mathcal{A}^{\perp}$ and $\mathcal{A}_{\top}$ have rank zero, $\Gamma[\mathcal{A}^{\perp}]$ and $\Gamma[\mathcal{A}_{\top}]$ are singletons, while $\mathcal{A}_{\perp} = \mathcal{A}^{\top} = \mathcal{A}$.

The first interesting case of the substitution product occurs in rank two when the flat X has rank one. This is illustrated below.

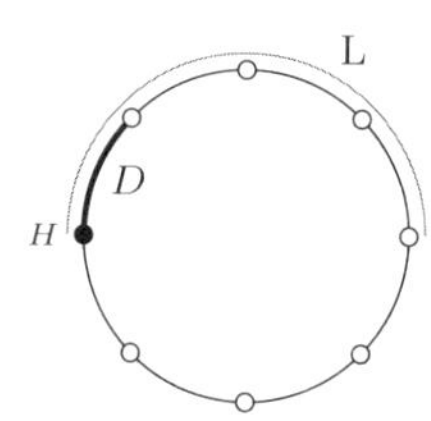

Here X supports the vertex H shown in black. The semicircle shown in gray is the top-lune L. The substitution product of H and L is the dark edge D.

In rank three, the interesting cases are when X has rank one or rank two. These are shown below.

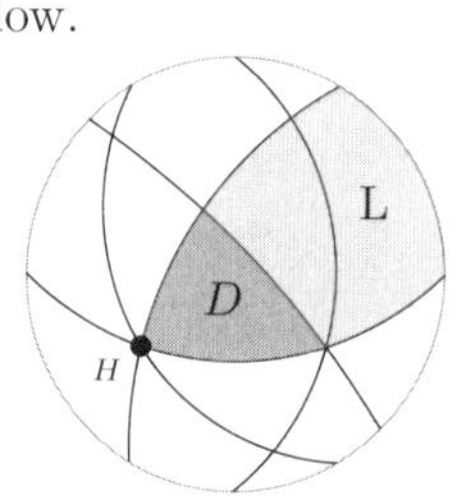

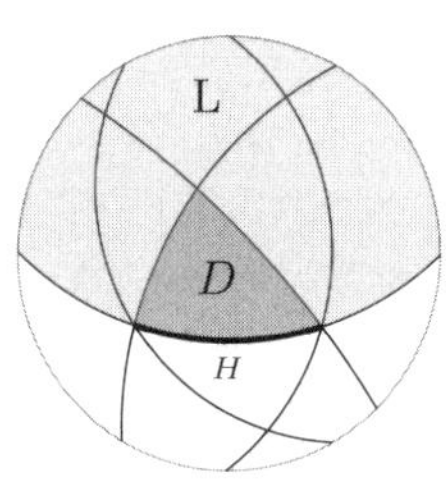

The flat X is the support of H. The shaded portion is the top-lune L. (It includes the chamber D.) In the first picture, H is a vertex and L is vertex-based, while in the second picture, H is an edge and L is edge-based. The substitution product of H and L is the chamber D.

4.8.2. Connection to category of lunes. Let $\mathrm{X} \le \mathrm{Y} \le \mathrm{Z}$ be flats and consider the arrangement $\mathcal{A}^{\mathrm{Z}}_{\mathrm{X}}$. Using Lemma 3.2, observe that the substitution product of chambers written in the general form

$$\Gamma[\mathcal{A}^{\mathrm{Y}}_{\mathrm{X}}] \times \Gamma[\mathcal{A}^{\mathrm{Z}}_{\mathrm{Y}}] \to \Gamma[\mathcal{A}^{\mathrm{Z}}_{\mathrm{X}}] \tag{4.19}$$

is equivalent to the composition product of lunes defined in Section 4.4.

Associativity of composition of lunes says that the diagram

$$\begin{array}{ccc} \Gamma[\mathcal{A}^{\mathrm{Y}}_{\mathrm{X}}] \times \Gamma[\mathcal{A}^{\mathrm{Z}}_{\mathrm{Y}}] \times \Gamma[\mathcal{A}^{\mathrm{W}}_{\mathrm{Z}}] & \longrightarrow & \Gamma[\mathcal{A}^{\mathrm{Y}}_{\mathrm{X}}] \times \Gamma[\mathcal{A}^{\mathrm{W}}_{\mathrm{Y}}] \\ \downarrow & & \downarrow \\ \Gamma[\mathcal{A}^{\mathrm{Z}}_{\mathrm{X}}] \times \Gamma[\mathcal{A}^{\mathrm{W}}_{\mathrm{Z}}] & \longrightarrow & \Gamma[\mathcal{A}^{\mathrm{W}}_{\mathrm{X}}] \end{array} \tag{4.20a}$$

commutes for any $\mathrm{X} \le \mathrm{Y} \le \mathrm{Z} \le \mathrm{W}$.

Unitality says that the maps

$$\Gamma[\mathcal{A}^{\mathrm{X}}_{\mathrm{X}}] \times \Gamma[\mathcal{A}^{\mathrm{Y}}_{\mathrm{X}}] \to \Gamma[\mathcal{A}^{\mathrm{Y}}_{\mathrm{X}}] \quad \text{and} \quad \Gamma[\mathcal{A}^{\mathrm{Y}}_{\mathrm{X}}] \times \Gamma[\mathcal{A}^{\mathrm{Y}}_{\mathrm{Y}}] \to \Gamma[\mathcal{A}^{\mathrm{Y}}_{\mathrm{X}}] \tag{4.20b}$$

are the canonical identifications (using that $\Gamma[\mathcal{A}^{\mathrm{X}}_{\mathrm{X}}]$ and $\Gamma[\mathcal{A}^{\mathrm{Y}}_{\mathrm{Y}}]$ are singletons).

4.8.3. Substitution product of chambers and faces. Recall the set of faces $\Sigma[\mathcal{A}]$. For any flat X, there is a map

$$\Gamma[\mathcal{A}^{\mathrm{X}}] \times \Sigma[\mathcal{A}_{\mathrm{X}}] \to \Sigma[\mathcal{A}]. \tag{4.21}$$

To define this map, pick any face F with support X, consider the map

$$\Gamma[\mathcal{A}^{\mathrm{X}}] \times \Sigma[\mathcal{A}_F] \to \Sigma[\mathcal{A}], \qquad (H, K/F) \mapsto HK,$$

and identify $\Sigma[\mathcal{A}_F]$ with $\Sigma[\mathcal{A}_{\mathrm{X}}]$. More directly,

$$\Gamma[\mathcal{A}^{\mathrm{X}}] \times \Sigma[\mathcal{A}_{\mathrm{X}}] \to \Sigma[\mathcal{A}], \qquad (H, \mathrm{L}) \mapsto G,$$

where L is a lune with base X which is the same as a face in $\mathcal{A}_{\mathrm{X}}$, H is a face with support X, and G is the unique top-dimensional in L which is greater than H (which exists by Proposition 3.12).

Under the identification

$$\Sigma[\mathcal{A}] = \bigsqcup_{\mathrm{Y}} \Gamma[\mathcal{A}^{\mathrm{Y}}],$$

the diagram

$$\begin{array}{ccc} \Gamma[\mathcal{A}^{\mathrm{X}}] \times \Gamma[\mathcal{A}^{\mathrm{Y}}_{\mathrm{X}}] & \longrightarrow & \Gamma[\mathcal{A}^{\mathrm{Y}}] \\ \downarrow & & \downarrow \\ \Gamma[\mathcal{A}^{\mathrm{X}}] \times \Sigma[\mathcal{A}_{\mathrm{X}}] & \longrightarrow & \Sigma[\mathcal{A}] \end{array} \tag{4.22}$$

commutes for any $\mathrm{X} \le \mathrm{Y}$.

4.8.4. Substitution product of top-lunes and chambers. Recall the set of top-lunes $\widehat{\Lambda}[\mathcal{A}]$. For any flat X, there is a map

$$\widehat{\Lambda}[\mathcal{A}^{\mathrm{X}}] \times \Gamma[\mathcal{A}_{\mathrm{X}}] \to \widehat{\Lambda}[\mathcal{A}]. \tag{4.23}$$

A top-lune in $\mathcal{A}^{\mathrm{X}}$ is a lune L in $\mathcal{A}$ with case X, while a chamber in $\mathcal{A}_{\mathrm{X}}$ is a top-lune M in $\mathcal{A}$ with base X. The above map, by definition, sends (L, M) to $\mathrm{L} \circ \mathrm{M}$. Note that the latter is a top-lune in $\mathcal{A}$ as required.

Under the identification

$$\widehat{\Lambda}[\mathcal{A}] = \bigsqcup_{\mathrm{X}} \Gamma[\mathcal{A}_{\mathrm{X}}],$$

the diagram

$$\begin{array}{ccc} \Gamma[\mathcal{A}^{\mathrm{Y}}_{\mathrm{X}}] \times \Gamma[\mathcal{A}_{\mathrm{Y}}] & \longrightarrow & \Gamma[\mathcal{A}_{\mathrm{X}}] \\ \downarrow & & \downarrow \\ \widehat{\Lambda}[\mathcal{A}^{\mathrm{Y}}] \times \Gamma[\mathcal{A}_{\mathrm{Y}}] & \longrightarrow & \widehat{\Lambda}[\mathcal{A}] \end{array} \tag{4.24}$$

commutes for any $\mathrm{X} \leq \mathrm{Y}$.

Notes

The second diagram in (4.9) for top-nested faces and top-lunes is given in [**8**, First diagram in (2.5)], and for the braid arrangement in [**9**, Diagram (10.49)]. The action of the Tits monoid on top-nested faces and of the Birkhoff monoid on top-lunes is considered in [**9**, Section 10.10.2].

Category of lunes. The category of lunes under the partial order opposite to $\preceq$ is a semiregular ordered category in the sense of Lawson [**277**]. More precisely, it satisfies the axioms (OC1), (OC2), (OC3), (OC4), (OC5), (OC6)(ii), (OC7), (OC8)(ii) stated in his Lemma 2.5. We elaborate briefly. (OC1), (OC2), (OC3) say that the category of lunes is internal to posets. (OC4) is contained in Corollary 3.17. (OC8)(ii) (which is stronger than (OC5)(ii) and (OC6)(ii)) is the same as Lemma 4.22. (OC7) is weaker than Lemma 4.33. (OC5)(i) is contained in Exercise 4.27.

Substitution product of chambers. We mention that the substitution product of chambers (4.18) generalizes to any LRB.

CHAPTER 5

Reflection arrangements

We review reflection arrangements. Roughly speaking, these are hyperplane arrangements equipped with reflection symmetries. The group generated by these symmetries is the Coxeter group of the arrangement. In addition to everything that comes with an arrangement, these symmetries allow us to define concepts such as face-types, flat-types, nested face-types and lune-types. These are orbits of faces, flats, nested faces and lunes under the Coxeter group action and display similar inter-relationships. Another interesting concept is that of the cycle-type function. One can also construct new objects like the Coxeter-Tits monoid by taking the semidirect product of the Coxeter group and Tits monoid.

Among reflection arrangements, there is a further subclass of good reflection arrangements which is closed under passage to arrangements over and under a flat. We recall the classification of reflection arrangements, and then list out those which are good.

5.1. Coxeter groups and reflection arrangements

5.1.1. Reflections. Let V be a finite-dimensional vector space over $\mathbb{R}$ equipped with an inner product. For any hyperplane H passing through the origin, reflection in H defines an orthogonal transformation of V. It fixes H pointwise, and sends any point on the line through the origin orthogonal to H to its negative. Let us denote this transformation by s_{H}.

A *Coxeter group* W on V is a finite group of orthogonal transformations of V generated by reflections in some finite set of hyperplanes through the origin. (The condition that the group generated by reflections be finite is very nontrivial. For instance, the group generated by reflections in two lines in $\mathbb{R}^2$ passing through the origin is finite iff the angle between them is a rational multiple of π.) For a Coxeter group W, the set of hyperplanes H such that $s_{\mathrm{H}} \in W$ is the *reflection arrangement* associated with W. This arrangement is central but not necessarily essential. Its ambient space is V.

Above we started with the group and constructed the arrangement from it. This procedure can also be reversed: We say that a hyperplane arrangement $\mathcal{A} = \{\mathrm{H}_i\}_{i \in I}$ with ambient space V is a *reflection arrangement* if for each i, the reflection s_{H_i} preserves $\mathcal{A}$. For a reflection arrangement $\mathcal{A}$, the group generated by the reflections s_{H_i} is called the *Coxeter group* of $\mathcal{A}$. (One can show that the Coxeter group does not have any more reflections than what we started with.)

5.1.2. Coxeter complex. A reflection arrangement $\mathcal{A}$ is necessarily simplicial. We will refer to elements of its Coxeter group as the Coxeter symmetries of $\mathcal{A}$. For a Coxeter group W, the regular cell complex Σ associated to its reflection arrangement is the *Coxeter complex* of W. It is a pure simplicial complex. Moreover, it is labeled. This fact will be elaborated later when we discuss the type map. (Let Δ be a pure simplicial complex and I be any set whose cardinality equals the rank of Δ. A *labeling* of Δ assigns to each vertex an element of I called its label such that any two vertices which share an edge have distinct labels. If such a labeling exists, then we say that Δ is a *labeled simplicial complex*.)

5.1.3. Examples. Arrangements of rank zero and rank one are reflection arrangements. The rank-two arrangement of n lines is a reflection arrangement precisely when the lines are equally spaced, that is, the arrangement is dihedral. Some important examples of reflection arrangements in $\mathbb{R}^n$ are shown in Table 5.1.

TABLE 5.1. Examples of reflection arrangements.

Reflection arrangement	Hyperplanes
Coordinate arrangement	$x_i = 0$ for $1 \leq i \leq n$
Braid arrangement or type A	$x_i = x_j$ for $1 \leq i, j \leq n$
Arrangement of type B	$x_i = \pm x_j$ and $x_i = 0$ for $1 \leq i, j \leq n$
Arrangement of type D	$x_i = \pm x_j$ for $1 \leq i, j \leq n$

The coordinate arrangement arises as the n-fold cartesian product of the rank-one arrangement with itself. The braid arrangement or type A arises from the symmetric group S_n on n letters acting on $\mathbb{R}^n$ by permuting coordinates. The reflections in this group are precisely transpositions. Similarly, the arrangement of type B arises from the signed symmetric group. These arrangements are discussed in detail in Chapter 6.

5.1.4. Coxeter group in action. Recall that we have associated many geometric objects such as faces, flats, cones, lunes, and so on to an arrangement. Some of these objects have extra structure such as that of a poset or monoid. Further, there are maps between these objects, for instance, the support map from faces to flats.

For a reflection arrangement, the Coxeter group W acts on each of these objects preserving whatever structure they may possess, and further is compatible with maps relating them. For instance,

$$w(FG) = w(F)w(G) \quad \text{and} \quad w(\mathrm{s}(F)) = \mathrm{s}(w(F)).$$

Some important properties of the action are given below.

- The action of W on the set of chambers is simply transitive. That is, given C and D, there is a unique $w \in W$ such that $w(C) = D$. It follows that there is a bijection $\Gamma \to W$ which preserves the left action of W on Γ and W. Thus, $\Gamma \cong W$ as left W-sets. In particular, the cardinality of W equals the number of chambers.
- If $w(F) = F$, then w fixes F (and in fact the entire flat $\mathrm{s}(F)$) pointwise.
- If $w(F) = G$, then $w(A) = A$ for any A smaller than both F and G.

Exercise 5.1. If F and G are faces of K and $w(F) = G$ for some w, then $F = G$. In other words, the action of W on the poset of faces Σ satisfies condition (C.46).

5.1.5. Pointed arrangements. A *pointed arrangement* is a pair $(\mathcal{A}, C)$, where $\mathcal{A}$ is a reflection arrangement and C is a chamber of $\mathcal{A}$. We will usually use the letter $\boldsymbol{\alpha}$ to denote a pointed arrangement. We will refer to C as the *reference chamber*. Let S be the set of reflections in the walls of C. It is a fact that S generates the Coxeter group of $\mathcal{A}$, which we are denoting by W. The pair (W, S) is called a *Coxeter system*.

5.2. Face-types, flat-types and lune-types

Face-types, flat-types and lune-types are the orbits of the action of the Coxeter group on faces, flats and lunes, respectively. We discuss these notions along with other similar notions which arise from the action of the Coxeter group.

5.2.1. Face-types. Define an equivalence relation on faces: $F \sim G$ if there is a Coxeter symmetry which sends F to G. An equivalence class is called a *face-type*. In other words, a face-type is an orbit of Σ under the action of W. We denote the set of face-types by Σ^W. The canonical map

$$\Sigma \to \Sigma^W, \qquad F \mapsto \mathrm{t}(F) \tag{5.1}$$

is called the *type map*. We refer to $\mathrm{t}(F)$ as the type of F.

Fix a chamber C. Then each face of C has a distinct type, and all face-types arise in this manner. Thus, Σ^W can be identified with the poset of faces of C. Since C is a simplex, this is a Boolean poset. Note that the type map is order-preserving.

It follows that the Coxeter complex Σ is a labeled simplicial complex: the label assigned to each vertex is its type.

Lemma 5.2. *Let A and B be faces of the same support. If F and G are faces greater than A and $w(F) = G$ for some $w \in W$, then $w(BF) = BG$. If F and G are faces greater than A and $\mathrm{t}(F) = \mathrm{t}(G)$, then $\mathrm{t}(BF) = \mathrm{t}(BG)$.*

PROOF. For the first claim: The hypothesis implies that $w(A) = A$. So w fixes the support of A pointwise, and hence $w(B) = B$. Thus $w(BF) = w(B)w(F) = BG$ as required. For the second claim, since F and G have the same type, there is a w such that $w(F) = G$. Now apply the first claim. □

Covering maps between posets are reviewed in Section C.4.5. The type map is an example of a covering map. This is a formal consequence of Exercise 5.1, see Proposition C.33, item (c),

Let $(\mathcal{A}, C)$ be a pointed arrangement with Coxeter system (W, S). Faces of C can be identified with subsets of S: Given a face F of C, any element $s \in S$ either fixes F or does not fix F. The required subset is obtained by picking those reflections that do not fix F. By this convention, the central face corresponds to the empty set, and C corresponds to S. Since face-types correspond to faces of C, we can identify them with subsets of S. The poset of face-types is then the Boolean poset of subsets of S. Viewed in this manner, we use the letters T, U, V for face-types. Note that $\mathrm{rk}(T) = |T|$.

For $\boldsymbol{\alpha} = (\mathcal{A}, C)$ and $T \leq S$, consider the pointed arrangement

$$\boldsymbol{\alpha}_T := (\mathcal{A}_F, C/F),$$

where F is the face of C of type T. The set of reflections in the walls of C/F identifies with the set $S \setminus T$. If G is a face greater than F of type U, then we will denote the type of G/F by U/T (instead of $U \setminus T$). By this convention, face-types of $\boldsymbol{\alpha}_T$ are subsets of S which contain T.

5.2.2. Group action on a set with an equivalence relation. Let G be a group acting on a set X. Suppose X has an equivalence relation $\sim$ compatible with the group action, that is, $x \sim y$ implies $g \cdot x \sim g \cdot y$. Then there is a commutative diagram

$$\begin{array}{ccc} X & \twoheadrightarrow & X_G \\ \downarrow & & \downarrow \\ X_\sim & \twoheadrightarrow & (X_\sim)_G = (X_G)_\sim . \end{array} \tag{5.2}$$

Here $X_\sim$ denotes the set of equivalence classes of X under $\sim$, while X_G denotes the set of orbits of X under G. Due to compatibility, there is an induced G-action on $X_\sim$, and an induced equivalence relation $\sim$ on X_G, and the orbits of the former correspond to classes of the latter yielding (5.2). Two elements $x, y \in X$ map to the same element of $(X_G)_\sim$ iff there exists an element $z \in X$ such that z and x lie in the same G-orbit, and $z \sim y$.

We will apply this construction to two situations. In both cases, G is the Coxeter group W. In the first case, X is the set of faces, and $\sim$ is (1.14), while in the second case, X is the set of nested faces, and $\sim$ is (3.13). Elements of $(X_G)_\sim$ will be called flat-types in the first case, and lune-types in the second case. Details follow.

5.2.3. Flat-types. A *flat-type* can be defined in two ways:

- Define an equivalence relation on flats: X $\sim$ Y if there is a Coxeter symmetry which sends X to Y. An equivalence class is a flat-type.
- Define an equivalence relation on face-types: $T \sim U$ if there is a flat X which supports both a face of type T and a face of type U. (If $T \sim U$ and $U \sim V$, then using transitivity of the group action on faces of type U, we deduce $T \sim V$.) An equivalence class is a flat-type.

There is a canonical bijection between the two sets of equivalence classes: In one direction, the class of T maps to the class of s(F), where F is any face of type T. In the other direction, the class of X maps to the class of t(F), where F is any face of support X.

Let Π^W denote the set of flat-types. We have the commutative diagram

$$\begin{array}{ccc} \Sigma & \xrightarrow{\ \mathrm{t}\ } & \Sigma^W \\ {\scriptstyle \mathrm{s}}\downarrow & & \downarrow{\scriptstyle \mathrm{s}} \\ \Pi & \xrightarrow[\ \mathrm{t}\ ]{} & \Pi^W \end{array} \qquad\qquad \begin{array}{ccc} \text{face} & \overset{\mathrm{t}}{\longmapsto} & \text{face-type} \\ {\scriptstyle \mathrm{s}}\downarrow & & \downarrow{\scriptstyle \mathrm{s}} \\ \text{flat} & \underset{\mathrm{t}}{\longmapsto} & \text{flat-type} \end{array} \tag{5.3}$$

with the support map and the type map on faces as before. This is a special case of (5.2). It is convenient to refer to both horizontal maps as the type maps, and to both vertical maps as the support maps. Thus, we can talk of the *support* of a face-type, and the *type* of a flat. We define the *support-type* of a face to be the type of its support, or equivalently, the support of its type.

In contrast to the type map on faces, the type map on flats is *not* a covering map in general.

Lemma 5.3. *Suppose F and G are two faces. Then F and G have the same support-type iff there exists a face H of the same type as F and of the same support as G.*

PROOF. Backward implication. By hypothesis, the support-type of H equals both the support-type of F and the support-type of G.

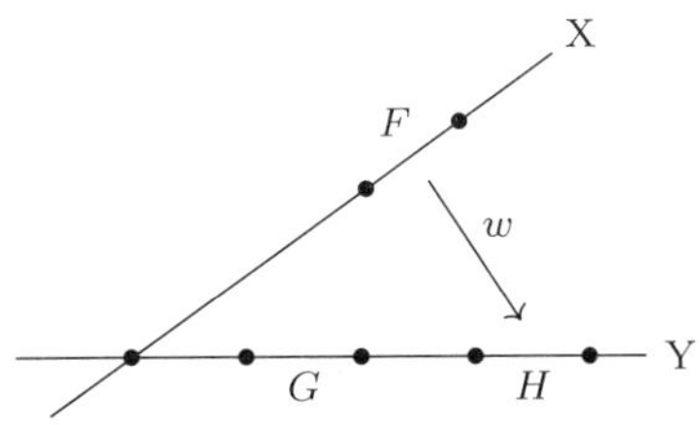

Forward implication. Let $\mathrm{X} = \mathrm{s}(F)$ and $\mathrm{Y} = \mathrm{s}(G)$. Then by hypothesis, there is an element w which sends X to Y. Set $H := wF$. Observe that H has the same type as F and the same support as G. □

Flat-types will usually be denoted by λ and μ. There is a partial order on flat-types. We say that $\lambda \leq \mu$ if there are faces F and G with $F \leq G$ such that the support-types of F and G are λ and μ, respectively. This partial order can also be described using face-types or flats. For instance, $\lambda \leq \mu$ if there are flats X and Y with $\mathrm{X} \leq \mathrm{Y}$ such that the types of X and Y are λ and μ, respectively. Note that the poset of flat-types has a unique minimum and maximum element. In addition, it is graded. The rank of λ equals the rank of F, where F is any face with support-type λ. Equivalently, it equals $|T|$ for any face-type T with support λ, and equals the rank of X for any flat X of type λ.

5.2.4. Nested face-types and lune-types. A *nested face-type* is a pair (T, U) of face-types such that $T \leq U$. Given a nested face (F, G), its *type* is defined to be the pair $(\mathrm{t}(T), \mathrm{t}(U))$. The latter is a nested face-type. The Coxeter group W acts on the set of nested faces, and the orbits of this action can be identified with nested face-types.

Let (T, U) and (T', U') be two nested face-types. We say $(T, U) \sim (T', U')$ if there exist nested faces (F, G) and (F', G') of types (T, U) and (T', U') such that $(F, G) \sim (F', G')$ in the sense of (3.13). Observe that (T, S) and (T', S) are equivalent iff T and T' have the same support.

A *lune-type* is an equivalence class under this equivalence relation on nested face-types. Alternatively, it is an orbit under the action of W on the set of lunes. We have the commutative diagram

$$\begin{array}{ccc} \text{nested face} & \xrightarrow{\ \mathrm{t}\ } & \text{nested face-type} \\ {\scriptstyle \mathrm{s}}\downarrow & & \downarrow{\scriptstyle \mathrm{s}} \\ \text{lune} & \xrightarrow[\ \mathrm{t}\ ]{} & \text{lune-type.} \end{array} \tag{5.4}$$

This is a special case of (5.2).

One may also consider the action of W on the set of top-nested faces under the equivalence relation (3.7). This yields the commutative diagram

$$
\begin{array}{ccc}
\text{top-nested face} & \xmapsto{\ \mathrm{t}\ } & \text{face-type} \\
\Big\downarrow{\scriptstyle \mathrm{s}} & & \Big\downarrow{\scriptstyle \mathrm{s}} \\
\text{top-lune} & \xmapsto[\ \mathrm{t}\ ]{} & \text{flat-type.}
\end{array} \tag{5.5}
$$

5.2.5. Opposition. The opposition map $F \mapsto \overline{F}$ clearly commutes with the action of W. In other words, $w\overline{F} = \overline{wF}$. More generally, it also preserves preserves opposition in stars:

Lemma 5.4. *Suppose C and D are chambers opposite to each other in the star of A, that is, $D = A\overline{C}$ and $C = A\overline{D}$. Then for any $w \in W$, wC and wD are opposite to each other in the star of wA.*

PROOF. The required calculation is $(wA)(\overline{wC}) = (wA)(w\overline{C}) = w(A\overline{C}) = wD$. □

Lemma 5.5. *Let F be any face and C be any chamber. Suppose u is the Coxeter symmetry defined by $u(FC) = C$. Then $u(\overline{F}C) = \overline{F_0}C$, where F_0 is the face of C of the same type as F.*

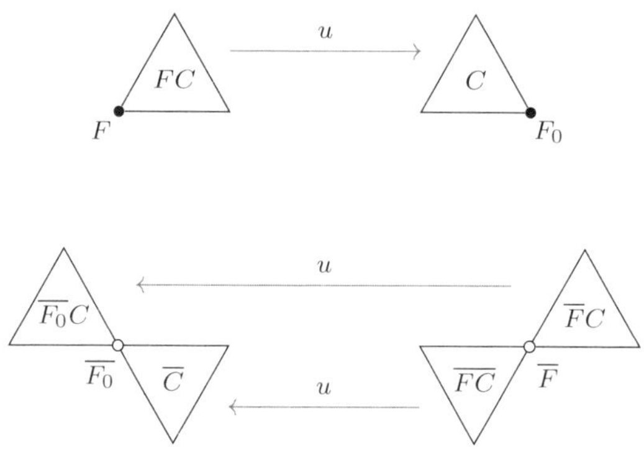

PROOF. Since u is type-preserving and $u(FC) = C$, we have $u(F) = F_0$. Further since u respects the opposition map, we deduce $u(\overline{FC}) = \overline{C}$ and $u(\overline{F}) = \overline{F_0}$. Now $\overline{FC}$ and $\overline{F}C$ are opposite chambers in the star of $\overline{F}$, while $\overline{C}$ and $\overline{F_0}C$ are opposite chambers in the star of $\overline{F_0}$. So by Lemma 5.4, it follows that $u(\overline{F}C) = \overline{F_0}C$, as required. □

5.3. Length, W-valued distance and weak order

In this discussion, we let C_0 denote the reference chamber.

The *length* of an element $w \in W$, denoted $l(w)$, is the smallest integer such that w can be expressed as a word in the generating set S whose length is that integer. Equivalently,

$$l(w) = \operatorname{dist}(C_0, wC_0). \tag{5.6}$$

Since the gallery metric is invariant under the diagonal action of W, we have

$$\operatorname{dist}(uC_0, vC_0) = l(u^{-1}v).$$

Define a W-valued gallery distance function by

$$d : \Gamma \times \Gamma \to W, \qquad d(uC_0, vC_0) = u^{-1}v.$$

Observe that $\text{dist}(C, D) = l(d(C, D))$. The function d is also invariant under the diagonal action of W. Further,

$$d(E, C) = d(E, D)d(D, C) \tag{5.7}$$

for any chambers C, D and E.

Let $u, v \in W$. We say that $u \leq v$ in the *weak order* on W if there is a minimal gallery $E \text{--} D \text{--} C$ with $d(D, C) = u$ and $d(E, C) = v$. Equivalently,

$$u \leq v \iff v^{-1}C_0 \text{--} u^{-1}C_0 \text{--} C_0 \iff C_0 \text{--} vu^{-1}C_0 \text{--} vC_0. \tag{5.8}$$

Alternatively, by letting $d(E, D) = w$,

$$u \leq v \iff v = wu \quad \text{and} \quad l(v) = l(w) + l(u).$$

The 'left' in the terminology refers to the fact that w appears to the left of u in the expression $v = wu$.

Exercise 5.6. Given $w \in W$, show that the parity of $\text{dist}(C, wC)$ is independent of C. In particular, the parity is odd when w is a reflection.

Exercise 5.7. Show that the chamber graph of a reflection arrangement of rank at least one is a balanced bipartite graph. (For definitions, see Section 1.10.4.) Equivalently, the number of $w \in W$ with $l(w)$ odd equals those with $l(w)$ even. For the braid arrangement, this says that the number of odd permutations equals the number of even permutations.

5.4. Subgroups of Coxeter groups

The Coxeter group W acts on the set of faces. For any face F, put

$$W_F := \{w \in W \mid w(F) = F\}.$$

This is the subgroup of W which leaves F invariant. It is called a *parabolic subgroup* of W. Elements of W_F fix the entire flat $\mathrm{s}(F)$ pointwise. Thus $W_F = W_G$ whenever F and G have the same support.

Similarly, the Coxeter group acts on the set of flats. For any flat X, put

$$\widehat{W}_{\mathrm{X}} := \{w \in W \mid w(\mathrm{X}) = \mathrm{X}\}.$$

Let W_{X} be the subgroup of $\widehat{W}_{\mathrm{X}}$ which fixes X pointwise. It is a normal subgroup of $\widehat{W}_{\mathrm{X}}$. This can be checked directly. It will also follow from the discussion below. Note that $W_{\mathrm{X}} = W_F$ whenever F has support X. We mention that W_{X} is a Coxeter group in its own right, the corresponding reflection arrangement is $\mathcal{A}_{\mathrm{X}}$.

For a top-lune L, let

$$W_{\mathrm{L}} := \{w \in W \mid w(\mathrm{L}) = \mathrm{L}\}.$$

For a top-lune L with base X, there is an exact sequence of groups

$$W_{\mathrm{X}} \hookrightarrow \widehat{W}_{\mathrm{X}} \twoheadrightarrow W_{\mathrm{L}}.$$

The first map is inclusion. The second map is defined as follows. Suppose $w \in \widehat{W}_{\mathrm{X}}$. Pick any top-nested face (F, C) with support L. Let u be the unique element such that $u(C) = w(F)C$. Thus, u sends (F, C) to $(w(F), w(F)C)$ which also has support L, and hence $u \in W_{\mathrm{L}}$. One can check that u does not depend on the particular choice of (F, C). Further, the map $w \mapsto u$ is a group homomorphism with kernel W_{X}, and the inclusion $W_{\mathrm{L}} \hookrightarrow \widehat{W}_{\mathrm{X}}$ is a section. As a consequence:

Proposition 5.8. *There is an isomorphism of groups*

$$\widehat{W}_{\mathrm{X}} \xrightarrow{\cong} W_{\mathrm{X}} \rtimes W_{\mathrm{L}}, \qquad w \mapsto (wu^{-1}, u). \tag{5.9}$$

Multiplication in the semidirect product is given by

$$(y, v) \cdot (x, u) = (yvxv^{-1}, vu).$$

Here is another way to think about the semidirect product. The group $\widehat{W}_{\mathrm{X}}$ acts on the set of top-lunes with base X. The subgroup W_{X} is normal and its action is simply transitive. That is, for top-lunes M and M′ with base X, there exists a unique $w \in W$ such that $w(\mathrm{M}) = \mathrm{M}'$ and w fixes X pointwise. This is because top-lunes with base X can be identified with chambers of $\mathcal{A}_{\mathrm{X}}$ (Lemma 3.2). Hence,

$$\widehat{W}_{\mathrm{X}} \xrightarrow{\cong} W_{\mathrm{X}} \rtimes W_{\mathrm{L}}, \qquad w \mapsto (x, x^{-1}w),$$

where $x \in W_{\mathrm{X}}$ is defined by $x(\mathrm{L}) = w(\mathrm{L})$.

Observe that W_F and W_G are conjugate subgroups if F and G have the same type. For a face-type T, we let W_T denote the conjugacy class of subgroups W_F with F of type T. Recall that face-types (as subsets of S) are defined using a reference chamber C. So we have a canonical representative for W_T, namely, W_{F_0}, where F_0 is the face of C of type T. To keep the notation flexible, we also use W_T to denote this particular subgroup.

Similarly, $\widehat{W}_{\mathrm{X}}$ and $\widehat{W}_{\mathrm{Y}}$ are conjugate subgroups if X and Y have the same type. For a flat-type λ, we let W_λ denote the conjugacy class of subgroups $\widehat{W}_{\mathrm{X}}$ with X of type λ. In contrast to face-types, there is no canonical representative for W_λ, but still we will allow ourselves to treat it as a subgroup (in which case it is assumed that some particular representative has been chosen).

Exercise 5.9. For any flat X and hyperplane H, show that:

- The reflection s_{H} leaves X invariant iff $\mathrm{X} = \mathrm{X} \cap \mathrm{H} + \mathrm{X} \cap \mathrm{H}^{\perp}$.
- s_{H} fixes X pointwise iff $\mathrm{X} \subseteq \mathrm{H}$.
- s_{H} leaves X invariant but not fixed iff $\mathrm{H}^{\perp} \subseteq \mathrm{X}$.

Exercise 5.10. For any top-lune L with base X, and hyperplane H, show that: The reflection s_{H} leaves L invariant iff $\mathrm{H}^{\perp} \subseteq \mathrm{X}$.

Exercise 5.11. For a top-lune L with base X, show that: If a reflection of W belongs to $\widehat{W}_{\mathrm{X}}$, then it either belongs to W_{X} or to W_{L}.

Exercise 5.12. Let L and L′ be top-lunes, and F and F' be faces supported by the bases of L and L′, respectively, such that $\mathrm{t}(F) = \mathrm{t}(F')$. Show that there exists a unique $u \in W$ such that $u(\mathrm{L}) = \mathrm{L}'$ and $u(F) = F'$. When L and L′ have the same base X, the element u belongs to $\widehat{W}_{\mathrm{X}}$, and when $F = F'$, it belongs to W_F.

Exercise 5.13. Fix a chamber C. Suppose X and X′ are flats and $w \in W$ is such that $w(\mathrm{X}) = \mathrm{X}'$. Show that w can be uniquely expressed in the form $w = vu$, with u and v subject to the following conditions: u sends the top-lune containing C with base X to the top-lune containing C with base X′, and v fixes X′ pointwise.

5.5. Cycle-type function and characteristic polynomial

We discuss some enumerative aspects of arrangements which are specific to reflection arrangements.

5.5.1. Numbers $|\lambda|$. Fix a flat-type λ. Suppose T is a face-type of support λ, and X is a flat of type λ. Let $|\lambda|$ denote the number of faces with type T and support X. Thus, it is the number of ways to complete the diagram

$$\begin{array}{ccc} ? & \longmapsto & T \\ \downarrow & & \downarrow \\ \mathrm{X} & \longmapsto & \lambda, \end{array}$$

as defined in (5.3). We show below that $|\lambda|$ only depends on λ, and not on the particular choice of T and X.

Lemma 5.14. *Let* X *and* X′ *be flats with type equal to the support of a face-type T. Then the number of faces with support* X *and type T equals the number of faces with support* X′ *and type T.*

Let T and T' be face-types with support equal to the type of a flat X. *Then the number of faces with support* X *and type T equals the number of faces with support* X *and type T'.*

PROOF. The first statement is clear since there is a Coxeter symmetry which takes X to X′, and it is type-preserving.

For the second statement: Let F be any face of support X. Then the subgroup $\widehat{W}_{\mathrm{X}}$ acts transitively on the faces of type $\mathrm{t}(F)$ and support X. So this number is the cardinality of $\widehat{W}_{\mathrm{X}}$ divided by the cardinality of W_F. But $W_F = W_{\mathrm{X}}$, so this number only depends on X and not on the particular F. □

Exercise 5.15. Check that: If λ is the maximum flat-type, then $|\lambda| = |W|$, the order of W.

Exercise 5.16. Let L be a top-lune whose base has type λ. Show that the group W_{L} has cardinality $|\lambda|$. In particular, $|\lambda|$ divides the order of W.

5.5.2. Cycle-type function. Recall the set of flats Π, and the set of flat-types Π^W. Consider the function

$$W \to \Pi \tag{5.10}$$

which sends $w \in W$ to the largest flat which is fixed pointwise by the action of w.

Lemma 5.17. *The inverse image of a flat* X *under* (5.10) *has cardinality* $|\mu(\mathcal{A}_{\mathrm{X}})|$.

PROOF. Let us temporarily denote this number by $f(\mathrm{X})$. Since W_{X} is the Coxeter group of $\mathcal{A}_{\mathrm{X}}$, its cardinality is the number of chambers of $\mathcal{A}_{\mathrm{X}}$ which we denote by c_{X}. Thus, we obtain

$$c_{\mathrm{X}} = \sum_{\mathrm{Y}:\, \mathrm{Y} \geq \mathrm{X}} f(\mathrm{Y}).$$

Now the Zaslavsky formula (1.48) applied to each $\mathcal{A}_X$ yields $f(X) = |\mu(\mathcal{A}_X)|$. □

The *cycle-type function*

$$W \to \Pi^W \tag{5.11}$$

is obtained by composing (5.10) with the type map from Π to Π^W. In other words, the cycle-type of w is the type of the largest flat which is fixed pointwise by w.

Lemma 5.17 yields the following.

Theorem 5.18. *Let W be a Coxeter group, and let $\mathcal{A}$ be its reflection arrangement. Then the number of elements of W with cycle-type λ is*

$$\sum_{X:\, t(X)=\lambda} |\mu(\mathcal{A}_X)|.$$

In particular, the number of elements of W whose cycle-type is the type of the minimum flat is $|\mu(\mathcal{A})|$.

5.5.3. Characteristic polynomial. We state below an important result concerning the factorization of the characteristic polynomial for reflection arrangements.

Theorem 5.19. *Given a reflection arrangement $\mathcal{A}$, there exist positive integers $e_1, \dots, e_n$ such that*

$$\chi(\mathcal{A}, t) = (t - e_1) \dots (t - e_n). \tag{5.12}$$

The integers $e_1, \dots, e_n$ are called the *exponents* of the Coxeter group of $\mathcal{A}$. The number 1 is always an exponent (assuming $\mathcal{A}$ has at least rank 1). This follows from (1.54b). Similarly, (1.54a) and (1.54c) yield:

$$\mu(\mathcal{A}) = \prod_{i=1}^{n} (-e_i) \quad \text{and} \quad c(\mathcal{A}) = \prod_{i=1}^{n} (e_i + 1). \tag{5.13}$$

Some interesting identities involving the Möbius function are given below.

Lemma 5.20. *In a reflection arrangement $\mathcal{A}$, for any chamber D,*

$$\sum_{G:\, G \leq D} \frac{\mu(\mathcal{A}_G)}{c^G c_G} = \begin{cases} 1 & \text{if } \mathcal{A} \text{ has rank } 0, \\ 0 & \text{otherwise,} \end{cases}$$

$$\sum_{\substack{G:\, G \leq D, \\ \mathrm{rk}(G)=k}} \frac{\mu(\mathcal{A}_G)}{c^G c_G} = \frac{\mathrm{wy}(\mathcal{A}, k)}{c},$$

$$\sum_{G:\, G \leq D} (-1)^{\mathrm{rk}(G)} \frac{\mu(\mathcal{A}_G)}{c^G c_G} = (-1)^{\mathrm{rk}(\mathcal{A})}.$$

PROOF. For the first identity: The result is clear if $\mathcal{A}$ has rank 0. So assume that $\mathcal{A}$ has rank at least 1. Note that the sum is only over faces of D. Using the symmetry in a reflection arrangement, it suffices to show that

$$\sum_{D} \sum_{G:\, G \leq D} \frac{\mu(\mathcal{A}_G)}{c^G c_G} = 0.$$

By definition, there are precisely c_G chambers greater than G. The above identity then follows from:

$$\sum_{G} \sum_{D:\, G \le D} \frac{\mu(\mathcal{A}_G)}{c^G c_G} = \sum_{G} \frac{\mu(\mathcal{A}_G)}{c^G} = \sum_{\mathrm{X}} \mu(\mathcal{A}_{\mathrm{X}}) = \sum_{\mathrm{X}} \mu(\mathrm{X}, \top) = 0.$$

The last step used (C.5b).

For the second identity: Repeating this calculation, we end up with a sum over all X of rank k which by (1.56) is $\mathrm{wy}(\mathcal{A}, k)$.

For the third identity: We may proceed as in the proof of the first identity and use the Zaslavsky formula in the last step. Alternatively, one can start with the second identity in Lemma 5.20, multiply it by $(-1)^k$, sum over k and use (1.57d). □

5.5.4. Counting chambers in a top-lune.

Lemma 5.21. *In a reflection arrangement $\mathcal{A}$, for any face F, the number of chambers in the top-lune $\mathrm{s}(F, D)$ is independent of D, and equals the number of chambers in $\mathcal{A}$ divided by the numbers of chambers in $\mathcal{A}_F$.*

PROOF. For any chambers D and E greater than F, there is a Coxeter symmetry which takes $\mathrm{s}(F, D)$ to $\mathrm{s}(F, E)$, and hence they contain the same number of chambers. To get the second statement, use the decomposition (3.16). □

Exercise 5.22. Show by an example that Lemma 5.21 fails for arbitrary arrangements.

5.6. Coxeter-Tits monoid

We introduce the Coxeter-Tits monoid. It is the semidirect product of the Coxeter group and the Tits monoid. Similarly, the semidirect product of the Coxeter group and the Birkhoff monoid yields the Coxeter-Birkhoff monoid. The support map relates the two monoids. By the same considerations, we define the Coxeter-Janus monoid.

5.6.1. Coxeter-Tits monoid. The Coxeter group W acts on the Tits monoid Σ. So we can form their semidirect product

$$\Sigma \rtimes W = \{(F, w) \mid F \in \Sigma,\, w \in W\}.$$

This is a monoid with product defined by

$$(G, v) \cdot (F, u) := (Gv(F), vu). \tag{5.14}$$

The unit element is (O, e). We refer to $\Sigma \rtimes W$ as the *Coxeter-Tits monoid* and abbreviate it to $\mathrm{W}\Sigma$. Note that Σ and W are submonoids of $\mathrm{W}\Sigma$ via the identifications $F \mapsto (F, e)$ and $w \mapsto (O, w)$.

Proposition 5.23. *The Coxeter-Tits monoid $\mathrm{W}\Sigma$ is freely generated by W and Σ subject to the relations*

$$w \cdot F = w(F) \cdot w \qquad \text{or equivalently} \qquad F \cdot w = w \cdot w^{-1}(F)$$

for $w \in W$ and $F \in \Sigma$.

PROOF. We have already remarked that W and Σ are submonoids of $\mathrm{W}\Sigma$. Further, we see from (5.14) that

$$(O, w) \cdot (F, e) = (w(F), e) \cdot (O, w),$$

with both sides equal to $(w(F), w)$. This yields the relation $w \cdot F = w(F) \cdot w$. Since it allows us to switch the order of elements of W and Σ, we obtain a presentation of $\mathrm{W}\Sigma$. $\square$

The discussion above applies to any group acting on any monoid. For instance, W also acts on the opposite of the Tits monoid. Set $\mathrm{W}\Sigma^{\mathrm{op}} := \Sigma^{\mathrm{op}} \rtimes W$. Explicitly, the product of $\mathrm{W}\Sigma^{\mathrm{op}}$ is given by

$$(G, v) \cdot (F, u) := (v(F)G, vu). \tag{5.15}$$

The order of the factors in the second coordinate is the same as in (5.14), but it has been reversed in the first coordinate. Thus, W and Σ^{op} are submonoids of $\mathrm{W}\Sigma^{\mathrm{op}}$. In fact, $\mathrm{W}\Sigma^{\mathrm{op}}$ is isomorphic to the opposite of $\mathrm{W}\Sigma$:

Proposition 5.24. *The map*

$$(\mathrm{W}\Sigma)^{\mathrm{op}} \to \mathrm{W}\Sigma^{\mathrm{op}}, \qquad (F, w) \mapsto (w^{-1}(F), w^{-1})$$

is an isomorphism of monoids.

This is straightforward to check.

5.6.2. Coxeter-Birkhoff monoid. Similarly, using the action of the Coxeter group W on the Birkhoff monoid Π, we can form the semidirect product

$$\Pi \rtimes W = \{(\mathrm{X}, w) \mid \mathrm{X} \in \Pi,\ w \in W\}.$$

This is a monoid under

$$(\mathrm{Y}, v) \cdot (\mathrm{X}, u) := (\mathrm{Y} \vee v(\mathrm{X}), vu). \tag{5.16}$$

The unit element is $(\perp, e)$. We refer to $\Pi \rtimes W$ as the *Coxeter-Birkhoff monoid* and abbreviate it to $\mathrm{W}\Pi$.

Note that W and Π are submonoids of $\mathrm{W}\Pi$ via the identifications $w \mapsto (\perp, w)$ and $\mathrm{X} \mapsto (\mathrm{X}, e)$. More precisely, the Coxeter-Birkhoff monoid $\mathrm{W}\Pi$ is freely generated by W and Π subject to the relations

$$w \cdot \mathrm{X} = w(\mathrm{X}) \cdot w \qquad \text{or equivalently} \qquad w \cdot w^{-1}(\mathrm{X}) = \mathrm{X} \cdot w$$

for $w \in W$ and $\mathrm{X} \in \Pi$.

Proposition 5.25. *The map*

$$(\mathrm{W}\Pi)^{\mathrm{op}} \to \mathrm{W}\Pi, \qquad (\mathrm{X}, w) \mapsto (w^{-1}(\mathrm{X}), w^{-1})$$

is an isomorphism of monoids.

5.6.3. Coxeter-Janus monoid. Finally, we consider the action of W on the Janus monoid J, and set

$$\mathrm{J} \rtimes W = \{(F, F', w) \mid (F, F') \in \mathrm{J},\, w \in W\}.$$

This is a monoid under

$$(G, G', v) \cdot (F, F', u) := (Gv(F), v(F')G', vu). \tag{5.17}$$

The unit element is (O, O, e). We refer to $\mathrm{J} \rtimes W$ as the *Coxeter-Janus monoid* and abbreviate it to WJ. It can also be obtained as the fiber product of the Coxeter-Tits monoid and its opposite over the Coxeter-Birkhoff monoid:

$$\mathrm{WJ} = \mathrm{W\Sigma} \times_{\mathrm{W\Pi}} \mathrm{W\Sigma}^{\mathrm{op}}.$$

In particular, we have a commutative diagram of monoids

$$\begin{array}{ccc} \mathrm{WJ} & \longrightarrow & \mathrm{W\Sigma}^{\mathrm{op}} \\ \big\downarrow & & \big\downarrow \mathrm{s} \\ \mathrm{W\Sigma} & \xrightarrow[\mathrm{s}]{} & \mathrm{W\Pi}, \end{array}$$

where s is the support map $(F, w) \mapsto (\mathrm{s}(F), w)$, the left-vertical map is projection on the first and third coordinates, while the top-horizontal map is projection on the second and third coordinates.

Proposition 5.26. *The map*

$$\mathrm{WJ} \to (\mathrm{WJ})^{\mathrm{op}}, \qquad (F, F', w) \mapsto (w^{-1}(F), w^{-1}(F'), w^{-1})$$

is an isomorphism of monoids.

This can also be seen as a formal consequence of Propositions 5.24 and 5.25, and the fiber product property.

5.6.4. Support map. The support map

$$\mathrm{s} : \mathrm{W\Sigma} \twoheadrightarrow \mathrm{W\Pi}, \qquad (F, w) \mapsto (\mathrm{s}(F), w)$$

can be understood algebraically as follows. Using (5.14),

$$(F, u) \cdot (G, v) \cdot (F, u) = (F, u) \iff v = u^{-1} \text{ and } Fu(G) = F. \tag{5.18}$$

In this situation, we say that (G, v) is a *pseudoinverse* of (F, u). Every element of $\mathrm{W\Sigma}$ has a pseudoinverse, so $\mathrm{W\Sigma}$ is a *regular semigroup*. Further, its set of idempotents form a subsemigroup, so it is an *orthodox semigroup*.

We say that (G, v) is an *inverse* of (F, u) if (G, v) is a pseudoinverse of (F, u) and vice-versa.

Lemma 5.27. *We have: (G, v) is an inverse of (F, u) iff $v = u^{-1}$ and F and $u(G)$ have the same support.*

PROOF. Employing (5.18), we see that (G, v) is an inverse of (F, u) iff $v = u^{-1}$, $Fu(G) = F$ and $Gv(F) = G$. Applying u to $Gv(F) = G$, we get $u(G)uv(F) = u(G)$. If $v = u^{-1}$, then this is the same as $u(G)F = u(G)$. □

Consider the quotient of $\mathrm{W\Sigma}$ obtained by identifying (F, u) and (F', u') whenever they have a common inverse. By Lemma 5.27, (F, u) and (F', u') get identified iff $u = u'$ and F and G have the same support. In other words, this quotient is precisely the Coxeter-Birkhoff monoid $\mathrm{W\Pi}$. The latter is an *inverse semigroup*, that is, every element has an unique inverse.

5.7. Good reflection arrangements

Reflection arrangements or Coxeter groups have been classified. They are usually listed using the notation of a Coxeter diagram. Here we introduce the notion of a good reflection arrangement. Then we go over each arrangement in the classification list, and say which one is good and which is not.

5.7.1. Irreducible reflection arrangements. If $\mathcal{A}$ and $\mathcal{A}'$ are reflection arrangements with Coxeter groups W and W', then the cartesian product $\mathcal{A} \times \mathcal{A}'$ is also a reflection arrangement and its Coxeter group is $W \times W'$. A reflection arrangement is *irreducible* if it cannot be expressed as a cartesian product of two reflection arrangements (both with nonzero ambient space). Such arrangements are necessarily essential. They have been completely classified up to gisomorphism. These are the arrangements of types A_n and B_n for $n \geq 0$, D_n for $n \geq 3$, H_3, H_4, F_4, E_6, E_7, E_8, and $I_2(m)$ for $m \geq 3$. The subscripts refer to the rank of the arrangement. There are some repetitions in this list, namely, $A_0 = B_0$, $A_1 = B_1$, $A_2 = I_2(3)$, $B_2 = I_2(4)$, and $A_3 = D_3$.

Types A, B and D have been briefly mentioned in Section 5.1.3. They are discussed in more detail in Chapter 6. Type $I_2(m)$ is the dihedral arrangement of m lines. We do not discuss the remaining types.

Warning. Braid arrangements are not irreducible, but their essentializations are. In this section, all references to type A are to the latter. In particular, the arrangement of type A_0 is the essential rank-zero arrangement.

5.7.2. Coxeter diagram. To each reflection arrangement $\mathcal{A}$, one can associate a graph with edges labeled by an integer greater than or equal to 3. The number of vertices is the rank of $\mathcal{A}$. This is called the *Coxeter diagram* of $\mathcal{A}$. (We omit the details.) The Coxeter diagrams of irreducible reflection arrangements are all trees (with very little branching).

5.7.3. Good reflection arrangements. Let $\mathcal{A}$ be a reflection arrangement with Coxeter group W. Then for any flat X, the arrangement $\mathcal{A}_{\mathrm{X}}$ is also a reflection arrangement with Coxeter group W_{X}. However, $\mathcal{A}^{\mathrm{X}}$ may not be a reflection arrangement. This motivates the following definition.

A reflection arrangement $\mathcal{A}$ is *good* if it has the property that for each flat X, the arrangement $\mathcal{A}^{\mathrm{X}}$ is cisomorphic to a reflection arrangement.

The class of good reflection arrangements is indeed closed under passage to arrangements under and over a flat, and under cartesian products. The list of good reflection arrangements is given below.

Theorem 5.28. *An irreducible reflection arrangement $\mathcal{A}$ is good iff $\mathcal{A}$ is of type A_{n-1} or B_n for $n \geq 1$ or $I_2(m)$ or H_3.*

PROOF. We provide a sketch. In this discussion, we will draw a tiny bit from Chapter 6.

- Type A and type B are good: in these types, an arrangement under a flat is again of the given type (Sections 6.3.11 and 6.7.11).
- All reflection arrangements of ranks 0, 1, 2 and 3 are good: The cases of rank 0 and 1 are clear. Any rank-two arrangement is cisomorphic to a reflection arrangement. It follows that rank-two and rank-three reflection arrangements are good. In particular, types $I_2(m)$ and H_3 are good.

- Type D_4 is not good: In this case, the arrangement under any of the hyperplanes is cisomorphic to the arrangement shown in Section 6.8.2. Observe that there are vertices, the boundary of whose top-stars are hexagons, but 6 does not divide 32, which is the number of chambers. So it cannot be a reflection arrangement (by Lemma 5.21).
- Type D_n for $n \geq 4$ and types E_6, E_7 and E_8 are not good: In each of these arrangements, there is face F such that $\mathcal{A}_F$ is cisomorphic to D_4. (This is because the Coxeter diagrams of these arrangements contain the Coxeter diagram of D_4.) Now pick a hyperplane X greater than $\mathrm{s}(F)$. Then $\mathcal{A}_F^{\mathrm{X}}$ is cisomorphic to the arrangement under a hyperplane of D_4, and hence is not a reflection arrangement. Therefore $\mathcal{A}^{\mathrm{X}}$ is also not a reflection arrangement (since reflection arrangements are closed under passage to arrangements over a flat).
- Types F_4 and H_4 are not good: An irreducible rank-three reflection arrangement has 24, 48 or 120 chambers. In type H_4, the arrangement under any hyperplane has 480 chambers, so it cannot be any of these. Further, the maximum boundary of a top-star of any vertex in this arrangement is a 12-gon, so it cannot be reducible either. The same argument works for type F_4, the arrangement under a hyperplane now has 96 chambers. □

It follows that:

Theorem 5.29. *An essential reflection arrangement $\mathcal{A}$ is good iff $\mathcal{A}$ is a cartesian product of arrangements of type A_{n-1} or B_n or $I_2(m)$ or H_3.*

We also note that a reflection arrangement is good iff its essentialization is good. In other words, a good reflection arrangement is a cartesian product of the above types possibly with an arbitrary rank-zero arrangement.

5.7.4. Enumerative aspects. By employing the defining property of a good reflection arrangement, the results of Lemmas 5.20 and 5.21 can be improved as follows.

Lemma 5.30. *In a good reflection arrangement $\mathcal{A}$, for any $F \leq H$,*

$$\sum_{G\colon F\leq G\leq H} \frac{\mu(\mathcal{A}_G^H)}{c_F^G c_G^H} = \begin{cases} 1 & \text{if } F = H, \\ 0 & \text{otherwise,} \end{cases}$$

and

$$\sum_{\substack{G\colon F\leq G\leq H,\\ \mathrm{rk}(G)=k}} \frac{\mu(\mathcal{A}_G^H)}{c_F^G c_G^H} = \frac{\mathrm{wy}(\mathcal{A}_F^H, k)}{c_F^H},$$

and for any face H,

$$\sum_{G\colon G\leq H} (-1)^{\mathrm{rk}(G)} \frac{\mu(\mathcal{A}_G^H)}{c^G c_G^H} = (-1)^{\mathrm{rk}(H)}.$$

Lemma 5.31. *In a good reflection arrangement $\mathcal{A}$, for any face F, the number of faces in the combinatorial lune $\mathrm{s}(F, G)$ only depends on the support of G, and equals the number of chambers in $\mathcal{A}^G$ divided by the numbers of chambers in $\mathcal{A}_F^G$.*

Exercise 5.32. Recall that the reflection arrangement of type D_4 is not good. Check that Lemma 5.31 fails for this arrangement.

Notes

Coxeter theory. The foundations of Coxeter theory were laid down by Tits [**415**]. Supplementary information is available in many places. See for instance the books by Abramenko and Brown [**2**], Björner and Brenti [**80**], Borovik and Borovik [**97**], Borovik, Gelfand and White [**98**, Chapters 5 and 7], Bourbaki [**99**], Davis [**137**], Grove and Benson [**212**], Humphreys [**238**] and Kane [**250**]. Short introductions can be found in [**8**, Sections 1.3 and 1.4], [**179**, Section 3], and [**339**, Chapter 11].

Flat-types. The partial order on flat-types in Section 5.2.3 is mentioned in [**8**, Section 2.6]. It agrees with the partial order considered by Saliola [**367**, Section 5, Remark 5.1]. The numbers $|\lambda|$ in Section 5.5.1 are defined in [**8**, Definition 5.7.4].

Coxeter-Tits monoid. The general construction of semidirect product of monoids is discussed in [**361**, Section 1.2.2]. Early references for regular and inverse semigroups are the papers by Green [**206**], Vagner [**417**] and Preston [**348**].

Characteristic polynomial. The factorization (5.12) of the characteristic polynomial is proved by Shephard and Todd [**382**, Theorem 5.3] by a case-by-case analysis. Warning: They work in the more general setting of unitary reflection groups, and the polynomial which they factorize is defined using dimension of the stabilizers of the group elements. For Coxeter groups, that is, in the real case, this polynomial agrees with the characteristic polynomial. (This is the content of Theorem 5.18.) A direct proof of their result is given by Solomon [**385**, Formula (1)]. The related second formula in (5.13) was observed by Coxeter [**126**] by a case-by-case analysis.

A similar factorization of the Poincare polynomial of the complement of a complexified reflection arrangement is given by Brieskorn [**102**, Theorem 6, part (ii)]. Orlik and Solomon show that the Poincare polynomial and the characteristic polynomial coincide for any complex arrangement [**326**, Theorem 5.2]. In a later paper, they obtain a factorization of the characteristic polynomial for unitary reflection groups [**327**, Theorem 4.8]. (Note that this differs from the factorization of Shephard and Todd.) This is generalized to free arrangements by Terao [**412**]. The latter result is discussed in [**328**, Theorem 6.60 and Corollary 6.62], and stated in [**400**, Theorem 4.14].

Good reflection arrangements. In [**1**, Proposition 5], Abramenko determined the class of finite Coxeter complexes with the property that their walls are again Coxeter complexes. His result is stated below.

Proposition 5.33. *Let Σ be a finite Coxeter complex. Every wall of Σ is a Coxeter complex iff the diagram of Σ does not contain a subdiagram of type D_4, F_4 or H_4. If Σ is of type A_n, B_n, H_3, then every wall of Σ is of type A_{n-1}, B_{n-1}, $I_2(6)$, respectively.*

Theorem 5.28 can be deduced from this result. The proof that we have sketched follows Abramenko's argument.

CHAPTER 6

Braid arrangement and related examples

We discuss some important examples of arrangements. The coordinate arrangement is treated first. It is the simplest example. We then focus on the braid arrangement. This is the reflection arrangement of type A and is a main example. Subsequently we treat the reflection arrangements of types B and D, more briefly. Finally we discuss graphic arrangements. They are associated to simple graphs and are the subarrangements of the braid arrangement. In all these examples, there is a rich interplay between geometry and combinatorics.

We employ the notation $[n] := \{1, 2, \ldots, n\}$.

6.1. Coordinate arrangement

The coordinate arrangement of rank n is the n-fold cartesian product of the arrangement of rank 1. We make explicit the notions of faces, flats, cones, lunes and so on for this arrangement, as summarized below.

Σ	$\{0, +, -\}^n$
Γ	$\{+, -\}^n$
$\Pi = \mathrm{G}$	$\{0, \pm\}^n$
$\Omega = \Lambda = \vec{\mathrm{G}}$	$\{0, +, -, \pm\}^n$
$\widehat{\Omega} = \widehat{\Lambda}$	$\{+, -, \pm\}^n$
W	$\mathbb{Z}_2^n$

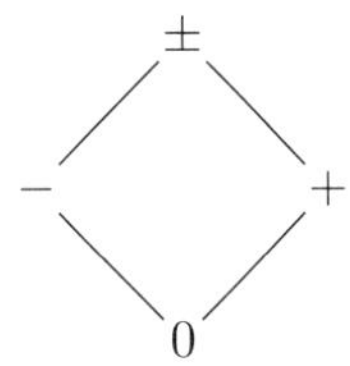

In this section, $\mathcal{A}$ denotes the coordinate arrangement of rank n, and W denotes its Coxeter group.

6.1.1. Coordinate arrangement.

6.1.1. Coordinate arrangement. The *coordinate arrangement* of rank n consists of the n hyperplanes

$$x_i = 0$$

for $1 \leq i \leq n$. It is the smallest arrangement of rank n in terms of number of hyperplanes. It is the n-fold cartesian product of the arrangement of rank 1. It is a reflection arrangement. Its Coxeter group is $\mathbb{Z}_2^n$, the product of n copies of $\mathbb{Z}_2$. The generator of the i-th copy of $\mathbb{Z}_2$ acts on $\mathbb{R}^n$ by changing the i-th coordinate to its negative.

Consider the vector space $\mathbb{R}^I$ consisting of functions from I to $\mathbb{R}$, with pointwise addition and scalar multiplication. It is useful to consider the coordinate arrangement in $\mathbb{R}^I$, where I is a finite set. It consists of the hyperplanes $x_a = 0$, with a varying over elements of I. It is gisomorphic to the coordinate arrangement of rank n, where $n = |I|$.

6.1.2. Small ranks. The linear and spherical models for $n = 1, 2, 3$ are shown below.

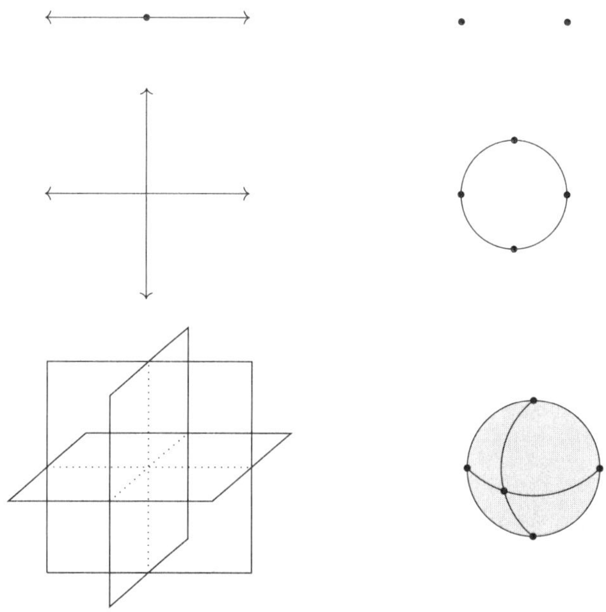

6.1.3. Faces and flats. Faces of $\mathcal{A}$ can be described as n-tuples in which each entry is either 0, $+$ or $-$. The rank of a face is the number of $+$ and $-$ in the n-tuple. Chambers are n-tuples in which each entry is either $+$ or $-$. A hyperplane separates two chambers if their n-tuples differ in the corresponding entry. The gallery distance between two chambers is the number of entries in which their n-tuples differ. The opposition map on faces interchanges $+$ and $-$. The Tits product on faces is given by (1.5), where $\epsilon_i(F)$ denotes the i-th entry in the tuple representating F. We have $F \lessdot G$ iff G is obtained from F by replacing exactly one 0 by either $+$ or $-$.

Flats of $\mathcal{A}$ can be described as n-tuples in which each entry is either 0 or $\pm$. The rank of a flat is the number of $\pm$ in the n-tuple. The support of a face is obtained by replacing each $+$ and each $-$ for $\pm$. There is a unique set of hyperplanes whose intersection is a given flat: they correspond to the 0 entries in the n-tuple. In other words, the maps (2.19) and (2.20) are inverse bijections. Thus, flats are the same as charts, and both may be identified with subsets of $[n]$. In these terms, the poset structure is given by reverse inclusion, and the support of a face is the subset of 0 entries in its n-tuple.

The lattice of flats is a Boolean poset. In particular, it is modular, that is, equality always holds in (1.20). Every flat is a factor, that is, every flat has a unique modular complement. In terms of subsets, modular complements are complementary subsets of $[n]$. The rank-one flats are the prime factors.

6.1.4. Cones and lunes. There is no distinction between cones, gallery intervals and lunes for the rank-one arrangement. Since these notions are compatible with cartesian product, the same is true for the coordinate arrangement. Cones (or lunes) of $\mathcal{A}$ can be described as n-tuples in which each entry is either 0, $+$, $-$ or $\pm$.

The case of a cone is obtained by replacing each $+$ and each $-$ for $\pm$. The base is obtained by replacing each $+$ and each $-$ for 0. The slack of a cone is the number of $+$ and $-$ in the n-tuple. Inclusion of cones corresponds to the coordinatewise partial order on n-tuples induced from the partial order on $\{0, +, -, \pm\}$ shown at the beginning of the section. There is a unique set of half-spaces whose intersection is a given cone. In other words, the maps (2.23) and (2.24) are inverse bijections. Thus, cones are also the same as dicharts.

Similarly, there is no distinction between top-stars, top-lunes, top-star-lunes and top-cones. They can be described as n-tuples in which each entry is either $+$, $-$, or $\pm$. Top-lunes with a base of rank i are the same as top-stars of faces of corank i, and consist of 2^i chambers. For instance, vertex-based top-lunes are the same as top-stars of panels, and consist of two adjacent chambers. The poset of top-stars or top-cones is dual to the poset of faces, with the maps (2.4) and (2.5) being inverse bijections.

Conjugate top-cones arise by taking cartesian products of conjugate top-cones in the rank-one arrangement. They can be completely classified as follows. Given faces F and G, the following are equivalent.

- The top-star of F and the top-star of G are conjugate.
- $F \wedge G = O$ and $F \vee G$ exists and is a chamber.
- F and G are complementary faces of a chamber.

In this situation, the top-star of G is the same as the top-lune $\mathrm{s}(F, C)$ where $C = F \vee G$. Thus, conjugates of the top-star of F are indexed by chambers greater than F. This is consistent with Lemma 3.43.

Exercise 6.1. Check that: For $H \leq G$, the lune $\mathrm{s}(H, G)$ defined by (3.8) has the following description.

- If 0 occurs in an entry of H, then $\mathrm{s}(H, G)$ equals G in that entry.
- If $+$ or $-$ occurs in an entry of H, then $\mathrm{s}(H, G)$ has $\pm$ in that entry.

Exercise 6.2. Recall the partial order (4.6) on lunes. Check that $\mathrm{L} \preceq \mathrm{M}$ iff the n-tuple of L is obtained from that of M by replacing some of the $\pm$ entries by either 0, $+$ or $-$.

6.1.5. Coxeter aspects. Face-types can be identified with subsets of $[n]$. All faces supported by a given flat have the same face-type. So diagram (5.3) specializes to

$$\begin{array}{ccc} \Sigma & \longrightarrow & \Sigma^W \\ \downarrow & & \| \\ \Pi & = & \Pi^W. \end{array}$$

It also follows that there is no distinction between nested face-types and lune-types.

Any given flat X is fixed by all Coxeter symmetries. In other words, $\widehat{W_{\mathrm{X}}} = W$ for all X. For any top-lune L with base X, we have $W = W_{\mathrm{X}} \times W_{\mathrm{L}}$. Note that this is a direct product.

6.1.6. Arrangements under and over a flat. Cartesian product. Recall that flats correspond to subsets of $[n]$. Let T be the subset corresponding to a flat X (indexing the 0 entries in the n-tuple) and let S be its complement. The arrangement $\mathcal{A}^{\mathrm{X}}$ is gisomorphic to the coordinate arrangement in $\mathbb{R}^S$, while the essentialization of $\mathcal{A}_{\mathrm{X}}$ is gisomorphic to the coordinate arrangement in $\mathbb{R}^T$. Similarly, the cartesian product of two coordinate arrangements is again a coordinate arrangement obtained by taking disjoint union of the two sets of coordinates.

Thus: The family of all coordinate arrangements is closed under passage to arrangements under and over a flat, and under cartesian products.

6.1.7. Substitution product of chambers. We describe the substitution product of chambers (4.18) in terms of n-tuples. Let X, S and T be as above. An element of $\Gamma[\mathcal{A}^{\mathrm{X}}]$ can be described as an n-tuple whose i-th entry is either $+$ or $-$ when $i \in S$, and 0 when $i \in T$. Similarly, an element of $\Gamma[\mathcal{A}_{\mathrm{X}}]$ can be described as an n-tuple whose i-th entry is either $+$ or $-$ when $i \in T$, and $\pm$ when $i \in S$. By combining the $+$ and $-$ in the two n-tuples, we obtain an n-tuple whose entries are either $+$ or $-$. This is the desired element of $\Gamma[\mathcal{A}]$. For example,

$$((0+-00-+), (-\pm\pm++\pm\pm)) \mapsto (-+-++-+).$$

Here $n = 7$ with $S = \{2, 3, 6, 7\}$ and $T = \{1, 4, 5\}$.

6.1.8. Category of lunes. Recall from Example 4.28 that the category of lunes for the rank-one arrangement is the category with two objects and two parallel arrows. It follows that the category of lunes for the coordinate arrangement of rank n is the n-fold cartesian power of this category.

The following describes composition of lunes in terms of n-tuples. Let L and M be two lunes. The following statements are equivalent.

- $\mathrm{b}(\mathrm{M}) = \mathrm{c}(\mathrm{L})$.
- 0 occurs in an entry of L iff $\pm$ does not occur in the corresponding entry of M.
- $\pm$ occurs in an entry of M iff 0 does not occur in the corresponding entry of L.

In this case, the lune $\mathrm{L} \circ \mathrm{M}$ is determined as follows.

- If 0 occurs in an entry of L, then $\mathrm{L} \circ \mathrm{M}$ equals M in that entry.
- If $\pm$ occurs in an entry of M, then $\mathrm{L} \circ \mathrm{M}$ equals L in that entry.

For example,

$$\mathrm{L} = (0, 0, 0, +, -, \pm) \quad \text{and} \quad \mathrm{M} = (0, +, -, \pm, \pm, \pm)$$

are composable lunes, and

$$\mathrm{L} \circ \mathrm{M} = (0, +, -, +, -, \pm).$$

Observe that when L has no $\pm$ entries and M has no 0 entries, we recover the substitution product of chambers discussed above.

Exercise 6.3. Check that:

(i) Given lunes $\mathrm{L} \leq \mathrm{N}$ with $\mathrm{b}(\mathrm{L}) = \mathrm{b}(\mathrm{N})$, there is a unique lune M such that $\mathrm{N} = \mathrm{L} \circ \mathrm{M}$.

(ii) Given lunes $\mathrm{N} \leq \mathrm{M}$ with $\mathrm{c}(\mathrm{N}) = \mathrm{c}(\mathrm{M})$, there is a unique lune L such that $\mathrm{N} = \mathrm{L} \circ \mathrm{M}$.

The first fact is true for any arrangement by Lemma 4.32, but the second one is not.

6.1.9. Möbius number and characteristic polynomial. For the coordinate arrangement $\mathcal{A}$ of rank n,

$$c(\mathcal{A}) = 2^n, \quad f(\mathcal{A}) = 3^n, \quad \mu(\mathcal{A}) = (-1)^n, \quad \chi(\mathcal{A}, t) = (t-1)^n.$$

This follows from (1.58) by taking n-th powers. These values can also be computed directly.

6.2. Rank-two arrangements

The arrangement of n lines in the plane, for $n \geq 2$, was introduced in Section 1.2.3. It has been used to illustrate many of the ideas and results in the preceding chapters. The cardinalities of the different geometric objects associated to an arrangement can be explicitly computed in this case. They are listed in Table 6.1. For objects such as face-types which are specific to reflection arrangements, we assume that the arrangement is dihedral, that is, the n lines are equally spaced.

TABLE 6.1. Enumeration in rank two.

Object	Cardinality
face	$4n+1$
chamber	$2n$
flat	$n+2$
cone	$2n^2+3n+2$
top-cone	$2n^2+1$
gallery interval	$2n^2+1$
top-nested face	$8n$
top-lune	$4n+1$
top-star	$4n+1$
top-star-lune	9 if $n=2$ and $6n+1$ if $n \geq 3$
nested face	$12n+1$
lune	$7n+2$
nested flat	$3n+3$
face-type	4
flat-type	3 if n is odd, and 4 if n is even
nested face-type	9
lune-type	7 if n is odd, and 9 if n is even

Note that:

- Top-stars and top-lunes have the same cardinality. Faces and top-stars have the same cardinality in any arrangement (Section 2.1.6).
- There is no distinction between top-cones and gallery intervals (Exercise 2.43).

- For n even, opposite vertices have the same type; as a result, there is no distinction between face-types and flat-types, and between nested face-type and lune-types.

We also recall from (1.59) that the Möbius number and characteristic polynomial are given by $\mu(\mathcal{A}) = n - 1$ and $\chi(\mathcal{A}, t) = t^2 - nt + n - 1$.

6.3. Braid arrangement. Compositions and partitions

The braid arrangement is a very important example of a reflection arrangement. It is a basic object for many considerations in algebra and combinatorics. The Coxeter group of the braid arrangement is the symmetric group. Many well-known combinatorial notions such as linear and partial orders, set and integer partitions, and so on correspond to geometric notions in the braid arrangement such as chambers, top-cones, flats, flat-types, and so on. In this section, we focus on compositions and partitions.

6.3.1. Braid arrangement. The *braid arrangement* on $[n]$ consists of the $\binom{n}{2}$ hyperplanes in $\mathbb{R}^n$ defined by

$$x_i = x_j$$

for $1 \leq i < j \leq n$. This is also called the *arrangement of type* A_{n-1}. It has rank $n-1$. It is not essential: The central face is one-dimensional and given by $x_1 = \cdots = x_n$. It is a reflection arrangement, the Coxeter group is the *symmetric group* on n letters, denoted S_n. An element of S_n, called a *permutation*, is a bijection from $[n]$ to itself. It acts by permuting the coordinates.

The canonical linear order of the set $[n]$ is not relevant to the definition of the arrangement. So it is also useful to proceed as follows. Let I be a finite set. The braid arrangement on I consists of the hyperplanes

$$x_a = x_b$$

in $\mathbb{R}^I$, as a and b vary over elements of I with $a \neq b$. Its Coxeter group is the group of all bijections from I to itself. We refer to elements of I as letters.

Recall that to any hyperplane arrangement, one can attach a variety of geometric notions such as faces, flats, and so on. For the braid arrangement, these geometric notions are equivalent to well-known combinatorial notions. The dictionary is given in Table 6.2. Some of these combinatorial objects are discussed below, the rest are discussed in Sections 6.4 and 6.5.

6.3.2. Set compositions and set partitions. Let I be a finite set. A *composition* of I is a finite sequence $(I_1, \dots, I_k)$ of disjoint nonempty subsets of I such that

$$I = \bigsqcup_{i=1}^{k} I_i.$$

The subsets I_i are the *blocks* of the composition. We write $F \vDash I$ to indicate that $F = (I_1, \dots, I_k)$ is a composition of I.

When the blocks are singletons, a composition of I amounts to a *linear order* on I. Two linear orders are *adjacent* if one is obtained from the other by switching two consecutive elements. A *minimal gallery* from one linear order to another is a minimal way to sort the first list by adjacent transpositions so as to obtain the second list.

TABLE 6.2. Geometry and combinatorics.

Geometry	Combinatorics
face	set composition
chamber	linear order
flat	set partition
cone	preorder
top-cone	partial order
gallery interval	partial order of order dimension 1 or 2
chart	simple graph
dichart	simple directed graph
top-nested face	set composition with a linear order on each block
top-lune	set partition with a linear order on each block, or parallel composition of linear orders
top-star	series composition of discrete partial orders
top-star-lune	series-parallel partial order
nested face	set composition with a composition of each block
lune	set partition with a composition of each block
face-type	integer composition
flat-type	integer partition
nested face-type	integer composition with a composition of each part
lune-type	integer partition with a composition of each part

Let F and G be compositions of I. We say G *refines* F if each block of F is a union of some contiguous set of blocks of G. In this case, we write $F \leq G$. This defines a partial order on the set of compositions of I. Maximal elements are linear orders. There is a unique minimum element, namely, the one-block composition of I.

A *partition* X of I is a collection X of disjoint nonempty subsets of I such that

$$I = \bigsqcup_{B \in \mathrm{X}} B.$$

The subsets B are the *blocks* of the partition. We write $\mathrm{X} \vdash I$ to indicate that X is a partition of I.

Let X and Y be partitions of I. We say that Y *refines* X if each block of X is a union of blocks of Y. In this case, we write $\mathrm{X} \leq \mathrm{Y}$. This defines a partial order on the set of partitions of I which is in fact a lattice. The top element is the partition into singletons and the bottom element is the partition whose only block is the whole set I.

6.3.3. Faces and flats. We now illustrate how faces correspond to set compositions, and flats to set partitions.

A face is defined by a system of equalities and inequalities which may be encoded by a composition of I: the equalities are used to define the blocks and the inequalities to order them. For example, for $I = \{a, b, c, d, e\}$,

$$x_a = x_c \leq x_b = x_d \leq x_e \qquad \longleftrightarrow \qquad ac|bd|e.$$

(The blocks have been separated by vertical bars and ordered from left to right. There is no order within each block.) Thus, faces correspond to compositions of the set I. In defining this correspondence, we have followed the convention that the values increase from left to right, that is, the value of the coordinates in the first block is smaller than the value of the coordinates in the second block, and so forth.

Under the above correspondence, chambers correspond to linear orders on I. For example, for $I = \{a, b, c, d, e\}$,

$$x_a \leq x_c \leq x_b \leq x_d \leq x_e \qquad \longleftrightarrow \qquad a|c|b|d|e.$$

A flat is defined by a system of equalities which may be encoded by a partition of I: the equalities are used to define the blocks. For example, for $I = \{a, b, c, d, e\}$,

$$x_a = x_c,\ x_b = x_d,\ x_e \qquad \longleftrightarrow \qquad \{ac, bd, e\}.$$

(The blocks have been separated by commas. There is no order within each block or among the blocks.) Thus, flats correspond to partitions of the set I.

6.3.4. Support map. The support map from faces to flats translates as follows. The *support* of a composition F of I is the partition $\mathrm{s}(F)$ of I obtained by forgetting the order among the blocks: if $F = (I_1, \ldots, I_k)$, then

$$\mathrm{s}(F) = \{I_1, \ldots, I_k\}.$$

6.3.5. Opposition map. The opposition map on faces translates as follows. The *opposite* of $F = (I_1, \ldots, I_k)$ is $\overline{F} = (I_k, \ldots, I_1)$. In other words, the opposition map reverses the order on the blocks of the set composition. This map restricts to chambers: the opposite of a linear order is a linear order.

6.3.6. Small ranks. The braid arrangement on $I = \{a\}$ is the rank-zero arrangement containing no hyperplanes.

The braid arrangement on $I = \{a, b\}$ consists of one hyperplane $x_a = x_b$. It is cisomorphic to the rank-one arrangement whose ambient space is one-dimensional. The latter is shown below on the left with the spherical model on the right.

$$a|b \longleftarrow\!\!-\!\!-\!\!-\; ab \;-\!\!-\!\!-\!\!\longrightarrow b|a \qquad\qquad \overset{a|b}{\bullet} \qquad \overset{b|a}{\bullet}$$

The central face corresponds to the one-block composition ab. It is not seen in the spherical model.

The braid arrangement on $I = \{a, b, c\}$ consists of the three hyperplanes $x_a = x_b$, $x_b = x_c$ and $x_a = x_c$. It is cisomorphic to the rank-two arrangement of three lines. The latter is shown below on the left with the spherical model on the right.

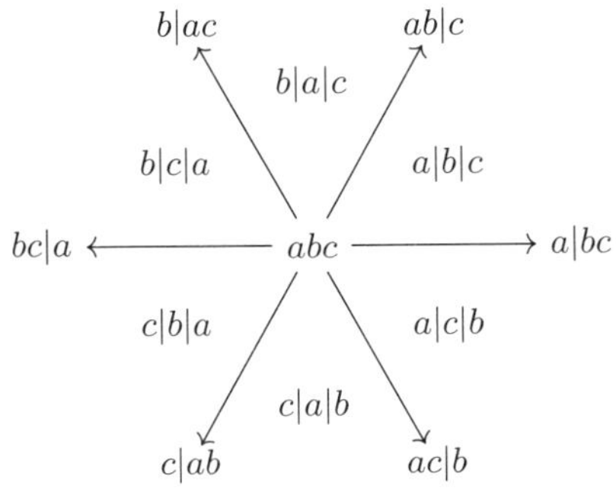

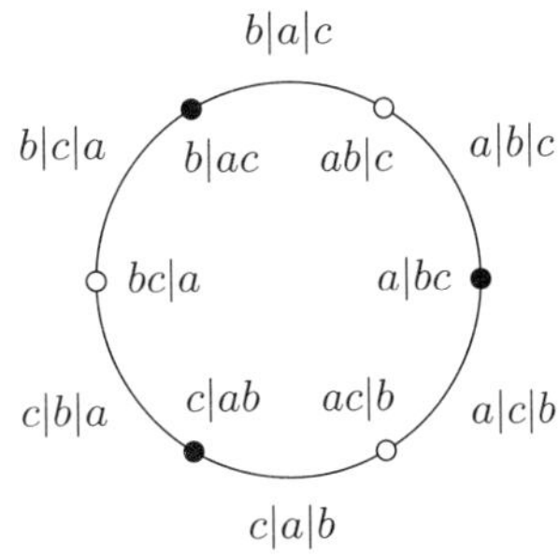

Faces are labeled by compositions of I. The central face corresponds to the one-block composition abc. It is not seen in the spherical model. There are two types of vertices shown in black and white, respectively.

The braid arrangement on $I = \{a, b, c, d\}$ consists of six hyperplanes. Its spherical model is shown below. The hyperplane $x_a = x_d$ is the outer circle, while $x_b = x_c$ is the horizontal line.

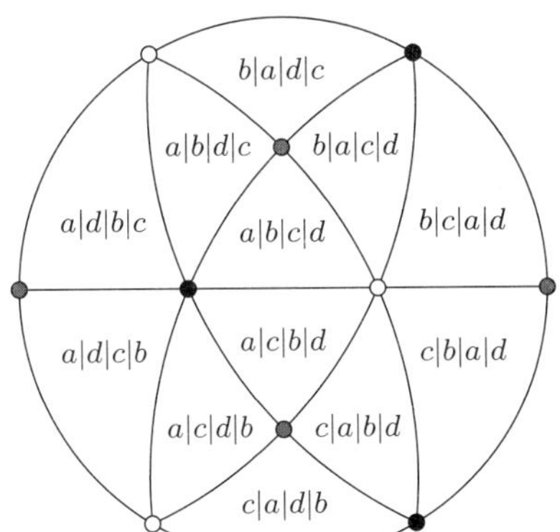

There are 24 triangles labeled by linear orders of which 12 are visible in the picture. The edges can be labeled by three-block compositions, and vertices by two-block compositions. There are three types of vertices shown in white, gray and black, respectively.

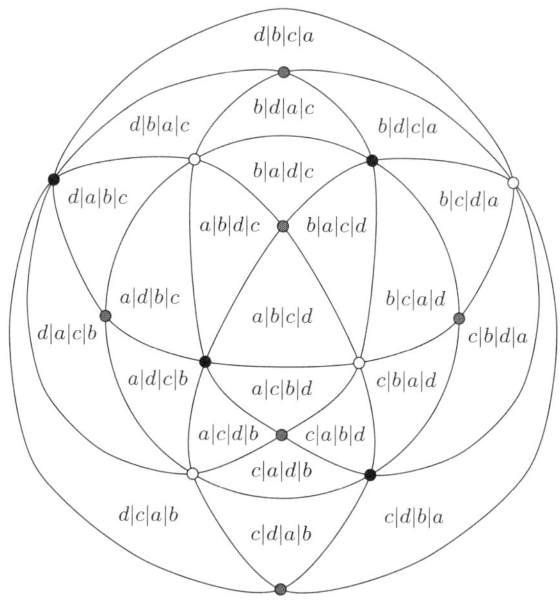

Here the spherical model has been flattened so that all triangles except $d|c|b|a$ are visible. The six hyperplanes can be seen in full as the six ovals.

The spherical model can also be visualized as a triangulated tetrahedron, with each of the four faces barycentrically subdivided into 6 triangles. This is illustrated below. In the picture, only two faces are visible. Each face has a white vertex in the center, gray vertices on its three edges, and black vertices as its vertices. The labels on some of the vertices have also been shown.

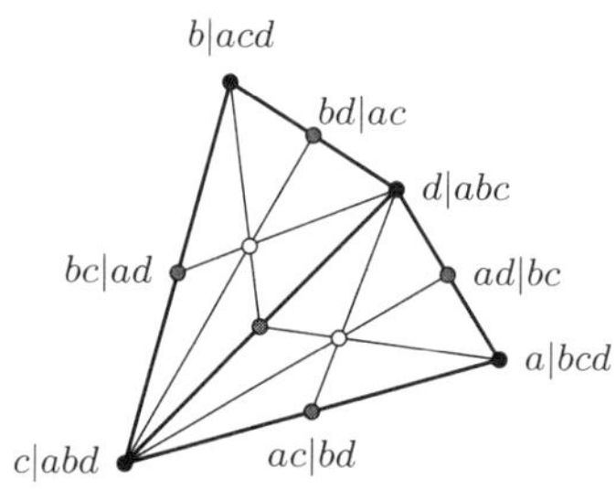

Another illustration is given below. Each of the two pictures shows two faces of the tetrahedron. To get the full tetrahedron, we identify the two pictures along their boundary, with one as the front side and the other as the back side.

Where are the six hyperplanes?

In another illustration shown below, three faces of the tetrahedron are fully visible, while one face is fully hidden. The three curved arcs on the boundary of the picture are the three sides of the hidden face. The visible faces have two straight sides and one curved side and share the black vertex in the center of the picture.

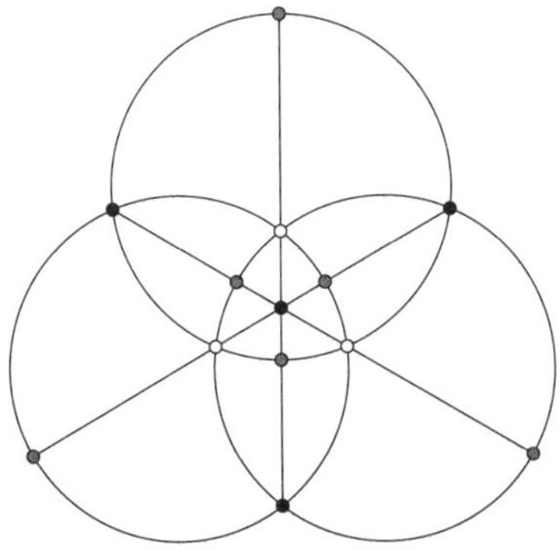

Three hyperplanes are seen in full as the three circles, while three are partly visible as the three straight lines.

6.3.7. Tits product. Let $F = (S_1, \ldots, S_p)$ and $G = (T_1, \ldots, T_q)$ be two compositions of I. Consider the pairwise intersections

$$A_{ij} := S_i \cap T_j$$

for $1 \leq i \leq p$, $1 \leq j \leq q$. A schematic picture is shown below.

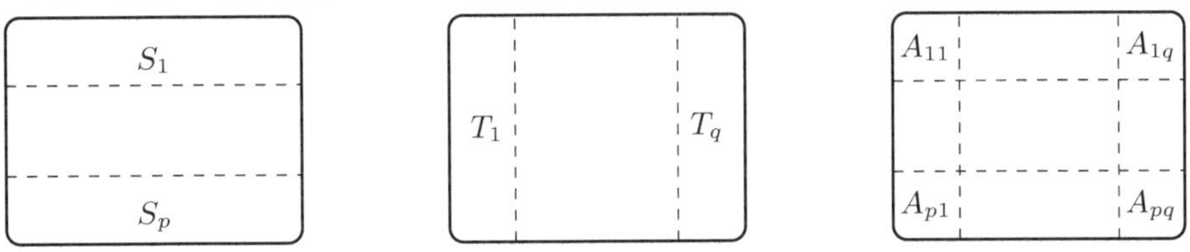

The *Tits product* FG is the composition obtained by listing the nonempty intersections A_{ij} in lexicographic order of the indices (i, j):

$$FG = (A_{11}, \ldots, A_{1q}, \ldots, A_{p1}, \ldots, A_{pq})^{\widehat{}}, \tag{6.1}$$

where the hat indicates that empty intersections are removed.

For example, to multiply $acde|bfg$ and $cdfg|b|ae$, we first compute the pairwise intersections.

$$\begin{bmatrix} acde \\ bfg \end{bmatrix} \qquad \begin{bmatrix} cdfg & b & ae \end{bmatrix} \qquad \begin{bmatrix} cd & \emptyset & ae \\ fg & b & \emptyset \end{bmatrix}$$

Now, from the last matrix, we read the nonempty entries in the first row followed by those in the second to obtain:

$$(acde|bfg)(cdfg|b|ae) = (cd|ae|fg|b).$$

There is a similar operation on set partitions. To multiply X and Y, intersect the blocks of X with the blocks of Y and remove empty intersections. This operation is commutative, and in fact agrees with the join X$\vee$Y, which is the smallest common refinement of X and Y.

Since the support map forgets the ordering on the blocks, it is easy to see that (1.11) holds.

6.3.8. Degeneracies in the Tits product of two vertices. Let us look at the Tits product of two vertices in detail. A vertex is a set composition with two blocks. Suppose $F = (S, T)$ and $G = (S', T')$ are vertices. Put

$$\begin{bmatrix} A & B \\ C & D \end{bmatrix} := \begin{bmatrix} S \cap S' & S \cap T' \\ T \cap S' & T \cap T' \end{bmatrix}.$$

(Note that $FG = (A, B, C, D)^\frown$ and $GF = (A, C, B, D)^\frown$.) Since S, T, S' and T' are nonempty, both entries in a row or column cannot be empty. The remaining possibilities are listed below.

Combinatorics	Geometry
All entries are nonempty	FG and GF are triangles
One diagonal entry is empty and the rest are nonempty	FG and GF are distinct edges
One off-diagonal entry is empty and the rest are nonempty	$FG = GF$ is an edge
Diagonal entries are empty and the rest are nonempty	$F = \overline{G}$
Off-diagonal entries are empty and the rest are nonempty	$F = G$

The first row is the generic case. The rest are degenerate cases of the generic case. Observe how the combinatorial and geometric degeneracies go hand-in-hand.

6.3.9. Integer compositions and integer partitions. Let n be a nonnegative integer. A *composition* $\alpha = (a_1, \ldots, a_k)$ of n is a finite sequence of positive integers such that

$$a_1 + a_2 + \cdots + a_k = n.$$

If the numbers a_i are allowed to be nonnegative, we say that α is a *weak composition* of n.

A *partition* $\lambda = (l_1, \ldots, l_k)$ of n is a finite sequence of positive integers such that

$$l_1 \geq l_2 \geq \cdots \geq l_k \quad \text{and} \quad l_1 + \cdots + l_k = n.$$

We write $\alpha \vDash n$ and $\lambda \vdash n$ to indicate that α is a composition of n, and λ is a partition of n. The numbers a_i and l_i are the *parts* of α and λ. The empty sequence is the only composition (and partition) of 0; it has no parts.

The *support* of a composition of n is the partition of n obtained by reordering the parts in decreasing order.

There is a partial order on the set of compositions given by refinement. We write $\alpha \leq \beta$ when β refines α. There is a similar partial order on the set of partitions. We say $\lambda \leq \mu$ if μ is obtained by refining each part in λ, and rearranging the parts in descending order.

6.3.10. Type map. The *type* of a composition F of I is the composition of $|I|$ whose parts are the sizes of the blocks of F. The *type* of a partition X of I is the partition of $|I|$ whose parts are the sizes of the blocks of X (listed in decreasing order).

The support and type maps commute with each other.

$$\begin{array}{ccc} \text{set composition} & \xmapsto{\ \mathrm{t}\ } & \text{composition} \\ \Big\downarrow{\scriptstyle \mathrm{s}} & & \Big\downarrow{\scriptstyle \mathrm{s}} \\ \text{set partition} & \xmapsto[\ \mathrm{t}\ ]{} & \text{partition} \end{array}$$

This is a specialization of (5.3).

6.3.11. Arrangements under and over a flat. Let $\mathcal{A}$ be any braid arrangement. The arrangement under a flat X of the braid arrangement $\mathcal{A}$ is again cisomorphic to a braid arrangement. More precisely, each block of X plays the role of one letter. For example, for $\mathrm{X} = \{cdf, ae, bg\}$, the arrangement $\mathcal{A}^{\mathrm{X}}$ is cisomorphic to the braid arrangement on the three letters cdf, ae and bg.

The arrangement over a flat X is cisomorphic to a cartesian product of braid arrangements. There is one braid arrangement for each block of X whose letters are the letters of that block. For example, for $\mathrm{X} = \{cdf, ae, bg\}$, the arrangement $\mathcal{A}_{\mathrm{X}}$ is cisomorphic to the cartesian product of the three braid arrangements on $\{c, d, f\}$, $\{a, e\}$ and $\{b, g\}$, respectively.

Exercise 6.4. Show that: Any braid arrangement $\mathcal{A}$ of rank at least one is prime. More generally, the arrangement $\mathcal{A}_{\mathrm{X}}$ is prime iff X has exactly one non-singleton block. (This provides examples for the second part of Exercise 1.53.)

Exercise 6.5. Characterize the set of faces F of the braid arrangement whose star Σ_F identifies with the set of faces of a coordinate arrangement.

6.3.12. Adjoint of the braid arrangement. For each subset S of $[n]$, let H_S denote the hyperplane in $\mathbb{R}^n$ with equation

$$\sum_{i \in S} x_i = 0.$$

The *all-subset arrangement* consists of the hyperplanes H_S, as S runs over all nonempty subsets of $[n]$. The *restricted all-subset arrangement* is the arrangement under the hyperplane $\mathrm{H}_{[n]}$ of the all-subset arrangement. There are $2^{n-1} - 1$ hyperplanes in this arrangement, one for each partition of $[n]$ into 2 blocks. The hyperplane corresponding to $\{S, T\}$ has equation

$$\sum_{i \in S} x_i = 0, \text{ or equivalently } \sum_{i \in T} x_i = 0.$$

This hyperplane is orthogonal to the line of the braid arrangement corresponding to $\{S, T\}$. Thus, the restricted all-subset arrangement is the adjoint of the braid arrangement.

6.3.13. The external product. In addition to the Tits product, there is another operation among faces of the braid arrangement. This is of a different nature: it combines faces from smaller arrangements to produce a face of a bigger one. It is combinatorially simpler.

Given a composition $F = (S_1, \dots, S_p)$ of I and a sequence of compositions H_i of S_i, $i = 1, \dots, p$, we set

$$\mu_F(H_1, \dots, H_p) = H, \tag{6.2}$$

where H is the ordered concatenation of the H_i. We refer to H as the *external product* of the H_i along F. It is a composition of I, finer than F. For example,

$$\mu_{abc|de}(ac|b, e|d) = (ac|b|e|d).$$

The external product restricts to linear orders: If each H_i is a linear order, then so is H, for any composition F.

The external product is unital: If F is the one-block composition, then μ_F is the identity. It is also associative, in the following sense. Let $G = (T_1, \dots, T_q)$ be a composition of I refining F. Let $(G/F)_i$ denote the composition of S_i consisting of those contiguous blocks of G which refine S_i. Similarly, given compositions H_j of T_j, $j = 1, \dots, q$, let $\underline{H}_i$ denote the subsequence of $H_1, \dots, H_q$ consisting of the H_j for which $T_j \subseteq S_i$. Thus $\underline{H}_i$ is a sequence of compositions of the blocks of $(G/F)_i$. Then we have

$$\mu_F(\mu_{(G/F)_1}(\underline{H}_1), \dots, \mu_{(G/F)_p}(\underline{H}_p)) = \mu_G(H_1, \dots, H_q).$$

The external product can also be defined for set partitions, integer compositions, and integer partitions, along the same lines.

6.4. Braid arrangement. Partial orders and graphs

We now focus on cones and charts for the braid arrangement. A key observation is that top-cones correspond to partial orders. Hence, results on top-cones from Chapter 2 specialize to interesting results on partial orders. We also mention that charts correspond to simple graphs, and dicharts to simple digraphs.

6.4.1. Preorders and partial orders. A *preorder* is a relation on a set which is reflexive and transitive. Let $x \sim y$ if $x \leq y$ and $y \leq x$. This is an equivalence relation whose classes are called the *blocks* of the preorder. A *preposet* is a set equipped with a preorder.

We say x and y are comparable if either $x \leq y$ or $y \leq x$. Let $x \sim y$ if there is a sequence of elements starting with x and ending at y such that adjacent elements in the sequence are comparable. This is an equivalence relation whose classes are called the *connected components* of the preorder.

A *partial order* is a preorder that is in addition anti-symmetric. A *poset* is a set equipped with a partial order.

A partial order is the same as a preorder whose blocks are singletons. Similarly, a preorder is the same as an equivalence relation together with a partial order on the set of blocks.

TABLE 6.3. Top-cones and partial orders.

Top-cone V	Partial order P
Half-spaces which contain V	Relations in P
Walls of V	Cover relations in P
Hyperplanes which cut V	Unordered pairs of incomparable elements of P
Chambers contained in V	Linear extensions of P
Top-cone opposite to V	Partial order opposite to P

6.4.2. Cones and top-cones. We turn to the correspondence between top-cones in the braid arrangement on I and partial orders on the set I stated in Table 6.2. Let V be a top-cone and P the corresponding partial order. The correspondence between geometric notions attached to V and combinatorial notions attached to P is given in Table 6.3. We elaborate on the first entry. For distinct $i, j \in I$, $i < j$ in P iff the half-space $x_i \leq x_j$ contains V.

Cones correspond to preorders. Recall that a cone is a top-cone in an arrangement under a flat. This ties to the fact that a preorder is a partial order on its blocks.

The fact that faces and flats are particular cones adopts the following form. A set composition is the same as a linear preorder: a preorder for which the partial order on the blocks is linear (the first block is smaller than the second block and so on). An equivalence relation is the same as a preorder for which the partial order on the blocks is discrete, and this amounts to a set partition.

Exercise 6.6. Notions of boundary, interior and closure for top-cones translate to partial orders. Check that: A two-block composition $S|T$ belongs to the closure of a partial order P iff the following condition holds.

$$\text{If } x < y \text{ in } P\text{, then either } x \text{ and } y \text{ both belong to } S\text{, or both belong to } T\text{,}\\ \text{or } x \text{ belongs to } S \text{ and } y \text{ belongs to } T.$$

Equivalently, no element of T is less than an element of S. (In this case, one says S is a lower set of P and T is an upper set of P.)

When does $S|T$ lie in the boundary or interior of P?

More generally, when does a set composition belong to the closure, boundary or interior of P?

6.4.3. Poset of partial orders. The partial order on top-cones translates to partial orders as follows. Fix a finite set I. For partial orders P and Q on I, we have

$$P \leq Q \iff x < y \text{ in } Q \text{ implies } x < y \text{ in } P \text{ for all } x, y \in I.$$

In other words, $P \leq Q$ iff Q is obtained from P by deleting some relations in P. This is the poset $\widehat{\Omega}$ of all partial orders on I. The discrete partial order on I is the maximum element, while linear orders on I are the minimal elements of $\widehat{\Omega}$.

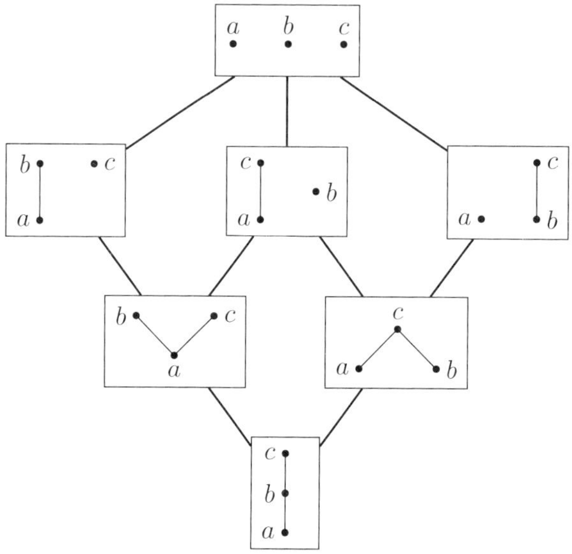

The picture shows the interval from a linear order to the discrete partial order on the set $I = \{a, b, c\}$.

Going back to the general case, $P \lessdot Q$ iff Q is obtained from P by deleting a cover relation of P. One can deduce that $\widehat{\Omega}$ is graded, with the rank of P equal to the number of unordered pairs of incomparable elements in P. In particular, the rank of $\widehat{\Omega}$ is equal to $\binom{n}{2}$, where n is the cardinality of I. This is a special case of Theorem 2.56. More generally, $\widehat{\Omega}$ is join-distributive, and in particular, every interval is upper semimodular. These are special cases of Theorems 2.58 and 2.60. The interval $[P, Q]$ is Boolean iff the relations in P and not in Q are cover relations in P. This is a special case of Proposition 2.59. The Möbius function of the poset of partial orders is

$$\mu(P, Q) = \begin{cases} \pm 1 & \text{if deleting some cover relations in } P \text{ yields } Q, \\ 0 & \text{otherwise.} \end{cases} \tag{6.3}$$

The sign is $+1$ if an even number of cover relations are deleted, and -1 if an odd number of cover relations are deleted. The value is 0 when P contains a noncover relation that is not present in Q. This is a special case of Corollary 2.64.

Exercise 6.7. Let P and Q be partial orders on I. If x and y are elements of I such that $x < y$ in P and $y < x$ in Q, then P and Q cannot have a common linear extension. Show that the converse is false. That is, find partial orders P and Q which do not have a common linear extension, and in which elements x and y of the above kind do not exist. This will also be an example for Exercise 2.31.

Exercise 6.8. The partial order on cones translates to preorders. Explicitly, for preorders P and Q on I, we have

$$P \leq Q \iff (x, y) \in Q \text{ implies } (x, y) \in P \text{ for all } x, y \in I.$$

Show that the poset of all preorders on a set of three elements is *not* graded. This is the case $n = 3$ of Exercise 2.57.

6.4.4. Case and base maps. The case and base maps from cones to flats translate as follows. The *case* of a preorder is the set partition whose blocks are the blocks of the preorder. The *base* of a preorder is the set partition whose blocks are the connected components of the preorder. In particular, the base of a preorder is the one-block partition iff the preorder is connected. Thus, a connected preorder corresponds to a salient cone (in the essentialization of the arrangement).

6.4.5. Order dimension. Convexity dimension of a top-cone translates to order dimension of a partial order. The *order dimension* of a partial order is the minimum number of linear orders whose join is the given partial order. In particular, a gallery interval is precisely a partial order of order dimension either 1 or 2. (A partial order of order dimension 1 is precisely a linear order.) Let us make this explicit.

Given two linear orders ℓ_1 and ℓ_2, one can associate a partial order by defining $x < y$ iff $x < y$ in both ℓ_1 and ℓ_2. Indeed, this partial order is the gallery interval $[\ell_1 : \ell_2]$. Thus, a partial order P is a gallery interval iff there exist linear extensions ℓ_1 and ℓ_2 with the following property.

$$x < y \text{ in both } \ell_1 \text{ and } \ell_2 \implies x < y \text{ in } P.$$

Equivalently:

For any $x, y \in P$ which are not related,
either $x < y$ in ℓ_1 and $y < x$ in ℓ_2, or vice versa.

All partial orders on 5 or fewer elements are of order dimension 1 or 2. On 6 elements, there are, up to isomorphism, exactly 3 partial orders of order dimension 3, as shown below. The remaining connected partial orders on 6 elements fall into one class of order dimension 1 (the linear orders) and 234 classes of order dimension 2.

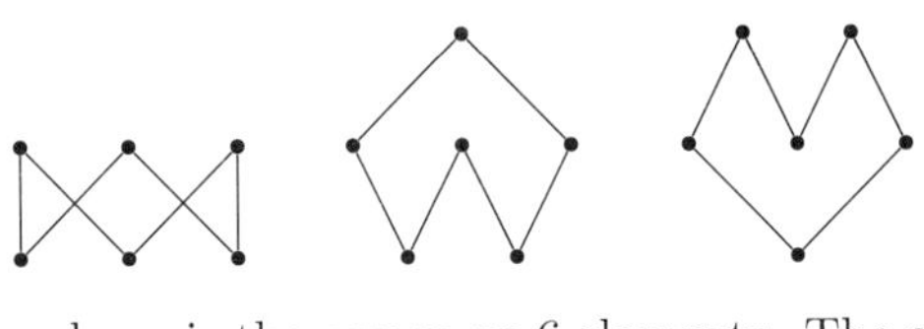

The first partial order above is the *crown* on 6 elements. The crown on $2n$ elements is of order dimension n.

6.4.6. Graphs and digraphs. We turn to the combinatorial side of charts and dicharts.

A *simple graph* on a finite set I is a subset g of $\binom{I}{2}$. An element $\{i, j\}$ is an *edge* between the *vertices* i and j of the graph g. By letting the edge $\{i, j\}$ represent the hyperplane $x_i = x_j$, simple graphs on I are in correspondence with charts in the braid arrangement on I.

The complete graph on I corresponds to the chart consisting of all hyperplanes, while the discrete graph corresponds to the chart with no hyperplanes. Connected graphs correspond to connected charts, and trees to coordinate charts.

A *simple directed graph* (simple digraph) on I is a subset r of $I \times I \setminus \{(a, a) \mid a \in I\}$. An element (i, j) of r is an *arrow* of the digraph from the vertex i to the vertex j. Simple digraphs on I are in correspondence with dicharts in the braid arrangement on I, with an arrow (i, j) representing the half-space $x_i \leq x_j$.

Recall the maps in (2.18):

$$\Pi[\mathcal{A}] \underset{\rho}{\overset{\lambda}{\rightleftarrows}} \mathrm{G}[\mathcal{A}] \underset{\rho'}{\overset{\rho''}{\leftleftarrows \!\!\xrightarrow{\lambda'}}} \vec{\mathrm{G}}[\mathcal{A}] \underset{\vec{\rho}}{\overset{\vec{\lambda}}{\leftrightarrows}} \Omega[\mathcal{A}].$$

They admit the following combinatorial descriptions. Let X be a partition of I, g a simple graph on I, r a simple digraph on I, and P a preorder on I. The simple graph $\lambda(\mathrm{X})$ is the disjoint union of the complete graphs on the blocks of X. The blocks of the partition $\rho(g)$ are the connected components of g. In the digraph $\lambda'(g)$ there are two arrows (i, j) and (j, i) for each edge $\{i, j\}$ in g. In the graph $\rho'(r)$ there is an edge $\{i, j\}$ if either arrow (i, j) or (j, i) is present in r. In the graph $\rho''(r)$ there is an edge $\{i, j\}$ only if both arrows (i, j) and (j, i) are present in r. For distinct i and j in I, we have (i, j) in the preorder $\vec{\rho}(r)$ if there is a directed path from i to j in r, and we have an arrow from i to j in the digraph $\vec{\lambda}(P)$ if $(i, j) \in P$.

Simple digraphs are the same as reflexive relations on the set I. Via λ', simple graphs correspond to reflexive symmetric relations, and partitions correspond to equivalence relations via $\lambda'\lambda$. The relation $\lambda'\rho'(r)$ is the symmetric closure of the relation r. The relation $\lambda\rho(g)$ is the transitive closure of the symmetric relation g. The relation $\vec{\lambda}\vec{\rho}(r)$ is the transitive closure of the relation r.

6.5. Braid arrangement. Linear compositions, partitions and shuffles

We now focus on nested faces and lunes for the braid arrangement. Top-nested faces and top-lunes correspond to linear compositions and linear partitions. There are similar descriptions for nested faces and lunes, as well as for nested face-types and lune-types. Combinatorial top-lunes relate to shuffles and quasi-shuffles. Top-star-lunes correspond to series-parallel partial orders (constructed from the operations of series and parallel composition), while conjugate top-cones corrrespond to conjugate partial orders. The substitution product of chambers specializes to the classical associative operad.

6.5.1. Series and parallel composition. Suppose I and J are disjoint sets, and P and Q are partial orders on I and J, respectively. There are two natural partial orders on $I \sqcup J$ as follows. The *disjoint union* or *parallel composition* of P and Q combines the relations in P and Q, with no additional relations. The *series composition* of P and Q combines the relations in P and Q, and in addition, imposes that every element of P be less than every element of Q.

6.5.2. Linear compositions and linear partitions. A *linear composition* of a set I is a composition in which each block is equipped with a linear order. A linear composition yields a linear order on I (the series composition of the linear orders on its blocks), but it cannot be recovered solely from it.

Similarly, a *linear partition* of I is a partition in which each block is equipped with a linear order. Equivalently, it is a partial order on I each of whose connected components is a linear order. This is the same as a partial order which can be obtained as a parallel composition of linear orders. Note that the base of a linear partition is the underlying set partition (obtained by forgetting the orders within each block).

6.5.3. Nested faces and lunes. Top-nested faces correspond to linear compositions: Given a top-nested face (H, D), the H specifies a set composition and the D specifies a linear order on each block of H. Using this along with (3.7) and (6.1), one can deduce that top-lunes correspond to linear partitions.

More generally, a nested face corresponds to a set composition in which each block is given the structure of a set composition. Similarly, a lune corresponds to a set partition is which each block is given the structure of a set composition.

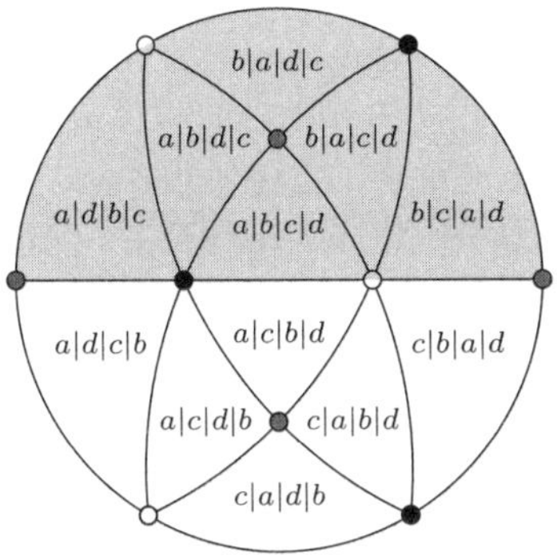

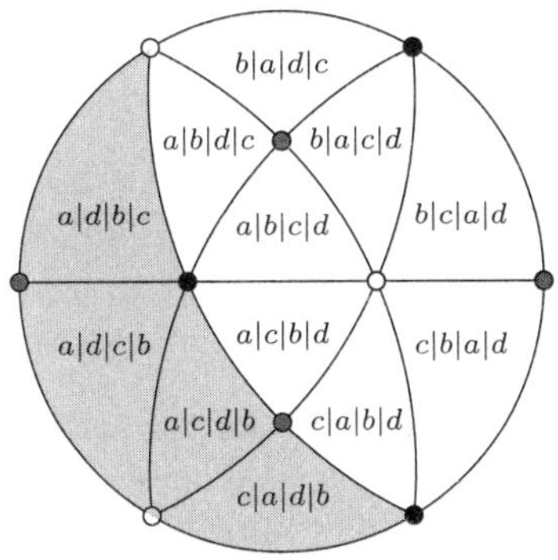

The top-lune on the left corresponds to the linear partition $\{a|d, b|c\}$, while the one on the right corresponds to $\{a|d|b, c\}$.

Exercise 6.9. Recall from Theorem 3.27 that every cone has an optimal decomposition into lunes. Check that: The optimal decomposition of a partial order P is the union of those linear partitions whose connected components are the same as those of P and within each component, we have a linear extension. Generalize to preorders.

6.5.4. Shuffles. Let $I = S \sqcup T$, ℓ and m linear orders on S and T, respectively. Let $\ell \cdot m$ be the concatenation of ℓ and m, a linear order on I. A *shuffle* of ℓ and m is any linear order on I whose restrictions to S and T are ℓ and m.

The shuffles of ℓ and m are precisely those linear orders C which satisfy $HC = D$, where $H = (S, T)$ and $D = \ell \cdot m$. This follows from (6.1). One may equally well choose $H = (T, S)$ and $D = m \cdot \ell$.

The linear orders ℓ and m give rise to two linear compositions, namely, (ℓ, m) and (m, ℓ). They correspond to the two choices of (H, D) given above. The support of either of them is the linear partition $\{\ell, m\}$. Its base is the two-block partition consisting of S and T. Viewing the linear partition as a partial order (parallel composition of ℓ and m), the shuffles of ℓ and m are precisely its linear extensions.

Thus, $\{\ell, m\}$ corresponds to a vertex-based top-lune, with shuffles corresponding to chambers.

Take for instance $I = \{a, b, c, d\}$. The shuffles of c and $a|d|b$ are

$$c|a|d|b, \; a|c|d|b, \; a|d|c|b, \; \text{and } a|d|b|c.$$

These are the linear extensions of the linear partition $\{c, a|d|b\}$ (viewed as a partial order). Similarly, the shuffles of $a|d$ and $b|c$ are

$$a|d|b|c, \; a|b|d|c, \; a|b|c|d, \; b|a|d|c, \; b|a|c|d, \; \text{and } b|c|a|d.$$

These are the linear extensions of $\{a|d, b|c\}$. See the pictures in Section 6.5.3.

This discussion generalizes to shuffles of a finite number of linear orders and to shuffles of set compositions. The latter correspond to solutions F of an equation $HF = G$ with $\mathrm{s}(F) = \mathrm{s}(G)$.

6.5.5. Quasi-shuffles. Let $I = S \sqcup T$, ℓ and m be as in Section 6.5.4. A *quasi-shuffle* of ℓ and m is any composition of I whose restrictions to S and T are ℓ and m, respectively. Every shuffle is a particular quasi-shuffle.

The quasi-shuffles of ℓ and m are precisely those compositions F which satisfy $HF = D$, where $H = (S, T)$ and $D = \ell \cdot m$. Other than the shuffles, the quasi-shuffles of $a|d$ and $b|c$ are

$$a|db|c,\ ab|d|c,\ ab|c|d,\ a|b|cd,\ b|a|cd,\ b|ac|d,\ \text{and } ab|cd.$$

The discussion generalizes to quasi-shuffles of a finite number of linear orders and to quasi-shuffles of set compositions. The latter correspond to solutions F of an equation $HF = G$.

6.5.6. Series composition of discrete partial orders. The notion of top-star of a face translates as follows.

A partial order is the top-star of the two-block composition $S|T$ iff elements of S are minimal, elements of T are maximal, and each element of S is less than each element of T.

More generally, a partial order is the top-star of a set composition F iff elements within any given block of F are incomparable, and all elements in a given block of F are less than all elements in a subsequent block of F. This is precisely a partial order which can be obtained as the series composition of discrete partial orders.

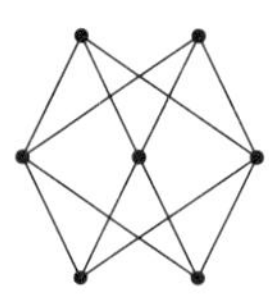

top-star

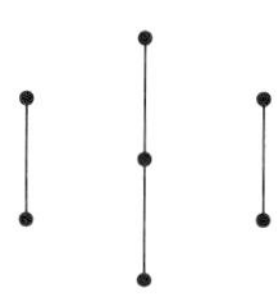

top-lune

The picture on the left shows a top-star obtained as series composition of three discrete partial orders, while the one on the right shows a top-lune obtained as parallel composition of three linear orders.

Exercise 6.10. Let P be a series composition of discrete partial orders. Check that: For any linear extension ℓ of P, there exists a unique linear extension ℓ' of P such that P is the gallery interval $[\ell : \ell']$. This is a specialization of Exercise 2.45.

Exercise 6.11. Let P be a parallel composition of linear orders. (It corresponds to a linear partition.) Check that: For any series composition ℓ of the given linear orders, P is of the form $[\ell : \ell']$ for a unique ℓ'. In fact, these are all the ways to realize P as a gallery interval. (In other words, the possibilities listed in Proposition 3.18 are the only ones for the braid arrangement.)

6.5.7. Series-parallel partial orders. A partial order is *series-parallel* if it can be obtained by a sequence of series compositions and parallel compositions starting with the partial order with one element.

The notion of top-star-lunes translates to the notion of series-parallel partial orders. Both definitions are inductive in nature; the series alternative in the former corresponds to series composition in the latter, while parallel alternative in the former corresponds to parallel composition in the latter. (Here we use that the arrangement over a flat of a braid arrangement is a cartesian product of smaller braid arrangements, and top-star-lunes are compatible with cartesian products. See Section 6.3.11.)

Series-parallel partial orders have order dimension either 1 or 2. This corresponds to the fact that top-star-lunes are gallery intervals (Lemma 3.42). It can also be directly checked by showing that partial orders of order dimension either 1 or 2 are preserved under series and parallel compositions.

6.5.8. Conjugate partial orders. The notion of conjugate top-cones translates as follows. Two partial orders on I are *conjugate* if any distinct x and y in I are incomparable in exactly one of the two partial orders. Conjugate partial orders have a common linear extension (obtained by combining the relations in the two). In other words, the braid arrangement satisfies the conjugate-meet property. Hence, by Corollary 3.47, we obtain: a partial order has order dimension either 1 or 2 iff it has a conjugate.

6.5.9. Nested face-types and lune-types. Compositions, partitions and the type map were reviewed in Sections 6.3.9 and 6.3.10. Nested faces, lunes, and their top-versions were described combinatorially in Section 6.5.2. Now let n be a positive integer. A nested face-type of n consists of a composition of n together with a composition of each of its parts. A lune-type of n consists of a partition of n together with a composition of each of its parts.

The type of a nested face is a nested face-type, and the type of a lune is a lune-type. They are obtained by replacing the involved sets by their cardinalities. One can explicitly see that diagrams (5.4) and (5.5) commute.

6.5.10. Classical associative operad. Let $\mathcal{A}$ be the braid arrangement on the set I. Arrangements under and over a flat of this arrangement are explained in Section 6.3.11. For a partition X of I, the substitution product of chambers (4.18) works as follows.

An element of $\Gamma[\mathcal{A}^{\mathrm{X}}]$ is a linear order on the blocks of X, while an element of $\Gamma[\mathcal{A}_{\mathrm{X}}]$ is a family of linear orders, one on each block of X. (The latter is the same as a linear partition whose base is X.) The two together specify a linear order on I, which is an element of $\Gamma[\mathcal{A}]$. For example, for the partition $\mathrm{X} = \{abd, fg, ce\}$,

$$((fg|ce|abd), \{a|d|b, g|f, c|e\}) \mapsto (g|f|c|e|a|d|b).$$

The substitution product in rank two is illustrated below.

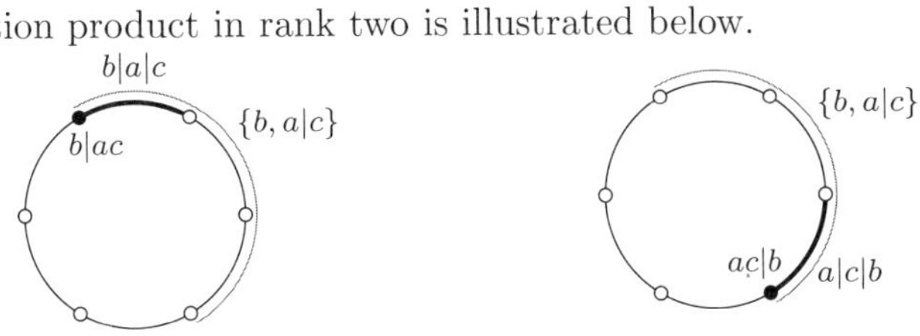

In the picture on the left, the substitution product of the vertex $b|ac$ (shown in black) with the top-lune $\{b, a|c\}$ (shown in gray) is the chamber $b|a|c$ (shown as the thick edge). Similarly, in the picture on the right, the substitution product of the vertex $ac|b$ with the top-lune $\{b, a|c\}$ is the chamber $a|c|b$.

Linear orders on finite sets with these structure maps constitute the *classical associative operad*.

TABLE 6.4. Enumeration in the braid arrangement.

Object	Sequence	OEIS
set composition	$1, 3, 13, 75, 541$	A000670
linear order	$1, 2, 6, 24, 120$	A000142
set partition	$1, 2, 5, 15, 52$	A000110
preorder	$1, 4, 29, 355, 6942$	A000798
partial order	$1, 3, 19, 219, 4231$	A001035
set composition with a linear order on each block	$1, 4, 24, 192, 1920$	A002866
set partition with a linear order on each block	$1, 3, 13, 73, 501$	A000262
series-parallel partial order	$1, 3, 19, 195, 2791$	A048172
set composition with a composition of each block	$1, 5, 37, 365, 4501$	A050351
set partition with a composition of each block	$1, 4, 23, 173, 1602$	A075729

6.6. Enumeration in the braid arrangement

We now discuss some enumerative features of the braid arrangement. This features in particular inversion sets, factorials, multinomial coefficients and formal power series.

6.6.1. Combinatorial objects. In Table 6.4, we list the first five terms in the sequences enumerating several combinatorial objects associated to the braid arrangement on $[n]$, starting at $n = 1$. The references are to the Online Encyclopedia of Integer Sequences [**384**]. The corresponding geometric objects are given in Table 6.2.

6.6.2. Gallery distance. Gallery distance is discussed in Section 1.10.3. It is related to inversion sets as follows.

Let C be a chamber in the braid arrangement on I. Write $C = C^1|\cdots|C^n$, where $n = |I|$. Define the inversion set of (C, D) to be

$$\mathrm{Inv}(C, D) := \{(i, j) \in [n] \times [n] \mid i < j \text{ and } C^i \text{ appears after } C^j \text{ in } D\}.$$

The gallery distance between C and D is then given by

$$\mathrm{dist}(C, D) = |\mathrm{Inv}(C, D)|. \tag{6.4}$$

Let F and G be faces with the same support. Write $F = F^1|\cdots|F^k$. Then G is a set composition obtained by permuting the F^i in some order. Define the inversion set of (F, G) to be

$$\mathrm{Inv}(F, G) := \{(i, j) \in [k] \times [k] \mid i < j \text{ and } F^i \text{ appears after } F^j \text{ in } G\}.$$

Then the distance between F and G is

$$\operatorname{dist}(F,G) = \sum_{(i,j)\in \operatorname{Inv}(F,G)} |F^i|\,|F^j|. \tag{6.5}$$

Consider now the general case, in which F and G are arbitrary faces. Here, we have

$$\operatorname{dist}(F,G) = \sum_{\substack{i<k\\ j>l}} |F^i \cap G^j|\,|F^k \cap G^l|, \tag{6.6}$$

where i and k index the blocks of F while j and l index the blocks of G.

6.6.3. Degrees and factorials. Recall that for set compositions $F \leq G$, the set composition $(G/F)_i$ consists of those contiguous blocks of G which refine the i-th block of F.

For any set composition G, let $\deg(G)$ denote the number of blocks of G. More generally, for $F \leq G$, let

$$\deg(G/F) = \prod_i \deg(G/F)_i. \tag{6.7}$$

In particular, $\deg(G/O) = \deg(G)$.

For F, a set composition consisting of two blocks, $\deg(G/F)$ is the product of two numbers, one for each block of F, as in the following example.

$$F = krish|na, \quad G = kr|i|sh|n|a, \quad \deg(G/F) = 3.2 = 6.$$

Here $kr|i|sh$ which refines $krish$ has 3 blocks, while $n|a$ which refines na has 2 blocks.

For any set composition G, let $\deg!(G)$ denote the factorial of the number of blocks of G. More generally, for $F \leq G$, let

$$\deg!(G/F) = \prod_i \deg!(G/F)_i. \tag{6.8}$$

In particular, $\deg!(G/O) = \deg!(G)$.

For example,

$$F = krish|na, \quad G = kr|i|sh|n|a, \quad \deg!(G/F) = 3!2! = 12.$$

In a similar manner, for integer compositions $\alpha \leq \beta$, one can define $\deg(\beta/\alpha)$ and $\deg!(\beta/\alpha)$.

Lemma 6.12. *For any integer m, and any set partition* X,

$$\sum_{F:\, \mathrm{s}(F)\leq \mathrm{X}} \binom{m}{\deg(F)} = m^{\deg(\mathrm{X})},$$

where $\deg(\mathrm{X})$ *denotes the number of blocks of* X.

PROOF. First assume m is positive. Take m boxes labeled 1 to m. The rhs counts the number of ways of putting each block of X in one of the m boxes. Each such assignment yields a set composition F with $\mathrm{s}(F) \leq \mathrm{X}$: each box is a block of F with empty boxes deleted. The number of assignments which yield the same F is precisely $\binom{m}{\deg(F)}$. This is the lhs. This proves the identity for m positive.

For the general case, note that both sides are polynomials in m. Since they agree for infinitely many values of m, they must agree for all values of m. □

6.6.4. Möbius number and characteristic polynomial. Let $\mathcal{A}$ denote the braid arrangement on $[n]$. It has rank $n-1$. Note that

$$\operatorname{rk}(F) = \deg(F) - 1 \quad \text{and} \quad \operatorname{rk}(G/F) = \deg(G) - \deg(F).$$

The number of chambers $c(\mathcal{A})$ is $n!$ and the number of faces $f(\mathcal{A})$ is the number of compositions of $[n]$. More generally,

$$c(\mathcal{A}^G) = \deg!(G) \quad \text{and} \quad c(\mathcal{A}_F^G) = \deg!(G/F).$$

The first one is the number of set compositions with the same support as G, while the second is the number of set compositions which are greater than F and have the same support as G. When F is the one-block composition, the second formula reduces to the first.

The Möbius number is given by

$$\mu(\mathcal{A}) = (-1)^{n-1}(n-1)!. \tag{6.9}$$

More generally, for $F \leq G$, we have

$$\mu(\mathcal{A}_F^G) = (-1)^{\operatorname{rk}(G/F)} \frac{\deg!(G/F)}{\deg(G/F)}. \tag{6.10}$$

This is because the arrangement $\mathcal{A}_F^G$ is cisomorphic to a cartesian product of braid arrangements, so the Möbius numbers multiply. It is also useful to note that

$$\frac{\mu(\mathcal{A}_F^G)}{c(\mathcal{A}_F^G)} = \frac{(-1)^{\operatorname{rk}(G/F)}}{\deg(G/F)}. \tag{6.11}$$

The characteristic polynomial of the braid arrangement is given by

$$\chi(\mathcal{A}, t) = \prod_{i=1}^{n-1} (t-i). \tag{6.12}$$

Putting $t = -1$, we see that the rhs up to sign equals the number of chambers. This is the Zaslavsky formula in the equivalent form (1.54c).

The *falling factorial* is the polynomial $t^{\underline{n}} = t(t-1)\cdots(t-n+1)$. By (6.12),

$$\chi(\mathcal{A}, t) = \frac{t^{\underline{n}}}{t}.$$

For $n \geq k$, let $s(n,k)$ to be the coefficient of t^k in $t^{\underline{n}}$. These are the *Stirling numbers of the first kind*. By Lemma 1.87, we have

$$\operatorname{wy}(\mathcal{A}, k) = s(n, k+1).$$

More generally, for any flat X, we have

$$\operatorname{wy}(\mathcal{A}^{\mathrm{X}}, k) = s(\deg(\mathrm{X}), k+1), \tag{6.13}$$

since $\mathcal{A}^{\mathrm{X}}$ is cisomorphic to the braid arrangement on $[d]$ where $d = \deg(\mathrm{X})$.

Exercise 6.13. Recall the adjoint of the braid arrangement from Section 6.3.12. Check that the characteristic polynomial of the adjoint of the rank-three braid arrangement is given by

$$\chi(\widehat{\mathcal{A}}, t) = -9 + 15t - 7t^2 + t^3 = (t-1)(t-3)^2,$$

and deduce that it contains 32 chambers.

6.6.5. Exponential and logarithm. For all $n \geq 1$, let

$$\zeta(n) = 1 \quad \text{and} \quad \mu(n) = (-1)^{n-1}(n-1)!.$$

Note that

$$e^{\chi} - 1 = \sum_{n \geq 1} \zeta(n) \frac{\chi^n}{n!} \quad \text{and} \quad \log(1+\chi) = \sum_{n \geq 1} \mu(n) \frac{\chi^n}{n!}.$$

This brings out the close relationship of ζ and μ with the exponential and logarithm. Let us understand this locally in $\Pi[I]$, the poset of partitions of I.

Let

$$v = \sum_{n \geq 1} v_n \frac{\chi^n}{n!}$$

be a formal power series. Let $e^v - 1$ denote the formal power series obtained by substituting v for χ in $e^{\chi} - 1$. A formal power series v gives rise to a function P_v on $\Pi[I]$ for every nonempty finite set I via

$$P_v(\mathrm{X}) := \prod_{B \in \mathrm{X}} v_{|B|}.$$

For instance, for $\mathrm{X} = \{ace, bf, dg\}$, we have $P_v(\mathrm{X}) = v_3 v_2 v_2$.

Lemma 6.14. *Suppose S and T are disjoint sets whose union is I. For any partition X of S, and Y of T,*

$$P_v(\mathrm{X} \sqcup \mathrm{Y}) = P_v(\mathrm{X}) P_v(\mathrm{Y}).$$

This multiplicative property follows from the definitions.

Proposition 6.15. *For a formal power series v and a partition X of I,*

$$P_{e^v - 1}(\mathrm{X}) = \sum_{\mathrm{Y}\colon \mathrm{Y} \geq \mathrm{X}} P_v(\mathrm{Y}). \tag{6.14}$$

The sum is over all Y greater than X in the poset $\Pi[I]$.

As functions on the poset $\Pi[I]$, we say that P_{e^v-1} is the exponential of P_v, and P_v is the logarithm of P_{e^v-1}. This is consistent with the usage in Section C.1.7.

PROOF. In view of Lemma 6.14, we may assume that $\mathrm{X} = \{I\}$ is the one-block partition. Let $|I| = n$. Put

$$\sum_i \frac{b_i}{i!} \chi^i := e^v - 1 = \sum_{m \geq 1} \frac{v^m}{m!} = \sum_{m \geq 1} \frac{1}{m!} \Big(\sum_{j > 0} v_j \frac{\chi^j}{j!} \Big)^m.$$

Note that $P_{e^v-1}(\{I\}) = b_n$. Each composition $(j_1, \dots, j_m)$ of n contributes to χ^n in the rhs, and the contribution is

$$\frac{1}{m!} \frac{v_{j_1}}{j_1!} \cdots \frac{v_{j_m}}{j_m!}.$$

Now

$$\sum_{\mathrm{Y}} P_v(\mathrm{Y}) = \sum_m \frac{1}{m!} \sum_{F\colon F \text{ has } m \text{ blocks}} P_v(F).$$

The inner sum is over all set compositions F with m blocks, and $P_v(F)$ is defined the same way as for set partitions. Given a composition $(j_1, \dots, j_m)$, there are

$$\binom{n}{j_1, \dots, j_m}$$

number of F whose type is that composition, and for each such F, $P_v(F)$ equals $v_{j_1} \dots v_{j_m}$. This completes the proof of (6.14). $\square$

6.6.6. Shuffles and quasi-shuffles. Recall shuffles and quasi-shuffles from Sections 6.5.4 and 6.5.5.

Exercise 6.16. Let $I = S_1 \sqcup \cdots \sqcup S_k$. For each $i = 1, \dots, k$, let G_i be a composition of S_i with p_i blocks. Show that shuffles of the G_i are in bijection with lattice paths in $\mathbb{N}^k$ from the origin to the point $(p_1, \dots, p_k)$ with steps of the form $(e_1, \dots, e_k)$ where exactly one e_i is 1 and all others are 0. (So the steps move from a point to any other vertex of the unit frame with origin at that point.) Deduce that the number of such shuffles with p blocks is the multinomial coefficient $\binom{p}{p_1,\dots,p_k}$.

Exercise 6.17. Let $I = S_1 \sqcup \cdots \sqcup S_k$. For each $i = 1, \dots, k$, let G_i be a composition of S_i with p_i blocks. Show that quasi-shuffles of the G_i are in bijection with lattice paths in $\mathbb{N}^k$ from the origin to the point $(p_1, \dots, p_k)$ with steps of the form $(e_1, \dots, e_k)$ where each e_i is either 0 or 1 and not all e_i are 0. (So the steps move from a point to any other vertex of the unit cube with origin at that point.)

In particular, the number of such quasi-shuffles only depends on $p_1, \dots, p_k$. Let $\binom{p}{p_1,\dots,p_k}_{\mathrm{qs}}$ denote the number of such quasi-shuffles with p blocks.

Exercise 6.18. Deduce that

$$\binom{p}{p_1, p_2}_{\mathrm{qs}} = \binom{p}{p-p_2,\ p-p_1,\ p_1+p_2-p}$$

(a multinomial coefficient) if $\max\{p_1, p_2\} \le p \le p_1 + p_2$, and is 0 otherwise.

Exercise 6.19. Show that

$$\binom{p}{p_1, \dots, p_k}_{\mathrm{qs}} = \sum_{i=0}^{p} (-1)^i \binom{p}{i} \binom{p-i}{p_1} \cdots \binom{p-i}{p_k}.$$

Exercise 6.20. Consider an equation of the form $\mathrm{Z} \vee \mathrm{X} = \mathrm{Y}$ among partitions of I. Suppose Z has k blocks respectively refined by $p_1, \dots, p_k$ blocks of Y. Show that the number of solutions X with p blocks is

$$\frac{p_1! \cdots p_k!}{p!} \binom{p}{p_1, \dots, p_k}_{\mathrm{qs}}.$$

Exercise 6.21. Employ (1.55) to deduce the identity

$$\binom{t}{p_1} \cdots \binom{t}{p_k} = \sum_{p=\max\{p_1,\dots,p_k\}}^{p_1+\cdots+p_k} \binom{p}{p_1, \dots, p_k}_{\mathrm{qs}} \binom{t}{p}.$$

In particular,

$$\binom{t}{p_1}\binom{t}{p_2} = \sum_{p=\max\{p_1,p_2\}}^{p_1+p_2} \binom{p}{p-p_2,\ p-p_1,\ p_1+p_2-p} \binom{t}{p}.$$

(Use (6.12) and Exercises 6.18 and 6.20.)

6.6.7. Numbers $|\lambda|$. Recall the numbers $|\lambda|$ from Section 5.5.1. For the braid arrangement, they are given as follows.

For a partition λ, let T be any composition of support λ, and X be any set partition of type λ. Then $|\lambda|$ is the number of set compositions of type T and support X. Explicitly,

$$|\lambda| = m_1!m_2!\ldots,$$

where in λ, the number 1 appears m_1 times, 2 appears m_2 times, and so on. This is because blocks of the same size can be permuted among themselves.

6.6.8. Cycle-type function. Recall the cycle-type function from Section 5.5.2. For the braid arrangement, it is given as follows.

Any permutation on n letters can be written in cycle notation. The lengths of the cycles form a partition of n. This is called the *cycle-type* of that permutation. Let S_n denote the symmetric group on n letters, and Par_n denote the set of partitions of n. The cycle-type function

$$\mathrm{S}_n \to \mathrm{Par}_n,$$

assigns to a permutation its cycle-type.

The number of permutations which consist of a single cycle, that is, whose cycle-type is the partition (n) (with a single part) is $(n-1)!$. This coincides with the absolute value of the Möbius number given in (6.9). This is in accordance with Theorem 5.18.

6.7. Arrangement of type B

We now discuss the arrangement of type B. Just as the braid arrangements correspond to the combinatorics of finite sets, the arrangements of type B correspond to the combinatorics of finite sets equipped with an involution (with a unique fixed point).

6.7.1. Arrangement of type B. The *arrangement of type B_n* consists of the n^2 hyperplanes in $\mathbb{R}^n$

$$x_i = x_j, \quad x_i = -x_j \quad \text{and} \quad x_k = 0$$

for $1 \le i < j \le n$ and $1 \le k \le n$. It is essential, and hence has rank n. It is a reflection arrangement, the Coxeter group is the *signed symmetric group* on n letters, denoted $\mathrm{S}_n^{\pm}$. It acts by permuting the coordinates, and changing some of them to their negative.

Similar to the braid arrangement, the canonical linear order of the set $[n]$ is not relevant to the definition of the arrangement. So we may proceed as follows. For a finite set I, the arrangement of type B on I consists of the hyperplanes

$$x_a = x_b, \quad x_a = -x_b \quad \text{and} \quad x_c = 0$$

in $\mathbb{R}^I$, as a, b and c vary over elements of I, with $a \neq b$. Put

$$\mathbf{I} := I \sqcup \overline{I} \sqcup \{0\},$$

where $\overline{I}$ is a copy of the set I and 0 is an element not in I or $\overline{I}$. For $a \in I$, write $\overline{a}$ for the corresponding element of $\overline{I}$ and set $\overline{0} = 0$. The set $\mathbf{I}$ is thus endowed with a canonical involution whose unique fixed point is 0 and which exchanges a and $\overline{a}$. The Coxeter group consists of all bijections of $\mathbf{I}$ which commute with the canonical involution. We refer to elements of $\mathbf{I}$ as letters. In case $I = [n]$, we may identify $\mathbf{I}$ with the set $[\mathbf{n}] := \{-n, \ldots, -1, 0, 1, \ldots, n\}$.

Geometric notions such as faces, flats, and so on can be described combinatorially in terms of type B set compositions, type B set partitions, and so on. In other words, there is a type B analogue of Table 6.2. There are two kinds of subsets of $\mathbf{I}$ which intervene in these combinatorial descriptions. They are as follows.

Let J be a subset of $\mathbf{I}$. We let $\overline{J}$ denote the image of J under the canonical involution. We say J is *involution-exclusive* if $a \in J$ implies $\overline{a} \notin J$, that is, if $J \cap \overline{J} = \emptyset$. We say J is *involution-inclusive* if $0 \in J$, and $a \in J$ implies $\overline{a} \in J$, that is, if $0 \in J$ and $J = \overline{J}$.

The collection of involution-exclusive subsets of $\mathbf{I}$ constitutes a simplicial complex: the set of proper faces of the n-dimensional *hyperoctahedron*, where $n = |I|$.

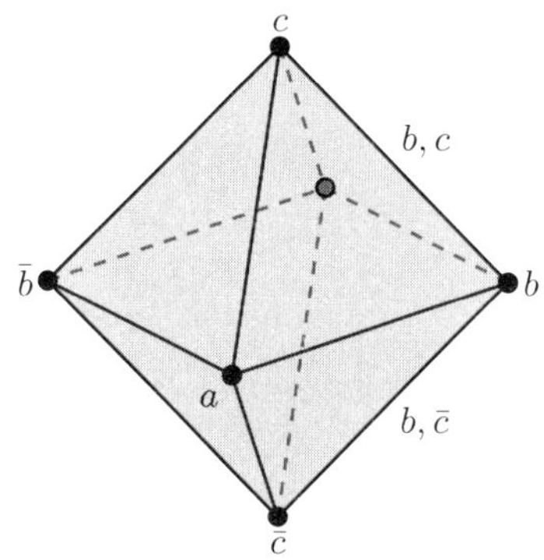

6.7.2. Type B set compositions and set partitions. A *type B composition* of I is a composition of the set $\mathbf{I}$ with the following two properties:

- If S is a block, then $\overline{S}$ is also a block.
- If a block S precedes a block T, then $\overline{T}$ precedes $\overline{S}$.

Such a composition is therefore of the form

$$(\overline{I_k}, \dots, \overline{I_1}, \mathbf{I}_0, I_1, \dots, I_k),$$

with $I_1, \dots, I_k$ involution-exclusive, $\mathbf{I}_0$ involution-inclusive, and

$$\mathbf{I} = \mathbf{I}_0 \sqcup \bigsqcup_{i=1}^{k} (I_i \sqcup \overline{I_i}).$$

We refer to $\mathbf{I}_0$ as the zero block, and to the remaining I_j and $\overline{I_j}$ as the nonzero blocks of the composition. Necessarily 0 lies in $\mathbf{I}_0$. Sometimes we specify a type B composition by displaying only the blocks $\mathbf{I}_0, I_1, \dots, I_k$, and omitting elements of $\overline{I}$ from the zero block. We refer to this as the *short notation*. The blocks are separated by vertical bars. The order within each block is irrelevant. For example, we may write either

$$\overline{e}|d\overline{b}|\overline{c}\overline{a}0ac|b\overline{d}|e \quad \text{or} \quad 0ac|b\overline{d}|e.$$

Refinement among type B set compositions is defined as it is among all set compositions. The resulting partial order on the set of type B compositions of I has a unique minimum element given by the one-block composition $(\mathbf{I})$. A maximal element is a *type B linear order*, that is, a type B composition in which all blocks are singletons (and in particular, $\mathbf{I}_0 = \{0\}$). Equivalently, a type B linear order is a linear order on $\mathbf{I}$ such that $a \le b$ implies $\overline{b} \le \overline{a}$.

A *type B partition* of I is a partition of $\mathbf{I}$ such that:

- If S is a block, then $\overline{S}$ is also a block.
- There is exactly one block S with $S = \overline{S}$.

It consists of an involution-inclusive subset $\mathbf{I}_0$ and a collection $\{I_j\}_{j\in J}$ of nonempty involution-exclusive subsets of $\mathbf{I}$, closed under the canonical involution and such that

$$\mathbf{I} = \mathbf{I}_0 \sqcup \bigsqcup_{j\in J} I_j.$$

An example is $\mathrm{X} = \{\overline{ce}, f\overline{g}, \overline{dba}0abd, \overline{f}g, ce\}$. We employ the same terminology as for type B compositions.

Type B partitions are partially ordered by refinement and the resulting poset is a lattice. The minimal element is the type B partition into one block and the maximal element is the type B partition in which all blocks are singletons.

For any type B set composition F, we let $z(F)$ denote the zero block of F. Similarly, for a type B set partition, we let $z(\mathrm{X})$ denote the zero block of X.

6.7.3. Support map. The *support* of a type B composition of I is the type B partition of I obtained by forgetting the order among the blocks.

Two type B compositions F and G with the same support have the same zero block and may only differ in the order in which the nonzero blocks are listed.

6.7.4. Opposition map. The *opposite* of a type B set composition is obtained by listing its blocks in reverse order. For example,

$$F = \bar{e}|d\bar{b}|\overline{ca}0ac|b\bar{d}|e, \qquad \overline{F} = e|b\bar{d}|\overline{ca}0ac|d\bar{b}|\bar{e}.$$

6.7.5. Tits product. The combinatorial description of the Tits product is similar to that for the braid arrangement. Let F and G be two type B compositions of I. The *Tits product* FG is the type B composition obtained by listing the pairwise intersections of the blocks in F with the blocks in G in lexicographic order and removing empty intersections from the nonzero blocks. The zero block of FG is the intersection of the zero blocks of F and G. Two small examples follow.

$$(a\bar{b}|0|\bar{a}b)(\bar{a}\bar{b}|0|ab) = (\bar{b}|a|0|\bar{a}|b), \qquad (a|\bar{b}0b|\bar{a})(\bar{b}|\bar{a}0a|b) = (a|\bar{b}|0|b|\bar{a}).$$

6.7.6. Small ranks. The arrangement of type B on $I = \{a\}$ consists of one hyperplane $x_a = 0$. The linear and spherical models are shown below.

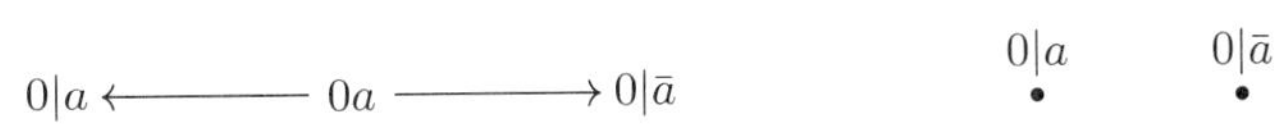

Compare and contrast with the braid arrangement on $I = \{a, b\}$. The models are identical, both being rank-one, but the way the faces are labeled is different.

The arrangement of type B on $I = \{a, b\}$ consists of the four hyperplanes $x_a = x_b$, $x_a = -x_b$, $x_a = 0$ and $x_b = 0$. It is the dihedral arrangement of four lines.

The linear and spherical models are shown below.

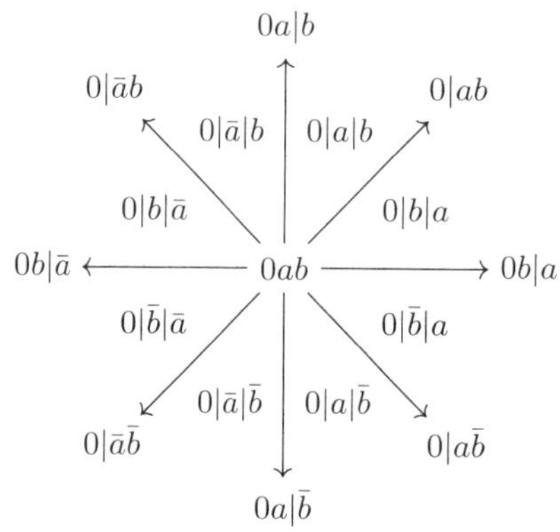

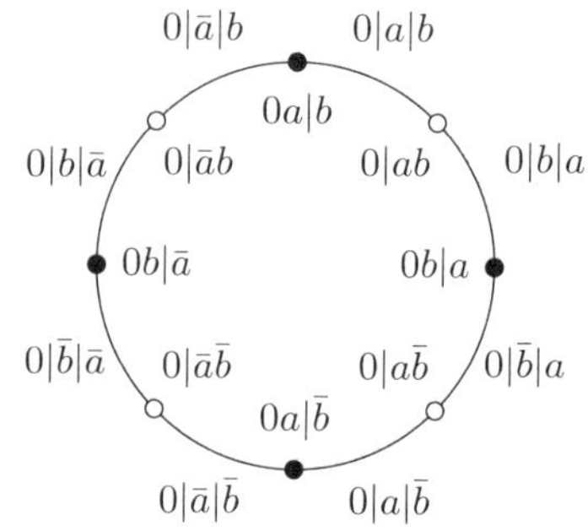

The faces are labeled by type B compositions of I using short notation.

The arrangement of type B_3 on $I = \{a, b, c\}$ consists of 9 hyperplanes. Two different spherical models are shown below. The set of hyperplanes splits into two orbits under the Coxeter group action. A representative from each orbit is chosen as the outer circle for the two pictures. Observe carefully the pattern of vertex-types on the outer circles; it is different in the two cases.

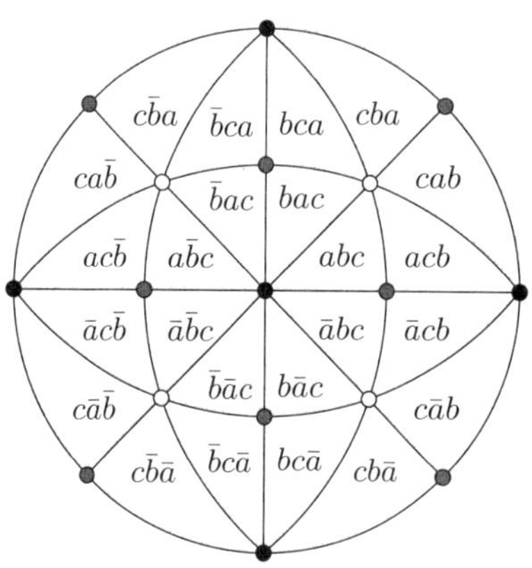

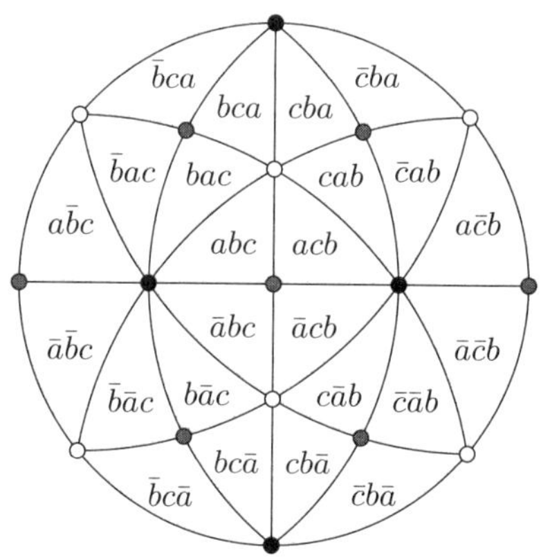

In the first picture, the horizontal line is $x_a = 0$, the vertical line is $x_b = 0$, while the outer circle is $x_c = 0$. In the second picture, the horizontal line is $x_a = 0$, the vertical line is $x_b = x_c$, while the outer circle is $x_b = -x_c$. The arrangement has 48 triangles of which 24 are visible in the pictures. The triangles are $0|a|b|c$, $0|b|a|c$, $0|\bar{a}|b|c$ and so on, but for convenience, we have shortened them to abc, bac, $\bar{a}bc$ and so on.

Other useful depictions follow.

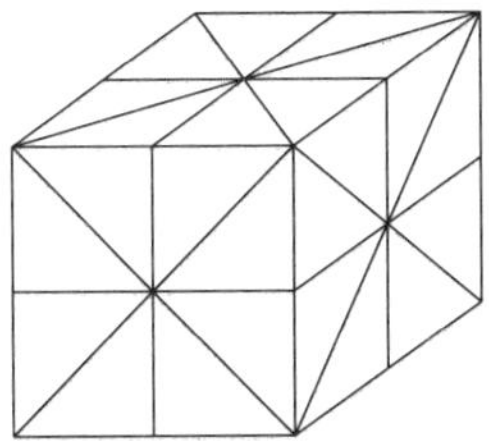 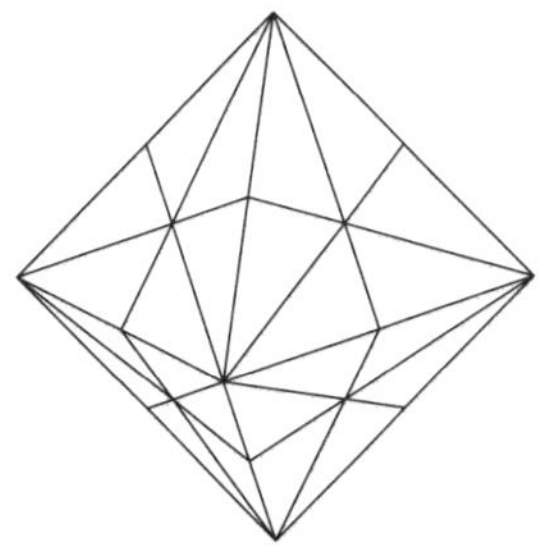

The arrangement has been intersected in one case with a centrally located cube each side of which is parallel to one of the three coordinate planes, in the other with the dual octahedron. (This is as the spherical model except that we have used a polytope instead of a sphere.) The 48 triangles can now be seen as 8 triangles on each of the six sides of the cube, or as 6 triangles on each of the eight sides of the octahedron.

6.7.7. Type B compositions and partitions. Let n be a nonnegative integer. A *type B composition* $\alpha = (a_0, a_1, \dots, a_k)$ of n is a finite sequence of integers with $a_0 \geq 0$, $a_i \geq 1$ for $i \geq 1$, and such that

$$a_0 + a_1 + \cdots + a_k = n.$$

A *type B partition* $\lambda = (l_0, l_1, \dots, l_k)$ of n is a finite sequence of integers with $l_0 \geq 0$, $l_i \geq 1$ for $i \geq 1$, and such that

$$l_1 \geq l_2 \geq \cdots \geq l_k \quad \text{and} \quad l_0 + l_1 + \cdots + l_k = n.$$

We refer to a_0 and l_0 as the *zero part*, and to the remaining a_i and l_i as the *nonzero parts* of α and λ.

The *support* of a type B composition of n is the type B partition of n obtained by reordering the nonzero parts in decreasing order.

6.7.8. Type map. The *type* of a type B composition $(\mathbf{I}_0, I_1, \dots, I_k)$ of I is the type B composition $\alpha = (a_0, a_1, \dots, a_k)$ of $|I|$ for which $|\mathbf{I}_0| = 2a_0 + 1$ and $a_j = |I_j|$ for $j \geq 1$. The *type* of a type B partition is defined similarly.

The support and type maps commute with each other.

$$\begin{array}{ccc} \text{type } B \text{ set composition} & \xrightarrow{\ \mathrm{t}\ } & \text{type } B \text{ composition} \\ \Big\downarrow{\scriptstyle \mathrm{s}} & & \Big\downarrow{\scriptstyle \mathrm{s}} \\ \text{type } B \text{ set partition} & \xrightarrow[\ \mathrm{t}\ ]{} & \text{type } B \text{ partition} \end{array}$$

This is a specialization of (5.3).

6.7.9. Type B graphs. A *type B simple graph* on I is a simple graph with vertex set $\mathbf{I}$ and with the following properties.

- If there is an edge between vertices a and b, then there is also an edge between $\overline{a}$ and $\overline{b}$.
- There is no edge between a vertex a and $\overline{a}$.

We may represent hyperplanes in the arrangement of type B on I as follows. Let $i, j \in I$. A hyperplane $x_i = x_j$ $(j \neq i)$ is represented by a pair of edges, one between i and j and the other between $\bar{i}$ and $\bar{j}$. Similarly, a hyperplane $x_i = -x_j$ $(j \neq i)$ is represented by an edge between i and $\bar{j}$ and another between $\bar{i}$ and j. Finally, a hyperplane $x_k = 0$ is represented by edges between 0 and k and between 0 and $\bar{k}$. In this manner, charts in the arrangement correspond to type B simple graphs.

Consider instead the following class of graphs: the vertex set is I, they may possess *half-edges* (but not loops), each edge between two distinct vertices i and j must be labeled $+$ or $-$, and parallel edges cannot receive the same label. A graph of this kind is said to be *simply signed* in the literature.

Removing the vertex 0 and identifying vertices and edges matched by the canonical involution turns a type B simple graph into a simply signed graph (edges within I or $\bar{I}$ are labeled $+$, edges between the two are labeled $-$). The type B graph can then be reconstructed by duplicating each edge and attaching the loose end of each half-edge to 0. Thus, type B simple graphs and simply signed graphs are equivalent notions. An illustration is provided below.

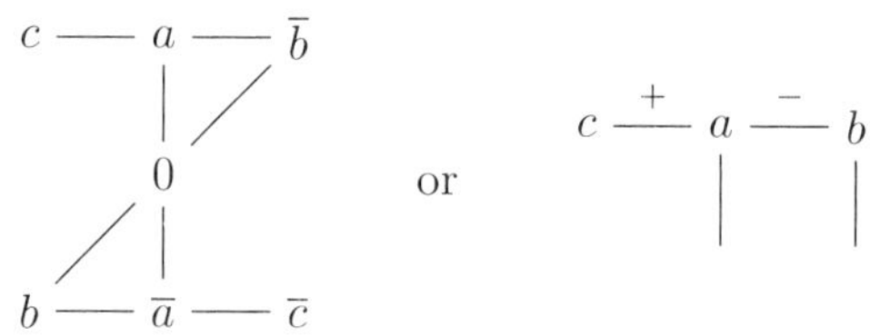

6.7.10. Type *B* partial orders. A *type B partial order* on I is a partial order on the set $\mathbf{I}$ with the following properties.

- If $a \leq b$, then also $\bar{b} \leq \bar{a}$.
- If $a \leq \bar{a}$, then in fact $a \leq 0 \leq \bar{a}$.

In such a poset, the connected component of 0 is involution-inclusive, while the other connected components are involution-exclusive and occur in pairs (with the two components in each pair being exchanged by the involution). An example follows.

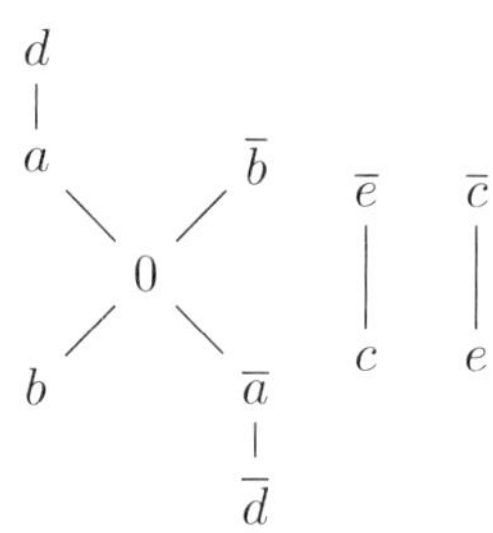

Top-cones in the arrangement of type B on I are in correspondence with such structures. Let $i, j \in I$. The top-cone corresponding to a partial order is defined by the inequalities $x_i \leq x_j$ if $i \leq j$, $x_i \leq -x_j$ if $i \leq \bar{j}$, $-x_i \leq x_j$ if $\bar{i} \leq j$, $0 \leq x_i$ if $0 \leq i$, and $x_i \leq 0$ if $i \leq 0$.

A type B partial order is also known as a *signed poset*.

6.7.11. Arrangements under and over a flat. Let $\mathcal{A}$ be an arrangement of type B. The arrangement under a flat X of the type B arrangement $\mathcal{A}$ is again cisomorphic to an arrangement of type B. The blocks of X play the role of the elements of $\mathbf{I}$, with the zero block of X playing the role of 0. The nonzero blocks appear in pairs and this defines the canonical involution.

The arrangement over a flat X is cisomorphic to a cartesian product of braid arrangements and an arrangement of type B. The arrangement of type B arises from the zero block of X, and its letters are the letters of that block. The nonzero blocks appear in pairs, and each pair gives one braid arrangement whose letters can be equivalently taken from either of the two blocks.

6.7.12. Substitution product of chambers. For a type B partition X of I, the substitution product of chambers (4.18) works as follows.

An element of $\Gamma[\mathcal{A}^{\mathrm{X}}]$ is a type B linear order on the blocks of X, while an element of $\Gamma[\mathcal{A}_{\mathrm{X}}]$ consists of a type B linear order on the zero block of X and linear orders, one on each nonzero block of X. (Recall that the nonzero blocks occur in pairs $(J, \overline{J})$. It is understood here that the linear order on $\overline{J}$ is obtained by reversing the linear order on J and applying the canonical involution to each letter.) This data together specifies a type B linear order on I, which is an element of $\Gamma[\mathcal{A}]$. For example, for the type B partition $\mathrm{X} = \{\overline{ce}, f\overline{g}, \overline{dba}0abd, \overline{f}g, ce\}$,

$$((\overline{ce}|\overline{f}g|\overline{dba}0abd|f\overline{g}|ce), \{\overline{d}|b|a|0|\overline{a}|\overline{b}|d, e|c, \overline{c}|\overline{e}, f|\overline{g}, g|\overline{f}\}) \\ \mapsto (\overline{c}|\overline{e}|g|\overline{f}|\overline{d}|b|a|0|\overline{a}|\overline{b}|d|f|\overline{g}|e|c).$$

In short notation, this can be rewritten as

$$((0abd|f\overline{g}|ce), \{0|\overline{a}|\overline{b}|d, e|c, f|\overline{g}\}) \mapsto (0|\overline{a}|\overline{b}|d|f|\overline{g}|e|c).$$

The substitution product in rank two is illustrated below.

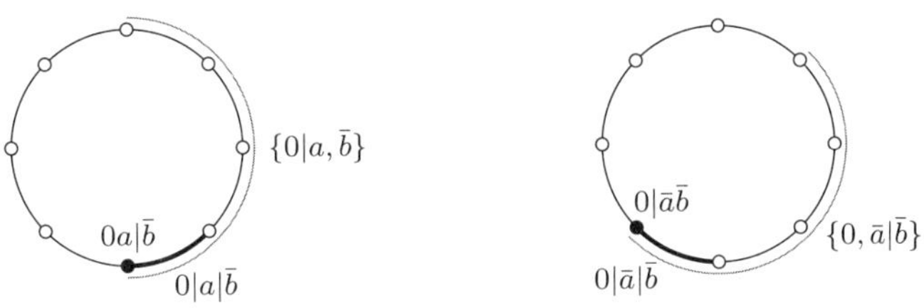

In both pictures, the substitution product of the vertex shown in black with the top-lune shown in gray is the thick edge. The top-lunes can also be denoted $\{0|a, b\}$ and $\{0, b|a\}$, respectively.

6.7.13. Type B degrees and double factorials. The *double factorials* are

$$(2k)!! = 2k(2k-2)\cdots 2 \quad \text{and} \quad (2k+1)!! = (2k+1)(2k-1)\cdots 1.$$

Let G be a type B set composition. If the number of blocks of G is $2k+1$, then $\mathrm{rk}(G) = k$. This is the number of nonzero blocks when G is written in short notation. Define

$$\mathbf{deg}(G) := \frac{(2k)!!}{(2k-1)!!} = \frac{4^k}{\binom{2k}{k}} \quad \text{and} \quad \mathbf{deg!}(G) := (2k)!! = k!\, 2^k.$$

Note that

$$\frac{(-1)^k}{\mathbf{deg}(G)} = \binom{-1/2}{k}. \tag{6.15}$$

For type B set compositions $F \leq G$, let $(G/F)_i$ denote the part of G which refines the i-th block of F, starting at $i = 0$ with the part that refines the block $z(F)$. Note that $(G/F)_0$ is a type B set composition, while the rest are set compositions. Define

$$\mathbf{deg}(G/F) := \mathbf{deg}(G/F)_0 \prod_{i \geq 1} \deg(G/F)_i, \tag{6.16}$$

$$\mathbf{deg!}(G/F) := \mathbf{deg!}(G/F)_0 \prod_{i \geq 1} \deg!(G/F)_i. \tag{6.17}$$

Recall that for a (type A) set composition, deg is the number of its blocks and deg! is the factorial of this number.

These definitions also apply to type B set partitions, type B integer compositions and type B integer partitions, and we employ similar notations for them.

Lemma 6.22. *For any integer m, and any type B set partition* X,

$$\sum_{F:\,\mathrm{s}(F) \leq \mathrm{X}} \binom{m}{\mathrm{rk}(F)} = (2m+1)^{\mathrm{rk}(\mathrm{X})},$$

and

$$\sum_{\substack{F:\,\mathrm{s}(F) \leq \mathrm{X} \\ z(F) = \{0\}}} \binom{m}{\mathrm{rk}(F)} = \begin{cases} (2m)^{\mathrm{rk}(\mathrm{X})} & \text{if } z(\mathrm{X}) = \{0\}, \\ 0 & \text{otherwise.} \end{cases}$$

Proof. We prove the first identity. First assume m is positive. Take $2m+1$ boxes labeled $\overline{m}, \dots, \overline{1}, 0, 1, \dots, m$. The rhs counts the number of ways of putting each block of X in one of the boxes with the condition that if a block goes in box i, then its opposite block goes in box $\overline{i}$. (In particular, the zero block of X must go in the box labeled 0.) Each such assignment yields a type B set composition F with $\mathrm{s}(F) \leq \mathrm{X}$: each box is a block of F with empty boxes deleted. The number of assignments which yield the same F is precisely $\binom{m}{\mathrm{rk}(F)}$. This is the lhs. This proves the first identity for m positive. For the general case, note that both sides are polynomials in m. Since they agree for infinitely many values of m, they must agree for all values of m. □

6.7.14. Möbius number and characteristic polynomial. Let $\mathcal{A}$ denote the arrangement of type B on $[n]$. The number of chambers $c(\mathcal{A})$ is $(2n)!!$, and the number of faces is the number of type B compositions of $[n]$. More generally,

$$c(\mathcal{A}^G) = \mathbf{deg!}(G) \quad \text{and} \quad c(\mathcal{A}^G_F) = \mathbf{deg!}(G/F).$$

The first one is the number of type B set compositions with the same support as G, while the second is the number of type B set compositions which are greater than F and have the same support as G. When F is the one-block type B composition, the second formula reduces to the first.

The Möbius number is given by

$$\mu(\mathcal{A}) = (-1)^n \, (2n-1)!!. \tag{6.18}$$

More generally, for $F \leq G$, we claim that

$$\mu(\mathcal{A}^G_F) = (-1)^{\mathrm{rk}(G/F)} \frac{\mathbf{deg!}(G/F)}{\mathbf{deg}(G/F)}. \tag{6.19}$$

When F is the central face, this reduces to (6.18). The general case can be deduced from this special case in conjunction with (6.10). The relevant fact is that $\mathcal{A}_F^G$ is cisomorphic to a cartesian product of braid arrangements and an arrangement of type B, so the Möbius numbers multiply. It is also useful to note that

$$\frac{\mu(\mathcal{A}_F^G)}{c(\mathcal{A}_F^G)} = \frac{(-1)^{\mathrm{rk}(G/F)}}{\mathbf{deg}(G/F)}. \tag{6.20}$$

The characteristic polynomial of the arrangement of type B is given by

$$\chi(\mathcal{A}, t) = \prod_{i=1}^{n} (t - (2i-1)). \tag{6.21}$$

Putting $t = -1$, we see that the rhs up to sign equals the number of chambers. This is the Zaslavsky formula in the equivalent form (1.54c). More generally,

$$\chi(\mathcal{A}, 2m+1) = (2n)!!\binom{m}{n}. \tag{6.22}$$

Let $s^{\pm}(m,k)$ denote the coefficient of x^k in the polynomial $(x-1)(x-3)\cdots(x-(2m-1))$. These are the *type B Stirling numbers*. By Lemma 1.87, we have

$$\mathrm{wy}(\mathcal{A}, k) = s^{\pm}(n,k).$$

More generally, for any flat X, we have

$$\mathrm{wy}(\mathcal{A}^{\mathrm{X}}, k) = s^{\pm}(\mathrm{rk}(\mathrm{X}), k). \tag{6.23}$$

This is because $\mathcal{A}^{\mathrm{X}}$ is also an arrangement of type B.

As a companion, one may also consider the coefficient of x^k in the polynomial $x(x-2)(x-4)\cdots(x-2(m-1))$. Observe that this equals $2^{m-k}s(m,k)$, where $s(m,k)$ is the Stirling number of type A.

6.8. Arrangement of type D

We discuss this reflection arrangement very briefly.

6.8.1. Arrangement of type *D*. For $n \geq 2$, the *arrangement of type* D_n consists of hyperplanes in $\mathbb{R}^n$, namely,

$$x_i = x_j \quad \text{and} \quad x_i = -x_j$$

for $1 \leq i < j \leq n$. It is a subarrangement of the arrangement of type B_n with the coordinate hyperplanes removed. It is also a reflection arrangement.

The arrangement of type D_2 consists of the two hyperplanes $x_1 = x_2$ and $x_1 = -x_2$. It is cisomorphic to the rank-two arrangement of two lines.

The arrangement of type D_3 is shown below. It has 24 chambers. Observe that it is cisomorphic to the arrangement of type A_3.

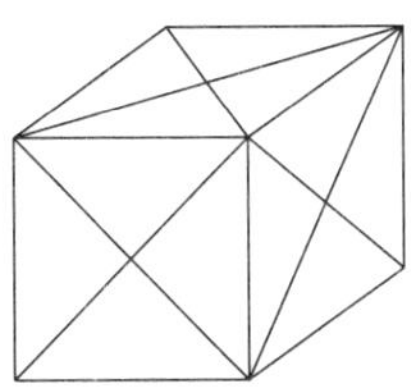

Just as for types A and B, for type D also, faces, flats, and other geometric objects admit combinatorial descriptions, but we do not discuss them here.

6.8.2. Arrangement under a flat. The arrangement under any of the hyperplanes of the arrangement of type D_4 is cisomorphic to the following rank-three arrangement:

$$x_1 = x_2,\ x_2 = x_3,\ x_1 = x_3,\ x_1 = -x_2,\ x_2 = -x_3,\ x_1 = -x_3 \text{ and } x_1 = 0.$$

Two ways to picture this arrangement are shown below.

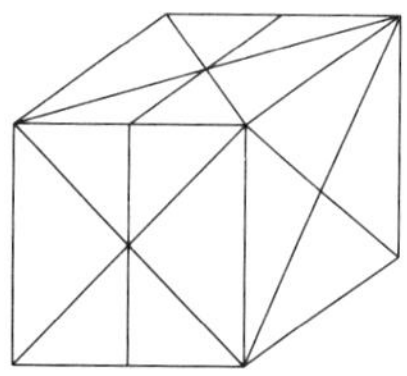

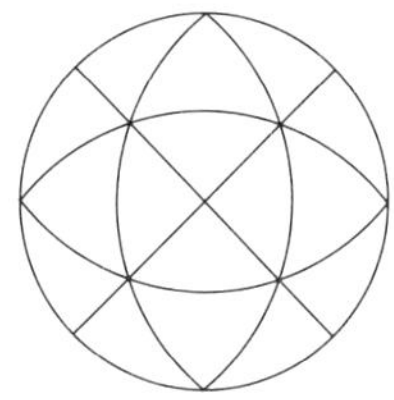

In the picture on the right, the outer circle is included in the arrangement.

In general, the arrangement under a flat of a type D arrangement sits "between" types B and D. That is, we have the hyperplanes $x_i = \pm x_j$ for all $i \neq j$, but $x_k = 0$ only for some k.

6.9. Graphic arrangements

We take a brief look at arrangements associated to simple graphs. They are subarrangements of the braid arrangement. The vertex set determines the ambient space and the edge set determines the hyperplanes. Every subarrangement of the braid arrangement arises in this manner from a unique simple graph.

6.9.1. Graphic arrangements. Let g be a simple graph on a finite set I. The elements of I are the vertices of g. An edge of g is a subset of I of cardinality 2. There are no repeated edges or loops. There are $2^{\binom{n}{2}}$ simple graphs on n vertices.

The arrangement $\mathcal{A}(g)$ in $\mathbb{R}^I$ consists of the hyperplanes $x_a = x_b$, one for each edge $\{a, b\}$ of g. This is the *graphic arrangement* of g. It is not essential: The dimension of the central face is the number of connected components of g. It follows that the rank of the graphic arrangement is

$$\operatorname{rk}(\mathcal{A}(g)) = |I| - c(g), \tag{6.24}$$

where $c(g)$ is the number of connected components of g. Combinatorial descriptions of some of the geometric objects associated to $\mathcal{A}(g)$ are summarized in Table 6.5 and detailed along the section.

TABLE 6.5. Combinatorics of graphic arrangements.

Geometry	Combinatorics
flat	bond or closed subgraph
chamber	acyclic orientation
face	bond with an acyclic orientation of the contraction
top-cone	partial order with covers among the graph edges
top-lune	bond and an acyclic orientation on each block

6.9.2. Examples. The *complete graph* k_I contains all possible edges among the elements of I. The arrangement $\mathcal{A}(k_I)$ is the braid arrangement on I. Graphic arrangements on I are precisely subarrangements of $\mathcal{A}(k_I)$. In the terminology of Section 2.6, graphic arrangements are the charts of braid arrangements.

The *discrete graph* d_I contains no edges (and has vertex set I). The arrangement $\mathcal{A}(d_I)$ is of rank 0, containing no hyperplanes.

A *tree* is a connected graph with no cycles. If g is a tree on n vertices, then $\mathcal{A}(g)$ is gisomorphic to the coordinate arrangement of rank $n-1$.

If g is a 4-cycle, then $\mathcal{A}(g)$ is cisomorphic to the smallest nonsimplicial arrangement (Section 1.2.5).

6.9.3. Product of graphic arrangements. Disjoint union of graphs results in product of arrangements. Suppose $I = S \sqcup T$. If g is a graph on S and h is a graph on T, then

$$\mathcal{A}(g \sqcup h) = \mathcal{A}(g) \times \mathcal{A}(h).$$

This is an arrangement in $\mathbb{R}^I = \mathbb{R}^S \times \mathbb{R}^T$.

6.9.4. Bonds and flats. Given a subset S of I, let $g|_S$ denote the graph *induced* by g on S: the set of vertices is S and the edges are the edges of g between elements of S.

A *bond* of g is a partition X of I into blocks B such that $g|_B$ is connected. Bonds correspond to flats of $\mathcal{A}(g)$ under the same correspondence as for the braid arrangement: the flat corresponding to X is defined by setting all variables in each block of X equal to each other.

The maximum flat corresponds to the partition of I into singletons. The minimum flat corresponds to the partition into connected components of g. The rank of the flat corresponding to X is the number of blocks of X minus the number of connected components of g.

6.9.5. Restriction and contraction. Let X be a bond of g. The *restriction* g_{X} is the disjoint union of the graphs induced on the blocks of X:

$$g_{\mathrm{X}} = \bigsqcup_{B \in \mathrm{X}} g|_B.$$

It is obtained by removing edges connecting distinct blocks of X. The vertex set is I. The *contraction* g^{X} is the simple graph obtained by identifying the vertices in each block of X. The vertex set is X and there is an edge between two blocks B and C if there is at least one edge in g connecting a vertex in B to a vertex in C. An example follows in which $I = \{a, b, c, d\}$ and $\mathrm{X} = a|bd|c$.

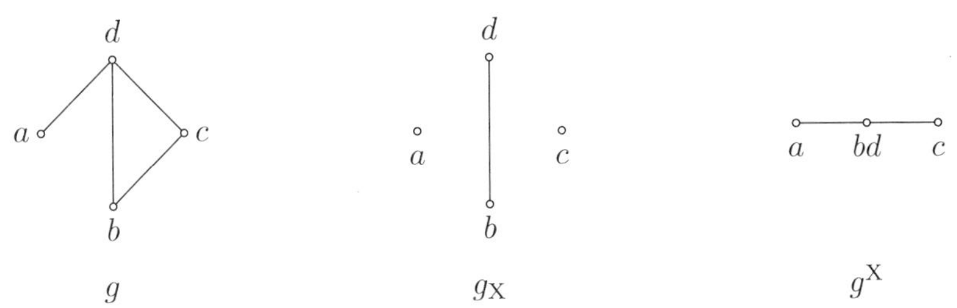

Let h be another graph on I. It is a *subgraph* of g if g contains every edge of h. A subgraph h of g is *closed* if it has the following property: if there is a sequence of edges $\{a, b\}, \{b, c\}, \ldots, \{y, z\}$ in h and $\{a, z\}$ is an edge of g, then $\{a, z\}$ is in h.

Closed subgraphs are in correspondence with bonds, with the bond X corresponding to the restriction g_{X} (which is a closed subgraph). The bond is recovered from the closed subgraph h as the partition of I into connected components of h. If h is the closed subgraph corresponding to X, g^{X} is obtained by contracting all edges in h.

The arrangements under and over a flat are again graphic. We have

$$\mathcal{A}(g)^{\mathrm{X}} = \mathcal{A}(g^{\mathrm{X}}) \quad \text{and} \quad \mathcal{A}(g)_{\mathrm{X}} = \mathcal{A}(g_{\mathrm{X}}) = \prod_{B \in \mathrm{X}} \mathcal{A}(g|_B).$$

6.9.6. Acyclic orientations, faces, and chambers. An *orientation* $\mathcal{O}$ of g is an assignment of a direction to each edge $\{a, b\}$ of g, denoted by $a \to b$ or $b \to a$, as the case may be. A sequence of the form $a \to b \to \cdots \to a$ with each $i \to j$ in $\mathcal{O}$ is a *cycle*. An orientation is *acyclic* if it contains no cycles. Acyclic orientations correspond to chambers: the chamber corresponding to $\mathcal{O}$ is defined by the inequalities $x_a \leq x_b$ for $a \to b$ in $\mathcal{O}$.

A face of $\mathcal{A}(g)$ is uniquely determined by a bond X together with an acyclic orientation of the contraction g^{X}. Note this corresponds to a chamber of $\mathcal{A}(g)^{\mathrm{X}}$. The flat corresponding to X is the support of the face.

The figure shows a path on 3 vertices and the corresponding arrangement (of rank 2).

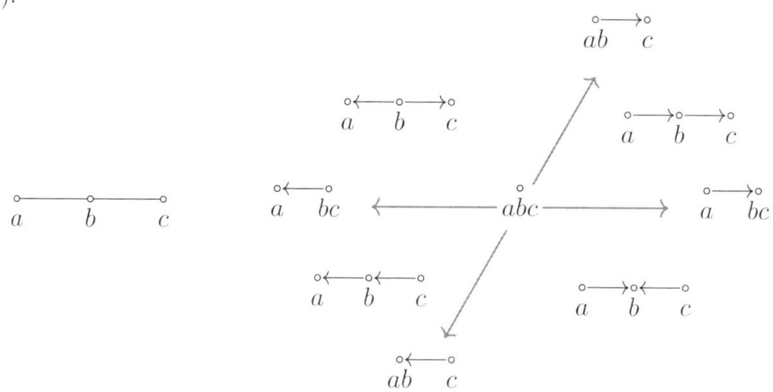

Exercise 6.23. Let g be a tree. Show that faces of a chamber of $\mathcal{A}(g)$ are in bijection with subsets of the edge set of g.

The figure shows the arrangement of the 4-cycle (the smallest nonsimplicial arrangement). There are four hyperplanes corresponding to the edges of the cycle. There are 14 chambers. Each visible chamber is labeled with an acyclic orientation. When crossing a hyperplane, the orientation of the corresponding edge is reversed.

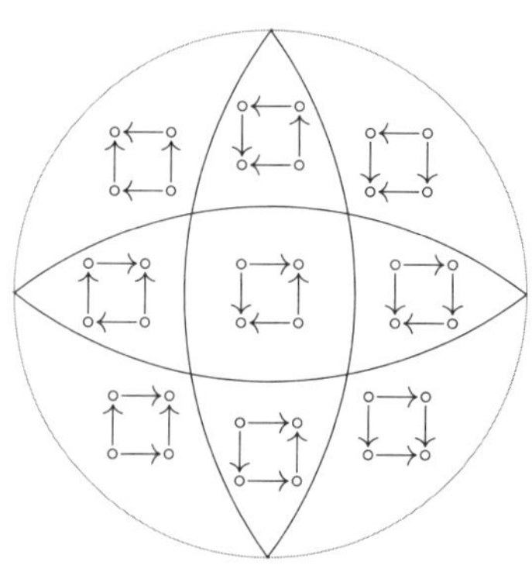

6.9.7. Top-cones and top-lunes. Consider a partial order p on I whose Hasse diagram is a subgraph of g. In other words, if $a \lessdot b$ is a cover relation in p, then $\{a, b\}$ is an edge of g. Top-cones of $\mathcal{A}(g)$ correspond to such structures. This follows from Exercise 2.21 applied to $\mathcal{A}(g)$ as a subarrangement of the braid arrangement. The base of the top-cone corresponds to the partition into connected components of p. The chambers contained in the top-cone correspond to the acyclic orientations $\mathcal{O}$ of g such that if $a \lessdot b$ in p, then $a \to b$ in $\mathcal{O}$. When the Hasse diagram of p equals g, the top-cone is a chamber. In this case, p is the transitive closure of the corresponding acyclic orientation of g.

A top-lune of $\mathcal{A}(g)$ is uniquely determined by a bond X together with acyclic orientations $\mathcal{O}_B$ of $g|_B$ for each block B of X. Note this corresponds to a chamber of $\mathcal{A}(g)_{\mathrm{X}}$. The flat corresponding to X is the base of the top-lune. The chambers contained in the top-lune correspond to the acyclic orientations $\mathcal{O}$ of g such that if $a \to b$ in $\mathcal{O}_B$ for some block B, then $a \to b$ in $\mathcal{O}$. A top-lune is a particular top-cone: the corresponding partial order is the transitive closure of $\bigsqcup_{B \in \mathrm{X}} \mathcal{O}_B$.

6.9.8. Chromatic and characteristic polynomials. An *n-coloring* of g is a function $f : I \to [n]$. An n-coloring is *proper* if $f(a) \neq f(b)$ whenever a and b are connected by an edge in g. The *chromatic function* $\chi(g, n)$ counts the number of proper n-colorings of g. It is a polynomial function of n. Moreover, it is related to the characteristic polynomial of the graphic arrangement by

$$\chi(\mathcal{A}(g), t) = \frac{1}{t^{c(g)}} \chi(g, t),$$

where $c(g)$ is the number of connected components of g. It follows from (1.54c) and (6.24) that

$$(-1)^{|I|} \chi(g, -1) \tag{6.25}$$

is the number of acyclic orientations of g. This is the *negative one color theorem* of Stanley.

6.9.9. Sinks and Möbius number. Let $\mathcal{O}$ be an orientation of g. If $\{a, b\}$ is an edge of g and $a \to b$ in $\mathcal{O}$, we say the edge is outgoing at a. A *sink* of $\mathcal{O}$ is a vertex a with no outgoing edges at a. Let $S(\mathcal{O})$ denote the set of sinks of $\mathcal{O}$. Note that if $\mathcal{O}$ is acyclic, then $S(\mathcal{O})$ contains at least one vertex in each connected component of g. Pick one vertex in each connected component of g. This yields a set S of vertices. It turns out that the number of acyclic orientations $\mathcal{O}$ with $S(\mathcal{O}) = S$ is independent of S. Moreover, it is equal to

$$(-1)^{|I| - c(g)} \mu(\mathcal{A}(g)) = |\mu(\mathcal{A}(g))|, \tag{6.26}$$

where $\mu(\mathcal{A}(g))$ is the Möbius number of $\mathcal{A}(g)$. In particular, when g is connected, $|\mu(\mathcal{A}(g))|$ counts the number of acyclic orientations of g with a unique fixed sink. This is [**210**, Theorem 7.3].

Notes

Coordinate arrangement. In the literature, the coordinate arrangement is also called the Boolean arrangement.

Braid arrangement. A detailed discussion on the braid arrangement is given in [**9**, Chapter 10]. The correspondence between geometry and combinatorics is discussed in [**9**, Section 13.5]. This is also discussed in the paper of Reiner, Postnikov and Williams [**346**, Section 3]; see in particular their Proposition 3.5. The correspondence between posets and top-cones goes back to Reiner's thesis [**356**], where similar correspondences for other reflection arrangements are also considered. The notions of order dimension of a poset and of conjugate posets were introduced by Dushnik and Miller [**165**]. They observed that conjugate posets have a common linear extension, and showed that a poset has order dimension either 1 or 2 iff the poset has a conjugate [**165**, Lemma 3.51 and Theorem 3.61]. For further results, see [**38, 39**]. The equivalence of order dimension and convexity dimension, and of convex closure of two permutations (which are related in the weak order) and gallery intervals is given by Björner and Wachs [**83**, Corollaries 6.7 and 6.4]. There is extensive literature on series-parallel posets, starting perhaps with [**275**]. Enumeration results on this class of posets are provided by Stanley [**394**], [**399**, Exercises 5.39 and 5.40] and Chapoton [**115**]. In the literature, the partial order on the poset of partial orders is opposite to the one given in Section 6.4.3, so upper semimodular becomes lower semimodular and join-distributive becomes meet-distributive. Dean and Keller [**139**] showed that every interval in the poset of partial orders is lower semimodular. Stanley [**393**] showed the stronger property that it is meet-distributive. This fact is also given in [**401**, Exercise 49]. The topology of the poset of partial orders is studied in [**85**, Section 2] and [**174**, Theorem 15]. Edelman and Klingsberg [**173**, Lemma 2.1] give the Möbius function of the poset of partial orders (6.3) when Q is the discrete poset.

Just as poset is short for partially ordered set, preposet is short for pre-partially ordered set. This terminology is used in [**9, 401, 346**]. Preposets are also sometimes called prosets [**176**].

Formulas (6.4), (6.5) and (6.6) are given in [**9**, (10.26), (10.37) and (10.38)]. The Zaslavsky formula (see text below (6.12)) is a special case of [**9**, Equation (10.4)].

The lattice of set partitions goes back to Birkhoff [**68**, Section 18]. His partial order is opposite to ours. Birkhoff also considered the poset of integer partitions. Again his partial order is opposite to the one given in Section 6.3.9. He discusses both posets in [**70**, Section I.8, Examples 9 and 10]. The poset of integer partitions with either one of the two conventions appears in [**8**, Definition 3.2.1], [**31**, page 546], [**360**, Section 9.2, page 227] and [**401**, Chapter 3, Exercise 135].

In the notation of [**399**, Equation (7.16)], $R_{\lambda\lambda} = |\lambda|$, with the latter as in Section 6.6.7.

The paper [**233**] introduces a family of semigroups that generalizes the Tits monoid of type A.

The description of the classical associative operad given in Section 6.5.10 can be found for instance in [**9**, Example B.2]. There are also other ways to formulate this operad, see for instance [**307**, Proposition 1.10 and Definition 1.12] and [**290**, Section 9.1].

The external product of set compositions and linear orders. These two are examples of monoids in the category of species. This and related structures are the main subject of [**9**, Part II]. See also [**10**, Section 10]. These considerations can be suitably extended to all real hyperplane arrangements. We plan to develop these ideas in a future work; some are implicit in this work already.

Adjoints and the all-subset arrangement. Adjoints of geometric lattices were considered by Crapo [**129**] and Cheung [**116**]. The all-subset arrangement (restricted or not) is considered in [**248, 249**]. Billera et al have shown that the number of chambers of the restricted all-subset arrangement is between the bounds $2^{\binom{n-1}{2}}$ and $2^{(n-1)^2}$. The exact number has been computed up to $n = 9$ [**65**, Theorem 1.2 and Proposition 1.1]. See [**65**] and [**79**] for additional information and references pertaining to the restricted all-subset arrangement.

Types B and D. Combinatorial descriptions of the arrangements of types B and D are given in [**296**, Sections B.5 and B.6]. For the combinatorics of their Coxeter groups, see [**80**, Sections 8.1 and 8.2] and [**339**, Chapter 13]. Also see [**42**, Section 4]. A short discussion is given in [**2**, Examples 1.82 and 1.83]. The arrangements between types B and D mentioned in Section 6.8 are considered by Zaslavsky [**443**] and later by Björner and Wachs [**84**, Section 9].

Work on signed graphs has been led by Zaslavsky, starting with [**443, 444, 445**]. For a comprehensive list of references, see [**446**]. Signed posets were introduced by Reiner [**356, 357**]. For recent work and additional references, see [**130**]. Stembridge studies cones in reflection arrangements [**407**].

Graphic arrangements. The study of these arrangements originates in work of Greene and Zaslavsky [**208, 210, 442**]. More recent references include [**328**, Section 2.4], [**321**, Section 5], [**400**, Section 2.3]. The study of the chromatic polynomial of graphs was initiated by George David Birkhoff [**71**] and Hassler Whitney [**432**]. The negative one color theorem (6.25) was originally obtained by Stanley by other methods [**392**, Corollary 1.3]. Its derivation from (1.54c) appears in [**208**, Section 1], [**210**, Theorem 7.1], and [**442**, Theorem G.1]. The expression for the Möbius number of graphic arrangements in terms of acyclic orientations (6.26) is given in [**210**, Theorem 7.3]. See [**150**, Section 1.1], [**175**, Section 3.2], [**272**], [**321**, Remark 5.5] and [**397**, Theorem 1.2(b)] for additional information.

A graphic arrangement $\mathcal{A}(g)$ is simplicial iff each biconnected component of the graph g is a complete graph. For this result and an alternative characterization of such graphs, see [**346**, Proposition 5.2 and Remark 5.3]. It follows that the simplicial graphic arrangements are precisely the cartesian products of braid arrangements.

Cographic arrangements constitute a different class of arrangements studied in [**208**, Section 3], [**210**, Section 8], and [**321**, Section 5]. They are associated to not necessarily simple graphs and are related to graphic arrangements by matroid duality. See also [**35, 230**] for recent related work.

Other work. Among vast interesting work on affine hyperplane arrangements related to the braid arrangement or other reflection arrangements, we mention [**20, 27, 28, 347, 398, 428**].

CHAPTER 7

Descent and lune equations

In any monoid, one may consider the equation $xy = a$, where a is fixed and either x or y or both are varying. We consider such equations for the Tits monoid. The main observation is that their solution sets (which consist of faces) are related to either topological balls or spheres. Hence the expression

$$\sum (-1)^{\mathrm{rk}(F)}$$

with F ranging over the solution set (which is an Euler characteristic), can be computed. This leads to numerous identities which are called descent identities, lune identities and the Witt identities.

More generally, one may consider such equations in the context of (left or right) modules over the Tits monoid. We attach a relative pair (X, A) of cell complexes to the solution set. The notation is meant to remind us of relative pairs in algebraic topology. By construction X is either a ball or sphere, but the topology of A can be complicated. We give examples where A is a ball or sphere or more generally a wedge of spheres. Since the Euler characteristic is known in these cases, we again obtain explicit identities. Left actions give descent identities while right actions give lune identities.

We also apply these considerations to the Birkhoff monoid. Since it is commutative, there is no distinction between left and right modules. So in this case we obtain descent-lune identities. We illustrate this for the example of charts.

Background information on cell complexes and Euler characteristics is given in Section A.1.

7.1. Descent equation

We introduce the descent equation. We study it first for chambers and then for faces. In the simplicial case, its solution set involves the notion of descent between chambers, and more generally, between faces. We discuss identities involving the Euler characteristic of the solution set.

7.1.1. Descent equation. Consider the equation $HC = D$, where C and D are given chambers and we want to find H. We call this the *descent equation for chambers*. For faces, there are two situations one may look at, namely, $HF = G$ and $HF \leq G$, where F and G are arbitrary fixed faces. We refer to either of these as the *descent equation for faces*.

Simplicial case. Let us first assume that $\mathcal{A}$ is a simplicial arrangement.

Proposition 7.1. *Let C and D be any chambers in a simplicial arrangement. If $H_1C = D$ and $H_2C = D$, then $(H_1 \wedge H_2)C = D$.*

PROOF. Since D is a simplex, its subfaces form a Boolean poset. Using the elementary exercise stated below, we deduce that: A wall of D contains $H_1 \wedge H_2$ iff it contains either H_1 or H_2. Now use (1.5) for all three products H_1C, H_2C and $(H_1 \wedge H_2)C$ with i indexing the walls of D. This is sufficient since any chamber is determined by its walls (Proposition 2.10). □

Exercise 7.2. Let A and B be subsets of a finite set I. Let J be a subset which is I minus a singleton. Check that J contains $A \cap B$ iff J contains either A or B.

Proposition 7.1 allows us to make the following definition. For chambers C and D, let $\mathrm{Des}(C, D)$ denote the smallest face H of D such that $HC = D$. In other words,

$$HC = D \iff \mathrm{Des}(C, D) \leq H \leq D. \tag{7.1}$$

We say that $\mathrm{Des}(C, D)$ is the *descent* of D wrt C.

Note that

$$\mathrm{Des}(C, D) = D \iff \overline{C} = D \quad \text{and} \quad \mathrm{Des}(C, D) = O \iff C = D. \tag{7.2}$$

To summarize:

Proposition 7.3. *Let C and D be chambers in a simplicial arrangement. Then the set of solutions to the equation $HC = D$ is a Boolean poset with minimum element $\mathrm{Des}(C, D)$ and maximum element D. A unique solution exists iff $\overline{C} = D$.*

Example 7.4. Let $\mathcal{A}$ be the coordinate arrangement of rank n. The face $F = \mathrm{Des}(C, D)$ admits the following description in terms of the n-tuples of Section 6.1. On entries where C and D differ, F takes the same sign as D. On entries where they agree, F is 0. The faces H as in (7.1) may only differ from D in the entries where C and D agree; those entries may be as in F or as in D. Thus those faces constitute a Boolean poset.

Example 7.5. Let $\mathcal{A}$ be the braid arrangement on I. The face $F = \mathrm{Des}(C, D)$ admits the following description in terms of compositions of I (Section 6.3). To obtain F, merge adjacent blocks of D as long as the linear orders induced by C and D on the merged block agree.

Now let F and G be any faces in a simplicial arrangement such that $GF = G$. Then FG and $F\overline{G}$ have the same support as G. So they are all chambers in the arrangement under this support. Define

$$\mathrm{Des}(F, G) := \mathrm{Des}(FG, G) \tag{7.3a}$$

and

$$\overline{\mathrm{Des}}(F, G) := \mathrm{Des}(F\overline{G}, G), \tag{7.3b}$$

where the rhs refer to the previous definition of $\mathrm{Des}(C, D)$. By definition, both are faces of G. When F and G are both chambers, the two coincide and agree with the previous definition.

Using (7.1) and (7.2) in conjunction with (1.9a) and (1.9b), one can deduce the following.

$$HF \leq G \iff \mathrm{Des}(F, G) \leq H \leq G, \tag{7.4a}$$

$$HF = G \iff \overline{\mathrm{Des}}(F, G) \leq H \leq G. \tag{7.4b}$$

Thus, $\mathrm{Des}(F,G)$ is the smallest face H of G such that $HF \leq G$, while $\overline{\mathrm{Des}}(F,G)$ is the smallest face H of G such that $HF = G$.

Further,

$$\mathrm{Des}(F,G) = G \iff \overline{F} = G \quad \text{and} \quad \mathrm{Des}(F,G) = O \iff F \leq G, \tag{7.5a}$$

$$\overline{\mathrm{Des}}(F,G) = G \iff \overline{F} \leq G \quad \text{and} \quad \overline{\mathrm{Des}}(F,G) = O \iff F = G. \tag{7.5b}$$

The two notions of descent are related by

$$\mathrm{Des}(F,G) \leq \overline{\mathrm{Des}}(F,G) \leq G.$$

This follows by taking $H = \overline{\mathrm{Des}}(F,G)$ in (7.4b) and substituting in (7.4a).

To summarize:

Proposition 7.6. *Let F and G be any faces in a simplicial arrangement, and consider the equations $HF = G$ and $HF \leq G$.*

- *If $GF \neq G$, then neither equation has any solution.*
- *Suppose $GF = G$. The set of solutions of $HF = G$ is a Boolean poset with minimum element $\overline{\mathrm{Des}}(F,G)$ and maximum element G. Further, there is a unique solution iff $\overline{F} \leq G$.*
- *Suppose $GF = G$. The set of solutions of $HF \leq G$ is a Boolean poset with minimum element $\mathrm{Des}(F,G)$ and maximum element G. Further, there is a unique solution iff $\overline{F} = G$.*

We now relate descents in the arrangement $\mathcal{A}$ to descents in the arrangement $\mathcal{A}_A$, where A is any face.

Let C and D be chambers, with D greater than A. Then

$$\mathrm{Des}(AC/A, D/A) = (\mathrm{Des}(C,D) \vee A)/A, \tag{7.6}$$

where the lhs is evaluated in the arrangement $\mathcal{A}_A$. This follows from the definition. Combining it with the first identity in (7.2), we deduce that

$$\mathrm{Des}(C,D) \vee A = D \iff AC = A\overline{D} \iff A\overline{C} = D. \tag{7.7}$$

More generally, let F and G be faces with $GF = G$ and G greater than A. Then

$$\mathrm{Des}(AF/A, G/A) = (\mathrm{Des}(F,G) \vee A)/A \tag{7.8a}$$

and

$$\overline{\mathrm{Des}}(AF/A, G/A) = (\overline{\mathrm{Des}}(F,G) \vee A)/A. \tag{7.8b}$$

This can be deduced from (7.6) by working in the arrangement under the support of G. Combining with the first identity in (7.5a) and (7.5b), respectively, we obtain

$$\mathrm{Des}(F,G) \vee A = G \iff AF = A\overline{G} \iff A\overline{F} = G, \tag{7.9a}$$

and

$$\overline{\mathrm{Des}}(F,G) \vee A = G \iff AF \leq A\overline{G} \iff A\overline{F} \leq G. \tag{7.9b}$$

Exercise 7.7. Suppose C and D are chambers which are both greater than A. Show that $\mathrm{Des}(C,D) \wedge A = O$.

General case. Now let us consider arbitrary arrangements (not necessarily simplicial). The set of solutions to the equation $HC = D$ may no longer be a Boolean poset. In fact, it may not have a unique minimal element. For example, take D to be a square, and C to be a triangle adjacent to D as shown below.

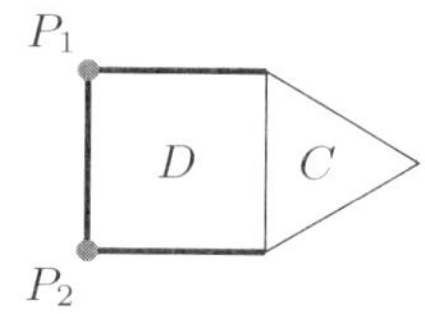

(For a concrete realization, see the arrangement in Section 1.2.4, or consider for instance the arrangement $x_1 = x_2$, $x_1 = -x_2$, $x_2 = x_3$, and $x_2 = -x_3$ in $\mathbb{R}^3$.) The two vertices P_1 and P_2 of D which are not on the edge shared by C and D both satisfy $HC = D$ but the central face does not since $C \neq D$. Thus P_1 and P_2 are the two minimal solutions. (Note that Proposition 7.1 fails.) An example of this kind is present in any nonsimplicial arrangement:

Lemma 7.8. *Let $\mathcal{A}$ be an arrangement. Then $\mathcal{A}$ is simplicial iff for any chambers C and D, the solution set of $HC = D$ is a Boolean poset.*

PROOF. The forward implication follows from Proposition 7.3. For the backward implication, suppose $\mathcal{A}$ is not simplicial. Pick a chamber D which is not a simplex. Then D has a panel such that at least two vertices of D are not on that panel. Let C be the chamber adjacent to D along this panel. Then the solution set of $HC = D$ cannot be a Boolean poset: The vertices of D not on the common panel are the minimal solutions, and there are at least two of them by construction. □

Exercise 7.9. Check that: For any chambers C and D in which D is a simplex, the solution set of $HC = D$ is a Boolean poset.

Let us understand the solution set in more detail in the nonsimplicial case. Let C and D be two fixed chambers. Divide the set of panels of D into two parts depending on whether the supporting wall of the panel separates or does not separate C and D. Label the panel $+$ if its supporting wall does not separate, and $-$ if its separates. This is illustrated below.

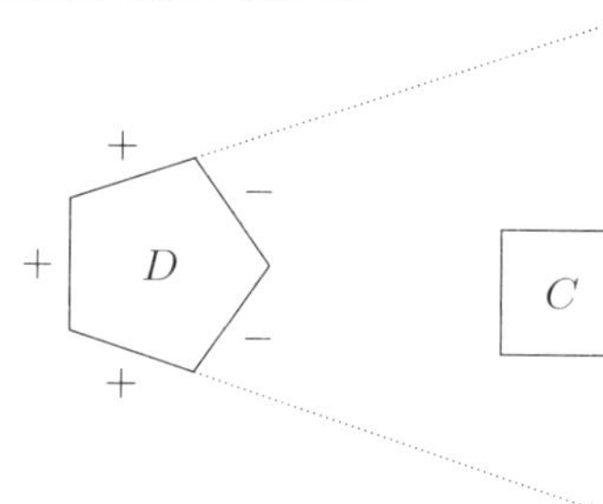

Observation 7.10. If $C = D$, then all panels of D are labeled $+$, and if $C = \overline{D}$, then all panels of D are labeled $-$. In all other cases, the set of panels with $+$ label along with all their faces forms a topological ball, the same happens for the set of panels with a $-$ label, and the two balls share a common boundary sphere.

This can be established by induction on $\operatorname{dist}(C, D)$. In the figure, the three edges labeled $+$ form a topological interval, as also the two edges labeled $-$, and the two intervals share two vertices, which is a 0-dimensional sphere.

Lemma 7.11. *Let H be a panel of D. Then $HC = D$ iff H is labeled $+$. Equivalently, $HC \neq D$ iff H is labeled $-$.*

This is contained in Lemma 1.55.

Lemma 7.12. *If a particular face of D does not solve $HC = D$, then no subface of it can solve this equation.*

PROOF. This follows from Lemma 1.6, item (2). $\square$

Proposition 7.13. *Let C and D be chambers in any arrangement. Then there is a unique solution to the equation $HC = D$ iff $\overline{C} = D$.*

PROOF. Note that $H = D$ is always a solution of $HC = D$. By Lemmas 7.11 and 7.12, this is the unique solution precisely when all panels of D are labeled $-$. This occurs iff $\overline{C} = D$. $\square$

A situation where the descent equation has a unique minimal solution is given below.

Proposition 7.14. *Suppose F is a face of C such that $\overline{F}C = D$. Then $HC = D$ iff $\overline{F} \leq H \leq D$.*

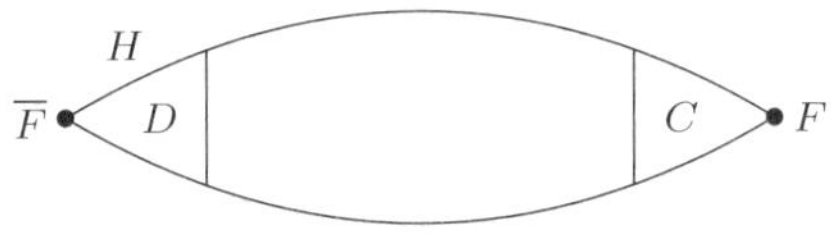

PROOF. The backward implication follows from Lemma 1.6, item (2). For the forward implication: By Lemma 1.6, item (1), H and $\overline{F}$ are faces of D. So they are joinable, and hence $H\overline{F} = \overline{F}H$ by Proposition 1.18. Also, $HFC = HC = D$, so HF is a face of D. Since HF and $H\overline{F}$ are faces of D, we deduce from Proposition 1.19 that $HF = H\overline{F} = H$. Together, we deduce that $\overline{F}H = H$, so $\overline{F} \leq H$ as required. $\square$

The following is a weaker form of Proposition 7.1 but it works for any arrangement.

Proposition 7.15. *Let H be a face of D. If $H_iC = D$ for all $H \leq H_i \lessdot D$, then $HC = D$.*

PROOF. Suppose $H_iC = D$ for all $H \leq H_i \lessdot D$. By Lemma 1.55, D is on the same side as C for the supports of all the H_i. (These are the walls of D which contain H.) By definition (1.5), HC also has the same property. Since a chamber is uniquely determined by its walls (Proposition 2.10), by working in the arrangement $\mathcal{A}_H$, we conclude that $HC = D$. $\square$

Exercise 7.16. Let $\{H_i\}$ be some set of panels of D whose meet is H. Show by an example that $H_iC = D$ for all i does not imply $HC = D$.

Exercise 7.17. Check that Proposition 7.15 holds with the chamber C replaced by any face F.

Let us now improve upon Lemma 7.11.

Proposition 7.18. *Consider the equation $HC = D$, with $\overline{C} \neq D$ and $C \neq D$. The faces which do not solve $HC = D$ are those in the $-$ ball, and O. The faces which solve $HC = D$ are those in the interior of the $+$ ball, and D.*

PROOF. It is clear that D solves the equation $HC = D$ while O does not. Let us look at the remaining faces of D. By Lemma 7.12, none of the faces in the $-$ ball solve this equation. The faces left are those in the interior of the $+$ ball. Any such face belongs only to the $+$ panels of D, so by Proposition 7.15, it solves the equation. □

Proposition 7.19. *Let F and G be any faces in any arrangement, and consider the equations $HF = G$ and $HF \leq G$.*

- *If $GF \neq G$, then neither equation has any solutions.*
- *If $GF = G$, then $HF = G$ has a unique solution iff $\overline{F} \leq G$.*
- *If $GF = G$, then $HF \leq G$ has a unique solution iff $\overline{F} = G$.*

PROOF. The first claim regarding $GF \neq G$ is clear. So we may assume $GF = G$. Then FG and $F\overline{G}$ have the same support as G. So they are all chambers in the arrangement under this support.

By (1.9b), the condition $HF = G$ is equivalent to $HF\overline{G} = G$, so the second claim follows by applying Proposition 7.13 with $D = G$ and $C = F\overline{G}$ and using (1.9a).

Similarly, by (1.9a), the condition $HF \leq G$ is equivalent to $HFG = G$, so the last claim follows by applying Proposition 7.13 with $D = G$ and $C = FG$ and using (1.9b). □

7.1.2. Descent identities.

Proposition 7.20. *In any arrangement, for any chambers C and D,*

$$\sum_{H\colon HC=D} (-1)^{\operatorname{rk}(H)} = \begin{cases} (-1)^{\operatorname{rk}(D)} & \text{if } \overline{C} = D, \\ 0 & \text{otherwise.} \end{cases} \tag{7.10}$$

PROOF. If the arrangement is simplicial, then we can apply Proposition 7.3. Since a Boolean poset is Eulerian (C.9), the lhs is up to sign an instance of (C.5a) for $\overline{C} = D$, and an instance of (C.5b) for $\overline{C} \neq D$.

In the general case, we can proceed as follows. We may assume that the arrangement has rank at least 1. We consider three cases. Let X denote the cell complex consisting of D and all its faces.

- $\overline{C} = D$. Then $H = D$ is the only solution.
- $C = D$. Then, all faces of D solve $HC = D$. The alternating sum (7.10) is the negative of the reduced Euler characteristic of X. Since X is a topological ball, the sum is zero.
- $\overline{C} \neq D$ and $C \neq D$. Equivalently, at least one but not all panels of D solve $HC = D$. Let A denote the cell complex consisting of those faces of D which do not solve $HC = D$. Then by Proposition 7.18, A is a topological ball of dimension one lower than that of X. Then

  $$\sum_{H\colon HC=D} (-1)^{\operatorname{rk}(H)} = -\chi(X) + \chi(A) = 0,$$

  since both X and A are topological balls. □

Let us see what is going on in a concrete example. Consider the nonsimplicial arrangement discussed in Section 1.2.4. It is redrawn below for convenience.

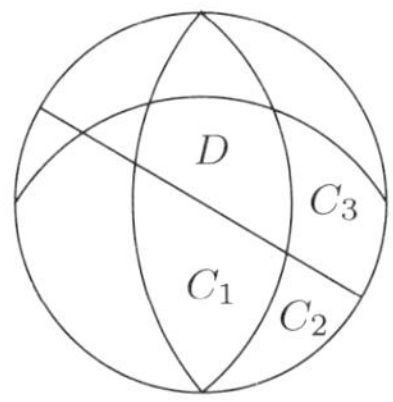

The solution set of $HC_1 = D$ consists of D itself, three (contiguous) edges of D, and the two vertices in-between. (These edges and vertices topologically form an open interval.) So the sum (7.10) is $-1+(1+1+1)-(1+1)=0$. The situation for $HC_3 = D$ is completely analogous. The solution set of $HC_2 = D$ consists of D itself, two (contiguous) edges of D, and the vertex in-between. So the sum (7.10) is $-1+(1+1)-1=0$.

Proposition 7.21. *In any arrangement, for any faces F and G,*

$$\sum_{H:\, HF=G} (-1)^{\operatorname{rk}(H)} = \begin{cases} (-1)^{\operatorname{rk}(G)} & \text{if } \overline{F} \le G, \\ 0 & \text{otherwise,} \end{cases} \tag{7.11a}$$

$$\sum_{H:\, HF\le G} (-1)^{\operatorname{rk}(H)} = \begin{cases} (-1)^{\operatorname{rk}(G)} & \text{if } \overline{F} = G, \\ 0 & \text{otherwise.} \end{cases} \tag{7.11b}$$

PROOF. If the arrangement is simplicial, then the result can be deduced from Proposition 7.6. In the general case, we proceed as follows. The argument will repeatedly make use of (1.9a) and (1.9b). We may assume $GF = G$, since otherwise there is no H satisfying $HF = G$ or $HF \le G$. Since all the action happens in the support of G, we may further assume that G is a chamber (and hence so is FG). The condition $HF \le G$ is equivalent to $HFG = G$, so (7.11b) is the same as (7.10) with $D = G$ and $C = FG$. Similarly, the condition $HF = G$ is equivalent to $HF\overline{G} = G$, so (7.11a) is the same as (7.10) with $D = G$ and $C = F\overline{G}$. □

Exercise 7.22. Use (7.11a) to deduce (7.11b), and vice-versa.

Exercise 7.23. Deduce Proposition 1.77 using Propositions 7.21 and 1.19.

7.2. Lune equation

The lune equation is a companion of the descent equation. Its solution set involves lunes which were studied in detail in Chapter 3. We discuss identities involving the Euler characteristic of the solution set.

7.2.1. Lune equation. Consider the equation $HC = D$, where H and D are fixed and we want to find C. We call this the *lune equation*. This is because its solutions form a combinatorial top-lune. This follows from (3.3) and Proposition 3.5.

The general case is to consider the equations $HF = G$ and $HF \le G$, where H and G are arbitrary fixed faces. The solution sets are, respectively, the interior and the closure of a combinatorial lune. This follows from (3.10) and (3.11).

7.2.2. Lune identities. We can use the topology of a lune (Proposition 3.11) to obtain identities involving the Euler characteristic of the solution sets of the lune equation.

Proposition 7.24. *In any arrangement, for any faces H and G,*

$$\sum_{F:\, HF=G} (-1)^{\mathrm{rk}(F)} = \begin{cases} (-1)^{\mathrm{rk}(G)} & \text{if } H \le G, \\ 0 & \text{otherwise,} \end{cases} \tag{7.12a}$$

$$\sum_{F:\, HF\le G} (-1)^{\mathrm{rk}(F)} = \begin{cases} (-1)^{\mathrm{rk}(G)} & \text{if } H = G, \\ 0 & \text{otherwise.} \end{cases} \tag{7.12b}$$

PROOF. We may assume $H \le G$, else both lhs are zero. We consider two further subcases.

- $H = G$. In this case, $HF = G$ holds for any face F whose support is smaller than $\mathrm{s}(G)$, and so the sum in (7.12a) as well as in (7.12b) is the negative of the reduced Euler characteristic of the sphere of dimension $\mathrm{rk}(G) - 1$, which is $(-1)^{\mathrm{rk}(G)}$.
- $H < G$. In this case, the sum in (7.12b) is the reduced Euler characteristic of a ball which is 0, while the sum in (7.12a) is the relative Euler characteristic of a pair consisting of a ball and its boundary sphere. Since the ball contributes 0, this sum is the reduced Euler characteristic of the sphere of dimension $\mathrm{rk}(G) - 2$ which is $(-1)^{\mathrm{rk}(G)}$. □

A more general result (with a different proof) is given below.

Proposition 7.25. *In any arrangement, for any faces H and G,*

$$\sum_{F:\, HF=G} (-1)^{\mathrm{rk}(F)} q^{\mathrm{dist}(H,F)} = \begin{cases} (-1)^{\mathrm{rk}(G)} q^{\mathrm{dist}(H,\overline{H})} & \text{if } H \le G, \\ 0 & \text{otherwise,} \end{cases} \tag{7.13a}$$

$$\sum_{F:\, HF\le G} (-1)^{\mathrm{rk}(F)} q^{\mathrm{dist}(H,F)} = \begin{cases} (-1)^{\mathrm{rk}(G)} q^{\mathrm{dist}(H,\overline{H})} & \text{if } H = G, \\ 0 & \text{otherwise.} \end{cases} \tag{7.13b}$$

PROOF. Let us look at (7.13b). We give two arguments. The idea is to introduce an intermediate indexing face K which is either FH or FG. The figure below shows faces H, F and G satisfying $HF \le G$ along with the two choices for K.

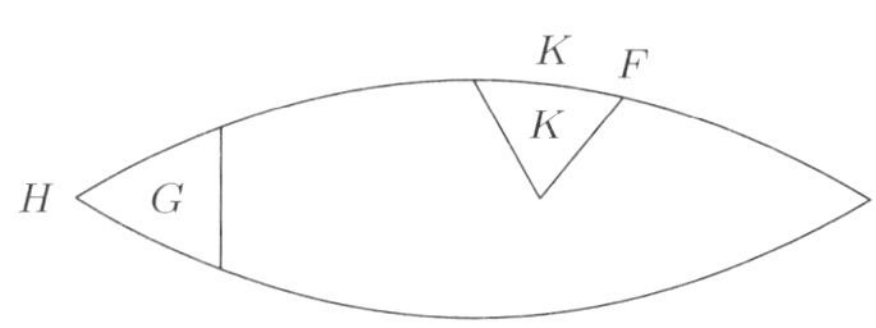

Using the first choice for K, the lhs of (7.13b) can be manipulated as follows.

$$\begin{aligned}\sum_{K:\, HK\le G} q^{\mathrm{dist}(H,K)} \sum_{F:\, FH=K} (-1)^{\mathrm{rk}(F)} &= \sum_{K:\, HK\le G,\, \overline{H}\le K} q^{\mathrm{dist}(H,K)} (-1)^{\mathrm{rk}(K)} \\ &= q^{\mathrm{dist}(H,\overline{H})} \sum_{K:\, \overline{H}\le K\le \overline{H}G} (-1)^{\mathrm{rk}(K)}.\end{aligned}$$

The first equality uses (7.11a). By the Eulerian property (1.43), the last sum is 0 unless $\overline{H} = \overline{H}G$, which is the same as $H = G$.

Alternatively, using the second choice for K, one may manipulate the lhs of (7.13b) as follows.

$$\sum_{K:\, HK=G} q^{\mathrm{dist}(H,K)} \sum_{F:\, FG=K} (-1)^{\mathrm{rk}(F)} = \sum_{K:\, HK=G,\, \overline{G}\leq K} q^{\mathrm{dist}(H,K)}(-1)^{\mathrm{rk}(K)}$$

If $H < G$, then there is no choice for K, hence the sum is 0. If $H = G$, then there is exactly one choice for K, namely, $K = \overline{G}$.

The identity (7.13a) can be proved similarly by summing over all K of the same support as G with $HK = G$, and then summing over all F with $FH = K$. □

Exercise 7.26. Use (7.12a) to deduce (7.12b), and vice-versa. More generally, do the same for (7.13a) and (7.13b).

Exercise 7.27. Deduce (7.12a) and (7.12b) by specializing (2.13) and (2.12), respectively, to $\mathrm{V} := \mathrm{s}(H, G)$. (Use (3.10) and (3.11).)

7.3. Witt identities

We now look at some special identities called the Witt identities which combine features of the descent and the lune identities.

7.3.1. Witt identity for chambers.

The Witt identity for chambers arises by considering the equation $HC = D$, where both H and C are regarded as variables.

Proposition 7.28. *In any arrangement, for a fixed chamber D, and scalars x^C indexed by chambers C,*

$$\sum_{H:\, H\leq D} (-1)^{\mathrm{rk}(H)} \Big(\sum_{C:\, HC=D} x^C \Big) = (-1)^{\mathrm{rk}(D)} x^{\overline{D}}. \tag{7.14}$$

We point out that the x^C instead of being scalars can more generally be taken to be elements of some fixed abelian group. Similar remark applies to the later identities.

PROOF. By interchanging the summations, the lhs of (7.14) can be written as

$$\sum_{C} \Big(\sum_{H:\, HC=D} (-1)^{\mathrm{rk}(H)} \Big) x^C.$$

The result now follows from (7.10). □

We refer to (7.14) as the *Witt identity*. A special case is given below.

Proposition 7.29. *In a reflection arrangement, for a fixed chamber D,*

$$\sum_{H:\, H\leq D} (-1)^{\mathrm{rk}(H)} \frac{1}{c_H} = (-1)^{\mathrm{rk}(D)} \frac{1}{c_O}, \tag{7.15}$$

where c_H is the number of chambers in the arrangement $\mathcal{A}_H$, and, in particular, c_O is the number of chambers in $\mathcal{A}$.

PROOF. In a reflection arrangement, by Lemma 5.21,

$$|\{C \mid HC = D\}| = \frac{c_O}{c_H}.$$

Now specialize all the x^C to 1 in (7.14) to deduce the result. □

A more general formulation of the Witt identity is given below.

Proposition 7.30. *In any arrangement, for a top-nested face (A, D), and scalars x^C indexed by chambers C,*

$$\sum_{H\colon A \leq H \leq D} (-1)^{\mathrm{rk}(H)} \Big(\sum_{C\colon HC = D} x^C \Big) = (-1)^{\mathrm{rk}(D)} \sum_{C\colon AC = A\overline{D}} x^C. \tag{7.16}$$

Setting A to be the central face recovers (7.14).

PROOF. For any chamber C' greater than A, put

$$x_A^{C'} := \sum_{C\colon AC = C'} x^C.$$

Then, for $A \leq H \leq D$,

$$\sum_{C\colon HC = D} x^C = \sum_{C'\colon HC' = D} \; \sum_{C\colon AC = C'} x^C = \sum_{C'\colon HC' = D, C' \geq A} x_A^{C'}.$$

(This is based on the lune decomposition (3.20).) Now apply (7.14) to the arrangement $\mathcal{A}_A$ and the scalars $x_A^{C'}$. □

One may generalize even further in the simplicial case:

Proposition 7.31. *In a simplicial arrangement, for a top-nested face (K, D), and scalars x^C indexed by chambers C,*

$$\sum_{H\colon H \leq K} (-1)^{\mathrm{rk}(H)} \Big(\sum_{C\colon HC = D} x^C \Big) = (-1)^{\mathrm{rk}(K)} \sum_{C\colon \mathrm{Des}(C, D) = K} x^C, \tag{7.17}$$

and more generally, for $A \leq K \leq D$,

$$\sum_{H\colon A \leq H \leq K} (-1)^{\mathrm{rk}(H)} \Big(\sum_{C\colon HC = D} x^C \Big) = (-1)^{\mathrm{rk}(K)} \sum_{C\colon \mathrm{Des}(C, D) \vee A = K} x^C. \tag{7.18}$$

For $K = D$, we see from (7.2) that (7.17) specializes to (7.14), and more generally from (7.7) that (7.18) specializes to (7.16).

PROOF. By interchanging the summations and using (7.1), the lhs of (7.17) can be written as

$$\sum_C \Big(\sum_{H\colon \mathrm{Des}(C,D) \leq H \leq K} (-1)^{\mathrm{rk}(H)} \Big) x^C.$$

The sum inside the parenthesis is zero unless $\mathrm{Des}(C, D) = K$, and (7.17) follows. Applying this identity to the arrangement $\mathcal{A}_A$ and the scalars $x_A^{C'}$ (as defined in the proof of Proposition 7.30), we deduce that up to the factor $(-1)^{\mathrm{rk}(K)}$, the lhs of (7.18) equals

$$\sum_{C'\colon \mathrm{Des}(C'/A,D/A)=K/A} x_A^{C'} = \sum_{C'\colon \mathrm{Des}(C'/A,D/A)=K/A} \sum_{C\colon AC=C'} x^C = \sum_{C\colon \mathrm{Des}(AC/A,D/A)=K/A} x^C.$$

Now apply (7.6). □

7.3.2. Witt identity for faces. There are two avatars of the Witt identity if we work with scalars indexed by faces instead of chambers. This is because instead of $HC = D$ we now have two equations to consider, namely, $HF \leq G$ and $HF = G$. By setting $G = D$ and the scalars x^F to be zero when F is not a chamber, any Witt identity for faces specializes to a Witt identity for chambers. In this sense, the results that we now discuss imply the previous ones. The proofs are similar, so we only indicate them briefly.

The analogue of Proposition 7.28 is as follows.

Proposition 7.32. *In any arrangement, for a fixed face G, and scalars x^F indexed by faces F,*

$$\sum_{H\colon H\leq G} (-1)^{\mathrm{rk}(H)}\Big(\sum_{F\colon HF\leq G} x^F\Big) = (-1)^{\mathrm{rk}(G)} x^{\overline{G}}, \tag{7.19a}$$

$$\sum_{H\colon H\leq G} (-1)^{\mathrm{rk}(H)}\Big(\sum_{F\colon HF=G} x^F\Big) = (-1)^{\mathrm{rk}(G)} \sum_{F\colon \overline{F}\leq G} x^F. \tag{7.19b}$$

PROOF. Interchange the summations, and use (7.11a) and (7.11b). □

The analogue of Proposition 7.30 is as follows.

Proposition 7.33. *In any arrangement, for a nested face (A, G), and scalars x^F indexed by faces F,*

$$\sum_{H\colon A\leq H\leq G} (-1)^{\mathrm{rk}(H)}\Big(\sum_{F\colon HF\leq G} x^F\Big) = (-1)^{\mathrm{rk}(G)} \sum_{F\colon AF=A\overline{G}} x^F, \tag{7.20a}$$

$$\sum_{H\colon A\leq H\leq G} (-1)^{\mathrm{rk}(H)}\Big(\sum_{F\colon HF=G} x^F\Big) = (-1)^{\mathrm{rk}(G)} \sum_{F\colon AF\leq A\overline{G}} x^F. \tag{7.20b}$$

This generalizes Proposition 7.32 and is proved by applying it to the arrangement $\mathcal{A}_A$. The symmetry between the two identities in the rhs is more manifest now.

The analogue of Proposition 7.31 is as follows.

Proposition 7.34. *In a simplicial arrangement, for a nested face (K, G), and scalars x^F indexed by faces F,*

$$\sum_{H\colon H\leq K} (-1)^{\mathrm{rk}(H)}\Big(\sum_{F\colon HF\leq G} x^F\Big) = (-1)^{\mathrm{rk}(K)} \sum_{F\colon GF=G,\, \mathrm{Des}(F,G)=K} x^F, \tag{7.21a}$$

(7.21b) $$\sum_{H:\, H\leq K} (-1)^{\operatorname{rk}(H)} \Big(\sum_{F:\, HF=G} x^F \Big) = (-1)^{\operatorname{rk}(K)} \sum_{F:\, GF=G,\, \overline{\operatorname{Des}}(F,G)=K} x^F.$$

and more generally, for $A \leq K \leq G$,

(7.22a) $$\sum_{H:\, A\leq H\leq K} (-1)^{\operatorname{rk}(H)} \Big(\sum_{F:\, HF\leq G} x^F \Big) = (-1)^{\operatorname{rk}(K)} \sum_{F:\, GF=G,\, \operatorname{Des}(F,G)\vee A=K} x^F,$$

(7.22b) $$\sum_{H:\, A\leq H\leq K} (-1)^{\operatorname{rk}(H)} \Big(\sum_{F:\, HF=G} x^F \Big) = (-1)^{\operatorname{rk}(K)} \sum_{F:\, GF=G,\, \overline{\operatorname{Des}}(F,G)\vee A=K} x^F.$$

PROOF. We explain the first two identities. By interchanging the summations and using (7.4a), the lhs of (7.21a) can be written as

$$\sum_{F:\, GF=G} \Big(\sum_{H:\, \operatorname{Des}(F,G)\leq H\leq K} (-1)^{\operatorname{rk}(H)} \Big) x^F.$$

The sum inside the parenthesis is zero unless $\operatorname{Des}(F,G) = K$. The first identity follows. The second identity can be deduced similarly from (7.4b). □

In the simplicial setting, (7.22a) and (7.22b) are the most general Witt identities from which all the earlier ones can be deduced.

There are specializations of Witt identities which are of interest. We mention one of them below.

Lemma 7.35. *For any face* G,

$$\sum_{H,F:\, HF=G} (-1)^{\operatorname{rk}(H)+\operatorname{rk}(F)} = \begin{cases} 1 & \text{if } G = O, \\ 0 & \text{otherwise.} \end{cases}$$

The sum is over both H *and* F.

PROOF. Put $x^F = (-1)^{\operatorname{rk}(F)}$ in (7.19b), and then use (1.44). Alternatively and more simply, we can first use either (7.11a) or (7.12a) followed by (1.44). □

Exercise 7.36. For faces F, H, H', G and G' with $\operatorname{s}(H) = \operatorname{s}(H')$,

$$\sum_{F:\, HF=G,\, FH'=G'} (-1)^{\operatorname{rk}(F)} = \begin{cases} (-1)^{\operatorname{rk}(G)} & \text{if } HG' = G \text{ and } \overline{H'}G = G', \\ 0 & \text{otherwise.} \end{cases}$$

Deduce (7.11a) and (7.12a) as consequences.

Exercise 7.37. For faces G and G',

$$\sum_{\substack{F,H:\\ HF=G,\, FH=G'}} (-1)^{\operatorname{rk}(H)+\operatorname{rk}(F)} = \begin{cases} (-1)^{\operatorname{rk}(K)+\operatorname{rk}(G)} & \text{if } G' = \overline{K}G \text{ for some } K \leq G, \\ 0 & \text{otherwise.} \end{cases}$$

The sum is over both H and F. (Use Exercise 7.36 with $H = H'$.) Deduce Lemma 7.35 as a consequence.

7.4. Descent-lune equation for flats

Interestingly, in the context of flats, the descent and lune equations merge into one equation, and so do the corresponding identities. In addition to alternating sums, the identities involve chamber counts in certain arrangements over and under flats. This is not surprising since these identities result from summing the identities for faces, and in doing so faces with the same support get lumped together.

7.4.1. Descent-lune equation. Consider the equation $\mathrm{Y} \vee \mathrm{X} = \mathrm{W}$, where X and W are fixed flats, and we want to solve for Y. Since $\mathrm{Y} \vee \mathrm{X} = \mathrm{X} \vee \mathrm{Y}$, this can be viewed as the flat-analogue of either the descent equation or the lune equation. Hence we call it the *descent-lune equation*.

7.4.2. Descent-lune identities. The analogue of Propositions 7.21 and 7.24 is given below. Recall that $c^{\mathrm{Y}}_{\mathrm{X}}$ denotes the number of chambers in the arrangement $\mathcal{A}^{\mathrm{Y}}_{\mathrm{X}}$.

Proposition 7.38. *In any arrangement, for any flats* X *and* W*,*

$$\sum_{\mathrm{Y}:\, \mathrm{X}\vee\mathrm{Y}=\mathrm{W}} (-1)^{\mathrm{rk}(\mathrm{Y})} c^{\mathrm{Y}} = \begin{cases} (-1)^{\mathrm{rk}(\mathrm{W})} c^{\mathrm{W}}_{\mathrm{X}} & \text{if } \mathrm{X} \leq \mathrm{W}, \\ 0 & \text{otherwise,} \end{cases} \tag{7.23a}$$

$$\sum_{\mathrm{Y}:\, \mathrm{X}\vee\mathrm{Y}\leq\mathrm{W}} (-1)^{\mathrm{rk}(\mathrm{Y})} c^{\mathrm{Y}} c^{\mathrm{W}}_{\mathrm{X}\vee\mathrm{Y}} = \begin{cases} (-1)^{\mathrm{rk}(\mathrm{W})} & \text{if } \mathrm{X} = \mathrm{W}, \\ 0 & \text{otherwise.} \end{cases} \tag{7.23b}$$

PROOF. Let us use Proposition 7.21. Pick any face F of support X. Sum (7.11a) over all G of support W, and we obtain (7.23a). If we sum (7.11b) instead, then we obtain (7.23b).

Alternatively, we may use Proposition 7.24. Pick any face H of support X. Sum (7.12a) over all G of support W, and we obtain (7.23a). If we sum (7.12b) instead, then we obtain (7.23b). □

Note that identity (1.42) can be recovered by setting $\mathrm{X} := \mathrm{W}$ in (7.23a) or (7.23b).

Exercise 7.39. Verify Proposition 7.38 explicitly for rank-two arrangements.

7.5. Descent and lune equations for partial-flats

Recall from Section 2.8 that partial-flats interpolate faces and flats. One can consider descent and lune equations and the corresponding identities in this more general context.

7.5.1. Descent and lune equations. Let $\sim$ be a partial-support relation. Consider the equation $\mathrm{yx} = \mathrm{w}$, where x and w are fixed partial-flats, and we want to find y. We call this the *descent equation for partial-flats*. If instead, y and w are fixed, and x is the unknown, then we call this the *lune equation for partial-flats*. One can also consider the variants resulting from $\mathrm{yx} \leq \mathrm{w}$.

For $\Sigma_\sim = \Sigma$, these equations specialize to the descent equation and lune equation, respectively, for faces. For $\Sigma_\sim = \Pi$, both equations specialize to the descent-lune equation for flats.

7.5.2. Descent and lune identities. For partial-flats $\mathrm{x} \leq \mathrm{y}$, let $c_{\mathrm{x}}^{\mathrm{y}}$ denote the number of faces with partial-support y which are greater than some fixed face in x. In particular, $c_{\mathrm{x}}^{\mathrm{x}} = 1$.

We now write down identities involving weighted sums over the solution sets of the descent and lune equations. They unify the corresponding identities for faces and flats.

Proposition 7.40. *In any arrangement, for partial-flats* x *and* w,

$$\sum_{\mathrm{y}:\, \mathrm{yx}=\mathrm{w}} (-1)^{\mathrm{rk}(\mathrm{y})} c^{\mathrm{y}} = \begin{cases} (-1)^{\mathrm{rk}(\mathrm{w})} c_{\overline{\mathrm{x}}}^{\mathrm{w}} & \textit{if } \overline{\mathrm{x}} \leq \mathrm{w}, \\ 0 & \textit{otherwise,} \end{cases} \tag{7.24a}$$

$$\sum_{\mathrm{y}:\, \mathrm{yx}\leq\mathrm{w}} (-1)^{\mathrm{rk}(\mathrm{y})} c^{\mathrm{y}} c_{\mathrm{yx}}^{\mathrm{w}} = \begin{cases} (-1)^{\mathrm{rk}(\mathrm{w})} & \textit{if } \overline{\mathrm{x}} = \mathrm{w}, \\ 0 & \textit{otherwise.} \end{cases} \tag{7.24b}$$

PROOF. Let us use Proposition 7.21. Fix a face F in x. Sum (7.11a) over all G in w, and we obtain (7.24a). If we sum (7.11b) instead, then we obtain (7.24b). □

Proposition 7.41. *In any arrangement, for partial-flats* y *and* w,

$$\sum_{\mathrm{x}:\, \mathrm{yx}=\mathrm{w}} (-1)^{\mathrm{rk}(\mathrm{x})} c^{\mathrm{x}} = \begin{cases} (-1)^{\mathrm{rk}(\mathrm{w})} c_{\mathrm{y}}^{\mathrm{w}} & \textit{if } \mathrm{y} \leq \mathrm{w}, \\ 0 & \textit{otherwise,} \end{cases} \tag{7.25a}$$

$$\sum_{\mathrm{x}:\, \mathrm{yx}\leq\mathrm{w}} (-1)^{\mathrm{rk}(\mathrm{x})} c^{\mathrm{x}} c_{\mathrm{yx}}^{\mathrm{w}} = \begin{cases} (-1)^{\mathrm{rk}(\mathrm{w})} & \textit{if } \mathrm{y} = \mathrm{w}, \\ 0 & \textit{otherwise.} \end{cases} \tag{7.25b}$$

PROOF. Here we use Proposition 7.24. Fix a face H in y. Sum (7.12a) over all G in w, and we obtain (7.23a). If we sum (7.12b) instead, then we obtain (7.23b). □

One may check that Propositions 7.40 and 7.41 specialize to Propositions 7.21 and 7.24 for $\Sigma_{\sim} = \Sigma$, and to Proposition 7.38 for $\Sigma_{\sim} = \Pi$.

Exercise 7.42. Show that: In any arrangement, for maximal partial-flats c and d,

$$\sum_{\mathrm{y}:\, \mathrm{yc}=\mathrm{d}} (-1)^{\mathrm{rk}(\mathrm{y})} c^{\mathrm{y}} = \begin{cases} (-1)^{\mathrm{rk}(\mathrm{d})} & \text{if } \overline{\mathrm{c}} = \mathrm{d}, \\ 0 & \text{otherwise.} \end{cases} \tag{7.26}$$

7.5.3. Weisner formula.

Proposition 7.43. *Suppose* $\sim$ *is a geometric partial-support relation. Then for partial-flats* $\mathrm{z} < \mathrm{x} \leq \mathrm{w}$,

$$\sum_{\mathrm{y}:\, \mathrm{y}\geq\mathrm{z},\, \mathrm{yx}=\mathrm{w}} \mu(\mathrm{z}, \mathrm{y}) = 0, \tag{7.27}$$

where μ *refers to the Möbius function of* $\Sigma_{\sim}$.

PROOF. The interval $[\mathrm{z}, \mathrm{w}]$ in the poset of partial-flats is a lattice. Apply the Weisner formula (C.7a) and use Lemma 2.74. □

One may check that Proposition 7.43 specializes to Proposition 1.77 for $\Sigma_{\sim} = \Sigma$.

7.6. Faces and flats for left Σ-sets

Let h be a set on which the Tits monoid Σ acts on the left. We introduce the notion of h-faces and h-flats and show that they carry the structure of a dimonoid. (A dimonoid is a set with two compatible associative operations, usually denoted $\vdash$ and $\dashv$. See [**289**].) For specific choices of h, one recovers familiar notions such as faces, flats, and lunes, as tabulated below.

left Σ-set h	h-face	h-flat
E	face	flat
Γ	top-nested face	top-lune
Σ	nested face	lune

There is a support map from h-faces to h-flats, which generalizes the support map from faces to flats, and from nested faces to lunes.

7.6.1. Left Σ-sets. A *left Σ-set* is a set h equipped with a rule which assigns to each face F and element $x \in \mathrm{h}$, an element $F \cdot x \in \mathrm{h}$ such that

$$F \cdot (G \cdot x) = FG \cdot x \quad \text{and} \quad O \cdot x = x.$$

In this situation, we say that the monoid Σ acts on h on the left. A map of left Σ-sets is a map $f : \mathrm{h} \to \mathrm{k}$ such that

$$f(F \cdot x) = F \cdot f(x).$$

This defines the category of left Σ-sets.

Some examples of left Σ-sets are given below.

- Let $\mathrm{E} := \{*\}$ denote any singleton set. The unique choice $F \cdot * = *$ defines an action of Σ on E. This is the terminal object in the category of left Σ-sets.
- The monoid Σ acts on itself, that is, $F \cdot G := FG$.
- Recall that the set of chambers Γ is a two-sided ideal of Σ. So Σ acts on Γ by restriction, that is, $F \cdot C := FC$. The inclusion $\Gamma \to \Sigma$ is a map of left Σ-sets.
- The monoid Σ acts on Π via $F \cdot \mathrm{X} := \mathrm{s}(F) \vee \mathrm{X}$.

7.6.2. Stars. Let h be a left Σ-set. For a face F, define

$$\mathrm{h}_F := \{F \cdot x \mid x \in \mathrm{h}\} = \{x \in \mathrm{h} \mid F \cdot x = x\}. \tag{7.28}$$

We call this set the star of F in h.

Note that Σ_F is the star of F, while Γ_F is the top-star of F. The notations are consistent with those in Section 1.7.3. Also Π_F consists of all flats X with $\mathrm{s}(F) \leq \mathrm{X}$.

7.6.3. h-faces. Let h be a left Σ-set. Let ${}^{\mathrm{h}}\Sigma$ denote the subset of $\Sigma \times \mathrm{h}$ consisting of pairs (F, x) such that

$$F \cdot x = x.$$

We refer to such a pair (F, x) as a h-*face*. Thus ${}^{\mathrm{h}}\Sigma$ is the set of h-faces.

- A E-face is the same as a face.
- A Σ-face is the same as a nested face, namely, a pair (H, F) with $H \leq F$. This follows from (1.9a).

- A Γ-face is the same as a top-nested face, namely, a pair (H, D) with $H \leq D$.
- A Π-face is a pair (H, X) with $\mathrm{s}(H) \leq \mathrm{X}$.

7.6.4. h-flats. Define an equivalence relation on ${}^{\mathrm{h}}\Sigma$ by

$$(F, x) \sim (G, y) \iff FG = F,\ GF = G,\ F \cdot y = x,\ G \cdot x = y. \tag{7.29}$$

Reflexivity follows from the definition of a h-face. Symmetry is clear. We omit the transitivity check.

Let ${}^{\mathrm{h}}\Pi$ denote the set of equivalence classes. We refer to an equivalence class as a h-*flat*. Thus ${}^{\mathrm{h}}\Pi$ is the set of h-flats.

- E-flats correspond to flats (1.14).
- Σ-flats correspond to lunes (3.13).
- Γ-flats correspond to top-lunes (3.7).
- A Π-flat is the same as a nested flat, namely, a pair (Y, X) with $\mathrm{Y} \leq \mathrm{X}$.

7.6.5. Support map. We call the canonical quotient map

$$\mathrm{s} : {}^{\mathrm{h}}\Sigma \to {}^{\mathrm{h}}\Pi$$

as the *support map*. Thus, the support of a h-face is a h-flat.

For h = E, this is the usual support map from faces to flats, for h = Γ, this is the support map from top-nested faces to top-lunes, and for h = Σ, this is the support map from nested faces to lunes.

7.6.6. Naturality. Suppose $f : \mathrm{h} \to \mathrm{k}$ is a map of left Σ-sets. If (F, x) is a h-face, then $(F, f(x))$ is a k-face:

$$F \cdot f(x) = f(F \cdot x) = f(x).$$

Thus, we obtain an induced map ${}^{\mathrm{h}}\Sigma \to {}^{\mathrm{k}}\Sigma$. Further, if $(F, x) \sim (G, y)$, then $(F, f(x)) \sim (G, f(y))$:

$$F \cdot f(y) = f(F \cdot y) = f(x) \quad \text{and} \quad G \cdot f(x) = f(G \cdot x) = f(y).$$

Thus, there is also an induced map ${}^{\mathrm{h}}\Pi \to {}^{\mathrm{k}}\Pi$, and a commutative diagram

$$\begin{array}{ccc} {}^{\mathrm{h}}\Sigma & \longrightarrow & {}^{\mathrm{k}}\Sigma \\ {\scriptstyle \mathrm{s}}\downarrow & & \downarrow{\scriptstyle \mathrm{s}} \\ {}^{\mathrm{h}}\Pi & \longrightarrow & {}^{\mathrm{k}}\Pi. \end{array} \tag{7.30}$$

To summarize, the assignments $\mathrm{h} \mapsto {}^{\mathrm{h}}\Sigma$ and $\mathrm{h} \mapsto {}^{\mathrm{h}}\Pi$ are functors, and the support map is a natural transformation between them.

Example 7.44. Take h := Σ and k := Π, and f to be the usual support map from faces to flats. Diagram (7.30) takes the following form.

$$\begin{array}{ccc} {}^{\Sigma}\Sigma & \longrightarrow & {}^{\Pi}\Sigma \\ \downarrow & & \downarrow \\ {}^{\Sigma}\Pi & \longrightarrow & {}^{\Pi}\Pi. \end{array} \qquad \begin{array}{ccc} (H, G) & \longmapsto & (H, \mathrm{s}(G)) \\ \downarrow & & \downarrow \\ \mathrm{s}(H, G) & \longmapsto & (\mathrm{s}(H), \mathrm{s}(G)). \end{array}$$

Here (H, G) is a nested face, and $\mathrm{s}(H, G)$ is a lune. Note from (4.5) that the bottom horizontal map is the base-case map.

Example 7.45. For any h, there is a unique morphism $\mathrm{h} \to \mathrm{E}$ of left Σ-sets. Since a E-face is a face and a E-flat is a flat, diagram (7.30) takes the form

$$\begin{array}{ccc} {}^{\mathrm{h}}\Sigma & \longrightarrow & \Sigma \\ {\scriptstyle s}\downarrow & & \downarrow{\scriptstyle s} \\ {}^{\mathrm{h}}\Pi & \longrightarrow & \Pi. \end{array} \tag{7.31}$$

The top horizontal map projects on the first coordinate. For $\mathrm{h} = \Gamma$, the bottom horizontal map specializes to the base map from top-lunes to flats (4.2), and for $\mathrm{h} = \Sigma$, it specializes to the base map from lunes to flats.

Exercise 7.46. Take $\mathrm{h} = \Gamma_{\sim}$, where $\sim$ is a partial-support relation on chambers. Condition (2.29) implies that $\Gamma_{\sim}$ is a left Σ-set. It interpolates between Γ and E. Describe $\Gamma_{\sim}$-faces and $\Gamma_{\sim}$-flats. Do the same for $\mathrm{h} = \Sigma_{\sim}$, where $\sim$ is a partial-support relation on faces. It is a left Σ-set due to condition (2.31c). It interpolates between Σ and Π.

Use Proposition 2.66 to recast the action of $\Sigma[\mathcal{A}]$ on $\Gamma_{\sim}[\mathcal{A}]$ as follows. For any subarrangement $\mathcal{A}'$, $\Sigma[\mathcal{A}]$ acts on $\Gamma[\mathcal{A}']$. Further, this action arises from the action of $\Sigma[\mathcal{A}']$ on $\Gamma[\mathcal{A}']$ via the morphism of monoids (1.15). As a special case, for any flat X, $\Sigma[\mathcal{A}]$ acts on $\Gamma[\mathcal{A}_{\mathrm{X}}]$. By Lemma 3.2, the latter is the set of top-lunes with base X. Explicitly, for a face F and top-lune L with base X, $F \cdot \mathrm{L}$ is the unique top-lune with base X which contains the chamber FC for any chamber C in L.

7.6.7. Dimonoids. The set of h-faces ${}^{\mathrm{h}}\Sigma$ is a bimodule over Σ, that is, Σ acts on ${}^{\mathrm{h}}\Sigma$ both on the left and the right, and the two actions commute with each other. The left and right actions are defined by

$$F \cdot (G, x) := (FG, F \cdot x) \qquad \text{and} \qquad (F, x) \cdot G := (FG, FG \cdot x).$$

Further observe that the map ${}^{\mathrm{h}}\Sigma \to \Sigma$ in (7.31) which projects on the first coordinate is a map of Σ-bimodules. It follows that ${}^{\mathrm{h}}\Sigma$ is a dimonoid [**289**, Example 2.2.d] under the operations

$$(F, x) \vdash (G, y) := F \cdot (G, y) = (FG, F \cdot y),$$
$$(F, x) \dashv (G, y) := (F, x) \cdot G = (FG, FG \cdot x).$$

Elements of the form (O, x) are the bar-units of this dimonoid.

The left and right actions of Σ on ${}^{\mathrm{h}}\Sigma$ are compatible with the equivalence relations (1.14) and (7.29): if $A \sim A'$ and $(F, x) \sim (F', x')$, then $A \cdot (F, x) \sim A' \cdot (F', x')$ and $(F, x) \cdot A \sim (F', x') \cdot A'$. Further, $A \cdot (F, x) \sim (F, x) \cdot A$. It follows that ${}^{\mathrm{h}}\Pi$ is a bimodule over Π with identical left and right actions. (The special case $\mathrm{h} = \Sigma$ recovers the action of the Birkhoff monoid on lunes discussed in Section 4.7.) Further, the map ${}^{\mathrm{h}}\Pi \to \Pi$ in (7.31) is a map of Π-bimodules. It follows that ${}^{\mathrm{h}}\Pi$ is a dimonoid.

Suppose $f : \mathrm{h} \to \mathrm{k}$ is a map of left Σ-sets. Then with the above structures, (7.30) is a commutative diagram of dimonoids. In particular, the support map ${}^{\mathrm{h}}\Sigma \to {}^{\mathrm{h}}\Pi$ is a morphism of dimonoids.

7.7. Descent equation for left Σ-sets

We now generalize the descent equation to any left Σ-set h. Setting the left Σ-set to be Γ or Σ recovers the situation in Section 7.1. In the descent equation, two elements, say x and y, in h are fixed. For each such x and y, we introduce a relative pair (X, A) of cell complexes, and formulate a general descent identity via the Euler characteristic of this pair. In the standard examples, the Euler characteristic can be computed explicitly, and we recover our earlier descent identities.

7.7.1. Descent equation. Let h be a left Σ-set. Consider the equation $F \cdot x = y$, where $x, y \in$ h are fixed, and we need to solve for F. This is the *descent equation for left Σ-sets.*

Let us introduce a notation for the solution set of the descent equation:

$$\Sigma_{x,y} := \{F \in \Sigma \mid F \cdot x = y\}.$$

We view $\Sigma_{x,y}$ as a subposet of the poset of faces Σ. This set may be empty. The stabilizer set of x is defined to be

$$\Sigma_{x,x} := \{F \in \Sigma \mid F \cdot x = x\}.$$

This is nonempty since it always contains the central face O.

Proposition 7.47. *$\Sigma_{x,x}$ is the closure of a combinatorial cone.*

PROOF. We verify the conditions of Proposition 2.7 for the stabilizer set.

- $O \cdot x = x$.
- If $G \cdot x = x$ and $F \leq G$, then $F \cdot x = F \cdot (G \cdot x) = FG \cdot x = G \cdot x = x$.
- If $F \cdot x = x$ and $G \cdot x = x$, then $FG \cdot x = F \cdot (G \cdot x) = x$. □

Proposition 7.48. *$\Sigma_{x,y}$ is a subset of $\Sigma_{y,y}$.*

PROOF. Suppose $F \cdot x = y$. Then

$$F \cdot y = F \cdot (F \cdot x) = (FF) \cdot x = F \cdot x = y.$$

The result follows. □

Proposition 7.49. *If $F \in \Sigma_{x,y}$ and $G \in \Sigma_{y,z}$, then $GF \in \Sigma_{x,z}$. Further, if $F \in \Sigma_{x,y}$, $G \in \Sigma_{y,y}$ and $GF = G$, then $G \in \Sigma_{x,y}$.*

PROOF. The first part follows from the definition of a left action:

$$GF \cdot x = G \cdot (F \cdot x) = G \cdot y = z.$$

The second part follows by taking $z = y$. □

Observe that there is a category whose objects are elements of h, and morphisms are $F : x \to y$ whenever $F \cdot x = y$.

Proposition 7.50. *If $\Sigma_{x,y}$ is nonempty, then its set of maximal elements is a combinatorial cone. This cone is the same as the set of top-dimensional faces of $\Sigma_{y,y}$. Thus, if $\Sigma_{x,y}$ is nonempty, then its closure is the stabilizer set $\Sigma_{y,y}$.*

PROOF. We deduce from Propositions 7.48 and 7.49 that the maximal elements in $\Sigma_{x,y}$ are the same as the top-dimensional faces of $\Sigma_{y,y}$. Now use Proposition 7.47 and the result follows. □

Proposition 7.51. *If $F, G \in \Sigma_{x,y}$ and $K \leq F$, then $KG \in \Sigma_{x,y}$.*

PROOF. This is a straightforward calculation.

$$KG \cdot x = K \cdot (G \cdot x) = K \cdot y = K \cdot (F \cdot x) = KF \cdot x = F \cdot x = y.$$ □

Proposition 7.52. *The set $\Sigma_{x,y}$ has the following properties.*

(1) *It is closed under taking products:*
If $F, G \in \Sigma_{x,y}$, then $FG \in \Sigma_{x,y}$.
(2) *It is a convex subposet of Σ:*
If $F, G \in \Sigma_{x,y}$ and $G \leq H \leq F$, then $H \in \Sigma_{x,y}$.
(3) *It is closed under taking gallery intervals:*
If $F, G \in \Sigma_{x,y}$, $\mathrm{s}(F) = \mathrm{s}(G)$ and $H \in [F\!:\!G]$, then $H \in \Sigma_{x,y}$.

PROOF. (1) Take $K := F$ in Proposition 7.51. Alternatively, specialize to $z = y$ in Proposition 7.49 and use Proposition 7.48.

(2) Take $K := H$ in Proposition 7.51 and use $HG = H$.

(3) By Proposition 7.48, $F, G \in \Sigma_{y,y}$. Since $\Sigma_{y,y}$ is the closure of a combinatorial cone, by Lemma 2.46, it is closed under taking gallery intervals. We deduce that $H \in \Sigma_{y,y}$. Now use the second part of Proposition 7.49. □

Exercise 7.53. Prove Proposition 7.52, item (3) using Proposition 7.51 and an induction on gallery distance between F and G.

Exercise 7.54. Suppose $F_1, F_2 \in \Sigma_{x,y}$, and F is such that $FF_1 = F$ and $F \leq F_2$. Then show that $F \in \Sigma_{x,y}$.

Exercise 7.55. Show that: If $\Sigma_{x,y}$ is nonempty, then its maximal elements have greater support than the maximal elements of $\Sigma_{x,x}$. If $\Sigma_{x,y}$ and $\Sigma_{y,x}$ are nonempty for $x \neq y$, then they are disjoint and their maximal elements have the same support.

7.7.2. Relative pair (X, A) of cell complexes. Fix $x, y \in \mathrm{h}$. Define X to be the cell complex obtained by taking closure of $\Sigma_{x,y}$. In other words, the cells of X are precisely those faces which are smaller than some face in $\Sigma_{x,y}$. If $\Sigma_{x,y}$ is empty, then X is the empty cell complex. If $\Sigma_{x,y}$ is nonempty, then by Proposition 7.50, X is the closure of a combinatorial cone, so it is either a topological sphere or ball. In particular, it is a pure cell complex.

Define A to be the complement of $\Sigma_{x,y}$ in X. A key observation is that A is a subcomplex of X, that is, if $G \in A$ and $F \leq G$, then $F \in A$. This follows from Proposition 7.52, item (2).

Lemma 7.56. *We have $x = y$ iff X is nonempty and A is the empty cell complex.*

PROOF. If $x = y$, then X contains the central face and A is the empty cell complex. Conversely, $\Sigma_{x,y}$ contains the central face, so $O \cdot x = y$ which means $x = y$. □

Thus, we have associated a pair (X, A) of cell complexes to $\Sigma_{x,y}$. Since X (if not the empty cell complex) is the closure of a combinatorial cone, it is also convenient to view it as a geometric cone. The structure of A is however more complicated in general. This is addressed below.

We say that a cone W (smaller than V) is full wrt V if $\mathrm{W} = \mathrm{V} \wedge \mathrm{c}(\mathrm{W})$. (In general, W will be smaller than $\mathrm{V} \wedge \mathrm{c}(\mathrm{W})$.)

An illustration is given below.

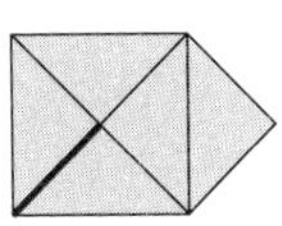
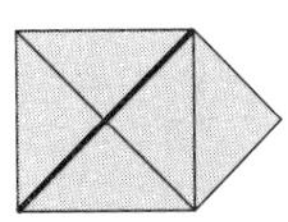

In both pictures, V is the shaded region and W is the thick line. In the first picture, W is not full wrt V, while in the second picture, W is full wrt V.

Going back to the general case, suppose V is a flat. Then W is full wrt V iff W is a flat smaller than V.

Lemma 7.57. *Suppose A is not the empty cell complex. Viewing A as a subset of the ambient space of the arrangement, A can be written as the union of certain geometric cones, each of which is full wrt X. In particular, if X is the ambient space, then A can be written as the union of certain geometric flats.*

PROOF. For any cone V contained in X, the top-dimensional faces of V either all belong to $\Sigma_{x,y}$, or all belong to A: Let F be a top-dimensional face of V. Suppose F belongs to $\Sigma_{x,y}$. Now, if G is another top-dimensional face of V, then pick a face $H \geq G$ which belongs to $\Sigma_{x,y}$, and by Proposition 7.51, $GF = G$ belongs to $\Sigma_{x,y}$.

Since A is a subcomplex, it follows that A is the union of all those cones V for which the top-dimensional faces of V all belong to A and which are full wrt X. □

7.7.3. Face-meet property. Let h be a left Σ-set, and $x, y \in$ h. We say that h satisfies the *face-meet property* wrt x and y if either of the following two equivalent conditions holds. They are formulated in terms of the relative pair (X, A).

- For any face H of the cell complex X, if $H_i \cdot x = y$ for all corank-one faces H_i of X which are greater than H, then $H \cdot x = y$.
- For any face H of the cell complex A, there exists a face K in A which is greater than H and has corank one in X.

This yields the following.

Lemma 7.58. *Suppose A is not the empty cell complex. Then* h *satisfies the face-meet property wrt x and y iff A can be written as a union of geometric cones, each of which is of corank one in X.*

A stronger form of the face-meet property is given in the two equivalent conditions below.

- For any maximal face G in $\Sigma_{x,y}$, and for any face H of G,
$$H_i \cdot x = y \text{ for all } H \leq H_i \lessdot G \text{ implies } H \cdot x = y.$$
- For any maximal face G in $\Sigma_{x,y}$, and for any face H of G,
$$\text{if } H \text{ is in } A, \text{ then } H_i \text{ is in } A \text{ for some } H \leq H_i \lessdot G.$$

Exercise 7.59. Suppose h satisfies the face-meet property wrt x and y, and A is nonempty. Deduce that if X is the ambient space, then A is the union of certain hyperplanes.

7.7.4. Descent identity. Let h be a left Σ-set. Observe that: For $x, y \in$ h,

$$\sum_{F:\, F\cdot x = y} (-1)^{\operatorname{rk}(F)} = -\chi(X) + \chi(A), \tag{7.32}$$

where χ is the reduced Euler characteristic (A.1). This is the descent identity.

If $\Sigma_{x,y}$ is empty, then X and A are both empty cell complexes, so $\chi(X) = \chi(A) = 0$. If $x = y$, then A is empty and $\chi(A) = 0$.

Since X is either the empty cell complex or a topological ball or sphere, its reduced Euler characteristic is either 0, 1 or -1. In the examples for which we have discussed the descent identity, A is either contractible or homotopy equivalent to a wedge of spheres (with all spheres of the same dimension), and one is able to calculate the number of these spheres explicitly. This is explained in more detail below.

7.7.5. Examples. Basic examples of left h-sets are given in Section 7.6. We go over them one by one.

Example 7.60. Let $\mathrm{h} = \mathrm{E}$. We have $\Sigma_{*,*} = \Sigma$, where $*$ is the unique element of E. This poset has a bottom element, namely, the central face, but no top element assuming that the arrangement has rank at least 1. The cell complex X is all of Σ while A is the empty cell complex. The descent identity is (1.41).

Example 7.61. Let $\mathrm{h} = \Gamma$. Fix chambers C and D. The poset $\Sigma_{C,D}$ has a top element, namely, the chamber D. If the arrangement $\mathcal{A}$ is simplicial, then $\Sigma_{C,D}$ has a bottom element as well, namely, the face $\mathrm{Des}(C, D)$, so in this case, $\Sigma_{C,D}$ is a Boolean poset (Proposition 7.3). Let us go back to the general case. The cell complex X consists of D and all its faces. By Proposition 7.15, Γ satisfies the stronger form of the face-meet property. So whenever A is not the empty cell complex, it is the union of certain panels of D. In fact, we have seen earlier that A is a topological ball or sphere. The descent identity is (7.10). The proof given there uses the technique of relative pairs and essentially evaluates the rhs of (7.32).

Exercise 7.62. Let $\mathrm{h} = \Gamma_{\sim}$, where $\sim$ is a partial-support relation on chambers (Exercise 7.46). Fix $\sim$-top-cones c and d. Describe the corresponding X and A. Deduce the descent identity (7.26). (A more general analysis for $\mathrm{h} = \Sigma_{\sim}$ is given in Example 7.66 below.)

Work out the special case when $\Gamma_{\sim}$ is the set of top-lunes with a fixed base X. Also check that the notation $\Sigma_{\mathrm{L},\mathrm{M}} := \{F \in \Sigma \mid F \cdot \mathrm{L} = \mathrm{M}\}$ agrees with the one introduced in Exercise 3.31.

Example 7.63. Now consider $\mathrm{h} = \Sigma$. Fix faces F and G. Observe that $\Sigma_{F,G}$ is nonempty iff $GF = G$. We assume this to be the case and proceed. By (1.9b),

$$\Sigma_{F,G} = \Sigma_{F\overline{G},G}.$$

Now G and $F\overline{G}$ have the same support, so by working in the arrangement under this support, we are back in the case $\mathrm{h} = \Gamma$. This yields the following. The poset $\Sigma_{F,G}$ has a top element, namely, G. If the arrangement $\mathcal{A}$ is simplicial, then $\Sigma_{F,G}$ has a bottom element, namely, $\overline{\mathrm{Des}}(F, G)$, so in this case, $\Sigma_{F,G}$ is a Boolean poset; also see Proposition 7.6. In the general case, X consists of G and all its faces, while A (if not the empty cell complex) is a topological ball or sphere. Also, Σ satisfies the stronger form of the face-meet property. The descent identity is (7.11a).

Example 7.64. Let $\mathrm{h} = \Pi$. Fix flats Y and W. We have

$$\Sigma_{\mathrm{Y},\mathrm{W}} = \{F \in \Sigma \mid \mathrm{s}(F) \vee \mathrm{Y} = \mathrm{W}\}.$$

This set is nonempty only if $\mathrm{Y} \leq \mathrm{W}$. So let us assume this to be the case and proceed. What is the relative pair (X, A)? Clearly, X is the closure of W, and A is the union of all hyperplanes in $\mathcal{A}^{\mathrm{W}}$ which contain Y. Topologically, A is the wedge of $c_{\mathrm{Y}}^{\mathrm{W}} - 1$ number of spheres. This is one less than the number of regions that the hyperplanes chop W into. (In this case, bear in mind that W is a topological sphere of one higher dimension.) We point out two extreme cases.

- $\mathrm{Y} = \bot$. The set $\Sigma_{\bot,\mathrm{W}}$ consists of faces F with support W, and A is the union of all hyperplanes in $\mathcal{A}^{\mathrm{W}}$.
- $\mathrm{Y} = \mathrm{W}$. The set $\Sigma_{\mathrm{W},\mathrm{W}}$ consists of faces F with support smaller than W, and A is the empty cell complex.

Evaluating the rhs of (7.32) yields the following.

$$\sum_{\mathrm{X}:\, \mathrm{X}\vee\mathrm{Y}=\mathrm{W}} (-1)^{\mathrm{rk}(\mathrm{X})} c^{\mathrm{X}} = \sum_{F\in\Sigma_{\mathrm{Y},\mathrm{W}}} (-1)^{\mathrm{rk}(F)} = -\chi(X) + \chi(A)$$
$$= (-1)^{\mathrm{rk}(\mathrm{W})} - (-1)^{\mathrm{rk}(\mathrm{W})-1}(c_{\mathrm{Y}}^{\mathrm{W}} - 1) = (-1)^{\mathrm{rk}(\mathrm{W})} c_{\mathrm{Y}}^{\mathrm{W}}.$$

This is the descent identity (7.23a).

Exercise 7.65. Show that $\mathrm{h} = \Pi$ satisfies the face-meet property. However, it does not satisfy the stronger form of the face-meet property. As a consequence, it also does not satisfy the property

$$H_1 \cdot x = H_2 \cdot x = y \implies (H_1 \wedge H_2) \cdot x = y.$$

(Recall from Proposition 7.1 that $\mathrm{h} = \Gamma$ satisfies this property when the arrangement is simplicial.)

Example 7.66. Let $\mathrm{h} = \Sigma_{\sim}$, where $\sim$ is a partial-support relation. Fix partial-flats y and w. First note that $\Sigma_{\mathrm{y},\mathrm{w}}$ is nonempty only if the support of y is contained in the support of w. So we proceed with this assumption. Clearly, X is the closure of w. Let $\mathcal{A}^{\mathrm{w}}$ denote the arrangement under the support of w. The subcomplex A is the union of $\mathrm{H} \wedge \mathrm{w}$ over those hyperplanes H in $\mathcal{A}^{\mathrm{w}}$ which either contain y, or separate y and w. (For the latter, we could use only those hyperplanes which are walls of w in view of Proposition 2.10.) The topology of A can be understood completely as follows.

First suppose w is a flat. Then A is the union of all hyperplanes in $\mathcal{A}^{\mathrm{w}}$ which contain y, and hence topologically A is a wedge of $c_{\mathrm{y}}^{\mathrm{w}} - 1$ spheres as in Example 7.64. (Recall that $c_{\mathrm{y}}^{\mathrm{w}}$ is the number of faces with partial-support w which are greater than some fixed face in y.)

Now suppose w is not a flat (and thus a topological ball). Then A is again homotopy equivalent to a wedge of spheres (all of dimension one less than that of w). The number of these spheres is exactly equal to the number of regions that the hyperplanes chop w into. (We do not consider regions which have some part on the boundary of w which is not in A.)

- Let $\bar{\mathrm{y}} \not\leq \mathrm{w}$. Then the number of these regions is zero: If there were such a region, $\bar{\mathrm{y}}$ must lie in it by construction, and hence also lie in w. But this contradicts the hypothesis.
- Let $\bar{\mathrm{y}} \leq \mathrm{w}$. In this case, all walls of w belong to A, and in addition, A contains the hyperplanes which pass through $\bar{\mathrm{y}}$. So the number of regions is $c_{\bar{\mathrm{y}}}^{\mathrm{w}}$.

This is illustrated below.

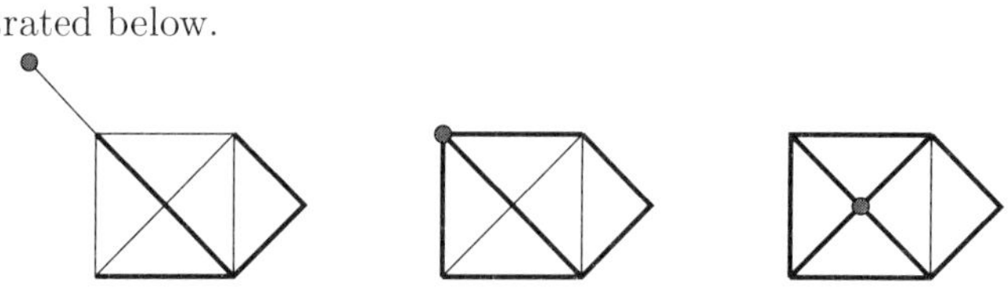

In each picture, w consists of the five triangles shown and X is its closure. The partial-flat $\bar{\mathrm{y}}$ is chosen differently in each picture. It is the vertex marked in gray. The resulting subcomplex A is marked by thick lines. In the first picture, $\bar{\mathrm{y}}$ is outside w, so A is contractible. (The number of regions is zero.) In the next two pictures, $\bar{\mathrm{y}}$ is inside w, and the number of regions that X is chopped into are two and four, respectively.

From the above analysis, we see that (7.32) specializes to (7.24a).

Exercise 7.67. Give an example of a left Σ-set which does not satisfy the face-meet property.

7.8. Lune equation for left Σ-sets

We now discuss the lune equation for left Σ-sets. The notions of h-faces and h-flats developed in Section 7.6 play a role in this discussion.

7.8.1. Lune equation. Let h be a left Σ-set. Consider the equation $F \cdot x = y$, where F and y are fixed, and we need to solve for x. This is the *lune equation for left Σ-sets*.

We introduce a notation for its solution set:

$$\ell(F, y) := \{x \in \mathrm{h} \mid F \cdot x = y\}. \tag{7.33}$$

If (F, y) is not a h-face, that is, if $F \cdot y \neq y$, then by Proposition 7.48, the lune equation has no solutions, that is, $\ell(F, y)$ is empty. So we may restrict to lune equations of h-faces.

For $\mathrm{h} = \Gamma$, solutions of the lune equation are combinatorial top-lunes (3.3). For $\mathrm{h} = \Sigma$, solutions are interiors of combinatorial lunes (3.11).

7.8.2. Lune decomposition. We saw in Section 3.3 how a lune can be decomposed into smaller lunes. This situation can be generalized as follows. The solution set $\ell(G, y)$ of a h-face (G, y) can be decomposed using a face $F \leq G$.

Proposition 7.68. *Let (G, y) be a h-face and $F \leq G$. Then (F, y) is a h-face, and $\ell(F, y) \subseteq \ell(G, y)$.*

PROOF. The first claim follows from Proposition 7.47. The second claim follows from the second statement of Proposition 7.49. □

Proposition 7.69. *Let (G, y) be a h-face and $F \leq G$. Then*

$$\ell(G, y) = \bigsqcup_{x:\, F\cdot x = x,\, G\cdot x = y} \ell(F, x), \tag{7.34}$$

with $\ell(F, y)$ being one of the summands.

PROOF. The main point to note is that if w belongs to the lhs, then it belongs to the summand in the rhs indexed by $x := F \cdot w$. All checks are straightforward. The last claim follows from Proposition 7.68. □

Recall from Lemma 1.36 that Σ_F is a monoid for a fixed face F. Consider $\{x \in \mathrm{h} \mid F \cdot x = x\}$. It consists of all h-faces of the form (F, x). It is a left Σ_F-set. Observe that the indexing set of the disjoint union in (7.34) is $\ell(G/F, y)$, which is a solution set of the lune equation for this Σ_F-set.

For $\mathrm{h} = \Gamma$ and $\mathrm{h} = \Sigma$, the decomposition (7.34) specializes to (3.20) and (3.22), respectively.

7.8.3. h-flats. Let us now understand the connection of the lune equation to h-flats.

Proposition 7.70. *Suppose (F, y) and (G, z) are* h*-faces which are equivalent in the sense of* (7.29). *Then the lune equations of (F, y) and (G, z) have the same solutions, that is, $\ell(F, y) = \ell(G, z)$.*

PROOF. Suppose $F \cdot x = y$. Then

$$G \cdot x = GF \cdot x = G \cdot (F \cdot x) = G \cdot y = z.$$

The result follows by symmetry. □

Recall that h-flats are equivalence classes under the relation (7.29). Thus, one can talk of the lune equation of a h-flat. It is now natural to ask whether distinct h-flats have distinct solution sets. This is not true in general. For instance, for h = E, all h-flats have the same solution set consisting of the unique element of E. However, the statement is true for Γ and Σ, see Propositions 3.9 and 3.13. Let us look at this problem in more detail.

Proposition 7.71. *Suppose the lune equations of the* h*-faces (F, y) and (G, z) have the same solutions. Then (FG, y) is a* h*-face and its lune equation also has the same solutions.*

PROOF. By hypothesis, $\ell(F, y) = \ell(G, z)$ and contains both y and z. Thus, $F \cdot z = y$ and $G \cdot y = z$. Let us first check that (FG, y) is a h-face:

$$FG \cdot y = F \cdot (G \cdot y) = F \cdot z = y.$$

Now we compare $\ell(F, y)$ and $\ell(FG, y)$. Suppose $F \cdot x = y$. Then $G \cdot x = z$. Hence $FG \cdot x = F \cdot z = y$. So $\ell(F, y) \subseteq \ell(FG, y)$. Conversely, suppose $FG \cdot x = y$, that is, $F \cdot (G \cdot x) = y$. Hence $G \cdot (G \cdot x) = z$. So $G \cdot x = z$, and hence $F \cdot x = y$. This proves the other inclusion. □

Proposition 7.72. *Distinct* h*-flats have distinct solution sets iff for any faces $F < G$ and* h*-face (G, y), we have $\ell(F, y) \subsetneq \ell(G, y)$.*

PROOF. The forward implication follows from Proposition 7.68. For the backward implication: Suppose the lune equations of the h-faces (F, y) and (G, z) have the same solutions. Then by Proposition 7.71, the lune equations of (F, y) and (FG, y) also have the same solutions. Since $F \leq FG$, the hypothesis implies $FG = F$. By symmetry, $GF = G$. Hence, F and G have the same support. We deduce from here that (F, y) and (G, z) are equivalent under (7.29) and determine the same h-flat. □

Exercise 7.73. Recall that a Π-flat is the same as a nested flat, that is, a pair (Y, Z) with $\mathrm{Y} \leq \mathrm{Z}$. The solutions of its lune equation consists of those flats X such that $\mathrm{Y} \vee \mathrm{X} = \mathrm{Z}$. Show that distinct Π-flats have distinct solutions sets.

Exercise 7.74. Take $\mathrm{h} = \Gamma_{\sim}$, where $\sim$ is a partial-support relation on chambers. Show that the solution set of the lune equation of a $\Gamma_{\sim}$-face is a combinatorial top-lune of the subarrangement corresponding to $\sim$ (Proposition 2.66). Further, all top-lunes of the subarrangement arise in this manner, so they index the distinct solution sets. Show that two $\Gamma_{\sim}$-faces (F, c) and (G, d) have the same solution set iff they have at least one solution in common and the set of hyperplanes in the subarrangement containing F is the same as those containing G. Deduce that distinct $\Gamma_\sim$-flats have distinct solution sets iff $\Gamma_\sim = \Gamma$.

Now take the left Σ-set $\mathrm{h} = \Sigma_\sim$, where $\sim$ is a partial-support relation on faces. Show that distinct $\Sigma_\sim$-flats have distinct solutions sets.

7.9. Lune equation for right Σ-sets

We now discuss the lune equation for right Σ-sets, that is, the Tits monoid acts on a set on the right. This complements the discussion in Section 7.7.

7.9.1. Right Σ-sets. A *right* Σ*-set* is a set h equipped with a rule which assigns to each face F and element $x \in \mathrm{h}$, an element $x \cdot F \in \mathrm{h}$ such that

$$(x \cdot F) \cdot G = x \cdot FG \quad \text{and} \quad x \cdot O = x.$$

In this situation, we say that Σ acts on h on the right. A map of right Σ-sets is a map $f : \mathrm{h} \to \mathrm{k}$ such that

$$f(x \cdot F) = f(x) \cdot F.$$

This defines the category of right Σ-sets.

For example, Σ acts on itself on the right. As a general construction, we can take quotient of Σ by any equivalence relation $\sim$ on Σ which satisfies (2.31b). For instance, for any partial-support relation $\sim$ on faces, $\Sigma_\sim$ is a right Σ-set. Among other examples of this construction, an extreme one is $\mathrm{h} = \mathrm{E}$ when all faces are identified into one equivalence class.

As another example, consider the coordinate arrangement of rank three.

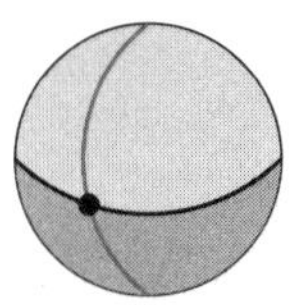

The black vertex and the two incident black edges belong to one class, the two light gray triangles and their common edge belong to one class, the two dark gray triangles and their common edge belong to one class, while all other classes are singletons. This defines a right Σ-set h.

7.9.2. Lune equation. Let h be a right Σ-set. Consider the equation $x \cdot F = y$, where $x, y \in \mathrm{h}$ are fixed, and we need to solve for F. This is the *lune equation for right* Σ*-sets.*

Let us introduce a notation for the solution set of the lune equation:

$${}_{x,y}\Sigma := \{F \in \Sigma \mid x \cdot F = y\}.$$

The stabilizer set of x is defined to be

$${}_{x,x}\Sigma := \{F \in \Sigma \mid x \cdot F = x\}.$$

Proposition 7.75. ${}_{x,x}\Sigma$ *is the closure of a combinatorial flat.*

PROOF. We verify the conditions of Proposition 1.16 for the stabilizer set.

- $x \cdot O = x$.
- If $x \cdot G = x$ and $GF = G$, then $x \cdot F = (x \cdot G) \cdot F = x \cdot GF = x \cdot G = x$.
- If $x \cdot F = x$ and $x \cdot G = x$, then $x \cdot FG = (x \cdot F) \cdot G = x$. □

Proposition 7.76. ${}_{x,y}\Sigma$ *is a subset of* ${}_{y,y}\Sigma$.

Proof. Suppose $x \cdot F = y$. Then
$$y \cdot F = (x \cdot F) \cdot F = x \cdot FF = x \cdot F = y.$$
The result follows. □

Proposition 7.77. *If $F \in {}_{x,y}\Sigma$ and $G \in {}_{y,z}\Sigma$, then $FG \in {}_{x,z}\Sigma$. Further, if $F \in {}_{x,y}\Sigma$, $G \in {}_{y,y}\Sigma$, and $FG = G$, then $G \in {}_{x,y}\Sigma$.*

Proof. The first part follows from the definition of a right action; the second part follows from the first by taking $z = y$. □

Proposition 7.78. *If $F, G \in {}_{x,y}\Sigma$ and $FK = F$, then $GK \in {}_{x,y}\Sigma$.*

Proof. This is a straightforward calculation.
$$x \cdot GK = (x \cdot G) \cdot K = y \cdot K = (x \cdot F) \cdot K = x \cdot FK = x \cdot F = y.$$ □

Proposition 7.79. *The set ${}_{x,y}\Sigma$ has the following properties.*

(1) *It is closed under taking products:*
If $F, G \in {}_{x,y}\Sigma$, then $FG \in {}_{x,y}\Sigma$.

(2) *It is a convex subposet of Σ:*
If $F, G \in {}_{x,y}\Sigma$ and $G \leq H \leq F$, then $H \in {}_{x,y}\Sigma$.

Proof. (1) Take $K := F$ in Proposition 7.78. Alternatively, specialize to $z = y$ in Proposition 7.77 and use the fact that ${}_{x,y}\Sigma$ is a subset of ${}_{y,y}\Sigma$.

(2) Take $K := H$ in Proposition 7.78 and use $GH = H$. □

Exercise 7.80. Show that any maximal element of ${}_{x,y}\Sigma$ is also a maximal element of ${}_{y,y}\Sigma$.

7.9.3. Lune decomposition. The relationship of the lune equation with lunes is brought forth by the following result.

Lemma 7.81. *Let* h *be a right Σ-set. For $x, y \in \mathrm{h}$,*
$$ {}_{x,y}\Sigma = \bigsqcup_{\substack{G \in {}_{x,y}\Sigma, \\ G \geq H}} \mathrm{s}(H, G)^o, \tag{7.35}$$
where H is some arbitrary but fixed maximal element of ${}_{x,x}\Sigma$, and $\mathrm{s}(H,G)^o$ is as in (3.11).

Proof. Let $F \in {}_{x,y}\Sigma$. Then by Proposition 7.77, $G := HF$ is also an element of ${}_{x,y}\Sigma$. Now let F' be any element of $\mathrm{s}(H,G)^o$, that is, $HF' = G$. Then
$$x \cdot F' = (x \cdot H) \cdot F' = x \cdot HF' = x \cdot G = y.$$
Hence F' belongs to ${}_{x,y}\Sigma$. This establishes the decomposition (7.35). □

The decomposition (7.35) may convey the wrong impression that ${}_{x,y}\Sigma$ is a nice geometric object such as a cone. The point is that the indexing set $\Sigma_H \cap {}_{x,y}\Sigma$ can be quite arbitrary. This is discussed below.

Lemma 7.82. *For any right Σ-set* h, *${}_{x,y}\Sigma$ is an upper set of ${}_{y,y}\Sigma$. Conversely: Given any upper set* U *of Σ, there is a right Σ-set* h *and $x, y \in \mathrm{h}$ such that ${}_{x,y}\Sigma = \mathrm{U}$. The same holds, more generally, if* U *is an upper set of* Cl(Y) *for some combinatorial flat* Y.

PROOF. The first statement follows from the second part of Proposition 7.77. For the converse: Consider the equivalence relation on Σ in which U is an equivalence class, and all remaining classes are singletons. This satisfies (2.31b), so the equivalence classes define a right Σ-set. Let x be the class of the central face, and y be U. Then ${}_{x,y}\Sigma = \mathrm{U}$.

For the more general claim, consider the equivalence relation on Σ in which U is an equivalence class, all other elements of Cl(Y) are singleton classes, and the complement of Cl(Y) is an equivalence class. This defines a right Σ-set and the rest is as above. □

Exercise 7.83. Give an example, where the set of maximal elements of ${}_{x,y}\Sigma$ is not a combinatorial cone. In other words, the analogues of Proposition 7.50 and Proposition 7.52, item (3) do not hold for the lune equation.

7.9.4. Partial order on a right Σ-set. Let h be a right Σ-set. We say that

$$x \leq y \text{ if there exists a face } F \text{ such that } x \cdot F = y. \tag{7.36}$$

Lemma 7.84. *For a right Σ-set,* (7.36) *defines a partial order. Further, any map $f : \mathrm{h} \to \mathrm{k}$ of right Σ-sets is order-preserving.*

PROOF. We check below that (7.36) defines a partial order.

- Reflexive. Since $x \cdot O = x$, we have $x \leq x$.
- Transitive. Suppose $x \leq y$ and $y \leq z$. So for some faces F and G, $x \cdot F = y$ and $y \cdot G = z$. Then $x \cdot FG = z$, so $x \leq z$.
- Antisymmetric. Suppose $x \leq y$ and $y \leq x$. Then, as above, for some faces F and G, $x \cdot F = y$ and $y \cdot G = x$, hence $x \cdot FG = x$. By Proposition 7.75, the stabilizer set of x is the closure of a combinatorial flat. Since it contains FG, it must also contain F (and G). Thus $x \cdot F = x$, and $x = y$.

For the second claim, take $x \leq y$ in h. Then for some F, $x \cdot F = y$. Applying f, we obtain $f(x \cdot F) = f(x) \cdot F = f(y)$. Hence, $f(x) \leq f(y)$ in k. □

Observe that ${}_{x,y}\Sigma$ is nonempty iff $x \leq y$. Further, for any $x \in \mathrm{h}$,

$$\Sigma = \bigsqcup_{y:\, y \geq x} {}_{x,y}\Sigma. \tag{7.37}$$

Exercise 7.85. Why does definition (7.36) (with F written on the left) not define a partial order on a left Σ-set?

Exercise 7.86. Suppose h is a right Σ-set. We know that $x \cdot F = y$ implies $y \cdot F = y$. Does $x \leq z \leq y$ and $x \cdot F = y$ imply $z \cdot F = y$?

7.9.5. Stars. Let h be a right Σ-set. For $x \in \mathrm{h}$, let

$$\mathrm{h}_x := \{y \in \mathrm{h} \mid y \geq x\}. \tag{7.38}$$

We call this set the *star* of x. It is a right Σ-set with action induced from h. Further, the map

$$\Sigma \twoheadrightarrow \mathrm{h}_x, \qquad F \mapsto x \cdot F \tag{7.39}$$

is a surjective map of right Σ-sets. In other words, h_x is the right Σ-submodule of h generated by x.

The inverse image of y under the map (7.39) is ${}_{x,y}\Sigma$. So the fibers of this map yield the decomposition (7.37).

Exercise 7.87. Suppose $f : \Sigma \twoheadrightarrow \mathrm{h}$ is a surjective map of right Σ-sets. Then show that h has a unique minimum element and it is given by $f(O)$. Give an example of a right Σ-set which is connected as a poset but has more than one minimum element.

7.9.6. Support map. Let h be a right Σ-set. Proposition 7.75 yields a *support map*

$$\mathrm{s} : \mathrm{h} \to \Pi, \qquad x \mapsto \mathrm{s}(x), \tag{7.40}$$

where $\mathrm{s}(x)$ is the flat determined by the maximal faces of ${}_{x,x}\Sigma$. For $\mathrm{h} := \Sigma$, this coincides with the usual support map.

Lemma 7.88. *The map* $\mathrm{s} : \mathrm{h} \to \Pi$ *is strictly order-preserving, that is,* $x < y$ *implies* $\mathrm{s}(x) < \mathrm{s}(y)$.

PROOF. Suppose $x < y$ with $x \cdot F = y$. Then, by hypothesis, F is not contained in ${}_{x,x}\Sigma$. Let G be any maximal element in ${}_{x,x}\Sigma$. Then, $x \cdot GF = y$, so GF belongs to ${}_{x,y}\Sigma$. Since ${}_{x,x}\Sigma$ is the closure of a combinatorial flat, $\mathrm{s}(GF) > \mathrm{s}(G)$. Also, since ${}_{x,y}\Sigma$ is a subset of ${}_{y,y}\Sigma$, GF belongs to the latter. Thus, $\mathrm{s}(x) < \mathrm{s}(y)$ as required. □

Exercise 7.89. Give an example of a right Σ-set which is not graded as a poset. Deduce that the map (7.40) does not preserve cover relations in general.

7.9.7. Relative pair (X, A) of cell complexes. Let h be a right Σ-set. To each $x, y \in \mathrm{h}$, we associate a relative pair of cell complexes (X, A) as follows. Let X be the closure of ${}_{x,y}\Sigma$, and A be the complement of ${}_{x,y}\Sigma$ in X. By Proposition 7.79, item (2), A is indeed a subcomplex of X.

Lemma 7.90. *For* $x, y \in \mathrm{h}$*:*

(1) *If* $x \not\leq y$, *then both* X *and* A *are the empty cell complexes.*
(2) *If* $x = y$, *then* X *is a nonempty cell complex while* A *is the empty cell complex.*
(3) *If* $x < y$, *then both* X *and* A *are nonempty cell complexes.*

PROOF. If $x \not\leq y$, then from the definition of the partial order ${}_{x,y}\Sigma$ is empty, and hence X and A are the empty cell complexes. If $x = y$, then the central face belongs to X, so it cannot belong to A, which means that A is the empty cell complex. If $x < y$, then X is nonempty but since $x \cdot O \neq y$, the central face does not belong to ${}_{x,y}\Sigma$, so it must belong to A which is then nonempty. □

Suppose (X, A) is a pair associated to some $x \leq y$. All maximal elements of ${}_{x,y}\Sigma$ have the same support. So X is a pure cell complex, and it has a well-defined support. Let Y denote the support of X. We say that F is a *boundary face* of X if F is contained in X but there is some face supported by Y which is greater than F but which is not in X. We deduce from Lemma 7.82 that A must necessarily contain the boundary faces of X. (This is in contrast to what happened for left actions.)

Exercise 7.91. Show that any lower set of Σ which does not contain any chambers can be realized for the subcomplex A for suitable choice of $x \leq y$.

Exercise 7.92. Show that the union of the closures of any nonempty subset of faces having the same support can be realized for the complex X for suitable choice of $x \leq y$. Given such a X, deduce that A could be any lower set of X containing all its boundary faces but none of its maximal faces.

Exercise 7.93. Give an example, where X is a topological cylinder, that is, a sphere with two holes, and A is the boundary of the cylinder, which is the disjoint union of two topological circles.

7.9.8. Lune identity. Let h be a right Σ-set. Observe that: For $x, y \in \mathrm{h}$,

$$\sum_{F:\, x\cdot F = y} (-1)^{\mathrm{rk}(F)} = -\chi(X) + \chi(A), \tag{7.41}$$

where χ is the reduced Euler characteristic (A.1). This is the lune identity.

If $x \not\leq y$, then both X and A are the empty cell complexes, and both sides of (7.41) are zero.

Example 7.94. Let $\mathrm{h} := \Sigma$. The partial order on Σ given by (7.36) coincides with the usual one. For $G \geq H$, note from (3.11) that

$$_{H,G}\Sigma = \mathrm{s}(H, G)^o.$$

In this case, the decomposition (7.35) assumes a trivial form with only one term in the rhs.

The star h_x given by (7.38), for $x := H$, coincides with the star of H. We have previously denoted this set by Σ_H, so the notations are consistent. The quotient map (7.39) is

$$\Sigma \to \Sigma_H, \qquad F \mapsto HF,$$

and Σ_H is the right ideal generated by H (Lemma 1.36). The decomposition of Σ given by (7.37) is precisely (3.17).

Now let us look at the relative pair (X, A). For $G \geq H$, X is the closure and A is the boundary of the lune $\mathrm{s}(H, G)$. For $G > H$, X is a topological ball and A is a topological sphere, while for $G = H$, X is a topological sphere and A is the empty cell complex. The lune identity (7.41) specializes to (7.12a), see the proof of the latter for more details.

Exercise 7.95. Generalize the above analysis to $\mathrm{h} := \Sigma_F$ for a fixed face F.

Example 7.96. Let $\mathrm{h} := \Sigma_\sim$, where $\sim$ is a partial-support relation on faces. The partial order on $\Sigma_\sim$ given by (7.36) coincides with the usual one. For $\mathrm{x} \leq \mathrm{w}$,

$$_{\mathrm{x},\mathrm{w}}\Sigma = \bigsqcup_{G:\, G\in \mathrm{w},\, G \geq H} \mathrm{s}(H, G)^o,$$

where H is some arbitrary but fixed element of x. This decomposition is a special case of (7.35). Since partial-flats are combinatorial cones, the G in the sum runs over a convex set of faces in the star of H (with the same support as w). As a consequence, the lunes $\mathrm{s}(H, G)$ together determine a cone. For a concrete instance, see the figure in Section 3.3.6. The interiors of the three shaded lunes make up $_{\mathrm{x},\mathrm{w}}\Sigma$. Their closure is X, and their boundaries (shown as thick lines) together make up A. In general, X is a topological ball or sphere, and A is homotopy equivalent to a wedge of spheres. This can be used to deduce the lune identity (7.25a). (Alternatively, one can work with each lune interior separately, and see that each of them contributes $(-1)^{\mathrm{rk}(\mathrm{w})}$.)

Exercise 7.97. Let $\mathrm{h} = \mathrm{E}$, the singleton Σ-set with the trivial right action. What are X and A in this case? Write down the lune identity. What happens for $\mathrm{h} = \Gamma$, with the right action induced from Σ?

7.10. Descent-lune equation for Π-sets

We consider the descent-lune equation for Π-sets. This generalizes the descent-lune equation for flats. It combines the features of the descent equation for left Σ-sets and the lune equation for right Σ-sets. We formulate a descent-lune identity using relative pairs (X, A), and work it out explicitly for the example of charts. We also give the descent-lune identity for the example of lunes.

7.10.1. Π-sets. A Π-*set* is a set h equipped with a rule which assigns to each flat X and element $x \in \mathrm{h}$, an element $\mathrm{X} \cdot x \in \mathrm{h}$ such that

$$\mathrm{X} \cdot (\mathrm{Y} \cdot x) = (\mathrm{X} \vee \mathrm{Y}) \cdot x \quad \text{and} \quad \bot \cdot x = x.$$

(Since Π is commutative, there is no distinction between left and right actions. As a convention, the action is written on the left.)

7.10.2. Descent-lune equation. Now suppose h is a Π-set. Consider the equation $\mathrm{X} \cdot x = y$, with x and y fixed and X variable. This is the *descent-lune equation*. Denote its solution set by

$$\Pi_{x,y} := \{\mathrm{X} \in \Pi \mid \mathrm{X} \cdot x = y\}.$$

Note that h is both a left and right Σ-set via

$$F \cdot x = x \cdot F = \mathrm{s}(F) \cdot x.$$

Clearly,

$$_{x,y}\Sigma = \Sigma_{x,y},$$

and taking supports of faces in this set yields $\Pi_{x,y}$. Thus, the analysis of both Sections 7.7 and 7.9 applies. We summarize the results below.

There is a partial order on h defined as follows. We say that

$$x \leq y \text{ if there exists a flat X such that } \mathrm{X} \cdot x = y. \tag{7.42}$$

Further, there is an order-preserving map

$$\mathrm{h} \to \Pi \tag{7.43}$$

which sends x to the largest flat which stabilizes x.

Lemma 7.98. *For any* Π*-set* h *and* $x, y \in \mathrm{h}$, $\Pi_{x,y}$ *is an upper set in* $\Pi_{y,y}$. *Conversely, given any flat* X *and an upper set* U *in the interval* $[\bot, \mathrm{X}]$ *of flats, there is a* Π*-set* h *such that* $\Pi_{x,y} = \mathrm{U}$ *for some* $x, y \in \mathrm{h}$.

PROOF. This can be deduced from Lemma 7.82. □

7.10.3. Descent-lune identity. Let h be a Π-set. Assume $x \leq y$. The relative pair (X, A) is defined in the same manner as before, that is, by applying the previous definitions to either $\Sigma_{x,y}$ or $_{x,y}\Sigma$. By definition,

$$\sum_{\mathrm{X}\colon \mathrm{X}\cdot x = y} (-1)^{\mathrm{rk}(\mathrm{X})} c^{\mathrm{X}} = -\chi(X) + \chi(A), \tag{7.44}$$

where χ is the reduced Euler characteristic (A.1). This is the descent-lune identity.

We have a complete understanding of what X and A could be. The cell complex X is the closure of a combinatorial flat, and A consists of faces supported by flats of some proper lower set of Π. Note from Lemma 7.98 that any proper lower set is possible for A. In particular, A may not be connected topologically. For instance, X could be the sphere and A could be the union of a great circle and a pair of antipodal points (disjoint from that circle). In Example 7.64, A was homotopy equivalent to a wedge of spheres.

7.10.4. Descent-lune identity for charts. Recall the poset of charts G from Section 2.6. It is a monoid under the join operation (intersection). The join-preserving map (2.19) turns G into a Π-set. Explicitly,

$$\mathrm{X} \cdot g = \lambda(\mathrm{X}) \vee g = g_{\mathrm{X}},$$

where recall that g_{X} consists of those hyperplanes in g which contain X. So, one can consider the descent-lune equation for charts. The descent-lune identity for charts is given below. Recall from (2.20) that for a chart h, the flat $\rho(h)$ is the intersection of all hyperplanes in h.

Proposition 7.99. *For any* $g, h \in \mathrm{G}$,

$$\sum_{\mathrm{X}:\, \mathrm{X}\cdot g=h} (-1)^{\mathrm{rk(X)}} c^{\mathrm{X}} = \begin{cases} (-1)^{\mathrm{rk}(\rho(h))} c(g^{\rho(h)}) & \text{if } g_{\rho(h)} = h, \\ 0 & \text{otherwise}, \end{cases} \tag{7.45}$$

where $c(g^{\rho(h)})$ *is the number of chambers in the arrangement* $g^{\rho(h)}$.

PROOF. The equation $\mathrm{X} \cdot g = h$ has a solution iff $\rho(h) \cdot g = h$. This is the same as the condition $g_{\rho(h)} = h$. So let us proceed under this assumption (else both sides are zero). The sum can be rewritten as $\sum_F (-1)^{\dim(F)}$, where F runs over all faces of $\mathcal{A}$ such that the intersection of the hyperplanes in g which contain F is $\rho(h)$. Equivalently, the sum is over all faces F in $\mathcal{A}^{\rho(h)}$ such that F is not on any hyperplane in $g^{\rho(h)}$. To calculate this sum: let us first sum over all faces F in $\mathcal{A}^{\rho(h)}$. This is the reduced Euler characteristic of a topological sphere. From this sum we substract the sum over all faces contained in the hyperplanes in $g^{\rho(h)}$. These form a space which is homotopy equivalent to a wedge of spheres, the number of spheres being the number of chambers in $g^{\rho(h)}$ minus 1. □

The above argument is indeed using the technique of relative pairs (X, A) by evaluating the rhs of (7.44). In this example, X is a sphere, and A is homotopy equivalent to a wedge of spheres.

7.10.5. Descent-lune identity for lunes. Recall from Section 4.7 that the set of lunes Λ is a Π-set. The partial order (7.42) coincides with the partial order $\preceq$ defined in (4.6). This follows from Exercise 4.52. Further, the map (7.43) specializes to the base map on lunes. This follows from Corollary 4.48.

Proposition 7.100. *For any lunes* L *and* M,

$$\sum_{\mathrm{X}:\, \mathrm{X}\cdot\mathrm{L}=\mathrm{M}} (-1)^{\mathrm{rk(X)}} c^{\mathrm{X}} = \begin{cases} (-1)^{\mathrm{rk(b(M))}} c^{\mathrm{b(M)}}_{\mathrm{b(L)}} & \text{if } \mathrm{L} \preceq \mathrm{M}, \\ 0 & \text{otherwise.} \end{cases} \tag{7.46}$$

PROOF. By Exercise 4.52, the equation $\mathrm{X} \cdot \mathrm{L} = \mathrm{M}$ has a solution iff $\mathrm{L} \preceq \mathrm{M}$. Under this assumption, X is a solution iff $\mathrm{X} \vee \mathrm{b(L)} = \mathrm{b(M)}$. This follows from Exercise 4.53. Now apply (7.23a). □

7.11. Flat-based lattices

We introduce the notion of a flat-based lattice. We generalize the descent-lune identity for flats to flat-based lattices under the hypothesis of supertightness.

7.11.1. Category of Π-based lattices. Recall the lattice of flats Π. A Π-based lattice is a finite lattice h equipped with a join-preserving map $\lambda : \Pi \to \mathrm{h}$. By our convention, a join-preserving map preserves all finite joins. In particular, it preserves minimum elements. A morphism between Π-based lattices is a join-preserving map $\mathrm{h} \to \mathrm{k}$ such that

$$\begin{array}{ccc} \mathrm{h} & \longrightarrow & \mathrm{k} \\ & \nwarrow \quad \nearrow & \\ & \Pi & \end{array}$$

commutes. This defines the category of Π-based lattices. This is an instance of a coslice category.

Exercise 7.101. Describe the terminal object and product in the category of Π-based lattices.

Any Π-based lattice h carries the structure of a Π-set. The action is defined by

$$\mathrm{X} \cdot x := \lambda(\mathrm{X}) \vee x.$$

It is easy to check that a morphism of Π-based lattices yields a map $\mathrm{h} \to \mathrm{k}$ of Π-sets. In other words, this construction yields a functor from the category of Π-based lattices to the category of Π-sets.

For Π, the action is simply given by the join operation of Π.

In view of (D.38), for a Π-based lattice h with structure map λ, the right adjoint $\rho : \mathrm{h} \to \Pi$ coincides with the map in (7.43).

7.11.2. Descent-lune equation for Π-based lattices. Recall the notion of tightness (Definition B.13).

Lemma 7.102. *Suppose $\lambda : \Pi \to \mathrm{h}$ is a supertight join-preserving map of lattices. Then for any $x, y \in \mathrm{h}$, and flat* X,

$$\mathrm{X} \cdot x = y \iff \mathrm{X} \vee \rho(x) = \rho(y) \text{ and } \lambda\rho(y) \vee x = y, \tag{7.47}$$

where ρ is the right adjoint of λ.

This is a special case of Lemma B.16. Also note that $\lambda\rho(y) \vee x = y$ is equivalent to $\rho(y) \cdot x = y$.

Consider the equation $\mathrm{X} \cdot x = y$ with x and y fixed and X variable. This is the descent-lune equation for a Π-based lattice. (When $\mathrm{h} = \Pi$, we recover the descent-lune equation for flats.) Let $\Pi_{x,y}$ denote its set of solutions. From (7.47), we deduce that

$$\Pi_{x,y} = \begin{cases} \Pi_{\rho(x),\rho(y)} & \text{if } \lambda\rho(y) \vee x = y, \\ \emptyset & \text{otherwise.} \end{cases}$$

In other words, under the hypothesis of supertightness, the solution set of the general descent-lune equation coincides with the solution set of the descent-lune equation for flats. Applying (7.23a) leads to the following descent-lune identity.

Proposition 7.103. *Suppose $\lambda : \Pi \to \mathrm{h}$ is a supertight join-preserving map of lattices. Then for any $x, y \in \mathrm{h}$,*

$$\sum_{\mathrm{X}:\, \mathrm{X}\cdot x = y} (-1)^{\mathrm{rk}(\mathrm{X})} c^{\mathrm{X}} = \begin{cases} (-1)^{\mathrm{rk}(\rho(y))} c^{\rho(y)}_{\rho(x)} & \text{if } \lambda\rho(y) \vee x = y, \\ 0 & \text{otherwise,} \end{cases} \tag{7.48}$$

where ρ is the right adjoint of λ.

More precisely, (7.48) holds whenever $\lambda_{x,y}$ is tight.

Recall the lattice of charts G. The map (2.19) turns G into a Π-based lattice. However this map is not supertight in general. Compare and contrast the descent-lune identity for charts (7.45) with the descent-lune identity (7.48) given above.

Exercise 7.104. Show that in the context of Π-based lattices, for any $x, y \in \mathrm{h}$ and flat X,

$$\mathrm{X} \cdot x \leq y \iff \mathrm{X} \leq \rho(y) \text{ and } x \leq y.$$

Use this to generalize (7.23b).

Notes

Descents. The notion of descent at a chamber D wrt another chamber C, denoted $\mathrm{Des}(C, D)$, occurs in [**296**, Section 1.6.2] and later in [**8**, Chapter 5]. (It generalizes in an appropriate sense the classical notion of descent of a permutation.) The more general notion of descent involving faces, denoted $\mathrm{Des}(F, G)$, dates back to the same time. Brown discusses chamber descents in the context of reflection arrangements [**103**, Section 9.3]. Observation (7.1) is given in [**103**, Proposition 4]. It is also explained in [**8**, Proposition 5.2.2]. Propositions 7.3 and 7.6 for the braid arrangement are given in [**9**, Propositions 10.11 and 10.12].

Witt formula. Formula (7.15) is due to Witt [**435**, Satz 3]. It is also given in [**238**, Section 1.11]. This identity is closely related to the flag f and flag h vectors associated to a labeled simplicial complex. If we multiply both sides of (7.15) by $(-1)^{\mathrm{rk}(D)} c_O$, the resulting identity can be understood in two steps as

$$\sum_{J \subseteq S} (-1)^{|S \setminus J|} f_J(D) = h_S(D) \quad \text{and} \quad h_S(D) = 1.$$

where $f_J(D)$ and $h_J(D)$ denote the components of the flag f and flag h vectors local to D. The former counts the number of chambers in the lune $\mathrm{s}(H, D)$ where H has type J, while the latter counts the number of chambers C such that the face $\mathrm{Des}(C, D)$ has type J. In particular, $h_S(D) = 1$ since $\overline{D}$ is the only chamber with the required property. For more details, see [**103**, Proposition 5] and [**296**, Section 1.7].

h-faces and h-flats. The material pertaining to dimonoids in Section 7.6 for the left Σ-set $\mathrm{h} = \Gamma$ is present in [**9**, Section 10.10]. Proposition 7.71 for $\mathrm{h} = \Gamma$ in the general context of LRBs is given in [**8**, (2.8)].

CHAPTER 8

Distance functions and Varchenko matrix

We introduce an abstract notion of distance function on chambers of an arrangement. The motivating example arises by assigning a weight to each half-space, and letting the distance between C and D to be the product of weights of all half-spaces that one has to move out of while going from C to D. An important special case is when all half-spaces have the same weight, say q, in which case, the distance between C and D is q power the number of hyperplanes separating C and D. (Recall that the latter is the gallery distance between C and D.)

A distance function gives rise to a matrix indexed by chambers whose entry in position (C, D) is the distance between C and D. This is the Varchenko matrix. For distance functions arising from weight functions on half-spaces, the determinant of this matrix has a nice factorization. The same is true, more generally, for the Varchenko matrix indexed by chambers of any top-cone.

8.1. Weights on half-spaces

Let $\mathcal{A}$ be an arrangement. We begin with distance functions on $\mathcal{A}$ which arise from weight functions on its half-spaces. This material builds on the discussion on separating hyperplanes, minimal galleries and their basic properties given in Section 1.10. An abstract approach to distance functions is given in Section 8.3.

8.1.1. Distance function on chambers. A *weight function* assigns a number (weight) to each half-space of $\mathcal{A}$. We write $\mathrm{wt}(\mathrm{h})$ for the weight on the half-space h. Given a weight function, for any chambers C and D, let

$$v_{C,D} := \prod_{\mathrm{h} \in r(C,D)} \mathrm{wt}(\mathrm{h}), \tag{8.1}$$

where recall that $r(C, D)$ consists of half-spaces h which contain C but do not contain D. This defines a function v on the set of pairs of chambers. We call it a (multiplicative) distance function.

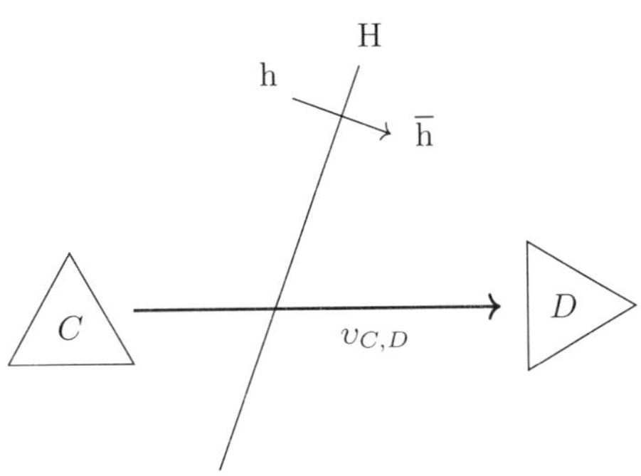

One may visualize $v_{C,D}$ as an arrow from C to D as shown in the figure. H is a typical hyperplane which separates C and D, and h and $\overline{\text{h}}$ are the two opposite half-spaces which it bounds. The weights of h and $\overline{\text{h}}$ are *not* required to be equal. By our convention, wt(h) will appear as a factor in $v_{C,D}$, while wt($\overline{\text{h}}$) will appear as a factor in $v_{D,C}$.

One may also interpret $v_{C,D}$ as the minimum (multiplicative) cost of going from C to D, the cost of going out of an half-space being the weight of that half-space.

The basic properties of v are listed below. They follow from (8.1) and the corresponding properties of the sets $r(C, D)$ listed in Proposition 1.66.

Proposition 8.1. *For any chamber C,*

$$v_{C,C} = 1. \tag{8.2a}$$

For faces F and G with the same support, and $F \leq C$ and $F \leq D$,

$$v_{C,GC} = v_{D,GD}. \tag{8.2b}$$

For any minimal gallery C -- D -- E,

$$v_{C,D}\, v_{D,E} = v_{C,E}. \tag{8.2c}$$

For any chambers C and D,

(8.2d) $v_{C,D}$ *is the product of the distances between adjacent chambers in any minimal gallery joining C and D.*

For any C, and G a face of D,

$$v_{C,D} = v_{C,GC}\, v_{GC,D}. \tag{8.2e}$$

For any D, and F a face of C,

$$v_{C,D} = v_{C,FD}\, v_{FD,D}. \tag{8.2f}$$

For faces F and G with the same support, and $F \leq C$ and $F \leq D$,

$$v_{C,D} = v_{GC,GD}. \tag{8.2g}$$

For any chambers C and D,

$$v_{C,D} = v_{\overline{D},\overline{C}}. \tag{8.2h}$$

Properties (8.2b) and (8.2g) may be rephrased as follows. For any chambers C and D, and faces F and G with the same support,

$$v_{FC,GC} = v_{FD,GD}, \tag{8.3}$$

and

$$v_{FC,FD} = v_{GC,GD}. \tag{8.4}$$

8.1.2. Symmetry and log-antisymmetry. A distance function v associated to a weight function wt is *symmetric* if

$$\text{wt}(\text{h}) = \text{wt}(\overline{\text{h}}) \qquad \text{or equivalently} \qquad v_{C,D} = v_{D,C},$$

and *log-antisymmetric* if

$$\text{wt}(\text{h})\,\text{wt}(\overline{\text{h}}) = 1 \qquad \text{or equivalently} \qquad v_{C,D}\, v_{D,C} = 1$$

for all half-spaces h and for all chambers C and D.

Observe that a symmetric distance function arises by assigning a weight to each hyperplane and letting $v_{C,D}$ be the product of weights of those hyperplanes which separate C and D.

Proposition 8.2. *For a log-antisymmetric distance function v,*

$$v_{C,D}\, v_{D,E} = v_{C,E}$$

for any chambers C, D and E.

In other words, the minimal gallery condition in (8.2c) can be dropped when v is log-antisymmetric.

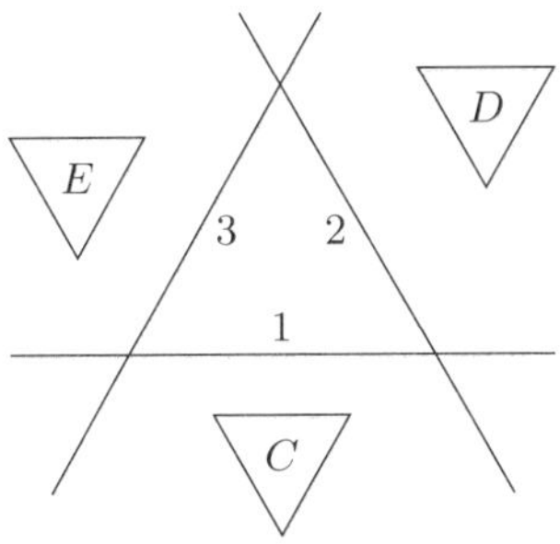

PROOF. We may assume that C, D and E are all distinct (the remaining cases are straightforward). Then there are three kinds of hyperplanes as shown in the figure whose associated half-spaces contribute to the weighted distances. The hyperplanes labeled 1 and 3 contribute once to both the lhs and rhs via the half-space which contains C. The hyperplane labeled 2 does not contribute to the rhs and contributes twice to the lhs via the two opposite half-spaces which it bounds; but since v is log-antisymmetric, the two contributions multiply to 1. □

8.1.3. Operations on distance functions. Let v be a distance function. Its transpose v^t is the distance function defined by

$$(v^t)_{C,D} := v_{D,C}. \tag{8.5}$$

In other words, the transpose interchanges the weights of opposite half-spaces.

Let v be a nowhere-zero distance function, that is, none of the weights are 0. This happens, for instance, if v is log-antisymmetric. In this situation, define the inverse distance function v^- by

$$(v^-)_{C,D} := v_{C,D}^{-1}. \tag{8.6}$$

In other words, the inverse replaces every weight by its inverse. Also let $v^{-t} := (v^t)^-$, that is,

$$(v^{-t})_{C,D} := v_{D,C}^{-1}.$$

Observe that v is symmetric iff $v = v^t$, and is log-antisymmetric iff $v = v^{-t}$.

Let v and v' be two distance functions. Their Hadamard product $v \times v'$ is the distance function defined by

$$(v \times v')_{C,D} := v_{C,D}\, v'_{C,D}.$$

In other words, the weight on a half-space for $v \times v'$ is the product of the weights on that half-space for v and for v'.

8.1.4. Distance function on faces. Property (8.3) can be used to extend any distance function v to the set of pairs of faces with the same support: For faces F and G with the same support, let

$$v_{F,G} := v_{FC,GC}, \tag{8.7}$$

where C is any chamber. The point is that the rhs does not depend on the specific choice of C.

The distance function further extends to the set of all pairs of faces as follows. For any faces F and G, let

$$v_{F,G} := v_{FG,GF}, \tag{8.8}$$

where for the rhs, (8.7) applies since FG and GF have the same support. Note that $v_{F,G} = v_{FG,G} = v_{F,GF}$.

Proposition 8.3. *For any face F,*

$$v_{F,F} = 1. \tag{8.9a}$$

For faces F and G with the same support, and $F \leq H$,

$$v_{F,G} = v_{H,GH}. \tag{8.9b}$$

For faces F, G and K with $FG \leq K$,

$$v_{F,G} = v_{K,G} \quad \text{and} \quad v_{G,F} = v_{G,K}. \tag{8.9c}$$

For faces F, G and H with the same support such that F -- G -- H,

$$v_{H,G}v_{G,F} = v_{H,F}. \tag{8.9d}$$

For faces F, H and K with $H \leq K$,

$$v_{K,F} = v_{K,HF}v_{HF,F}. \tag{8.9e}$$

For faces F, H and K with $H \leq F$,

$$v_{K,F} = v_{K,HK}v_{HK,F}. \tag{8.9f}$$

For faces F and G with the same support, and $F \leq H$ and $F \leq K$,

$$v_{H,K} = v_{GH,GK}. \tag{8.9g}$$

For faces F and G,

$$v_{F,G} = v_{\overline{G},\overline{F}}. \tag{8.9h}$$

Properties (8.9b) and (8.9g) may be rephrased as follows. For faces F, G, H and K with $\mathrm{s}(F) = \mathrm{s}(G)$,

$$v_{FH,GH} = v_{F,G} = v_{FK,GK}, \tag{8.10}$$

and

$$v_{FH,FK} = v_{GH,GK}. \tag{8.11}$$

PROOF. Identity (8.9a) follows from (8.2a).

For (8.9b): If H is a chamber, then we are reduced to (8.7). For the general case, we can choose a chamber C greater than H, and observe that $GC = GHC$.

For (8.9c): Using (8.9b),

$$v_{F,G} = v_{FG,GF} = v_{K,GFK} = v_{K,GK} = v_{K,G}.$$

The second identity is similar.

Identity (8.9d) follows from (8.2c) and (1.35).

When F and K are chambers, (8.9e) specializes to (8.2f). The general case can be reduced to this case as follows. Note that KF, FK and HFK have the same support, and $KF \text{--} HFK \text{--} FK$ by the gate property. So by (8.9d),

$$v_{KF,FK} = v_{KF,HFK} v_{HFK,FK}.$$

Further, since $FK \geq FH$, we have $v_{HFK,FK} = v_{HF,FH}$. Substituting this in the above identity yields the result.

Identity (8.9f) is similar or can also be deduced from (8.9e) and (8.9h). The latter follows from (8.2h).

Identity (8.9g) can be deduced from (8.2g). □

Proposition 8.4. *For a log-antisymmetric distance function* v,

$$v_{H,G}\, v_{G,F} = v_{H,F}$$

for any faces F, G *and* H *with the same support.*

PROOF. This follows by picking any chamber C, and applying Proposition 8.2 to the chambers HC, GC and FC. □

Proposition 8.5. *For a distance function* v, *and faces* F', G, G' *and* H *with* $\mathrm{s}(G) = \mathrm{s}(G')$,

$$v_{F',G} v_{G'F',H} = v_{F',GH} v_{G',H}$$

This identity reduces to a tautology when G and G' are both chambers.

PROOF. Using (8.8), (8.9e) and (8.10), the lhs is

$$v_{F',G} v_{G'F',H} = v_{F'G,GF'} v_{G'F'H,HG'F'} = v_{F'GH,GF'H} v_{G'F'H,G'HF'} v_{G'HF',HG'F'}.$$

Similarly, using (8.8), (8.9f) and (8.10), the rhs is

$$v_{F',GH} v_{G',H} = v_{F'GH,GHF'} v_{G'H,HG'} = v_{F'GH,GF'H} v_{GF'H,GHF'} v_{G'HF',HG'F'}.$$

So the check boils down to the identity

$$v_{G'F'H,G'HF'} = v_{GF'H,GHF'}$$

which holds by (8.11). This uses that G and G' have the same support. □

8.1.5. Weights on half-flats. Suppose wt is a weight function on the set of half-spaces in $\mathcal{A}$. Then it extends to a weight function on the set of half-flats: The weight on a half-flat h is the product of weights of all half-spaces which contain h but do not contain the support of h. For any flat X, a half-space of $\mathcal{A}^{\mathrm{X}}$ is precisely a half-flat whose support is X. In particular, wt induces a weight function on half-flats with support X. Let us denote it by wt^{X}.

Let v^{X} denote the distance function on chambers of $\mathcal{A}^{\mathrm{X}}$ which arises from wt^{X}. We claim that

$$(v^{\mathrm{X}})_{F,G} = v_{F,G}, \tag{8.12}$$

with the latter as in (8.7). It is easy to see that both sides are equal to the product of weights of all half-spaces in $\mathcal{A}$ which contain F but do not contain G.

8.1.6. Distance from a fixed chamber. Let v be a log-antisymmetric distance function. For each chamber C, define

$$v_C := \sum_D v_{C,D}. \tag{8.13}$$

Then for any chambers C and C', we have

$$v_C = v_{C,C'} v_{C'}. \tag{8.14}$$

This follows from Proposition 8.2. More generally, for each face F, define

$$v_F := \sum_{G\colon \mathrm{s}(G)=\mathrm{s}(F)} v_{F,G}.$$

Then for any faces F and F' with the same support, we have

$$v_F = v_{F,F'} v_{F'}.$$

This follows from Proposition 8.4.

8.2. Sampling weights from a matrix

Let $\mathcal{A}$ be any arrangement. We discuss weight functions on $\mathcal{A}$ that arise by sampling entries of a fixed square matrix. The special case when the matrix has size 1 is also nontrivial and of interest. We begin the discussion with this case.

8.2.1. Uniform weights. Fix a scalar q. Define a distance function v_q on $\mathcal{A}$ by assigning the weight q to each half-space. By (8.1),

$$(v_q)_{C,D} := q^{\mathrm{dist}(C,D)}, \tag{8.15}$$

where $\mathrm{dist}(C, D)$ is the gallery distance between C and D, or equivalently, it is the number of hyperplanes which separate C and D. This is an example of a symmetric distance function. It is log-antisymmetric precisely when $q = \pm 1$.

The induced function on pairs of faces (8.8) is given by

$$(v_q)_{F,G} = q^{\mathrm{dist}(F,G)}, \tag{8.16}$$

where $\mathrm{dist}(F, G)$ is the number of hyperplanes which separate F and G.

The cases $q = \pm 1$ and $q = 0$ are of special interest. The distance function v_1 is identically 1,

$$(v_{-1})_{C,D} = (-1)^{\mathrm{dist}(C,D)} \tag{8.17}$$

keeps track of the parity of the gallery distance, while v_0 is the delta function

$$(v_0)_{C,D} = \begin{cases} 1 & \text{if } C = D, \\ 0 & \text{otherwise.} \end{cases} \tag{8.18}$$

Since $v_{\pm 1}$ are log-antisymmetric, by Proposition 8.4,

$$(v_{\pm 1})_{H,G} (v_{\pm 1})_{G,F} = (v_{\pm 1})_{H,F} \tag{8.19}$$

whenever F, G and H have the same support.

8.2.2. Odd-even invariant. We apply the discussion in Section 8.1.6 to the log-antisymmetric distance functions $\upsilon_{\pm 1}$. For υ_1, we have

$$(\upsilon_1)_C = c(\mathcal{A}),$$

the number of chambers in the arrangement. This does not depend on the choice of C. More generally, $(\upsilon_1)_F$ equals the number of faces of the same support as F.

For υ_{-1}, we have

$$(\upsilon_{-1})_C = \sum_D (-1)^{\operatorname{dist}(C,D)}.$$

Up to sign, the rhs is independent of C. (This follows from (8.14).) Its absolute value is called the *odd-even invariant* of the arrangement. Recall from Section 1.10.4 that the chamber graph of an arrangement is bipartite. The absolute difference in the sizes of the two parts of this bipartite graph is precisely the odd-even invariant. In particular, this invariant is zero iff the bipartite graph is balanced. By Exercises 1.60 and 5.7, this invariant is zero for reflection arrangements of rank at least one, and for arrangements with an odd number of hyperplanes. More generally, we have

$$(\upsilon_{-1})_F = \sum_{G:\, \mathrm{s}(G)=\mathrm{s}(F)} (-1)^{\operatorname{dist}(F,G)}.$$

Up to sign, the rhs only depends on the support of F, say X. Taking absolute value, we get the *odd-even invariant* of the flat X. When X is the maximum flat, we recover the odd-even invariant of the arrangement.

For the rank-two arrangement of n lines, the odd-even invariants are 0 for the maximum flat, $1 + (-1)^{n-1}$ for any of the lines, and 1 for the minimum flat.

Exercise 8.6. Show that: The odd-even invariants for the braid arrangement are as follows. The invariant is 0 unless X has at most one block of odd size, and in this case, the invariant is deg!(X), the factorial of the number of blocks of X. (Employ (6.5) for the distance between two faces with the same support.)

In particular, for the rank-three braid arrangement, the odd-even invariants are 0 for the maximum flat, 0 for each hyperplane, 0 for flats of the form $\{a, bcd\}$, 2 for flats of the form $\{ab, cd\}$, and 1 for the minimum flat.

8.2.3. Square matrices. We set up some terminology to deal with square matrices of size r. A general square matrix of size r is denoted

$$Q := (q_{ij})_{1 \le i,j \le r}.$$

Given square matrices P and Q of size r, let $P \times Q$ denote the matrix obtained by multiplying the corresponding entries of P and Q. This is the Hadamard product on matrices [**231**, Chapter 5]. The unit element for this product is the matrix all of whose entries are 1. A matrix Q is invertible wrt this product iff each entry of Q is nonzero. In this case, the inverse is obtained by inverting each entry of Q. We denote the inverse by Q^-. The transpose of Q is given the usual meaning and denoted Q^t. When Q is invertible, we let $Q^{-t} := (Q^t)^-$, that is, take transpose of Q and invert each entry.

We say Q is *symmetric* if $Q = Q^t$. We say Q is *log-antisymmetric* if $Q = Q^{-t}$, that is,

$$q_{ij} q_{ji} = 1 \quad \text{for } 1 \le i, j \le r.$$

In particular, the diagonal entries of a log-antisymmetric matrix are either 1 or -1. A log-antisymmetric matrix is symmetric iff the matrix entries are either 1 or -1.

8.2.4. Matrix weights. Fix a matrix Q of size r. Suppose $f : \mho \to [r]$ with $\mho$ being the set of half-spaces in $\mathcal{A}$. With this data, define a distance function v^f_Q on $\mathcal{A}$ by assigning to the half-space h the weight

$$\mathrm{wt}^f_Q(\mathrm{h}) := q_{f(\mathrm{h})f(\bar{\mathrm{h}})}.$$

By (8.1),

$$(v^f_Q)_{C,D} := \prod \mathrm{wt}^f_Q(\mathrm{h}),$$

where the product is over all half-spaces h which contain C but do not contain D.

Observe that

$$(v^f_Q)^t = v^f_{Q^t}, \quad (v^f_Q)^- = v^f_{Q^-} \quad \text{and} \quad v^f_P \times v^f_Q = v^f_{P\times Q}.$$

In particular, if Q is symmetric, then v^f_Q is symmetric, and if Q is log-antisymmetric, then v^f_Q is log-antisymmetric.

Now suppose $r = 1$, so Q is of the form (q) for some scalar q. Note that there is a unique function of the form $f : \mho \to [1]$; thus f provides no additional information in this case. It is clear that $v^f_Q = v_q$ and we recover the distance function of uniform weights.

8.3. Distance functions

We have discussed distance functions on arrangements arising from weight functions on half-spaces. We now approach them through the axiomatic treatment of Section E.2 which defines (left, right) distance functions on bands. The band under consideration here is the Tits monoid.

8.3.1. Distance functions. By definition, a distance function on chambers is a function v on pairs of chambers which satisfies (8.2a), (8.2b), (8.2e) and (8.2f). In particular, weight functions on half-spaces give rise to distance functions.

Proposition 8.7. *Suppose v is any function on pairs of chambers which satisfies (8.2a). Then conditions (8.2c), (8.2d), (8.2e) and (8.2f) are equivalent to one another.*

PROOF. Clearly (8.2c) and (8.2d) are equivalent. By the gate property, we have $C \text{--} GC \text{--} D$ and $C \text{--} FD \text{--} D$ whenever $G \leq D$ and $F \leq C$. Thus, (8.2e) and (8.2f) are both special cases of (8.2c). Conversely, either (8.2e) or (8.2f) implies (8.2d), and hence also (8.2c). □

Thus, a distance function on chambers is a function v on pairs of chambers which satisfies (8.2a), (8.2b), and any of the equivalent conditions (8.2c), (8.2d), (8.2e) and (8.2f). The latter says that distance functions, left distance functions, and right distance functions are all equivalent notions.

Proposition 8.8. *Suppose v is a nowhere-zero distance function on chambers. Then it necessarily arises from a choice of a nonzero weight for each half-space.*

PROOF. We first show that (8.2g) holds. Applying (8.2c) to $C \text{--} GC \text{--} GD$ and to $C \text{--} D \text{--} GD$, we obtain

$$v_{C,GC}\, v_{GC,GD} = v_{C,GD} = v_{C,D}\, v_{D,GD}.$$

Now by (8.2b),

$$v_{C,GC} = v_{D,GD}.$$

Canceling this off from the previous equation, we obtain $v_{C,D} = v_{GC,GD}$ as required. Finally, any distance function satisfies (8.2d) which then implies (8.1). $\square$

Exercise 8.9. Give an example of a distance function on chambers which does not arise from a weight function on half-spaces.

Exercise 8.10. Suppose in the definition of distance function, axiom (8.2b) is replaced by (8.2g). (The latter corresponds to the additional axiom (E.10).) Then show that distance functions necessarily arise from weight functions on half-spaces. In particular, axiom (8.2b) is then a consequence.

By definition, a distance function on faces is a function v on pairs of faces with the same support which satisfies (8.9a), (8.9b), (8.9e), and (8.9f). This is equivalent to a distance function on chambers.

8.3.2. Partial-flats. Let $\sim$ be a partial-support relation on faces (Section 2.8). The monoid $\Sigma_\sim$ is a LRB. For chambers c and d, that is, for partial-flats whose support is the maximum flat, define

$$v_{\mathrm{c},\mathrm{d}} := \begin{cases} 1 & \text{if } \mathrm{c} = \mathrm{d}, \\ 0 & \text{otherwise}. \end{cases} \tag{8.20}$$

We claim that this is a distance function on $\Sigma_\sim$. The nontrivial part is to check property (E.9b). This follows from Corollary 2.68. The extension of this function to faces with the same support (E.12) is given by

$$v_{\mathrm{x},\mathrm{y}} := \begin{cases} 1 & \text{if x and y have an upper bound}, \\ 0 & \text{otherwise}. \end{cases}$$

If the partial-support relation is geometric, then by Lemma 2.74, this simplifies to

$$v_{\mathrm{x},\mathrm{y}} := \begin{cases} 1 & \text{if } \mathrm{x} = \mathrm{y}, \\ 0 & \text{otherwise}. \end{cases}$$

8.3.3. Invariant distance functions. Now assume that $\mathcal{A}$ is a reflection arrangement. A distance function is *invariant* if it is preserved by the action of the Coxeter group W. That is,

$$v_{C,D} = v_{wC,wD}$$

for all w, C and D. An invariant distance function is necessarily symmetric: When C and D are adjacent, there is a reflection which interchanges C and D, hence $v_{C,D} = v_{D,C}$. The general case can be deduced from this one by using (8.2d).

Now assume that the distance function arises from a weight function on half-spaces. In this case, invariance is equivalent to $\mathrm{wt}(\mathrm{h}) = \mathrm{wt}(w\mathrm{h})$ for all w and h. Further, since the distance function is symmetric, this condition can be rewritten as $\mathrm{wt}(\mathrm{H}) = \mathrm{wt}(w\mathrm{H})$ for all w and H. To summarize, an invariant distance function amounts to assigning a weight to each hyperplane such that hyperplanes in the same W-orbit have the same weight. For type A, there is a single orbit, so all hyperplanes must have the same weight, say q, and $v = v_q$. This is also the case for other irreducible types except type B, type F_4 and $I_2(m)$ with m even where there are two orbits.

8.4. Varchenko matrix

Fix an arrangement $\mathcal{A}$. Any weight function on its half-spaces gives rise to a distance function v on chambers. The matrix $(v_{C,D})$ is the Varchenko matrix. Its determinant has a nice factorization. Each factor is of the form $1-\alpha$, where α is a product of weights of some of the half-spaces. In particular, the Varchenko matrix is invertible whenever no nontrivial product of the weights is 1. The Witt identity plays a key role in the proof.

8.4.1. Varchenko matrix. Let $\Bbbk$ be any field. Assign to each half-space h a formal variable denoted wt(h), and put

$$R := \Bbbk[\mathrm{wt}(\mathrm{h}), \mathrm{h} \text{ is a half-space}].$$

This is the polynomial ring over $\Bbbk$ in the variables wt(h). For any chambers C and D, let

$$v_{C,D} := \prod_{\mathrm{h} \in r(C,D)} \mathrm{wt}(\mathrm{h}), \tag{8.21}$$

where recall that $r(C, D)$ is the set of half-spaces h which contain C but do not contain D. Each $v_{C,D}$ is a monomial in the ring R. If $C = D$, then $v_{C,D} = 1$.

Note that if we specialize the variables to scalars, then we have a weight function, and the monomials $v_{C,D}$ specialize to (8.1). The properties listed in Proposition 8.1 continue to hold in the present formal setting.

Consider the matrix

$$A := (v_{C,D})$$

indexed by chambers with entries in the ring R. The entry in row C and column D is $v_{C,D}$. Note that the diagonal entries of A are all 1, while the off-diagonal entries are monomials of degree at least 1. We refer to A as the *Varchenko matrix* of the arrangement $\mathcal{A}$.

8.4.2. Determinant of Varchenko matrix. Let F be any face which is not a chamber. Define b_F to be the product of weights of all half-spaces whose base contains F. Thus,

$$b_F = \prod_{\mathrm{H} \supseteq F} \mathrm{wt}(\mathrm{H}^+)\,\mathrm{wt}(\mathrm{H}^-) \tag{8.22}$$

with H running over the hyperplanes that contain F and $\mathrm{H}^{\pm}$ denoting the two opposite half-spaces bounded by H. This is a monomial in R. Pick any hyperplane H which contains F. Define β_F to be half the number of chambers C such that $C \wedge \mathrm{H} = F$. The number β_F does not depend on the particular choice of H; this will be a consequence of the discussion below.

Let X be any non-maximum flat. Put $b_{\mathrm{X}} := b_F$ and $\beta_{\mathrm{X}} := \beta_F$, where F is any face with support X. More directly, b_{X} is the product of weights of all half-spaces which contain X. Note that $\mathrm{X} \neq \mathrm{Y}$ implies $b_{\mathrm{X}} \neq b_{\mathrm{Y}}$.

Theorem 8.11. *The determinant of the Varchenko matrix A is given by*

$$\det(A) = \prod_{\mathrm{X}:\, \mathrm{X} \neq \top} (1 - b_{\mathrm{X}})^{c^{\mathrm{X}} \beta_{\mathrm{X}}}, \tag{8.23}$$

where c^{X} is the number of faces with support X. *The product is over all non-maximum flats.*

The exponent can also be directly described as follows. Pick any hyperplane H which contains X. Then $c^{\mathrm{X}}\beta_{\mathrm{X}}$ is half the number of chambers C which have the property that X is the support of $C \wedge \mathrm{H}$.

The above result can be recast in the language of bilinear forms: Let $R\Gamma$ denote the linearization of the set of chambers Γ over the ring R. Define a bilinear form on $R\Gamma$ by letting

$$\langle C, D \rangle := v_{C,D} \tag{8.24}$$

on the basis elements. Theorem 8.11 says that the determinant of this bilinear form (in the canonical basis of chambers) is given by formula (8.23).

A more general result is given below.

Theorem 8.12. *Let* V *be a combinatorial top-cone. Let* $A|_{\mathrm{V}}$ *denote the restriction of the Varchenko matrix* A *to the chambers in* V*. Then*

$$\det(A|_{\mathrm{V}}) = \prod_{F\colon F \in \mathrm{V}^o, F \notin \Gamma} (1 - b_F)^{\beta_F}. \tag{8.25}$$

The product is over all faces in the interior of V *which are not chambers. Equivalently,*

$$\det(A|_{\mathrm{V}}) = \prod_{\mathrm{X}\colon \mathrm{X} \neq \top} (1 - b_{\mathrm{X}})^{c^{\mathrm{X},\mathrm{V}}\beta_{\mathrm{X}}}, \tag{8.26}$$

where $c^{\mathrm{X},\mathrm{V}}$ *is the number of faces with support* X *which belong to the interior of* V.

It is possible that X does not support any faces belonging to the interior of V. In this case $c^{\mathrm{X},\mathrm{V}} = 0$, and X does not contribute to the factorization. More interestingly, it is also possible that X does support faces belonging to the interior of V but it still does not contribute to the factorization because $\beta_{\mathrm{X}} = 0$.

When V is the ambient space, we have $c^{\mathrm{X},\mathrm{V}} = c^{\mathrm{X}}$, and Theorem 8.12 specializes to Theorem 8.11. As further illustrations, consider the following two top-cones in rank three.

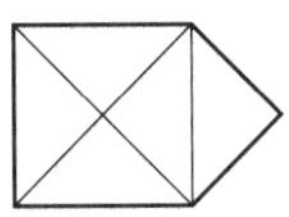

The determinant for the first top-cone is $(1 - xx')(1 - yy')(1 - zz')$. There are three edges in the interior of the top-cone. Here x, x' are weights of the two half-spaces whose base supports one edge, similarly, y, y' are weights for the second edge, and z, z' for the third edge. The determinant for the second top-cone is $(1 - xx')^2(1 - yy')^2(1 - zz')$. Here z, z' correspond to the vertical edge in the interior, while x, x', y, y' correspond to the slanting interior edges. There is a vertex P in the interior, but $\beta_P = 0$, so it does not contribute.

We now work towards proving Theorem 8.12.

Let S denote the ring obtained by localizing R at the set of all polynomials whose constant term is nonzero. In other words, S consists of fractions whose numerator is any polynomial, and denominator is a polynomial whose constant term is nonzero.

Proposition 8.13. *The Varchenko matrix A is invertible over the ring S. Further, each entry of A^{-1} is of the form*

$$\frac{p}{\prod_{\mathrm{X}:\,\mathrm{X}\neq\top}(1-b_{\mathrm{X}})^{k_{\mathrm{X}}}}, \tag{8.27}$$

for some $p \in R$ and nonnegative integers k_{X}. (These depend on the matrix entry.)

More generally, the same is true for the matrix $A|_{\mathrm{V}}$, where in addition, the denominator in (8.27) can be restricted to factors of the form $1 - b_F$, where F varies over faces in the interior of V which are not chambers.

Proof. The determinant of A is a polynomial with constant term 1. So it is invertible in S, and we deduce that A is invertible over S. We now prove the entries in the inverse have the form (8.27).

Let $\gamma : S\Gamma \to S\Gamma^*$ be the S-linear map induced by the bilinear form (8.24). Explicitly,

$$\gamma(D) = \sum_C v_{D,C}\, C^*.$$

For any top-nested face (H, D), define

$$m(H, D) := \sum_{C:\,HC=D} v_{D,C}\, C^*.$$

We claim that each $m(H, D)$ can be written as a linear combination

$$\sum_C a_C \gamma(C), \tag{a}$$

where each coefficient a_C is of the form (8.27). The proof proceeds by backward induction on the rank of H. Note that $m(D, D) = \gamma(D)$, so the claim holds if H is a chamber. This is the induction base.

For the induction step, we proceed as follows. The Witt identity (7.16) applied to the vectors $x^C := v_{D,C}C^*$ yields

$$\begin{aligned}\sum_{K:H\leq K\leq D} (-1)^{\mathrm{rk}(K)}\, m(K, D) &= (-1)^{\mathrm{rk}(D)} \sum_{C:\,HC=H\overline{D}} v_{D,C}\, C^* \\ &= (-1)^{\mathrm{rk}(D)} v_{D,H\overline{D}}\, m(H, H\overline{D}).\end{aligned}$$

The second equality used

$$v_{D,H\overline{D}} v_{H\overline{D},C} = v_{D,C},$$

which holds by (8.2f). Rearranging the terms of the above identity, we obtain

$$m(H, D) - (-1)^{\mathrm{rk}(D/H)} v_{D,H\overline{D}}\, m(H, H\overline{D}) = \sum_{K:H<K\leq D} (-1)^{\mathrm{rk}(K/H)+1}\, m(K, D).$$

By induction hypothesis, the rhs can be expressed in the form (a). Interchanging the roles of D and $H\overline{D}$, we note that

$$m(H, H\overline{D}) - (-1)^{\mathrm{rk}(H\overline{D}/H)} v_{H\overline{D},D}\, m(H, D)$$

is also of the form (a). Now observe that

$$1 - v_{D,H\overline{D}} v_{H\overline{D},D} = 1 - b_{\mathrm{X}},$$

where X is the support of H. It follows that both $m(H, D)$ and $m(H, H\overline{D})$ can be expressed in the form (a). (This necessitates dividing by $1 - b_{\mathrm{X}}$.) This completes the induction step and the claim is proved.

To finish the argument, we note that for any chamber D, we have $m(O, D) = D^*$. So each D^* is of the form (a). Applying γ^{-1} to both sides, we see that $\gamma^{-1}(D^*)$ is a linear combination of chambers whose coefficients are of the form (8.27).

Now let us deal with the general case of a combinatorial top-cone V. For that, specialize the above discussion by setting $\mathrm{wt}(\mathrm{h}) = 0$ for all half-spaces h whose base is a wall of V. Hence $v_{C,D} = 0$ whenever one of C or D belongs to V but the other does not. Now take $D \in \mathrm{V}$. Then $m(H, D)$ can be written in the form (a), with C varying over chambers in V. To understand a_C, we proceed as before. Note that $H \in \mathrm{V}^o \iff H\overline{D} \in \mathrm{V}$ (Exercise 2.24). In this situation, we need to divide by $1 - b_H$ to obtain $m(H, D)$ (exactly as before). On the other hand, if H belongs to the boundary of V, then $v_{D,H\overline{D}} = 0$, and hence

$$m(H, D) = \sum_{K: H < K \leq D} (-1)^{\mathrm{rk}(K/H)+1} m(K, D),$$

and no division is required. □

PROOF OF THEOREM 8.12. Proposition 8.13 provides us the factors of the determinant, namely,

$$\det(A|_{\mathrm{V}}) = \prod_{F:\, F \in \mathrm{V}^o, F \notin \Gamma} (1 - b_F)^{\beta'_F}$$

for some nonnegative integers β'_F. We now show that $\beta'_F = \beta_F$. We proceed by induction on the number of chambers in V. Take any hyperplane H which cuts V, say into V_1 and V_2. By assigning weight 0 to the two opposite half-spaces bounded by H, and applying induction hypothesis to both V_1 and V_2, we see that $\beta'_F = \beta_F$ for all F not contained in H. This takes care of all interior faces of V, except those that lie on every hyperplane that cuts V. There can be at most one such face, say H, with V being the top-star of H. In this case, by passing to $\mathcal{A}_H$, we may assume that H is the central face and V is the ambient space. We want to show that $\beta'_O = \beta_O$. Observe from the definition that the leading monomial in $\det(A)$ is

$$(-1)^{c/2} \prod_C v_{C,\overline{C}} = \big(- \prod_{\mathrm{h}} \mathrm{wt}(\mathrm{h}) \big)^{c/2},$$

where c is the number of chambers. Further, by our factorization, this equals $\prod_{F \notin \Gamma} (-b_F)^{\beta'_F}$. Now fix a half-space h. Comparing the exponent of $\mathrm{wt}(\mathrm{h})$ in the two expressions for the leading monomial, we obtain

$$\sum_{F:\, F \leq \mathrm{H}} \beta'_F = c/2,$$

where H is the base of h. Grouping chambers C according to $C \wedge \mathrm{H}$, we obtain

$$\sum_{F:\, F \leq \mathrm{H}} \beta_F = c/2.$$

Since the l' and l values for non-minimum faces are known to be equal, we conclude from the above two identities that $\beta'_O = \beta_O$. □

8.4.3. Varchenko matrix in an arrangement under a flat. Fix a flat X. Consider the matrix

$$A^{\mathrm{X}} := (v_{F,G}),$$

indexed by faces with support X, with $v_{F,G}$ as in (8.7). The matrix entries are in the ring R with weights treated as formal variables. This is the Varchenko matrix associated to the flat X. When X is the maximum flat, we recover A. The discussion leading to (8.12) shows that in fact, A^{X} is the Varchenko matrix of $\mathcal{A}^{\mathrm{X}}$ for the weight function wt^{X}. Hence, by specializing (8.23), we obtain

$$\det(A^{\mathrm{X}}) = \prod_{\mathrm{Y}:\,\mathrm{Y}<\mathrm{X}} (1 - b_{\mathrm{Y}}^{\mathrm{X}})^{c^{\mathrm{Y}}\beta_{\mathrm{Y}}^{\mathrm{X}}}, \tag{8.28}$$

where $b_{\mathrm{Y}}^{\mathrm{X}}$ and $\beta_{\mathrm{Y}}^{\mathrm{X}}$ are the same as b_{Y} and β_{Y} but in $\mathcal{A}^{\mathrm{X}}$ and for the weight function wt^{X}. Explicitly, $b_{\mathrm{Y}}^{\mathrm{X}}$ is the product of weights of all half-spaces in $\mathcal{A}$ which contain Y but do not contain X. Pick a hyperplane H which contains Y but not X. Then $c^{\mathrm{Y}}\beta_{\mathrm{Y}}^{\mathrm{X}}$ is the number of faces K of support X with the property that Y is the support of $K \wedge \mathrm{H}$.

8.4.4. Formal inverse of Varchenko matrix and non-stuttering paths. Let S' denote the ring of formal power series in the variables $\mathrm{wt}(\mathrm{h})$. Then $R \hookrightarrow S \hookrightarrow S'$. Since the Varchenko matrix is invertible over S, it is also invertible over S'.

To be completely general, let us fix a flat X, and work with the associated Varchenko matrix $A^{\mathrm{X}} = (v_{F,G})$. Denote its inverse by $(v^{F,G})$. Explicitly,

$$\sum_{G:\,\mathrm{s}(G)=\mathrm{X}} v^{F,G} v_{G,K} = \begin{cases} 1 & \text{if } F = K, \\ 0 & \text{if } F \neq K. \end{cases} \tag{8.29}$$

Here F and K are faces of support X, and the sum is over all faces G of support X.

Example 8.14. When X is the minimum flat, $A^{\mathrm{X}} = (1)$, the matrix of size 1 whose only entry is 1, and hence $v^{O,O} = 1$. When X is a rank-one flat with faces F and $\overline{F}$,

$$A^{\mathrm{X}} = \begin{pmatrix} 1 & v_{F,\overline{F}} \\ v_{\overline{F},F} & 1 \end{pmatrix} \quad \text{and} \quad (A^{\mathrm{X}})^{-1} = \frac{1}{1 - v_{F,\overline{F}}v_{\overline{F},F}} \begin{pmatrix} 1 & -v_{F,\overline{F}} \\ -v_{\overline{F},F} & 1 \end{pmatrix},$$

and thus

$$v^{F,F} = v^{\overline{F},\overline{F}} = \frac{1}{1 - v_{F,\overline{F}}v_{\overline{F},F}}, \qquad v^{F,\overline{F}} = \frac{-v^{F,\overline{F}}}{1 - v_{F,\overline{F}}v_{\overline{F},F}}, \qquad v^{\overline{F},F} = \frac{-v^{\overline{F},F}}{1 - v_{F,\overline{F}}v_{\overline{F},F}}.$$

The coefficients of the inverse of the Varchenko matrix can be expressed as formal power series, that is, as elements of S'. The coefficients of these power series are related to non-stuttering paths in X. This is explained below.

A *path* α in X is a finite sequence of faces $F_0 - F_1 - \cdots - F_n$ all of support X. We say that α is a path of length n from F_0 to F_n, and write $s(\alpha) = F_0$, $t(\alpha) = F_n$ and $l(\alpha) = n$. The path α is *non-stuttering* if any two consecutive faces in its sequence are distinct, that is, $F_i \neq F_{i+1}$ for all i. In the present discussion, we will only be considering non-stuttering paths. Note that a face by itself is a non-stuttering path of length 0. For a path $\alpha = F_0 - \cdots - F_n$, define

$$v_\alpha := \prod_{i=1}^{n} v_{F_{i-1},F_i}.$$

If α has length 0, then $v_\alpha = 1$.

Lemma 8.15. *For faces F and G with support* X, *we have*

$$v^{F,G} = \sum_{\alpha:\, s(\alpha)=F, t(\alpha)=G} (-1)^{l(\alpha)} v_\alpha. \tag{8.30}$$

This is an identity in the ring S' of formal power series. The sum is over all non-stuttering paths α in X from F to G. The rhs makes sense because there are only finitely many non-stuttering paths which contribute to a given monomial.

PROOF. We use the formal identity

$$D^{-1} = (I + (D - I))^{-1} = \sum_{i \geq 0} (-1)^i (D - I)^i.$$

Let D be the associated Varchenko matrix. The diagonal entries of $D - I$ are all zero. So the entry in row F and column G of $(D - I)^i$ is v_α summed over all non-stuttering paths of length i from F to G. □

We now complement the discussion in Example 8.14 using the above result.

Example 8.16. Let X be the minimum flat. Then there is only one non-stuttering path in X from the central face O to itself, and it has length 0. Thus, we have $v^{O,O} = 1$.

Let X be a rank-one flat with faces F and $\overline{F}$. Then the non-stuttering paths in X from F to itself are

$$F, \quad F - \overline{F} - F, \quad F - \overline{F} - F - \overline{F} - F,$$

and so on. Similarly, the non-stuttering paths from F to $\overline{F}$ are

$$F - \overline{F}, \quad F - \overline{F} - F - \overline{F}, \quad F - \overline{F} - F - \overline{F} - F - \overline{F},$$

and so on. Thus,

$$v^{F,F} = 1 + v_{F,\overline{F}} v_{\overline{F},F} + (v_{F,\overline{F}} v_{\overline{F},F})^2 + \dots$$

and

$$v^{F,\overline{F}} = -v_{F,\overline{F}} - v_{\overline{F},F}(v_{F,\overline{F}})^2 - (v_{\overline{F},F})^2 (v_{F,\overline{F}})^3 \dots.$$

The power series for $v^{\overline{F},F}$ and $v^{\overline{F},\overline{F}}$ can be computed similarly.

8.4.5. Assembly of Varchenko matrices. Define a matrix B indexed by faces as follows.

$$B_{F,G} := \begin{cases} v_{F,G} & \text{if } GF = G, \\ 0 & \text{otherwise,} \end{cases} \tag{8.31}$$

with $v_{F,G}$ as in (8.8). Then

$$\det(B) = \prod_{\mathrm{Y} < \mathrm{X}} (1 - b_{\mathrm{Y}}^{\mathrm{X}})^{c^{\mathrm{Y}} \beta_{\mathrm{Y}}^{\mathrm{X}}}. \tag{8.32}$$

The product is over all Y and X such that Y is properly contained in X.

PROOF. Write B as a block-matrix indexed by flats, with the (Y, X)-block consisting of the entries $B_{F,G}$ with F of support Y and G of support X. This block-matrix is triangular, with the diagonal (X, X)-block being the associated Varchenko matrix A^{X}. Thus

$$\det(B) = \prod_{\mathrm{X}} \det(A^{\mathrm{X}}),$$

where the product is over all flats X. Now apply (8.28). □

Example 8.17. Let $\mathcal{A}$ be the rank-one arrangement with chambers C and $\overline{C}$. Then

$$B = \left(\begin{array}{c|cc} 1 & 1 & 1 \\ \hline 0 & 1 & v_{C,\overline{C}} \\ 0 & v_{\overline{C},C} & 1 \end{array}\right)$$

with rows and columns indexed in the order $(O, C, \overline{C})$. There are two flats in $\mathcal{A}$, and hence B splits into four blocks. Multiplying the determinants of the two diagonal blocks, we see that $\det(B) = 1 - v_{C,\overline{C}}v_{\overline{C},C}$.

8.4.6. A linear system of equations.

Consider the linear system over the ring S

$$\sum_{F:\, HF=G} x^F v_{F,G} = 0, \tag{8.33}$$

indexed by $O < H \leq G$. The variables are the x^F and the coefficients are the $v_{F,G}$.

By a triangularity argument, one can see that the linear system

$$\sum_{F:\, HF\leq G} x^F v_{F,G} = 0,$$

indexed by $O < H \leq G$, has the same solution space as (8.33).

Both linear systems are in fact defined over the ring R, but it is more convenient to solve the systems over the ring S. The first step is to get rid of the linear dependencies among the equations:

Lemma 8.18. *Consider the linear system*

$$\sum_{F:\, GF=G} x^F v_{F,G} = 0, \tag{8.34}$$

indexed by faces $G > O$. The solution space of (8.34) *coincides with that of* (8.33).

PROOF. Note that (8.34) is a smaller system than (8.33) consisting of the equations indexed by $O < H \leq G$ with $H = G$. So we need to show that any solution of (8.34) also solves (8.33). We apply induction on $\mathrm{rk}(G) - \mathrm{rk}(H)$. The base case is when $H = G$. These equations are in the smaller system, so the base case holds. Let $O < A \leq G$. Start with the Witt identity (7.20b) with x^F replaced by $x^F v_{F,G}$. By induction, the summands inside the parenthesis for $H > A$ are zero. Thus,

$$\begin{aligned}(-1)^{\mathrm{rk}(A)} \sum_{F:\, AF=G} x^F v_{F,G} &= (-1)^{\mathrm{rk}(G)} \sum_{F:\, AF\leq A\overline{G}} x^F v_{F,G} \\ &= (-1)^{\mathrm{rk}(G)} v_{A\overline{G},G} \sum_{F:\, AF\leq A\overline{G}} x^F v_{F,A\overline{G}} \\ &= (-1)^{\mathrm{rk}(G)} v_{A\overline{G},G} \sum_{F:\, AF=A\overline{G}} x^F v_{F,A\overline{G}}.\end{aligned}$$

The second step used (8.9f). The last step used induction and (8.9c). Now interchanging the roles of G and $A\overline{G}$ yields a similar identity. Combining the two yields

$$(1 - v_{G,A\overline{G}} v_{A\overline{G},G}) \sum_{F\colon AF=G} x^F v_{F,G} = 0.$$

Since the coefficient is invertible in the ring S, it can be canceled off. This completes the induction step. □

Theorem 8.19. *The solution space of the linear system (8.33) is one-dimensional. Explicitly, starting with an arbitrary value of x^O, there is a unique solution which can be computed recursively by the formula*

$$x^G = \frac{-1}{1 - v_{G,\overline{G}} v_{\overline{G},G}} \sum_{F\colon F<G} (x^F + (-1)^{\mathrm{rk}(G)} x^{\overline{F}} v_{\overline{G},G}) \tag{8.35a}$$

or by the formula

$$x^G = \frac{1}{1 - v_{G,\overline{G}} v_{\overline{G},G}} \sum_{F\colon F<G} (v_{\overline{G},G} x^F v_{F,\overline{F}} + (-1)^{\mathrm{rk}(G)} x^{\overline{F}} v_{\overline{F},F}). \tag{8.35b}$$

PROOF. To the linear system (8.34), add the equation

$$x^O = \alpha,$$

where α is fixed but arbitrary. The matrix of this linear system is precisely the square matrix B defined in (8.31). By (8.32), the determinant of B is invertible in the ring S, so this system has a unique solution. This shows that the solution space of (8.34) is one-dimensional. Now apply Lemma 8.18. This proves the first part.

To derive (8.35a), we proceed as follows. Applying the Witt identity (7.19a) with x^F replaced by $x^F v_{F,G}$, we obtain

$$x^G - (-1)^{\mathrm{rk}(G)} x^{\overline{G}} v_{\overline{G},G} = - \sum_{F\colon F<G} x^F.$$

Interchanging the roles of G and $\overline{G}$, and solving for x^G and $x^{\overline{G}}$ yields (8.35a). Formula (8.35b) can be deduced in a similar manner by applying (7.19b). □

Example 8.20. Let $\mathcal{A}$ be the rank-one arrangement with chambers C and $\overline{C}$. The linear system (8.33) consists of two equations:

$$x^O v_{O,C} + x^C v_{C,C} + x^{\overline{C}} v_{\overline{C},C} = 0 \quad \text{and} \quad x^O v_{O,\overline{C}} + x^C v_{C,\overline{C}} + x^{\overline{C}} v_{\overline{C},\overline{C}} = 0.$$

Its solution is given by

$$x^C = -\frac{1 - v_{\overline{C},C}}{1 - v_{C,\overline{C}} v_{\overline{C},C}} x^O \quad \text{and} \quad x^{\overline{C}} = -\frac{1 - v_{C,\overline{C}}}{1 - v_{C,\overline{C}} v_{\overline{C},C}} x^O,$$

where x^O is arbitrary.

8.5. Symmetric Varchenko matrix

We now specialize to the case when the Varchenko matrix is symmetric. This happens when the weights of the two associated half-spaces of any hyperplane are equal. In this case, we can forget half-spaces and work with weights on hyperplanes.

8.5.1. Weights on hyperplanes. Assign to each hyperplane H, a formal variable denoted wt(H). For any chambers C and D, let

$$\upsilon_{C,D} := \prod_{\mathrm{H} \in g(C,D)} \mathrm{wt}(\mathrm{H}),$$

where recall that $g(C,D)$ is the set of all hyperplanes H which separate C and D. Put $A := (\upsilon_{C,D})$. Since this matrix is symmetric, we refer to it as the *symmetric Varchenko matrix*.

Theorem 8.21. *The determinant of the symmetric Varchenko matrix A is given by*

$$\det(A) = \prod_{\mathrm{X}:\, \mathrm{X} \neq \top} (1 - a_{\mathrm{X}}^2)^{c^{\mathrm{X}} \beta_{\mathrm{X}}}, \tag{8.36}$$

where a_{X} is the product of weights of all hyperplanes which contain X, *and β_{X} is as in* (8.23).

PROOF. This follows from Theorem 8.11 by equating the variables wt(h) and wt($\overline{\mathrm{h}}$) and calling this new variable wt(H), where H is the base of h and $\overline{\mathrm{h}}$. □

More generally, specializing Theorem 8.12, we obtain:

Theorem 8.22. *Let* V *be a combinatorial top-cone. Let $A|_{\mathrm{V}}$ denote the restriction of the symmetric Varchenko matrix A to the chambers in* V. *Then*

$$\det(A|_{\mathrm{V}}) = \prod_{F:\, F \in \mathrm{V}^{\circ},\, F \notin \Gamma} (1 - a_F^2)^{\beta_F}, \tag{8.37}$$

where a_F is the product of weights of all hyperplanes which contain F. Equivalently,

$$\det(A|_{\mathrm{V}}) = \prod_{\mathrm{X}:\, \mathrm{X} \neq \top} (1 - a_{\mathrm{X}}^2)^{c^{\mathrm{X},\mathrm{V}} \beta_{\mathrm{X}}}, \tag{8.38}$$

with a_{X} and β_{X} as in (8.36) *and $c^{\mathrm{X},\mathrm{V}}$ as in* (8.26).

Similarly, one can define the symmetric Varchenko matrix A^{X} for any flat X. Its determinant can be obtained by specializing (8.28):

$$\det(A^{\mathrm{X}}) = \prod_{\mathrm{Y}:\, \mathrm{Y} < \mathrm{X}} (1 - (a_{\mathrm{Y}}^{\mathrm{X}})^2)^{c^{\mathrm{Y}} \beta_{\mathrm{Y}}^{\mathrm{X}}}, \tag{8.39}$$

where $a_{\mathrm{Y}}^{\mathrm{X}}$ is the product of weights of all hyperplanes in $\mathcal{A}$ which contain Y but do not contain X, and $\beta_{\mathrm{Y}}^{\mathrm{X}}$ is as before.

Let us now consider the linear system (8.33). One can show by induction that the solution has antipodal invariance, that is, $x^F = x^{\overline{F}}$. Formulas (8.35a) and (8.35b) then simplify to

$$x^G = \frac{-1}{1 - (-1)^{\mathrm{rk}(G)} \upsilon_{G,\overline{G}}} \sum_{F:\, F < G} x^F \tag{8.40a}$$

and

$$x^G = \frac{(-1)^{\mathrm{rk}(G)}}{1 - (-1)^{\mathrm{rk}(G)} \upsilon_{G,\overline{G}}} \sum_{F:\, F < G} x^F \upsilon_{F,\overline{F}} \tag{8.40b}$$

respectively. Solving (8.40a) yields

$$x^G = \Big(\sum_{O<F_1 \cdots <F_n = G} \prod_{i=1}^{n} \frac{-1}{1-(-1)^{\mathrm{rk}(F_i)} v_{F_i,\overline{F_i}}} \Big) x^O. \tag{8.40c}$$

The sum is over all strict chains of faces from O to G. (A similar formula can be obtained from the other recursion.) For $G = O$, there is only one chain, and the formula is a tautology. For $G = P$ a vertex, again there is only one chain, and the formula gives

$$x^P = \frac{-1}{1+v_{P,\overline{P}}} x^O.$$

Note that in the symmetric case, the formulas given in Example 8.20 specialize to this formula.

8.5.2. Equal weights. Let q be a formal variable. For any chambers C and D, let

$$q_{C,D} := q^{\mathrm{dist}(C,D)}.$$

Put $A := (q_{C,D})$. We call it the q-*Varchenko matrix*.

Theorem 8.23. *The determinant of the q-Varchenko matrix is given by*

$$\det(A) = \prod_{\mathrm{X}:\, \mathrm{X} \neq \top} (1-q^{2\, n_{\mathrm{X}}})^{c^{\mathrm{X}} \beta_{\mathrm{X}}}, \tag{8.41}$$

where n_{X} is the number of hyperplanes which contain X, *and β_{X} is as in* (8.23).

Proof. This follows from Theorem 8.11 by equating all the variables wt(h) and calling this new variable q, or similarly from Theorem 8.21 by equating the wt(H) to q. □

Since (8.41) is a formal identity, we may substitute any scalar value for q. We conclude that $\det(A)$ is nonzero when q is not a root of unity.

More generally: Fix a flat X. For any faces F and G with support X, let

$$q_{F,G} := q^{\mathrm{dist}(F,G)}.$$

Put $A^{\mathrm{X}} := (q_{F,G})$. Then, by (8.28),

$$\det(A^{\mathrm{X}}) = \prod_{\mathrm{Y}:\, \mathrm{Y}<\mathrm{X}} (1-q^{2\, n_{\mathrm{Y}}^{\mathrm{X}}})^{c^{\mathrm{Y}} \beta_{\mathrm{Y}}^{\mathrm{X}}}, \tag{8.42}$$

where $n_{\mathrm{Y}}^{\mathrm{X}}$ is the number of hyperplanes which contain Y but do not contain X, and $\beta_{\mathrm{Y}}^{\mathrm{X}}$ is as before. Note that the determinant of A^{X} is nonzero when q is not a root of unity.

Let $R\Gamma[\mathcal{A}^{\mathrm{X}}]$ denote the linearization of the set of chambers of $\mathcal{A}^{\mathrm{X}}$ (or equivalently faces with support X). Define a bilinear form on this space by

$$\langle F, G \rangle := q^{\mathrm{dist}(F,G)}.$$

Its determinant in the canonical basis is given by (8.42). In particular, this bilinear form is non-degenerate when q is not a root of unity.

Denote the inverse of A^{X} by $(q^{F,G})$. Explicitly,

$$\sum_{G:\, \mathrm{s}(G)=\mathrm{X}} q^{F,G} q^{\mathrm{dist}(G,K)} = \begin{cases} 1 & \text{if } F = K, \\ 0 & \text{if } F \neq K. \end{cases} \tag{8.43}$$

Here F and K are faces of support X, and the sum is over all faces G of support X. By Cramer's rule, each $q^{F,G}$ is a rational function in q with the denominator factorizing as a product of terms of the form $(1-q^n)$. One can formally invert the denominator and express each $q^{F,G}$ as a formal power series in q, whose coefficients are related to non-stuttering paths in X: For a path $\alpha = F_0 - \cdots - F_n$, define

$$q_\alpha := \prod_{i=1}^{n} q^{\operatorname{dist}(F_{i-1},F_i)} = q^{\sum_{i=1}^{n} \operatorname{dist}(F_{i-1},F_i)}.$$

Then

$$q^{F,G} = \sum_{\alpha:\, s(\alpha)=F, t(\alpha)=G} (-1)^{l(\alpha)} q_\alpha. \tag{8.44}$$

This is a specialization of (8.30). The sum is over all non-stuttering paths α in X from F to G. The rhs makes sense because there are only finitely many non-stuttering paths which contribute to a given power of q.

Example 8.24. When X is the minimum flat, $A^{\mathrm{X}} = (1)$, the matrix of size 1 whose only entry is 1, and hence $q^{O,O} = 1$.

Let X be a rank-one flat with faces F and $\overline{F}$. Put $d := \operatorname{dist}(F, \overline{F})$. Then

$$A^{\mathrm{X}} = \begin{pmatrix} 1 & q^d \\ q^d & 1 \end{pmatrix} \quad \text{and} \quad (A^{\mathrm{X}})^{-1} = \frac{1}{1-q^{2d}} \begin{pmatrix} 1 & -q^d \\ -q^d & 1 \end{pmatrix}.$$

Hence

$$q^{F,F} = q^{\overline{F},\overline{F}} = \frac{1}{1-q^{2d}} = 1 + q^{2d} + q^{4d} + \dots$$

and

$$q^{F,\overline{F}} = q^{\overline{F},F} = \frac{-q^d}{1-q^{2d}} = -q^d - q^{3d} - q^{5d} - \dots.$$

One may readily check that the power series expressions are consistent with (8.44).

Suppose $\Bbbk = \mathbb{C}$. Then each $q^{F,G}$ is a meromorphic function with poles possible only at roots of unity. If X is not the minimum flat, then $q^{F,G}$ has at least one pole, so the power series in the rhs of (8.44) has radius of convergence 1, and (8.44) holds for $|q| < 1$.

Consider the linear system

$$\sum_{F:\, HF=G} x^F q^{\operatorname{dist}(F,G)} = 0, \tag{8.45}$$

indexed by $O < H \leq G$. The variables are the x^F and the coefficients are the $q^{\operatorname{dist}(F,G)}$.

Theorem 8.25. *Given x^O, the linear system (8.45) has a unique solution which can be computed recursively by*

$$x^G = \frac{-1}{1-(-1)^{\operatorname{rk}(G)} q^{\operatorname{dist}(G,\overline{G})}} \sum_{F:\, F<G} x^F \tag{8.46a}$$

or

$$x^G = \frac{(-1)^{\operatorname{rk}(G)}}{1-(-1)^{\operatorname{rk}(G)} q^{\operatorname{dist}(G,\overline{G})}} \sum_{F:\, F<G} x^F q^{\operatorname{dist}(F,\overline{F})}. \tag{8.46b}$$

Explicitly,

$$x^G = \Big(\sum_{O < F_1 \cdots < F_n = G} \prod_{i=1}^{n} \frac{-1}{1-(-1)^{\mathrm{rk}(F_i)} q^{\mathrm{dist}(F_i, \overline{F_i})}} \Big) x^O. \tag{8.46c}$$

The sum is over all strict chains of faces from O to G.

This is a special case of Theorem 8.19. The formulas are readily seen from formulas (8.40a), (8.40b) and (8.40c). In the last formula, when $G = P$ is a vertex, there is only one chain, namely, $O < P$, and we obtain

$$x^P = \frac{-1}{1+q^{\mathrm{dist}(P,\overline{P})}} x^O.$$

Example 8.26. Let $\mathcal{A}$ be the rank-one arrangement with chambers C and $\overline{C}$. The linear system (8.45) consists of two equations:

$$x^O + x^C + q\, x^{\overline{C}} = 0 \quad \text{and} \quad x^O + q\, x^C + x^{\overline{C}} = 0.$$

Its solution is given by

$$x^C = x^{\overline{C}} = -\frac{1}{1+q} x^O,$$

where x^O is arbitrary.

Now let $\mathcal{A}$ be the rank-two arrangement of d lines. The solution to the linear system (8.45) is given by

$$x^P = -\frac{1}{1+q^{d-1}} x^O \quad \text{and} \quad x^C = \frac{1-q^{d-1}}{(1-q^d)(1+q^{d-1})} x^O,$$

where x^O is arbitrary. Here P is any vertex, and C is any chamber. These formulas can be obtained from (8.46c).

Example 8.27. Let q be a scalar. Then given x^O, (8.45) has a unique solution if q is not a root of unity. This applies to $q = 0$. In this case, the solution to (8.45) is explicitly given by

$$x^G = (-1)^{\mathrm{rk}(G)} x^O. \tag{8.47}$$

This follows directly from (8.46b) since except the term corresponding to $F = O$, the remaining terms in the sum become zero. Alternatively, (8.46a) yields $\sum_{F:\, F \leq G} x^F = 0$ for each $G > O$. Thus $x^G = \mu(O,G) x^O$, where $\mu(O,G)$ is the Möbius function of the poset of faces. Now apply (1.43).

For $q = 1$, the solution is not determined by the value of x^O. In this case, the solutions are precisely the Zie elements, whose study we will begin in Chapter 10.

8.5.3. An application.

We give an interesting application of formula (8.41).

Lemma 8.28. *Let W be a finite Coxeter group. For q not a root of unity, the element*

$$\sum_{w \in W} q^{l(w)} w$$

is invertible in the group algebra of W.

PROOF. We will make use of Lemma D.25, item (3), which is an elementary fact from algebra.

Let $\mathsf{\Gamma}[\mathcal{A}]$ denote the linearization of the set of chambers $\Gamma[\mathcal{A}]$. It is a faithful left module over W (since it is isomorphic to the left regular representation of W). The action of $\sum_{w\in W} q^{l(w)}w$ on $\mathsf{\Gamma}[\mathcal{A}]$ is given by the linear map

$$D \mapsto \sum_{w\in W} q^{l(w)} wD,$$

where $l(w)$ is taken wrt the reference chamber C. We want to show that this map is an isomorphism. The matrix of this map in the canonical basis of chambers agrees with the matrix of the following bilinear form on chambers

$$\langle uC, vC\rangle = q^{l(vu^{-1})} = q^{\mathrm{dist}(u^{-1}C, v^{-1}C)}.$$

By the change of basis $uC \mapsto u^{-1}C$, this bilinear form agrees with

$$\langle D, D'\rangle = q^{\mathrm{dist}(D,D')}.$$

Formula (8.41) implies that the determinant of this bilinear form is nonzero if q is not a root of unity. So the linear map is an isomorphism as required. □

8.6. Braid arrangement

The braid arrangement was reviewed in Sections 6.3–6.6. Recall that a top-cone of the braid arrangement corresponds to a partial order. Chambers in the top-cone correspond to linear extensions of the partial order. We now specialize Theorem 8.12 to this setting. We give an explicit combinatorial description for the Varchenko matrix indexed by linear extensions of a partial order, followed by a formula for its determinant. As special cases, we consider the Varchenko matrix indexed by permutations, and by Catalan paths.

8.6.1. Varchenko matrix indexed by linear extensions of a partial order. Let P be a partial order on a finite set I. For each ordered pair (i, j) of incomparable elements in P, let a_{ij} be a parameter. (We do not assume $a_{ij} = a_{ji}$.) Given two linear extensions ℓ_1 and ℓ_2 of P, define

$$v_{\ell_1,\ell_2} := \prod_{\substack{(i,j)\in I^2\\ i<j \text{ in } \ell_1\\ i>j \text{ in } \ell_2}} a_{ij}. \tag{8.48}$$

(If ℓ_1 and ℓ_2 disagree on $\{i, j\}$, then i and j are incomparable in P.) We obtain a matrix

$$A_P := (v_{\ell_1,\ell_2})$$

indexed by linear extensions of P.

We say that a composition F of I is a *prelinear extension* of P if whenever $i < j$ in P, the block of i in F strictly precedes the block of j. We say it is an *almost-linear extension* of P if in addition F contains exactly one non-singleton block, say B. For such F, we let $c(F)$ denote the size of B, and set

$$b_F := \prod_{\substack{(i,j)\in B^2\\ i\neq j}} a_{ij}. \tag{8.49}$$

Since F extends P, all pairs (i, j) occurring in the product are incomparable in P.

Theorem 8.29. *We have*

$$\det(A_P) = \prod_F (1 - b_F)^{(c(F)-2)!}. \tag{8.50}$$

The product is over the set of almost-linear extensions F of P.

PROOF. We employ the dictionary between geometry and combinatorics of the braid arrangement on I from Table 6.2. For each pair $(i,j) \in I^2$ with $i \neq j$ there is one half-space, namely, h_{ij} given by the equation $x_i < x_j$. We set

$$\mathrm{wt}(\mathrm{h}_{ij}) := a_{ij}$$

for each pair of incomparable elements in P.

Chambers correspond to linear orders, and a chamber is contained in h_{ij} precisely when $i < j$ in the corresponding order. The partial order P gives rise to a top-cone whose chambers correspond to the linear extensions of P. It follows that (8.48) is a special case of (8.21).

Let i and j be two elements in I and let C be a chamber. The corresponding linear order is of the form

$$C = a_1|\cdots|a_p|i|b_1|\cdots|b_q|j|c_1|\cdots|c_r \quad \text{or} \quad C = a_1|\cdots|a_p|j|b_1|\cdots|b_q|i|c_1|\cdots|c_r$$

Let H be the hyperplane $x_i = x_j$. In either case, the face $C \wedge \mathrm{H}$ then corresponds to the composition

$$a_1|\cdots|a_p|\{i, b_1, \ldots, b_q, j\}|c_1|\cdots|c_r.$$

Faces F in the interior of the cone correspond to prelinear extensions of P. For F to satisfy $C \wedge \mathrm{H} = F$, F must contain exactly one non-singleton block, and in fact be an almost-linear extension of P. Conversely, given such an extension F of P, the chambers C that satisfy $C \wedge \mathrm{H} = F$ only differ in the ordering of the elements $i, b_1, \ldots, b_q, j$, with the proviso that i and j must be the endpoints of this segment; the remaining elements must be ordered as in F. In conclusion, if F is not an almost-linear extension of P, $\beta_F = 0$, and if F is such an extension, $\beta_F = q!$.

The hyperplanes that contain F are those of equation $x_h = x_k$ where h and k belong to the non-singleton block. Therefore, b_F as defined by (8.22) agrees with (8.49). Finally, (8.50) follows from (8.25). □

Example 8.30. We illustrate the result with a simple example in which $I = \{i, j, k\}$.

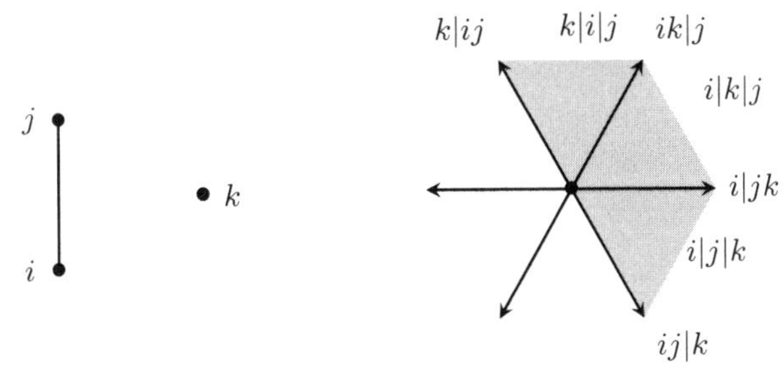

partial order   top-cone

The linear extensions are $i|j|k$, $i|k|j$ and $k|i|j$. Listing them in that order, the matrix is

$$\begin{bmatrix} 1 & a_{jk} & a_{ik}a_{jk} \\ a_{kj} & 1 & a_{ik} \\ a_{ki}a_{kj} & a_{ki} & 1 \end{bmatrix}.$$

The almost-linear extensions are $i|jk$ and $ik|j$. The determinant is

$$(1 - a_{ik}a_{ki})(1 - a_{jk}a_{kj}).$$

8.6.2. Varchenko matrix indexed by permutations. For each ordered pair (i, j) of distinct elements in $[n]$, let a_{ij} be a parameter. The set of *inversions* of a permutation $\sigma \in \mathrm{S}_n$ is

$$\mathrm{Inv}(\sigma) := \{(h, k) \in [n]^2 \mid h < k,\, \sigma(h) > \sigma(k)\}.$$

Given two permutations σ and $\tau \in \mathrm{S}_n$, let

$$v_{\sigma,\tau} := \prod_{(h,k)\in \mathrm{Inv}(\sigma^{-1}\tau)} a_{\tau(k)\tau(h)}. \tag{8.51}$$

We obtain a matrix $A := (v_{\sigma,\tau})$ indexed by permutations in S_n. For each subset B of $[n]$, let

$$b_B := \prod_{\substack{(i,j)\in B^2 \\ i\neq j}} a_{ij}.$$

Also, let $c(B)$ denote the size of B.

Proposition 8.31. *We have*

$$\det(A) = \prod_{B} (1 - b_B)^{(c(B)-2)!(n-c(B)+1)!}. \tag{8.52}$$

The product is over all subsets B of $[n]$ of size at least 2.

PROOF. We apply Theorem 8.29 to the discrete partial order on $[n]$. The matrix entries (8.48) are indexed by all linear orders on $[n]$. Identify the linear order $i_1|\cdots|i_n$ with the permutation in S_n given by $h \mapsto i_h$. If ℓ_1 corresponds to σ and ℓ_2 corresponds to τ, then a pair (i, j) satisfies $i < j$ in ℓ_1 and $i > j$ in ℓ_2 iff $(\tau^{-1}(j), \tau^{-1}(i)) \in \mathrm{Inv}(\sigma^{-1}\tau)$. It follows that (8.51) agrees with (8.48). The almost-linear extensions of the discrete partial order are the compositions F with exactly one non-singleton block B. For such F, we have $b_F = b_B$ and $c(F) = c(B)$. Given B, the number of such F is $(n - c(B) + 1)!$. Formula (8.52) follows from (8.50). $\square$

Let q be a fixed parameter and set $a_{ij} = q$ for all i, j. Then $v_{\sigma,\tau} = q^{\mathrm{inv}(\sigma^{-1}\tau)}$ where inv denotes number of inversions. Formula (8.52) specializes to

$$\det(A) = \prod_{i=2}^{n} \left(1 - q^{i(i-1)}\right)^{\binom{n}{i}(i-2)!(n-i+1)!}. \tag{8.53}$$

We call this the *Zagier formula*. It is also a special case of (8.41).

8.6.3. Varchenko matrix indexed by Catalan paths. Let $\mathcal{C}_n$ be the set of *Catalan paths* of length n. They run from $(0,0)$ to (n,n) and do not go below the diagonal. The number of such paths is the Catalan number $C_n = \frac{1}{n+1}\binom{2n}{n}$ [**399**, Exercise 6.19.h].

We identify each pair (i,j) with the cell in the integer lattice whose top right corner is (i,j). Let a_{ij}^+ and a_{ij}^- be two sets of parameters, with $1 \leq i < j \leq n$. In other words, we have two parameters for each cell (i,j) located above the diagonal and between 0 and n.

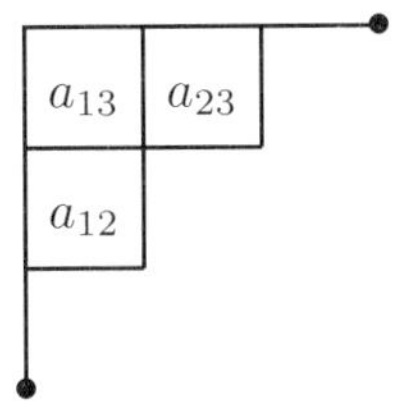

Given two paths p and q in $\mathcal{C}_n$, let $\mathcal{D}(p,q)$ denote the region between the two paths. Given a cell $(i,j) \in \mathcal{D}(p,q)$, let

$$\epsilon(i,j) := \begin{cases} + & \text{if } p \text{ is above } q \text{ on that cell,} \\ - & \text{if } p \text{ is below } q \text{ on that cell.} \end{cases}$$

Define

$$v_{p,q} := \prod_{(i,j)\in\mathcal{D}(p,q)} a_{ij}^{\epsilon(i,j)}. \tag{8.54}$$

In other words, we pick one parameter for each cell between p and q; we choose it from one set of parameters $(+)$ or the other $(-)$ depending on the relative position of the paths. Then we multiply them all. An example follows.

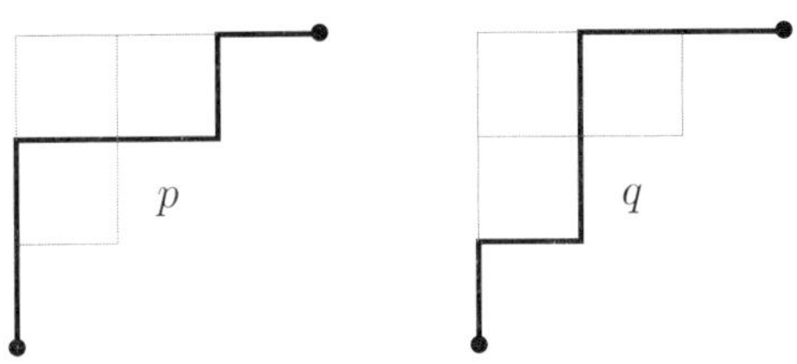

$$v_{p,q} = a_{12}^+ a_{23}^-$$

Proposition 8.32. *We have*

$$\det(v_{p,q})_{p,q\in\mathcal{C}_n} = \prod_{1\leq i<j\leq n} (1 - a_{ij}^+ a_{ij}^-)^{\binom{i+j-3}{i-1}\binom{2n-i-j-1}{n-j}}. \tag{8.55}$$

Let q be a fixed parameter. If $a_{ij}^+ = a_{ij}^- = q$ for all i,j, then $v_{p,q} = q^{a(p,q)}$ where $a(p,q)$ is the area of $\mathcal{D}(p,q)$. On the other hand, if $a_{ij}^+ = q$ and $a_{ij}^- = q^{-1}$, then $v_{p,q} = q^{a(p,q)}$ where $a(p,q)$ is the signed area of the region (the area below p and above q, minus the area above p and below q). According to (8.55), in this case the matrix is singular.

PROOF. We apply (8.50) to the cartesian product poset $[2] \times [n]$. Its linear extensions can be identified with Catalan paths of length n as follows. If the i-th element of the linear extension occurs in $1 \times [n]$, the i-th step of the path is vertical. If the element occurs in $2 \times [n]$, the step is horizontal. An example follows.

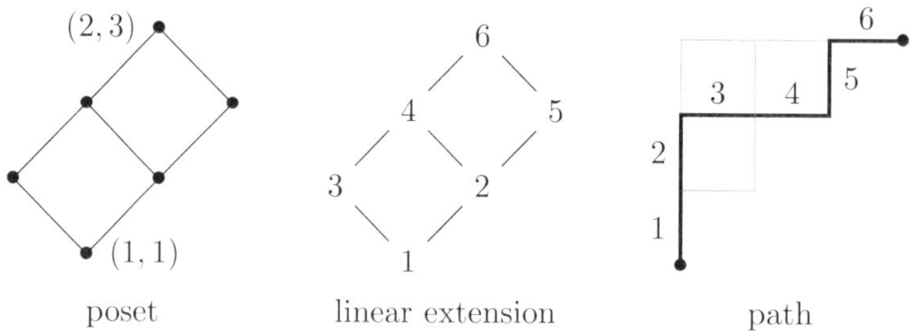

One verifies that $(i, j) \in \mathcal{D}(p, q)$ iff $(2, i)$ and $(1, j)$ occur in opposite orders in the linear extensions corresponding to p and q. Thus, $(v_{p,q})$ is the matrix (8.48) of the poset $[2] \times [n]$. The relevant set compositions of $[2] \times [n]$ contain one block B of size 2 and all other blocks are of size 1. The former block consists of pairs $(2, i)$ and $(1, j)$ with $i < j$. The first block A must consist of the pair $(1, 1)$ and the last block C of $(2, n)$. The segment between A and B is a shuffle of

$$(2, 1)|\cdots|(2, i-1) \quad \text{and} \quad (1, 2)|\cdots|(1, j-1).$$

The segment between B and C is a shuffle of

$$(2, i+1)|\cdots|(2, n-1) \quad \text{and} \quad (1, j+1)|\cdots|(1, n).$$

For any such composition F, $b_F = a_{ij}^+ a_{ij}^-$ and $\beta_F = 1$ ($c(F) = 2$). Formula (8.55) now follows from (8.50). □

Exercise 8.33. Let $\mathcal{P}_{m,n}$ be the set of all lattice paths from $(0, 0)$ to (m, n). The number of such paths is the binomial coefficient $\binom{m+n}{n}$. Let a_{ij}^+ and a_{ij}^- be two sets of parameters, with $1 \leq i \leq m$, $1 \leq j \leq n$. Given two paths p and q in $\mathcal{P}_{m,n}$, define $v_{p,q}$ by the same formula as in (8.54). Show that

$$\det(v_{p,q})_{p,q \in \mathcal{P}_{m,n}} = \prod_{\substack{1 \leq i \leq m \\ 1 \leq j \leq n}} (1 - a_{ij}^+ a_{ij}^-)^{\binom{i+j-3}{i-1}\binom{m+n-i-j-1}{n-j}}. \tag{8.56}$$

(Example 8.30 addresses the case $m = 2$, $n = 1$.)

8.6.4. Equal weights. Let P be a partial order on a set I. We specialize Theorem 8.29 by setting $a_{ij} = q$ for all $i, j \in I$, where q is a fixed parameter. We obtain a matrix $A_P = (q^{\langle \ell_1, \ell_2 \rangle})$ with entries indexed by linear extensions ℓ_1, ℓ_2 of P, and

$$\langle \ell_1, \ell_2 \rangle = |\{(i, j) \in I^2 \mid i < j \text{ in } \ell_1,\ i > j \text{ in } \ell_2\}|.$$

Given an antichain B of P, let $I/_B$ denote the quotient of the set I in which all elements of B are identified (and the remaining elements are kept). The partial order P induces a relation on $I/_B$. Since B is an antichain, this relation is acyclic. Its transitive closure is then a partial order $P/_B$. Let $L(B)$ denote the number of linear extensions of $P/_B$.

Proposition 8.34. *We have*

$$\det(A_P) = \prod_B (1 - q^{|B|(|B|-1)})^{(|B|-2)!\,L(B)}. \tag{8.57}$$

The product is over all antichains B of P other than the singletons.

PROOF. An almost-linear extension F of P yields an antichain B of P (its non-singleton block) and a linear extension of $P/_B$, and conversely. We have $b_F = q^{|B|(|B|-1)}$ and $c(F) = |B|$. The number of F giving rise to B is $L(B)$. Thus, (8.57) is a special case of (8.50). □

8.7. Type B arrangement

The type B arrangement was reviewed in Section 6.7. Top-cones of this arrangement correspond to type B partial orders (signed posets). Chambers in the top-cone correspond to linear extensions of the signed poset which are type B linear orders. We specialize Theorem 8.12 to this setting.

Let the sets I and $\mathbf{I}$ be as in Section 6.7.1. Let P be a signed poset on I. It is a particular partial order on the set $\mathbf{I}$. For each ordered pair (i, j) of incomparable elements in P and with $i \neq \bar{j}$, let a_{ij} be a parameter. We require that for all such pairs,

$$a_{ij} = a_{\bar{j}\,\bar{i}}. \tag{8.58}$$

Given two type B linear extensions ℓ_1 and ℓ_2 of P, define

$$v_{\ell_1,\ell_2} := \prod_{\substack{(i,j)\in\mathbf{I}^2,\, i\neq\bar{j}\\ i<j \text{ in } \ell_1\\ i>j \text{ in } \ell_2}} \sqrt{a_{ij}}. \tag{8.59}$$

Contributing indices occur in pairs (i, j) and $(\bar{j}, \bar{i})$. By (8.58), v_{ℓ_1,ℓ_2} is a monomial in the variables a_{ij}. We obtain a matrix $A_P := (v_{\ell_1,\ell_2})$ indexed by type B linear extensions of P.

Let F be a type B set composition that is a prelinear extension of P (whenever $i < j$ in P, the block of i in F strictly precedes the block of j). Given $(i, j) \in \mathbf{I}^2$, write $i \sim_F j$ when i and j lie in the same block of F and also $i \neq j, \bar{j}$. Note that if $i \sim_F j$, then i and j are incomparable in P. Define

$$b_F := \prod_{\substack{(i,j)\in\mathbf{I}^2\\ i\sim_F j}} \sqrt{a_{ij}}. \tag{8.60}$$

Suppose now that F has either exactly one or exactly two non-singleton blocks. In the former case, the non-singleton block must be the zero block of F and its size be odd and at least 3. Let $c(F)$ be this size. In the latter, the non-singleton blocks must be paired by the involution. Let $c(F)$ be their common size (at least 2).

An analysis similar to that in Theorem 8.29 yields the following result as a special case of Theorem 8.12.

Theorem 8.35. *We have*

$$\det(A_P) = \prod_F (1 - b_F)^{(c(F)-3)!!} \cdot \prod_F (1 - b_F)^{(c(F)-2)!}. \tag{8.61}$$

The first product is over the set of type B prelinear extensions F of P with exactly one non-singleton block, the second over those with exactly two.

Exercise 8.36. Formulate the type B analogue of Proposition 8.31 and deduce it from Theorem 8.35. The matrix is now indexed by signed permutations.

Exercise 8.37. Formulate the type B analogue of Proposition 8.34. Deduce it as the case of equal weights in Theorem 8.35.

Exercise 8.38. Formulate the type B analogue of the Zagier formula (8.53) and deduce it from either Exercise 8.36 or 8.37. The determinant in this case equals

$$\prod_{i=1}^{n}(1-q^{2i^2})^{\frac{(2n)!!}{2i}} \cdot \prod_{i=2}^{n}(1-q^{i(i-1)})^{(2n)!!\frac{n-i+1}{i(i-1)}}.$$

Notes

Odd-even invariant. Proposition 8.2 improves upon [**9**, Proposition 10.22]; the figure shown in its proof is the same as [**9**, Figure 10.12]. The odd-even invariant is considered by Lawrence in the more general context of oriented matroids [**276**]; see also [**167, 254**].

Varchenko matrix. The factorization (8.36) is due to Varchenko [**420**, Theorem (1.1)], [**421**, Theorem 2.6.2]. He obtained it in the greater generality of affine arrangements. This result can be deduced as a special case of (8.25) by the method of coning (see Section 14.3.1). Varchenko worked with a weight function on hyperplanes. The more general factorization for a weight function on half-spaces (8.23) appears to be new. It was stated in [**9**, Equation (10.131)]. The proof that we have given here follows Varchenko's ideas; working with arbitrary top-cones allowed us to simplify his inductive argument.

The number β_X which appears in the exponent in (8.36) is the Crapo invariant of $\mathcal{A}_X$. This follows from [**440**, Theorem D]. The invariant was introduced in [**127**] for arbitrary matroids. Fine information on this invariant of arrangements is provided in [**210**]. See [**272, 273**] and [**441**] for the case of oriented matroids.

Our results on Varchenko matrices overlap to an extent with those in the thesis of Gente [**200**], although they were obtained independently and by different methods. Gente considers the Varchenko matrix for top-cones (in the case of symmetric weights) and obtains the factorization of its determinant [**200**, Theorem 4.5]. Her formula is equivalent to (8.38). Proposition 8.34 is precisely [**200**, Theorem 5.6]. Theorems 8.12 and 8.29, and most of the results in Section 8.4, appear to be new.

Formula (8.53) was first proved by Zagier [**438**, Theorem 2]. The connection to the Varchenko factorization was pointed out in [**163**, Sections 4 and 6] and [**223**]. Lemma 8.28 for the symmetric group is in Zagier's paper [**438**, Section 2], and the general case is in [**223**, Section 4]. Formula (8.52) for symmetric weights is given in [**163**, Formula (43)]. This and similar formulas for types B and D are obtained in [**351**].

Some other papers dealing with the Varchenko bilinear form are [**142, 144, 145, 373**]. Additional references can be found in Krattenthaler's surveys [**262**, Theorem 55] and [**263**, Section 5.7]. A "quasiclassical" version of Varchenko's result appears in his earlier paper with Schechtman [**372**]. Brylawski and Varchenko extended this result from arrangements to matroids [**106**, Theorem 4.16].

Part II.

CHAPTER 9

Birkhoff algebra and Tits algebra

We study the Birkhoff algebra and the Tits algebra. They are obtained by linearizing the Birkhoff monoid of flats and the Tits monoid of faces, respectively. Since the poset of flats is a lattice, the Birkhoff algebra is a split-semisimple commutative algebra. Thus, its simple modules are one-dimensional, and there is one for each flat. Further, any module is a direct sum of simple modules. The Tits algebra is a non-commutative elementary algebra. Its split-semisimple quotient is precisely the Birkhoff algebra (under the support map). Thus, its simple modules are also one-dimensional and indexed by flats. However, an arbitrary module (for instance, the module of chambers) is not a direct sum of simple modules. A consequence of knowing the simple modules is that one can compute the eigenvalues and multiplicities of the action of any given element of the Tits algebra on any module. For the module of chambers, this yields the Bidigare-Hanlon-Rockmore (BHR) theorem.

A general framework for studying the Birkhoff algebra is given by algebras obtained by linearizing lattices. We give three more examples of this nature, namely, the algebra of charts, the algebra of dicharts and the algebra of cones. These are obtained by linearizing the lattices of charts, dicharts and cones, respectively. Further, the four algebras relate to one another by linearizations of join-preserving maps. In the same vein, a general framework for studying the Tits algebra would be a "noncommutative lattice" such as a left regular band, but we do not pursue this idea in detail.

Modules over the Birkhoff algebra and Tits algebra have a primitive part. For instance, the algebra of charts can be viewed as a module over the Birkhoff algebra, and its primitive part is spanned by connected charts. The primitive part of the algebra of cones is spanned by cones whose base is the minimum flat.

The Janus algebra is obtained by linearizing the Janus monoid of bi-faces. Just like the Tits algebra, it is elementary and its split-semisimple quotient is the Birkhoff algebra. It admits a deformation by a parameter q which we call the q-Janus algebra. When q is not a root of unity, this algebra is split-semisimple and Morita equivalent to the Birkhoff algebra.

Background material on algebras and modules is given in Appendix D. This includes split-semisimple commutative algebras, simple modules, characters, complete systems of idempotents, radical of an algebra, diagonalizable elements and elementary algebras. Algebras obtained by linearizing lattices are treated in Section D.9.

We will continue to follow Convention 1.1. For instance, $\Sigma[\mathcal{A}]$ denotes the Tits algebra of $\mathcal{A}$ and $\sqcap[\mathcal{A}]$ denotes the Birkhoff algebra of $\mathcal{A}$. When $\mathcal{A}$ is understood from the context, we may simply write Σ, $\sqcap$, and so on.

Convention 9.1. All modules over the Birkhoff algebra and the Tits algebra are assumed to be finite-dimensional.

9.1. Birkhoff algebra

The Birkhoff algebra is the linearization of the Birkhoff monoid. It has a basis indexed by flats, with the product given by the join operation. It is a split-semisimple commutative algebra; the primitive idempotents can be written explicitly using the Möbius function of the lattice of flats. It is self-dual as a module over itself, and the self-duality isomorphism can be explicitly described by choosing a family of nonzero scalars indexed by flats. This is a specialization of the results of Section D.9.

9.1.1. Birkhoff algebra. Recall the set of flats $\Pi[\mathcal{A}]$. Since it is a lattice under the partial order of inclusion, it carries a monoid structure given by the join operation. Let $\mathsf{\Pi}[\mathcal{A}]$ denote its linearization over a field $\Bbbk$, with canonical basis $\mathtt{H}$. It is a commutative $\Bbbk$-algebra:

$$\mathtt{H}_{\mathrm{X}} \cdot \mathtt{H}_{\mathrm{Y}} := \mathtt{H}_{\mathrm{X} \vee \mathrm{Y}}.$$

We call this the *Birkhoff algebra*.

9.1.2. $\mathtt{Q}$-basis and split-semisimplicity. Define the $\mathtt{Q}$-basis of $\mathsf{\Pi}[\mathcal{A}]$ by

$$\mathtt{H}_{\mathrm{X}} = \sum_{\mathrm{Y}:\,\mathrm{Y} \geq \mathrm{X}} \mathtt{Q}_{\mathrm{Y}} \quad \text{or equivalently} \quad \mathtt{Q}_{\mathrm{X}} = \sum_{\mathrm{Y}:\,\mathrm{Y} \geq \mathrm{X}} \mu(\mathrm{X}, \mathrm{Y})\, \mathtt{H}_{\mathrm{Y}}. \tag{9.1}$$

In particular, the unit element is

$$\mathtt{H}_{\bot} = \sum_{\mathrm{Y}} \mathtt{Q}_{\mathrm{Y}}. \tag{9.2}$$

Specializing Theorem D.47, we obtain:

Theorem 9.2. *The Birkhoff algebra is a split-semisimple commutative algebra. Its dimension equals the number of flats in $\mathcal{A}$. The unique complete system of primitive orthogonal idempotents is given by the $\mathtt{Q}$-basis:*

$$\mathtt{Q}_{\mathrm{X}} \cdot \mathtt{Q}_{\mathrm{Y}} = \begin{cases} \mathtt{Q}_{\mathrm{X}} & \text{if } \mathrm{X} = \mathrm{Y}, \\ 0 & \text{otherwise.} \end{cases} \tag{9.3}$$

By (D.25), we have:

$$\mathtt{H}_{\mathrm{Y}} \cdot \mathtt{Q}_{\mathrm{X}} = \begin{cases} \mathtt{Q}_{\mathrm{X}} & \text{if } \mathrm{X} \geq \mathrm{Y}, \\ 0 & \text{otherwise.} \end{cases} \tag{9.4}$$

From now on, whenever convenient, we will abbreviate $\Pi[\mathcal{A}]$ to Π.

9.1.3. Rank one. Let $\mathcal{A}$ be the arrangement of rank one. It has two flats, namely, the minimum flat $\bot$ and the maximum flat $\top$. The $\mathtt{Q}$-basis elements are given by

$$\mathtt{Q}_{\bot} = \mathtt{H}_{\bot} - \mathtt{H}_{\top}, \qquad \mathtt{Q}_{\top} = \mathtt{H}_{\top}.$$

One can readily check that they define a complete system.

9.1.4. Characters of modules. Let h be a module over the Birkhoff algebra, and Ψ_{h} the associated representation. For any element w of the Birkhoff algebra, $\Psi_{\mathsf{h}}(w)$ denotes the linear operator on h given by multiplication by w, and $w \cdot \mathsf{h}$ denotes its image. In other words, $w \cdot \mathsf{h}$ consists of all elements of the form $w \cdot h$, as h varies over elements of h. Following (D.1), the character of h is the linear functional

$$\chi_{\mathsf{h}} : \Pi \to \Bbbk, \qquad \chi_{\mathsf{h}}(w) = \operatorname{tr}(\Psi_{\mathsf{h}}(w)),$$

where tr denotes trace.

9.1.5. Exponential and logarithm. Let (ξ_{X}) and (η_{X}) be two families of scalars indexed by flats which are related by

$$\xi_{\mathrm{X}} = \sum_{\mathrm{Y}:\, \mathrm{Y} \geq \mathrm{X}} \eta_{\mathrm{Y}} \quad \text{and} \quad \eta_{\mathrm{X}} = \sum_{\mathrm{Y}:\, \mathrm{Y} \geq \mathrm{X}} \mu(\mathrm{X}, \mathrm{Y})\, \xi_{\mathrm{Y}}. \tag{9.5}$$

In other words, they are the exponential and logarithm of each other in the lattice of flats. They correspond to the linear functional $f : \Pi \to \Bbbk$ by

$$\xi_{\mathrm{X}} = f(\mathtt{H}_{\mathrm{X}}) \qquad \text{and} \qquad \eta_{\mathrm{X}} = f(\mathtt{Q}_{\mathrm{X}}). \tag{9.6}$$

See (D.26) and (D.27).

Some choices for these families are given below.

Example 9.3. For each flat X, put

$$\xi_{\mathrm{X}} = \begin{cases} 1 & \text{if } \mathrm{X} = \top, \\ 0 & \text{otherwise} \end{cases} \quad \text{and} \quad \eta_{\mathrm{X}} = \mu(\mathcal{A}_{\mathrm{X}}). \tag{9.7}$$

This choice is a specialization of (D.28).

Example 9.4. For each flat X, put

$$\xi_{\mathrm{X}} = c(\mathcal{A}_{\mathrm{X}}) \quad \text{and} \quad \eta_{\mathrm{X}} = |\mu(\mathcal{A}_{\mathrm{X}})|, \tag{9.8}$$

where $c(\mathcal{A}_{\mathrm{X}})$ is the number of chambers in $\mathcal{A}_{\mathrm{X}}$. The validity of this choice is equivalent to the Zaslavsky formula (1.48).

Similarly, for each flat X, put

$$\xi_{\mathrm{X}} = f(\mathcal{A}_{\mathrm{X}}) \quad \text{and} \quad \eta_{\mathrm{X}} = \sum_{\mathrm{Y}:\, \mathrm{Y} \geq \mathrm{X}} |\mu(\mathcal{A}_{\mathrm{X}}^{\mathrm{Y}})|, \tag{9.9}$$

where $f(\mathcal{A}_{\mathrm{X}})$ is the number of faces in $\mathcal{A}_{\mathrm{X}}$. The validity of this choice is equivalent to formula (1.49).

Example 9.5. Let h be a module over Π. For each flat X, put

$$\xi_{\mathrm{X}}(\mathsf{h}) := \dim(\mathtt{H}_{\mathrm{X}} \cdot \mathsf{h}) \quad \text{and} \quad \eta_{\mathrm{X}}(\mathsf{h}) := \dim(\mathtt{Q}_{\mathrm{X}} \cdot \mathsf{h}). \tag{9.10}$$

This choice is a specialization of (D.29). It is illustrated below.

- For $\mathsf{h} = \Pi$ (viewed as a module over itself), $\xi_{\mathrm{X}}(\Pi)$ is the number of flats greater than X, and $\eta_{\mathrm{X}}(\Pi) \equiv 1$.
- Let E denote the module $\Bbbk$, with each basis element $\mathtt{H}_{\mathrm{X}}$ acting by the identity. Then $\xi_{\mathrm{X}}(\mathsf{E}) \equiv 1$, and $\eta_{\mathrm{X}}(\mathsf{E})$ is 1 if $\mathrm{X} = \top$ and 0 otherwise.
- Let $\mathsf{\Lambda}$ denote the linearization of the set of lunes Λ. Recall from Section 4.7 that the Birkhoff monoid acts on Λ. Hence, by linearization, the Birkhoff algebra acts on $\mathsf{\Lambda}$. By Corollary 4.48, $\xi_{\mathrm{X}}(\mathsf{\Lambda})$ is the number of lunes whose base is greater than X. It then follows that $\eta_{\mathrm{X}}(\mathsf{\Lambda})$ is the number of lunes whose base is equal to X.

The linear functional $f : \Pi \to \Bbbk$ associated to $\xi_X(\mathsf{h})$ (or to $\eta_X(\mathsf{h})$) is the character χ_{h} of h.

Exercise 9.6. For each flat X, put

$$\xi_X = c(\mathcal{A}_X)^2 \quad \text{and} \quad \eta_X = \sum_{Y \wedge Z = X} |\mu(\mathcal{A}_Y)||\mu(\mathcal{A}_Z)|.$$

The sum is over both Y and Z. Check that (9.5) holds. Also compare with (9.8).

9.1.6. Simple modules and diagonalizability.

Theorem 9.7. *The Birkhoff algebra Π has one simple module (up to isomorphism) for each flat X. It is one-dimensional and defined by the multiplicative character*

$$\chi_X : \Pi \to \Bbbk, \qquad \sum_Y w^Y \mathtt{H}_Y \mapsto \sum_{Y:\, Y \leq X} w^Y.$$

Proof. This is a special case of Theorem D.51. □

Let $\eta_X(\mathsf{h})$ be as in (9.10). Then: For any $w \in \Pi$,

$$\chi_{\mathsf{h}}(w) = \sum_X \chi_X(w)\, \eta_X(\mathsf{h}). \tag{9.11}$$

This is a special case of (D.30).

Theorem 9.8. *Any module h is a direct sum of simple modules with $\eta_X(\mathsf{h})$ being the multiplicity of the simple module corresponding to the flat X.*

Proof. This is a special case of Theorem D.52. □

Theorem 9.9. *Let h be a module over the Birkhoff algebra. For $w = \sum_X w^X \mathtt{H}_X$, the linear operator $\Psi_{\mathsf{h}}(w)$ is diagonalizable. It has an eigenvalue*

$$\lambda_X(w) = \chi_X(w) = \sum_{Y:\, Y \leq X} w^Y \tag{9.12}$$

for each X, with multiplicity $\eta_X(\mathsf{h})$.

Proof. This is a special case of Theorem D.53. □

9.1.7. Self-duality. Let $\Pi[\mathcal{A}]^*$ denote the linear dual of $\Pi[\mathcal{A}]$. Since $\Pi[\mathcal{A}]$ is an algebra, it is a module over itself. We view $\Pi[\mathcal{A}]^*$ as a $\Pi[\mathcal{A}]$-module with the dual action (D.2). Let $\mathtt{M}$ be the basis of $\Pi[\mathcal{A}]^*$ dual to $\mathtt{H}$, and $\mathtt{P}$ be the basis dual to $\mathtt{Q}$. The action on these bases is given by

$$\mathtt{H}_Y \cdot \mathtt{M}_W = \sum_{X:\, X \vee Y = W} \mathtt{M}_X \quad \text{and} \quad \mathtt{H}_Y \cdot \mathtt{P}_X = \begin{cases} \mathtt{P}_X & \text{if } X \geq Y, \\ 0 & \text{otherwise.} \end{cases} \tag{9.13}$$

This is a specialization of (D.34).

Theorem 9.10. *For any ξ and η as in (9.5), the map $\Pi[\mathcal{A}] \to \Pi[\mathcal{A}]^*$ given by*

$$\mathtt{H}_X \mapsto \sum_Y \xi_{X \vee Y}\, \mathtt{M}_Y \qquad \text{or equivalently} \qquad \mathtt{Q}_X \mapsto \eta_X\, \mathtt{P}_X$$

is a morphism of $\Pi[\mathcal{A}]$-modules. In particular, if $\eta_X \neq 0$ for all X, then this map is an isomorphism.

Proof. This is a special case of Theorem D.59. □

Theorem 9.11. *The maps*

$$\psi, \varphi : \Pi[\mathcal{A}] \to \Pi[\mathcal{A}]^*$$

defined by

$$\psi(\mathtt{H}_{\mathrm{Y}}) := \sum_{\mathrm{X}} c_{\mathrm{X}\vee\mathrm{Y}}\, \mathtt{M}_{\mathrm{X}} \qquad \textit{and} \qquad \varphi(\mathtt{H}_{\mathrm{Y}}) := \sum_{\mathrm{X}:\, \mathrm{X}\vee\mathrm{Y}=\top} \mathtt{M}_{\mathrm{X}}$$

are isomorphisms of $\Pi[\mathcal{A}]$*-modules.*

PROOF. This follows by specializing Theorem 9.10 to the choices of ξ given in (9.7) and (9.8). □

9.1.8. Subarrangements. Given a subarrangement $\mathcal{A}'$ of $\mathcal{A}$, recall the morphism of monoids in (1.16). Its linearization is a morphism of algebras

$$\Pi[\mathcal{A}] \to \Pi[\mathcal{A}']. \tag{9.14}$$

Exercise 9.12. Show that the morphism (9.14) sends $\mathtt{Q}_{\mathrm{Y}}$ to itself if Y is a flat of $\mathcal{A}'$, and to 0 otherwise.

9.1.9. Over and under a flat. Cartesian product. We briefly discuss how the Birkhoff algebra behaves under passage to arrangements over and under a flat, and under taking cartesian product of arrangements.

For any flat X of $\mathcal{A}$, the map

$$\Delta_{\mathrm{X}} : \Pi[\mathcal{A}] \to \Pi[\mathcal{A}_{\mathrm{X}}], \qquad \mathtt{H}_{\mathrm{Y}} \mapsto \mathtt{H}_{\mathrm{X}\vee\mathrm{Y}/\mathrm{X}} \tag{9.15}$$

is an algebra homomorphism. For flats $\mathrm{X} \le \mathrm{Y}$, the diagram

$$\begin{array}{ccc} & \Pi[\mathcal{A}] & \\ {}^{\Delta_{\mathrm{X}}}\swarrow & & \searrow^{\Delta_{\mathrm{Y}}} \\ \Pi[\mathcal{A}_{\mathrm{X}}] & \xrightarrow[\Delta_{\mathrm{Y}/\mathrm{X}}]{} & \Pi[\mathcal{A}_{\mathrm{Y}}] \end{array} \tag{9.16}$$

commutes, where $\Delta_{\mathrm{Y}/\mathrm{X}}(\mathtt{H}_{\mathrm{Z}/\mathrm{X}}) = \mathtt{H}_{\mathrm{Y}\vee\mathrm{Z}/\mathrm{Y}}$.

For any flat X of $\mathcal{A}$, let

$$\mu_{\mathrm{X}} : \Pi[\mathcal{A}_{\mathrm{X}}] \to \Pi[\mathcal{A}], \qquad \mathtt{H}_{\mathrm{Z}/\mathrm{X}} \mapsto \mathtt{H}_{\mathrm{Z}}. \tag{9.17}$$

These maps satisfy a diagram similar to (9.16). The map μ_{X} is a section of the map Δ_{X}, that is, $\Delta_{\mathrm{X}}\mu_{\mathrm{X}} = \mathrm{id}$. Composing in the other direction yields

$$\mu_{\mathrm{X}}\Delta_{\mathrm{X}}(z) = \mathtt{H}_{\mathrm{X}} \cdot z. \tag{9.18}$$

Note that μ_{X} preserves products, that is, $\mu_{\mathrm{X}}(z \cdot w) = \mu_{\mathrm{X}}(z) \cdot \mu_{\mathrm{X}}(w)$, but it does not preserve the unit, so it is not an algebra homomorphism.

Exercise 9.13. Check that for any flats X and Y,

$$\Delta_{\mathrm{Y}}\mu_{\mathrm{X}} = \mu_{\mathrm{Y}\vee\mathrm{X}/\mathrm{Y}}\Delta_{\mathrm{X}\vee\mathrm{Y}/\mathrm{X}}.$$

We call this the bimonoid axiom for flats. It links the Birkhoff algebras of $\mathcal{A}$, $\mathcal{A}_{\mathrm{X}}$, $\mathcal{A}_{\mathrm{Y}}$ and $\mathcal{A}_{\mathrm{X}\vee\mathrm{Y}}$.

For any flat X, the linear map

$$\Pi[\mathcal{A}] \to \Pi[\mathcal{A}^{\mathrm{X}}], \qquad \sum_{\mathrm{Y}} x^{\mathrm{Y}}\mathtt{H}_{\mathrm{Y}} \mapsto \sum_{\mathrm{Y}:\, \mathrm{Y}\le\mathrm{X}} x^{\mathrm{Y}}\mathtt{H}_{\mathrm{Y}} \tag{9.19}$$

is an algebra homomorphism.

For any arrangements $\mathcal{A}$ and $\mathcal{A}'$, there is an algebra isomorphism

$$\Pi[\mathcal{A} \times \mathcal{A}'] \to \Pi[\mathcal{A}] \otimes \Pi[\mathcal{A}'], \qquad \mathtt{H}_{(\mathrm{X},\mathrm{X}')} \mapsto \mathtt{H}_{\mathrm{X}} \otimes \mathtt{H}_{\mathrm{X}'}. \tag{9.20}$$

This follows from (1.22).

Exercise 9.14. Check that the map (9.15) is the special case of the algebra homomorphism (9.61) for $\mathcal{A}' := \mathcal{A}_{\mathrm{X}}$. Deduce that (9.15) sends $\mathtt{Q}_{\mathrm{Y}}$ to $\mathtt{Q}_{\mathrm{Y}/\mathrm{X}}$ if $\mathrm{Y} \geq \mathrm{X}$, and to 0 otherwise.

9.2. Algebras of charts, dicharts and cones

In Section 2.6, we defined charts and dicharts for an arrangement, and related them to flats and cones. These are all lattices linked by join-preserving maps. Recall that the linearization of the lattice of flats is the Birkhoff algebra. In a similar manner, we linearize the other three lattices to obtain the algebra of charts, the algebra of dicharts and the algebra of cones. They can be studied in much the same way as the Birkhoff algebra by using the general theory of Section D.9. They are split-semisimple and self-dual. Further, the self-duality isomorphisms can be chosen in a mutually consistent way.

9.2.1. Algebras of charts, dicharts and cones. Let $\mathsf{G}[\mathcal{A}]$ and $\vec{\mathsf{G}}[\mathcal{A}]$ denote the linearizations of the set of charts $\mathrm{G}[\mathcal{A}]$ and dicharts $\vec{\mathrm{G}}[\mathcal{A}]$, respectively. We write $\mathtt{H}$ for the canonical basis. Both spaces are linearizations of a (Boolean) lattice, and hence commutative algebras, with product induced by the join operation (intersection). For instance, the product in $\mathsf{G}[\mathcal{A}]$ is given by

$$\mathtt{H}_g \cdot \mathtt{H}_h := \mathtt{H}_{g\cap h}.$$

Similarly, let $\Omega[\mathcal{A}]$ denote the linearization of the set of cones $\Omega[\mathcal{A}]$. It is a commutative algebra via

$$\mathtt{H}_{\mathrm{V}} \cdot \mathtt{H}_{\mathrm{W}} := \mathtt{H}_{\mathrm{V}\vee\mathrm{W}}.$$

By Theorem D.47, the algebras $\mathsf{G}[\mathcal{A}]$, $\vec{\mathsf{G}}[\mathcal{A}]$ and $\Omega[\mathcal{A}]$ are split-semisimple. Let $\mathtt{Q}$ denote the basis of primitive idempotents in each case. Explicitly, for charts, by (D.22),

$$\mathtt{H}_g = \sum_{h:\, h\subseteq g} \mathtt{Q}_h \qquad \text{or equivalently} \qquad \mathtt{Q}_g = \sum_{h:\, h\subseteq g} (-1)^{|g\setminus h|}\, \mathtt{H}_h.$$

For dicharts, the formulas are very similar. For cones,

$$\mathtt{H}_{\mathrm{V}} = \sum_{\mathrm{W}:\, \mathrm{W}\geq \mathrm{V}} \mathtt{Q}_{\mathrm{W}} \qquad \text{or equivalently} \qquad \mathtt{Q}_{\mathrm{V}} = \sum_{\mathrm{W}:\, \mathrm{W}\geq \mathrm{V}} \mu(\mathrm{V},\mathrm{W})\, \mathtt{H}_{\mathrm{W}},$$

where μ is the Möbius function of the lattice of cones.

The maps in the first diagram in (2.18) are join-preserving. So by linearizing, we obtain the following commutative diagram of algebras.

$$\begin{array}{ccc} \mathsf{G}[\mathcal{A}] & \xrightarrow{\lambda'} & \vec{\mathsf{G}}[\mathcal{A}] \\ {\scriptstyle\lambda}\uparrow & & \uparrow{\scriptstyle\vec{\lambda}} \\ \Pi[\mathcal{A}] & \xrightarrow[i]{} & \Omega[\mathcal{A}] \end{array} \tag{9.21}$$

The maps on the $\mathtt{H}$-basis are straightforward. On the $\mathtt{Q}$-basis, they can be described by employing the general formula (D.36) and using the maps in the second diagram in (2.18). This is elaborated below.

The map λ sends $\mathtt{H}_{\mathrm{X}}$ to $\mathtt{H}_g$, where g is the set of hyperplanes containing X, see (2.19). On the $\mathtt{Q}$-basis, it is given by

$$\mathtt{Q}_{\mathrm{X}} \mapsto \sum_{g:\, \rho(g)=\mathrm{X}} \mathtt{Q}_g, \tag{9.22}$$

where ρ is the right adjoint of λ given in (2.20). Explicitly, the sum is over all charts g whose center is X.

The map i sends $\mathtt{H}_{\mathrm{X}}$ to $\mathtt{H}_{\mathrm{X}}$, and on the $\mathtt{Q}$-basis,

$$\mathtt{Q}_{\mathrm{X}} \mapsto \sum_{\mathrm{V}:\, \mathrm{b}(\mathrm{V})=\mathrm{X}} \mathtt{Q}_{\mathrm{V}}. \tag{9.23}$$

The sum is over all cones V whose base is X.

The map λ' sends $\mathtt{H}_g$ to $\mathtt{H}_r$, where r is the set of those half-spaces whose base is in g. On the $\mathtt{Q}$-basis, it is given by

$$\mathtt{Q}_g \mapsto \sum_{r:\, \rho'(r)=g} \mathtt{Q}_r, \tag{9.24}$$

with ρ' is the right adjoint of λ' given in (2.22). Explicitly, the sum is over all dicharts r such that g is the set of hyperplanes bounding the half-spaces in r.

The map $\vec{\lambda}$ sends $\mathtt{H}_{\mathrm{V}}$ to $\mathtt{H}_r$, where r is the set of half-spaces containing V, see (2.23). On the $\mathtt{Q}$-basis, it is given by

$$\mathtt{Q}_{\mathrm{V}} \mapsto \sum_{r:\, \vec{\rho}(r)=\mathrm{V}} \mathtt{Q}_r, \tag{9.25}$$

where $\vec{\rho}$ is the right adjoint of $\vec{\lambda}$ given in (2.24). Explicitly, the sum is over all dicharts r such that the intersection of the half-spaces in r is V.

9.2.2. Modules. View $\mathsf{G}[\mathcal{A}]$, $\vec{\mathsf{G}}[\mathcal{A}]$ and $\Omega[\mathcal{A}]$ as modules over $\Pi[\mathcal{A}]$ via the algebra homomorphisms in (9.21). For these modules, the numbers $\xi_{\mathrm{X}}(\mathsf{h})$ and $\eta_{\mathrm{X}}(\mathsf{h})$ defined in (9.10) can be described using Lemma D.62. For instance, $\xi_{\mathrm{X}}(\mathsf{G})$ is the number of charts whose center is greater than X which is the same as 2 power the number of hyperplanes which contain X, while $\eta_{\mathrm{X}}(\mathsf{G})$ is the number of charts whose center is X. Similarly, $\xi_{\mathrm{X}}(\Omega)$ is the number of cones whose base is greater than X, while $\eta_{\mathrm{X}}(\Omega)$ is the number of cones whose base is X. (Compare with the module Λ in Example 9.5.)

Any module over $\Pi[\mathcal{A}]$ has a primitive part (Section D.9.5). By Lemma D.54,

$$\mathcal{P}(\mathsf{h}) = \mathtt{Q}_{\perp} \cdot \mathsf{h}. \tag{9.26}$$

More generally,

$$\mathcal{P}_{\mathrm{X}}(\mathsf{h}) = \bigoplus_{\mathrm{Y}:\, \mathrm{Y} \leq \mathrm{X}} \mathtt{Q}_{\mathrm{Y}} \cdot \mathsf{h}, \tag{9.27}$$

with $\mathrm{X} = \perp$ recovering the primitive part.

For the modules under consideration, a basis for these spaces can be given using Lemma D.61. For instance, the primitive part of G, denoted $\mathcal{P}(\mathsf{G})$, has a basis of $\mathtt{Q}_g$, as g varies over all connected charts, and more generally, $\mathcal{P}_{\mathrm{X}}(\mathsf{G})$ has a basis of $\mathtt{Q}_g$, as g varies over all charts whose center is smaller than X. Similarly, the primitive part of Ω, denoted $\mathcal{P}(\Omega)$, has a basis of $\mathtt{Q}_{\mathrm{V}}$, as V varies over all cones whose base is the minimum flat, and more generally, $\mathcal{P}_{\mathrm{X}}(\Omega)$ has a basis of $\mathtt{Q}_{\mathrm{V}}$, as V varies over all cones whose base is smaller than X.

9.2.3. Self-duality. Let (ξ_g) and (η_g) be two families of scalars indexed by charts such that

$$\xi_g = \sum_{h:\, h\subseteq g} \eta_h \qquad \text{or equivalently} \qquad \eta_g = \sum_{h:\, h\subseteq g} (-1)^{|g\setminus h|}\, \xi_h. \tag{9.28}$$

As a special case of Theorem D.59, we obtain:

Theorem 9.15. *For any ξ and η as in (9.28), the map $\mathsf{G}[\mathcal{A}] \to \mathsf{G}[\mathcal{A}]^*$ given by*

$$\mathtt{H}_g \mapsto \sum_h \xi_{g\cap h}\, \mathtt{M}_h \qquad \textit{or equivalently} \qquad \mathtt{Q}_g \mapsto \eta_g\, \mathtt{P}_g$$

is a morphism of $\mathsf{G}[\mathcal{A}]$-modules. In particular, if $\eta_g \neq 0$ for all g, then this map is an isomorphism.

In view of the algebra homomorphism from flats to charts, the above map is also a morphism of $\Pi[\mathcal{A}]$-modules.

Theorem 9.15 also holds with charts replaced by dicharts or cones (with appropriate changes in the notation). The analogous result for flats is given in Theorem 9.10. There is a nice choice for η in all four cases which is consistent with the maps in (9.21). We elaborate on this below. For a flat X, cone V, chart g and dichart r, let

$$\eta_{\mathrm{X}} := \mu(\mathrm{X}, \top), \quad \eta_{\mathrm{V}} := \mu(\mathrm{V}, \top), \quad \eta_g := (-1)^{|g|} \quad \text{and} \quad \eta_r := (-1)^{|r|}. \tag{9.29}$$

We point out that the latter two are also Möbius functions of their respective lattices. So we are in the setup of (D.28), with a simple formula for the corresponding ξ. For instance, for the lattice of charts,

$$\xi_g = \begin{cases} 1 & \text{if } g \text{ has no hyperplanes,} \\ 0 & \text{otherwise.} \end{cases} \tag{9.30}$$

Lemma 9.16. *Let $\mathcal{A}$ be an arrangement. Then for any flat X,*

$$\mu(\mathrm{X}, \top) = \sum_{g:\, \rho(g)=\mathrm{X}} (-1)^{|g|}, \qquad \mu(\mathrm{X}, \top) = \sum_{\mathrm{V}:\, \mathrm{b}(\mathrm{V})=\mathrm{X}} \mu(\mathrm{V}, \top),$$

for any chart g,

$$(-1)^{|g|} = \sum_{r:\, \rho'(r)=g} (-1)^{|r|},$$

and for any cone V,

$$\mu(\mathrm{V}, \top) = \sum_{r:\, \vec{\rho}(r)=\mathrm{V}} (-1)^{|r|}.$$

PROOF. All identities are special cases of (D.40). (One has to check that the hypothesis of Lemma D.63 are met.) Also, the first identity is the same as formula (1.51). □

It follows from (9.29) that

$$\sum_{g:\, \rho(g)=\mathrm{X}} \eta_g = \eta_{\mathrm{X}}, \qquad \sum_{\mathrm{V}:\, \mathrm{b}(\mathrm{V})=\mathrm{X}} \eta_{\mathrm{V}} = \eta_{\mathrm{X}}, \qquad \sum_{r:\, \rho'(r)=g} \eta_r = \eta_g, \qquad \sum_{r:\, \vec{\rho}(r)=\mathrm{V}} \eta_r = \eta_{\mathrm{V}}.$$

A comparison with (9.22), (9.23), (9.24) and (9.25) yields a commutative diagram

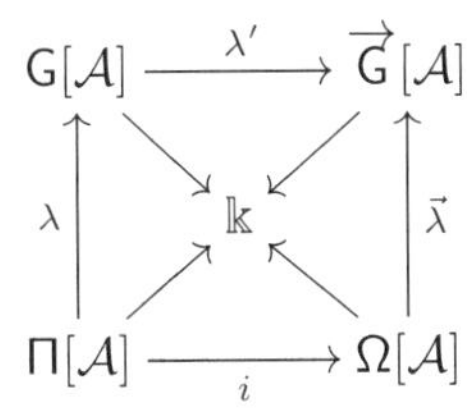

where the four maps to $\Bbbk$ all send $\mathtt{Q}_x$ to η_x. This can also be seen as a special case of Lemma D.64.

9.3. Tits algebra

The Tits algebra is the linearization of the Tits monoid. It is an example of an elementary algebra. Its split-semisimple quotient is precisely the Birkhoff algebra (and the kernel of the support map is its radical). We also introduce the projective Tits algebra.

Background information on elementary algebras is given in Section D.8.

9.3.1. Tits algebra. Recall the set of faces $\Sigma[\mathcal{A}]$. It carries the structure of a monoid under the product (1.5). This is the Tits monoid. Let $\Sigma[\mathcal{A}]$ denote its linearization over a field $\Bbbk$, with canonical basis $\mathtt{H}$. It is an algebra:

$$\mathtt{H}_F \cdot \mathtt{H}_G := \mathtt{H}_{FG}.$$

We call this the *Tits algebra.*

The linearization of the support map (1.2) yields

$$\text{(9.31)} \qquad \mathrm{s} : \Sigma[\mathcal{A}] \twoheadrightarrow \Pi[\mathcal{A}], \qquad \mathtt{H}_F \mapsto \mathtt{H}_{\mathrm{s}(F)}.$$

We continue to call this the support map. It is a surjective morphism of algebras, that is,

$$\text{(9.32)} \qquad \mathrm{s}(z \cdot w) = \mathrm{s}(z) \cdot \mathrm{s}(w).$$

This is a formal consequence of (1.11).

9.3.2. Radical of the Tits algebra. Let N denote the kernel of the support map (9.31). We set out to prove that

$$\mathsf{N} = \mathrm{rad}(\Sigma[\mathcal{A}]),$$

the radical of the Tits algebra.

Since the support map is an algebra homomorphism, N is an ideal of the Tits algebra. Let $z = \sum_F x^F \mathtt{H}_F$ be any element of $\Sigma[\mathcal{A}]$. Then observe that

$$\text{(9.33)} \qquad z \in \mathsf{N} \iff \sum_{F:\, \mathrm{s}(F)=\mathrm{X}} x^F = 0 \text{ for all flats X}.$$

In particular, for this to occur, $x^O = 0$. Note that for any faces F and G with the same support, $\mathtt{H}_F - \mathtt{H}_G$ belongs to N, and elements of this form linearly span N.

Exercise 9.17. Check that for any face F, the element $\mathtt{H}_F - \mathtt{H}_{\overline{F}}$ is nilpotent of order 2.

An element of the Tits algebra is homogeneous if it is a linear combination of faces with the same support. For any such element x, let us denote this common support by $\mathrm{s}(x)$. By convention, $\mathrm{s}(0) = \top$, the maximum flat. Note that an arbitrary element of the Tits algebra can be written as a linear combination of homogeneous elements. Further, the product of homogeneous elements is again homogeneous. Also, $\mathrm{s}(x), \mathrm{s}(y) \leq \mathrm{s}(x \cdot y)$ for any homogeneous elements x and y.

Lemma 9.18. *If $x \in \mathsf{N}$ is homogeneous and F is a face such that $\mathrm{s}(x) \leq \mathrm{s}(F)$, then $\mathtt{H}_F \cdot x = 0$. More generally, if x and y are homogeneous, $x \in \mathsf{N}$ and $\mathrm{s}(x) \leq \mathrm{s}(y)$, then $y \cdot x = 0$.*

PROOF. The second statement follows from the first. To prove the first: Write $x = \sum_{K:\,\mathrm{s}(K)=\mathrm{X}} a^K \mathtt{H}_K$, where $\mathrm{X} = \mathrm{s}(x)$. By hypothesis, $FK = F$ for all K of support X. Thus,

$$\mathtt{H}_F \cdot x = \sum_{K:\,\mathrm{s}(K)=\mathrm{X}} a^K \mathtt{H}_F \cdot \mathtt{H}_K = \Big(\sum_{K:\,\mathrm{s}(K)=\mathrm{X}} a^K \Big) \mathtt{H}_F = 0,$$

by (9.33). □

Lemma 9.19. *For any nonnegative integer k, the ideal N^k only contains elements which are linear combinations of faces of rank at least k.*

PROOF. Consider $x_1 \cdot x_2 \cdot \ldots \cdot x_k \in \mathsf{N}^k$, where each x_i is a homogeneous element of N. Then

$$\bot \leq \mathrm{s}(x_1) \leq \mathrm{s}(x_1 \cdot x_2) \leq \cdots \leq \mathrm{s}(x_1 \cdot \ldots \cdot x_k).$$

If equality holds in any place, say $\mathrm{s}(x_1 \cdot \ldots \cdot x_{i-1}) = \mathrm{s}(x_1 \cdot \ldots \cdot x_i)$, then $\mathrm{s}(x_i) \leq \mathrm{s}(x_1 \cdot \ldots \cdot x_{i-1})$, and hence $x_1 \cdot \ldots \cdot x_i = 0$ by Lemma 9.18. Thus we may assume

$$\bot < \mathrm{s}(x_1) < \mathrm{s}(x_1 \cdot x_2) < \cdots < \mathrm{s}(x_1 \cdot \ldots \cdot x_k)$$

from which we deduce that $x_1 \cdot x_2 \cdot \ldots \cdot x_k$ is a linear combination of faces of rank at least k. □

Lemma 9.20. *For $0 \leq k \leq \mathrm{rk}(\mathcal{A})$, given a face K of rank k, there exists an element of N^k which is a linear combination of faces of the same support as K in which $\mathtt{H}_K$ has a nonzero coefficient.*

PROOF. Pick any chain of faces $O \lessdot F_1 \lessdot F_2 \lessdot \cdots \lessdot F_k = K$. For $1 \leq i \leq k$, let G_i be the face opposite to F_i in the star of F_{i-1} (with the convention $F_0 = O$). Equivalently, $G_i = F_{i-1}\overline{F_i}$. Then the element

$$(\mathtt{H}_{F_1} - \mathtt{H}_{G_1}) \cdot \ldots \cdot (\mathtt{H}_{F_k} - \mathtt{H}_{G_k})$$

is a sum of 2^k distinct faces all of the same support, with coefficients ± 1, with $\mathtt{H}_{F_k}$ appearing with coefficient $+1$.

Alternatively: If the arrangement is simplicial and the field characteristic is 0, then one can also argue as follows. Let z be the element of N obtained by adding all vertices of K and subtracting all vertices of $\overline{K}$. Then the coefficient of $\mathtt{H}_K$ in z^k is $k!$: The only way $\mathtt{H}_K$ can arise is by multiplying the vertices of K in different orders. In particular, $z^k \neq 0$ and z^k is a linear combination of faces of the same support as K. □

As a consequence of Lemmas 9.19 and 9.20:

Proposition 9.21. *The ideal* N *is nilpotent. Its nilpotency index is* $\operatorname{rk}(\mathcal{A})+1$, *that is,* $\operatorname{rk}(\mathcal{A})+1$ *is the smallest index* k *such that* $\mathsf{N}^k = 0$.

Proposition 9.22. *The Tits algebra is elementary. Its split-semisimple quotient is the Birkhoff algebra, with the support map as the quotient map. In particular, the radical of the Tits algebra is the kernel of the support map:* $\operatorname{rad}(\Sigma[\mathcal{A}]) = \mathsf{N}$ *and it consists precisely of the nilpotent elements of the Tits algebra.*

PROOF. Apply Proposition D.22 to the nilpotent ideal N, and use Theorem 9.2. All claims follow. □

In conjunction with Lemma 1.10, we deduce:

Corollary 9.23. *The quotient of the Tits algebra by its radical is commutative, and equals its abelianization.*

This is indeed special since, in general, there are commutative algebras with nontrivial radical, and noncommutative algebras whose radical is zero.

Corollary 9.24. *The dimension of the radical of the Tits algebra is equal to the number of faces minus the number of flats.*

9.3.3. Rank one. Let $\mathcal{A}$ be the arrangement of rank one with chambers C and $\overline{C}$. The Tits algebra of $\mathcal{A}$ is isomorphic to the algebra of upper triangular 2 by 2 matrices (assuming $\Bbbk$ does not have characteristic 2). An explicit isomorphism is

$$\mathtt{H}_O \mapsto \begin{pmatrix} 1 & 0 \\ 0 & 1 \end{pmatrix} \qquad \mathtt{H}_C \mapsto \begin{pmatrix} 0 & 1 \\ 0 & 1 \end{pmatrix} \qquad \mathtt{H}_{\overline{C}} \mapsto \begin{pmatrix} 0 & -1 \\ 0 & 1 \end{pmatrix}.$$

In this case, the radical of the Tits algebra is one-dimensional and spanned by $\mathtt{H}_C - \mathtt{H}_{\overline{C}}$. This can be seen directly from Exercise 9.17. The radical identifies with the one-dimensional space of strictly upper triangular 2 by 2 matrices.

Exercise 9.25. By Example D.36, the algebra of upper triangular matrices is elementary. Show that the algebra of upper triangular n by n matrices for $n \geq 3$ cannot be isomorphic to the Tits algebra of any arrangement. (Compare the respective dimensions of the split-semisimple quotients and the nilpotency indices of the radicals.)

Exercise 9.26. Check by direct calculation that the idempotents in the Tits algebra of the rank-one arrangement are given by

$$0, \quad \mathtt{H}_O, \quad \beta\,\mathtt{H}_C + (1-\beta)\,\mathtt{H}_{\overline{C}}, \quad \mathtt{H}_O + \beta\,\mathtt{H}_C + (-1-\beta)\,\mathtt{H}_{\overline{C}},$$

with β an arbitrary scalar.

9.3.4. Projective Tits algebra. Consider the linearization of the opposition map. It sends z to $\overline{z}$, where

$$z = \sum_F x^F \mathtt{H}_F \quad \text{and} \quad \overline{z} = \sum_F x^F \mathtt{H}_{\overline{F}}. \tag{9.34}$$

It is an algebra isomorphism by (1.7), that is,

$$\overline{z \cdot w} = \overline{z} \cdot \overline{w}. \tag{9.35}$$

An element z of the Tits algebra is called projective if $z = \overline{z}$, that is, the coefficients of $\mathtt{H}_F$ and $\mathtt{H}_{\overline{F}}$ in z are equal for all F. Observe that the set of all projective elements is a subalgebra of the Tits algebra. We call it the *projective Tits algebra*. It has a basis indexed by projective faces, namely,

$$\mathtt{H}_{\{O,O\}} := \mathtt{H}_O \quad \text{and} \quad \mathtt{H}_{\{F,\overline{F}\}} := \mathtt{H}_F + \mathtt{H}_{\overline{F}} \text{ for } F \neq O. \tag{9.36}$$

This is the H-basis of the projective Tits algebra. Observe that

$$\mathtt{H}_{\{F,\overline{F}\}} \cdot \mathtt{H}_{\{G,\overline{G}\}} = \mathtt{H}_{\{FG,\overline{FG}\}} + \mathtt{H}_{\{F\overline{G},\overline{F}G\}}.$$

Proposition 9.27. *The projective Tits algebra is elementary. Its radical consists of the projective elements in the radical of the Tits algebra (or equivalently, projective elements in the kernel of the support map). Further, if the field characteristic is not* 2*, then the split-semisimple quotient is the Birkhoff algebra.*

PROOF. This can be deduced from Proposition 9.22 using either Lemma D.45 or Lemma D.46, the latter for the group of two elements consisting of the identity and the opposition map. □

Exercise 9.28. Assume that the field characteristic is not 2. Let r denote the rank of the arrangement. Show that the nilpotency index of the radical of the projective Tits algebra is $r/2+1$ if r is even, and $(r+1)/2$ if r is odd. (Formulate analogues of (9.33) and Lemmas 9.18, 9.19 and 9.20.)

9.4. Left module of chambers

The space of chambers is the linearization of the set of chambers. It is a two-sided ideal of the Tits algebra. It has a trivial centralizer. This implies that the Tits algebra has a trivial center. We focus particularly on the left ideal structure of chambers. We refer to it as the left module of chambers. It is faithful. Thus, there is an injective algebra homomorphism from the Tits algebra to the endomorphism algebra of chambers. It factors through the space of top-nested faces.

9.4.1. Left module of chambers. Let $\mathsf{\Gamma}[\mathcal{A}]$ denote the linearization of the set of chambers $\Gamma[\mathcal{A}]$ over a field $\Bbbk$, with canonical basis H. It is a two-sided ideal of the Tits algebra. In particular, it is a left module over the Tits algebra:

$$\mathtt{H}_F \cdot \mathtt{H}_C := \mathtt{H}_{FC}.$$

We call this the *left module of chambers*.

Lemma 9.29. *The left module of chambers is faithful. That is, for any* $z \in \Sigma[\mathcal{A}]$, $z \cdot \mathtt{H}_C = 0$ *for all* C *implies that* $z = 0$.

PROOF. Let $z = \sum_H x^H \mathtt{H}_H$. Then, for any chambers C and D,

$$\sum_{H:\, HC=D} x^H = 0.$$

(The lhs is the coefficient of $\mathtt{H}_D$ in $z \cdot \mathtt{H}_C$.) We deduce from Proposition 7.14 that for any $G \leq D$,

$$\sum_{H:\, G \leq H \leq D} x^H = 0.$$

This implies that $x^H = 0$ for all H, that is, $z = 0$. □

9.4.2. Center of the Tits algebra. We show that the Tits algebra has a trivial center. In fact, the following stronger result holds.

Lemma 9.30. *The centralizer of the two-sided ideal of chambers is trivial. That is, for any $z \in \Sigma[\mathcal{A}]$, $\mathtt{H}_C \cdot z = z \cdot \mathtt{H}_C$ for all C implies that z is a scalar multiple of $\mathtt{H}_O$.*

PROOF. The proof is similar to that of Lemma 9.29. Let $z = \sum_H x^H \mathtt{H}_H$. The condition $\mathtt{H}_C \cdot z = z \cdot \mathtt{H}_C$ implies that for any chambers $C \neq D$,

$$\sum_{H:\, HC=D} x^H = 0.$$

We deduce from Proposition 7.14 that for any $O < G \leq D$,

$$\sum_{H:\, G \leq H \leq D} x^H = 0.$$

This implies that $x^H = 0$ for all $H \neq O$. So z is a scalar multiple of $\mathtt{H}_O$. □

Corollary 9.31. *The Tits algebra has a trivial center. That is, the center consists of scalar multiples of the unit element $\mathtt{H}_O$.*

9.4.3. Radical. Recall from Section D.5.4 that the radical of a module is the intersection of its maximal submodules. Let us compute the radical of $\Gamma[\mathcal{A}]$.

Proposition 9.32. *The radical of $\Gamma[\mathcal{A}]$ is the unique maximal submodule of $\Gamma[\mathcal{A}]$. Its dimension is one less than the number of chambers. Explicitly,*

$$\sum_C x^C \mathtt{H}_C \in \operatorname{rad}(\Gamma[\mathcal{A}]) \iff \sum_C x^C = 0. \tag{9.37}$$

In other words, the radical is linearly spanned by elements of the form $\mathtt{H}_C - \mathtt{H}_D$.

PROOF. By (D.12), the radical equals $\operatorname{rad}(\Sigma[\mathcal{A}]) \cdot \Gamma[\mathcal{A}]$. Using (9.33), we deduce (9.37). The claim about dimension then follows. The uniqueness claim follows next by the definition of the radical. □

As a consequence of (D.13):

Corollary 9.33. *The module $\Gamma[\mathcal{A}]$ is semisimple iff $\mathcal{A}$ has rank zero.*

Exercise 9.34. Suppose $\mathcal{A}$ has rank at least one. Check directly using (9.37) that every one-dimensional submodule of $\Gamma[\mathcal{A}]$ lies inside $\operatorname{rad}(\Gamma[\mathcal{A}])$. In particular, $\operatorname{rad}(\Gamma[\mathcal{A}])$ does not have any complementary submodule in $\Gamma[\mathcal{A}]$.

Exercise 9.35. Check that $\operatorname{rad}(\Gamma[\mathcal{A}]) = \Gamma[\mathcal{A}] \cap \operatorname{rad}(\Sigma[\mathcal{A}])$.

Exercise 9.36. Use the fact that $\mathcal{A}$ is gallery connected to deduce that $\operatorname{rad}(\Gamma[\mathcal{A}])$ is linearly spanned by elements of the form $\mathtt{H}_C - \mathtt{H}_D$, where C and D are adjacent chambers.

9.4.4. Endomorphism algebra of chambers. Recall from Section D.1.1 that a left module M over an algebra A gives rise to an algebra homomorphism from A to the endomorphism algebra of M. In particular, the left module of chambers $\Gamma[\mathcal{A}]$ yields an algebra homomorphism

$$\Sigma[\mathcal{A}] \to \operatorname{End}_{\Bbbk}(\Gamma[\mathcal{A}]), \tag{9.38}$$

the latter being the algebra of endomorphisms of $\Gamma[\mathcal{A}]$. Since the left module of chambers is faithful (Lemma 9.29), the map (9.38) is injective. By comparing dimensions, we deduce that

$$f(\mathcal{A}) \leq c(\mathcal{A})^2,$$

where $f(\mathcal{A})$ is the number of faces and $c(\mathcal{A})$ is the number of chambers of $\mathcal{A}$.

9.4.5. Rank one. Let $\mathcal{A}$ be the arrangement of rank one with chambers C and $\overline{C}$. Identifying the endomorphism algebra of chambers with 2 by 2 matrices, the map (9.38) is given by

$$\alpha\,\mathtt{H}_O + \beta\,\mathtt{H}_C + \gamma\,\mathtt{H}_{\overline{C}} \mapsto \begin{pmatrix} \alpha+\beta & \beta \\ \gamma & \alpha+\gamma \end{pmatrix}. \tag{9.39}$$

Observe directly that this map is injective. Its image consists of those matrices whose column sums are equal. The image of the radical of the Tits algebra consists of matrices whose columns are identical with column sum 0. (That is, $\alpha = \beta + \gamma = 0$.)

Consider the matrix in (9.39). It has eigenvalues α and $\alpha + \beta + \gamma$ with eigenvectors $\mathtt{H}_C - \mathtt{H}_{\overline{C}}$ and $\beta\,\mathtt{H}_C + \gamma\,\mathtt{H}_{\overline{C}}$, respectively. If $\beta + \gamma = 0$, then α is a repeated eigenvalue. One can check that the matrix is diagonalizable iff either $\beta = \gamma = 0$ or $\beta + \gamma \neq 0$.

We also deduce from these calculations that the module $\Gamma[\mathcal{A}]$ has a unique one-dimensional submodule, namely, the subspace spanned by $\mathtt{H}_C - \mathtt{H}_{\overline{C}}$. In fact, this is the radical of $\Gamma[\mathcal{A}]$ by (9.37). The uniqueness of this submodule shows that $\Gamma[\mathcal{A}]$ does *not* decompose as a direct sum of simple modules. These observations are consistent with Proposition 9.32 and Corollary 9.33. Following (D.1), taking the trace of the matrix in (9.39), we see that the character of $\Gamma[\mathcal{A}]$ is the linear functional

$$\chi_{\Gamma[\mathcal{A}]} : \Sigma[\mathcal{A}] \to \Bbbk, \qquad \alpha\,\mathtt{H}_O + \beta\,\mathtt{H}_C + \gamma\,\mathtt{H}_{\overline{C}} \mapsto 2\alpha + \beta + \gamma. \tag{9.40}$$

9.4.6. Another viewpoint on the endomorphism algebra. Let us go back to the general case. Consider the canonical identification

$$\mathrm{End}_{\Bbbk}(\Gamma[\mathcal{A}]) \cong \Gamma[\mathcal{A}]^* \otimes \Gamma[\mathcal{A}]$$

by viewing a basis element

$$\mathtt{M}_D \otimes \mathtt{H}_C \in \Gamma[\mathcal{A}]^* \otimes \Gamma[\mathcal{A}]$$

as the endomorphism

$$\mathtt{H}_E \mapsto \begin{cases} \mathtt{H}_C & \text{if } D = E, \\ 0 & \text{otherwise.} \end{cases}$$

Here $\mathtt{M}$ denotes the basis of $\Gamma[\mathcal{A}]^*$ dual to $\mathtt{H}$. This induces a product on the algebra $\Gamma[\mathcal{A}]^* \otimes \Gamma[\mathcal{A}]$ which is given by

$$(\mathtt{M}_{D_2} \otimes \mathtt{H}_{C_2})\,(\mathtt{M}_{D_1} \otimes \mathtt{H}_{C_1}) = \begin{cases} \mathtt{M}_{D_1} \otimes \mathtt{H}_{C_2} & \text{if } D_2 = C_1, \\ 0 & \text{otherwise,} \end{cases} \tag{9.41}$$

and the injective algebra homomorphism (9.38) takes the form

$$\Sigma[\mathcal{A}] \to \Gamma[\mathcal{A}]^* \otimes \Gamma[\mathcal{A}], \qquad \mathtt{H}_F \mapsto \sum_{(D,C):\, FD=C} \mathtt{M}_D \otimes \mathtt{H}_C. \tag{9.42}$$

The sum is over both C and D.

9.4.7. Space of top-nested faces. Let $\widehat{\mathsf{Q}}[\mathcal{A}]$ denote the linearization of the set of top-nested faces. We consider two bases on it, namely, $\mathtt{H}$ and $\mathtt{K}$, related by

$$\mathtt{H}_{H,D} = \sum_{G:\, G \leq H} \mathtt{K}_{G,D} \quad \text{or equivalently} \quad \mathtt{K}_{H,D} = \sum_{G:\, G \leq H} (-1)^{\mathrm{rk}(H/G)} \mathtt{H}_{G,D}. \tag{9.43}$$

For this equivalence, we used (1.43).

The map (9.42) factors as

$$\Sigma[\mathcal{A}] \hookrightarrow \widehat{\mathsf{Q}}[\mathcal{A}] \hookrightarrow \Gamma[\mathcal{A}]^* \otimes \Gamma[\mathcal{A}] \tag{9.44}$$

with

$$\mathtt{H}_H \mapsto \sum_{D:\, D \geq H} \mathtt{H}_{H,D} \quad \text{and} \quad \mathtt{H}_{H,D} \mapsto \sum_{C:\, HC = D} \mathtt{M}_C \otimes \mathtt{H}_D.$$

For a simplicial arrangement, using (7.1), the second map can be rewritten as

$$\mathtt{K}_{H,D} \mapsto \sum_{C:\, \mathrm{Des}(C,D) = H} \mathtt{M}_C \otimes \mathtt{H}_D. \tag{9.45}$$

In terms of the endomorphism algebra, the second map can be expressed as

$$\mathtt{H}_{H,D} \cdot \mathtt{H}_C = \begin{cases} \mathtt{H}_D & \text{if } HC = D, \\ 0 & \text{otherwise,} \end{cases}$$

or, in the simplicial case, as

$$\mathtt{K}_{H,D} \cdot \mathtt{H}_C = \begin{cases} \mathtt{H}_D & \text{if } \mathrm{Des}(C,D) = H, \\ 0 & \text{otherwise.} \end{cases} \tag{9.46}$$

Exercise 9.37. Verify that the map $\widehat{\mathsf{Q}}[\mathcal{A}] \to \Gamma[\mathcal{A}]^* \otimes \Gamma[\mathcal{A}]$ is injective as implicitly claimed in (9.44). (Follow the proof of Lemma 9.29.)

9.4.8. Left module of projective chambers. Let us use the term chamber element to mean a linear combination of chambers. A chamber element z is *projective* if $z = \overline{z}$, that is, the coefficients of $\mathtt{H}_D$ and $\mathtt{H}_{\overline{D}}$ in z are equal for all D. The notation $\overline{z}$ is as in (9.34). The space of all projective chamber elements has a basis indexed by projective chambers. It is a left module over the projective Tits algebra. We call it the *left module of projective chambers*.

9.5. Modules over the Tits algebra

Recall that the Tits algebra is elementary and its split-semisimple quotient is the Birkhoff algebra. This allows us to classify its simple modules by multiplicative characters indexed by flats with simple explicit formulas. An arbitrary module does not break as a direct sum of simple modules. However, by employing the technique of composition series, one can write down explicit formulas for the eigenvalues and multiplicities of the action of any element of the Tits algebra on it. For the module of chambers, this yields the Bidigare-Hanlon-Rockmore (BHR) theorem. Invertible elements in the Tits algebra are precisely those whose eigenvalues are all nonzero.

9.5.1. Multiplicative characters.

Theorem 9.38. *The simple modules over $\Sigma[\mathcal{A}]$ are one-dimensional and indexed by flats. Let χ_{X} denote the multiplicative character corresponding to the flat* X. *It is specified by*

$$\mathrm{s}(z) = \sum_{\mathrm{X}} \chi_{\mathrm{X}}(z)\, \mathtt{Q}_{\mathrm{X}}. \tag{9.47}$$

On a H*-basis element, it is given by*

$$\chi_{\mathrm{X}}(\mathtt{H}_F) = \begin{cases} 1 & \text{if } \mathrm{s}(F) \leq \mathrm{X}, \\ 0 & \text{otherwise.} \end{cases} \tag{9.48}$$

Equivalently, for $w = \sum_F w^F \mathtt{H}_F$,

$$\chi_{\mathrm{X}}(w) = \sum_{F:\, \mathrm{s}(F) \leq \mathrm{X}} w^F. \tag{9.49}$$

In particular, the multiplicative characters corresponding to the minimum and maximum flats are given by

$$\chi_{\perp}(\mathtt{H}_F) = \begin{cases} 1 & \text{if } F = O, \\ 0 & \text{otherwise,} \end{cases} \qquad \text{and} \qquad \chi_{\top}(\mathtt{H}_F) = 1 \text{ for all } F. \tag{9.50}$$

PROOF. Apply Theorem D.35. This yields the first two statements. In particular, $\chi_{\mathrm{X}}(\mathtt{H}_F)$ is the coefficient of $\mathtt{Q}_{\mathrm{X}}$ in $\mathtt{H}_{\mathrm{s}(F)}$. Now use (9.1) to first get (9.48) and then (9.49). □

The multiplicative characters of the Tits algebra obtained in Theorem 9.38 can also be directly computed as follows. Let $\chi : \Sigma \to \Bbbk$ be multiplicative. Since $\mathtt{H}_F$ is an idempotent, $\chi(\mathtt{H}_F)$ is either 0 or 1. Let X be the join of the supports of all faces F for which $\chi(\mathtt{H}_F) = 1$. We claim that $\chi = \chi_{\mathrm{X}}$, with the latter as in (9.48). For this, let G be the product of all faces F with $\chi(\mathtt{H}_F) = 1$ taken in some order. Then G has support X and $\chi(\mathtt{H}_G) = 1$. Further, for any face F with $\mathrm{s}(F) \leq \mathrm{X}$, we have $GF = G$ and hence $\chi(\mathtt{H}_{GF}) = \chi(\mathtt{H}_G)$ which implies $\chi(\mathtt{H}_F) = 1$. This proves the claim.

Note very carefully that this calculation does not prove Theorem 9.38 since it does not show that all simple modules arise from multiplicative characters.

9.5.2. Characters. Let h be a left module over the Tits algebra, and Ψ_{h} the associated representation. For any element w of the Tits algebra, $\Psi_{\mathsf{h}}(w)$ denotes the linear operator on h given by multiplication by w, and $w \cdot \mathsf{h}$ denotes its image. Thus,

$$\Psi_{\mathsf{h}}(w) : \mathsf{h} \to \mathsf{h}, \qquad \Psi_{\mathsf{h}}(w)(h) := w \cdot h.$$

Following (D.1), the character of h is the linear functional

$$\chi_{\mathsf{h}} : \Sigma \to \Bbbk, \qquad \chi_{\mathsf{h}}(w) = \mathrm{tr}(\Psi_{\mathsf{h}}(w)),$$

where tr denotes trace.

Recall from Proposition 9.22 that the Tits algebra is elementary. We now apply the general discussion in Section D.8.4 to the module h. The character χ_{h} factors through the support map yielding the commutative diagram

$$\begin{array}{ccc} \Sigma & & \\ {\scriptstyle s}\downarrow & \searrow^{\chi_{\mathsf{h}}} & \\ \mathsf{\Pi} & \xrightarrow[\chi_{\mathsf{h}}]{\quad} & \Bbbk. \end{array}$$

The induced linear functional on $\mathsf{\Pi}$ is also denoted χ_{h}. Following (9.6), for each flat X, put

$$\xi_{\mathrm{X}}(\mathsf{h}) = \chi_{\mathsf{h}}(\mathtt{H}_{\mathrm{X}}) \qquad \text{and} \qquad \eta_{\mathrm{X}}(\mathsf{h}) = \chi_{\mathsf{h}}(\mathtt{Q}_{\mathrm{X}}). \tag{9.51}$$

Thus,

$$\xi_{\mathrm{X}}(\mathsf{h}) = \sum_{\mathrm{Y}:\,\mathrm{Y}\geq\mathrm{X}} \eta_{\mathrm{Y}}(\mathsf{h}) \quad \text{or equivalently} \quad \eta_{\mathrm{X}}(\mathsf{h}) = \sum_{\mathrm{Y}:\,\mathrm{Y}\geq\mathrm{X}} \mu(\mathrm{X},\mathrm{Y})\,\xi_{\mathrm{Y}}(\mathsf{h}). \tag{9.52}$$

The integer $\eta_{\mathrm{X}}(\mathsf{h})$ agrees with (D.19). It is the number of times the simple module with multiplicative character χ_{X} appears as a composition factor in a composition series of h. By (D.20), for $w \in \Sigma$,

$$\chi_{\mathsf{h}}(w) = \sum_{\mathrm{X}} \chi_{\mathrm{X}}(w)\,\eta_{\mathrm{X}}(\mathsf{h}). \tag{9.53}$$

Recall from (D.4) that the trace of an idempotent operator is the dimension of its image. Applying this to the idempotent $\mathtt{H}_F$, we get

$$\xi_{\mathrm{X}}(\mathsf{h}) = \dim(\mathtt{H}_F \cdot \mathsf{h}), \tag{9.54}$$

where F is any face with support X. The fact that this number does not depend on the particular choice of F can also be seen directly:

Lemma 9.39. *Let F and G be faces of the same support. For a left Σ-module h, there is an isomorphism*

$$\mathtt{H}_F \cdot \mathsf{h} \xrightarrow{\cong} \mathtt{H}_G \cdot \mathsf{h}$$

given by multiplication by $\mathtt{H}_G$, with inverse given by multiplication by $\mathtt{H}_F$.

PROOF. This follows from the second equality in (1.6), and (1.13). □

Similarly, we have

$$\eta_{\mathrm{X}}(\mathsf{h}) = \dim(\mathtt{Q}_F \cdot \mathsf{h}) \tag{9.55}$$

for any idempotent $\mathtt{Q}_F$ which lifts $\mathtt{Q}_{\mathrm{X}}$. Such idempotents will be constructed in Chapter 11.

If the action of Σ on h factors through the support map, then h becomes a module over $\mathsf{\Pi}$, and $\xi_{\mathrm{X}}(\mathsf{h})$ and $\eta_{\mathrm{X}}(\mathsf{h})$ coincide with (9.10).

Example 9.40. For the left module of chambers $\mathsf{\Gamma}$,

$$\xi_{\mathrm{X}}(\mathsf{\Gamma}) = c(\mathcal{A}_{\mathrm{X}}) \quad \text{and} \quad \eta_{\mathrm{X}}(\mathsf{\Gamma}) = |\mu(\mathcal{A}_{\mathrm{X}})|, \tag{9.56}$$

where $c(\mathcal{A}_{\mathrm{X}})$ is the number of chambers in $\mathcal{A}_{\mathrm{X}}$. This can be understood as follows. The space $\mathtt{H}_F \cdot \mathsf{\Gamma}$ has a basis consisting of all chambers greater than F. This yields the formula for $\xi_{\mathrm{X}}(\mathsf{\Gamma})$. The formula for $\eta_{\mathrm{X}}(\mathsf{\Gamma})$ then follows from (9.8) (in view of (9.5) and (9.52)).

Similarly, for Σ as a left module over itself,

$$\xi_{\mathrm{X}}(\Sigma) = f(\mathcal{A}_{\mathrm{X}}) \quad \text{and} \quad \eta_{\mathrm{X}}(\Sigma) = \sum_{\mathrm{Y}:\,\mathrm{Y}\geq\mathrm{X}} |\mu(\mathcal{A}_{\mathrm{X}}^{\mathrm{Y}})|, \tag{9.57}$$

where $f(\mathcal{A}_X)$ is the number of faces in $\mathcal{A}_X$. The space $\mathtt{H}_F \cdot \Sigma$ has a basis consisting of all faces greater than F. This yields the formula for $\xi_X(\Sigma)$. The formula for $\eta_X(\Sigma)$ then follows from (9.9).

For $\mathsf{h} = \Gamma$ and $\mathsf{h} = \Sigma$, the isomorphism in Lemma 9.39 is the linearization of the bijections in Lemma 1.37.

The character of the left module of chambers Γ is given by

$$\chi_\Gamma(w) = \sum_X \chi_X(w)\, |\mu(\mathcal{A}_X)|. \tag{9.58}$$

This follows from (9.53) and (9.56). Similarly, by enploying (9.57), we obtain a character formula for the left module of faces Σ.

Exercise 9.41. Give an example of a left module h over the Tits algebra such that $\xi_X(\mathsf{h})$ and $\eta_X(\mathsf{h})$ match those given in Exercise 9.6.

Exercise 9.42. Let h be a left module over Σ obtained by linearizing a left Σ-set h. Check that $\xi_X(\mathsf{h})$ is the cardinality of the set h_F defined in (7.28), where F is any face with support X.

We mention that the above discussion can also be carried out for right modules. If h is a left Σ-module, then h^* is a right Σ-module, and vice versa. We note that $w \cdot \mathsf{h}$ and $\mathsf{h}^* \cdot w$ have the same dimension (Lemma D.3). Applying this to $w = \mathtt{H}_F$, we conclude:

$$\xi_X(\mathsf{h}) = \xi_X(\mathsf{h}^*) \quad \text{and} \quad \eta_X(\mathsf{h}) = \eta_X(\mathsf{h}^*). \tag{9.59}$$

Recall that a linear functional on Σ is called a character of Σ if it is the character of some Σ-module h. Multiplicative characters are those which arise from one-dimensional modules.

Proposition 9.43. *The characters of the Tits algebra correspond to families* (η_X) *of nonnegative integers indexed by flats, with the multiplicative ones corresponding to those families in which exactly one* η_X *is* 1 *and the rest are* 0.

PROOF. This follows from Proposition D.39. □

The characters also correspond to certain families (ξ_X) of nonnegative integers. The point is that the ξ_X cannot be arbitrary. They are such that the η_X defined from them via (9.52) are nonnegative.

9.5.3. Eigenvalue-multiplicity theorem. Theorem D.38 gives the eigenvalues and multiplicities of the action of any element of an elementary algebra on a module. Applying it to the Tits algebra and using (9.49), we obtain:

Theorem 9.44. *Let* h *be a (left or right) module over the Tits algebra* Σ. *Then all elements of* Σ *are simultaneously triangularizable on* h. *For* $w = \sum_F w^F \mathtt{H}_F$, *the linear operator* $\Psi_{\mathsf{h}}(w)$ *has an eigenvalue*

$$\lambda_X(w) := \chi_X(w) = \sum_{F:\, \mathrm{s}(F) \leq X} w^F \tag{9.60}$$

for each $X \in \Pi$, *with multiplicity* $\eta_X(\mathsf{h})$ *given by* (9.51).

This is the *eigenvalue-multiplicity theorem.*

Exercise 9.45. Deduce Theorem 9.9 from Theorem 9.44 using the fact that any module over the Birkhoff algebra is a module over the Tits algebra via the support map.

9.5.4. Bidigare-Hanlon-Rockmore. By specializing Theorem 9.44 to the left module of chambers $\mathsf{h} = \Gamma$ and using formula (9.56), we obtain:

Theorem 9.46. *For* $w = \sum_F w^F \mathtt{H}_F$, *the linear operator* $\Psi_\Gamma(w)$ *has an eigenvalue* $\lambda_{\mathrm{X}}(w)$ *defined by* (9.60) *for each* $\mathrm{X} \in \Pi$, *with multiplicity* $|\mu(\mathcal{A}_{\mathrm{X}})|$.

This is the *Bidigare-Hanlon-Rockmore theorem*, or BHR for short. Note very carefully that this result makes no claim about the diagonalizability of $\Psi_\Gamma(w)$.

Example 9.47. Let $\mathcal{A}$ be the rank-one arrangement with chambers C and $\overline{C}$. It has two flats, namely, $\perp$ and $\top$. Let $w = \alpha\,\mathtt{H}_O + \beta\,\mathtt{H}_C + \gamma\,\mathtt{H}_{\overline{C}}$. By BHR, the eigenvalues of $\Psi_\Gamma(w)$ are

$$\lambda_\perp(w) = \alpha \quad \text{and} \quad \lambda_\top(w) = \alpha + \beta + \gamma,$$

and both have multiplicity one. This is consistent with the explicit calculations done in Section 9.4.5.

The module Γ has a unique composition series, namely, $0 \lessdot \mathrm{rad}(\Gamma) \lessdot \Gamma$, where $\mathrm{rad}(\Gamma)$ denotes the radical of Γ. It is the submodule of Γ spanned by $\mathtt{H}_C - \mathtt{H}_{\overline{C}}$. In fact, this composition series coincides with the radical series of Γ. (See (D.14) for the definition of radical series.) The eigenvalue $\lambda_\perp$ corresponds to the composition factor $\mathrm{rad}(\Gamma)$, while $\lambda_\top$ corresponds to the composition factor $\Gamma/\mathrm{rad}(\Gamma)$. The calculation for the latter goes as follows.

$$(\alpha\,\mathtt{H}_O + \beta\,\mathtt{H}_C + \gamma\,\mathtt{H}_{\overline{C}}) \cdot \mathtt{H}_C = \alpha\,\mathtt{H}_C + \beta\,\mathtt{H}_C + \gamma\,\mathtt{H}_{\overline{C}} = (\alpha + \beta + \gamma)\,\mathtt{H}_C$$

since $\mathtt{H}_C$ and $\mathtt{H}_{\overline{C}}$ represent the same element of $\Gamma/\mathrm{rad}(\Gamma)$.

Exercise 9.48. Recall the notion of uniserial modules from Section D.5.9. Show that the left module of chambers $\Gamma[\mathcal{A}]$ is uniserial iff $\mathcal{A}$ has rank either 0 or 1.

9.5.5. Subarrangements. Given a subarrangement $\mathcal{A}'$ of $\mathcal{A}$, recall the morphism of monoids in (1.15). Let us denote it by f. If F is a face of $\mathcal{A}$, then $f(F)$ is the face of $\mathcal{A}'$ whose sign sequence is that of F restricted to $\mathcal{A}'$. Its linearization is a morphism of algebras

$$\Sigma[\mathcal{A}] \to \Sigma[\mathcal{A}'] \tag{9.61}$$

that we still denote by f. If $w = \sum_{F \in \Sigma[\mathcal{A}]} w^F\, \mathtt{H}_F$, then $f(w) = \sum_{G \in \Sigma[\mathcal{A}']} f(w)^G\, \mathtt{H}_G$, where

$$f(w)^G = \sum_{F:\, f(F) = G} w^F.$$

Exercise 9.49. Let $\mathcal{A}'$ be a subarrangement of $\mathcal{A}$. Let X be a flat of $\mathcal{A}'$ (and hence also a flat of $\mathcal{A}$). Check that $\chi_{\mathrm{X}} f = \chi_{\mathrm{X}}$, where f denotes the map (9.61). (Use (9.47) and Exercise 9.12.)

Exercise 9.50. For any subarrangement $\mathcal{A}'$ of $\mathcal{A}$, there is a left action of $\Sigma[\mathcal{A}]$ on $\Gamma[\mathcal{A}']$ obtained by linearizing the action given in Exercise 7.46. Give explicit formulas for the eigenvalues and multiplicities for the left module $\Gamma[\mathcal{A}']$. (Use (9.61) to reduce the problem to the BHR theorem for the Tits algebra $\Sigma[\mathcal{A}']$.)

9.5.6. Invertible elements and zero divisors. Let $A^{\times}$ denote the set of invertible elements in an algebra A. It is a group under multiplication. Invertible elements and zero divisors of the Tits algebra are characterized below.

Proposition 9.51. *Let w be an element of the Tits algebra. Then*

$$w \in \Sigma^{\times} \iff \mathrm{s}(w) \in \Pi^{\times} \iff \lambda_{\mathrm{X}}(w) \neq 0 \text{ for all flats } \mathrm{X},$$

with $\lambda_{\mathrm{X}}(w)$ as is (9.60). Similarly,

$$w \text{ is a zero divisor} \iff \lambda_{\mathrm{X}}(w) = 0 \text{ for some flat } \mathrm{X}.$$

PROOF. This is a special case of Proposition D.43. (Recall that $\lambda_{\mathrm{X}}(w) = \chi_{\mathrm{X}}(w)$.) □

There is an algorithm to compute the inverse of $u \in \Sigma^{\times}$. By the proposition,

$$\mathrm{s}(u) = \sum_{\mathrm{X}} \lambda_{\mathrm{X}}(u)\, \mathtt{Q}_{\mathrm{X}} \ \in \Pi^{\times}.$$

The inverse in the Birkhoff algebra is given by

$$\sum_{\mathrm{X}} \lambda_{\mathrm{X}}(u)^{-1}\, \mathtt{Q}_{\mathrm{X}}.$$

Choose any lift of this element to the Tits algebra. Call it v. Then $uv = \mathtt{H}_O + z$, where z belongs to the radical of Σ. Since z is nilpotent, $\mathtt{H}_O + z$ is invertible with inverse $\mathtt{H}_O - z + z^2 - \dots$. Multiplying v on the right by this element yields the inverse of u. (After choosing v, one may instead compute vu and proceed as before.)

9.5.7. Bilinear forms. Let f be any linear functional on Σ which factors through the support map. Define the families of scalars ξ and η by (9.6). Define a symmetric bilinear form on Σ by

$$\Sigma \times \Sigma \to \Bbbk, \qquad \langle x, y\rangle = f(x \cdot y). \tag{9.62}$$

It is clearly associative, that is, (D.3) holds. By hypothesis, it induces a symmetric bilinear form on Π.

Lemma 9.52. *The radical of the bilinear form (9.62) contains the radical of the Tits algebra. Equality holds iff the induced bilinear form on Π is nondegenerate iff $\eta_{\mathrm{X}} \neq 0$ for all* X.

PROOF. Recall from Proposition 9.22 that the radical of the Tits algebra is the kernel of the support map. This proves everything except the last claim which follows from Lemma D.57 specialized to $P = \Pi$. □

Some examples are given below.

- The character χ_{h} of any Σ-module h is a linear functional of the above kind. In this case, equality holds iff $\eta_{\mathrm{X}}(\mathsf{h}) \neq 0$ for all X iff the associated graded module of any composition series of h is a faithful Π-module.
- Let $f(\mathtt{H}_F)$ be 1 if F is a chamber, and 0 otherwise. In this case, by (9.7), $\eta_{\mathrm{X}} = \mu(\mathrm{X}, \top)$ which is nonzero, so equality holds. The induced bilinear form on Π is the one discussed in the beginning of Example D.58 specialized to $P = \Pi$.

Exercise 9.53. Frobenius algebras are reviewed in Section D.1.5. The linear functional f considered above is not Frobenius since (9.62) has a nontrivial radical. Is the Tits algebra Frobenius?

9.6. Filtration by flats of a right module

Any right module over the Tits algebra has a filtration indexed by the poset of flats. It can be used to give another proof of the eigenvalue-multiplicity theorem. More perspective on this filtration is given later in Chapter 13 when we discuss decomposable series of right modules.

9.6.1. Filtration. For a right Σ-module, the isomorphism in Lemma 9.39 gets replaced by an equality:

Lemma 9.54. *Let F and G be faces of the same support. For a right Σ-module h, we have $\mathsf{h} \cdot \mathtt{H}_F = \mathsf{h} \cdot \mathtt{H}_G$.*

PROOF. This follows from (1.13). □

Let h be a right Σ-module. For each flat X, put

$$\mathcal{D}_{\mathrm{X}}(\mathsf{h}) := \mathsf{h} \cdot \mathtt{H}_F, \tag{9.63}$$

where F is any face of support X. This is well-defined by Lemma 9.54. Also

$$\dim \mathcal{D}_{\mathrm{X}}(\mathsf{h}) = \xi_{\mathrm{X}}(\mathsf{h}) \tag{9.64}$$

by (9.54) (for a right Σ-module). Note that

$$\mathcal{D}_{\perp}(\mathsf{h}) = \mathsf{h}.$$

Further,

$$\mathrm{X} \leq \mathrm{Y} \implies \mathcal{D}_{\mathrm{X}}(\mathsf{h}) \supseteq \mathcal{D}_{\mathrm{Y}}(\mathsf{h}). \tag{9.65}$$

To see this, pick $F \leq G$ such that F has support X, and G has support Y. Any element of $\mathcal{D}_{\mathrm{Y}}(\mathsf{h})$ is of the form $x \cdot \mathtt{H}_G$. Call this element z. Then

$$z \cdot \mathtt{H}_F = (x \cdot \mathtt{H}_G) \cdot \mathtt{H}_F = x \cdot \mathtt{H}_{GF} = x \cdot \mathtt{H}_G = z,$$

so z belongs to $\mathcal{D}_{\mathrm{X}}(\mathsf{h})$, proving (9.65). Finally, we claim that each $\mathcal{D}_{\mathrm{X}}(\mathsf{h})$ is a submodule of h. To see this, let $z = x \cdot \mathtt{H}_F$ be any element of $\mathcal{D}_{\mathrm{X}}(\mathsf{h})$ (with F of support X). Then for any G, $z \cdot \mathtt{H}_G = x \cdot \mathtt{H}_{FG}$. This is an element of $\mathcal{D}_{\mathrm{s}(FG)}(\mathsf{h})$ which is a subspace of $\mathcal{D}_{\mathrm{X}}(\mathsf{h})$ by (9.65). In conclusion:

Lemma 9.55. *For a right Σ-module h, the submodules $\mathcal{D}_{\mathrm{X}}(\mathsf{h})$ defined by* (9.63) *define a filtration of h indexed by flats, that is,* (9.65) *holds.*

The following is a key property of this filtration.

Lemma 9.56. *Let h be a right Σ-module. For any flats X and Y,*

$$\mathcal{D}_{\mathrm{X}}(\mathsf{h}) \cap \mathcal{D}_{\mathrm{Y}}(\mathsf{h}) = \mathcal{D}_{\mathrm{X} \vee \mathrm{Y}}(\mathsf{h}).$$

For any flats X, Y and Z,

$$\mathcal{D}_{\mathrm{X}}(\mathsf{h}) \cap (\mathcal{D}_{\mathrm{Y}}(\mathsf{h}) + \mathcal{D}_{\mathrm{Z}}(\mathsf{h})) = \mathcal{D}_{\mathrm{X} \vee \mathrm{Y}}(\mathsf{h}) + \mathcal{D}_{\mathrm{X} \vee \mathrm{Z}}(\mathsf{h}).$$

This generalizes to a finite sum.

PROOF. For the first part: The rhs is contained in the lhs by (9.65). For the reverse containment: Take F and G of supports X and Y. Let x belong to the lhs. Then x is fixed by both F and G, and hence by FG (and GF). So x belongs to the rhs.

For the second part: The rhs is contained in the lhs by the first part. For the reverse containment: Take F, G and H of supports X, Y and Z. Let x belong to the lhs. Then $x = x \cdot \mathtt{H}_F$ and there are x_1 and x_2 such that $x = x_1 \cdot \mathtt{H}_G + x_2 \cdot \mathtt{H}_H$.

Then $x = x \cdot \mathtt{H}_F = (x_1 \cdot \mathtt{H}_G + x_2 \cdot \mathtt{H}_H) \cdot \mathtt{H}_F = x_1 \cdot \mathtt{H}_{GF} + x_2 \cdot \mathtt{H}_{HF}$. This element is in the rhs. The same argument works for more than two summands. □

Example 9.57. We make some of the above ideas more explicit in the case when the right Σ-module arises from a left Σ-set. This makes use of the discussion in Section 7.8.

Let h be a left Σ-set. Then its linearization h is a left Σ-module. We write

$$\mathtt{H}_F \cdot \mathtt{H}_x = \mathtt{H}_{F \cdot x}.$$

Observe that the set $\{\mathtt{H}_y \mid F \cdot y = y\}$ is a basis for $\mathtt{H}_F \cdot \mathsf{h}$.

The dual h^* is a right Σ-module. Writing $\mathtt{M}$ for the basis dual to the $\mathtt{H}$-basis,

$$\mathtt{M}_y \cdot \mathtt{H}_F = \sum_{x:\, F \cdot x = y} \mathtt{M}_x. \tag{9.66}$$

The indexing set is $\ell(F, y)$, see (7.33). The set $\{\mathtt{M}_y \cdot \mathtt{H}_F \mid F \cdot y = y\}$ is a basis for $\mathsf{h}^* \cdot \mathtt{H}_F$. Note that there is a bijection between bases of $\mathtt{H}_F \cdot \mathsf{h}$ and $\mathsf{h}^* \cdot \mathtt{H}_F$. In particular, the two spaces have the same dimension, as expected by (9.59).

For $F \leq G$ and $G \cdot y = y$, using (7.34), we have

$$\mathtt{M}_y \cdot \mathtt{H}_G = \sum_{z \in \ell(G,y)} \mathtt{M}_z = \sum_{x:\, F \cdot x = x,\, G \cdot x = y} \Big(\sum_{z \in \ell(F,x)} \mathtt{M}_z \Big).$$

This expresses a basis element of $\mathsf{h}^* \cdot \mathtt{H}_G$ as a sum of basis elements of $\mathsf{h}^* \cdot \mathtt{H}_F$. In particular, $\mathsf{h}^* \cdot \mathtt{H}_G$ is a subspace of $\mathsf{h}^* \cdot \mathtt{H}_F$ consistent with (9.65).

9.6.2. Associated graded module. Let $\mathsf{k} = \oplus_{\mathrm{X}} \mathsf{k}_{\mathrm{X}}$ denote the associated graded module of the filtration in Lemma 9.55. That is, k_{X} is the quotient of $\mathcal{D}_{\mathrm{X}}(\mathsf{h})$ by the sum of all $\mathcal{D}_{\mathrm{Y}}(\mathsf{h})$ for $\mathrm{Y} > \mathrm{X}$.

Note very carefully that we are dealing here with a filtration indexed by the poset of flats and *not* a usual filtration indexed by a chain. This extra level of difficulty can be handled by the technical property given in Lemma 9.56 as follows.

Lemma 9.58. *For any upper set* U *in the lattice of flats, there is a vector space isomorphism*

$$\sum_{\mathrm{Y} \in \mathrm{U}} \mathcal{D}_{\mathrm{Y}}(\mathsf{h}) = \bigoplus_{\mathrm{Y} \in \mathrm{U}} \mathsf{k}_{\mathrm{Y}}.$$

In particular: There is a vector space isomorphism

$$\mathcal{D}_{\mathrm{X}}(\mathsf{h}) \cong \bigoplus_{\mathrm{Y}:\, \mathrm{Y} \geq \mathrm{X}} \mathsf{k}_{\mathrm{Y}}.$$

In the special case $\mathrm{X} = \bot$, *we have* $\mathsf{h} \cong \mathsf{k}$.

PROOF. The second part follows by applying the first part to the upper set $[\mathrm{X}, \top]$. To prove the first part, we proceed by induction on the cardinality of U. For convenience, put

$$\mathsf{h}_{\mathrm{U}} := \sum_{\mathrm{Y} \in \mathrm{U}} \mathcal{D}_{\mathrm{Y}}(\mathsf{h}).$$

Pick a minimal element of U. Call it X. Then

$$\mathsf{h}_{\mathrm{U}}/\mathsf{h}_{\mathrm{U} \setminus \mathrm{X}} \cong \mathcal{D}_{\mathrm{X}}(\mathsf{h})/(\mathcal{D}_{\mathrm{X}}(\mathsf{h}) \cap \mathsf{h}_{\mathrm{U} \setminus \mathrm{X}}) = \mathcal{D}_{\mathrm{X}}(\mathsf{h})/(\sum_{\mathrm{Y} > \mathrm{X}} \mathcal{D}_{\mathrm{Y}}(\mathsf{h})) = \mathsf{k}_{\mathrm{X}}.$$

The second step used Lemma 9.56. In conjunction with the induction hypothesis, we obtain

$$\mathsf{h}_{\mathrm{U}} \cong \mathsf{h}_{\mathrm{U}\setminus\mathrm{X}} \oplus \mathsf{h}_{\mathrm{U}}/\mathsf{h}_{\mathrm{U}\setminus\mathrm{X}} \cong \Big(\bigoplus_{\mathrm{Y}\in\mathrm{U}\setminus\mathrm{X}} \mathsf{k}_{\mathrm{Y}}\Big) \oplus \mathsf{k}_{\mathrm{X}} \cong \bigoplus_{\mathrm{Y}\in\mathrm{U}} \mathsf{k}_{\mathrm{Y}}.$$

This completes the induction step. □

9.6.3. Second proof of the eigenvalue-multiplicity theorem. We now give a second proof of Theorem 9.44.

Let us assume that h is a right Σ-module. Consider the filtration of h defined in Lemma 9.55. It is indexed by flats, with the X-component given by $\mathcal{D}_{\mathrm{X}}(\mathsf{h})$. Its associated graded module is $\mathsf{k} = \oplus_{\mathrm{X}}\mathsf{k}_{\mathrm{X}}$. We employ Lemma 9.58. The eigenvalues of $\Psi_{\mathsf{h}}(w)$ are the same as those of $\Psi_{\mathsf{k}}(w)$. Using the definition of k_{X}, we see that $\mathtt{H}_F$ acts on k_{X} by the identity if $\mathrm{s}(F) \leq \mathrm{X}$, and by zero otherwise. Equivalently, by (9.48), w acts on k_{X} by scalar multiplication by $\chi_{\mathrm{X}}(w)$. The multiplicity of this eigenvalue is the dimension of k_{X}, and by (9.52), this must equal $\eta_{\mathrm{X}}(\mathsf{h})$ since the dimension of $\mathcal{D}_{\mathrm{X}}(\mathsf{h})$ (which is $\xi_{\mathrm{X}}(\mathsf{h})$ by (9.64)) is the sum of the dimensions of k_{Y} for $\mathrm{Y} \geq \mathrm{X}$. This completes the argument.

Now suppose h were a left Σ-module. Then we can apply the above argument to its dual h^* which is a right Σ-module. Dualizing (that is, looking at the transpose matrix) does not affect eigenvalues and multiplicities. Also recall from (9.59) that $\eta_{\mathrm{X}}(\mathsf{h}) = \eta_{\mathrm{X}}(\mathsf{h}^*)$. This concludes the proof.

9.7. Primitive part and decomposable part

We introduce the notion of primitive part of a left module and the decomposable part of a right module over the Tits algebra. They are related to each other by duality.

9.7.1. Primitive part of a left module. For a left Σ-module h, the *primitive part* of h is the subspace defined by

$$\mathcal{P}(\mathsf{h}) = \bigcap_{F>O} \ker(\Psi_{\mathsf{h}}(\mathtt{H}_F) : \mathsf{h} \to \mathsf{h}).$$

In other words,

$$x \in \mathcal{P}(\mathsf{h}) \iff \mathtt{H}_F \cdot x = 0 \text{ for all } F > O.$$

Exercise 9.59. Check that: $x \in \mathcal{P}(\mathsf{h}) \iff \mathtt{H}_P \cdot x = 0$ for all vertices P.

The definition of primitive part can be rephrased as follows.

Lemma 9.60. *The primitive part $\mathcal{P}(\mathsf{h})$ is the invariant subspace of h for the multiplicative character $\chi_\perp$ defined in* (9.50). *In other words,*

$$x \in \mathcal{P}(\mathsf{h}) \iff z \cdot x = \chi_\perp(z)x \text{ for all } z \in \Sigma. \tag{9.67}$$

Equivalently, $\mathcal{P}(\mathsf{h})$ is the sum of the simple submodules of h with multiplicative character $\chi_\perp$. In particular, $\mathcal{P}(\mathsf{h})$ is semisimple.

9.7.2. Decomposable part of a right module. For a right Σ-module h, the *decomposable part* of h is the subspace defined by

$$\mathcal{D}(\mathsf{h}) = \sum_{F>O} \mathsf{h} \cdot \mathtt{H}_F = \sum_{\mathrm{X}>\perp} \mathcal{D}_{\mathrm{X}}(\mathsf{h}), \tag{9.68}$$

with $\mathcal{D}_{\mathrm{X}}(\mathsf{h})$ as in (9.63). It is a submodule of h.

Decomposable part and primitive part are dual notions:

Proposition 9.61. *For a left Σ-module h, the spaces $\mathcal{P}(\mathsf{h})$ and $\mathcal{D}(\mathsf{h}^*)$ are orthogonal to each other under the canonical pairing between h and h^*.*

9.8. Over and under a flat. Cartesian product

We briefly discuss how the Tits algebra behaves under passage to arrangements over and under a flat, and under taking cartesian product of arrangements.

9.8.1. Over a flat. Let us linearize the isomorphisms in Lemmas 1.37 and 1.38. For faces F and G with the same support, there is an algebra isomorphism

$$\beta_{G,F} : \Sigma[\mathcal{A}_F] \to \Sigma[\mathcal{A}_G], \qquad \mathtt{H}_{K/F} \mapsto \mathtt{H}_{GK/G}. \tag{9.69}$$

Its inverse is $\beta_{F,G}$. Similarly, for any face with support X, there are canonical inverse algebra isomorphisms

$$\beta_{\mathrm{X},F} : \Sigma[\mathcal{A}_F] \to \Sigma[\mathcal{A}_{\mathrm{X}}] \qquad \text{and} \qquad \beta_{F,\mathrm{X}} : \Sigma[\mathcal{A}_{\mathrm{X}}] \to \Sigma[\mathcal{A}_F]. \tag{9.70}$$

Identities such as

$$\beta_{\mathrm{X},F} = \beta_{\mathrm{X},G}\beta_{G,F} \quad \text{and} \quad \beta_{G,F} = \beta_{G,\mathrm{X}}\beta_{\mathrm{X},F}$$

always hold.

For any face H of $\mathcal{A}$, the map

$$\Delta_H : \Sigma[\mathcal{A}] \to \Sigma[\mathcal{A}_H], \qquad \mathtt{H}_F \mapsto \mathtt{H}_{HF/H} \tag{9.71}$$

is an algebra homomorphism. For faces F and G with the same support, the diagram

$$\begin{array}{ccc} & \Sigma[\mathcal{A}] & \\ {\scriptstyle\Delta_F}\swarrow & & \searrow{\scriptstyle\Delta_G} \\ \Sigma[\mathcal{A}_F] & \xrightarrow[\beta_{G,F}]{} & \Sigma[\mathcal{A}_G] \end{array} \tag{9.72}$$

commutes. For faces $F \leq G$, the diagram

$$\begin{array}{ccc} & \Sigma[\mathcal{A}] & \\ {\scriptstyle\Delta_F}\swarrow & & \searrow{\scriptstyle\Delta_G} \\ \Sigma[\mathcal{A}_F] & \xrightarrow[\Delta_{G/F}]{} & \Sigma[\mathcal{A}_G] \end{array} \tag{9.73}$$

commutes, where $\Delta_{G/F}(\mathtt{H}_{K/F}) := \mathtt{H}_{GK/G}$.

For any face F of $\mathcal{A}$, let

$$\mu_F : \Sigma[\mathcal{A}_F] \to \Sigma[\mathcal{A}], \qquad \mathtt{H}_{K/F} \mapsto \mathtt{H}_K. \tag{9.74}$$

These maps satisfy a diagram similar to (9.73) with $\mu_{G/F}$ defined in the obvious manner. The map μ_F is a section of the map Δ_F, that is, $\Delta_F\mu_F = \mathrm{id}$. Composing in the other direction yields

$$\mu_F\Delta_F(z) = \mathtt{H}_F \cdot z. \tag{9.75}$$

Note that μ_F preserves products, that is, $\mu_F(z \cdot w) = \mu_F(z) \cdot \mu_F(w)$, but it does not preserve the unit, so it is not an algebra homomorphism.

Exercise 9.62. Check that for any faces F and G,

$$\Delta_G \mu_F = \mu_{GF/G} \beta_{GF,FG} \Delta_{FG/F}.$$

We call this the bimonoid axiom for faces. It links the Tits algebras of $\mathcal{A}$, $\mathcal{A}_F$, $\mathcal{A}_G$, $\mathcal{A}_{FG}$ and $\mathcal{A}_{FG}$.

Exercise 9.63. Recall the maps μ_X and Δ_X from Section 9.1.9. For any face F with support X, check that the diagrams

$$\begin{array}{ccc} \Sigma[\mathcal{A}_X] \xrightarrow{\beta_{F,X}} \Sigma[\mathcal{A}_F] & \xrightarrow{\mu_F} & \Sigma[\mathcal{A}] \\ \downarrow{\scriptstyle s} & & \downarrow{\scriptstyle s} \\ \Pi[\mathcal{A}_X] & \xrightarrow[\mu_X]{\hspace{6em}} & \Pi[\mathcal{A}] \end{array} \qquad \begin{array}{ccc} \Sigma[\mathcal{A}] \xrightarrow{\Delta_F} \Sigma[\mathcal{A}_F] & \xrightarrow{\beta_{X,F}} & \Sigma[\mathcal{A}_X] \\ \downarrow{\scriptstyle s} & & \downarrow{\scriptstyle s} \\ \Pi[\mathcal{A}] & \xrightarrow[\Delta_X]{\hspace{6em}} & \Pi[\mathcal{A}_X] \end{array}$$

commute.

Exercise 9.64. Let X be a flat containing a face F. Let $\chi_{X/F}$ denote the multiplicative character of $\Sigma[\mathcal{A}_F]$ indexed by the flat X/F. The composite $\chi_{X/F}\Delta_F$ is an algebra homomorphism, and hence determines a multiplicative character of $\Sigma[\mathcal{A}]$. Check that $\chi_{X/F}\Delta_F = \chi_X$.

Exercise 9.65. Check that Δ_H induces a map from the projective Tits algebra of $\mathcal{A}$ to the projective Tits algebra of $\mathcal{A}_H$. On the H-basis defined in (9.36), it sends $\mathtt{H}_{\{F,\overline{F}\}}$ to $\mathtt{H}_{\{HF/H, H\overline{F}/H\}}$.

Exercise 9.66. Check that: For any flat X, there is an algebra homomorphism

$$\Sigma[\mathcal{A}] \to \Sigma[\mathcal{A}_X]. \tag{9.76}$$

It is defined as $\beta_{X,F}\Delta_F$, where F is any face with support X. This does not depend on the particular choice of F.

Exercise 9.67. Check that the map (9.76) is the special case of the algebra homomorphism (9.61) for $\mathcal{A}' := \mathcal{A}_X$. Deduce the result in Exercise 9.64 as a special case of the result in Exercise 9.49.

9.8.2. Under a flat. For any flat X, the linear map

$$\Sigma[\mathcal{A}] \to \Sigma[\mathcal{A}^X], \qquad \sum_F x^F \mathtt{H}_F \mapsto \sum_{F:\, s(F) \leq X} x^F \mathtt{H}_F \tag{9.77}$$

is an algebra homomorphism.

9.8.3. Cartesian product. For any arrangements $\mathcal{A}$ and $\mathcal{A}'$, there is an algebra isomorphism

$$\Sigma[\mathcal{A} \times \mathcal{A}'] \to \Sigma[\mathcal{A}] \otimes \Sigma[\mathcal{A}'], \qquad \mathtt{H}_{(F,F')} \mapsto \mathtt{H}_F \otimes \mathtt{H}_{F'}. \tag{9.78}$$

This follows from (1.21). Similarly, there is an isomorphism

$$\Gamma[\mathcal{A} \times \mathcal{A}'] \to \Gamma[\mathcal{A}] \otimes \Gamma[\mathcal{A}'], \qquad \mathtt{H}_{(C,C')} \mapsto \mathtt{H}_C \otimes \mathtt{H}_{C'}. \tag{9.79}$$

9.9. Janus algebra and its one-parameter deformation

Recall the Janus monoid from Section 1.5. Its linearization is the Janus algebra. It is an elementary algebra, with the Birkhoff algebra as its split-semisimple quotient. Further, it admits a one-parameter deformation. When the parameter is not a root of unity, the deformed algebra is in fact a split-semisimple (noncommutative) algebra which is Morita equivalent to the Birkhoff algebra. The proof uses the factorization of the determinant of the Varchenko matrices.

9.9.1. Janus algebra. The linearization of the Janus monoid yields an algebra. We call this the *Janus algebra*, and denote it by $\mathsf{J}[\mathcal{A}]$. Using $\mathtt{H}$ for the canonical basis, we write

$$\mathtt{H}_{(F,F')} \cdot \mathtt{H}_{(G,G')} = \mathtt{H}_{(FG,G'F')}. \tag{9.80}$$

Diagram (1.18) yields a commutative diagram of algebras

$$\begin{array}{ccc} \mathsf{J}[\mathcal{A}] & \longrightarrow & \Sigma^{\mathrm{op}}[\mathcal{A}] \\ \downarrow & & \downarrow \mathrm{s} \\ \Sigma[\mathcal{A}] & \xrightarrow[\mathrm{s}]{} & \Pi[\mathcal{A}]. \end{array} \tag{9.81}$$

Proposition 9.68. *The Janus algebra $\mathsf{J}[\mathcal{A}]$ is elementary. Its radical is the kernel of the map $\mathsf{J}[\mathcal{A}] \to \Pi[\mathcal{A}]$. The dimension of the radical is equal to the number of bi-faces minus the number of flats.*

This can be proved in a manner similar to Proposition 9.22 by showing that the kernel of the map $\mathsf{J}[\mathcal{A}] \twoheadrightarrow \Pi[\mathcal{A}]$ is a nilpotent ideal. We omit the details.

9.9.2. Projective Janus algebra. Analogous to the projective Tits algebra, one can define the *projective Janus algebra* as follows. It is the subalgebra of the Janus algebra consisting of elements which are invariant under the map which sends z to $\overline{z}$, where

$$z = \sum_{(F,F')} x^{F,F'} \mathtt{H}_{(F,F')} \quad \text{and} \quad \overline{z} = \sum_{(F,F')} x^{F,F'} \mathtt{H}_{(\overline{F},\overline{F'})}. \tag{9.82}$$

It is an elementary algebra, with the Birkhoff algebra as its split-semisimple quotient.

9.9.3. Modules over the Janus algebra. Since the Janus algebra is elementary, its module theory is similar to that of the Tits algebra which was discussed in Section 9.5. In particular, one can define the numbers $\xi_{\mathrm{X}}(\mathsf{h})$ and $\eta_{\mathrm{X}}(\mathsf{h})$ for any module h over the Janus algebra.

Exercise 9.69. Formulate Theorems 9.38 and 9.44 for the Janus algebra.

The Janus algebra is isomorphic to its opposite algebra under the map which interchanges the two coordinates. It follows that the categories of left modules and right modules over the Janus algebra are isomorphic. Explicitly, if h is a left module, then it is also a right module via

$$x \cdot \mathtt{H}_{(F,F')} := \mathtt{H}_{(F',F)} \cdot x$$

for $x \in \mathsf{h}$. In particular, the dual of a left module can again be viewed as a left module. Hence, the notion of self-duality makes sense for modules over the Janus algebra.

Now let k be a left module over the Tits algebra. Then k^* is a right module over its opposite algebra, and as a result $\mathsf{k} \otimes \mathsf{k}^*$ is a left module over the Janus algebra. In addition, we deduce that the latter is self-dual.

Exercise 9.70. Consider the left module

$$\mathsf{h} := \Gamma[\mathcal{A}] \otimes \Gamma[\mathcal{A}]^*$$

over the Janus algebra. Explicitly, using (9.66), the action is given by

$$\mathtt{H}_{(F,F')} \cdot (\mathtt{H}_C \otimes \mathtt{M}_{C'}) = \sum_{D:\, F'D = C'} \mathtt{H}_{FC} \otimes \mathtt{M}_D. \tag{9.83}$$

Note that the rhs is zero if F' is not a face of C'. Check that $\xi_{\mathrm{X}}(\mathsf{h})$ and $\eta_{\mathrm{X}}(\mathsf{h})$ match those given in Exercise 9.6.

9.9.4. *q*-Janus algebra. Fix a scalar q. The binary operation

$$\mathtt{H}_{(F,F')} \cdot \mathtt{H}_{(G,G')} := q^{\mathrm{dist}(F',G)}\, \mathtt{H}_{(FG,G'F')} \tag{9.84}$$

is associative. To see this, take three bi-faces (F,F'), (G,G') and (H,H'). Their associativity boils down to the identity

$$q^{\mathrm{dist}(F',G)} q^{\mathrm{dist}(G'F',H)} = q^{\mathrm{dist}(F',GH)} q^{\mathrm{dist}(G',H)}.$$

This holds by Proposition 8.5 applied to the distance function v_q of Section 8.2.1. Thus, the product (9.84) indeed defines an algebra. We call it the *q-Janus algebra* and denote it by $\mathsf{J}_q[\mathcal{A}]$. Setting $q = 1$ recovers the Janus algebra.

Exercise 9.71. Show that the q-Janus algebra is isomorphic to its opposite algebra.

Exercise 9.72. Show that the Janus algebra and the (-1)-Janus algebra are isomorphic.

Exercise 9.73. Check that: The space of chambers has a left action of the q-Janus algebra given by

$$\mathtt{H}_{(F,F')} \cdot \mathtt{H}_C := q^{\mathrm{dist}(C,F'C)}\, \mathtt{H}_{FC}. \tag{9.85}$$

Note very carefully the different roles played by F and F'. We denote this module by $\Gamma_q[\mathcal{A}]$. When $q = 1$, the action (9.85) factors through the quotient map $\mathsf{J}[\mathcal{A}] \to \Sigma[\mathcal{A}]$ to recover the usual left module of chambers $\Gamma[\mathcal{A}]$.

Dualizing $\Gamma_q[\mathcal{A}]$ yields $\Gamma_q[\mathcal{A}]^*$ which we again view as a left module. Write down the action explicitly.

Exercise 9.74. Let p and q be scalars. Check that: If h is a left module over $\mathsf{J}_p[\mathcal{A}]$ and k is a left module over $\mathsf{J}_q[\mathcal{A}]$, then $\mathsf{h} \otimes \mathsf{k}$ is a left module over $\mathsf{J}_{pq}[\mathcal{A}]$ via

$$\mathtt{H}_{(F,F')} \cdot (x \otimes y) := (\mathtt{H}_{(F,F')} \cdot x) \otimes (\mathtt{H}_{(F,F')} \cdot y)$$

for $x \in \mathsf{h}$ and $y \in \mathsf{k}$.

Apply this to the examples in Exercise 9.73 to deduce that

$$\Gamma_p[\mathcal{A}] \otimes \Gamma_q[\mathcal{A}]^*$$

is a left module over $\mathsf{J}_{pq}[\mathcal{A}]$. When $p = q = 1$, we recover the module of Exercise 9.70.

9.9.5. Split-semisimplicity for q not a root of unity. The behavior of the q-Janus algebra for generic values of q is quite different from that of the Janus algebra. The precise result is as follows.

Theorem 9.75. *Suppose q is not a root of unity. The q-Janus algebra over $\Bbbk$ is split-semisimple, that is, isomorphic to a product of matrix algebras over $\Bbbk$. There is one matrix algebra for each flat* X*, with the size of the matrix being the number of faces with support* X.

To prove this result, we will construct a Q-basis of $\mathsf{J}_q[\mathcal{A}]$ on which the product is given by

$$\mathtt{Q}_{(F,F')} \cdot \mathtt{Q}_{(G,G')} = \begin{cases} \mathtt{Q}_{(F,G')} & \text{if } F' = G, \\ 0 & \text{otherwise.} \end{cases} \tag{9.86}$$

Then for any flat X, the Q-basis elements indexed by bi-faces with support X form the basis of a matrix algebra. Further, Q-basis elements for different flats are orthogonal, so $\mathsf{J}_q[\mathcal{A}]$ breaks as a product of these matrix algebras.

9.9.6. Morita equivalence. Two algebras are *Morita equivalent* if their module categories are equivalent.

Theorem 9.76. *The q-Janus algebra, for q not a root of unity, and the Birkhoff algebra are Morita equivalent.*

PROOF. The algebra of $n \times n$ matrices over $\Bbbk$ is Morita equivalent to $\Bbbk$, see for instance [**268**, Theorem (17.20)]. Now combine Theorems 9.2 and 9.75. □

9.9.7. 0-Janus algebra. Since the construction of the Q-basis is a little involved, let us first tackle the case $q = 0$. In this situation, the product (9.84) simplifies to

$$\mathtt{H}_{(F,F')} \cdot \mathtt{H}_{(G,G')} = \begin{cases} \mathtt{H}_{(FG,G'F')} & \text{if } F'G = GF', \\ 0 & \text{otherwise.} \end{cases} \tag{9.87}$$

This follows from (1.32).

Define a partial order on the set of bi-faces by

$$(F, F') \leq (G, G') \iff F \leq G,\ F' \leq G',\ FG' = G,\ F'G = G'. \tag{9.88}$$

Reflexivity and antisymmetry are clear. Transitivity requires a small check, which we omit. Note that any bi-face greater than (F, F') is obtained by arbitrarily picking a face G greater than F, and then setting $G' = F'G$ (or by picking a face G' greater than F', and setting $G = FG'$). See Lemma 1.37 in this regard.

Warning. The partial order (9.88) on bi-faces is related to but different from the partial order on bi-faces in Lemma 1.27.

Now define the Q-basis of $\mathsf{J}_0[\mathcal{A}]$ by

$$\mathtt{H}_{(F,F')} = \sum_{(G,G') \geq (F,F')} \mathtt{Q}_{(G,G')}. \tag{9.89}$$

The sum is over all bi-faces (G, G') which are greater than (F, F').

Lemma 9.77. *Formula* (9.86) *holds.*

PROOF. One way to prove this is to assume it and then deduce (9.87) from it. The calculation goes as follows.

$$\begin{aligned}\mathtt{H}_{(F,F')}\cdot\mathtt{H}_{(H,H')} &= \Big(\sum_{G'\colon G'\geq F'} \mathtt{Q}_{(FG',G')}\Big)\cdot\Big(\sum_{K\colon K\geq H}\mathtt{Q}_{(K,H'K)}\Big)\\ &= \sum_{K\colon K\geq F',K\geq H}\mathtt{Q}_{(FK,H'K)}.\end{aligned}$$

This sum is zero unless F' and H are joinable, that is, unless $F'H = HF'$ (Proposition 1.18). Assuming this, the calculation proceeds as follows.

$$\mathtt{H}_{(F,F')}\cdot\mathtt{H}_{(H,H')} = \sum_{K\colon K\geq F'H}\mathtt{Q}_{(FK,H'K)} = \mathtt{H}_{(FH,H'F')}.$$

For the last step, observe that FH, $H'F'$ and $F'H = HF'$ all have the same support. So by Lemma 1.37, faces K greater than $F'H$ are in correspondence with faces $FHK = FK$ greater than FH, and also with faces $H'F'K = H'K$ greater than $H'F'$. □

This completes the proof of Theorem 9.75 when $q = 0$.

9.9.8. Back to the general case. Now suppose q is any scalar which is not a root of unity. Define the Q-basis of $\mathtt{J}_q[\mathcal{A}]$ by

(9.90)
$$\mathtt{H}_{(F,F')} = \sum_{G\colon G\geq F}\ \sum_{\substack{G'\colon FG'=G,\\ \mathrm{s}(G')=\mathrm{s}(G)}} q^{\mathrm{dist}(F',G')}\,\mathtt{Q}_{(G,G')}.$$

An illustration of how the bi-faces (F,F') and (G,G') relate to each other is shown below.

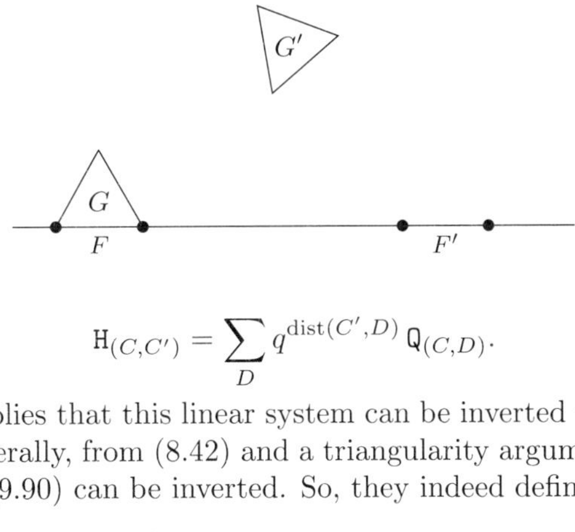

In particular,

(9.91)
$$\mathtt{H}_{(C,C')} = \sum_{D} q^{\mathrm{dist}(C',D)}\,\mathtt{Q}_{(C,D)}.$$

Formula (8.41) implies that this linear system can be inverted when q is not a root of unity. More generally, from (8.42) and a triangularity argument, we deduce that the linear system (9.90) can be inverted. So, they indeed define the Q-basis.

Lemma 9.78. *Formula* (9.86) *holds.*

PROOF. We first illustrate the case of pairs of chambers. Again we assume (9.86) to deduce (9.84).

$$\begin{aligned}\mathtt{H}_{(C,C')}\cdot\mathtt{H}_{(E,E')} &= \Big(\sum_{D} q^{\mathrm{dist}(C',D)}\,\mathtt{Q}_{(C,D)}\Big)\cdot\Big(\sum_{D'} q^{\mathrm{dist}(E',D')}\,\mathtt{Q}_{(E,D')}\Big)\\ &= q^{\mathrm{dist}(C',E)}\Big(\sum_{D'} q^{\mathrm{dist}(E',D')}\,\mathtt{Q}_{(C,D')}\Big)\\ &= q^{\mathrm{dist}(C',E)}\,\mathtt{H}_{(C,E')}.\end{aligned}$$

The first and last steps used (9.91). In the second step, we used $D = E$ to eliminate D. The general calculation goes as follows.

$$\begin{aligned}
\mathtt{H}_{(F,F')} \cdot \mathtt{H}_{(H,H')} &= \Big(\sum_{G:\, G\geq F} \sum_{\substack{G':\, FG'=G,\\ \mathrm{s}(G')=\mathrm{s}(G)}} q^{\mathrm{dist}(F',G')}\, \mathtt{Q}_{(G,G')} \Big) \\
&\quad \cdot \Big(\sum_{K:\, K\geq H} \sum_{\substack{K':\, HK'=K,\\ \mathrm{s}(K')=\mathrm{s}(K)}} q^{\mathrm{dist}(H',K')}\, \mathtt{Q}_{(K,K')} \Big) \\
&= \sum_{G:\, G\geq F} \sum_{\substack{K:\, FK=G,\\ \mathrm{s}(K)=\mathrm{s}(G)}} \sum_{\substack{K':\, HK'=K,\\ \mathrm{s}(K')=\mathrm{s}(K)}} q^{\mathrm{dist}(F',K)} q^{\mathrm{dist}(H',K')}\, \mathtt{Q}_{(G,K')} \\
&= \sum_{G:\, G\geq FH} \sum_{\substack{K':\, FHK'=G,\\ \mathrm{s}(K')=\mathrm{s}(G)}} q^{\mathrm{dist}(F',HK')} q^{\mathrm{dist}(H',K')}\, \mathtt{Q}_{(G,K')} \\
&= \sum_{G:\, G\geq FH} \sum_{\substack{K':\, FHK'=G,\\ \mathrm{s}(K')=\mathrm{s}(G)}} q^{\mathrm{dist}(F',H)} q^{\mathrm{dist}(H'F',K')}\, \mathtt{Q}_{(G,K')} \\
&= q^{\mathrm{dist}(F',H)}\, \mathtt{H}_{(FH,H'F')}.
\end{aligned}$$

The first and last steps used (9.90). In the second step, we used $G' = K$ to eliminate G'. In the next step, we used $HK' = K$ to eliminate K. In the third step, we applied Proposition 8.5 to the distance function v_q of Section 8.2.1. □

This completes the proof of Theorem 9.75.

Exercise 9.79. Check that

$$\mathtt{H}_{(F,F')} \cdot \mathtt{Q}_{(G,G')} = \begin{cases} q^{\mathrm{dist}(F',G)}\, \mathtt{Q}_{(FG,G'F')} & \text{if } \mathrm{s}(G) \geq \mathrm{s}(F), \\ 0 & \text{otherwise.} \end{cases} \tag{9.92}$$

In particular, $\mathtt{H}_{(F,F)} \cdot \mathtt{Q}_{(O,O)} = 0$ for any $F > O$. Compare (9.92) with (9.4).

Exercise 9.80. Show that: For q not a root of unity, the $\mathtt{Q}_{(F,F)}$, as F varies over all faces, form a complete system of primitive orthogonal idempotents of $\mathsf{J}_q[\mathcal{A}]$. In particular,

$$\mathtt{H}_{(O,O)} = \sum_F \mathtt{Q}_{(F,F)}.$$

The sum is over all faces F. This is a special case of (9.90).

Exercise 9.81. For q not a root of unity and h a left J_q-module, the *primitive part* of h is the subspace $\mathcal{P}(\mathsf{h})$ defined by

$$x \in \mathcal{P}(\mathsf{h}) \iff \mathtt{H}_{(F,F)} \cdot x = 0 \text{ for all } F > O.$$

Show that

$$\mathcal{P}(\mathsf{h}) = \mathtt{Q}_{(O,O)} \cdot \mathsf{h}. \tag{9.93}$$

The rhs consists of all elements of the form $\mathtt{Q}_{(O,O)} \cdot x$ for $x \in \mathsf{h}$. Compare (9.93) with (9.26).

9.9.9. Rank one. Let $\mathcal{A}$ be the rank-one arrangement with chambers C and $\overline{C}$. The linear system (9.90) is as follows.

$$\mathtt{H}_{(C,C)} = \mathtt{Q}_{(C,C)} + q\,\mathtt{Q}_{(C,\overline{C})}, \qquad \mathtt{H}_{(C,\overline{C})} = \mathtt{Q}_{(C,\overline{C})} + q\,\mathtt{Q}_{(C,C)},$$

$$\mathtt{H}_{(\overline{C},C)} = \mathtt{Q}_{(\overline{C},C)} + q\,\mathtt{Q}_{(\overline{C},\overline{C})}, \qquad \mathtt{H}_{(\overline{C},\overline{C})} = \mathtt{Q}_{(\overline{C},\overline{C})} + q\,\mathtt{Q}_{(\overline{C},C)},$$

$$\mathtt{H}_{(O,O)} = \mathtt{Q}_{(O,O)} + \mathtt{Q}_{(C,C)} + \mathtt{Q}_{(\overline{C},\overline{C})}.$$

For $q \neq \pm 1$, this linear system can be inverted yielding:

$$\mathtt{Q}_{(C,C)} = \frac{1}{1-q^2}(\mathtt{H}_{(C,C)} - q\,\mathtt{H}_{(C,\overline{C})}), \quad \mathtt{Q}_{(C,\overline{C})} = \frac{1}{1-q^2}(\mathtt{H}_{(C,\overline{C})} - q\,\mathtt{H}_{(C,C)}),$$

$$\mathtt{Q}_{(\overline{C},C)} = \frac{1}{1-q^2}(\mathtt{H}_{(\overline{C},C)} - q\,\mathtt{H}_{(\overline{C},\overline{C})}), \quad \mathtt{Q}_{(\overline{C},\overline{C})} = \frac{1}{1-q^2}(\mathtt{H}_{(\overline{C},\overline{C})} - q\,\mathtt{H}_{(\overline{C},C)}),$$

$$\mathtt{Q}_{(O,O)} = \mathtt{H}_{(O,O)} - \frac{1}{1-q^2}(\mathtt{H}_{(C,C)} - q\,\mathtt{H}_{(C,\overline{C})} - q\,\mathtt{H}_{(\overline{C},C)} + \mathtt{H}_{(\overline{C},\overline{C})}).$$

These are the $\mathtt{Q}$-basis elements of the q-Janus algebra for $q \neq \pm 1$.

9.9.10. Diagonal 0-Janus algebra. Let $\Sigma_0[\mathcal{A}]$ denote the commutative algebra, indexed by faces, with product defined by

$$\mathtt{H}_F \cdot \mathtt{H}_G := \begin{cases} \mathtt{H}_{FG} & \text{if } FG = GF, \\ 0 & \text{otherwise.} \end{cases} \tag{9.94}$$

Define the $\mathtt{Q}$-basis by

$$\mathtt{H}_F = \sum_{G:\, G \geq F} \mathtt{Q}_G \qquad \text{or equivalently} \qquad \mathtt{Q}_F = \sum_{G:\, G \geq F} (-1)^{\mathrm{rk}(G/F)}\, \mathtt{H}_G.$$

The two statements are equivalent by (1.43).

Theorem 9.82. *The commutative algebra $\Sigma_0[\mathcal{A}]$ is split-semisimple, with $\mathtt{Q}$ as its basis of primitive idempotents.*

PROOF. The usual way to proceed is to use the product on the $\mathtt{Q}$-basis, and deduce the one on the $\mathtt{H}$-basis using Proposition 1.18. Another approach is to add a top element to the poset of faces making it a lattice. The algebra of this lattice is split-semisimple by Theorem D.47, and $\Sigma_0[\mathcal{A}]$ is the quotient of it by the span of the top element (which is a primitive idempotent). □

Observe that $\Sigma_0[\mathcal{A}]$ is the diagonal subalgebra of $\mathrm{J}_0[\mathcal{A}]$ under the map

$$\mathtt{H}_F \mapsto \mathtt{H}_{(F,F)} \qquad \text{or equivalently} \qquad \mathtt{Q}_F \mapsto \mathtt{Q}_{(F,F)}.$$

Formula (9.86) then gives another proof of Theorem 9.82. We refer to $\Sigma_0[\mathcal{A}]$ as the *diagonal* 0-*Janus algebra*.

9.9.11. v-Janus algebra. The q-Janus algebra can be generalized further as follows. Let v be any distance function arising from a weight function. By Proposition 8.5, the binary operation

$$\mathtt{H}_{(F,F')} \cdot \mathtt{H}_{(G,G')} := v_{F',G}\, \mathtt{H}_{(FG,G'F')} \tag{9.95}$$

is associative. The resulting algebra is the *v-Janus algebra* which we denote by $\mathsf{J}_v[\mathcal{A}]$. Setting the distance function to be v_q recovers the q-Janus algebra.

We say that a distance function v is *generic* if the associated Varchenko matrices A^{X} are invertible for all flats X, that is, the $b^{\mathrm{X}}_{\mathrm{Y}}$ which appear in the factorization (8.28) are never equal to 1. Observe that the distance function v_q is generic if q is not a root of unity.

When v is generic, the linear system

$$\mathtt{H}_{(F,F')} = \sum_{G:\, G \geq F} \sum_{\substack{G':\, FG'=G,\\ \mathrm{s}(G')=\mathrm{s}(G)}} v_{F',G'}\, \mathtt{Q}_{(G,G')} \tag{9.96}$$

is invertible and defines the $\mathtt{Q}$-basis of $\mathsf{J}_v[\mathcal{A}]$. By the same calculation as before, we see that (9.86) holds. This yields the following generalization of Theorem 9.75.

Theorem 9.83. *For a generic distance function v, the v-Janus algebra is split-semisimple.*

9.10. Coxeter-Tits algebra

We now consider the Coxeter analogues of the Tits, Birkhoff and Janus monoids defined in Section 5.6, and linearize them as shown below.

Monoid	Linearized algebra
Coxeter-Tits monoid $W\Sigma$	Coxeter-Tits algebra $\mathsf{W\Sigma}$
$W\Sigma^{\mathrm{op}}$	$\mathsf{W\Sigma}^{\mathrm{op}}$
Coxeter-Birkhoff monoid $W\Pi$	Coxeter-Birkhoff algebra $\mathsf{W\Pi}$
Coxeter-Janus monoid WJ	Coxeter-Janus algebra WJ

Since each one of them is a semidirect product, their modules can be described in terms of an action of the Coxeter group and an action of the Tits or Birkhoff or Janus monoid with a compatibility condition. Let us make this explicit.

Let W denote the group algebra of W. A left $\mathsf{W\Sigma}$-module is the same as a vector space M which is a left W-module and a left Σ-module, and the two actions are compatible via

$$w(\mathtt{H}_F \cdot m) = w(\mathtt{H}_F) \cdot w(m)$$

for $m \in M$, $w \in W$ and $F \in \Sigma$. This follows from Proposition 5.23. Note that we write $\mathtt{H}_F \cdot m$ for the action of Σ but $w(m)$ for the action of W. The latter is consistent with the notation $w(F)$ that we use for the action W on Σ.

Similarly, a left $\mathsf{W\Sigma}^{\mathrm{op}}$-module is the same as a vector space M which is a left W-module and a right Σ-module, and the two actions are compatible via

$$w(m \cdot \mathtt{H}_F) = w(m) \cdot w(\mathtt{H}_F)$$

for $m \in M$, $w \in W$ and $F \in \Sigma$. In view of Proposition 5.24, this is also the same as a right $\mathsf{W\Sigma}$-module.

A left $\mathsf{W}\Pi$-module is the same as a vector space M which is a Π-module and a left W-module, and the two actions are compatible via

$$w(\mathtt{H}_{\mathrm{X}} \cdot m) = w(\mathtt{H}_{\mathrm{X}}) \cdot w(m)$$

for $m \in M$, $w \in W$ and $\mathrm{X} \in \Pi$. (We do not need to specify left or right for a Π-module since Π is commutative.) A right $\mathsf{W}\Pi$-module admits a similar description with W acting on the right. We do not need to consider these explicitly in view of Proposition 5.25.

A left WJ-module is the same as a vector space M which is a left W-module and a left J-module, and the two actions are compatible via

$$w(\mathtt{H}_{F,F'} \cdot m) = w(\mathtt{H}_{F,F'}) \cdot w(m)$$

for $m \in M$, $w \in W$ and $(F, F') \in \mathrm{J}$. We can also consider right WJ-modules, but it is not required in view of Proposition 5.26.

Notes

Exponential and logarithm. In certain situations, the families (ξ_{X}) and (η_{X}) linked by (9.5) can be interpreted as the moments and cumulants of a random variable (possibly noncommutative). For more details, see the remark at the end of [**10**, Section 13.5].

Tits algebra. Lemma 9.29 is due to Bidigare [**62**, Theorem 2.4.2]. Propositions 9.21 and 9.22 are also due to him [**62**, Section 2.3.3, and in particular, Theorem 2.3.5]. Theorem 9.38 (which is a consequence of these results) was noted by Brown [**103**, Theorem 3]. It is implicit in Bidigare's work [**62**, Sections 2.3 and 2.4]; formula (9.48) is given at the bottom of page 21 of his thesis. Brown worked in the wider context of LRBs. Theorem 9.38 in more general contexts even beyond LRBs is given in [**195**, Corollary 9] and [**16**, Theorem 4.3]. These references also provide pointers to the older semigroup literature. See also [**377**].

The maps (9.44) in the generality of LRBs are contained in [**8**, Diagram (5.8)]; also see [**9**, Diagram (12.14)]. Formula (9.45) is contained in [**8**, Lemma 5.6.1]. The bilinear form (9.62) for the linear functional $f := \chi_{\Gamma}$, namely, the character of Γ, is studied in [**8**, Section 2.5]; Lemma 9.52 corresponds to [**8**, Corollary 2.5.2].

BHR theorem. Bidigare, Hanlon and Rockmore [**61**, Theorem 4.1] or [**62**, Section 2.5] proved Theorem 9.46. Our second proof of Theorem 9.44 given in Section 9.6 is inspired by their method. They worked in the setting of Example 9.57. In hindsight, Bidigare, Hanlon and Rockmore worked essentially with the decomposable series of $\mathsf{h} = \Gamma^*$. In their Theorem 4.4, they state and prove the first part of Lemma 9.56 for $\mathsf{h} = \Gamma^*$. Similarly, their Proposition 4.8 is our Lemma 9.58 for $\mathsf{h} = \Gamma^*$; however, the short inductive proof that they give makes no mention of the key property of the filtration given in the second part of Lemma 9.56.

Later, another proof of the BHR theorem (which we have not discussed here) was given by Brown and Diaconis [**105**, Theorem 1]. More information on their work is given in the notes to Chapter 12.

Still later, a third proof of the BHR theorem was given by Brown based on the fact that the Tits algebra is elementary [**103**, Theorem 4]. Our first proof of Theorem 9.44 follows Brown's approach. His proof, though written for the module of chambers Γ, works for any h. As noted earlier, Brown worked in the wider context of LRBs. For generalizations beyond that, see work of Steinberg [**402**, Theorems 6.3 and 6.4], [**403**, Formulas (8.4) and (8.5)] and [**405**, Theorems 14.11 and 14.12].

Exercise 9.50 is an algebraic reformulation of a result of Athanasiadis and Diaconis [**29**, Theorem 3.3, part (i)]. Details on the solution to this exercise as well as numerous examples can be found in this reference.

Janus algebra. The algebra obtained by linearizing a band is elementary, with the split-semisimple quotient being the linearization of its support lattice. This result is given by Brown [**104**, Theorem B.1]. In view of Lemma 1.27, Proposition 9.68 becomes a special case of this result. More results on the Janus algebra will be given in future work.

Finite-dimensional algebras. For references on the general theory of finite-dimensional algebras, see the notes to Appendix D.

Bialgebras. There are important connections between the theory of modules over the Birkhoff algebra, Tits algebra and Janus algebra and the classical theory of connected bialgebras. The connection is made through the braid arrangement and the Hopf monoids in species of [**9**, Part II]. The ingredients are present in [**10**, Section 13]. This inspires much of our work throughout this monograph, although the connections are often implicit. For instance, the maps μ and Δ occurring in Sections 9.1.9 and 9.8.1 correspond to certain higher product and coproduct operations, while β corresponds to the braiding. We plan to provide details in future work.

CHAPTER 10

Lie and Zie elements

We introduce Lie elements. Recall that left modules over the Tits algebra have a primitive part. Lie elements are the primitive part of the left module of chambers. These elements can also be expressed as solutions to a linear system of equations whose variables are chambers. The set of chambers involved in any given equation form a top-lune. We pay special attention to Lie elements of rank-one and rank-two arrangements; the antisymmetry relation appears in rank-one and the Jacobi identity in rank-two arrangements.

We also introduce Zie elements. They are the primitive part of the Tits algebra viewed as a left module over itself. They can also be expressed as solutions to a linear system of equations whose variables are faces. In fact, we consider two such linear systems. The set of faces involved in any given equation form either the interior or the closure of a lune. A Zie element is special if the central face appears in it with coefficient 1. A special Zie element projects any left module over the Tits algebra onto its primitive part. Using this principle, we derive formulas for the dimensions of the spaces of Lie elements and Zie elements. Both formulas involve the Möbius function of the lattice of flats.

Lie elements carry a substitution product which specifies a way to multiply Lie elements in arrangements under a flat with Lie elements in arrangements over a flat. It is obtained by restricting the substitution product of chambers. Similarly, Lie can be multiplied with Zie on the right and with chambers on the left. This is the restriction of the substitution product of chambers with faces on the right and with top-lunes on the left.

10.1. Lie elements

We introduce Lie elements as solutions to a linear system of equations whose variables are chambers. We then discuss various characterizations for it named after Friedrichs, Ree and Garsia. They involve the notions of primitive part, top-lunes and descents, respectively. The Garsia criterion is also intimately connected to the Witt identity.

10.1.1. Lie elements. Recall the left module of chambers $\Gamma[\mathcal{A}]$. We write a typical element as

$$z = \sum_C x^C \mathtt{H}_C.$$

An element $z \in \Gamma[\mathcal{A}]$ is a *Lie element* if

$$\sum_{C:\, HC=D} x^C = 0 \text{ for all } O < H \leq D. \tag{10.1}$$

This is a linear system in the variables x^C.

We denote the set of Lie elements by $\mathsf{Lie}[\mathcal{A}]$. It is a subspace of $\Gamma[\mathcal{A}]$.

- Note very carefully that $H = O$ is excluded from (10.1): If not, then $z = 0$ would be the only solution since all its coefficients x^C would be forced to be zero.
- Since the condition (10.1) is in terms of the Tits product, cisomorphic arrangements have the "same" Lie elements. In particular, to understand $\mathsf{Lie}[\mathcal{A}]$, one may replace $\mathcal{A}$ by its essentialization.

Lemma 10.1. *If $\mathcal{A}$ has rank zero, then* $\mathsf{Lie}[\mathcal{A}] = \Gamma[\mathcal{A}] = \Bbbk$.

PROOF. Suppose $\mathcal{A}$ has rank zero. Then, it has only one chamber which is the central face, so (10.1) is vacuously true. Hence $\mathsf{Lie}[\mathcal{A}] = \Gamma[\mathcal{A}]$, spanned by $\mathtt{H}_O$. □

Lemma 10.2. *If $\mathcal{A}$ has rank at least one, then the sum of the coefficients of any Lie element is zero. That is, $z \in \mathsf{Lie}[\mathcal{A}]$ implies*

$$\sum_C x^C = 0. \tag{10.2}$$

PROOF. Let D be any chamber. Since $\mathcal{A}$ has rank at least one, $D > O$. So we may choose $H = D$ in (10.1). This yields (10.2). □

Equivalently, using (9.37):

Lemma 10.3. *If $\mathcal{A}$ has rank at least one, then* $\mathsf{Lie}[\mathcal{A}] \subseteq \mathrm{rad}(\Gamma[\mathcal{A}])$.

A more precise result is given below. For any face F, view $\mathsf{Lie}[\mathcal{A}_F]$ as a subspace of $\Gamma[\mathcal{A}]$ via the composite of inclusion maps

$$\mathsf{Lie}[\mathcal{A}_F] \to \Gamma[\mathcal{A}_F] \to \Gamma[\mathcal{A}].$$

Lemma 10.4. *We have*

$$\sum_{F \in \Sigma[\mathcal{A}] \setminus \Gamma[\mathcal{A}]} \mathsf{Lie}[\mathcal{A}_F] = \mathrm{rad}(\Gamma[\mathcal{A}]).$$

The sum is over all faces F which are not chambers.

PROOF. By Lemma 10.2, the lhs is a subset of the rhs. We need to show that equality holds. If C and D are adjacent chambers with common panel H, then $\mathtt{H}_{C/H} - \mathtt{H}_{D/H}$ is an element of $\mathsf{Lie}[\mathcal{A}_H]$. For any chambers C' and D', by picking a gallery joining them, $\mathtt{H}_{C'} - \mathtt{H}_{D'}$ can be written as a sum of $(\mathtt{H}_C - \mathtt{H}_D)$'s, with C and D adjacent; hence it is an element of the lhs. Since such elements span the rhs, equality holds. □

10.1.2. Friedrichs primitive part criterion. Recall from Section 9.7 that left modules over the Tits algebra have a primitive part.

Lemma 10.5. *The space of Lie elements is the primitive part of the left module of chambers:*

$$\mathcal{P}(\Gamma[\mathcal{A}]) = \mathsf{Lie}[\mathcal{A}].$$

Explicitly,

$$z \in \mathsf{Lie}[\mathcal{A}] \iff \mathtt{H}_H \cdot z = 0 \textit{ for all } H > O.$$

PROOF. Let H be any face of $\mathcal{A}$. Then

$$\mathtt{H}_H \cdot \big(\sum_C x^C \mathtt{H}_C\big) = \sum_C x^C \mathtt{H}_{HC} = \sum_{D\colon H \leq D} \Big(\sum_{C\colon HC=D} x^C \Big) \mathtt{H}_D.$$

This equals 0 iff

$$\sum_{C\colon HC=D} x^C = 0 \text{ for all } D \geq H.$$

The result follows from (10.1). □

We refer to the characterization of Lie elements given by Lemma 10.5 as the *Friedrichs criterion.*

Exercise 10.6. Check that: $z \in \mathsf{Lie}[\mathcal{A}] \iff \mathtt{H}_P \cdot z = 0$ for all vertices P. (See Exercise 9.59 in this regard.)

10.1.3. Ree top-lune criterion. Recall from Section 3.2.1 that any top-nested face (H, D) gives rise to a top-lune

$$\mathrm{s}(H, D) = \{C \mid HC = D\}.$$

Note that D always belongs to this top-lune. Further, this top-lune is a singleton (consisting of D) iff $H = O$. All top-lunes arise in this manner from top-nested faces. The definition of a Lie element may now be rewritten as follows.

Lemma 10.7. *We have* $z \in \mathsf{Lie}[\mathcal{A}]$ *iff*

$$\sum_{C \in \mathrm{L}} x^C = 0 \tag{10.3}$$

for all non-singleton combinatorial top-lunes L *in* $\mathcal{A}$.

When L is the maximum flat, equation (10.3) specializes to (10.2).

Lemma 10.8. *We have* $z \in \mathsf{Lie}[\mathcal{A}]$ *iff* (10.3) *holds for all vertex-based combinatorial top-lunes* L *in* $\mathcal{A}$, *or equivalently,* (10.1) *holds for all vertices* H.

PROOF. It suffices to consider only vertex-based top-lunes, since by Corollary 3.25, any non-singleton top-lune can be written as a disjoint union of vertex-based top-lunes. □

We refer to the description of Lie elements given by Lemma 10.7 or Lemma 10.8 as the *Ree criterion.* A Lie element may be visualized as a scalar assigned to each chamber such that the sum of the scalars in every vertex-based top-lune is zero. (The scalar assigned to C is x^C.)

Exercise 10.9. Show that $z \in \mathsf{Lie}[\mathcal{A}]$ iff the sum of the coefficients of z of chambers in any non-singleton top-cone of $\mathcal{A}$ is zero.

10.1.4. Orthogonality with decomposable part. Let $\Gamma[\mathcal{A}]^*$ denote the vector space dual to $\Gamma[\mathcal{A}]$. We use the letter $\mathtt{M}$ to denote the basis of $\Gamma[\mathcal{A}]^*$ which is dual to the $\mathtt{H}$-basis of $\Gamma[\mathcal{A}]$. Since $\Sigma[\mathcal{A}]$ acts on $\Gamma[\mathcal{A}]$ on the left, it acts on $\Gamma[\mathcal{A}]^*$ on the right. Using the definition

$$\langle \mathtt{M}_D \cdot \mathtt{H}_H, \mathtt{H}_C \rangle = \langle \mathtt{M}_D, \mathtt{H}_H \cdot \mathtt{H}_C \rangle,$$

we see that the right action is given by

$$\mathtt{M}_D \cdot \mathtt{H}_H = \sum_{C:HC=D} \mathtt{M}_C. \tag{10.4}$$

Note that the rhs is zero if $H \not\leq D$. This formula is an instance of (9.66).

Let $\mathsf{Lie}[\mathcal{A}]^{\perp}$ denote the decomposable part of $\Gamma[\mathcal{A}]^*$. By definition (9.68), it is the span of the elements (10.4) as (H, D) varies over top-nested faces with $H > O$. Equivalently, it is spanned by the elements

$$c_{\mathrm{L}} = \sum_{C \in \mathrm{L}} \mathtt{M}_C$$

as L varies over non-singleton top-lunes. (It suffices to take vertex-based top-lunes.) Lemma 10.7 says that z is a Lie element iff $\langle c_{\mathrm{L}}, z \rangle = 0$ for every non-singleton top-lune L. Equivalently:

Lemma 10.10. *The spaces* $\mathsf{Lie}[\mathcal{A}]$ *and* $\mathsf{Lie}[\mathcal{A}]^{\perp}$ *are orthogonal to each other under the canonical pairing between* $\Gamma[\mathcal{A}]$ *and* $\Gamma[\mathcal{A}]^*$.

Since $\mathsf{Lie}[\mathcal{A}]$ is the primitive part of $\Gamma[\mathcal{A}]$, what we are seeing above is an illustration of the duality between primitive and decomposable parts (Proposition 9.61).

10.1.5. Opposition map. The opposition map preserves the space of Lie elements. This can be deduced from either the Friedrichs criterion or the Ree criterion. In fact, the opposition map sends a Lie element either to itself or its negative:

Lemma 10.11. *For* $z \in \mathsf{Lie}[\mathcal{A}]$, *the coefficients* x^D *and* $x^{\overline{D}}$ *differ at most by a sign. More precisely,*

$$x^D = (-1)^{\mathrm{rk}(\mathcal{A})} x^{\overline{D}}. \tag{10.5}$$

PROOF. Suppose z is a Lie element. Now apply the Witt identity (7.14) to the scalars x^C. By definition of a Lie element (10.1), the sum inside the parenthesis in (7.14) is zero for $H > O$. So the lhs of (7.14) reduces to x^D (coming from the term $H = O$), and we get (10.5). □

We say a Lie element z is *projective* if it is a projective chamber element, that is, the coefficients of $\mathtt{H}_D$ and $\mathtt{H}_{\overline{D}}$ in z are equal for all D.

Lemma 10.12. *If* $\mathcal{A}$ *has even rank, then all Lie elements are projective. If* $\mathcal{A}$ *has odd rank, then there are no nonzero projective Lie elements (assuming that the field characteristic is not* 2*).*

PROOF. This follows from (10.5). □

10.1.6. Garsia descent criterion. Consider the top-lune $s(H, D)$ in a rank-three simplicial arrangement shown below. (Only those three hyperplanes in the arrangement which are relevant to the discussion are shown.)

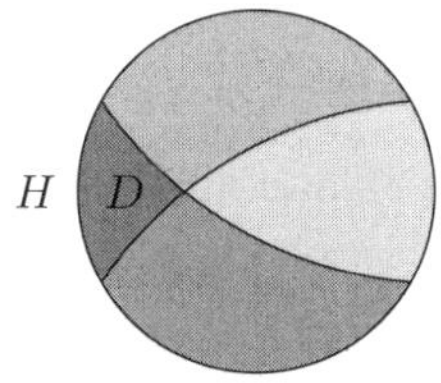

The top-lune $s(H, D)$ consists of chambers C such that $HC = D$. This set can be split into four parts (shown in dark, dark-medium, light-medium and light shades) depending on the value of $\mathrm{Des}(C, D)$ (which is a face of H), see (7.1). When this is the central face, we get only the chamber D shown in dark shade, when this is a vertex of H, we get chambers either in the dark-medium or the light-medium shades, and when this is H, we get chambers in the light shade.

Suppose we are given a Lie element $z \in \mathsf{Lie}[\mathcal{A}]$. By the Ree criterion, the sum of the coefficients of z of chambers in all four regions together in zero. Since the dark shaded and dark-medium shaded regions together form a top-lune, the sum of the coefficients of z of chambers in these two regions in also zero. Similarly, the sum in the dark shaded and light-medium shaded regions is zero. By putting these facts together, we deduce that the coefficient of D in z is equal to the sum of the coefficients of z of chambers in the light shaded region. This property of Lie elements is formalized below.

Lemma 10.13. *Suppose $\mathcal{A}$ is a simplicial arrangement. Then $z \in \mathsf{Lie}[\mathcal{A}]$ iff for any top-nested face (K, D),*

$$x^D = (-1)^{\mathrm{rk}(K)} \sum_{C\colon \mathrm{Des}(C,D)=K} x^C, \tag{10.6}$$

or equivalently, by (9.46),

$$\mathtt{K}_{K,D} \cdot z = (-1)^{\mathrm{rk}(K)} x^D \, \mathtt{H}_D.$$

The lhs of (10.6) is a specialization of the rhs when K is the central face. This follows from (7.2).

PROOF. Suppose $z \in \mathsf{Lie}[\mathcal{A}]$. Now apply the Witt identity (7.17) to the scalars x^C. By (10.1), the sum inside the parenthesis in (7.17) is zero for $H > O$. So the lhs of (7.17) reduces to x^D (coming from the term $H = O$). This proves (10.6).

Conversely, suppose (10.6) holds for all top-nested faces (K, D). Let $H \leq D$ and $H > O$. Then, by (7.1),

$$\sum_{C\colon HC=D} x^C = \sum_{C\colon \mathrm{Des}(C,D)\leq H} x^C = \sum_{K\colon K\leq H} \; \sum_{C\colon \mathrm{Des}(C,D)=K} x^C$$
$$= \sum_{K\colon K\leq H} (-1)^{\mathrm{rk}(K)} x^D = 0.$$

(The second equality is the decomposition illustrated in the preceding figure.) Since x^D does not depend on K, the sum over K is 0 by (1.44). This verifies (10.1). Hence $z \in \mathsf{Lie}[\mathcal{A}]$. □

We refer to the description of Lie elements given by Lemma 10.13 as the *Garsia criterion*.

Exercise 10.14. For a simplicial arrangement, deduce Lemma 10.11 as a special case of Lemma 10.13 by setting $K = D$.

10.1.7. Cartesian product. For any arrangements $\mathcal{A}$ and $\mathcal{A}'$, the isomorphism (9.79) restricts to the space of Lie elements:

$$\mathsf{Lie}[\mathcal{A} \times \mathcal{A}'] \xrightarrow{\cong} \mathsf{Lie}[\mathcal{A}] \otimes \mathsf{Lie}[\mathcal{A}']. \tag{10.7}$$

This can be checked using the Friedrichs criterion.

10.2. Lie in small ranks. Antisymmetry and Jacobi identity

We discuss Lie elements of arrangements of small rank. The rank-zero case was treated in Lemma 10.1. Lie elements of rank-one and rank-two arrangements play an important role in Lie theory. The antisymmetry relation appears in rank one and the Jacobi identity in rank two.

10.2.1. Rank one and antisymmetry. Consider the rank-one arrangement in which the ambient space has dimension one, and there is only one hyperplane consisting of the origin.

There is only one non-singleton top-lune consisting of the two chambers. It follows that $\mathsf{Lie}[\mathcal{A}]$ is one-dimensional. The coefficients of the two chambers are a and $-a$. The simplest choices are $a = 1$ and $a = -1$. Either of them spans $\mathsf{Lie}[\mathcal{A}]$, and their sum is zero. This can be shown as follows.

$$\Big(\overset{1}{\bullet} \qquad \overset{\overline{1}}{\bullet}\Big) + \Big(\overset{\overline{1}}{\bullet} \qquad \overset{1}{\bullet}\Big) = 0. \tag{10.8}$$

This is the *antisymmetry relation*. (By convention, $\overline{1}$ denotes -1.)

10.2.2. Rank two and Jacobi identity. Now consider the rank-two arrangement of 3 lines.

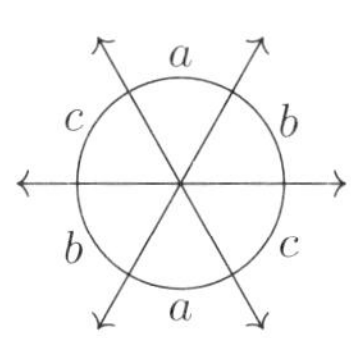

There are six chambers. A non-singleton top-lune is either one of the six half-spaces or the full ambient space. It follows that $\mathsf{Lie}[\mathcal{A}]$ is two-dimensional. The coefficients of the chambers (read in clockwise cyclic order) are a, b, c, a, b and c subject to the condition $a + b + c = 0$. For example, one may take $a = 1$, $b = -1$, and $c = 0$.

Other similar choices are $a = 0$, $b = 1$, and $c = -1$, or $a = -1$, $b = 0$, and $c = 1$. Any two of these yield a basis for $\mathsf{Lie}[\mathcal{A}]$, and the sum of all three is 0. This can be shown as follows.

(10.9) 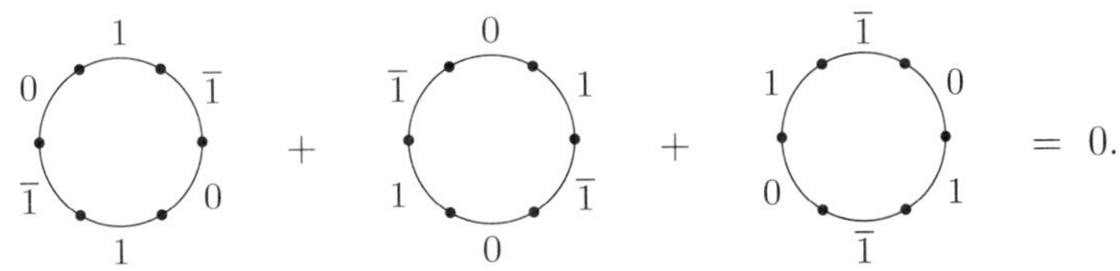

This is the *Jacobi identity* for the hexagon. (By convention, $\overline{1}$ denotes -1.)

The above analysis readily generalizes to the rank-two arrangement of n lines: The hexagon gets replaced by a $2n$-gon, and $\mathsf{Lie}[\mathcal{A}]$ is $(n-1)$-dimensional. The coefficients of the chambers (read in clockwise cyclic order) are $a_1, \ldots, a_n, a_1, \ldots, a_n$ subject to the condition $a_1 + \cdots + a_n = 0$. Jacobi identity consists of n terms adding up to 0. Each term is a $2n$-gon whose two adjacent sides (and their opposites) have coefficients 1 and $\overline{1}$, and the remaining sides have coefficient 0. For instance:

(10.10) 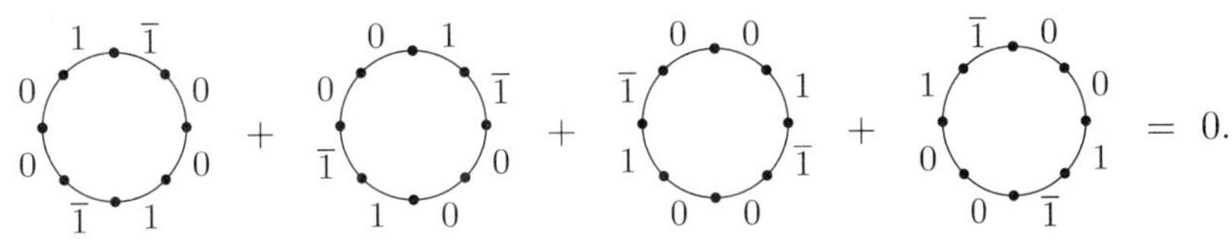

This is the *Jacobi identity* for the octagon.

Exercise 10.15. Verify directly the Garsia criterion (Lemma 10.13) for the rank-two arrangement of n lines. The outline is as follows. For $K = O$, (10.6) is a tautology; for $K = D$, it says that the coefficient of D equals that of $\overline{D}$; for K a vertex of D, it says that $a_1 = -(a_2 + \cdots + a_n)$, where $a_1, \ldots, a_n$ are the chamber coefficients in cyclic order starting at D.

10.3. Zie elements

We introduce Zie elements, and then discuss various characterizations for it named after Friedrichs, Ree and Garsia. This parallels the discussion of Lie elements given in Section 10.1. We are essentially replacing chambers with faces in the discussion. This has the effect of replacing top-lunes with lunes (and having to deal with their interiors and closures), and replacing descents between chambers with descents between faces.

A Zie element is special if the central face appears in it with coefficient 1. Special Zie elements are precisely those idempotents of the Tits algebra whose support is the primitive idempotent of the Birkhoff algebra corresponding to the minimum flat.

This section can be read in conjunction with Section 10.5 where Zie elements in small ranks are worked out.

10.3.1. Zie elements. Consider the Tits algebra $\Sigma[\mathcal{A}]$. Write a typical element as

$$z = \sum_F x^F \mathtt{H}_F.$$

An element $z \in \Sigma[\mathcal{A}]$ is a *Zie element* if

$$\sum_{F\colon HF=G} x^F = 0 \text{ for all } O < H \leq G. \tag{10.11a}$$

This is a linear system in the variables x^F.

Equivalently, $z \in \Sigma[\mathcal{A}]$ is a Zie element if

$$\sum_{F\colon HF\leq G} x^F = 0 \text{ for all } O < H \leq G. \tag{10.11b}$$

To see this: Note that the equation (10.11b) for $O < H \leq G$ is the sum of the equations (10.11a) for $O < H \leq G'$ with G' running between H and G. Hence the claim follows from a triangularity argument on the poset of faces.

We denote the set of Zie elements by $\mathsf{Zie}[\mathcal{A}]$. It is a subspace of $\Sigma[\mathcal{A}]$. As for Lie elements, note that $H = O$ is excluded from the defining equations. Also, cisomorphic arrangements have the "same" Zie elements.

Lemma 10.16. *If $\mathcal{A}$ has rank zero, then* $\mathsf{Zie}[\mathcal{A}] = \Sigma[\mathcal{A}] = \Bbbk$.

PROOF. Suppose $\mathcal{A}$ has rank zero. Then, it has only one face, namely, the central face, so (10.11a) is vacuously true. Hence $\mathsf{Zie}[\mathcal{A}] = \Sigma[\mathcal{A}]$, spanned by $\mathtt{H}_O$. □

Lemma 10.17. *Suppose z is a Zie element. Then*

$$\sum_{F\colon \mathrm{s}(F)\leq \mathrm{X}} x^F = 0 \textit{ for all non-minimum flats } \mathrm{X}. \tag{10.12}$$

In particular, if $\mathcal{A}$ has rank at least one, then

$$\sum_{F} x^F = 0. \tag{10.13}$$

The sum is over all faces F.

PROOF. Consider the special case of (10.11a) in which $O < H = G$. Let $\mathrm{X} := \mathrm{s}(H) = \mathrm{s}(G)$. Recalling from (1.12) that $GF = G$ iff $\mathrm{s}(F) \leq \mathrm{X}$, we obtain (10.12). Letting X be the maximum flat yields (10.13). □

10.3.2. Special Zie elements. A Zie element z is *special* if the coefficient in z of the central face is 1, that is, if $x^O = 1$. Such elements do exist; examples are given in Chapters 11 and 14. Special Zie elements form an affine space of dimension one less than the dimension of $\mathsf{Zie}[\mathcal{A}]$.

Lemma 10.18. *For $z \in \Sigma[\mathcal{A}]$, the following conditions are equivalent.*

$$x^O = 1 \quad \textit{and} \quad \sum_{F\colon \mathrm{s}(F)\leq \mathrm{X}} x^F = 0 \textit{ for all non-minimum flats } \mathrm{X}. \tag{10.14}$$

$$\sum_{F\colon \mathrm{s}(F)= \mathrm{X}} x^F = \mu(\bot, \mathrm{X}) \textit{ for all flats } \mathrm{X}. \tag{10.15}$$

$$\text{(10.16)} \qquad \mathrm{s}(z) = \mathtt{Q}_\perp, \text{ the } \mathtt{Q}\text{-basis element defined in (9.1).}$$

When z is a special Zie element, all conditions hold.

PROOF. For the equivalence between the first two conditions: Denote the lhs of (10.15) by $f(\mathrm{X})$. In (10.14), the condition $x^O = 1$ is the same as $f(\perp) = 1$, while the equations say: for any $\mathrm{Y} > \perp$,

$$\sum_{\mathrm{X}:\, \mathrm{X} \leq \mathrm{Y}} f(\mathrm{X}) = 0.$$

By (C.5a) and (C.5b), this linear system has a unique solution, namely, $f(\mathrm{X}) = \mu(\perp, \mathrm{X})$ for all X.

For the equivalence between the last two conditions: Note that

$$\mathrm{s}(z) = \sum_F x^F \mathtt{H}_{\mathrm{s}(F)} = \Big(\sum_{F:\, \mathrm{s}(F) = \mathrm{X}} x^F \Big) \mathtt{H}_{\mathrm{X}}.$$

By (9.1), this equals $\mathtt{Q}_\perp$ iff the term in parenthesis is $\mu(\perp, \mathrm{X})$.

By Lemma 10.17, a special Zie element satisfies condition (10.14), and hence the other two conditions as well. ☐

10.3.3. Friedrichs primitive part criterion. The space of Zie elements is the primitive part of the Tits algebra (as a left module over itself). This is the *Friedrichs criterion*. It is elaborated below.

Lemma 10.19. *We have*

$$\mathcal{P}(\Sigma[\mathcal{A}]) = \mathsf{Zie}[\mathcal{A}].$$

Explicitly,

$$z \in \mathsf{Zie}[\mathcal{A}] \iff \mathtt{H}_H \cdot z = 0 \text{ for all } H > O.$$

PROOF. Let H be any face of $\mathcal{A}$. Then

$$\mathtt{H}_H \cdot \Big(\sum_F x^F \mathtt{H}_F \Big) = \sum_F x^F \mathtt{H}_{HF} = \sum_{G:\, H \leq G} \Big(\sum_{F:\, HF = G} x^F \Big) \mathtt{H}_G.$$

This equals 0 iff

$$\sum_{F:\, HF=G} x^F = 0 \text{ for all } G \geq H.$$

The result follows from (10.11a). ☐

We now discuss some consequences of the Friedrichs criterion.

Lemma 10.20. *Every Lie element is a Zie element. Conversely, any Zie element which is a linear combination of chambers is a Lie element. In other words,*

$$\mathsf{Lie}[\mathcal{A}] = \mathsf{Zie}[\mathcal{A}] \cap \Gamma[\mathcal{A}].$$

Lemma 10.21. *The subspace $\mathsf{Zie}[\mathcal{A}]$ is a right ideal of $\Sigma[\mathcal{A}]$. More precisely: If z is a special Zie element, then $\mathsf{Zie}[\mathcal{A}]$ is the right ideal of $\Sigma[\mathcal{A}]$ generated by z.*

PROOF. Let z be a special Zie element. For any element w of the Tits algebra, $z \cdot w$ is a Zie element since by Lemma 10.19,

$$\mathtt{H}_F \cdot (z \cdot w) = (\mathtt{H}_F \cdot z) \cdot w = 0$$

whenever $F > O$. Thus the right ideal generated by z is contained in $\mathsf{Zie}[\mathcal{A}]$. Equality holds since for any Zie element z',

$$z \cdot z' = \big(\sum_F x^F \mathtt{H}_F\big) \cdot z' = \sum_F x^F \mathtt{H}_F \cdot z' = x^O z' = z'$$

again using Lemma 10.19. □

Lemma 10.22. *Any Zie element is a quasi-idempotent. More precisely, any Zie element z satisfies $z^2 = x^O z$. A nonzero Zie element is an idempotent iff it is special.*

PROOF. Let z be a Zie element. By Lemma 10.19,

$$z \cdot z = \big(\sum_F x^F \mathtt{H}_F\big) \cdot z = \sum_F x^F (\mathtt{H}_F \cdot z) = x^O z.$$

This proves the first claim. Note that z is an idempotent iff $x^O z = z$. Assuming z to be nonzero, this happens precisely when $x^O = 1$, that is, when z is special. □

Lemma 10.23. *Conjugation of a special Zie element by an invertible element of the Tits algebra produces another special Zie element.*

PROOF. Let u be an invertible element and z be a special Zie element. We want to show that $u \cdot z \cdot u^{-1}$ is also a special Zie element. We may assume that the coefficient of $\mathtt{H}_O$ in u is 1. By Lemma 10.19, $u \cdot z = z$. By Lemma 10.21, $z \cdot u^{-1}$ is a Zie element, and it is special because the coefficient of $\mathtt{H}_O$ in u^{-1} is also 1. □

Lemma 10.24. *Let z be an element of the Tits algebra. Then z is a special Zie element iff z is an idempotent and $\mathrm{s}(z) = \mathtt{Q}_\perp$.*

PROOF. The forward implication follows from Lemmas 10.18 and 10.22. For the backward implication: Pick a special Zie element, say z'. (We use here that such an element exists.) In particular, z' is an idempotent. Now $\mathrm{s}(z) = \mathrm{s}(z') = \mathtt{Q}_\perp$. Hence, by Theorem D.33, z and z' are conjugate by an element of $\mathtt{H}_O + \mathrm{rad}(\Sigma)$, where $\mathrm{rad}(\Sigma)$ is the radical of the Tits algebra. So, by Lemma 10.23, z is also a special Zie element. □

Exercise 10.25. Show that: If z is a Zie element, then $\mathrm{s}(z) = x^O \mathtt{Q}_\perp$. Deduce that a Zie element z belongs to the radical of the Tits algebra iff $x^O = 0$.

10.3.4. Ree lune criterion. Recall from Section 3.2.2 that any nested face (H, G) gives rise to a lune

$$\mathrm{s}(H, G) = \{F \mid HF = G \text{ and } \mathrm{s}(F) = \mathrm{s}(G)\}.$$

Note that G always belongs to this lune. Further, this lune is a singleton (consisting of G) iff $H = O$. All lunes arise in this manner from nested faces. The closure, interior and boundary of $\mathrm{s}(H, G)$ are given by

$$\{F \mid HF \leq G\}, \quad \{F \mid HF = G\} \quad \text{and} \quad \{F \mid HF < G\}$$

respectively. This lune $\mathrm{s}(H, G)$ is a flat precisely when $H = G$, in which case its closure equals its interior. The definition of a Zie element may now be rewritten as follows.

Lemma 10.26. *We have* $z \in \mathsf{Zie}[\mathcal{A}]$ *iff*

$$\sum_{F \in \mathrm{L}^o} x^F = 0 \text{ for all non-singleton combinatorial lunes } \mathrm{L} \tag{10.17a}$$

iff

$$\sum_{F \in \mathrm{Cl}(\mathrm{L})} x^F = 0 \text{ for all non-singleton combinatorial lunes } \mathrm{L}, \tag{10.17b}$$

where L^o *and* $\mathrm{Cl}(\mathrm{L})$ *denote the interior and closure of* L*, respectively.*

When L runs over non-minimum flats, both statements specialize to (10.12).

Lemma 10.27. *We have* $z \in \mathsf{Zie}[\mathcal{A}]$ *iff* (10.17a) *holds for all vertex-based combinatorial lunes iff* (10.17b) *holds for all vertex-based combinatorial lunes.*

PROOF. By Corollary 3.26, the interior of any non-singleton combinatorial lune can be written as a disjoint union of the interiors of vertex-based combinatorial lunes. This proves the first equivalence. The second equivalence then follows from the usual triangularity argument. □

This is the *Ree criterion*. A Zie element may be visualized as a scalar assigned to each face such that the sum of the scalars in the interior (closure) of every vertex-based lune is 0. (The scalar assigned to F is x^F.)

10.3.5. Orthogonality with decomposable part. Let $\Sigma[\mathcal{A}]^*$ denote the vector space dual to $\Sigma[\mathcal{A}]$. We use the letter $\mathtt{M}$ to denote the basis of $\Sigma[\mathcal{A}]^*$ which is dual to the $\mathtt{H}$-basis of $\Sigma[\mathcal{A}]$. We view $\Sigma[\mathcal{A}]$ as a left module over itself. Since $\Sigma[\mathcal{A}]$ acts on itself on the left, it acts on $\Sigma[\mathcal{A}]^*$ on the right. Using the definition

$$\langle \mathtt{M}_G \cdot \mathtt{H}_H, \mathtt{H}_F \rangle = \langle \mathtt{M}_G, \mathtt{H}_H \cdot \mathtt{H}_F \rangle,$$

we see that the right action is given by

$$\mathtt{M}_G \cdot \mathtt{H}_H = \sum_{F:\, HF=G} \mathtt{M}_F. \tag{10.18}$$

Note that the rhs is zero if $H \not\leq G$. This formula is an instance of (9.66).

Let $\mathsf{Zie}[\mathcal{A}]^{\perp}$ denote the decomposable part of $\Sigma[\mathcal{A}]^*$. By definition (9.68), it is the span of the elements (10.18) as (H, G) varies over nested faces with $H > O$. Equivalently, it is spanned by the elements

$$f_{\mathrm{L}} = \sum_{F \in \mathrm{Cl}(\mathrm{L})} \mathtt{M}_F$$

as L varies over non-singleton lunes. (We can also take the interior L^o instead of the closure $\mathrm{Cl}(\mathrm{L})$.) Lemma 10.26 says that z is a Zie element iff $\langle f_{\mathrm{L}}, z \rangle = 0$ for every non-singleton lune L. Equivalently:

Lemma 10.28. *The spaces* $\mathsf{Zie}[\mathcal{A}]$ *and* $\mathsf{Zie}[\mathcal{A}]^{\perp}$ *are orthogonal to each other under the canonical pairing between* $\Sigma[\mathcal{A}]$ *and* $\Sigma[\mathcal{A}]^*$.

Since $\mathsf{Zie}[\mathcal{A}]$ is the primitive part of $\Sigma[\mathcal{A}]$, what we are then seeing above is an illustration of the duality between primitive and decomposable parts (Proposition 9.61).

Exercise 10.29. Show that $z \in \mathsf{Zie}[\mathcal{A}]$ iff the sum of the coefficients of z of faces in the interior of any non-singleton cone is zero.

Exercise 10.30. Use the Ree criterion to deduce Lemma 10.20.

10.3.6. Opposition map. The opposition map preserves the space of Zie elements. This can be deduced from either the Friedrichs criterion or the Ree criterion. We say a Zie element is *projective* if it belongs to the projective Tits algebra, that is, if it is fixed by the opposition map. A related result is given below.

Lemma 10.31. *If $z \in \mathsf{Zie}[\mathcal{A}]$, then for any face G,*

$$\sum_{F:\, F \leq G} x^F = (-1)^{\mathrm{rk}(G)} x^{\overline{G}}. \tag{10.19}$$

PROOF. Suppose z is a Zie element. Now apply the Witt identity (7.19a) to the scalars x^F. By definition of a Zie element (10.11b), the sum inside the parenthesis in (7.19a) is zero for $H > O$. So the lhs of (7.19a) reduces to $\sum_{F:\, F \leq G} x^F$ (coming from the term $H = O$), and we get (10.19).

Alternatively, one can use (7.19b) and (10.11a). □

10.3.7. Garsia descent criterion. Consider the lune $\mathrm{s}(H, G)$ in a rank-three simplicial arrangement shown below. (Only those three hyperplanes in the arrangement which are relevant to the discussion are shown.)

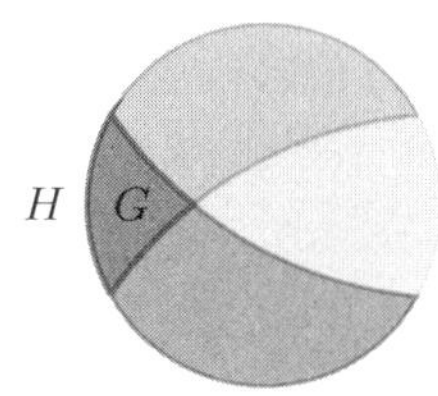

The closure of the lune $\mathrm{s}(H, G)$ consists of faces F such that $HF \leq G$. This set can be split into four parts (shown in dark, dark-medium, light-medium and light shades) depending on the value of $\mathrm{Des}(F, G)$ (which is necessarily a face of H), see (7.3a). When this is the central face, we get faces of G shown in dark shade, when this is a vertex of H, we get faces either in the dark-medium or light-medium shaded regions, and when this is H, we get faces in the light shaded region. Note very carefully the shades on the boundaries of these regions; compare and contrast with the figure in Section 10.1.6.

Suppose we are given a Zie element $z \in \mathsf{Zie}[\mathcal{A}]$. By applying the Ree criterion, arguing as in the Lie case, we deduce that the sum of the coefficients of z of faces in the dark shaded region (that is, faces of G) is equal to the sum in the light shaded region. This leads to the *Garsia criterion*:

Lemma 10.32. *Suppose $\mathcal{A}$ is a simplicial arrangement. Then $z \in \mathsf{Zie}[\mathcal{A}]$ iff for any nested face (K, G),*

$$\sum_{F:\, F \leq G} x^F = (-1)^{\mathrm{rk}(K)} \sum_{\substack{F:\, GF = G,\\ \mathrm{Des}(F,G) = K}} x^F. \tag{10.20}$$

The lhs is a specialization of the rhs when K is the central face. This follows from (7.5a).

PROOF. Suppose $z \in \mathsf{Zie}[\mathcal{A}]$. Now apply the Witt identity (7.21a) to the scalars x^F. By (10.11b), the sum inside the parenthesis in (7.21a) is zero for $H > O$. So the lhs of (7.21a) reduces to $\sum_{F:\, F \leq G} x^G$ (coming from the term $H = O$). This proves (10.20).

Conversely, suppose (10.20) holds for any pair of faces $K \leq G$. Let $H \leq G$ and $H > O$. Then, by (7.4a),

$$\sum_{F:\, HF \leq G} x^F = \sum_{\substack{F:\, GF=G,\\ \mathrm{Des}(F,G) \leq H}} x^F = \sum_{K:\, K \leq H} \sum_{\substack{F:\, GF=G,\\ \mathrm{Des}(F,G)=K}} x^F$$

$$= \sum_{K:\, K \leq H} (-1)^{\mathrm{rk}(K)} \Big(\sum_{F:\, F \leq G} x^F \Big) = 0.$$

(The second equality is the decomposition illustrated in the preceding figure.) The term inside the parenthesis does not depend on K. So the sum over K is 0 by (1.44). This verifies (10.11b). Hence $z \in \mathsf{Zie}[\mathcal{A}]$. □

There is another way to formulate the Garsia criterion by decomposing the interior of a lune rather than its closure. It is illustrated below.

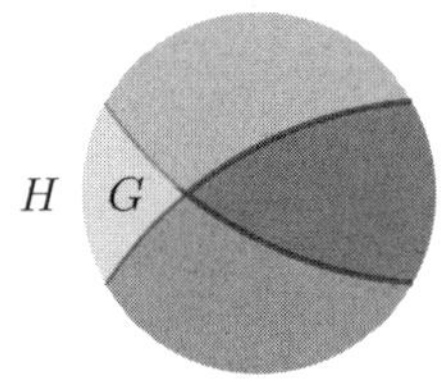

The interior of the lune $\mathrm{s}(H,G)$ consists of faces F such that $HF = G$. This set can be split into four parts (shown as the four shaded regions) depending on the value of $\overline{\mathrm{Des}}(F,G)$ (which is necessarily a face of H), see (7.3b). Note very carefully how the shades on the boundaries have changed from before. Now if $z \in \mathsf{Zie}[\mathcal{A}]$, then the coefficient of the face G in z equals the sum of the coefficients of z of faces in the dark shaded region. (The light shaded region contains only one face, namely, G.)

Lemma 10.33. *Suppose $\mathcal{A}$ is a simplicial arrangement. Then $z \in \mathsf{Zie}[\mathcal{A}]$ iff for any nested face (K,G),*

$$x^G = (-1)^{\mathrm{rk}(K)} \sum_{\substack{F:\, GF=G,\\ \overline{\mathrm{Des}}(F,G)=K}} x^F. \tag{10.21}$$

The lhs is a specialization of the rhs when K is the central face. This follows from (7.5b).

The proof is similar to the previous case, the relevant identity is (7.21b).

Exercise 10.34. For a simplicial arrangement, deduce Lemma 10.31 as a special case of both Lemma 10.32 and Lemma 10.33 by setting $K = G$.

10.3.8. Cartesian product. Induced element under a flat. For any arrangements $\mathcal{A}$ and $\mathcal{A}'$, the isomorphism (9.78) restricts to the space of Zie elements:

$$\mathsf{Zie}[\mathcal{A} \times \mathcal{A}'] \xrightarrow{\cong} \mathsf{Zie}[\mathcal{A}] \otimes \mathsf{Zie}[\mathcal{A}']. \tag{10.22}$$

This can be checked using the Friedrichs criterion.

Similarly, the map (9.77) restricts to the space of Zie elements, that is, the truncation of a Zie element of $\mathcal{A}$ to faces under a flat X yields a Zie element of $\mathcal{A}^{\mathrm{X}}$.

10.4. Zie elements and primitive part of modules

Zie elements can be used to study the primitive part of any left module over the Tits algebra. More precisely, a special Zie element projects a left module onto its primitive part. This allows us to derive formulas for the dimensions of the space of Lie and Zie elements.

10.4.1. Zie elements and primitive part of modules. Let z be an element of the Tits algebra Σ and h a left Σ-module. Recall that $\Psi_{\mathsf{h}}(z)$ denotes the linear operator on h given by left multiplication by z.

Proposition 10.35. *If z is a Zie element, then the image of $\Psi_{\mathsf{h}}(z)$ is contained in $\mathcal{P}(\mathsf{h})$. Moreover, $\Psi_{\mathsf{h}}(z)$ acts on $\mathcal{P}(\mathsf{h})$ by scalar multiplication by the coefficient of the central face in z. If z is a special Zie element, then $\Psi_{\mathsf{h}}(z)$ projects h onto $\mathcal{P}(\mathsf{h})$.*

Conversely: If $\Psi_{\mathsf{h}}(z)$ maps h to $\mathcal{P}(\mathsf{h})$ for every left module h, then z is a Zie element.

Proof. Let $z = \sum_F x^F \mathtt{H}_F$. Let $h \in \mathsf{h}$. By Lemma 10.19,

$$\mathtt{H}_H \cdot (z \cdot h) = (\mathtt{H}_H \cdot z) \cdot h = 0$$

for all $H > O$. Thus $z \cdot h \in \mathcal{P}(\mathsf{h})$ as required. If h itself is primitive, then

$$z \cdot h = \sum_F x^F \mathtt{H}_F \cdot h = x^O h.$$

For the converse, choose $\mathsf{h} := \Sigma$. Then $z \cdot \mathtt{H}_F$ is a Zie element for any face F. Hence $z = z \cdot \mathtt{H}_O$ is a Zie element. □

Remark 10.36. For each $z \in \Sigma$, the linear map $\Psi_{\mathsf{h}}(z)$ is natural in h. Let $\mathcal{F}$ denote the forgetful functor from the category of left Σ-modules to the category of vector spaces. One may recover the Tits algebra as the algebra of endomorphisms (natural transformations) of the functor $\mathcal{F}$: the map

$$\Sigma \to \mathrm{End}(\mathcal{F}), \quad z \mapsto \Psi(z)$$

is an isomorphism. (This is a general fact for the category of modules over an algebra.) Proposition 10.35 shows that one may similarly recover the space of Zie elements as

$$\mathsf{Zie} \cong \mathrm{Hom}(\mathcal{F}, \mathcal{P}).$$

10.4.2. Dimensions of Lie and Zie.

Proposition 10.37. *For any left Σ-module h,*

$$\dim(\mathcal{P}(\mathsf{h})) = \eta_\perp(\mathsf{h}) = \sum_{\mathrm{Y}} \mu(\perp, \mathrm{Y})\, \xi_{\mathrm{Y}}(\mathsf{h}), \tag{10.23}$$

with $\xi_{\mathrm{X}}(\mathsf{h})$ and $\eta_{\mathrm{X}}(\mathsf{h})$ as in (9.51).

Proof. We use that special Zie elements z do exist. By Proposition 10.35, $\mathcal{P}(\mathsf{h}) = z \cdot \mathsf{h}$. By Lemma 10.24, z is an idempotent which lifts $\mathtt{Q}_\perp$. Now apply (9.55). □

Let us apply (10.23) to $\mathsf{h} = \Gamma$ and $\mathsf{h} = \Sigma$. Combining the Friedrichs criterion (Lemmas 10.5 and 10.19) with (9.56) and (9.57), we obtain formulas for the dimensions of the space of Lie and Zie elements:

Theorem 10.38. *For any arrangement $\mathcal{A}$,*

$$\dim(\mathsf{Lie}[\mathcal{A}]) = \sum_{\mathrm{X}} \mu(\perp, \mathrm{X})\, c_{\mathrm{X}} = |\mu(\mathcal{A})| \tag{10.24}$$

where c_{X} is the number of chambers in $\mathcal{A}_{\mathrm{X}}$, and

$$\dim(\mathsf{Zie}[\mathcal{A}]) = \sum_{\mathrm{X}} \mu(\perp, \mathrm{X})\, d_{\mathrm{X}} = \sum_{\mathrm{X}} |\mu(\mathcal{A}^{\mathrm{X}})|, \tag{10.25}$$

where d_{X} is the number of faces in $\mathcal{A}_{\mathrm{X}}$. In each sum, X varies over all flats in the arrangement.

10.5. Zie in small ranks

Let us try to understand Zie elements of arrangements of rank one and two. The rank-zero case was treated in Lemma 10.16.

10.5.1. Rank one. Let $\mathcal{A}$ be the rank-one arrangement consisting of the central face, and chambers C and $\overline{C}$. Then, the ambient space is the only non-singleton lune of $\mathcal{A}$. Hence,

$$x^O\, \mathtt{H}_O + x^C\, \mathtt{H}_C + x^{\overline{C}}\, \mathtt{H}_{\overline{C}} \in \mathsf{Zie}[\mathcal{A}] \iff x^O + x^C + x^{\overline{C}} = 0. \tag{10.26}$$

Thus, $\mathsf{Zie}[\mathcal{A}]$ is two-dimensional. This can be double-checked from (10.25):

$$\dim(\mathsf{Zie}[\mathcal{A}]) = 1 \cdot 3 - 1 \cdot 1 = 2.$$

Observe that any special Zie element is of the form

$$\mathtt{H}_O - p\,\mathtt{H}_C - (1-p)\,\mathtt{H}_{\overline{C}},$$

where p is an arbitrary scalar. Let us compute the action of this element on $\Gamma[\mathcal{A}]$. For instance,

$$\begin{aligned}(\mathtt{H}_O - p\,\mathtt{H}_C - (1-p)\,\mathtt{H}_{\overline{C}}) \cdot \mathtt{H}_C &= \mathtt{H}_C - p\,\mathtt{H}_C - (1-p)\,\mathtt{H}_{\overline{C}} \\ &= (1-p)\,\mathtt{H}_C - (1-p)\,\mathtt{H}_{\overline{C}},\end{aligned}$$

which is a Lie element. Further,

$$(\mathtt{H}_O - p\,\mathtt{H}_C - (1-p)\,\mathtt{H}_{\overline{C}}) \cdot (\mathtt{H}_C - \mathtt{H}_{\overline{C}}) = \mathtt{H}_C - \mathtt{H}_{\overline{C}}.$$

So its action on a Lie element gives back the same Lie element. This is consistent with Proposition 10.35.

Compare the above calculations with those in Section 9.4.5. In this case, $\mathsf{Lie}[\mathcal{A}] = \mathrm{rad}(\Gamma[\mathcal{A}])$.

Exercise 10.39. Verify Lemma 10.24 for the rank-one arrangement. (Use the description of idempotents given in Exercise 9.26.)

10.5.2. Rank two. Let $\mathcal{A}$ be the rank-two arrangement of n lines. Then, applying (10.25),

$$\dim(\mathsf{Zie}[\mathcal{A}]) = (4n+1) - 3n + (n-1) = 2n.$$

Let us look at the case $n = 3$. A Zie element is shown in the diagram below.

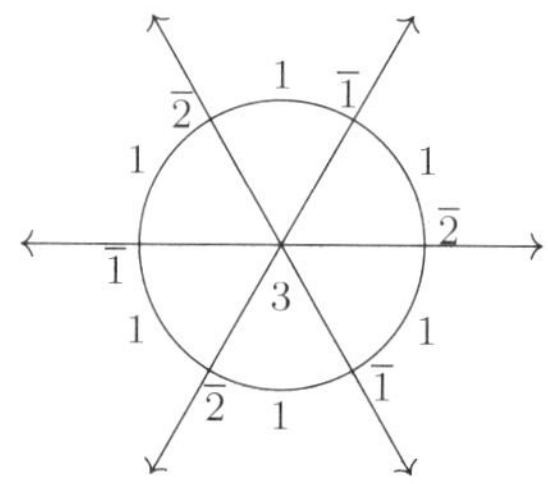

This is a quasi-idempotent element with $x^O = 3$.

The Ree criterion says that the sum of the scalars in the closure (or interior) of every non-singleton lune is zero. The three lines, the six half-spaces, and the ambient space are the closures of non-singleton lunes. Among them, the lines and half-spaces are vertex-based lunes. For a line, $-2+3-1=0$; for a half-space, $(-2+3-1)+(1-1+1-2+1)=0$; and for the ambient space, $(-2+3-1)+(1-1+1-2+1)+(1-1+1-2+1)=0$. The ambient space is not a vertex-based lune, so, as is evident, the last calculation is a consequence of the previous two. The origin, the 6 rays, and the six sectors are the closures of singleton lunes, thus there is no condition on the sum of their coefficients.

Now let us check the Garsia criterion. Take G to be any one of the six edges. Then

$$\sum_{F:\, F \leq G} x^G = 3 - 2 - 1 + 1 = 1 = x^{\overline{G}}.$$

This verifies (10.19). Now take K to be a vertex of G. Then the lhs of (10.20) is as above, while the rhs is

$$(-1)(1+1-2-1) = 1$$

as required.

Exercise 10.40. Check that the action of the above Zie element on any chamber produces a Lie element, while the action on a Lie element produces the same Lie element multiplied by 3.

Exercise 10.41. For the arrangement $\mathcal{A}$ of n lines, check directly by solving the linear system (10.11a) that $\dim(\mathsf{Zie}[\mathcal{A}]) = 2n$.

10.6. Substitution product of Lie

We introduce the substitution product of Lie. It is a procedure to obtain Lie elements of an arrangement by "multiplying" Lie elements of the arrangements under and over its flats. This can be viewed as a restriction of the substitution product of chambers (Section 4.8.1). The connection to the classical Lie operad is given in Section 14.9.6.

Recall that every Lie element is a Zie element. The substitution product of Lie can be generalized as follows. One can multiply a Lie element of the arrangement under a flat with a Zie element of the arrangement over the same flat to obtain a Zie element of the arrangement. This product can be viewed as a restriction of the substitution product of chambers and faces (Section 4.8.3).

Similarly, one can multiply a chamber element of the arrangement under a flat with a Lie element of the arrangement over the same flat to obtain a chamber element of the arrangement. This product can be viewed as a restriction of the substitution product of top-lunes and chambers (Section 4.8.4).

10.6.1. Substitution product of Lie. Recall the substitution product of chambers (4.18). By linearizing, we obtain a map

$$\Gamma[\mathcal{A}^{\mathrm{X}}] \otimes \Gamma[\mathcal{A}_{\mathrm{X}}] \to \Gamma[\mathcal{A}] \tag{10.27}$$

for any flat X. We now show that this map restricts to the space of Lie elements.

Proposition 10.42. *For any flat* X, *there is a unique linear map*

$$\mathsf{Lie}[\mathcal{A}^{\mathrm{X}}] \otimes \mathsf{Lie}[\mathcal{A}_{\mathrm{X}}] \to \mathsf{Lie}[\mathcal{A}] \tag{10.28}$$

such that the diagram

$$\begin{array}{ccc} \Gamma[\mathcal{A}^{\mathrm{X}}] \otimes \Gamma[\mathcal{A}_{\mathrm{X}}] & \longrightarrow & \Gamma[\mathcal{A}] \\ \uparrow & & \uparrow \\ \mathsf{Lie}[\mathcal{A}^{\mathrm{X}}] \otimes \mathsf{Lie}[\mathcal{A}_{\mathrm{X}}] & \longrightarrow & \mathsf{Lie}[\mathcal{A}] \end{array} \tag{10.29}$$

commutes.

We call (10.28) the substitution product of Lie.

PROOF. An element of $\mathsf{Lie}[\mathcal{A}^{\mathrm{X}}]$ is a scalar assigned to each face F of support X such that the Ree criterion is satisfied in $\mathcal{A}^{\mathrm{X}}$, while an element of $\mathsf{Lie}[\mathcal{A}_{\mathrm{X}}]$ is a scalar assigned to each chamber greater than F (where F is a fixed face of support X) such that the Ree criterion is satisfied in $\mathcal{A}_F$.

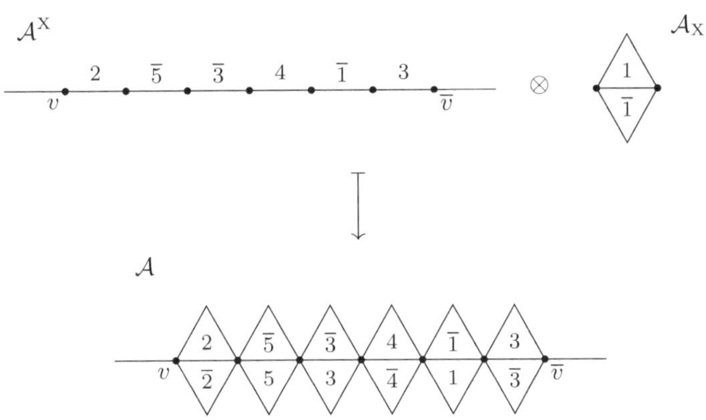

An illustration is provided in the diagram. The arrangement $\mathcal{A}$ has rank 3. The flat X has rank 2, so it is a great circle. The arrangement $\mathcal{A}^{\mathrm{X}}$ is a 12-gon, six of whose sides have been marked out. They comprise a top-lune of $\mathcal{A}^{\mathrm{X}}$; note that the assigned values add up to zero: $2 - 5 - 3 + 4 - 1 + 3 = 0$. (For practical purposes, the full 12-gon is not shown. But the six values shown determine the remaining six values, since opposite sides have the same value by (10.5).) The arrangement $\mathcal{A}_{\mathrm{X}}$ consists of two chambers (the hemispheres), with values say 1 and -1. Substitution (10.27) assigns labels to $24 = 12 * 2$ faces, 12 of which are shown above. For convenience, we assume that they are triangles.

We check that this assignment satisfies the Ree criterion. Accordingly, let L be a non-singleton top-lune of $\mathcal{A}$. If L does not contain any of the 24 triangles, then the Ree criterion clearly holds. So suppose that L contains at least one of the 24 triangles.

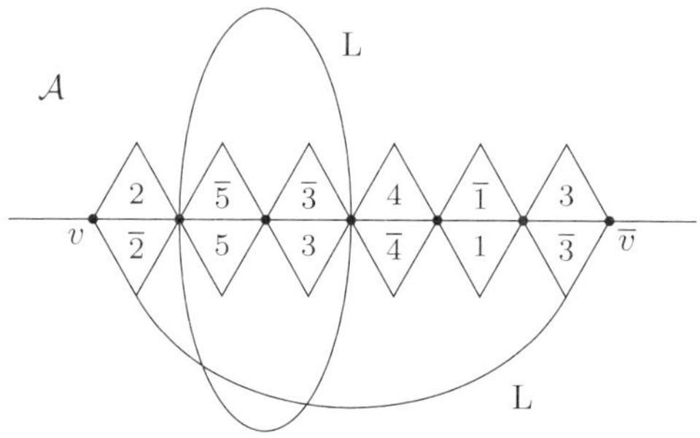

There are two cases.

- $\mathrm{b}(\mathrm{L}) \leq \mathrm{X}$.

  In the picture, L is the semi-circular ball with $\mathrm{b}(\mathrm{L}) = \{v, \overline{v}\}$. Note that 6 of the 24 triangles belong to L, and the sum of their values is zero. The reason is that in this case $\mathrm{L} \wedge \mathrm{X}$ is a non-singleton top-lune of $\mathcal{A}^{\mathrm{X}}$, and the values in $\mathcal{A}^{\mathrm{X}}$ satisfy the Ree criterion.
- $\mathrm{b}(\mathrm{L}) \not\leq \mathrm{X}$.

  In the picture, L is the oval. Note that 4 of the 24 triangles belong to L, and the sum of their values is zero. In fact, the sum can be broken into two parts and each part sums to zero ($5+\overline{5}=0$ and $3+\overline{3}=0$). The reason is that in this case, for any face F with support X which belongs to the closure of L, $\mathrm{L} \wedge \Gamma_F$ is a non-singleton top-lune of $\mathcal{A}_F$, and the values in $\mathcal{A}_F$ satisfy the Ree criterion.

The general case works in exactly the same manner as this illustration. □

Exercise 10.43. Prove Proposition 10.42 using the Friedrichs criterion. The outline is as follows. To check that $\mathtt{H}_H$ acts by zero for $H > O$, split the analysis into two cases depending on whether $\mathrm{s}(H) \leq \mathrm{X}$ or not. In the first case, use that the first tensor factor is a Lie element and the fact that $H > O$. In the second case, use that the second tensor factor is a Lie element and the fact that $FH > F$ for any F with support X.

Example 10.44. Let us understand how substitution works in rank two. Each term in any Jacobi identity arises from the substitution product (10.28) on a rank-one flat. For instance, the middle term in (10.9) arises as follows.

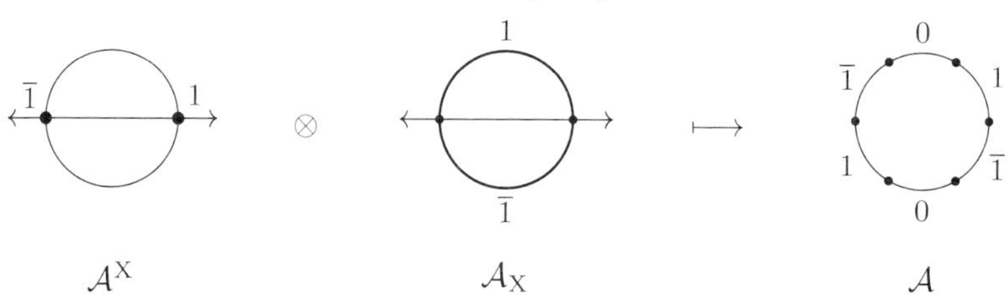

Example 10.45. Let us now go to rank three. The spherical model of the braid arrangement $\mathcal{A}$ on [4] is shown below. We illustrate how the substitution product can be used to generate elements of $\mathsf{Lie}[\mathcal{A}]$.

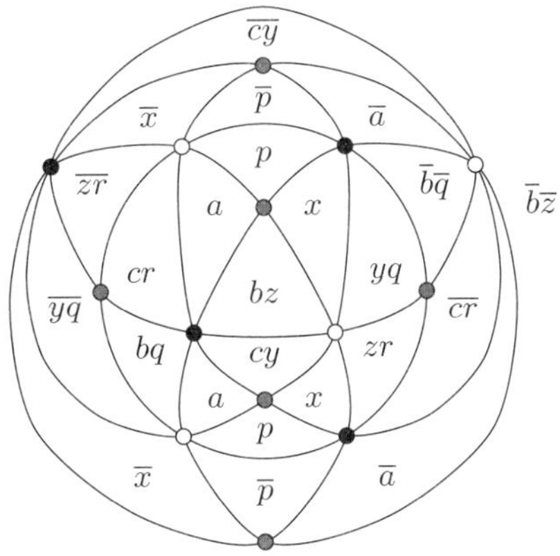

Pick any a, b and c subject to $a+b+c=0$, any x, y and z subject to $x+y+z=0$, and any p, q and r subject to $p+q+r=0$. Using this data, we assign labels to chambers as shown in the figure: bq means $b+q$, $\overline{cr}$ means $-c-r$, and so on. (We use multiplicative notation for typesetting reasons.) Why is this a Lie element? To see this, use (10.28) three times for three different flats and take their sum. In each case $\mathcal{A}^{\mathrm{X}}$ and $\mathcal{A}_{\mathrm{X}}$ are cisomorphic to either the rank-one arrangement or the rank-two arrangement of three lines. One flat is the central circle which is surrounded by p's, q's and r's. The remaining two flats are of rank 1. For that, locate a pair of opposite vertices which are surrounded by a's, b's and c's, and another pair of opposite vertices which are surrounded by x's, y's and z's.

One can produce other Lie elements by substituting at other flats. We will see later that Lie elements which arise in this manner provide a spanning set for all Lie elements.

Exercise 10.46. Use Exercise 1.39 to check that: If $w \in \Gamma[\mathcal{A}^{\mathrm{X}}]$ and $z \in \Gamma[\mathcal{A}_{\mathrm{X}}]$ are both nonzero, then the substitution product (10.27) applied to $w \otimes z$ yields a nonzero element of $\Gamma[\mathcal{A}]$. Deduce that the same is true with Γ replaced by Lie.

Exercise 10.47. Let X and Y be modular complements in the lattice of flats. Each chamber F in $\mathcal{A}^{\mathrm{X}}$ is contained in a unique chamber of $\mathcal{A}_{\mathrm{Y}}$, which we denote by $\mathrm{Y}F$. (Also see Exercise 3.10.) Show that if $\sum_F x^F \mathtt{H}_F$ is a Lie element of $\mathcal{A}^{\mathrm{X}}$, then $\sum_F x^F \mathtt{H}_{\mathrm{Y}F}$ is a Lie element of $\mathcal{A}_{\mathrm{Y}}$. This fact may fail if Y is a complement of X which is not modular.

10.6.2. Iterated substitution product. The substitution product of Lie can be written in the general form

$$\mathsf{Lie}[\mathcal{A}^{\mathrm{Y}}_{\mathrm{X}}] \otimes \mathsf{Lie}[\mathcal{A}^{\mathrm{Z}}_{\mathrm{Y}}] \to \mathsf{Lie}[\mathcal{A}^{\mathrm{Z}}_{\mathrm{X}}] \tag{10.30}$$

for $\mathrm{X} \leq \mathrm{Y} \leq \mathrm{Z}$. Compare with (4.19). Diagrams (4.20a) and (4.20b) yield the following.

The diagram

$$
\begin{array}{ccc}
\mathsf{Lie}[\mathcal{A}_{\mathrm{X}}^{\mathrm{Y}}] \otimes \mathsf{Lie}[\mathcal{A}_{\mathrm{Y}}^{\mathrm{Z}}] \otimes \mathsf{Lie}[\mathcal{A}_{\mathrm{Z}}^{\mathrm{W}}] & \longrightarrow & \mathsf{Lie}[\mathcal{A}_{\mathrm{X}}^{\mathrm{Y}}] \otimes \mathsf{Lie}[\mathcal{A}_{\mathrm{Y}}^{\mathrm{W}}] \\
\downarrow & & \downarrow \\
\mathsf{Lie}[\mathcal{A}_{\mathrm{X}}^{\mathrm{Z}}] \otimes \mathsf{Lie}[\mathcal{A}_{\mathrm{Z}}^{\mathrm{W}}] & \longrightarrow & \mathsf{Lie}[\mathcal{A}_{\mathrm{X}}^{\mathrm{W}}]
\end{array}
\tag{10.31a}
$$

commutes for any $\mathrm{X} \leq \mathrm{Y} \leq \mathrm{Z} \leq \mathrm{W}$.

The maps

$$
\mathsf{Lie}[\mathcal{A}_{\mathrm{X}}^{\mathrm{X}}] \otimes \mathsf{Lie}[\mathcal{A}_{\mathrm{X}}^{\mathrm{Y}}] \to \mathsf{Lie}[\mathcal{A}_{\mathrm{X}}^{\mathrm{Y}}] \quad \text{and} \quad \mathsf{Lie}[\mathcal{A}_{\mathrm{X}}^{\mathrm{Y}}] \otimes \mathsf{Lie}[\mathcal{A}_{\mathrm{Y}}^{\mathrm{Y}}] \to \mathsf{Lie}[\mathcal{A}_{\mathrm{X}}^{\mathrm{Y}}]
\tag{10.31b}
$$

are the canonical identifications.

For any chain of flats $\mathrm{X}_0 < \mathrm{X}_1 < \cdots < \mathrm{X}_{k-1} < \mathrm{X}_k$, there is a linear map

$$
\mathsf{Lie}[\mathcal{A}_{\mathrm{X}_0}^{\mathrm{X}_1}] \otimes \mathsf{Lie}[\mathcal{A}_{\mathrm{X}_1}^{\mathrm{X}_2}] \otimes \cdots \otimes \mathsf{Lie}[\mathcal{A}_{\mathrm{X}_{k-1}}^{\mathrm{X}_k}] \to \mathsf{Lie}[\mathcal{A}_{\mathrm{X}_0}^{\mathrm{X}_k}]
\tag{10.32}
$$

obtained by repeated application of (10.30). It is well-defined in view of (10.31a). We call (10.32) the iterated substitution product of Lie.

Exercise 10.48. Show that: For any arrangement $\mathcal{A}$, the space $\mathsf{Lie}[\mathcal{A}]$ is nonzero. (Recall that $\mathsf{Lie}[\mathcal{A}_{\mathrm{X}}^{\mathrm{Y}}]$ is one-dimensional when $\mathrm{X} \lessdot \mathrm{Y}$. Now use iterated substitution (10.32) and Exercise 10.46. Alternatively, use that dimemnsion of $\mathsf{Lie}[\mathcal{A}]$ is $|\mu(\mathcal{A})|$ which is nonzero by (1.47).)

10.6.3. Substitution product of Lie and Zie. By linearizing the substitution product of chambers and faces (4.21), we obtain a map

$$
\Gamma[\mathcal{A}^{\mathrm{X}}] \otimes \Sigma[\mathcal{A}_{\mathrm{X}}] \to \Sigma[\mathcal{A}]
\tag{10.33}
$$

for any flat X.

Proposition 10.49. *For any flat* X*, there is a unique linear map*

$$
\mathsf{Lie}[\mathcal{A}^{\mathrm{X}}] \otimes \mathsf{Zie}[\mathcal{A}_{\mathrm{X}}] \to \mathsf{Zie}[\mathcal{A}]
\tag{10.34}
$$

such that the diagram

$$
\begin{array}{ccc}
\Gamma[\mathcal{A}^{\mathrm{X}}] \otimes \Sigma[\mathcal{A}_{\mathrm{X}}] & \longrightarrow & \Sigma[\mathcal{A}] \\
\uparrow & & \uparrow \\
\mathsf{Lie}[\mathcal{A}^{\mathrm{X}}] \otimes \mathsf{Zie}[\mathcal{A}_{\mathrm{X}}] & \longrightarrow & \mathsf{Zie}[\mathcal{A}]
\end{array}
\tag{10.35}
$$

commutes.

This result can be proved exactly like Proposition 10.42 using the Ree criterion both for Lie and Zie elements. Alternatively, one may also use the Friedrichs criterion along the lines of Exercise 10.43.

We call (10.34) the substitution product of Lie and Zie. One may check that: For any $\mathrm{X} \leq \mathrm{Y}$, the diagram

$$\begin{array}{ccc} \mathsf{Lie}[\mathcal{A}^{\mathrm{X}}] \otimes \mathsf{Lie}[\mathcal{A}^{\mathrm{Y}}_{\mathrm{X}}] \otimes \mathsf{Zie}[\mathcal{A}_{\mathrm{Y}}] & \longrightarrow & \mathsf{Lie}[\mathcal{A}^{\mathrm{Y}}] \otimes \mathsf{Zie}[\mathcal{A}_{\mathrm{Y}}] \\ \downarrow & & \downarrow \\ \mathsf{Lie}[\mathcal{A}^{\mathrm{X}}] \otimes \mathsf{Zie}[\mathcal{A}_{\mathrm{X}}] & \longrightarrow & \mathsf{Zie}[\mathcal{A}] \end{array} \tag{10.36}$$

commutes. The top horizontal map involves (10.30) while the rest involve (10.34).

10.6.4. Substitution product of chambers and Lie. Let $\widehat{\Lambda}[\mathcal{A}]$ denote the space obtained by linearizing $\widehat{\Lambda}[\mathcal{A}]$ with canonical basis $\mathtt{H}$. Define an injective linear map

$$\Gamma[\mathcal{A}] \to \widehat{\Lambda}[\mathcal{A}], \qquad \sum_C x^C\, \mathtt{H}_C \mapsto \sum_{\mathrm{L}} x^{\mathrm{L}}\, \mathtt{H}_{\mathrm{L}}, \tag{10.37}$$

where $x^{\mathrm{L}} := \sum_{C \in \mathrm{L}} x^C$. In other words, x^{L} is obtained by summing x^C over all chambers C contained in L.

Each chamber is a top-lune. However, note very carefully that (10.37) is *not* the inclusion map of chambers into top-lunes. For example, for the rank-one arrangement with chambers C and $\overline{C}$, the map (10.37) is given by

$$a\, \mathtt{H}_C + b\, \mathtt{H}_{\overline{C}} \mapsto a\, \mathtt{H}_C + b\, \mathtt{H}_{\overline{C}} + (a+b)\, \mathtt{H}_{\top}.$$

By linearizing the substitution product of top-lunes and chambers (4.23), we obtain a map

$$\widehat{\Lambda}[\mathcal{A}^{\mathrm{X}}] \otimes \Gamma[\mathcal{A}_{\mathrm{X}}] \to \widehat{\Lambda}[\mathcal{A}] \tag{10.38}$$

for any flat X.

Proposition 10.50. *For any flat* X, *there is a unique linear map*

$$\Gamma[\mathcal{A}^{\mathrm{X}}] \otimes \mathsf{Lie}[\mathcal{A}_{\mathrm{X}}] \to \Gamma[\mathcal{A}] \tag{10.39}$$

such that the diagram

$$\begin{array}{ccc} \widehat{\Lambda}[\mathcal{A}^{\mathrm{X}}] \otimes \Gamma[\mathcal{A}_{\mathrm{X}}] & \longrightarrow & \widehat{\Lambda}[\mathcal{A}] \\ \uparrow & & \uparrow \\ \Gamma[\mathcal{A}^{\mathrm{X}}] \otimes \mathsf{Lie}[\mathcal{A}_{\mathrm{X}}] & \longrightarrow & \Gamma[\mathcal{A}] \end{array} \tag{10.40}$$

commutes. The vertical maps are induced from (10.37).

We omit the proof. We mention that only the top-lunes $\mathrm{s}(H, D)$ with $\mathrm{X} = \mathrm{s}(G)$ for some $H \leq G \leq D$ play a part in the commutativity of (10.40).

We call (10.39) the substitution product of Γ and Lie. One may check that it is indeed the restriction of the map (10.27). For any $\mathrm{X} \leq \mathrm{Y}$, the diagram

$$\begin{array}{ccc} \Gamma[\mathcal{A}^{\mathrm{X}}] \otimes \mathsf{Lie}[\mathcal{A}^{\mathrm{Y}}_{\mathrm{X}}] \otimes \mathsf{Lie}[\mathcal{A}_{\mathrm{Y}}] & \longrightarrow & \Gamma[\mathcal{A}^{\mathrm{X}}] \otimes \mathsf{Lie}[\mathcal{A}_{\mathrm{X}}] \\ \downarrow & & \downarrow \\ \Gamma[\mathcal{A}^{\mathrm{Y}}] \otimes \mathsf{Lie}[\mathcal{A}_{\mathrm{Y}}] & \longrightarrow & \Gamma[\mathcal{A}] \end{array} \tag{10.41}$$

commutes. The top horizontal map involves (10.30) while the rest involve (10.39).

We now record a significant property of the substitution product of chambers and Lie.

Lemma 10.51. *The map*

$$\bigoplus_{\mathrm{H}} \Gamma[\mathcal{A}^{\mathrm{H}}] \otimes \mathsf{Lie}[\mathcal{A}_{\mathrm{H}}] \twoheadrightarrow \mathrm{rad}(\Gamma[\mathcal{A}])$$

is surjective. The sum is over all hyperplanes H.

PROOF. For any adjacent chambers C and D whose common panel, say F, has support Z,

$$\mathtt{H}_F \otimes \beta_{\mathrm{Z},F}(\mathtt{H}_{C/F} - \mathtt{H}_{D/F}) \mapsto \mathtt{H}_C - \mathtt{H}_D.$$

By Proposition 9.32, the element $\mathtt{H}_C - \mathtt{H}_D$ belongs to the radical of $\Gamma[\mathcal{A}]$, and further by Exercise 9.36, the radical is linearly spanned by such elements. The result follows. □

A companion result for the substitution product of Lie and Zie is given later in Exercise 13.65.

10.6.5. Projective case. Recall from Section 9.4.8 that $\Gamma[\mathcal{A}]$ has a subspace consisting of projective chamber elements. The map (10.27) restricts to these subspaces. Similarly, the spaces $\Sigma[\mathcal{A}]$ and $\widehat{\Lambda}[\mathcal{A}]$ have projective analogues, and the maps (10.33) and (10.38) restrict to those. A simple way to deduce this is as follows.

Linearize the category of lunes (Section 4.4). By Lemma 4.39, the opposition map acts on this category. So we have the linear subcategory which is invariant under this action. Explicitly, its linear space of morphisms has a basis indexed by projective lunes, and they compose as

$$(\mathtt{H}_{\mathrm{L}} + \mathtt{H}_{\overline{\mathrm{L}}}) \circ (\mathtt{H}_{\mathrm{M}} + \mathtt{H}_{\overline{\mathrm{M}}}) = (\mathtt{H}_{\mathrm{L}\circ\mathrm{M}} + \mathtt{H}_{\overline{\mathrm{L}\circ\mathrm{M}}}) + (\mathtt{H}_{\overline{\mathrm{L}}\circ\mathrm{M}} + \mathtt{H}_{\mathrm{L}\circ\overline{\mathrm{M}}}).$$

As explained in Section 4.8, the maps (10.27), (10.33) and (10.38) arise from the composition operation on lunes, so they indeed have projective analogues.

As a consequence, (10.28), (10.34) and (10.39) also restrict to projective Lie, projective Zie and projective chamber elements.

Notes

Classical Lie elements. Lie elements for the braid arrangement (that is, the arrangement of type A) have been extensively studied. Standard references are Reutenauer's book [**360**] and Garsia's paper [**196**]. We discuss these Lie elements in Section 14.9. In modern terms, they are elements of the classical Lie operad. In the older literature, results about the classical Lie operad are often phrased in terms of the free Lie algebra. The term Lie polynomials is used in this setting. Different characterizations of Lie polynomials are given in [**360**, Theorem 1.4]. Similar discussion for Lie series is given in [**360**, Theorem 3.1]. Reutenauer's criteria (iii) in Theorems 1.4 and 3.1 corresponds to our Friedrichs criterion (or more precisely to Exercise 10.6). This originated in work of Friedrichs [**188**, footnote on page 19]. Early papers related to this criterion are those of Cohn [**122**], Magnus [**295**, Theorem I], Lyndon [**293**] and Finkelstein [**183**]. Reutenauer's criteria (iv) in Theorem 3.1 corresponds to our Ree criterion (Lemma 10.8). (The equivalent Lemma 10.10 specialized to type A is stated later in Lemma 14.62.) This originated in work of Ree [**354**, Theorem 2.2]. Lemma 10.13 for type A is due to Garsia [**196**, Theorem 2.1, (i) and (iii)].

In the discussion in Section 10.2.2: Suppose $\Bbbk = \mathbb{C}$. Let w denote the primitive cube-root of unity. Then $a = 1$, $b = w$, and $c = w^2$, and $a = 1$, $b = w^2$, and $c = w$ yields a basis for $\mathsf{Lie}[\mathcal{A}]$. This is a special case of a result of Klyachko [**258**].

Type B Lie elements. Lie elements for the arrangement of type B are treated in Section 14.10. They have appeared in the literature in a slightly different guise in the work of Bergeron, Gottlieb and Wachs. For details, see the notes to Chapter 14.

Lie elements for arrangements. Lie elements for arbitrary arrangements do not seem to have been explicitly considered in the literature. In fact, we mention that the linear system (10.1) can be used to define Lie elements for any LRB. The Friedrichs and Ree criteria continue to hold. Moreover, the substitution product of chambers restricts to Lie elements yielding (10.28). The same proof works in this general context.

CHAPTER 11

Eulerian idempotents

We saw that the Tits algebra is elementary and its split-semisimple quotient is the Birkhoff algebra. The quotient map is the support map. Let us refer to a complete system of primitive orthogonal idempotents of the Tits algebra as an Eulerian family. (In the exposition, an Eulerian family is defined slightly differently and later shown to be equivalent to a complete sytem.) Any Eulerian family lifts the primitive idempotents of the Birkhoff algebra. There are two theoretically significant methods to construct and characterize Eulerian families. The first method starts with a homogeneous section of the support map. This is the Saliola construction. The second method starts with a family of special Zie elements. Each Eulerian family gives rise to a $\mathtt{Q}$-basis of the Tits algebra. This is in contrast to the Birkhoff algebra which has a unique complete system which also serves as the unique $\mathtt{Q}$-basis. These ideas are further developed in Chapter 15 through consideration of the lune-incidence algebra with a summary given in Section 15.5.

The Saliola construction is recursive in nature and involves alternating sums. Hence it is nontrivial to write down closed formulas for the Eulerian idempotents in general. For a good reflection arrangement, we give cancelation-free formulas for the Eulerian idempotents associated to the uniform section.

As an application, we discuss the extension problem for chambers. Any chamber element of $\mathcal{A}$ induces a chamber element of $\mathcal{A}_F$ by Tits projection on the face F. The extension problem is to start with chamber elements of $\mathcal{A}_F$ for each noncentral face $F > O$ which are "mutually compatible", and construct chamber elements of $\mathcal{A}$ whose Tits projections are these given elements. We show that the solution space is a translate of the space of Lie elements.

11.1. Homogeneous sections of the support map

Fix an arrangement $\mathcal{A}$. We define a class of (linear) sections of the support map of $\mathcal{A}$. We call them homogeneous sections.

11.1.1. Homogeneous sections. Recall the Birkhoff algebra $\Pi[\mathcal{A}]$, the Tits algebra $\Sigma[\mathcal{A}]$ and the support map (9.31) relating them. Let

$$\mathtt{u} : \Pi[\mathcal{A}] \to \Sigma[\mathcal{A}]$$

be any section of the support map. (The section is only required to be a linear map, not an algebra map.) For each flat X, let $\mathtt{u}_\mathrm{X} := \mathtt{u}(\mathtt{H}_\mathrm{X})$ denote the value of $\mathtt{u}$ on $\mathtt{H}_\mathrm{X}$. Thus

$$\mathrm{s}(\mathtt{u}_\mathrm{X}) = \mathtt{H}_\mathrm{X}. \tag{11.1}$$

We say that a section $\mathtt{u}$ of the support map is *homogeneous* if each $\mathtt{u}_X$ only involves faces of support X. That is,

$$\mathtt{u}_X = \sum_{F:\, \mathrm{s}(F)=X} \mathtt{u}^F \mathtt{H}_F \tag{11.2}$$

for scalars $\mathtt{u}^F$. Applying the support map and using (11.1), we obtain

$$\sum_{F:\, \mathrm{s}(F)=X} \mathtt{u}^F = 1. \tag{11.3}$$

Note that $\mathtt{u}_\perp = \mathtt{H}_O$. Conversely, a choice of elements $\mathtt{u}_X$ of the form (11.2) with property (11.3) determines a homogeneous section $\mathtt{u}$.

Remark 11.1. For an arbitrary section, $\mathtt{u}_X$ may also contain terms from flats other than X. The condition for $\mathtt{u}$ to be a section is: For any X, the sum of the coefficients of faces in $\mathtt{u}_X$ with support Y is 1 if Y = X and 0 if Y $\neq$ X.

Lemma 11.2. *Let* X *be a flat, and* G *be a face with support* X. *Let* $\mathtt{u}_X$ *be an element of* $\Sigma[\mathcal{A}]$ *of the form* (11.2). *Then*

$$\mathtt{H}_G \cdot \mathtt{u}_X = \mathtt{H}_G \iff \sum_{F:\, \mathrm{s}(F)=X} \mathtt{u}^F = 1 \iff \mathtt{u}_X \cdot \mathtt{u}_X = \mathtt{u}_X \text{ and } \mathtt{u}_X \text{ is nonzero.}$$

PROOF. By (1.13), $\mathrm{s}(F) = \mathrm{s}(G)$ implies $GF = G$. Hence,

$$\mathtt{H}_G \cdot \mathtt{u}_X = \Big(\sum_{F:\, \mathrm{s}(F)=X} \mathtt{u}^F\Big)\, \mathtt{H}_G \quad \text{and} \quad \mathtt{u}_X \cdot \mathtt{u}_X = \Big(\sum_{F:\, \mathrm{s}(F)=X} \mathtt{u}^F\Big)\, \mathtt{u}_X.$$

Both equivalences follow. Note the relevance of requiring $\mathtt{u}_X$ to be nonzero. □

The preceding discussion yields the following.

Lemma 11.3. *The following are equivalent.*

(1) *A homogeneous section* $\mathtt{u}$ *of* $\mathcal{A}$.
(2) *A family of scalars* $(\mathtt{u}^F)$ *indexed by faces* F, *which satisfy* (11.3) *for each flat* X.
(3) *A family of nonzero elements* $\{\mathtt{u}_X\}_{X\in\Pi}$ *indexed by flats of the form* (11.2) *with*

$$\mathtt{u}_X \cdot \mathtt{u}_X = \mathtt{u}_X. \tag{11.4}$$

In particular, $\mathtt{u}_X$ *is a nonzero idempotent of the Tits algebra.*

Lemma 11.4. *The dimension of the affine space of all homogeneous sections is equal to the number of faces minus the number of flats.*

PROOF. Apply Lemma 11.3, item (2). For each flat we get the number of faces with that support minus one. □

Suppose the base field is the real numbers and all scalars $\mathtt{u}^F$ are nonnegative. In this case, by Lemma 11.3, item (2), a homogeneous section constitutes a family of probability distributions: a distribution on the set of faces supported on X, one for each flat X.

11.1.2. Set-theoretic sections. Consider the (set-theoretic) support map (1.2) relating the Birkhoff monoid $\Pi[\mathcal{A}]$ and the Tits monoid $\Sigma[\mathcal{A}]$. Let

$$\sec : \Pi[\mathcal{A}] \to \Sigma[\mathcal{A}]$$

be any section of the support map. Note that $\sec(\top)$ is an arbitrarily chosen chamber. Linearizing sec yields a homogeneous section u. Explicitly, the scalars u^F are given by

$$\mathrm{u}^F := \begin{cases} 1 & \text{if } F \text{ is in the image of sec,} \\ 0 & \text{otherwise.} \end{cases}$$

In this case, we say that the homogeneous section u is *set-theoretic.*

11.1.3. The uniform section. Suppose that the field characteristic is 0. We say that a homogeneous section u is *uniform* if $\mathrm{u}^F = \mathrm{u}^G$ whenever F and G have the same support. Equivalently, u is uniform if

$$\mathrm{u}^F = \frac{1}{c^F},$$

where c^F is the number of faces with support $\mathrm{s}(F)$.

11.1.4. Projective sections. We say that a homogeneous section u is *projective* if $\mathrm{u}^F = \mathrm{u}^{\overline{F}}$ for all faces F.

Lemma 11.5. *Assume the rank of the arrangement to be at least one. Let $\Bbbk$ be any field. A projective section exists iff the characteristic of $\Bbbk$ is not 2.*

PROOF. A projective section is the same as a choice of scalars $\mathrm{u}^{\{F,\overline{F}\}}$, one for each projective face $\{F, \overline{F}\}$, such that $\mathrm{u}^{\{O,O\}} = 1$, and for each non-minimum flat X,

$$2 \sum_{\{F,\overline{F}\}:\, \mathrm{s}(F)=\mathrm{X}} \mathrm{u}^{\{F,\overline{F}\}} = 1.$$

In particular, $2\,\mathrm{u}^{\{P,\overline{P}\}} = 1$ for any vertex P. Clearly, these equations can be solved iff the field characteristic is not 2. □

The uniform section (assuming characteristic 0) is clearly projective. In contrast, a set-theoretic section of an arrangement of rank at least one can never be projective.

11.1.5. Example. Consider the rank-one arrangement with chambers C and $\overline{C}$. Homogeneous sections are characterized by an arbitrary scalar p via

$$\mathrm{u}^O = 1,\ \mathrm{u}^C = p,\ \mathrm{u}^{\overline{C}} = 1 - p.$$

There are two set-theoretic sections, namely,

$$\mathrm{u}^O = 1,\ \mathrm{u}^C = 1,\ \mathrm{u}^{\overline{C}} = 0 \qquad \text{and} \qquad \mathrm{u}^O = 1,\ \mathrm{u}^C = 0,\ \mathrm{u}^{\overline{C}} = 1.$$

These are the cases $p = 1$ and $p = 0$, respectively. There is only one projective section and it is the uniform section. It is given by

$$\mathrm{u}^O = 1,\ \mathrm{u}^C = \mathrm{u}^{\overline{C}} = \frac{1}{2}.$$

This is the case $p = 1/2$.

Now consider the rank-two arrangement of 3 lines. The figure on the left shows a set-theoretic section, while the one on the right shows a projective section. The figure in the middle is the uniform section. The number written on the face F stands for the coefficient $\mathtt{u}^F$.

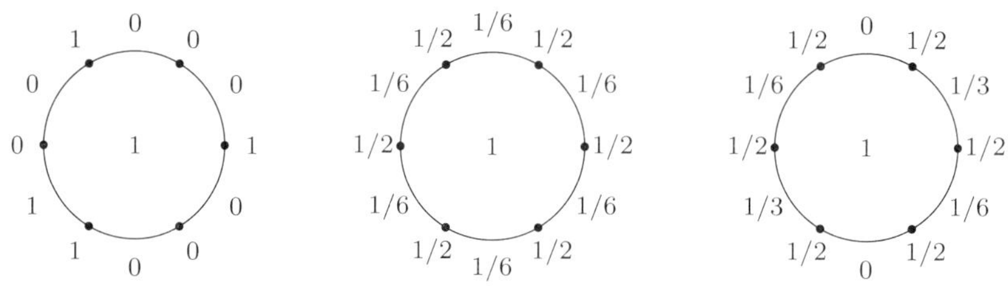

11.1.6. Induced section over a flat. Recall that for any face H of $\mathcal{A}$, the flats of $\mathcal{A}_H$ correspond to the flats of $\mathcal{A}$ which contain H. Whenever a flat X of $\mathcal{A}$ contains H, we write X/H for the corresponding flat of $\mathcal{A}_H$.

Suppose $\mathtt{u}$ is a homogeneous section of $\mathcal{A}$. For each $G \geq H$, define

$$\mathtt{u}_H^G := \sum_{\substack{F:\, HF=G,\\ \mathrm{s}(F)=\mathrm{s}(G)}} \mathtt{u}^F. \tag{11.5}$$

The sum is over all faces in the combinatorial lune $\mathrm{s}(H,G)$, see (3.8). In particular, for any chamber $D \geq H$,

$$\mathtt{u}_H^D := \sum_{C:\, HC=D} \mathtt{u}^C. \tag{11.6}$$

The sum is over all chambers in the combinatorial top-lune $\mathrm{s}(H,D)$, see (3.3).

Lemma 11.6. *A homogeneous section* $\mathtt{u}$ *of* $\mathcal{A}$ *induces a homogeneous section* $\mathtt{u}_H$ *of* $\mathcal{A}_H$*, with the scalar associated to the face* G/H *being* $\mathtt{u}_H^G$.

PROOF. Let X be any flat containing H. Then

$$\begin{aligned}\sum_{G/H:\,\mathrm{s}(G/H)=\mathrm{X}/H} \mathtt{u}_H^G &= \sum_{\substack{G:\, G\geq H,\\ \mathrm{s}(G)=\mathrm{X}}} \sum_{\substack{F:\, HF=G,\\ \mathrm{s}(F)=\mathrm{s}(G)}} \mathtt{u}^F \\ &= \sum_{F:\,\mathrm{s}(F)=\mathrm{X}} \mathtt{u}^F \\ &= 1.\end{aligned}$$

The first step used (11.5), while the last step used (11.3). By Lemma 11.3, $\mathtt{u}_H$ is a homogeneous section. □

For the three homogeneous sections $\mathtt{u}$ shown in the pictures in Section 11.1.5, the respective induced homogeneous sections $\mathtt{u}_F$ for F a vertex are shown below.

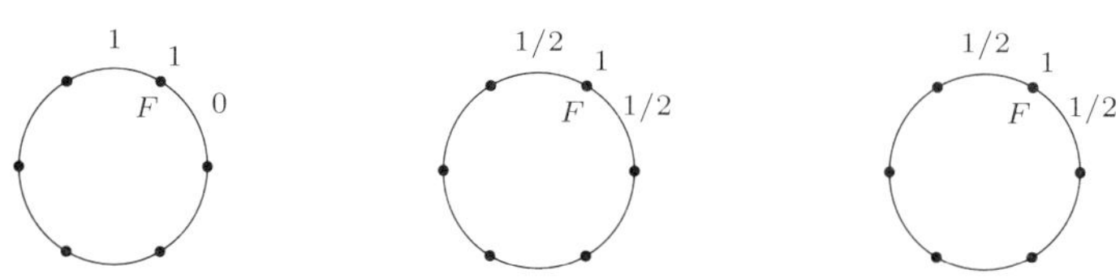

In the middle picture, the value $1/2$ on an edge containing F was obtained as $1/6+1/6+1/6$, while in the last picture, the value $1/2$ was obtained as $1/3+1/6+0$.

Going back to the general case: Consistent with (11.2), for any face H, and flat X containing H, put

$$\mathrm{u}_{\mathrm{X}/H} = \sum_{G\colon G\geq H, \mathrm{s}(G)=\mathrm{X}} \mathrm{u}_H^G \, \mathrm{H}_{G/H}. \tag{11.7}$$

This is an element of $\Sigma[\mathcal{A}_H]$. It is clear from (11.5) and (11.7) that

$$\mathrm{u}_{\mathrm{X}/H} = \Delta_H(\mathrm{u}_\mathrm{X}), \tag{11.8}$$

with Δ_H as in (9.71). Now let $\beta_{G,F}$ be as in (9.69). Using (9.72) and (9.73), we deduce that $\beta_{G,F}(\mathrm{u}_{\mathrm{X}/F}) = \mathrm{u}_{\mathrm{X}/G}$ for any F and G with the same support, and $\Delta_{G/H}(\mathrm{u}_{\mathrm{X}/H}) = \mathrm{u}_{\mathrm{X}/G}$ for any $G \geq H$. This can be expressed succintly as follows.

Lemma 11.7. *The homogeneous sections* u_H *of* $\mathcal{A}_H$ *induced from a homogeneous section* u *of* $\mathcal{A}$ *satisfy the following compatibility conditions. For any* $G \geq H$,

$$(\mathrm{u}_H)_{G/H} = \mathrm{u}_G, \tag{11.9}$$

and for any F *and* G *with the same support,*

$$\beta_{G,F}(\mathrm{u}_F) = \mathrm{u}_G. \tag{11.10}$$

The following is an equivalent form of the identity (11.9). For all $A \leq H \leq G$,

$$\mathrm{u}_H^G = \sum_{\substack{F\colon F\geq A, HF=G,\\ \mathrm{s}(F)=\mathrm{s}(G)}} \mathrm{u}_A^F. \tag{11.11}$$

(The notations in the two identities do not correspond.)

Exercise 11.8. Let u be a homogeneous section of $\mathcal{A}$. Check that:

- If u is set-theoretic arising from sec : $\Pi[\mathcal{A}] \to \Sigma[\mathcal{A}]$, then the induced homogeneous section u_H on $\mathcal{A}_H$ is also set-theoretic and arises from the section
$$\mathrm{sec}_H : \Pi[\mathcal{A}_H] \to \Sigma[\mathcal{A}_H], \qquad \mathrm{sec}_H(\mathrm{X}/H) := H\,\mathrm{sec}(\mathrm{X})/H.$$
- If u is projective, then so is the induced homogeneous section u_H on $\mathcal{A}_H$, that is, $\mathrm{u}_H^F = \mathrm{u}_H^{H\overline{F}}$ for all $H \leq F$.

However, this property is not true in general for the uniform section. For instance, it fails in the smallest nonsimplicial arrangement in rank three (Section 1.2.5). It also fails for the simplicial arrangement shown in Section 6.8.2.

11.1.7. Induced section under a flat. Suppose u is a homogeneous section of $\mathcal{A}$. Then, for a fixed flat X, restricting u to flats $\mathrm{Y} \leq \mathrm{X}$ yields a homogeneous section of the arrangement $\mathcal{A}^\mathrm{X}$, which we denote by u^X.

11.1.8. Cartesian product. Suppose u is a homogeneous section of $\mathcal{A}$, and u′ is a homogeneous section of $\mathcal{A}'$. Then we obtain an induced homogeneous section $\mathrm{u} \times \mathrm{u}'$ on $\mathcal{A} \times \mathcal{A}'$:

$$(\mathrm{u} \times \mathrm{u}')^{(F,F')} := \mathrm{u}^F {\mathrm{u}'}^{F'}.$$

Equivalently,

$$(\mathrm{u} \times \mathrm{u}')_{(\mathrm{X},\mathrm{X}')} = \mathrm{u}_\mathrm{X} \otimes \mathrm{u}'_{\mathrm{X}'}.$$

If u and u′ are set-theoretic (uniform, projective), then so is $\mathrm{u} \times \mathrm{u}'$.

11.2. Eulerian idempotents

We now relate homogeneous sections to Eulerian families. The latter are families of orthogonal idempotents in the Tits algebra. Connection of Eulerian families with complete systems and algebra sections of the support map is treated in the next section.

11.2.1. Eulerian families. Fix an arrangement $\mathcal{A}$. An *Eulerian family* of $\mathcal{A}$ is a set $\mathtt{E} := \{\mathtt{E}_{\mathrm{X}}\}_{\mathrm{X}\in\Pi}$ indexed by flats, where each $\mathtt{E}_{\mathrm{X}}$ is an element of the Tits algebra $\Sigma[\mathcal{A}]$ of the form

$$\mathtt{E}_{\mathrm{X}} = \sum_{F:\,\mathrm{s}(F)\geq \mathrm{X}} a^F \mathtt{H}_F \tag{11.12}$$

with a nonzero base term, that is, $a^G \neq 0$ for at least one face G of support X. These elements are required to be idempotent and mutually orthogonal:

$$\mathtt{E}_{\mathrm{X}} \cdot \mathtt{E}_{\mathrm{Y}} = \begin{cases} \mathtt{E}_{\mathrm{X}} & \text{if } \mathrm{X} = \mathrm{Y}, \\ 0 & \text{if } \mathrm{X} \neq \mathrm{Y}. \end{cases} \tag{11.13}$$

We refer to the $\mathtt{E}_{\mathrm{X}}$ as the *Eulerian idempotents*. The element $\mathtt{E}_\bot$ associated to the minimum flat $\bot$ is the *first Eulerian idempotent*. The nonzero base term condition says that the coefficient of $\mathtt{H}_O$ in $\mathtt{E}_\bot$ is nonzero.

We will prove the following result.

Proposition 11.9. *Eulerian families of $\mathcal{A}$ are in correspondence with homogeneous sections of the support map of $\mathcal{A}$.*

In particular, such families always exist.

11.2.2. From a homogeneous section to an Eulerian family. Saliola construction. Suppose we are given a homogeneous section $\mathtt{u}$. Define elements of $\Sigma[\mathcal{A}]$ indexed by flats recursively by the formula

$$\mathtt{E}_{\mathrm{X}} := \mathtt{u}_{\mathrm{X}} - \sum_{\mathrm{Y}:\mathrm{Y}>\mathrm{X}} \mathtt{u}_{\mathrm{X}} \cdot \mathtt{E}_{\mathrm{Y}}, \tag{11.14}$$

beginning with the maximum flat and proceeding down. Thus, $\mathtt{E}_\top = \mathtt{u}_\top$, and for a hyperplane X,

$$\mathtt{E}_{\mathrm{X}} = \mathtt{u}_{\mathrm{X}} - \mathtt{u}_{\mathrm{X}} \cdot \mathtt{E}_\top = \mathtt{u}_{\mathrm{X}} - \mathtt{u}_{\mathrm{X}} \cdot \mathtt{u}_\top,$$

and so on till we reach $\mathtt{E}_\bot$ indexed by the minimum flat. In $\mathtt{E}_{\mathrm{X}}$, the term $\mathtt{u}_{\mathrm{X}}$ involves faces with support X, while the remaining terms involve faces with support strictly greater than X.

Example 11.10. Suppose the homogeneous section $\mathtt{u}$ is set-theoretic arising from $\mathrm{sec} : \Pi[\mathcal{A}] \to \Sigma[\mathcal{A}]$. Then formula (11.14) takes the simpler form:

$$\mathtt{E}_{\mathrm{X}} := \mathtt{H}_{\mathrm{sec}(\mathrm{X})} - \sum_{\mathrm{Y}:\mathrm{Y}>\mathrm{X}} \mathtt{H}_{\mathrm{sec}(\mathrm{X})} \cdot \mathtt{E}_{\mathrm{Y}}.$$

The first two steps are as follows.

- $\mathtt{E}_\top = \mathtt{H}_{\mathrm{sec}(\top)}$. This is a chamber. Call it $\mathtt{H}_C$.
- For any panel F in the image of sec, we have $\mathtt{E}_{\mathrm{s}(F)} = \mathtt{H}_F - \mathtt{H}_{FC}$.

In general, $\mathtt{E}_{\mathrm{X}}$ only contains faces in the star of $\mathrm{sec}(\mathrm{X})$.

Lemma 11.11. *We have*

$$\mathtt{u}_X \cdot \mathtt{E}_X = \mathtt{E}_X, \tag{11.15}$$

$$\mathtt{u}_X \cdot \Big(\sum_{Y:Y \geq X} \mathtt{E}_Y \Big) = \mathtt{u}_X, \tag{11.16}$$

$$\mathtt{H}_F \cdot \Big(\sum_{Y:\, Y \geq s(F)} \mathtt{E}_Y \Big) = \mathtt{H}_F. \tag{11.17}$$

PROOF. Formula (11.15) follows by premultiplying (11.14) with $\mathtt{u}_X$ and then using (11.4). Substituting it in the lhs of (11.14) and rearranging terms yields (11.16). Premultiplying this with $\mathtt{H}_F$ (where $X = s(F)$) and using Lemma 11.2 yields (11.17). □

In particular, by setting $X = \bot$ in (11.16) and using $\mathtt{u}_\bot = \mathtt{H}_O$, or by setting $F = O$ in (11.17), we obtain:

$$\mathtt{H}_O = \sum_X \mathtt{E}_X. \tag{11.18}$$

Lemma 11.12. *For any face F and flat X, if $s(F) \not\leq X$, then $\mathtt{H}_F \cdot \mathtt{E}_X = 0$. In particular, $\mathtt{H}_F \cdot \mathtt{E}_\bot = 0$ for $F > O$.*

PROOF. We do a backward induction on the rank of X. If $X = \top$, then the statement is vacuously true. This is the induction base. The induction step is shown below. Put $Z = s(F)$.

$$\begin{aligned} \mathtt{H}_F \cdot \mathtt{E}_X &= \mathtt{H}_F \cdot \mathtt{u}_X - \sum_{Y:Y>X} \mathtt{H}_F \cdot \mathtt{u}_X \cdot \mathtt{E}_Y \\ &= \mathtt{H}_F \cdot \mathtt{u}_X - \sum_{Y:Y \geq Z \vee X} \mathtt{H}_F \cdot \mathtt{u}_X \cdot \mathtt{E}_Y - \sum_{Y:Y>X,\, Z \not\leq Y} \mathtt{H}_F \cdot \mathtt{u}_X \cdot \mathtt{E}_Y. \end{aligned} \tag{a}$$

Each term in $\mathtt{H}_F \cdot \mathtt{u}_X$ has support $Z \vee X$. It follows from Lemma 11.2 that

$$\mathtt{H}_F \cdot \mathtt{u}_X = \mathtt{H}_F \cdot \mathtt{u}_X \cdot \mathtt{u}_{Z \vee X}.$$

This along with the induction hypothesis implies that for any $Y > X$, $Z \not\leq Y$,

$$\mathtt{H}_F \cdot \mathtt{u}_X \cdot \mathtt{E}_Y = \mathtt{H}_F \cdot \mathtt{u}_X \cdot \mathtt{u}_{Z \vee X} \cdot \mathtt{E}_Y = 0,$$

and hence the last term in (a) is zero. The first two terms are manipulated as follows.

$$\mathtt{H}_F \cdot \mathtt{u}_X - \sum_{Y:Y \geq Z \vee X} \mathtt{H}_F \cdot \mathtt{u}_X \cdot \mathtt{E}_Y = \mathtt{H}_F \cdot \mathtt{u}_X \cdot \Big(\mathtt{u}_{Z \vee X} - \sum_{Y:Y \geq Z \vee X} \mathtt{u}_{Z \vee X} \cdot \mathtt{E}_Y \Big) = 0.$$

In the last step, we used (11.16) to deduce that the term inside the parenthesis is zero. □

We refer to Lemma 11.12 as the *Saliola lemma*. It is an important result which will be repeatedly used in the text.

Lemma 11.13. *Given a homogeneous section $\mathtt{u}$, the elements $\mathtt{E}_X$ defined by* (11.14) *yield an Eulerian family.*

PROOF. It is clear that each $\mathtt{E}_X$ is of the form (11.12) and has a nonzero base term. We need to check (11.13). We do a backward induction on the rank of X. By (11.4), $\mathtt{E}_\top = \mathtt{u}_\top$ is idempotent, and by the Saliola lemma, $\mathtt{E}_\top \cdot \mathtt{E}_Y = 0$ for any $Y < \top$. This is the induction base. The induction step is completed below.

- $X \not\leq Y$.

  In this case, $\mathtt{E}_X \cdot \mathtt{E}_Y = 0$ by the Saliola lemma.
- $X < Y$.

  In this case, by multiplying (11.14) on the right by $\mathtt{E}_Y$, we obtain

  $$\mathtt{E}_X \cdot \mathtt{E}_Y = \mathtt{u}_X \cdot \mathtt{E}_Y - \sum_{Z:Z>X} \mathtt{u}_X \cdot \mathtt{E}_Z \cdot \mathtt{E}_Y = \mathtt{u}_X \cdot \mathtt{E}_Y - \mathtt{u}_X \cdot \mathtt{E}_Y = 0.$$

  In the second step, by the induction hypothesis, only the summand for $Z = Y$ contributed.
- $X = Y$.

  In this case, by multiplying (11.14) on the right by $\mathtt{E}_X$, we obtain

  $$\mathtt{E}_X \cdot \mathtt{E}_X = \mathtt{u}_X \cdot \mathtt{E}_X - \sum_{Z:Z>X} \mathtt{u}_X \cdot \mathtt{E}_Z \cdot \mathtt{E}_X = \mathtt{u}_X \cdot \mathtt{E}_X = \mathtt{E}_X.$$

  In the second step, by the induction hypothesis or by the Saliola lemma, the sum is zero. In the last step, we used (11.15).

□

We refer to this construction of an Eulerian family starting from a homogeneous section as the *Saliola construction*.

Exercise 11.14. Show that $\mathtt{E}_\perp \cdot \mathtt{u}_\top = 0$ whenever $\perp \neq \top$.

11.2.3. From an Eulerian family to a homogeneous section. The base term. Suppose $\mathtt{E}$ is an Eulerian family. Using (11.12), write

$$\mathtt{E}_X = \mathtt{u}_X + \sum_{F:\, \mathrm{s}(F)>X} a^F \mathtt{H}_F. \tag{11.19}$$

The element $\mathtt{u}_X$ is the part of $\mathtt{E}_X$ consisting of faces of support X. This is the base term of $\mathtt{E}_X$. It is *nonzero* by hypothesis. The remaining terms are the higher terms consisting of faces of support strictly greater than X. Since $\mathtt{E}_X$ is an idempotent,

$$\mathtt{u}_X + \text{ higher terms} = \mathtt{E}_X = \mathtt{E}_X \cdot \mathtt{E}_X = \mathtt{u}_X \cdot \mathtt{u}_X + \text{ higher terms}.$$

Thus $\mathtt{u}_X \cdot \mathtt{u}_X = \mathtt{u}_X$, and hence by Lemma 11.3 the $\mathtt{u}_X$ yield a homogeneous section.

11.2.4. Equivalence between homogeneous sections and Eulerian families. We claim that the two constructions discussed above are inverse to each other. One direction is clear. The nontrivial direction is proved below.

Lemma 11.15. *Suppose* $\mathtt{E}$ *is an Eulerian family. Then* (11.14) *holds with the* $\mathtt{u}_X$ *defined by* (11.19).

PROOF. Write

$$\mathtt{E}_X = \mathtt{u}_X - \Big(\sum_{Y:Y>X} \mathtt{u}_X \cdot \mathtt{E}_Y \Big) + \mathrm{err}_X .$$

We view err_X as the error term and would like to show it to be 0 for all X. We do this by a backward induction on the rank of X.

Clearly, $\mathtt{E}_\top = \mathtt{u}_\top$ and hence $\mathrm{err}_\top = 0$. This is the induction base. For the induction step: Suppose $\mathrm{err}_\mathrm{Y} = 0$ for all $\mathrm{Y} > \mathrm{X}$. In other words, suppose the $\mathtt{E}_\mathrm{Y}$ for $\mathrm{Y} > \mathrm{X}$ are given by the Saliola construction (11.14). For $\mathrm{Y} > \mathrm{X}$,

$$0 = (\mathtt{E}_\mathrm{X} - \mathrm{err}_\mathrm{X}) \cdot \mathtt{E}_\mathrm{Y} = \mathtt{E}_\mathrm{X} \cdot \mathtt{E}_\mathrm{Y} - \mathrm{err}_\mathrm{X} \cdot \mathtt{E}_\mathrm{Y} = - \mathrm{err}_\mathrm{X} \cdot \mathtt{E}_\mathrm{Y}.$$

The first step used Lemma 11.13 while the last step used orthogonality of the Eulerian idempotents (11.13). Hence, $\mathrm{err}_\mathrm{X} \cdot \mathtt{E}_\mathrm{Y} = 0$ for all $\mathrm{Y} > \mathrm{X}$. Also, by construction, err_X only contains faces with support strictly greater than X. Write $\mathrm{err}_\mathrm{X} = \sum x^F \mathtt{H}_F$. Suppose $\mathrm{err}_\mathrm{X} \neq 0$. Then there exists a face F such that $x^F \neq 0$ but $x^G = 0$ for all $G < F$. In particular, $\mathrm{s}(F) > \mathrm{X}$. Let us calculate the coefficient of $\mathtt{H}_F$ in $\mathrm{err}_\mathrm{X} \cdot \mathtt{E}_{\mathrm{s}(F)}$.

$$\begin{aligned}\langle \mathrm{err}_\mathrm{X} \cdot \mathtt{E}_{\mathrm{s}(F)}, \mathtt{H}_F \rangle = \langle x^F \mathtt{H}_F \cdot \mathtt{E}_{\mathrm{s}(F)}, \mathtt{H}_F \rangle &= \langle x^F \mathtt{H}_F \cdot \mathtt{u}_{\mathrm{s}(F)}, \mathtt{H}_F \rangle \\ &= \langle x^F \mathtt{H}_F, \mathtt{H}_F \rangle = x^F \neq 0.\end{aligned}$$

Thus $\mathrm{err}_\mathrm{X} \cdot \mathtt{E}_{\mathrm{s}(F)} \neq 0$, which is a contradiction. Hence $\mathrm{err}_\mathrm{X} = 0$ as required. □

This completes the proof of Proposition 11.9.

11.2.5. Visualizing an Eulerian idempotent. One also deduces from (11.14) that Eulerian idempotents have a more rigid form than what is specified by (11.12), namely,

$$\mathtt{E}_\mathrm{X} = \sum_{F:\, \mathrm{s}(F) = \mathrm{X}} \; \sum_{G:\, G \geq F} a^G \mathtt{H}_G. \tag{11.20}$$

In other words, we need to sum only over those faces G which have a face F with support X. Further, if F and F' both have support X, and $G \geq F$ and $G' \geq F'$ are such that $F'G = G'$, then $a^G a^{F'} = a^{G'} a^F$. In other words, the coefficients of faces in the star of F are in proportion to those in the star of F'.

An illustration in rank three is given below. For the flat X shown as the thick line, the Eulerian idempotent $\mathtt{E}_\mathrm{X}$ involves edges on that line and the chambers in the shaded region. (Only the front half is visible in the picture.)

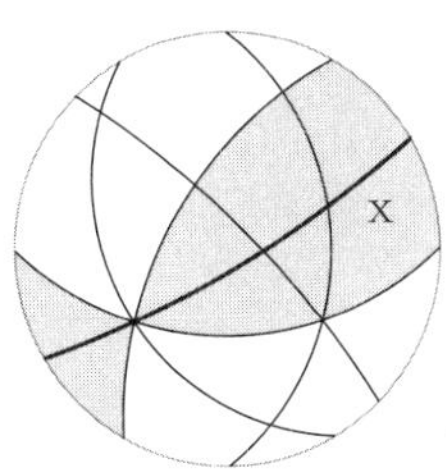

11.2.6. Projective Eulerian families. Recall the projective Tits algebra from Section 9.3.4. An Eulerian family $\mathtt{E}$ is *projective* if each Eulerian idempotent $\mathtt{E}_\mathrm{X}$ belongs to the projective Tits algebra.

Proposition 11.16. *Projective Eulerian families of $\mathcal{A}$ are in correspondence with projective sections of the support map of $\mathcal{A}$.*

This is obtained by restricting the correspondence in Proposition 11.9.

11.2.7. Over and under a flat. Cartesian product. We now see how the correspondence between Eulerian families and homogeneous sections behaves under passage to arrangements over and under a flat, and under taking cartesian product of arrangements.

Suppose $\mathtt{E}$ is an Eulerian family of $\mathcal{A}$. For any face H, and flat X containing H, define

$$\mathtt{E}_{\mathrm{X}/H} := \Delta_H(\mathtt{E}_{\mathrm{X}}), \tag{11.21}$$

with Δ_H as in (9.71). Since it is an algebra homomorphism, it follows from (11.13) that

$$\mathtt{E}_H := \{\mathtt{E}_{\mathrm{X}/H}\}_{\mathrm{X} \geq \mathrm{s}(H)}$$

is a family of mutually orthogonal idempotents of $\Sigma[\mathcal{A}_H]$.

Lemma 11.17. *Let* $\mathtt{u}$ *be a homogeneous section of* $\mathcal{A}$, *and let* $\mathtt{E}$ *be its associated Eulerian family. Then for any face* H, $\mathtt{E}_H$ *is the Eulerian family of* $\mathcal{A}_H$ *associated to the homogeneous section* $\mathtt{u}_H$. *In particular,* $\mathtt{E}_{\mathrm{s}(H)/H}$ *is the first Eulerian idempotent of* $\mathcal{A}_H$.

PROOF. Let $\mathtt{E}'_{\mathrm{X}/H}$ denote the Eulerian idempotents associated to $\mathtt{u}_H$. We want to show $\mathtt{E}'_{\mathrm{X}/H} = \mathtt{E}_{\mathrm{X}/H}$. We do a backward induction on the rank of X. By (11.8), $\mathtt{u}_{\top/H} = \Delta_H(\mathtt{u}_\top)$, so the result holds for $\mathrm{X} = \top$. This is the induction base. The induction step is as follows.

$$\begin{aligned}
\mathtt{E}'_{\mathrm{X}/H} &:= \mathtt{u}_{\mathrm{X}/H} - \sum_{\mathrm{Y}:\mathrm{Y}>\mathrm{X}} \mathtt{u}_{\mathrm{X}/H} \cdot \mathtt{E}'_{\mathrm{Y}/H} \\
&= \Delta_H(\mathtt{u}_{\mathrm{X}}) - \sum_{\mathrm{Y}:\mathrm{Y}>\mathrm{X}} \Delta_H(\mathtt{u}_{\mathrm{X}}) \cdot \Delta_H(\mathtt{E}_{\mathrm{Y}}) \\
&= \Delta_H(\mathtt{u}_{\mathrm{X}}) - \sum_{\mathrm{Y}:\mathrm{Y}>\mathrm{X}} \Delta_H(\mathtt{u}_{\mathrm{X}} \cdot \mathtt{E}_{\mathrm{Y}}) \\
&= \Delta_H(\mathtt{E}_{\mathrm{X}}).
\end{aligned}$$

The definition is used in the first step, and the induction hypothesis is used in the next step. The remaining steps used (11.8) and the fact that Δ_H is an algebra homomorphism. □

Suppose $\mathtt{E} := \{\mathtt{E}_{\mathrm{Y}}\}$ is an Eulerian family of $\mathcal{A}$. For $\mathrm{Y} \leq \mathrm{X}$, let $\mathtt{E}^{\mathrm{X}}_{\mathrm{Y}}$ denote the image of $\mathtt{E}_{\mathrm{Y}}$ under (9.77). Since this map is an algebra homomorphism, it follows from (11.13) that

$$\mathtt{E}^{\mathrm{X}} := \{\mathtt{E}^{\mathrm{X}}_{\mathrm{Y}}\}_{\mathrm{Y} \leq \mathrm{X}}$$

is a family of mutually orthogonal idempotents of $\Sigma[\mathcal{A}^{\mathrm{X}}]$, that is,

$$\mathtt{E}^{\mathrm{X}}_{\mathrm{Y}} \cdot \mathtt{E}^{\mathrm{X}}_{\mathrm{Z}} = \begin{cases} \mathtt{E}^{\mathrm{X}}_{\mathrm{Y}} & \text{if } \mathrm{Y} = \mathrm{Z}, \\ 0 & \text{if } \mathrm{Y} \neq \mathrm{Z}. \end{cases} \tag{11.22}$$

Lemma 11.18. *Let* $\mathtt{u}$ *be a homogeneous section of* $\mathcal{A}$, *and let* $\mathtt{E}$ *be its associated Eulerian family. Then for any flat* X, $\mathtt{E}^{\mathrm{X}}$ *is the Eulerian family of* $\mathcal{A}^{\mathrm{X}}$ *associated to the homogeneous section* $\mathtt{u}^{\mathrm{X}}$.

PROOF. This can be shown by induction on the rank of X similar to the proof of Lemma 11.17. □

Suppose $\mathtt{E} := \{\mathtt{E}_{\mathrm{X}}\}$ is an Eulerian family of $\mathcal{A}$, and $\mathtt{E}' := \{\mathtt{E}'_{\mathrm{X}'}\}$ is an Eulerian family of $\mathcal{A}'$. Then

$$\mathtt{E} \times \mathtt{E}' := \{\mathtt{E}_{\mathrm{X}} \otimes \mathtt{E}'_{\mathrm{X}'}\}_{(\mathrm{X},\mathrm{X}')}$$

is an Eulerian family of $\mathcal{A} \times \mathcal{A}'$. More precisely:

Lemma 11.19. *Let* $\mathtt{u}$ *and* $\mathtt{u}'$ *be homogeneous sections of* $\mathcal{A}$ *and* $\mathcal{A}'$, *and let* $\mathtt{E}$ *and* $\mathtt{E}'$ *be their associated Eulerian families. Then* $\mathtt{E}\times\mathtt{E}'$ *is the Eulerian family of* $\mathcal{A}\times\mathcal{A}'$ *associated to the homogeneous section* $\mathtt{u} \times \mathtt{u}'$.

11.3. Eulerian families, complete systems and algebra sections

We now show that any Eulerian family is a complete system of primitive orthogonal idempotents of the Tits algebra, and all complete systems are of this form. Recall that for any elementary algebra, there is a correspondence between complete systems and algebra sections (Theorem D.32). Hence, this result can be stated as follows.

Theorem 11.20. *The following pieces of data are equivalent.*

- *An Eulerian family* $\mathtt{E}$ *of* $\mathcal{A}$.
- *A complete system of primitive orthogonal idempotents of* $\Sigma[\mathcal{A}]$.
- *An algebra section* $\Pi[\mathcal{A}] \to \Sigma[\mathcal{A}]$ *of the support map.*

11.3.1. Eulerian families and algebra sections. Let $\mathtt{E}$ be an Eulerian family associated to the homogeneous section $\mathtt{u}$. Applying the support map to (11.14) and using (9.32) and (11.1) yields

$$\mathrm{s}(\mathtt{E}_{\mathrm{X}}) = \mathtt{H}_{\mathrm{X}} - \sum_{\mathrm{Y}:\mathrm{Y}>\mathrm{X}} \mathtt{H}_{\mathrm{X}} \cdot \mathrm{s}(\mathtt{E}_{\mathrm{Y}}).$$

Applying induction and using (9.1) and (9.4), we deduce that

$$\mathrm{s}(\mathtt{E}_{\mathrm{X}}) = \mathtt{Q}_{\mathrm{X}}. \tag{11.23}$$

Thus, the Eulerian idempotents map to the primitive idempotents of the Birkhoff algebra. Further, the map

$$\Pi[\mathcal{A}] \hookrightarrow \Sigma[\mathcal{A}], \qquad \mathtt{Q}_{\mathrm{X}} \mapsto \mathtt{E}_{\mathrm{X}} \tag{11.24}$$

is an algebra section of the support map. This is because the $\mathtt{E}_{\mathrm{X}}$ are idempotent, mutually orthogonal, and by (11.18), they add up to $\mathtt{H}_{O}$ (the unit element). Since algebra sections correspond to complete systems, we have:

Theorem 11.21. *Any Eulerian family of* $\mathcal{A}$ *is a complete system of primitive orthogonal idempotents of* $\Sigma[\mathcal{A}]$.

For the converse: Recall from Theorem D.31 that any two algebra sections of the support map are conjugate by an element of $\mathtt{H}_{O} + \mathrm{rad}(\Sigma)$. (The latter is a subgroup of the group of invertible elements of the Tits algebra.) Now use the result below.

Lemma 11.22. *Conjugation of any Eulerian family by an invertible element of the Tits algebra produces another Eulerian family.*

PROOF. Let $\mathtt{E}$ be an Eulerian family, and v be an invertible element. Put $\mathtt{E}'_{\mathrm{X}} := v \cdot \mathtt{E}_{\mathrm{X}} \cdot v^{-1}$. Then clearly, the $\mathtt{E}'_{\mathrm{X}}$ are idempotent and mutually orthogonal, and each $\mathtt{E}'_{\mathrm{X}}$ only involves faces of support greater than X. Further, $\mathrm{s}(\mathtt{E}'_{\mathrm{X}}) = \mathrm{s}(v) \cdot \mathrm{s}(\mathtt{E}_{\mathrm{X}}) \cdot$

$\mathtt{s}(v^{-1}) = \mathtt{s}(\mathtt{E}_{\mathrm{X}}) = \mathtt{Q}_{\mathrm{X}}$, so $\mathtt{E}'_{\mathrm{X}}$ has a nonzero base term. Thus, $\mathtt{E}'$ is an Eulerian family. $\square$

This completes the proof of Theorem 11.20.

Exercise 11.23. Check that formula (11.23) can be equivalently expressed as

$$\chi_{\mathrm{X}}(\mathtt{E}_{\mathrm{Y}}) = \begin{cases} 1 & \text{if } \mathrm{X} = \mathrm{Y}, \\ 0 & \text{otherwise}, \end{cases} \tag{11.25}$$

where χ_{X} are the multiplicative characters of $\Sigma[\mathcal{A}]$ given by (9.48).

Exercise 11.24. Assume that the field characteristic is not 2. Recall from Proposition 9.27 that the projective Tits algebra is elementary. Formulate the analogue of Theorem 11.20 for the projective Tits algebra.

11.3.2. Conjugation is simply transitive. We know from Lemma 11.22 that the group $\mathtt{H}_O + \mathrm{rad}(\Sigma)$ acts by conjugation on the set of all Eulerian families. By looking at the base terms, one can write down the corresponding action on the set of homogeneous sections. It is as follows.

Lemma 11.25. *For $z \in \mathrm{rad}(\Sigma)$, the action of $\mathtt{H}_O + z$ on $\mathtt{u}$ is*

$$\mathtt{u}_{\mathrm{X}} \mapsto (\mathtt{H}_O + z^{\mathrm{X}}) \cdot \mathtt{u}_{\mathrm{X}},$$

where z^{X} denotes the part of z involving faces of support smaller than X. (Note that $z^{\perp} = 0$.)

PROOF. Let us write $(\mathtt{H}_O + z) \cdot \mathtt{u}_{\mathrm{X}} \cdot (\mathtt{H}_O + z)^{-1}$ as

$$(\mathtt{H}_O + z^{\mathrm{X}} + \text{ higher terms}) \cdot \mathtt{u}_{\mathrm{X}} \cdot (\mathtt{H}_O + \text{ terms under X} + \text{ higher terms}).$$

'Terms under X' refer to terms involving faces whose support is smaller than X. Since we only want terms involving faces of support X, the higher terms can be safely ignored. The 'terms under X' taken together belong to the radical, so by (9.33), left multiplication by $\mathtt{u}_{\mathrm{X}}$ yields 0. Thus, we are left with $(\mathtt{H}_O + z^{\mathrm{X}}) \cdot \mathtt{u}_{\mathrm{X}}$. $\square$

Lemma 11.26. *The action of $\mathtt{H}_O + \mathrm{rad}(\Sigma)$ on the set of Eulerian families (or homogeneous sections) is simply transitive: Given homogeneous sections $\mathtt{u}$ and $\mathtt{u}'$, there is a unique $z \in \mathrm{rad}(\Sigma)$ such that*

$$\mathtt{u}'_{\mathrm{X}} = (\mathtt{H}_O + z^{\mathrm{X}}) \cdot \mathtt{u}_{\mathrm{X}}$$

for all flats X.

PROOF. Let z_{X} denote the part of z consisting of faces with support X. To construct z, we need to construct z_{X} for each flat X. We do that by induction on the rank of X. Note that $z_{\perp} = 0$ is the unique solution for $\mathrm{X} = \perp$. Now suppose that z_{Y} are constructed for all $\mathrm{Y} < \mathrm{X}$, and they are unique. To construct z_{X}, we need to solve the equation

$$\Big(\mathtt{H}_O + \sum_{\mathrm{Y}:\, \mathrm{Y}<\mathrm{X}} z_{\mathrm{Y}} + z_{\mathrm{X}}\Big) \cdot \mathtt{u}_{\mathrm{X}} = \mathtt{u}'_{\mathrm{X}}.$$

By Lemma 11.2, $z_{\mathrm{X}} \cdot \mathtt{u}_{\mathrm{X}}$ equals z_{X}. Thus,

$$z_{\mathrm{X}} := \mathtt{u}'_{\mathrm{X}} - \Big(\mathtt{H}_O + \sum_{\mathrm{Y}:\, \mathrm{Y}<\mathrm{X}} z_{\mathrm{Y}}\Big) \cdot \mathtt{u}_{\mathrm{X}}$$

is the unique solution. (By (9.33), $\mathtt{u}'_{\mathrm{X}} - \mathtt{u}_{\mathrm{X}}$ and hence z_{X} indeed belongs to the radical.) This completes the induction step. $\square$

11.3.3. Idempotents in the Tits algebra.

Lemma 11.27. *For an idempotent e in the Tits algebra, the following are equivalent.*

(1) *e is primitive.*
(2) *$e = \mathtt{E}_{\mathrm{X}}$ for some Eulerian family $\mathtt{E}$ and some flat X.*
(3) *$\mathrm{s}(e) = \mathtt{Q}_{\mathrm{X}}$ for some flat X.*

In particular, any primitive idempotent is necessarily an Eulerian idempotent.

Proof. (1) implies (2). By applying Lemma D.4 to $1-e$, we see that e belongs to a complete system, and hence by Theorem 11.20, it belongs to an Eulerian family

(2) implies (3). This is the same as (11.23).

(3) implies (1). e is an idempotent which lifts the primitive idempotent $\mathtt{Q}_{\mathrm{X}}$, so it itself must be primitive by Lemma D.28. □

Lemma 11.28. *An element e of the Tits algebra is idempotent iff $e = \mathtt{E}_{\mathrm{X}_1} + \dots + \mathtt{E}_{\mathrm{X}_k}$ for some Eulerian family $\mathtt{E}$ and distinct flats X_i.*

Proof. The backward implication is clear. For the forward implication: Applying Lemma D.4 to both e and $1-e$, we see that e can be written as the sum of mutually orthogonal primitive idempotents which belong to a complete system. By Theorem 11.20, any such complete system is an Eulerian family. □

11.4. Q-bases of the Tits algebra

We introduce Q-bases of the Tits algebra. There is one such basis for every homogeneous section $\mathtt{u}$, or equivalently, one for every Eulerian family $\mathtt{E}$. We compare the Q-bases with the (unique) Q-basis of the Birkhoff algebra.

An abstract approach to Q-bases is given later in Section 11.5.5.

11.4.1. Q-basis of the Tits algebra.

Let $\mathtt{E}$ be an Eulerian family. For any face F, put

$$\mathtt{Q}_F := \mathtt{H}_F \cdot \mathtt{E}_{\mathrm{s}(F)}. \tag{11.26}$$

In particular, $\mathtt{Q}_O = \mathtt{E}_\perp$, and $\mathtt{Q}_C = \mathtt{H}_C$ for any chamber C.

Lemma 11.29. *Each $\mathtt{Q}_F$ is a primitive idempotent, with*

$$\mathrm{s}(\mathtt{Q}_F) = \mathtt{Q}_{\mathrm{s}(F)}. \tag{11.27}$$

Further, the set $\{\mathtt{Q}_F\}$ is a basis of the Tits algebra $\Sigma[\mathcal{A}]$.

Proof. The following calculation shows that $\mathtt{Q}_F$ is an idempotent.

$$\mathtt{Q}_F \cdot \mathtt{Q}_F = \mathtt{H}_F \cdot \mathtt{E}_{\mathrm{s}(F)} \cdot \mathtt{H}_F \cdot \mathtt{E}_{\mathrm{s}(F)} = \mathtt{H}_F \cdot \mathtt{E}_{\mathrm{s}(F)} \cdot \mathtt{E}_{\mathrm{s}(F)} = \mathtt{H}_F \cdot \mathtt{E}_{\mathrm{s}(F)} = \mathtt{Q}_F.$$

Using (9.4) and (11.23), we deduce

$$\mathrm{s}(\mathtt{Q}_F) = \mathrm{s}(\mathtt{H}_F \cdot \mathtt{E}_{\mathrm{s}(F)}) = \mathrm{s}(\mathtt{H}_F) \cdot \mathrm{s}(\mathtt{E}_{\mathrm{s}(F)}) = \mathtt{H}_{\mathrm{s}(F)} \cdot \mathtt{Q}_{\mathrm{s}(F)} = \mathtt{Q}_{\mathrm{s}(F)}.$$

By Lemma 11.27, item (1), $\mathtt{Q}_F$ is primitive.

The element $\mathtt{Q}_F$ written in the H-basis only involves faces greater than F, and further by Lemma 11.2, $\mathtt{H}_F$ appears with coefficient 1. By triangularity, the set $\{\mathtt{Q}_F\}$ is a basis of $\Sigma[\mathcal{A}]$. □

We refer to $\{\mathtt{Q}_F\}$ as the $\mathtt{Q}$-basis. It follows from (11.2) and (11.15) that

$$\mathtt{E}_\mathrm{X} = \sum_{F:\, \mathrm{s}(F)=\mathrm{X}} \mathtt{u}^F \mathtt{Q}_F. \tag{11.28}$$

This is a more precise way of writing (11.20).

Exercise 11.30. Let $\mathtt{u}$ be a homogeneous section with associated Eulerian family $\mathtt{E}$ and $\mathtt{Q}$-basis. Fix a specific $\mathtt{Q}$-basis element, say $\mathtt{Q}_F$. Give an example of a homogeneous section $\mathtt{u}'$ whose associated Eulerian family $\mathtt{E}'$ satisfies $\mathtt{Q}_F = \mathtt{E}'_{\mathrm{s}(F)}$. (Since $\mathtt{Q}_F$ is a primitive idempotent, such a $\mathtt{u}'$ and $\mathtt{E}'$ will exist by Lemma 11.27, item (2).)

11.4.2. Visualizing a $\mathtt{Q}$-basis element. An illustration of a $\mathtt{Q}$-basis element is shown below.

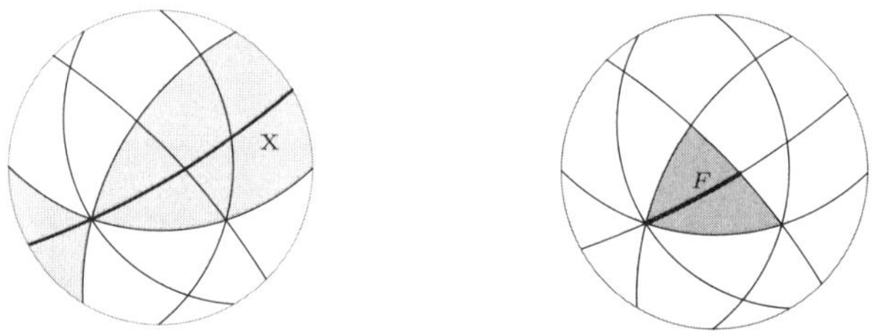

On the left, we have redrawn the picture of the Eulerian idempotent $\mathtt{E}_\mathrm{X}$ from Section 11.2.5. The picture on the right shows $\mathtt{Q}_F$ with $\mathrm{s}(F) = \mathrm{X}$. It involves the edge F and the two shaded chambers. In a sense, $\mathtt{E}_\mathrm{X}$ is local to X, while $\mathtt{Q}_F$ is local to F. The passage between the two is governed by (11.26) and (11.28).

11.4.3. Rank one. Consider the rank-one arrangement with chambers C and $\overline{C}$. Fix an arbitrary scalar p. Recall from Section 11.1.5 that any homogeneous section $\mathtt{u}$ is of the form $\mathtt{u}^O = 1$, $\mathtt{u}^C = p$, $\mathtt{u}^{\overline{C}} = 1 - p$. The associated Eulerian family $\mathtt{E}$ is

$$\mathtt{E}_\top = p\,\mathtt{H}_C + (1-p)\,\mathtt{H}_{\overline{C}} \quad \text{and} \quad \mathtt{E}_\bot = \mathtt{H}_O - p\,\mathtt{H}_C - (1-p)\,\mathtt{H}_{\overline{C}}.$$

The $\mathtt{Q}$-basis is

$$\mathtt{Q}_C = \mathtt{H}_C, \quad \mathtt{Q}_{\overline{C}} = \mathtt{H}_{\overline{C}}, \quad \mathtt{Q}_O = \mathtt{H}_O - p\,\mathtt{H}_C - (1-p)\,\mathtt{H}_{\overline{C}}.$$

Note very carefully that $\mathtt{Q}_O$ is not orthogonal to either $\mathtt{Q}_C$ or $\mathtt{Q}_{\overline{C}}$ in general.

By Lemma 11.27, we conclude that $\mathtt{E}_\bot$ and $\mathtt{E}_\top$, as p varies, yield all the primitive idempotents. Further by Lemma 11.28, the only other idempotents are 0 and $\mathtt{E}_\bot + \mathtt{E}_\top = \mathtt{H}_O$. This agrees with the claim made in Exercise 9.26.

11.4.4. Product of $\mathtt{H}$- and $\mathtt{Q}$-bases elements. The following is a noncommutative analogue of (9.4).

Lemma 11.31. *For any faces F and G,*

$$\mathtt{H}_F \cdot \mathtt{Q}_G = \begin{cases} \mathtt{Q}_{FG} & \text{if } GF = G, \\ 0 & \text{if } GF > G. \end{cases} \tag{11.29}$$

In particular, if F and G have the same support, then

$$\mathtt{H}_F \cdot \mathtt{Q}_G = \mathtt{Q}_F. \tag{11.30}$$

PROOF. For any F and G,

$$\mathtt{H}_F \cdot \mathtt{Q}_G = \mathtt{H}_F \cdot \mathtt{H}_G \cdot \mathtt{E}_{\mathrm{s}(G)} = \mathtt{H}_{FG} \cdot \mathtt{E}_{\mathrm{s}(G)}.$$

If $GF = G$, then $\mathrm{s}(FG) = \mathrm{s}(G)$, and the above quantity equals $\mathtt{Q}_{FG}$. If $GF > G$, then $\mathrm{s}(FG) > \mathrm{s}(G)$, and the above quantity equals 0 by the Saliola lemma (Lemma 11.12). □

11.4.5. Change of basis formulas. Consider the matrix $(\mathtt{u}_H^G)$ with entries defined by (11.5). This matrix can be inverted. More precisely: For $F \leq G$, let $\mathtt{a}_F^G$ be the unique scalars which satisfy $\mathtt{a}_F^F = 1$ for all F, and

$$\sum_{K:\, F \leq K \leq G} \mathtt{u}_F^K \mathtt{a}_K^G = 0 = \sum_{K:\, F \leq K \leq G} \mathtt{a}_F^K \mathtt{u}_K^G \tag{11.31}$$

for all $F < G$. Explicitly,

$$\mathtt{a}_F^G = -\mathtt{u}_F^G + \sum_{F<H<G} \mathtt{u}_F^H \mathtt{u}_H^G - \sum_{F<H<K<G} \mathtt{u}_F^H \mathtt{u}_H^K \mathtt{u}_K^G + \dots \tag{11.32}$$

for all $F < G$. The first sum is over H, the second sum is over H and K, and so on. Note that some of the $\mathtt{a}_F^G$ could be negative even when all the $\mathtt{u}_F^G$ are nonnegative.

Lemma 11.32. *The* H- *and* Q-*bases of the Tits algebra are related by*

$$\mathtt{H}_F = \sum_{K:\, F \leq K} \mathtt{u}_F^K\, \mathtt{Q}_K \quad \text{and} \quad \mathtt{Q}_F = \sum_{G:\, F \leq G} \mathtt{a}_F^G\, \mathtt{H}_G. \tag{11.33}$$

In particular,

$$\mathtt{H}_O = \sum_K \mathtt{u}^K \mathtt{Q}_K. \tag{11.34}$$

PROOF. First note that (11.34) follows from (11.18) and (11.28). Now multiply both sides of this identity on the left by $\mathtt{H}_F$, and then use (11.5) and (11.29). This proves the first formula in (11.33), and the second formula then follows. □

We deduce from (11.28) and the second formula in (11.33) that for an Eulerian family E associated to the homogeneous section u,

$$\mathtt{E}_{\mathrm{X}} = \sum_{F:\, \mathrm{s}(F) = \mathrm{X}} \; \sum_{G:\, F \leq G} \mathtt{u}^F \mathtt{a}_F^G\, \mathtt{H}_G. \tag{11.35}$$

Exercise 11.33. Suppose the homogeneous section u is set-theoretic arising from sec : $\Pi[\mathcal{A}] \to \Sigma[\mathcal{A}]$. Check that

$$\mathtt{H}_F = \sum_K \mathtt{Q}_K,$$

where the sum is over all faces K for which there exists a face G in the image of sec such that $FG = K$ and $GF = G$. In particular, $\mathtt{H}_O$ is obtained by summing over all faces in the image of sec. (Also see Exercise 11.8).

Exercise 11.34. Let E be any Eulerian family. Establish the following identities using the outline given below.

$$\mathtt{E}_{\mathrm{s}(F)} \cdot \mathtt{Q}_F = \mathtt{E}_{\mathrm{s}(F)}, \quad \mathtt{Q}_F \cdot \mathtt{E}_{\mathrm{s}(F)} = \mathtt{E}_{\mathrm{s}(F)} \quad \text{and} \quad \mathtt{Q}_F \cdot \mathtt{Q}_G = \mathtt{Q}_F \text{ if } \mathrm{s}(F) = \mathrm{s}(G).$$

Use (11.26) to obtain the first two identities and use either of them to deduce the third. Alternatively, use (11.29) to obtain the third identity and then use (11.28) to deduce the first two.

Exercise 11.35. Fix a flat X. Let h_{X} denote the linear span of $\mathtt{Q}_F$, as F varies over all faces with support X. Use (11.26) and (11.29) to deduce that:

- h_{X} is a left ideal of Σ. Moreover, it is the left ideal of Σ generated by $\mathtt{E}_{\mathrm{X}}$.
- The radical of h_{X} is linearly spanned by elements of the form $\mathtt{Q}_F - \mathtt{Q}_G$, where F and G both have support X.
- The quotient of h_{X} by its radical is one-dimensional with multiplicative character χ_{X}. (The latter is defined in (9.48).)

For X $= \top$, the left ideal $\mathsf{h}_\top$ generated by $\mathtt{E}_\top$ coincides with the left ideal of chambers Γ. (For further perspective, see the discussion in Section 13.6.3.)

11.4.6. Over and under a flat. Cartesian product. The discussion of Section 11.2.7 implies:

Lemma 11.36. *A* $\mathtt{Q}$*-basis of* $\Sigma[\mathcal{A}]$ *induces a* $\mathtt{Q}$*-basis of* $\Sigma[\mathcal{A}_H]$ *for every face* H *and a* $\mathtt{Q}$*-basis of* $\Sigma[\mathcal{A}^{\mathrm{X}}]$ *for every flat* X*. Similarly, a* $\mathtt{Q}$*-basis of* $\Sigma[\mathcal{A}]$ *and a* $\mathtt{Q}$*-basis of* $\Sigma[\mathcal{A}']$ *induce a* $\mathtt{Q}$*-basis of* $\Sigma[\mathcal{A} \times \mathcal{A}']$.

Aspects of the passage to the arrangement over a flat are discussed in more detail below.

Lemma 11.37. *For* $F \geq H$,

$$\mathtt{H}_{F/H} = \sum_{K:\, F \leq K} \mathtt{u}_F^K \, \mathtt{Q}_{K/H} \quad \textit{and} \quad \mathtt{Q}_{F/H} = \sum_{G:\, F \leq G} \mathtt{a}_F^G \, \mathtt{H}_{G/H}. \tag{11.36}$$

These are the change of basis formulas for the $\mathtt{H}$- and $\mathtt{Q}$-bases of $\Sigma[\mathcal{A}_H]$.

PROOF. By either (11.9) or equivalently (11.11), the first formula is a restatement of the first formula in (11.33). The second formula follows by inverting the matrix of coefficients $(\mathtt{u}_F^K)$ with $H \leq F \leq K$. The coefficients of the inverse will match the $\mathtt{a}_F^G$. □

Lemma 11.38. *We have*

$$\Delta_G(\mathtt{Q}_K) = \begin{cases} \mathtt{Q}_{GK/G} & \textit{if } KG = K, \\ 0 & \textit{otherwise,} \end{cases} \qquad \textit{and} \qquad \mu_F(\mathtt{Q}_{K/F}) = \mathtt{Q}_K, \tag{11.37}$$

with Δ_G *as in* (9.71) *and* μ_F *as in* (9.74)*. In particular,*

$$\mathtt{Q}_F = \mu_F(\mathtt{Q}_{F/F}). \tag{11.38}$$

PROOF. For the first formula: By (11.26) and the fact that Δ_G is an algebra homomorphism,

$$\Delta_G(\mathtt{Q}_K) = \Delta_G(\mathtt{H}_K \cdot \mathtt{E}_{\mathrm{s}(K)}) = \Delta_G(\mathtt{H}_K) \cdot \Delta_G(\mathtt{E}_{\mathrm{s}(K)}).$$

If $KG \neq K$, then the rhs is zero by the Saliola lemma (Lemma 11.12). If $KG = K$, then the calculation continues using (11.21) as follows.

$$\Delta_G(\mathtt{H}_K) \cdot \Delta_G(\mathtt{E}_{\mathrm{s}(K)}) = \mathtt{H}_{GK/G} \cdot \mathtt{E}_{\mathrm{s}(K)/G} = \mathtt{Q}_{GK/G}.$$

The last step again used (11.26) but for $\mathcal{A}_G$.

For the second formula:

$$\mu_F(\mathtt{Q}_{K/F}) = \mu_F(\Delta_F(\mathtt{Q}_K)) = \mathtt{H}_F \cdot \mathtt{Q}_K = \mathtt{Q}_K.$$

We used the first formula, (9.75) and (11.29) in that order. Alternatively, one may also use (9.74) and the second formula in (11.36). □

Now $\mathtt{Q}_{F/F} = \mathtt{E}_{\mathrm{s}(F)/F}$ is the first Eulerian idempotent of $\Sigma[\mathcal{A}_F]$ and μ_F is the inclusion map. Thus, by (11.38), each $\mathtt{Q}$-basis element of $\mathcal{A}$ is interpretable as the first Eulerian idempotent of an arrangement over a flat of $\mathcal{A}$.

Lemma 11.39. *If F and G have the same support, then*

$$\beta_{G,F}(\mathtt{Q}_{K/F}) = \mathtt{Q}_{GK/G}, \tag{11.39}$$

with $\beta_{G,F}$ as in (9.69). *In particular,*

$$\beta_{G,F}(\mathtt{Q}_{F/F}) = \mathtt{Q}_{G/G}. \tag{11.40}$$

PROOF. This follows from (9.72) and the first formula in (11.37). □

11.4.7. Projective Tits algebra. Let $\mathtt{E}$ be a projective Eulerian family with associated $\mathtt{Q}$-basis. The Eulerian idempotents $\mathtt{E}_{\mathrm{X}}$ belong to the projective Tits algebra, and hence so do the elements

$$\mathtt{Q}_{\{O,O\}} := \mathtt{Q}_O \quad \text{and} \quad \mathtt{Q}_{\{F,\overline{F}\}} := \mathtt{Q}_F + \mathtt{Q}_{\overline{F}} \text{ for } F \neq O. \tag{11.41}$$

These elements relate by triangularity to the $\mathtt{H}$-basis elements defined in (9.36). So they constitute a basis of the projective Tits algebra, which we refer to as the $\mathtt{Q}$-basis.

11.5. Families of Zie idempotents

Zie elements were introduced in Section 10.3. Recall that a Zie element is special if its coefficient of the central face is 1. We now show that an element of the Tits algebra is a special Zie element iff it is the first Eulerian idempotent of some Eulerian family. In particular, such elements exist. More generally, we consider special Zie families made up of special Zie elements in arrangements over flats and extend Proposition 11.9 as follows.

Theorem 11.40. *The following pieces of data are equivalent.*

- *A homogeneous section* $\mathtt{u}$ *of* $\mathcal{A}$.
- *An Eulerian family* $\mathtt{E}$ *of* $\mathcal{A}$.
- *A special Zie family* P *of* $\mathcal{A}$.

Homogeneous sections are trivial to construct. All one needs to do is to assign scalars to each face such that the sum in each flat is 1. In contrast, construction of Eulerian families or of special Zie elements is completely nontrivial.

11.5.1. Special Zie families. A *Zie family* of $\mathcal{A}$ is a set $\mathsf{P} := \{\mathsf{P}_{\mathrm{X}}\}_{\mathrm{X}\in\Pi}$ indexed by flats, where each P_{X} is a Zie element of the arrangement $\mathcal{A}_{\mathrm{X}}$ over X. A Zie family P is *special* if each P_{X} is a special Zie element of $\mathcal{A}_{\mathrm{X}}$, and, in particular, $\mathsf{P}_{\perp}$ is a special Zie element of $\mathcal{A}$.

11.5.2. From an Eulerian family to a special Zie family. Suppose $\mathtt{E}$ is an Eulerian family. Recall: Using this family, one can define the $\mathtt{Q}$-basis of $\Sigma[\mathcal{A}]$, with basis element $\mathtt{Q}_F$ given by (11.26). Written in the $\mathtt{H}$-basis, $\mathtt{Q}_F$ only involves faces greater than F. The element $\mathtt{Q}_O$ equals the first Eulerian idempotent $\mathtt{E}_{\perp}$. Similarly, the induced Eulerian family of $\mathcal{A}_F$ yields a $\mathtt{Q}$-basis of $\Sigma[\mathcal{A}_F]$ for any face F. The element $\mathtt{Q}_{F/F}$ equals the first Eulerian idempotent $\mathtt{E}_{\mathrm{s}(F)/F}$.

Now, for each flat X, define an element of $\Sigma[\mathcal{A}_{\mathrm{X}}]$ by

$$\mathsf{P}_{\mathrm{X}} := \beta_{\mathrm{X},F}(\mathtt{Q}_{F/F}), \tag{11.42}$$

where F is any face with support X, and $\beta_{\mathrm{X},F}$ is as in (9.70). (It follows from (11.40) that the rhs does not depend on the specific choice of F.) In particular,

$$\mathsf{P}_{\perp} = \mathtt{Q}_O = \mathtt{E}_{\perp}.$$

The Saliola lemma (Lemma 11.12) and the Friedrichs criterion (Lemma 10.19) imply that the first Eulerian idempotent $\mathtt{E}_{\perp}$ is a Zie element of $\mathcal{A}$. Also it is clearly special. Similarly, for each flat X, P_{X} is a special Zie element of $\mathcal{A}_{\mathrm{X}}$ since $\mathtt{Q}_{F/F}$ is the first Eulerian idempotent of $\Sigma[\mathcal{A}_F]$ and $\beta_{\mathrm{X},F}$ is an algebra isomorphism. Thus, we have constructed a special Zie family $\{\mathsf{P}_{\mathrm{X}}\}$.

11.5.3. From a special Zie family to an Eulerian family. We saw how to go from an Eulerian family to a special Zie family. Now we show that this procedure can be reversed. Accordingly:

- Suppose we are given a special Zie element P_{X} of $\mathcal{A}_{\mathrm{X}}$, one for each flat X of $\mathcal{A}$.

Starting from this data, we construct a homogeneous section $\mathtt{u}$, and its associated Eulerian family $\mathtt{E}$ and the $\mathtt{Q}$-basis. In fact, we will first get hold of the $\mathtt{Q}$-basis, and use it to define $\mathtt{E}$ and $\mathtt{u}$. Details follow.

For any face F with support X, define

$$\mathtt{Q}_{F/F} := \beta_{F,\mathrm{X}}(\mathsf{P}_{\mathrm{X}}) \quad \text{and} \quad \mathtt{Q}_F := \mu_F(\mathtt{Q}_{F/F}), \tag{11.43}$$

where μ_F is as in (9.74). Observe that $\mathtt{Q}_F$ is of the form

$$\mathtt{Q}_F = \mathtt{H}_F + \sum_{G:\, G>F} \mathtt{a}_F^G\, \mathtt{H}_G \tag{11.44}$$

for some scalars $\mathtt{a}_F^G$. (Since P_{X} is special, the coefficient of $\mathtt{H}_F$ in $\mathtt{Q}_F$ is 1.) By triangularity, the set $\{\mathtt{Q}_F\}$ as F varies is a basis of $\Sigma[\mathcal{A}]$. We next claim that Lemma 11.31 holds.

- If F and G have the same support, then $\mathtt{H}_F \cdot \mathtt{Q}_G = \mathtt{Q}_F$. This is (11.30). ($\mathtt{H}_F \cdot \mathtt{Q}_G = \mu_F \Delta_F \mu_G(\mathtt{Q}_{G/G}) = \mu_F \beta_{F,G} \Delta_G \mu_G(\mathtt{Q}_{G/G}) = \mu_F \beta_{F,G}(\mathtt{Q}_{G/G}) = \mu_F(\mathtt{Q}_{F/F}) = \mathtt{Q}_F$.)
- If $GF = G$, then $\mathtt{H}_F \cdot \mathtt{Q}_G = \mathtt{Q}_{FG}$. This is the first alternative in (11.29). (Using the previous case, $\mathtt{H}_F \cdot \mathtt{Q}_G = \mathtt{H}_F \cdot \mathtt{H}_G \cdot \mathtt{Q}_G = \mathtt{H}_{FG} \cdot \mathtt{Q}_G = \mathtt{Q}_{FG}$.)
- If $GF > G$, then $\mathtt{H}_F \cdot \mathtt{Q}_G = 0$. This is the second alternative in (11.29). (Similarly, $\mathtt{H}_F \cdot \mathtt{Q}_G = \mathtt{H}_F \cdot \mathtt{H}_G \cdot \mathtt{Q}_G = \mathtt{H}_{FG} \cdot \mathtt{Q}_G = \mathtt{H}_{FG} \cdot \mathtt{H}_{GF} \cdot \mathtt{Q}_G = \mathtt{H}_{FG} \cdot \mu_G(\mathtt{H}_{GF/G}) \cdot \mu_G(\mathtt{Q}_{G/G}) = \mathtt{H}_{FG} \cdot \mu_G(\mathtt{H}_{GF/G} \cdot \mathtt{Q}_{G/G}) = 0$ by the Friedrichs criterion since $\mathtt{Q}_{G/G}$ is a Zie element of $\Sigma[\mathcal{A}_G]$.)

Further, for any face F,

$$\mathrm{s}(\mathtt{Q}_F) = \mathtt{Q}_{\mathrm{s}(F)}. \tag{11.45}$$

The required calculation is

$$\mathrm{s}(\mathtt{Q}_F) = \mathrm{s}(\mu_F \beta_{F,\mathrm{X}}(\mathsf{P}_{\mathrm{X}})) = \mu_{\mathrm{X}}\, \mathrm{s}(\mathsf{P}_{\mathrm{X}}) = \mu_{\mathrm{X}}(\mathtt{Q}_{\mathrm{X}/\mathrm{X}}) = \mathtt{Q}_{\mathrm{X}},$$

with $\mathrm{X} = \mathrm{s}(F)$. The second step used the first diagram in Exercise 9.63, while the third step used the forward implication of Lemma 10.24 for the arrangement $\mathcal{A}_{\mathrm{X}}$.

Now we proceed to the construction of $\mathtt{E}$ and $\mathtt{u}$. Since $\mathtt{Q}$ is a basis, there exist unique scalars $\mathtt{u}^F$ such that

$$\mathtt{H}_O = \sum_F \mathtt{u}^F \mathtt{Q}_F. \tag{11.46}$$

Now set

$$\mathtt{E}_{\mathrm{X}} := \sum_{F:\, \mathrm{s}(F)=\mathrm{X}} \mathtt{u}^F \mathtt{Q}_F \quad \text{and} \quad \mathtt{u}_{\mathrm{X}} := \sum_{F:\, \mathrm{s}(F)=\mathrm{X}} \mathtt{u}^F \mathtt{H}_F. \tag{11.47}$$

By construction, (11.18) holds, that is, the sum of the $\mathtt{E}_{\mathrm{X}}$ is $\mathtt{H}_O$. We now claim that (11.3), (11.14) and (11.26) hold with $\mathtt{Q}$ as in (11.43) and $\mathtt{E}_{\mathrm{X}}$ and $\mathtt{u}_{\mathrm{X}}$ as in (11.47).

Applying the support map to (11.46) and using (11.45), we obtain

$$\mathtt{H}_{\perp} = \sum_{F} \mathtt{u}^F \mathtt{Q}_{\mathrm{s}(F)} = \sum_{\mathrm{X}} \Big(\sum_{F:\, \mathrm{s}(F)=\mathrm{X}} \mathtt{u}^F \Big) \mathtt{Q}_{\mathrm{X}}.$$

A comparison with (9.2) shows that the sums in parenthesis are all 1. This proves (11.3). Thus, the scalars $\mathtt{u}^F$ indeed determine a homogeneous section $\mathtt{u}$. Formulas (11.30) and (11.3) imply (11.26). Using these along with the second alternative in (11.29), we deduce:

$$\mathtt{u}_{\mathrm{X}} \cdot \mathtt{E}_{\mathrm{X}} = \mathtt{E}_{\mathrm{X}} \qquad \text{and} \qquad \mathtt{u}_{\mathrm{X}} \cdot \mathtt{E}_{\mathrm{Y}} = 0 \ \text{ if } \mathrm{X} \not\leq \mathrm{Y}.$$

Now multiply both sides of (11.18) on the left by $\mathtt{u}_{\mathrm{X}}$ and use the above two identities to obtain

$$\mathtt{u}_{\mathrm{X}} = \mathtt{E}_{\mathrm{X}} + \sum_{\mathrm{Y}:\mathrm{X}<\mathrm{Y}} \mathtt{u}_{\mathrm{X}} \cdot \mathtt{E}_{\mathrm{Y}}.$$

This proves (11.14), and the claim is established.

Thus, we obtain an Eulerian family starting with a special Zie family. This construction is clearly inverse to the previous construction. This completes the proof of Theorem 11.40.

Exercise 11.41. List all special Zie families of the rank-one arrangement using (10.26). Compute the corresponding Eulerian families using the above procedure and check that we indeed get all of them. (Eulerian families of the rank-one arrangement are listed in Section 11.4.3.)

11.5.4. Special Zie elements.

We record the following important characterization of special Zie elements.

Lemma 11.42. *For an element z of the Tits algebra, the following are equivalent.*

(1) *z is a special Zie element.*
(2) *$z = \mathtt{E}_{\perp}$, the first Eulerian idempotent of some Eulerian family $\mathtt{E}$.*
(3) *z is an idempotent which lifts $\mathtt{Q}_{\perp}$.*

PROOF. For the equivalence between (2) and (3), set $\mathrm{X} = \perp$ in Lemma 11.27. The equivalence between (1) and (2) can be viewed as the first step in the equivalence between special Zie families and Eulerian families. More directly, (1) implies (3) by the forward implication of Lemma 10.24, while (2) implies (1) by the Saliola lemma and the Friedrichs criterion. □

This result subsumes Lemma 10.24. (Recall that the proof of the backward implication of the latter made use of the existence of a special Zie element.) Also observe that Lemma 10.23 is contained in Lemma 11.22.

11.5.5. Abstract approach to Q-bases. In Section 11.4, we attached a Q-basis to an Eulerian family. However, one can also define Q-bases directly without any reference to Eulerian families. This is done as follows.

We say that a basis of the Tits algebra is a Q-basis if each basis element $\mathtt{Q}_F$ is of the form (11.44) and Lemma 11.31 holds.

Proposition 11.43. *Special Zie families of $\mathcal{A}$ are in correspondence with* Q*-bases of the Tits algebra of $\mathcal{A}$.*

PROOF. Starting with a special Zie family, we define the Q-basis via (11.43). Subsequent analysis shows that (11.44) and (11.29) hold. Conversely, suppose we are given a Q-basis. Using the second alternative in (11.29), we deduce by the Friedrichs criterion that $\mathtt{Q}_O$ is a special Zie element. More generally, the element $\mathtt{Q}_{F/F}$ defined via $\mathtt{Q}_F = \mu_F(\mathtt{Q}_{F/F})$ is a special Zie element of $\mathcal{A}_F$. Now define the special Zie family P via (11.42). Formula (11.30) shows that P_{X} does not depend on the specific choice of F. □

It follows that abstract Q-bases are in correspondence with Eulerian families. To go from an Eulerian family to a Q-basis, we use (11.26). To go from a Q-basis to an Eulerian family, we use (11.46) and (11.47).

Exercise 11.44. Starting with the above definition of a Q-basis, deduce that each $\mathtt{Q}_F$ is an idempotent.

11.5.6. Projective Zie families. A Zie family P is *projective* if each Zie element P_{X} belongs to the projective Tits algebra of $\mathcal{A}_{\mathrm{X}}$.

Proposition 11.45. *Projective Eulerian families of $\mathcal{A}$ are in correspondence with projective special Zie families of $\mathcal{A}$.*

This is obtained by restricting the correspondence in Theorem 11.40.

PROOF. Suppose each $\mathtt{E}_{\mathrm{X}}$ is projective. Then by Exercise 9.65, $\mathtt{Q}_{F/F} = \Delta_F(\mathtt{E}_{\mathrm{s}(F)})$ is projective. So, by definition (11.42), each P_{X} is projective. Conversely, suppose P is a projective special Zie family. Then, with the Q-basis defined as in (11.43), the opposition map sends each $\mathtt{Q}_F$ to $\mathtt{Q}_{\overline{F}}$. Now applying the opposition map to (11.46), we see that $\mathtt{u}^F = \mathtt{u}^{\overline{F}}$. So $\mathtt{u}$ is projective from which we deduce that $\mathtt{E}$ is projective using Proposition 11.16. □

As with the Tits algebra, one can also take an abstract approach to Q-bases of the projective Tits algebra. We say that a Q-basis of the Tits algebra is *projective* if the opposition map sends $\mathtt{Q}_F$ to $\mathtt{Q}_{\overline{F}}$ for each F (or equivalently, if each $\mathtt{Q}_{F/F}$ belongs to the projective Tits algebra of $\mathcal{A}_F$). Further, we say that a basis of the projective Tits algebra is a Q-basis if it is obtained from a projective Q-basis of the Tits algebra via (11.41).

Proposition 11.46. *Projective special Zie families of $\mathcal{A}$ are in correspondence with* Q*-bases of the projective Tits algebra of $\mathcal{A}$.*

This is obtained by restricting the correspondence in Proposition 11.43.

11.5.7. Over and under a flat. Cartesian product. The compatibility between homogeneous sections and Eulerian families given in Lemmas 11.17, 11.18 and 11.19 can be extended to include special Zie families as follows.

Suppose P_X is a Zie element of $\mathcal{A}_X$. For a flat X containing a face H, let $\mathsf{P}_{X/H}$ denote the Zie element of $\mathcal{A}_{X/H}$ corresponding to P_X. A special Zie family $\mathsf{P} = \{\mathsf{P}_X\}$ of $\mathcal{A}$ induces a special Zie family of $\mathcal{A}_H$, namely,

$$\mathsf{P}_H := \{\mathsf{P}_{X/H}\}.$$

Lemma 11.47. *Suppose for an arrangement $\mathcal{A}$,*

$$\mathtt{u} \longleftrightarrow \mathtt{E} \longleftrightarrow \mathsf{P}$$

under the equivalences of Theorem 11.40. *Then for the arrangement $\mathcal{A}_H$,*

$$\mathtt{u}_H \longleftrightarrow \mathtt{E}_H \longleftrightarrow \mathsf{P}_H.$$

Suppose P_Y is a Zie element of $\mathcal{A}_Y$. Let $X \geq Y$. Let P^X_Y denote the Zie element of $\mathcal{A}^X_Y$ obtained by truncating to faces contained in X. A special Zie family $\mathsf{P} = \{\mathsf{P}_Y\}$ of $\mathcal{A}$ induces a special Zie family of $\mathcal{A}^X$, namely,

$$\mathsf{P}^X := \{\mathsf{P}^X_Y\}.$$

Lemma 11.48. *Suppose for an arrangement $\mathcal{A}$,*

$$\mathtt{u} \longleftrightarrow \mathtt{E} \longleftrightarrow \mathsf{P}$$

under the equivalences of Theorem 11.40. *Then for the arrangement $\mathcal{A}^X$,*

$$\mathtt{u}^X \longleftrightarrow \mathtt{E}^X \longleftrightarrow \mathsf{P}^X.$$

Suppose P_X is a Zie element of $\mathcal{A}_X$ and $\mathsf{P}'_{X'}$ is a Zie element of $\mathcal{A}_{X'}$. Then $\mathsf{P}_X \otimes \mathsf{P}'_{X'}$ is a Zie element of $(\mathcal{A} \times \mathcal{A}')_{(X,X')}$. A special Zie family $\mathsf{P} = \{\mathsf{P}_X\}$ of $\mathcal{A}$ and a special Zie family $\mathsf{P}' = \{\mathsf{P}'_{X'}\}$ of $\mathcal{A}'$ induces a special Zie family of $\mathcal{A} \times \mathcal{A}'$, namely,

$$\mathsf{P} \times \mathsf{P}' := \{\mathsf{P}_X \otimes \mathsf{P}'_{X'}\}_{(X,X')}.$$

Lemma 11.49. *Suppose for arrangements $\mathcal{A}$ and $\mathcal{A}'$,*

$$\mathtt{u} \longleftrightarrow \mathtt{E} \longleftrightarrow \mathsf{P} \qquad \textit{and} \qquad \mathtt{u}' \longleftrightarrow \mathtt{E}' \longleftrightarrow \mathsf{P}'$$

under the equivalences of Theorem 11.40. *Then for the arrangement $\mathcal{A} \times \mathcal{A}'$,*

$$\mathtt{u} \times \mathtt{u}' \longleftrightarrow \mathtt{E} \times \mathtt{E}' \longleftrightarrow \mathsf{P} \times \mathsf{P}'.$$

Exercise 11.50. Make Lemma 11.36 explicit for abstract $\mathtt{Q}$-bases. Further, extend Lemmas 11.47, 11.48 and 11.49 to include abstract $\mathtt{Q}$-bases (by making use of Proposition 11.43).

11.5.8. Naturality under partial-support maps. All preceding constructions and results generalize to a left regular band. It is of interest to understand how constructions for different LRBs relate under a morphism of monoids. For a partial-support map, the situation is as follows.

Recall from Section 2.8 that for any partial-support relation $\sim$, the set of partial-flats $\Sigma_\sim$ is a left regular band. Let $\mathrm{s}_\sim : \Sigma \to \Sigma_\sim$ denote the partial-support map. It is a morphism of monoids. We use the same notation for its linearization. Then a homogeneous section $\mathtt{u}$, an Eulerian family $\mathtt{E}$ and a special Zie family P

on Σ induce, respectively, a homogeneous section $\mathtt{u}_\sim$, an Eulerian family $\mathtt{E}_\sim$ and a special Zie family $\mathtt{P}_\sim$ on $\Sigma_\sim$. Further,

$$\mathtt{u} \longleftrightarrow \mathtt{E} \longleftrightarrow \mathtt{P} \text{ implies } \mathtt{u}_\sim \longleftrightarrow \mathtt{E}_\sim \longleftrightarrow \mathtt{P}_\sim.$$

In this setup, we have

$$\mathrm{s}_\sim(\mathtt{Q}_F) = \mathtt{Q}_{\mathrm{s}_\sim(F)}. \tag{11.48}$$

Now consider the specialization $\Sigma_\sim = \Pi$. In this case, there is a unique homogeneous section, resulting in a unique Eulerian family, which is nothing but the $\mathtt{Q}$-basis of Π. Hence the support of any Eulerian family (and the associated $\mathtt{Q}$-basis) of Σ is the $\mathtt{Q}$-basis of Π. This was the content of (11.23) and (11.27).

11.6. Eulerian idempotents for good reflection arrangements

Recall from Section 5.7 that a good reflection arrangement is a reflection arrangement in which every arrangement under a flat is also a reflection arrangement. We give cancelation-free formulas for Eulerian idempotents of good reflection arrangements associated to the uniform section (assuming characteristic 0). Important specializations of these formulas, namely, to the braid arrangement and the arrangement of type B are discussed later in Sections 12.5 and 12.6.

11.6.1. The uniform section. Let $\mathtt{u}$ be the uniform section of a good reflection arrangement $\mathcal{A}$, that is, faces with the same support are assigned the same scalar. In other words,

$$\mathtt{u}^F := \frac{1}{c^F},$$

where c^F is the number of faces of support $\mathrm{s}(F)$. Then, for any face H, the induced section $\mathtt{u}_H$ on $\mathcal{A}_H$ is also uniform. This follows from Lemma 5.31. This property is indeed special, see the last part of Exercise 11.8 in this regard.

Lemma 11.51. *For the uniform section of a good reflection arrangement $\mathcal{A}$,*

$$\mathtt{u}^G_F = \frac{1}{c^G_F} \quad \text{and} \quad \mathtt{a}^G_F = \frac{\mu(\mathcal{A}^G_F)}{c^G_F}, \tag{11.49}$$

with $\mathtt{a}^G_F$ as in (11.31). *(Recall that c^G_F denotes the number of chambers in $\mathcal{A}^G_F$.)*

PROOF. The first formula follows from the preceding discussion. The second formula then follows from Lemma 5.30. ☐

11.6.2. $\mathtt{H}$- and $\mathtt{Q}$-bases.

Lemma 11.52. *For the uniform section of a good reflection arrangement $\mathcal{A}$, the $\mathtt{H}$- and $\mathtt{Q}$-bases are related by*

$$\mathtt{H}_F = \sum_{G:\, F \leq G} \frac{1}{c^G_F} \mathtt{Q}_G \quad \text{and} \quad \mathtt{Q}_F = \sum_{G:\, F \leq G} \frac{\mu(\mathcal{A}^G_F)}{c^G_F} \mathtt{H}_G. \tag{11.50}$$

PROOF. This is obtained by specializing (11.33) using (11.49). ☐

In the sums, faces with the same support can be lumped together. For instance,

$$\mathtt{Q}_F = \sum_{\mathrm{X}:\, \mathrm{X} \geq \mathrm{s}(F)} \frac{\mu(\mathcal{A}^{\mathrm{X}}_F)}{c^{\mathrm{X}}_F} \sum_{G:\, F \leq G,\, \mathrm{s}(G) = \mathrm{X}} \mathtt{H}_G.$$

11.6.3. Eulerian idempotents.

Theorem 11.53. *For a good reflection arrangement $\mathcal{A}$, the Eulerian idempotents for the uniform section are given by*

$$\mathtt{E}_{\mathrm{X}} = \frac{1}{c^{\mathrm{X}}} \sum_{F:\, \mathrm{s}(F)=\mathrm{X}} \mathtt{Q}_F = \frac{1}{c^{\mathrm{X}}} \sum_{F:\, \mathrm{s}(F)=\mathrm{X}} \ \sum_{G:\, F \leq G} \frac{\mu(\mathcal{A}_F^G)}{c_F^G} \mathtt{H}_G, \tag{11.51}$$

where c^{X} is the number of faces with support X.

In particular, the first Eulerian idempotent is

$$\mathtt{E}_\perp = \mathtt{Q}_O = \sum_F \frac{\mu(\mathcal{A}^F)}{c^F} \mathtt{H}_F = \sum_{\mathrm{X}} \frac{\mu(\perp, \mathrm{X})}{c^{\mathrm{X}}} \sum_{F:\, \mathrm{s}(F)=\mathrm{X}} \mathtt{H}_F. \tag{11.52}$$

PROOF. This follows from (11.28) and the second formula in (11.50). □

Here is another way to deduce the above formulas. Since $\mathcal{A}$ is a reflection arrangement, the uniform section and hence $\mathtt{E}_\perp$ is invariant under the action of the Coxeter group. In particular, the coefficients of all chambers in $\mathtt{E}_\perp$ must be equal. Since $\mathtt{E}_\perp$ is a special Zie element, the sum of the chamber coefficients is $\mu(\perp, \top)$ by (10.15), so each coefficient is $\frac{\mu(\perp,\top)}{c^\top}$. Let $\mathtt{E}_\perp^{\mathrm{X}}$ denote the part of $\mathtt{E}_\perp$ consisting of faces of support smaller than X. By Lemma 11.18, $\mathtt{E}_\perp^{\mathrm{X}}$ is the first Eulerian idempotent of $\mathcal{A}^{\mathrm{X}}$, which is also a reflection arrangement and $\mathtt{u}^{\mathrm{X}}$ is uniform. Hence applying the result just proved, the coefficient of any face of support X in $\mathtt{E}_\perp$ is $\frac{\mu(\perp,\mathrm{X})}{c^{\mathrm{X}}}$. This proves (11.52). This formula in turn can be used to deduce the second formula in (11.50): Use that $\mathtt{Q}_{F/F}$ is the first Eulerian idempotent of $\mathcal{A}_F$ for $\mathtt{u}_F$, and $\mathcal{A}_F$ is good and $\mathtt{u}_F$ is uniform, and finally apply (11.38).

Corollary 11.54. *For any arrangement, the element*

$$\sum_{\mathrm{X}} \frac{\mu(\perp, \mathrm{X})}{c^{\mathrm{X}}} \sum_{F:\, \mathrm{s}(F)=\mathrm{X}} \mathtt{H}_F \tag{11.53}$$

satisfies the equivalent conditions of Lemma 10.18. *In addition, for a good reflection arrangement, it is a special Zie element.*

PROOF. It is easiest to see that the element (11.53) satisfies the condition (10.15). Moreover, for a good reflection arrangement, this element is the first Eulerian idempotent for the uniform section by Theorem 11.53, and hence is a special Zie element by Lemma 11.42. □

Exercise 11.55. Show directly using the Friedrichs criterion (Lemma 10.19) that (11.53) is a special Zie element for a good reflection arrangement. (Use the Weisner formula (1.46a).) This result is false in general for reflection arrangements. A counterexample is given in Corollary 11.75.

11.6.4. Lumping Eulerian idempotents. Let $\mathtt{E}_{\mathrm{X}}$ be the Eulerian idempotents defined in (11.51). For $k \geq 0$, set

$$\mathtt{E}_k := \sum_{\mathrm{X}:\, \mathrm{rk}(\mathrm{X})=k} \mathtt{E}_{\mathrm{X}}. \tag{11.54}$$

Note that the $\mathtt{E}_k$ form a family of orthogonal idempotents, but they are no longer primitive. Using (11.51) and Lemma 5.30, we obtain:

Theorem 11.56. *For a good reflection arrangement,*

$$\mathtt{E}_k = \sum_{\mathrm{X}} \frac{\mathrm{wy}(\mathcal{A}^{\mathrm{X}}, k)}{c^{\mathrm{X}}} \sum_{F:\, \mathrm{s}(F)=\mathrm{X}} \mathtt{H}_F,$$

where $\mathrm{wy}(\mathcal{A}, k)$ *are the Whitney numbers of the first kind defined in* (1.56).

An alternative proof of this result is given after Theorem 12.71.

11.7. Extension problem and dimension of Lie

Recall the left module of chambers $\Gamma[\mathcal{A}]$. Given elements $v_H \in \Gamma[\mathcal{A}_H]$, one for every noncentral face H, which are mutually compatible in a certain sense, can one find an element of $\Gamma[\mathcal{A}]$ whose Tits projection on each H is precisely v_H? This is the extension problem. We show that this problem always has a solution and further the solutions form an affine subspace which is a translate of $\mathsf{Lie}[\mathcal{A}]$, the space of Lie elements. Moreover, when the v_H are projective, we can say the following: When $\mathcal{A}$ has odd rank, there is a unique projective solution, and when $\mathcal{A}$ has even rank, all solutions are necessarily projective.

As a consequence, we rederive that the dimension of $\mathsf{Lie}[\mathcal{A}]$ is $|\mu(\mathcal{A})|$, the absolute value of the Möbius number of $\mathcal{A}$. The Zaslavsky formula also emerges as a byproduct.

11.7.1. Extension problem. For any function u on the set of chambers of $\mathcal{A}$, let u^C denote the value of u on the chamber C. (Such a function may be identified with the chamber element $\sum_C u^C \mathtt{H}_C$.) The *content* of u is defined to be $\sum_C u^C$. For any face H, such a function u induces a function u_H on the set of chambers of $\mathcal{A}_H$, whose value on a chamber D/H is given by

$$u_H^D := \sum_{C:\, HC=D} u^C. \tag{11.55}$$

(Compare with (11.6).) Note that the content of u_H (which is the sum of u_H^D over D) is the same as the content of u. Further note by (10.1) that u is a Lie element iff $u_H = 0$ for all $H > O$.

Consider the following linear system in the variables u^C:

$$\sum_{C:\, HC=D} u^C = v_H^D.$$

There is one equation for each $O < H \leq D$, and the v_H^D are specified constants.

This may be rephrased as follows. Given a function v_H on the chambers of $\mathcal{A}_H$ for each $H > O$, find all functions u on the chambers of $\mathcal{A}$ such that for any face $H > O$,

$$u_H = v_H. \tag{11.56}$$

We call this the *extension problem* since it involves finding u by knowing its induced values in all proper stars.

For the extension problem to have a solution, it is clear that the v_H must satisfy the following compatibility conditions. For any $G \geq H$,

$$(v_H)_{G/H} = v_G, \tag{11.57}$$

and for any F and G with the same support,

$$\beta_{G,F}(v_F) = v_G. \tag{11.58}$$

(Compare with Lemma 11.7.) We show below that these conditions are also sufficient to have a solution.

Theorem 11.57. *Let $\mathcal{A}$ be any arrangement. Suppose for each face $H > O$, we are given a function v_H on the chambers of $\mathcal{A}_H$. Then, there exists a function u on the chambers of $\mathcal{A}$ satisfying* (11.56) *for each $H > O$ iff the v_H satisfy the compatibility conditions* (11.57) *and* (11.58). *Further, the affine space of all solutions for u has dimension equal to the dimension of* $\mathsf{Lie}[\mathcal{A}]$.

PROOF. Suppose u is a solution. Then

$$\begin{aligned} u' \text{ is a solution } &\iff u - u' \text{ satisfies } (u-u')_H = 0 \text{ for each } H > O \\ &\iff u - u' \text{ is a Lie element.} \end{aligned}$$

Thus, the second part follows from the first.

We now show that the compatibility conditions imply the existence of u. Any v_H which satisfy (11.57) and (11.58) must each have the same content. So we can normalize the content to be 1, unless the content of each v_H is zero. Let us suppose that this is not the case and proceed. We first give the gist of the construction. Theorem 11.40 plays a key role.

- Pick any homogeneous section u' of $\mathcal{A}$.
- Construct a special Zie family P using u' and the v_H.
- Construct a homogeneous section u of $\mathcal{A}$ from the special Zie family P. Due to the tweaking done using the v_H, it will be different from u' in general. The scalars $u^D := \mathsf{u}^D$ yield a solution.

The construction of P goes as follows. Let $\mathsf{P}_\perp$ be the first Eulerian idempotent of $\mathcal{A}$ associated to u'. (We ignore the higher Eulerian idempotents.) For each $H > O$, define a homogeneous section v_H on $\mathcal{A}_H$:

$$(\mathsf{v}_H)^K := \begin{cases} v_H^K & \text{if } K \text{ is a chamber,} \\ (\mathsf{u}')_H^K & \text{otherwise.} \end{cases}$$

The first case uses the given functions v_H of content 1, while the second case uses the arbitrarily chosen homogeneous section u'. Let $\mathsf{Q}_{H/H}$ be the first Eulerian idempotent of $\mathcal{A}_H$ associated to v_H. The compatibility relation (11.58) implies that for any F and G with the same support,

$$\beta_{G,F}(\mathsf{Q}_{F/F}) = \mathsf{Q}_{G/G}.$$

This defines a special Zie element P_{X} of $\mathcal{A}_{\mathrm{X}}$, namely,

$$\mathsf{P}_{\mathrm{X}} := \beta_{\mathrm{X},F}(\mathsf{Q}_{F/F}),$$

where F is any face of support F. Now define u to be the homogeneous section of $\mathcal{A}$ corresponding to the special Zie family P. Fix a $H > O$. By Lemma 11.47, the special Zie family P_H corresponds to the homogeneous section u_H of $\mathcal{A}_H$. However, by the compatibility relation (11.57), the same family also corresponds to the homogeneous section v_H. Hence, $\mathsf{u}_H = \mathsf{v}_H$ for all $H > O$. This says that $u^D := \mathsf{u}^D$ is a solution.

Now let us deal with case when each v_H has content zero. Take any function p on chambers with content 1. Put $v'_H := v_H + p_H$. These have content 1 and are mutually compatible. So applying the above result, there is a q such that $q_H = v'_H$ for all $H > O$. Then $u := q - p$ is the desired solution. Even more explicitly, fix a chamber C. Take p to be the function which is 1 on C and 0 on all other chambers. Put

$$(v'_H)^D := \begin{cases} v_H^D + 1 & \text{if } D = HC, \\ v_H^D & \text{otherwise.} \end{cases}$$

Let q be such that $q_H = v'_H$ for all $H > O$. Then u is obtained from q by subtracting 1 from the value of C. □

A more direct argument is given after Theorem 15.42.

11.7.2. Projective case. We now consider the extension problem for projective functions on chambers. Interestingly, the solution depends on the parity of $\operatorname{rk}(\mathcal{A})$. We assume that the field characteristic is not 2.

Lemma 11.58. *Let u be a projective function on the chambers of an arrangement $\mathcal{A}$. Then, for any chamber D and face $H \leq D$,*

$$\frac{1}{2} \sum_{G:\, H < G \leq D} (-1)^{\operatorname{rk}(G)+1} u_G^D = \begin{cases} (-1)^{\operatorname{rk}(H)} u_H^D & \text{if } \operatorname{rk}(\mathcal{A}_H) \text{ is odd,} \\ 0 & \text{if } \operatorname{rk}(\mathcal{A}_H) \text{ is even.} \end{cases} \tag{11.59}$$

PROOF. Apply the Witt identity (7.16) to obtain:

$$u_H^D + \sum_{G:\, H < G \leq D} (-1)^{\operatorname{rk}(G/H)} u_G^D = (-1)^{\operatorname{rk}(D/H)} u_H^{H\overline{D}}.$$

Since u is projective, $u_H^D = u_H^{H\overline{D}}$. Analyzing further according to the parity of $\operatorname{rk}(D/H)$ yields (11.59). □

Theorem 11.59. *Let $\mathcal{A}$ be any arrangement of odd rank. Suppose for each face $H > O$, we are given a projective function v_H on the chambers of $\mathcal{A}_H$ which satisfy the compatibility conditions* (11.57) *and* (11.58)*. Then there is a unique projective function u on the chambers of $\mathcal{A}$ satisfying* (11.56) *for each $H > O$. Explicitly, for a chamber C, the scalar u^C is given by*

$$u^C = \frac{1}{2} \sum_{F:\, O < F \leq C} (-1)^{\operatorname{rk}(F)+1} v_F^C. \tag{11.60}$$

PROOF. Formula (11.60) is obtained by setting $H = O$ in (11.59). This proves the uniqueness assertion. The same argument can be given in a different guise: Suppose u and u' are two projective solutions. Then $u - u'$ is a projective Lie element. However, by Lemma 10.12, the only such element in odd rank is 0. So $u = u'$ which proves uniqueness.

By Theorem 11.57, there exists a solution u. Since the v_H are projective, $\overline{u}$ defined by $\overline{u}^C := u^{\overline{C}}$ is also a solution. Hence $(u + \overline{u})/2$ is also a solution and it is projective. This proves existence. □

Exercise 11.60. Check directly that (11.60) is a solution (without appealing to Theorem 11.57).

Theorem 11.61. *Let $\mathcal{A}$ be any arrangement of even rank. Suppose for each face $H > O$, we are given a projective function v_H on the chambers of $\mathcal{A}_H$ which satisfy the compatibility conditions* (11.57) *and* (11.58)*. Then any function u on the* *chambers of $\mathcal{A}$ satisfying* (11.56) *for each $H > O$ is necessarily projective, and the affine space of all such u has dimension equal to the dimension of* $\mathsf{Lie}[\mathcal{A}]$.

PROOF. Using Theorem 11.57 and arguing as in the second para of the previous proof, a projective solution u exists. By (10.5), all Lie elements of even rank are projective. Further, any solution is of the form u plus a Lie element. So it must be projective. □

11.7.3. Dimension of Lie and Zaslavsky formula. We discuss some consequences of the extension problem.

Lemma 11.62. *For any arrangement $\mathcal{A}$,*

$$\sum_{\mathrm{X}\in\Pi} \dim(\mathsf{Lie}[\mathcal{A}_{\mathrm{X}}]) = \dim(\mathsf{\Gamma}[\mathcal{A}]). \tag{11.61}$$

Note that the rhs is the same as the number of chambers in $\mathcal{A}$.

PROOF. The space of all functions u on chambers has dimension equal to the number of chambers. We construct this space inductively. By Theorem 11.57, the dimension of the space of choices for defining u_G assuming u_K has been defined for all $K > G$, is $\dim(\mathsf{Lie}[\mathcal{A}_G])$. Also if F and G have the same support, then $u_F = u_G$. So the dimension of the space of all functions u on chambers is the lhs of (11.61). The result follows. □

Lemma 11.63. *For any arrangement $\mathcal{A}$ of rank at least* 1,

$$\sum_{\substack{\mathrm{X}\in\Pi \\ \mathrm{rk}(\mathrm{X}) \textit{ is even}}} \dim(\mathsf{Lie}[\mathcal{A}_{\mathrm{X}}]) = \frac{1}{2}\dim(\mathsf{\Gamma}[\mathcal{A}]) = \sum_{\substack{\mathrm{X}\in\Pi \\ \mathrm{rk}(\mathrm{X}) \textit{ is odd}}} \dim(\mathsf{Lie}[\mathcal{A}_{\mathrm{X}}]). \tag{11.62}$$

PROOF. Proceeding as in the previous proof, we count the dimension of the space of all projective functions u on chambers of $\mathcal{A}$. In view of Theorems 11.59 and 11.61, we obtain:

$$\sum_{\substack{\mathrm{X}\in\Pi \\ \mathrm{rk}(\mathcal{A}_{\mathrm{X}}) \text{ is even}}} \dim(\mathsf{Lie}[\mathcal{A}_{\mathrm{X}}]) = \frac{1}{2}\dim(\mathsf{\Gamma}[\mathcal{A}]).$$

This in conjunction with (11.61) implies (11.62). □

Theorem 11.64. *For any arrangement $\mathcal{A}$,*

$$\dim(\mathsf{Lie}[\mathcal{A}]) = (-1)^{\mathrm{rk}(\mathcal{A})}\mu(\mathcal{A}) = |\mu(\mathcal{A})|. \tag{11.63}$$

PROOF. We proceed by induction on the rank of $\mathcal{A}$. We know from (11.62) that for rank at least 1,

$$\sum_{\mathrm{X}\in\Pi} (-1)^{\mathrm{rk}(\mathcal{A}_{\mathrm{X}})} \dim(\mathsf{Lie}[\mathcal{A}_{\mathrm{X}}]) = 0.$$

Rewriting,

$$\begin{aligned}(-1)^{\mathrm{rk}(\mathcal{A})}\dim(\mathsf{Lie}[\mathcal{A}]) &= -\sum_{\mathrm{X}:\,\mathrm{X}>\bot} (-1)^{\mathrm{rk}(\mathcal{A}_{\mathrm{X}})}\dim(\mathsf{Lie}[\mathcal{A}_{\mathrm{X}}]) \\ &= -\sum_{\mathrm{X}:\,\mathrm{X}>\bot} \mu(\mathcal{A}_{\mathrm{X}}) \\ &= \mu(\mathcal{A}).\end{aligned}$$

The second step used the induction hypothesis, and the last step used (C.5b). This completes the induction step and proves the first equality. The second equality follows since this number must be nonnegative. □

The second equality in (11.63) is a nontrivial result, see Proposition 1.80. Interpreting the Möbius number as the dimension of $\mathsf{Lie}[\mathcal{A}]$ up to sign is a nice way of understanding this result. We further note that in view of (11.61), the Zaslavsky formula (1.48) is a consequence of (11.63).

11.8. Rank-two arrangements

Let $\mathcal{A}$ be the rank-two arrangement of n lines. In the spherical model, this is a polygon with $2n$ sides. We compute the Eulerian idempotents starting with an arbitrary homogeneous section, and later specialize to the projective and uniform cases.

Let $\mathtt{u}$ be any homogeneous section of $\mathcal{A}$. This defines the $\mathtt{Q}$-basis. The expressions for the $\mathtt{Q}$-basis elements in terms of the $\mathtt{H}$-basis elements can be computed using the second formula in (11.33). They are as follows.

$$\mathtt{Q}_C = \mathtt{H}_C.$$

For any vertex P,

$$\mathtt{Q}_P = \mathtt{H}_P - \mathtt{u}_P^C\,\mathtt{H}_C - \mathtt{u}_P^D\,\mathtt{H}_D,$$

where C and D are the two edges which are greater than P.

$$\mathtt{Q}_O = \mathtt{H}_O - \sum_P \mathtt{u}^P\,\mathtt{H}_P + \sum_C (-\mathtt{u}^C + \mathtt{u}^P\mathtt{u}_P^C + \mathtt{u}^Q\mathtt{u}_Q^C)\,\mathtt{H}_C.$$

The first sum is over all vertices P, while the second sum is over all edges C, with P and Q being its two vertices.

Using (11.28), we obtain the following formulas for the Eulerian idempotents.

$$\mathtt{E}_\top = \mathtt{u}_\top = \sum_C \mathtt{u}^C H_C.$$

For the line X supporting the vertices P and $\overline{P}$,

$$\begin{aligned}\mathtt{E}_{\mathrm{X}} &= \mathtt{u}^P\,\mathtt{Q}_P + \mathtt{u}^{\overline{P}}\,\mathtt{Q}_{\overline{P}}\\ &= \mathtt{u}^P\,\mathtt{H}_P + \mathtt{u}^{\overline{P}}\,\mathtt{H}_{\overline{P}} - (\mathtt{u}^P\mathtt{u}_P^C\,\mathtt{H}_C + \mathtt{u}^P\mathtt{u}_P^D\,\mathtt{H}_D + \mathtt{u}^{\overline{P}}\mathtt{u}_{\overline{P}}^{\overline{C}}\mathtt{H}_{\overline{C}} + \mathtt{u}^{\overline{P}}\mathtt{u}_{\overline{P}}^{\overline{D}}\,\mathtt{H}_{\overline{D}}),\end{aligned}$$

where C and D are the two edges which are greater than P.

$$\mathtt{E}_\perp = \mathtt{Q}_O = \mathtt{H}_O - \sum_P \mathtt{u}^P\,\mathtt{H}_P + \sum_C (-\mathtt{u}^C + \mathtt{u}^P\mathtt{u}_P^C + \mathtt{u}^Q\mathtt{u}_Q^C)\,\mathtt{H}_C.$$

The first sum is over all vertices P, while the second sum is over all edges C, with P and Q being its two vertices.

Proposition 11.65. *For the rank-two arrangement of n lines, the Eulerian idempotents associated to any projective section $\mathtt{u}$ are given by*

$$\mathtt{E}_\top = \sum_C \mathtt{u}^C H_C,$$

$$\mathtt{E}_{\mathrm{X}} = \frac{1}{2}\,\mathtt{H}_P + \frac{1}{2}\,\mathtt{H}_{\overline{P}} - \frac{1}{4}\,(\mathtt{H}_C + \mathtt{H}_D + \mathtt{H}_{\overline{C}} + \mathtt{H}_{\overline{D}}),$$

where X *is the support of the vertices* P *and* $\overline{P}$, *and* C *and* D *are the two edges which are greater than* P,

$$\mathtt{E}_\perp = \mathtt{H}_O - \frac{1}{2}\sum_P \mathtt{H}_P + \sum_C (-\mathtt{u}^C + \frac{1}{2})\,\mathtt{H}_C. \tag{11.64}$$

PROOF. In this case, $\mathtt{u}^P = 1/2$ for any vertex P, and $\mathtt{u}^C_P = 1/2$ for any edge $C \geq P$. Substitute these in the previous formulas. □

Proposition 11.66. *For the rank-two arrangement of* n *lines, the Eulerian idempotents associated to the uniform section are given by*

$$\mathtt{E}_\top = \frac{1}{2n}\sum_C H_C,$$

$$\mathtt{E}_\mathrm{X} = \frac{1}{2}\mathtt{H}_P + \frac{1}{2}\mathtt{H}_{\overline{P}} - \frac{1}{4}(\mathtt{H}_C + \mathtt{H}_D + \mathtt{H}_{\overline{C}} + \mathtt{H}_{\overline{D}}),$$

where X *is the support of the vertices* P *and* $\overline{P}$, *and* C *and* D *are the two edges which are greater than* P,

$$\mathtt{E}_\perp = \mathtt{H}_O - \frac{1}{2}\sum_P \mathtt{H}_P + \frac{1}{2}\sum_C \frac{n-1}{n}\mathtt{H}_C. \tag{11.65}$$

PROOF. The additional ingredient from the projective case is that $\mathtt{u}^C = 1/2n$ for each edge C. So formula (11.64) simplifies. □

Exercise 11.67. Any rank-two arrangement is a good reflection arrangement. Check that the formulas in Proposition 11.66 agree with those given by Theorem 11.53.

11.9. Rank-three arrangements

For rank-three arrangements, we compute the first Eulerian idempotent of any projective section. (The higher idempotents can be computed using the rank-two case which we have discussed.) As a fun application, we show that the Möbius number of any rank-three arrangement equals the number of hyperplanes minus half the number of chambers. We also show that there is a unique projective section whose induced sections both over and under any proper flat are uniform.

11.9.1. First Eulerian idempotent.

Let $\mathcal{A}$ be any rank-three arrangement. Let $\mathtt{u}$ be any homogeneous section of $\mathcal{A}$. By the Witt identity (7.14), for any chamber D,

$$\mathtt{u}^D - \Big(\sum_{P<D} \mathtt{u}^D_P\Big) + \Big(\sum_{E<D} \mathtt{u}^D_E\Big) - 1 = -\mathtt{u}^{\overline{D}}.$$

The first sum is over all vertices P of D, while the second sum is over all edges E of D. If $\mathtt{u}$ is projective, then

$$\mathtt{u}^D = \frac{1}{2}\Big(\sum_{P<D} \mathtt{u}^D_P - \frac{|E|}{2} + 1\Big), \tag{11.66}$$

where $|E|$ is the number of edges of D, and further for any $P < E < D$, $\mathtt{u}^P = \mathtt{u}^E_P = \mathtt{u}^D_E = 1/2$.

By (11.35), the first Eulerian idempotent of $\mathcal{A}$ is given by

$$\mathtt{E}_\perp = \mathtt{H}_O - \sum_P \mathtt{u}^P\, \mathtt{H}_P + \sum_E (-\mathtt{u}^E + \mathtt{u}^P \mathtt{u}^E_P + \mathtt{u}^Q \mathtt{u}^E_Q)\, \mathtt{H}_E$$
$$+ \sum_D \Big(- \mathtt{u}^D + \sum_{P<D} \mathtt{u}^P \mathtt{u}^D_P + \sum_{E<D} \mathtt{u}^E \mathtt{u}^D_E - \sum_{P<E<D} \mathtt{u}^P \mathtt{u}^E_P \mathtt{u}^D_E \Big)\, \mathtt{H}_D.$$

The first sum is over all vertices P. The second sum is over all edges E, with P and Q being the vertices of E. The third sum is over all chambers D, the inside sums being all over vertices P of D, all edges E of D and all incident vertex-edge pairs $P < E$ of D. If $\mathtt{u}$ is projective, then the above formula simplifies as follows.

Proposition 11.68. *For any rank-three arrangement, the first Eulerian idempotent of any projective section* $\mathtt{u}$ *is given by*

$$\mathtt{E}_\perp = \mathtt{H}_O - \frac{1}{2} \sum_P \mathtt{H}_P + \sum_E (-\mathtt{u}^E + \frac{1}{2})\, \mathtt{H}_E - \sum_D \Big(\frac{1}{2} - \frac{1}{2} \sum_{E<D} \mathtt{u}^E \Big)\, \mathtt{H}_D. \tag{11.67}$$

11.9.2. Möbius number and the number of chambers. Let $h(\mathcal{A})$ denote the number of hyperplanes in $\mathcal{A}$. Recall that $\mu(\mathcal{A})$ denotes the Möbius number of $\mathcal{A}$, and $c(\mathcal{A})$ denotes the number of chambers in $\mathcal{A}$.

Proposition 11.69. *For any rank-three arrangement* $\mathcal{A}$,

$$\mu(\mathcal{A}) = h(\mathcal{A}) - \frac{1}{2} c(\mathcal{A}). \tag{11.68}$$

PROOF. Let $\mathtt{u}$ be any projective section of $\mathcal{A}$. Its first Eulerian idempotent is given by (11.67). By Lemma 11.42, this is a special Zie element. Hence by (10.15) applied to the maximum flat $\top$,

$$\mu(\mathcal{A}) = \sum_C \Big(-\frac{1}{2} + \frac{1}{2} \sum_{E:\, E<C} \mathtt{u}^E \Big) = -\frac{1}{2} c(\mathcal{A}) + \frac{1}{2} \sum_{E<C} \mathtt{u}^E$$
$$= -\frac{1}{2} c(\mathcal{A}) + \sum_E \mathtt{u}^E = -\frac{1}{2} c(\mathcal{A}) + h(\mathcal{A}).$$

In the third step, we used that each edge E is contained in exactly two chambers. For the last step, we used that the sum of $\mathtt{u}^E$ for all edges supported in a hyperplane is 1. □

Exercise 11.70. Deduce (11.68) directly by combining (1.47), the Zaslavsky formula (1.48) and (C.5b) (applied to the lattice of flats).

Proposition 11.71. *Suppose* $\mathcal{A}_n$ *is a sequence of simplicial rank-three arrangements with the property that the number of edges in any rank-two flat grows indefinitely with* n. *Then*

$$\lim_n \frac{\mu(\mathcal{A}_n)}{c(\mathcal{A}_n)} = -\frac{1}{2}.$$

PROOF. For any edge E of $\mathcal{A}_n$, let $\mathtt{u}^E_n$ be the reciprocal of the number of edges in the support of E. Then given ϵ, for n sufficiently large,

$$\frac{h(\mathcal{A}_n)}{c(\mathcal{A}_n)} = \frac{1}{c(\mathcal{A}_n)} \sum_E \mathtt{u}^E_n = \frac{1}{2c(\mathcal{A}_n)} \sum_C \sum_{E<C} \mathtt{u}^E_n < \frac{\epsilon}{2}.$$

For the last step, we note that since chambers are triangles, the inside sum always has three terms, and each term can be made arbitrarily small by hypothesis. Thus the ratio of the number of hyperplanes to the number of chambers goes to zero. Now apply (11.68). □

Example 11.72. Let us look at the ratio of the Möbius number to the number of chambers in some rank-three arrangements.

For type $A_3 = D_3$, the ratio is $-1/4$, for any wall of D_4, the ratio is $-9/32$, for type B_3, the ratio is $-5/16$, and for type H_3, the ratio is $-3/8$.

For $n \geq 3$, let $\mathcal{A}_n$ be the arrangement of n lines, one of which is the equator, and all the remaining $n-1$ lines pass through the north and south poles and are equally spaced. In the language of projective geometry, this is called a near-pencil. It is a good reflection arrangement, obtained as the cartesian product of the dihedral arrangement of $n-1$ lines with the rank-one arrangement. An explicit calculation shows that

$$h(\mathcal{A}_n) = n, \quad \mu(\mathcal{A}_n) = 2 - n \qquad \text{and} \qquad c(\mathcal{A}_n) = 4n - 4$$

(thus verifying (11.68)). The limit of the ratio of $\mu(\mathcal{A}_n)$ to $c(\mathcal{A}_n)$ is $-1/4$ and not $-1/2$. Note that except the equator, all other lines always have 4 edges, so the hypothesis of Proposition 11.71 is violated.

In the survey article [**216**], Grünbaum describes all known rank-three simplicial arrangements. For instance, for his arrangement $\mathcal{A}(37,1)$, $\mu(\mathcal{A}) = -323$ and $c(\mathcal{A}) = 720$, and their ratio is approximately -0.4486111. The arrangement $\mathcal{A}(37,1)$ is a part of an infinite family denoted $\mathcal{A}(4m+1,1)$ for $m \geq 2$. This family does satisfy the hypothesis of Proposition 11.71, and so the ratio tends to $-1/2$.

A companion to Proposition 11.71 is given below. It refers to the Takeuchi element defined in Section 12.3.

Proposition 11.73. *Suppose $\mathcal{A}_n$ is a sequence of simplicial rank-three arrangements with the property that the number of edges in any rank-two flat grows indefinitely with n. Then for the uniform section* $\mathtt{u}$,

$$\lim_n \mathtt{E}_\perp - \frac{1}{2}(\mathtt{H}_O + \mathtt{Tak}) = 0,$$

where $\mathtt{Tak}$ *refers to the Takeuchi element* (12.15).

The lhs means that given $\epsilon > 0$, for sufficiently large n, the coefficient of every face for the element inside the limit is smaller than ϵ.

PROOF. This follows from (11.67). □

Note from (11.65) that the formula in Proposition 11.73 also holds for rank-two arrangements. These observations can be understood from the first identity in (12.25).

11.9.3. Canonical homogeneous section. Recall from Section 11.1 that a homogeneous section $\mathtt{u}$ of $\mathcal{A}$ induces a homogeneous section $\mathtt{u}_F$ of $\mathcal{A}_F$ for any face F, and a homogeneous section $\mathtt{u}^{\mathrm{X}}$ of $\mathcal{A}^{\mathrm{X}}$ for any flat X.

Since $\mathcal{A}$ has rank 3, an arrangement under or over a flat of $\mathcal{A}$ has rank between 0 and 3. This is tabulated below.

F	rank of $\mathcal{A}_F$	X	rank of $\mathcal{A}^{\mathrm{X}}$
central face	3	$\perp$	0
vertex	2	antipodal points	1
edge	1	great circle	2
chamber	0	$\top$	3

Proposition 11.74. *There is a unique projective section* $\mathtt{u}$ *of* $\mathcal{A}$ *such that the induced sections* $\mathtt{u}_F$ *of* $\mathcal{A}_F$ *for* $F > O$ *and* $\mathtt{u}^{\mathrm{X}}$ *of* $\mathcal{A}^{\mathrm{X}}$ *for* $\mathrm{X} \neq \top$ *are uniform. Explicitly, for any vertex* P, $\mathtt{u}^P = 1/2$, *for any edge* E, $\mathtt{u}^E$ *is the reciprocal of the number of edges in the great circle supporting* E, *and for any chamber* D,

$$\mathtt{u}^D = \frac{1}{2}\Big(\sum_P \frac{1}{c_P^D} - \frac{|E|}{2} + 1\Big),$$

where the sum is over all vertices P *of* D, *and* $|E|$ *is the number of edges of* D. *In particular, for any triangle* D,

$$\mathtt{u}^D = \frac{1}{2}\Big(\sum_P \frac{1}{c_P^D} - \frac{1}{2}\Big).$$

We call this the *canonical homogeneous section.*

PROOF. Uniformity of the sections $\mathtt{u}^{\mathrm{X}}$ for $\mathrm{X} \neq \top$ implies the claims about $\mathtt{u}^P$ and $\mathtt{u}^E$. The nontrivial part is to understand the coefficients of chambers. For this, we apply Theorem 11.59. It guarantees existence. Further, for any chamber D, $\mathtt{u}_P^D = \frac{1}{c_P^D}$ for any vertex P of D, and $\mathtt{u}_E^D = \frac{1}{2}$ for any edge E of D. Substitute these in (11.60). Alternatively, one may also substitute in (11.66). □

11.9.4. Arrangement of type D_4. Suppose $\mathcal{A}$ is any wall of the reflection arrangement of type D_4. Then $\mathcal{A}$ is of rank three and cisomorphic to the arrangement shown in Section 6.8.2 which is reproduced below.

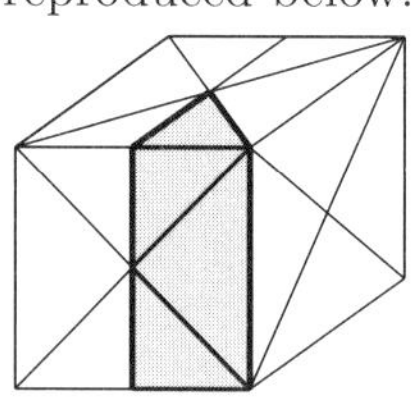

Let us compute the canonical homogeneous section. For any of the small triangles D, the boundary of the stars of the three vertices are polygons of sizes 4, 6 and 6. So by Proposition 11.74,

$$\mathtt{u}^D = \frac{1}{2}\Big(\frac{1}{4} + \frac{1}{6} + \frac{1}{6} - \frac{1}{2}\Big) = \frac{1}{24}.$$

For any of the 8 large triangles D adjacent to the small triangles, the polygon sizes are 6, 6 and 6, and so $\mathtt{u}^D = 0$. For the remaining 8 large triangles D, the polygon sizes are again 4, 6 and 6, and so $\mathtt{u}^D = 1/24$.

The first Eulerian idempotent given by (11.67) specializes to:

$$\mathtt{E}_\perp = \mathtt{H}_O - \frac{1}{2}\sum_P \mathtt{H}_P + \frac{1}{3}\sum_E \mathtt{H}_E + \frac{3}{8}\sum_{E'} \mathtt{H}_{E'} - \frac{1}{4}\sum_D \mathtt{H}_D - \frac{7}{24}\sum_{D'} \mathtt{H}_{D'}. \tag{11.69}$$

The first sum is over all 18 vertices P, the second sum is over the 24 edges E whose support is a hexagon, the third sum is over the 24 edges E' whose support is an octagon, the fourth sum is over the 8 large triangles D adjacent to the small triangles, and the last sum is over the remaining 24 triangles D'.

It follows that (11.69) is a special Zie element of $\mathcal{A}$. Let us use this to illustrate some properties of Zie elements. To illustrate (10.11a), consider the top-lune shown in the figure. The interior of this lune consists of four edges and five triangles (one of which is on the bottom side and not visible). The sum of their coefficients is zero:

$$\frac{1}{3}(2) + \frac{3}{8}(2) - \frac{1}{4}(1) - \frac{7}{24}(4) = 0.$$

The sum of the coefficients of all chambers is

$$-\frac{1}{4}(8) - \frac{7}{24}(24) = -9,$$

which is the Möbius number of the lattice of flats of $\mathcal{A}$. This is consistent with (10.15).

Let us contrast this with the element (11.53). This is the Fulman element of parameter 0 and is given by

$$\mathtt{H}_O - \frac{1}{2}\sum_P \mathtt{H}_P + \frac{1}{3}\sum_E \mathtt{H}_E + \frac{3}{8}\sum_{E'} \mathtt{H}_{E'} - \frac{9}{32}\sum_D \mathtt{H}_D. \tag{11.70}$$

It differs from (11.69) only in the chamber coefficients, all 32 triangles now have the coefficient $-9/32$. We explain the calculation needed to arrive at (11.70): The number c^{X} for X of rank 0 is 1, of rank 1 is 2, of rank 2 is either 6 or 8, and of rank 3 is 32. The Möbius functions $\mu(\bot, \mathrm{X})$ are 1, -1, 2, 3 and -9, respectively.

We claim that (11.70) is *not* a Zie element of $\mathcal{A}$. There are many ways to check this. For instance, (10.11a) fails. The sum of the coefficients of faces in the interior of the lune shown in the figure is not zero:

$$\frac{1}{3}(2) + \frac{3}{8}(2) - \frac{9}{32}(5) \approx 0.0104167 \neq 0.$$

However the equivalent conditions of Lemma 10.18 hold. In particular, the sum of the coefficients of all faces is zero:

$$1 - \frac{1}{2}(18) + \frac{1}{3}(24) + \frac{3}{8}(24) - \frac{9}{32}(32) = 0.$$

Corollary 11.75. *The element* (11.53) *for the arrangement of type* D_4 *is not a Zie element, and hence cannot be the first Eulerian idempotent of any homogeneous section.*

PROOF. Let X be a wall of the D_4 arrangement and let $\mathcal{A}$ denote the arrangement under X of the D_4 arrangement. The truncation of the element (11.53) to faces smaller than X will yield the element (11.70). We noted above that this is not a Zie element of $\mathcal{A}$. Since truncations preserve Zie elements, the element we started with cannot be a Zie element. □

Notes

Eulerian idempotents. Primitive idempotents in the wider context of LRBs appeared in work of Brown [**103**, Sections 8.5 and 8.7]; see also the notes to Chapter 12. The recursive construction (11.14) of Eulerian idempotents starting with a homogeneous section is due to Saliola [**365**, Section 1.5.1], [**368**, Section 5.1], [**367**, Section 5], [**366**, Section 4]. In

[**367**], his ℓ_X corresponds to our $\mathtt{u}_X$. In [**366**], he works in the generality of LRBs. His main result says that the Eulerian idempotents arising from his construction form a complete system of primitive orthogonal idempotents of the Tits algebra. The proof that we have given closely follows his exposition, the key observation being Lemma 11.12. For related work, see [**303**, Section 5.1], [**304**, Section 4.3] and [**52**, Section 3].

The second construction of Eulerian idempotents from a special Zie family is new. However, the analogous construction for the invariant Tits algebra of the braid arrangement in a special case is present in the literature. These are the Krob-Leclerc-Thibon idempotents discussed later in Section 16.11.6.

Q-basis. The Q-basis for the Tits algebra was introduced by Saliola [**365**, Section 1.5.2], [**368**, Section 5.2], [**366**, Corollary 4.4] via the equation (11.26). He did not use any specific notation for this basis. See also [**303**, Corollary 5.2] and [**304**, Corollary 4.4].

For the braid arrangement, Schocker [**376**, Corollary 4.4 and Proposition 5.1] constructed a Q-basis starting with a special Zie family. He did not use the latter terminology. That he indeed starts out with special Zie elements can be seen from his Corollary 4.4 and the Friedrichs criterion for Zie elements. His formula (5.2) is our (11.29). The result of Exercise 11.35 for the braid arrangement (except for the connection with $\mathtt{E}_X$) is given in his Theorem 5.4 and Corollary 5.5. The precise setting of Schocker's construction is explained later in Section 12.5.5. It is analogous to the discussion of the Krob-Leclerc-Thibon idempotents in Section 16.11.6. However, Schocker did not give the general construction (11.47) of Eulerian idempotents. He only dealt with a special case arising from Dynkin elements which is given in Exercise 14.70.

CHAPTER 12

Diagonalizability and characteristic elements

For any element of the Tits algebra, the sum of all its coefficients is an eigenvalue. When the element satisfies an additional hypothesis which we call the top-separating condition, this eigenvalue has a unique eigenvector (up to scaling) in the left module of chambers. There is an explicit formula for this eigenvector which we call the Brown-Diaconis formula. While working over the reals, the eigenvector can be interpreted as the stationary distribution of a random walk associated to the given element. From this point of view, elements with nonnegative coefficients are of interest. We also discuss the Billera-Brown-Diaconis formula which treats a special case in rank-three arrangements.

Next, we consider the general problem of diagonalizability of elements of the Tits algebra. An important sufficient condition for diagonalizability is that the element be separating. The separating condition is similar to but stronger than the top-separating condition. The key step in the proof is to determine the homogeneous section, called the eigensection, whose associated Eulerian family will diagonalize the given element. The eigensection can be built by a separate computation in each flat, and in each case, one can employ the Brown-Diaconis formula. This leads to the Brown formulas for the eigensection of a separating element, and also for the associated Eulerian family.

We study in detail the Takeuchi element. It is defined via an "alternating" sum of faces (similar to the Euler characteristic). Its square is the unit of the Tits algebra. The Takeuchi element is projective and commutes with every projective element. It is neither separating nor does it have nonnegative coefficients, yet it is diagonalizable. Its eigenvalues are ± 1, and any projective section is an eigensection. Further, it acts on the left module of chambers by sending a chamber to its opposite (up to sign). We also consider a more general class of elements called the Fulman elements. For a good reflection arrangement, the uniform section is an eigensection for the Fulman elements which then leads to an explicit diagonalization (using the cancelation-free formulas for the corresponding Eulerian idempotents). We specialize to the braid arrangement and discuss in particular the Adams elements. We also give a similar discussion for the arrangement of type B.

Takeuchi and Fulman elements are examples of characteristic elements. The eigenvalues of such elements are powers of a fixed parameter. This interesting class of elements is intimately connected to the characteristic polynomial of the arrangement.

12.1. Stationary distribution

Let w be an element of the Tits algebra. Theorem 9.44 gives the eigenvalues of w for the action on any module h. We would now like to work towards finding the eigenvectors of w. It is particularly important to understand the eigenvectors of $\lambda_\top$

(the eigenvalue associated to the maximum flat) for the module $\mathsf{h} = \Gamma$. The main result is as follows. If $\lambda_{\mathrm{X}} \neq \lambda_{\top}$ for all $\mathrm{X} \neq \top$, then $\lambda_{\top}$ has a unique eigenvector (normalized so that the sum of its coefficients is 1). If in addition w is a probability distribution (assuming the base field to be $\mathbb{R}$), then $\lambda_{\top}$ is the largest eigenvalue of w and its unique normalized eigenvector can be interpreted as the stationary distribution of a random walk on chambers associated to w.

Notation 12.1. In this section, w, w_F are elements of the Tits algebra, while w^F, w_F^G are scalars. Similarly, u, u_F are chamber elements, while u^F, u_F^C are scalars.

12.1.1. Top-eigenvectors. Fix an element $w = \sum_F w^F\, \mathtt{H}_F$ of the Tits algebra. The content of w is defined to be the sum of the coefficients w^F. For each flat X, put

$$\lambda_{\mathrm{X}} = \sum_{F:\, \mathrm{s}(F) \leq \mathrm{X}} w^F. \tag{12.1}$$

In particular, $\lambda_{\perp} = w^O$, while $\lambda_{\top}$ is the content of w. We refer to the λ_{X} as the *eigenvalues* of w.

Let $u = \sum_C u^C\, \mathtt{H}_C$ be a chamber element. We say that u is a *top-eigenvector* for w if u is of content 1 and

$$w \cdot u = \lambda\, u \tag{12.2}$$

for some scalar λ, the *eigenvalue*. Thus, the sum of the u^C is 1, and by taking the content in (12.2), we see that $\lambda = \lambda_{\top}$ necessarily. Note in passing that an eigenvector u of w with eigenvalue different from $\lambda_{\top}$ must have content 0.

Lemma 12.2. *Let w be an element of the Tits algebra, and u be a chamber element of content* 1*. Then u is a top-eigenvector of w iff for each chamber C,*

$$(\lambda_{\top} - \lambda_{\perp})\, u^C = \sum_{F:\, O < F \leq C} w^F u_F^C, \tag{12.3}$$

with u_F^C as in (11.55).

PROOF. This is a straightforward calculation.

$$\begin{aligned} w \cdot u &= \Big(\sum_H w^H\, \mathtt{H}_H\Big) \cdot \Big(\sum_C u^C\, \mathtt{H}_C\Big) \\ &= \sum_{H,C} w^H u^C\, \mathtt{H}_{HC} \\ &= \sum_D \Bigg(\sum_{H:\, H \leq D} w^H \Big(\sum_{C:\, HC = D} u^C \Big) \Bigg) \mathtt{H}_D \\ &= \sum_D \Big(\sum_{H:\, H \leq D} w^H u_H^D \Big)\, \mathtt{H}_D \end{aligned}$$

Comparing with the coefficient of $\mathtt{H}_D$ in $\lambda_{\top} u$, we obtain

$$\lambda_{\top}\, u^D = \sum_{H:\, H \leq D} w^H u_H^D.$$

The summand in the rhs corresponding to $H = O$ is $w^O u_O^D = \lambda_{\perp} u^D$. Bringing it to the lhs yields (12.3). $\square$

Given w, for any face F, define $w_F := \Delta_F(w)$, with Δ_F as in (9.71). It is an element of the Tits algebra of $\mathcal{A}_F$. Explicitly,

$$w_F = \sum_{G:\, G \geq F} w_F^G \, \mathtt{H}_{G/F}, \quad \text{where} \quad w_F^G := \sum_{K:\, FK=G} w^K. \tag{12.4}$$

Observe that w and w_F have the same content. In fact, the eigenvalues of w_H are those eigenvalues of w which are indexed by flats X containing H. An equivalent result is given in Exercise 9.64.

Similarly, given u, for any face F, define $u_F := \Delta_F(u)$. Explicitly,

$$u_F = \sum_{C:\, C \geq F} u_F^C \, \mathtt{H}_{C/F}.$$

It has the same content as u. Since Δ_F is an algebra homomorphism, we have the key fact:

Lemma 12.3. *If u is a top-eigenvector for w, then u_F is a top-eigenvector for w_F, with the same eigenvalue.*

Note very carefully that the rhs of (12.3) only involves the coefficients of u_F for $F > O$. In conjunction with Lemma 12.3, this suggests that one can construct a top-eigenvector u of w by an inductive procedure by first constructing it in the stars of all faces $F > O$. This is similar to the extension problem discussed in Section 11.7, and we will make use of the basic setup there in what follows.

12.1.2. Brown-Diaconis formula. We say an element w of the Tits algebra is *top-separating* if $\lambda_{\mathrm{X}} \neq \lambda_\top$ for any $\mathrm{X} \neq \top$. Note that if w is top-separating, then so is w_F for any face F.

Our goal now is to show that a top-separating element has a unique top-eigenvector.

Lemma 12.4. *Let w be an element of the Tits algebra such that $\lambda_\top \neq \lambda_\perp$. Suppose for each face $H > O$, we are given a top-eigenvector v_H of w_H such that the v_H satisfy the compatibility conditions* (11.57) *and* (11.58). *Then there is a unique top-eigenvector u of w satisfying* (11.56) *for each $H > O$. Explicitly, for a chamber C, the scalar u^C is given by*

$$u^C = \frac{1}{\lambda_\top - \lambda_\perp} \sum_{F:\, O<F\leq C} w^F v_F^C. \tag{12.5}$$

Proof. Suppose u is a top-eigenvector of w satisfying (11.56) for each $H > O$. Since $\lambda_\top \neq \lambda_\perp$, (12.3) can be rewritten as in (12.5). This proves uniqueness of u. For existence of u, we check below that the u defined by (12.5) satisfies (11.56).

$$\begin{aligned} u_H^D &= \sum_{C:\, HC=D} u^C \\ &= \frac{1}{\lambda_\top - \lambda_\perp} \sum_{C:\, HC=D} \; \sum_{F:\, O<F\leq C} w^F \, v_F^C \\ &= \frac{1}{\lambda_\top - \lambda_\perp} \sum_{F:\, O<F,\, HF\leq D} w^F \sum_{C:\, F\leq C,\, HC=D} v_F^C. \end{aligned}$$

The condition $HC = D$ in the inside sum can be replaced by $FHC = FD$, so by (11.57), the inside sum equals v^{FD}_{FH}, and by (11.58), this further equals v^D_{HF}. This is illustrated below.

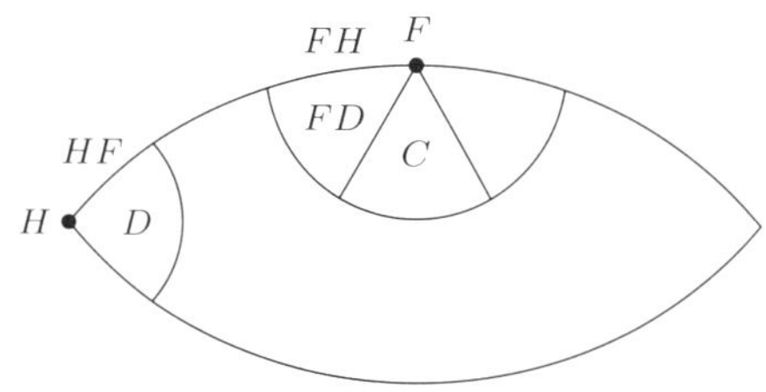

(For some context on this picture, see the discussion in Section 3.4.7.)

Substituting, the calculation continues as follows.

$$\begin{aligned} u^D_H &= \frac{1}{\lambda_\top - \lambda_\bot} \sum_{F:\, O<F,\, HF\leq D} w^F\, v^D_{HF} \\ &= \frac{1}{\lambda_\top - \lambda_\bot} \sum_{G:\, H\leq G\leq D} v^D_G \sum_{F:\, O<F,\, HF=G} w^F \\ &= \frac{1}{\lambda_\top - \lambda_\bot} \Big((w^H_H - \lambda_\bot)\, v^D_H + \sum_{G:\, H<G\leq D} w^G_H\, v^D_G \Big) \\ &= \frac{1}{\lambda_\top - \lambda_\bot} \Big((w^H_H - \lambda_\bot)\, v^D_H + (\lambda_\top - w^H_H)\, v^D_H \Big) \\ &= v^D_H. \end{aligned}$$

In the third step, we broke the first sum depending on whether $G = H$ or $G > H$, and used (12.4). In the second-last step, we used (12.3) for the eigenvector v_H of w_H.

Finally, to see that u has content 1, we recall that u and u_H have the same content, and $u_H = v_H$ and v_H has content 1. ☐

Theorem 12.5. *Suppose w is a top-separating element of the Tits algebra. Then w has a unique top-eigenvector u. Its eigenvalue is $\lambda_\top$. Explicitly, in rank at least one, for a chamber C, the scalar u^C is given by*

$$(12.6)\quad u^C = \frac{w^C}{\lambda_\top - \lambda_\bot} + \sum_{O<F<C} \frac{w^F w^C_F}{(\lambda_\top - \lambda_\bot)(\lambda_\top - \lambda_{\mathrm{s}(F)})} + \sum_{O<F<G<C} \frac{w^F w^G_F w^C_G}{(\lambda_\top - \lambda_\bot)(\lambda_\top - \lambda_{\mathrm{s}(F)})(\lambda_\top - \lambda_{\mathrm{s}(G)})} + \dots.$$

The first sum is over F, the second sum is over F and G, and so on. (The top-separating condition ensures that the denominators are nonzero.)

We refer to (12.6) as the *Brown-Diaconis formula.*

PROOF. We show that w has a unique top-eigenvector u by induction on the rank of $\mathcal{A}$. For rank 0, $C = O$ and clearly $u^C = 1$ is the unique eigenvector. This is the induction base. Since w is top-separating, so is w_H for any face H. Hence by the induction hypothesis, for each $H > O$, the element w_H has a unique top-eigenvector, say v_H. By uniqueness, the v_H must satisfy the compatibility conditions (11.57) and (11.58). Therefore by Lemma 12.4, w has a unique top-eigenvector u satisfying $u_H = v_H$. In conjunction with Lemma 12.3, this proves both existence and uniqueness of u.

Formula (12.6) follows by inductively applying (12.5) to each u^C_F. □

Example 12.6. Let $\mathcal{A}$ be the rank-one arrangement with chambers C and $\overline{C}$. An element of the Tits algebra w is top-separating if $w^C + w^{\overline{C}} \neq 0$. If this happens, then w has a unique top-eigenvector u whose coefficients are

$$u^C = \frac{w^C}{w^C + w^{\overline{C}}} \quad \text{and} \quad u^{\overline{C}} = \frac{w^{\overline{C}}}{w^C + w^{\overline{C}}}.$$

Only the first term in (12.6) contributed.

For any rank-two arrangement, the unique top-eigenvector u of a top-separating element w has coefficients

$$u^C = \frac{w^C}{\lambda_\top - \lambda_\perp} + \frac{w^P w^C_P}{(\lambda_\top - \lambda_\perp)(\lambda_\top - \lambda_{\mathrm{s}(P)})} + \frac{w^Q w^C_Q}{(\lambda_\top - \lambda_\perp)(\lambda_\top - \lambda_{\mathrm{s}(Q)})},$$

where P and Q are the two vertices of C.

Exercise 12.7. In a rank-two arrangement, consider the element $w = \frac{1}{f_0}\sum_P \mathtt{H}_P$, where the sum is over all vertices P, and f_0 is the number of vertices. Check that w is a top-separating element, and its unique top-eigenvector is $\frac{1}{f_0}\sum_C \mathtt{H}_C$.

12.1.3. Billera-Brown-Diaconis formula. Let $\mathcal{A}$ be any rank-three arrangement. Let f_0 denote the number of vertices in $\mathcal{A}$. Recall from Exercise 1.5 that f_0 is at least 6. Consider the element of the Tits algebra defined by

$$w = \frac{1}{f_0}\sum_P \mathtt{H}_P, \tag{12.7}$$

where the sum is over all vertices P. This element is clearly top-separating. Hence by Theorem 12.5, it has a unique top-eigenvector.

Theorem 12.8. *The unique top-eigenvector of* (12.7) *is given by*

$$u^C = \frac{i-2}{2(f_0-2)}, \tag{12.8}$$

where i is the number of sides of the chamber C.

We refer to (12.8) as the *Billera-Brown-Diaconis formula.*

PROOF. Let us apply (12.6). In this case: $\lambda_\top = 1$, $\lambda_\perp = 0$. For a vertex P, an edge E and a chamber C, $w^P = \frac{1}{f_0}$, $w^E = w^C = 0$,

$$\lambda_\top - \lambda_{\mathrm{s}(P)} = 1 - \frac{2}{f_0} \quad \text{and} \quad \frac{w^C_E}{\lambda_\top - \lambda_{\mathrm{s}(E)}} = \frac{1}{2}.$$

Substituting these, we obtain

$$u^C = \frac{1}{(f_0-2)}\Big(\sum_{O<P<C} w^C_P + \sum_{O<P<E<C} \frac{1}{2} w^E_P\Big).$$

The term in parenthesis can be understood as follows. For any vertex K, let us mark a vertex P of C with a circle if $PK = C$ and with a semicircle if PK is an edge E of C. Counting 1 for each circle and 1/2 for each semicircle and dividing the sum by f_0 gives the contribution of the vertex K to the term in parenthesis. Now let us combine the contributions of a pair of opposite vertices K and $\overline{K}$. This is illustrated below when C is a hexagonal face, with the contribution of K shown in black, and that of $\overline{K}$ shown in white.

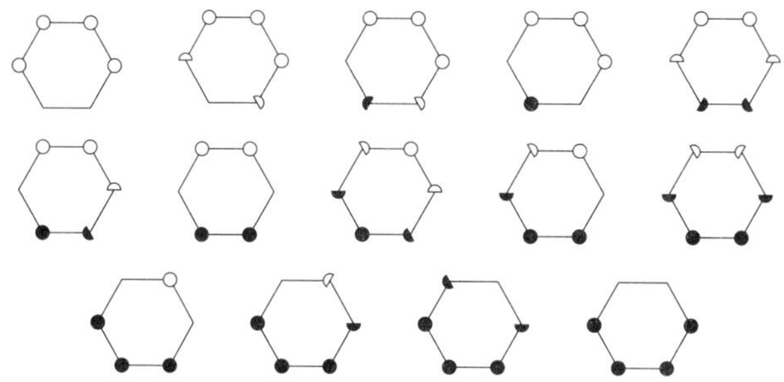

Observe that the contribution of either vertex is contiguous and their sum is always $4/f_0$ (though their individual contributions may differ resulting in the various possibilities shown above). In general, if C is an i-gon, then the total contribution is $(i-2)/f_0$. Since there are $f_0/2$ pairs of opposite vertices, the term in parenthesis is $(i-2)/2$ as required. □

12.1.4. Eigenvectors for nonnegative elements. Assume that the base field is $\mathbb{R}$. An element of the Tits algebra is *nonnegative* if all its coefficients are nonnegative. We now show that a nonnegative element has a top-eigenvector but it may not be unique. This is in contrast to the previous result for top-separating elements. Note that for a nonnegative element, $\lambda_{\mathrm{X}} \leq \lambda_{\mathrm{Y}}$ when $\mathrm{X} \leq \mathrm{Y}$. Therefore, such an element is top-separating precisely if $\lambda_{\mathrm{X}} \neq \lambda_{\top}$ for every hyperplane X.

Proposition 12.9. *Every nonnegative element w in the Tits algebra has a top-eigenvector.*

PROOF. Represent the action of w on $\Gamma[\mathcal{A}]$ by a matrix T whose entries are indexed by pairs of chambers:

$$T_{D,C} := \sum_{F:\, FC=D} w^F,$$

with D as the row index, and C as the column index. The column sums of T are all equal to $\lambda_{\top}$, which is the content of w. So $\lambda_{\top}$ is an eigenvalue of T. (The row vector with all entries 1 is an eigenvector). Hence it must have at least one column-eigenvector with real entries. The problem with such an eigenvector is that its content may be zero. However, if the entries of the matrix are nonnegative, then a generalization of the Perron-Frobenius Theorem guarantees that we can find a nonzero column-eigenvector with nonnegative entries [**232**, Lemma 8.3.1] (which we can then normalize to have content 1). □

12.1.5. Stationary distribution. Suppose the scalars w^F are nonnegative and add up to 1. Then w can be interpreted as a probability distribution on the set of faces. It induces a random walk on the set of chambers: Suppose we are currently in chamber C. Then pick a face F at random (with probability w^F) and move to FC. With this interpretation, a top-eigenvector u for w is the same as a stationary distribution for this random walk (provided all coefficients of u are nonnegative).

Theorem 12.10. *Suppose w is a top-separating probability distribution on the set of faces. Then the associated random walk has a unique stationary distribution u given by the Brown-Diaconis formula* (12.6).

This is essentially a restatement of Theorem 12.5 with a small additional observation. We need to know that the coefficients of the eigenvector u are nonnegative, but this is clear from Brown-Diaconis formula. Alternatively, one may also apply the Perron-Frobenius Theorem.

12.2. Diagonalizability and eigensections

Diagonalizable elements in an algebra are discussed in Section D.4. By definition, an element of the Tits algebra is diagonalizable if it can be expressed as a linear combination of mutually orthogonal idempotents. We show that elements which satisfy a separating condition or a nonnegativity condition are diagonalizable. The key idea is to choose an appropriate homogeneous section $\mathtt{u}$ so that the given element w can be expressed using the Eulerian family associated to $\mathtt{u}$. We refer to such a $\mathtt{u}$ as an eigensection of w.

Notation 12.11. In this section, w, w_F, w^{X}, w_{X} are elements of the Tits algebra, while w^F, w^G_F are scalars. Similarly, $\mathtt{u}$, $\mathtt{u}_F$ are homogeneous sections, while $\mathtt{u}^F$, $\mathtt{u}^G_F$ are scalars.

12.2.1. Eigensections. For any element of the Tits algebra $w = \sum_F w^F \mathtt{H}_F$, set

$$w^{\mathrm{X}} := \sum_{F:\, \mathrm{s}(F) \leq \mathrm{X}} w^F \mathtt{H}_F \quad \text{and} \quad w_{\mathrm{X}} := \sum_{F:\, \mathrm{s}(F) = \mathrm{X}} w^F \mathtt{H}_F. \tag{12.9}$$

By definition,

$$w^{\mathrm{X}} = \sum_{\mathrm{Y}:\, \mathrm{Y} \leq \mathrm{X}} w_{\mathrm{Y}}.$$

Definition 12.12. Let w be an element of the Tits algebra and $\mathtt{u}$ be a homogeneous section. We say that $\mathtt{u}$ is an *eigensection* for w if there exist scalars $\lambda = (\lambda_{\mathrm{X}})$ indexed by flats X, such that for any flat X,

$$w^{\mathrm{X}} \cdot \mathtt{u}_{\mathrm{X}} = \lambda_{\mathrm{X}}\, \mathtt{u}_{\mathrm{X}}, \tag{12.10}$$

with $\mathtt{u}_{\mathrm{X}}$ as in (11.2).

Observe that an eigensection of w is the same as a family $(\mathtt{u}_{\mathrm{X}})$, where each $\mathtt{u}_{\mathrm{X}}$ is a top-eigenvector of w^{X} in the arrangement $\mathcal{A}^{\mathrm{X}}$. Since $\mathtt{u}_{\mathrm{X}}$ has content 1, taking the content of both sides of (12.10), we note that λ_{X} is given by (12.1). In particular, it depends only on w and not on the choice of $\mathtt{u}$.

Proposition 12.13. *Given a homogeneous section* $\mathtt{u}$ *and* $\lambda = (\lambda_{\mathrm{X}})$, *there exists a unique w with eigenvalues λ and eigensection* $\mathtt{u}$.

PROOF. To construct w, we need to construct w_{X} for each flat X. We do that by induction on the rank of X. Setting $\mathrm{X} := \bot$ in (12.10) and using $\mathtt{u}_\bot = \mathtt{H}_O$ yields

$$w^\bot = w_\bot = \lambda_\bot \mathtt{H}_O.$$

Now suppose that w_{Y} are uniquely constructed for all $\mathrm{Y} < \mathrm{X}$. To construct w_{X}, we need to solve the equation

$$\Big(w_{\mathrm{X}} + \sum_{\mathrm{Y}:\, \mathrm{Y} < \mathrm{X}} w_{\mathrm{Y}}\Big) \cdot \mathtt{u}_{\mathrm{X}} = \lambda_{\mathrm{X}}\, \mathtt{u}_{\mathrm{X}}.$$

(This is a reformulation of (12.10).) By Lemma 11.2, $w_X \cdot \mathtt{u}_X = w_X$ always holds. Thus

$$w_X := \lambda_X \mathtt{u}_X - \Big(\sum_{Y:\, Y<X} w_Y \Big) \cdot \mathtt{u}_X$$

is the unique solution. This completes the induction step. □

A more precise result is given below.

Proposition 12.14. *Given a triple* $(w, \lambda, \mathtt{u})$,

$$\text{(12.11)} \qquad \mathtt{u} \textit{ is an eigensection of } w \textit{ with eigenvalues } \lambda \iff w = \sum_X \lambda_X \mathtt{E}_X,$$

where $\mathtt{E}$ *is the Eulerian family associated to* $\mathtt{u}$.

Proof. Forward implication. Since the sum of the $\mathtt{E}_X$ is $\mathtt{H}_O$, it suffices to show that $w \cdot \mathtt{E}_X = \lambda_X \mathtt{E}_X$. This follows from:

$$w \cdot \mathtt{E}_X = w^X \cdot \mathtt{E}_X = w^X \cdot \mathtt{u}_X \cdot \mathtt{E}_X = \lambda_X \mathtt{u}_X \cdot \mathtt{E}_X = \lambda_X \mathtt{E}_X.$$

The first equality used the Saliola lemma (Lemma 11.12). The remaining ones used (11.15) and (12.10).

Backward implication. We provide two arguments. Applying Proposition 12.13, let v be the unique element with eigenvalues λ and eigensection $\mathtt{u}$. Now apply the forward implication to $(v, \lambda, \mathtt{u})$ to obtain

$$v = \sum_X \lambda_X \mathtt{E}_X.$$

Therefore $v = w$. Alternatively: Truncating $w = \sum_Y \lambda_Y \mathtt{E}_Y$ to faces of support smaller than X,

$$w^X = \sum_{Y:\, Y \leq X} \lambda_Y \mathtt{E}_Y^X,$$

where $\mathtt{E}_Y^X$ is the part of $\mathtt{E}_Y$ consisting of faces of support smaller than X. In particular, $\mathtt{E}_X^X = \mathtt{u}_X$. Hence

$$w^X \cdot \mathtt{u}_X = \sum_{Y:\, Y \leq X} \lambda_Y \mathtt{E}_Y^X \cdot \mathtt{E}_X^X = \lambda_X \mathtt{E}_X^X = \lambda_X \mathtt{u}_X.$$

The third equality used (11.22). □

We know from Theorems D.34 and 11.20 that every diagonalizable element can be diagonalized using an Eulerian family. In conjunction with Proposition 12.14, we obtain:

Corollary 12.15. *An element of the Tits algebra is diagonalizable iff it has an eigensection.*

Exercise 12.16. Recall from Lemma 11.6 that a homogeneous section $\mathtt{u}$ of $\mathcal{A}$ induces a homogeneous section $\mathtt{u}_F$ of $\mathcal{A}_F$. Check that: If $\mathtt{u}$ is an eigensection for w, then $\mathtt{u}_F$ is an eigensection for w_F.

12.2.2. Diagonalizability for separating elements and Brown formulas. We say an element w of the Tits algebra is *separating* if for any $\mathrm{X} < \mathrm{Y}$, we have $\lambda_{\mathrm{X}} \neq \lambda_{\mathrm{Y}}$. This condition is stronger than the top-separating condition. More precisely, w is separating iff w^{X} (viewed as an element of the Tits algebra of $\mathcal{A}^{\mathrm{X}}$) is top-separating for each X. Also, if w is separating, then so is w_F for any face F.

Theorem 12.17. *Suppose w is a separating element of the Tits algebra. Then w has a unique eigensection* $\mathtt{u}$. *Explicitly,* $\mathtt{u}_F^F = 1$ *and for* $F < G$,

$$\mathtt{u}_F^G = \frac{w_F^G}{\lambda_{\mathrm{s}(G)} - \lambda_{\mathrm{s}(F)}} + \sum_{F<H<G} \frac{w_F^H w_H^G}{(\lambda_{\mathrm{s}(G)} - \lambda_{\mathrm{s}(F)})(\lambda_{\mathrm{s}(G)} - \lambda_{\mathrm{s}(H)})} + \sum_{F<H<K<G} \frac{w_F^H w_H^K w_K^G}{(\lambda_{\mathrm{s}(G)} - \lambda_{\mathrm{s}(F)})(\lambda_{\mathrm{s}(G)} - \lambda_{\mathrm{s}(H)})(\lambda_{\mathrm{s}(G)} - \lambda_{\mathrm{s}(K)})} + \dots, \tag{12.12}$$

and $\mathtt{u}^G = \mathtt{u}_O^G$. *The first sum is over H, the second sum is over H and K, and so on. The scalars w_F^G are as in* (12.4).

Moreover, w is diagonalizable, with

$$w = \sum_{\mathrm{X}} \lambda_{\mathrm{X}}\, \mathtt{E}_{\mathrm{X}}$$

for a unique Eulerian family $\mathtt{E}$.

PROOF. The last claim follows from the first by (12.11). For the first claim: To construct $\mathtt{u}$, we need to construct each $\mathtt{u}_{\mathrm{X}}$. This is a top-eigenvector of w^{X} in $\mathcal{A}^{\mathrm{X}}$, and we can apply Theorem 12.5. The special case $F = O$ of formula (12.12) follows from the Brown-Diaconis formula (12.6); for the case of arbitrary F, we employ in addition Exercise 12.16. □

We would now like to invert the matrix $(\mathtt{u}_F^G)$ defined by (12.12) by using formula (11.32). The following identity is useful for that purpose.

Lemma 12.18. *Let $x_0, x_1, \dots, x_n$ be distinct scalars. Then*

$$(-1)^n \prod_{i=1}^{n} \frac{1}{x_i - x_0} = \sum_{(a_1,\dots,a_k)\models n} (-1)^k \prod_{j=1}^{k} \frac{1}{(x_{b_j} - x_{b_j - 1}) \dots (x_{b_j} - x_{b_{j-1}})},$$

where $b_j = a_1 + \dots + a_j$ and $b_0 = 0$. The sum is over all compositions $(a_1, \dots, a_k)$ of n.

PROOF. Note that $x_n - x_{n-1}$ appears in all terms in the rhs. Split the rhs into two depending on whether $a_k = 1$ or $a_k > 1$. Denoting the rhs by $f(x_0, \dots, x_n)$, this yields the recursion

$$f(x_0, \dots, x_n) = \frac{1}{x_n - x_{n-1}} \big(- f(x_0, \dots, x_{n-1}) + f(x_0, \dots, x_{n-2}, x_n) \big).$$

Note that in the second term, the variable x_{n-1} is absent. Solving this recursion yields the result. □

Theorem 12.19. *Let w be a separating element in the Tits algebra, and* $\mathtt{u}$ *be its unique eigensection. Let* $\mathtt{E}$ *be the associated Eulerian family, and* $\mathtt{Q}$ *the associated basis. Then*

$$\mathtt{E}_{\mathrm{X}} = \sum_{F:\,\mathrm{s}(F)=\mathrm{X}} \; \sum_{G:\, F\leq G} \mathtt{u}^F \mathtt{a}_F^G\, \mathtt{H}_G \quad \textit{and} \quad \mathtt{Q}_F = \sum_{G:\, F\leq G} \mathtt{a}_F^G\, \mathtt{H}_G, \tag{12.13}$$

where

$$\mathtt{a}_F^G = -\frac{w_F^G}{\lambda_{\mathrm{s}(G)} - \lambda_{\mathrm{s}(F)}} + \sum_{F<H<G} \frac{w_F^H w_H^G}{(\lambda_{\mathrm{s}(H)} - \lambda_{\mathrm{s}(F)})(\lambda_{\mathrm{s}(G)} - \lambda_{\mathrm{s}(F)})} - \sum_{F<H<K<G} \frac{w_F^H w_H^K w_K^G}{(\lambda_{\mathrm{s}(H)} - \lambda_{\mathrm{s}(F)})(\lambda_{\mathrm{s}(K)} - \lambda_{\mathrm{s}(F)})(\lambda_{\mathrm{s}(G)} - \lambda_{\mathrm{s}(F)})} + \dots. \tag{12.14}$$

The first sum is over H, the second sum is over H and K, and so on.

PROOF. Formulas (12.13) are the same as the second formula in (11.33) and (11.35). The nontrivial part is to obtain the formula for $\mathtt{a}_F^G$. For this, substitute (12.12) in (11.32), collect together the terms involving w_F^G, $w_F^H w_H^G$, and so on, and simplify each coefficient using Lemma 12.18. This yields (12.14). □

We refer to (12.13), with $\mathtt{u}^F$ and $\mathtt{a}_F^G$ given by (12.12) and (12.14), as the *Brown formulas* for the Eulerian idempotents of a separating element. Note that (12.14) is *not* a cancelation-free formula because of the alternating signs.

12.2.3. Diagonalizability for nonnegative elements. We now work over $\mathbb{R}$ and show that a nonnegative element has a (not necessarily unique) eigensection. Note that a nonnegative element is separating provided $\lambda_{\mathrm{X}} \neq \lambda_{\mathrm{Y}}$ whenever X has codimension 1 in Y.

Theorem 12.20. *Every nonnegative element w in the Tits algebra has an eigensection* $\mathtt{u}$, *and hence is diagonalizable.*

PROOF. This can be deduced from Proposition 12.9 by passing to arrangements under different flats. □

Remark 12.21. Theorem 12.17 generalizes to left regular bands (the same arguments work). It may then be used to deduce Theorem 12.20 (again for LRBs) as follows. Given any element w, let Σ_w denote the submonoid of Σ generated by those faces F for which $w^F \neq 0$. In particular, w belongs to the algebra Σ_w obtained by linearizing Σ_w. The point is that Σ_w is a LRB but it may not come from any arrangement.

Now suppose the base field is $\mathbb{R}$ and w is nonnegative. Then w is a separating element of Σ_w, so Theorem 12.17 applies showing that w is diagonalizable in Σ_w and hence in Σ. Note that this argument does not require the Perron-Frobenius Theorem.

If w is nonnegative but not separating, then w may have more than one eigensection. However some of the eigenvalues λ_{X} will then coincide. Lumping together the Eulerian idempotents with the same eigenvalue will give the diagonalization of w and it will be independent of the chosen eigensection by uniqueness. For the same reason, this diagonalization will coincide with the one constructed from Σ_w.

12.2.4. Jordan-Chevalley decomposition in rank 1. Consider the rank-one arrangement with chambers C and $\overline{C}$. Let w be an element of the Tits algebra. Then w is nilpotent iff the coefficient of $\mathtt{H}_O$ is zero and the sum of the coefficients of $\mathtt{H}_C$ and $\mathtt{H}_{\overline{C}}$ is zero. This is because nilpotent elements are precisely those whose support is zero (Proposition 9.22).

Similarly, w is diagonalizable iff w is either a scalar multiple of $\mathtt{H}_O$ or the sum of the coefficients of $\mathtt{H}_C$ and $\mathtt{H}_{\overline{C}}$ is nonzero. To see this, recall from Section 11.4.3 that the Eulerian idempotents are

$$\mathtt{E}_\top = p\,\mathtt{H}_C + (1-p)\,\mathtt{H}_{\overline{C}} \quad \text{and} \quad \mathtt{E}_\perp = \mathtt{H}_O - p\,\mathtt{H}_C - (1-p)\,\mathtt{H}_{\overline{C}}$$

with p arbitrary. Hence the diagonalizable elements have the form

$$a\,\mathtt{E}_\perp + b\,\mathtt{E}_\top = a\,\mathtt{H}_O + (b-a)p\,\mathtt{H}_C + (b-a)(1-p)\,\mathtt{H}_{\overline{C}}$$

from which the claim follows. Alternatively: Recall from Lemma 9.29 that the left module of chambers Γ is faithful. Hence, by Corollary D.18, w is diagonalizable iff $\Psi_\Gamma(w)$ is diagonalizable. The latter condition was analyzed in Section 9.4.5 and using it the claim follows.

In general, w can be uniquely expressed as a sum of a diagonalizable and a nilpotent element which commute with each other. This is the Jordan-Chevalley decomposition (Proposition D.44).

Lemma 12.22. *For the rank-one arrangement with chambers C and $\overline{C}$, the Jordan-Chevalley decomposition is given by*

$$a\mathtt{H}_O + b\,\mathtt{H}_C + c\,\mathtt{H}_{\overline{C}} = \begin{cases} (a\mathtt{H}_O + b\,\mathtt{H}_C + c\,\mathtt{H}_{\overline{C}}) + 0 & \text{if } b+c \neq 0, \\ a\mathtt{H}_O + (b\,\mathtt{H}_C + c\,\mathtt{H}_{\overline{C}}) & \text{if } b+c = 0. \end{cases}$$

This follows from (and contains) the above classification of diagonalizable and nilpotent elements. As a consequence, we note that $w = \mathtt{H}_O + \alpha\,\mathtt{H}_C - \alpha\,\mathtt{H}_{\overline{C}}$, for $\alpha \neq 0$, is an element which is neither nilpotent nor diagonalizable.

Exercise 12.23. Show that no nonzero nilpotent element of the Tits algebra (of any arrangement) commutes with a separating element.

12.2.5. Minimum polynomial. Recall from Theorem D.15 that an element w is diagonalizable if its minimum polynomial factorizes into distinct linear factors. Related ideas are presented below.

Lemma 12.24. *Let w be any element of the Tits algebra. Then for any face F of support X, the element $\mathtt{H}_F \cdot (w - \lambda_\mathrm{X})$ can only contain faces strictly greater than F.*

PROOF. This follows by noting that $\mathtt{H}_F \cdot (w^\mathrm{X} - \lambda_\mathrm{X}) = 0$. □

Lemma 12.25. *For all elements w of the Tits algebra, we have $\prod_\mathrm{X}(w - \lambda_\mathrm{X}) = 0$.*

PROOF. Recall the poset of flats Π. Observe that $w - \lambda_\perp$ does not contain the central face. More generally, we claim that the product $\prod(w - \lambda_\mathrm{Y})$, where Y varies over some lower set of Π, cannot contain any face whose support occurs in this lower set. This can be established by induction using Lemma 12.24. Applying the claim to the full poset Π yields what we want.

Alternatively, one can show that $\mathtt{H}_F \cdot \prod(w - \lambda_\mathrm{Y}) = 0$, the product being over any upper set of Π which contains $\mathrm{s}(F)$. Setting F equal to the central face yields what we want. □

By a similar argument, one can show:

Lemma 12.26. *For all separating elements w, we have $(w - \lambda_1)\dots(w - \lambda_k) = 0$, where $\lambda_1, \dots, \lambda_k$ are the distinct eigenvalues of w.*

This gives a simple direct proof of the fact that separating elements are diagonalizable.

12.3. Takeuchi element

We define the Takeuchi element of the Tits algebra by taking an "alternating" sum of faces (the coefficient of each face F is either $+1$ or -1 depending on the parity of the rank of F). The Takeuchi element is diagonalizable (though it is neither separating nor nonnegative in general). Its eigenvalues only take values ± 1, and its eigensections are precisely the projective sections. Thus, it can be expressed as a linear combination of Eulerian idempotents with coefficients ± 1. The Takeuchi element has a commutative counterpart which belongs to the Birkhoff algebra. Even more interestingly, we have the two-sided Takeuchi element, which is an element of the Janus algebra.

In this discussion, we will make use of some identities from Chapter 7. Also we assume that the field characteristic is not 2.

12.3.1. Takeuchi element. Fix an arrangement $\mathcal{A}$. Consider the element of the Tits algebra $\Sigma[\mathcal{A}]$ defined by

$$\mathtt{Tak}[\mathcal{A}] := \sum_{F} (-1)^{\mathrm{rk}(F)}\, \mathtt{H}_F. \tag{12.15}$$

The sum is over all faces F. We call this the *Takeuchi element* of $\mathcal{A}$. If the arrangement is clear from context, then we may only write $\mathtt{Tak}$.

Lemma 12.27. *The Takeuchi element has order* 2*:*

$$\mathtt{Tak} \cdot \mathtt{Tak} = \mathtt{H}_O. \tag{12.16}$$

In particular, it is invertible, with its inverse being itself.

PROOF. The required calculation is done below.

$$\begin{aligned}\mathtt{Tak} \cdot \mathtt{Tak} &= \Big(\sum_{H} (-1)^{\mathrm{rk}(H)}\, \mathtt{H}_H\Big) \cdot \Big(\sum_{F} (-1)^{\mathrm{rk}(F)}\, \mathtt{H}_F\Big) \\ &= \sum_{G} \Big(\sum_{H,F:\, HF=G} (-1)^{\mathrm{rk}(H)+\mathrm{rk}(F)}\Big)\, \mathtt{H}_G \\ &= \mathtt{H}_O\end{aligned}$$

The last step used Lemma 7.35. □

Lemma 12.28. *The Takeuchi element is diagonalizable. Its minimum polynomial is* $(x+1)(x-1)$. *Explicitly,*

$$\mathtt{Tak} = \Big(\frac{\mathtt{H}_O + \mathtt{Tak}}{2}\Big) - \Big(\frac{\mathtt{H}_O - \mathtt{Tak}}{2}\Big). \tag{12.17}$$

The elements in parenthesis are orthogonal idempotents which add up to $\mathtt{H}_O$.

PROOF. We see from (12.16) that $(x+1)(x-1)$ is the minimum polynomial of $\mathtt{Tak}$. It follows from Theorem D.15 that $\mathtt{Tak}$ is diagonalizable. □

Lemma 12.29. *For all* $z \in \Sigma[\mathcal{A}]$,

$$z \cdot \mathtt{Tak} = \mathtt{Tak} \cdot \overline{z},$$

with notation as in (9.34).

PROOF. This can be deduced from identities (7.11a) and (7.12a). □

As a consequence:

Lemma 12.30. *The Takeuchi element belongs to the center of the projective Tits algebra. Further, the elements* $\frac{1}{2}(\mathtt{H}_O + \mathtt{Tak})$ *and* $\frac{1}{2}(\mathtt{H}_O - \mathtt{Tak})$ *are central idempotents of the projective Tits algebra (that is, idempotents which belong to the center).*

An improvement of this result is given later in Propositions 13.60 and 13.62.

12.3.2. Commutative Takeuchi element. Applying the support map (9.31) to (12.15), we obtain:

$$\mathrm{s}(\mathtt{Tak}) = \sum_{\mathrm{X}} (-1)^{\mathrm{rk}(\mathrm{X})} c^{\mathrm{X}}\, \mathtt{H}_{\mathrm{X}}. \tag{12.18}$$

The sum is over all flats X and c^{X} is the number of faces with support X. We call this the *commutative Takeuchi element* of $\mathcal{A}$. It is an element of the Birkhoff algebra. On the Q-basis,

$$\mathrm{s}(\mathtt{Tak}) = \sum_{\mathrm{X}} (-1)^{\mathrm{rk}(\mathrm{X})}\, \mathtt{Q}_{\mathrm{X}}. \tag{12.19}$$

This follows from (9.1) and (1.42).

12.3.3. Under and over operations. Recall that for any element w of the Tits algebra $\Sigma[\mathcal{A}]$, one can define elements w^{X} and w_H by (12.9) and (12.4). These are the under and over operations, respectively. They preserve the Takeuchi element (up to sign). More precisely:

Lemma 12.31. *For any face H and flat* X,

$$(-1)^{\mathrm{rk}(H)}\mathtt{Tak}[\mathcal{A}]_H = \mathtt{Tak}[\mathcal{A}_H] \quad \textit{and} \quad \mathtt{Tak}[\mathcal{A}]^{\mathrm{X}} = \mathtt{Tak}[\mathcal{A}^{\mathrm{X}}]. \tag{12.20}$$

PROOF. The first claim is a restatement of (7.12a). The second claim is clear. □

Exercise 12.32. Let $\mathcal{A}'$ be a subarrangement of $\mathcal{A}$. Show that the linearization of the map (1.15) sends $\mathtt{Tak}[\mathcal{A}]$ to $(-1)^{\dim(O')-\dim(O)}\mathtt{Tak}[\mathcal{A}']$, where O' denotes the central face of $\mathcal{A}'$. Deduce the first identity in (12.20) by specializing to $\mathcal{A}' := \mathcal{A}_H$.

12.3.4. Action on chambers. The eigenvalues and eigenspaces for the action of the Takeuchi element on the left module of chambers can be computed directly as follows.

Lemma 12.33. *We have*

$$\mathtt{Tak} \cdot \Big(\sum_C x^C\, \mathtt{H}_C \Big) = (-1)^{\mathrm{rk}(\mathcal{A})} \sum_C x^{\overline{C}}\, \mathtt{H}_C. \tag{12.21}$$

In particular: For any chamber C,

$$\mathtt{Tak} \cdot \mathtt{H}_C = (-1)^{\mathrm{rk}(C)}\, \mathtt{H}_{\overline{C}}.$$

PROOF. The necessary calculation is shown below.

$$\begin{aligned}
\mathtt{Tak} \cdot \Big(\sum_C x^C\, \mathtt{H}_C \Big) &= \sum_{H,C} (-1)^{\mathrm{rk}(H)} x^C\, \mathtt{H}_{HC} \\
&= \sum_D \Bigg(\sum_{H:\, H \le D} (-1)^{\mathrm{rk}(H)} \Big(\sum_{C:\, HC=D} x^C \Big) \Bigg) \mathtt{H}_D \\
&= \sum_D (-1)^{\mathrm{rk}(D)} x^{\overline{D}}\, \mathtt{H}_D.
\end{aligned}$$

The last step used the Witt identity (7.14). It is also possible to sum H and C in the opposite order and use the descent identity (7.10). □

Thus, for any chamber C, the subspace spanned by $\mathtt{H}_C$ and $\mathtt{H}_{\overline{C}}$ is invariant under the action of $\mathtt{Tak}$. Analyzing the action on each of these subspaces, we obtain:

Lemma 12.34. *The action of the Takeuchi element on the left module of chambers* $\Gamma[\mathcal{A}]$ *is diagonalizable, with eigenvalues* $+1$ *and* -1, *and with the elements*

$$\mathtt{H}_C + (-1)^{\mathrm{rk}(C)}\, \mathtt{H}_{\overline{C}} \quad \text{and} \quad \mathtt{H}_C - (-1)^{\mathrm{rk}(C)}\, \mathtt{H}_{\overline{C}},$$

as C varies, yielding a basis for the $+1$ *and* -1 *eigenspaces, respectively.*

Exercise 12.35. Generalize (12.21) as follows.

$$\mathtt{Tak} \cdot \Big(\sum_F x^F\, \mathtt{H}_F \Big) = \sum_G \Big((-1)^{\mathrm{rk}(G)} \sum_{F:\, \overline{F} \le G} x^F \Big)\, \mathtt{H}_G.$$

(Use either (7.19b) or (7.11a).)

12.3.5. Eigenvalues and eigensections. Recall from Definition 12.12 that for any element of the Tits algebra, there is a notion of eigenvalues and eigensections.

Lemma 12.36. *The eigenvalues of the Takeuchi element are given by*

$$\lambda_{\mathrm{X}} = (-1)^{\mathrm{rk}(\mathrm{X})}. \tag{12.22}$$

PROOF. This follows from (12.1) and (1.42). Alternatively, recall that the Tits algebra is elementary with the Birkhoff algebra as its split-semisimple quotient under the support map. As a consequence, λ_{X} is the coefficient of $\mathtt{Q}_{\mathrm{X}}$ in $\mathrm{s}(\mathtt{Tak})$, see (9.47) and (9.60). Now use (12.19). □

In particular, $\lambda_{\mathrm{X}} = \lambda_{\mathrm{Y}}$ whenever the ranks of X and Y differ by an even integer. Thus the Takeuchi element is not separating if $\mathcal{A}$ has rank greater than 2. It is interesting to note that

$$\lambda_{\top} - \lambda_{\bot} = \begin{cases} -2 & \text{if } \mathcal{A} \text{ has odd rank,} \\ 0 & \text{if } \mathcal{A} \text{ has even rank.} \end{cases}$$

In particular, Lemma 12.4 applies to the Takeuchi element only if $\mathcal{A}$ has odd rank.

We now turn to the computation of the eigensections. Recall that a homogeneous section $\mathtt{u}$ is projective if $\mathtt{u}^F = \mathtt{u}^{\overline{F}}$ for all faces F.

Lemma 12.37. *Let* $\mathtt{u}$ *be any homogeneous section. Then* $\mathtt{u}$ *is an eigensection for the Takeuchi element iff* $\mathtt{u}$ *is projective.*

In particular, the uniform section (defined in characteristic 0) is an eigensection of the Takeuchi element.

PROOF. We need to solve the equations (12.10). First consider $\mathtt{u}_\top = \sum_C \mathtt{u}^C \mathtt{H}_C$. Observe from (12.21) that

$$\mathtt{Tak} \cdot \mathtt{u}_\top = (-1)^{\mathrm{rk}(\mathcal{A})} \mathtt{u}_\top \iff \mathtt{u}^C = \mathtt{u}^{\overline{C}} \text{ for all chambers } C.$$

Recall from (12.20) that for any flat X, the element $\mathtt{Tak}^{\mathrm{X}}$ can be identified with the Takeuchi element of $\mathcal{A}^{\mathrm{X}}$. Hence by the above calculation, for any flat X,

$$\mathtt{Tak}^{\mathrm{X}} \cdot \mathtt{u}_{\mathrm{X}} = (-1)^{\mathrm{rk}(\mathrm{X})} \mathtt{u}_{\mathrm{X}} \iff \mathtt{u}^F = \mathtt{u}^{\overline{F}} \text{ for all faces } F \text{ of support X.}$$

The result follows. □

Combining with Proposition 12.14, we obtain:

Lemma 12.38. *For an Eulerian family* E *of a homogeneous section* u,

$$\mathtt{u} \text{ is projective} \iff \mathtt{Tak} = \sum_{\mathrm{X}} (-1)^{\mathrm{rk}(\mathrm{X})} \mathtt{E}_{\mathrm{X}}. \tag{12.23}$$

Thus:

Lemma 12.39. *For any projective Eulerian family* E,

$$\sum_F (-1)^{\mathrm{rk}(F)} \mathtt{H}_F = \sum_{\mathrm{X}} (-1)^{\mathrm{rk}(\mathrm{X})} \mathtt{E}_{\mathrm{X}}. \tag{12.24}$$

In particular, this holds for the Eulerian family associated to the uniform section.

Lemma 12.40. *For any projective Eulerian family* E,

$$\frac{1}{2}(\mathtt{H}_O + \mathtt{Tak}) = \sum_{\mathrm{X}:\, \mathrm{rk}(\mathrm{X}) \text{ is even}} \mathtt{E}_{\mathrm{X}} \quad \text{and} \quad \frac{1}{2}(\mathtt{H}_O - \mathtt{Tak}) = \sum_{\mathrm{X}:\, \mathrm{rk}(\mathrm{X}) \text{ is odd}} \mathtt{E}_{\mathrm{X}}. \tag{12.25}$$

PROOF. The rhs of (12.24) gives a diagonalization of the Takeuchi element. Now lump together the Eulerian idempotents with eigenvalue +1 and those with eigenvalue −1, and compare with (12.17) to obtain (12.25). □

We remark that (12.24) is a nontrivial identity. Recall from (11.32) and (11.35) that the coefficients of the Eulerian idempotents are given by an alternating sum; so a lot of delicate cancelations are taking place in the rhs of (12.24). In fact, the lhs is giving a cancelation-free formula for the rhs (though we did not start out with this motivation).

Exercise 12.41. Check identity (12.24) for rank-two arrangements using the formulas given in Proposition 11.65.

Exercise 12.42. For the Eulerian family associated to the uniform section of a good reflection arrangement, check (12.24) explicitly using formula (11.51) and the last identity in Lemma 5.30.

Exercise 12.43. Deduce Theorem 11.59 by specializing Lemma 12.4 to $w = \mathtt{Tak}$.

12.3.6. Two-sided Takeuchi element. Fix an arrangement $\mathcal{A}$. Consider the element of the Janus algebra $\mathsf{J}[\mathcal{A}]$ defined by

$$\mathbf{Tak}[\mathcal{A}] := \sum_{F} (-1)^{\operatorname{rk}(F)} \mathtt{H}_{(F,F)}. \tag{12.26}$$

The sum is over all faces F. We call this the *two-sided Takeuchi element* of $\mathcal{A}$. Projecting it on either coordinate yields the Takeuchi element. Thus, in diagram (9.81):

$$\begin{array}{ccc} \mathsf{J}[\mathcal{A}] & \longrightarrow & \Sigma^{\mathrm{op}}[\mathcal{A}] \\ \big\downarrow & & \big\downarrow \mathrm{s} \\ \Sigma[\mathcal{A}] & \xrightarrow[\mathrm{s}]{} & \Pi[\mathcal{A}]. \end{array} \qquad \begin{array}{ccc} \mathbf{Tak} & \longmapsto & \mathtt{Tak} \\ \big\downarrow & & \big\downarrow \\ \mathtt{Tak} & \longmapsto & \mathrm{s}(\mathtt{Tak}). \end{array}$$

Recall from Proposition 9.68 that the Janus algebra is elementary with the Birkhoff algebra as its split-semisimple quotient. The quotient map is the composite in the diagram above. We now deduce from (12.19) that the possible eigenvalues of the two-sided Takeuchi element on any module (over the Janus algebra) are ± 1, and formula (12.22) holds.

Lemma 12.44. *We have*

$$\mathbf{Tak} \cdot \mathbf{Tak} = \sum_{K \le G} (-1)^{\operatorname{rk}(K)+\operatorname{rk}(G)} \mathtt{H}_{(G,\overline{K}G)} = \sum_{K \le G} (-1)^{\operatorname{rk}(K)+\operatorname{rk}(G)} \mathtt{H}_{(\overline{K}G,G)}. \tag{12.27}$$

The sum is over both K and G. These expressions are cancelation-free.

Proof. This can be deduced using Exercise 7.37. Also note that the two expressions imply each other since **Tak**, and hence **Tak** $\cdot$ **Tak**, is invariant under the map which switches the two coordinates. □

Lemma 12.45. *For all $z \in \mathsf{J}[\mathcal{A}]$,*

$$z \cdot \mathbf{Tak} = \mathbf{Tak} \cdot \overline{z},$$

with notation as in (9.82). *In particular, the two-sided Takeuchi element belongs to the center of the projective Janus algebra.*

Proof. This can be deduced using Exercise 7.36. □

Exercise 12.46. Deduce Lemmas 12.27 and 12.29 as consequences of Lemmas 12.44 and 12.45, respectively.

Exercise 12.47. Show that for any bi-chamber (C, D),

$$(\mathtt{H}_{(O,O)} - \mathbf{Tak}^2) \cdot \mathtt{H}_{(C,D)} = 0 = \mathtt{H}_{(C,D)} \cdot (\mathtt{H}_{(O,O)} - \mathbf{Tak}^2).$$

Exercise 12.48. Show that

$$\mathbf{Tak}^3 = \sum_{K \le H \le G} (-1)^{\operatorname{rk}(K)+\operatorname{rk}(H)+\operatorname{rk}(G)} \mathtt{H}_{(G,K\overline{H}G)}.$$

The sum is over K, H and G. (Use formula (12.27) and Exercise 7.36.) Generalize to $\mathbf{Tak}^i$ by summing over chains involving i faces.

Exercise 12.49. Formulate the analogue of (12.20) for the two-sided Takeuchi element. Also, for any subarrangement $\mathcal{A}'$ of $\mathcal{A}$, show that the linearization of the map (1.19) sends $\mathbf{Tak}[\mathcal{A}]$ to $(-1)^{\dim(O')-\dim(O)}\, \mathbf{Tak}[\mathcal{A}']$, where O' denotes the central face of $\mathcal{A}'$.

Lemma 12.50. *Let r denote the rank of the arrangement. For $r \geq 1$, the element $(\mathtt{H}_{(O,O)} - \mathbf{Tak})^r(\mathtt{H}_{(O,O)} + \mathbf{Tak})^{r-1}$ is a linear combination of bi-chambers. For $r \geq 0$,*

$$(\mathtt{H}_{(O,O)} - \mathbf{Tak})^{r+1}(\mathtt{H}_{(O,O)} + \mathbf{Tak})^r = 0,$$

that is, the minimum polynomial of **Tak** *divides* $(1-x)^{r+1}(1+x)^r$.

PROOF. The first fact implies the second in view of Exercise 12.47. To obtain the first fact, we do an induction on r, and apply the second fact on arrangements under different hyperplanes. □

Lemma 12.51. *For a subarrangement $\mathcal{A}'$ of $\mathcal{A}$ of the same rank as $\mathcal{A}$, the minimum polynomial of* $\mathbf{Tak}[\mathcal{A}']$ *divides the minimum polynomial of* $\mathbf{Tak}[\mathcal{A}]$.

PROOF. This is a consequence of Exercise 12.49. □

Theorem 12.52. *Assume that the field characteristic does not divide $r!$. For an arrangement of rank r, the minimum polynomial of* **Tak** *equals* $(1-x)^{r+1}(1+x)^r$. *In particular,* **Tak** *is not diagonalizable for $r \geq 1$.*

PROOF. We can always remove hyperplanes from the arrangement until we have a subarrangement that is gisomorphic to the coordinate arrangement of rank r. The minimum polynomial of the latter is $(1-x)^{r+1}(1+x)^r$ provided the field characteristic does not divide $r!$. (This requires an argument which we omit.) By Lemma 12.51, this polynomial must divide the the minimum polynomial of **Tak**. Combining with Lemma 12.50, the result follows. □

Exercise 12.53. Using the expression for $\mathbf{Tak}^i$ given in Exercise 12.48, check by direct calculation that the minimum polynomial of **Tak** is $(1-x)^2(1+x)$ for rank-one and $(1-x)^3(1+x)^2$ for rank-two arrangements.

12.4. Characteristic elements

Fix an arrangement $\mathcal{A}$. A characteristic element of parameter t is an element of the Tits algebra whose eigenvalues are powers of t, more precisely, the eigenvalue for X is $t^{\mathrm{rk(X)}}$. The Takeuchi element, for instance, is a characteristic element of parameter -1. These elements are closely linked to the characteristic polynomial of arrangements: For a characteristic element w of parameter t, the sum of the coefficients of faces of support X in w equals the characteristic polynomial of $\mathcal{A}^{\mathrm{X}}$ (in the variable t).

The Fulman element of parameter t is the characteristic element of parameter t for which faces with the same support have the same coefficient. For a good reflection arrangement, the uniform section is an eigensection for the Fulman element, thus giving an explicit diagonalization. The product of the Fulman elements of parameters s and t is the Fulman element of parameter st. Further, the algebra generated by all the Fulman elements is a split-semisimple commutative algebra of dimension equal to the rank of the arrangement plus 1.

12.4.1. Characteristic elements.

An element w of the Tits algebra $\Sigma[\mathcal{A}]$ is called a *characteristic element* if there exists a scalar t such that

$$\lambda_{\mathrm{X}}(w) = t^{\mathrm{rk(X)}}, \tag{12.28}$$

for each flat X, with λ_{X} as in (12.1). The dependence of λ_{X} on w is made explicit by writing it in parenthesis.

Observe that:

- If w is a characteristic element, then $\lambda_{\perp}(w) = 1$, that is, the coefficient of the central face in w is 1.
- If $\mathcal{A}$ has rank 0, then $\mathtt{H}_O$ is a characteristic element of parameter t for all t, and there are no other characteristic elements.
- If $t \neq 0$ and t is not a root of unity, then any characteristic element of parameter t is separating, and in particular, has a unique eigensection.

Lemma 12.54. *For any elements u, v of the Tits algebra, and for any flat* X*,*

$$\lambda_{\mathrm{X}}(u \cdot v) = \lambda_{\mathrm{X}}(u)\lambda_{\mathrm{X}}(v).$$

In particular, if u is a characteristic element of parameter s and v is a characteristic element of parameter t, then uv is a characteristic element of parameter st.

PROOF. The linear functional λ_{X} is the same as the multiplicative character χ_{X} defined in (9.49), so the above formula holds. This may also be checked directly. The main observation is $(u \cdot v)_{\mathrm{X}} = u_{\mathrm{X}} \cdot v_{\mathrm{X}}$. Taking the sum of coefficients in the H-basis on both sides yields the formula. □

Using Proposition 9.51, we deduce:

- For $t \neq 0$, any characteristic element of parameter t is invertible in the Tits algebra, and its inverse is a characteristic element of parameter t^{-1}.
- Any characteristic element of parameter 0 is a zero divisor (assuming rank of $\mathcal{A}$ to be at least 1).

12.4.2. Relation to characteristic polynomial. Recall the characteristic polynomial of an arrangement from Section 1.13.4.

Lemma 12.55. *Let $w = \sum_F w^F \mathtt{H}_F$ be any element of the Tits algebra. Then w is a characteristic element of parameter t iff for every flat* X*,*

$$\sum_{F:\, \mathrm{s}(F)=\mathrm{X}} w^F = \chi(\mathcal{A}^{\mathrm{X}}, t), \tag{12.29}$$

where $\chi(\mathcal{A}^{\mathrm{X}}, t)$ is the characteristic polynomial of $\mathcal{A}^{\mathrm{X}}$.

PROOF. Denote the lhs of (12.29) by $f(\mathrm{X})$. Note that

w is a characteristic element of parameter t

$$\iff \sum_{\mathrm{Y}:\, \mathrm{Y} \le \mathrm{X}} f(\mathrm{Y}) = t^{\mathrm{rk}(\mathrm{X})} \iff f(\mathrm{X}) = \sum_{\mathrm{Y}:\, \mathrm{Y} \le \mathrm{X}} \mu(\mathrm{Y}, \mathrm{X})\, t^{\mathrm{rk}(\mathrm{Y})}.$$

Now apply definition (1.53). □

It follows that the set of characteristic elements of parameter t is an affine subspace of $\Sigma[\mathcal{A}]$ of dimension equal to the number of faces minus the number of flats, for any scalar t. In particular, there exist characteristic elements of parameter t. We now consider the special cases $t = 0, 1, -1$.

Lemma 12.56. *Let $w = \sum_F w^F \mathtt{H}_F$ be any element of the Tits algebra. Then: w is a characteristic element of parameter* 0 *iff for every flat* X*,*

$$\sum_{F:\, \mathrm{s}(F)=\mathrm{X}} w^F = \mu(\perp, \mathrm{X}); \tag{12.30a}$$

w is a characteristic element of parameter 1 iff $w^O = 1$ and for every non-minimum flat X,

$$\sum_{F:\,\mathrm{s}(F)=\mathrm{X}} w^F = 0; \tag{12.30b}$$

w is a characteristic element of parameter -1 iff for every flat X,

$$\sum_{F:\,\mathrm{s}(F)=\mathrm{X}} w^F = (-1)^{\mathrm{rk}(\mathrm{X})} c^{\mathrm{X}}, \tag{12.30c}$$

where c^{X} is the number of faces with support X.

PROOF. We employ (12.29). The three cases follow from (1.54a), (1.54b) and (1.54c), respectively. □

Corollary 12.57. *We have:*

(1) *Any special Zie element is a characteristic element of parameter* 0.
(2) *The unit $\mathtt{H}_O$ is a characteristic element of parameter* 1.
(3) *The Takeuchi element is a characteristic element of parameter* -1.

PROOF. For (1), use (12.30a) and (10.15). For (2), use (12.30b). For (3), use (12.30c). This can also be seen directly by comparing (12.22) and (12.28). □

Exercise 12.58. Let $\mathcal{A}$ be an arrangement of rank n. Its *Whitney polynomial* is

$$\omega(\mathcal{A}, t, u) := \sum_{\mathrm{Y} \leq \mathrm{Z}} \mu(\mathrm{Y}, \mathrm{Z})\, t^{\mathrm{rk}(\mathrm{Y})} u^{n-\mathrm{rk}(\mathrm{Z})}. \tag{12.31}$$

(i) Note that $\omega(\mathcal{A}, t, 0) = \chi(\mathcal{A}, t)$.
(ii) Let w be a characteristic element of parameter t. Show that

$$\sum_F u^{n-\mathrm{rk}(F)} w^F = \omega(\mathcal{A}, t, u). \tag{12.32}$$

(iii) Employ the Takeuchi element to deduce that

$$\omega(\mathcal{A}, -1, -u) = (-1)^n \sum_{i=0}^{n} f_{n-i}(\mathcal{A})\, u^i, \tag{12.33}$$

where $f_i(\mathcal{A})$ denotes the number of rank i faces of $\mathcal{A}$. Note this is equivalent to (1.50).

We say that a scalar t is *critical* for $\mathcal{A}$ if it is a root of $\chi(\mathcal{A}^{\mathrm{X}}, t)$ for some flat X. The set of critical scalars is finite. The scalar 0 is not critical, in view of (1.47) and (1.54a).

Let t be a non-critical scalar. Given a characteristic element w of parameter t, set

$$\mathtt{u}_{\mathrm{X}} = \frac{1}{\chi(\mathcal{A}^{\mathrm{X}}, t)} \sum_{F:\,\mathrm{s}(F)=\mathrm{X}} w^F \mathtt{H}_F.$$

In view of (12.29), $\mathtt{u}_{\mathrm{X}}$ is a homogeneous section as in (11.3). We recover w from $\mathtt{u}$ by multiplying by the scalars $\chi(\mathcal{A}^{\mathrm{X}}, t)$. Thus:

Proposition 12.59. *Let t be a non-critical scalar for $\mathcal{A}$. The set of characteristic elements of parameter t is in bijection with the set of homogeneous sections of the support map.*

12.4.3. Uniqueness modulo the radical. It follows from (12.29) that two elements w and w' are characteristic of the same parameter iff $\lambda_X(w-w') = 0$ for each flat X. Since the λ_X are the characters of the simple Σ-modules (Theorem 16.5), the preceding holds precisely when $w - w' \in \mathrm{rad}(\Sigma)$. (This can also be seen from (9.33) and (12.29).) Thus:

Proposition 12.60. *The space of characteristic elements of parameter t is*

$$w_0 + \mathrm{rad}(\Sigma),$$

where w_0 is any such element.

It also follows that any two characteristic elements of the same parameter differ by a nilpotent element.

12.4.4. Multiplicativity. Suppose for each scalar t, w_t is a characteristic element of parameter t. By Lemma 12.54, both $w_s \cdot w_t$ and w_{st} are characteristic elements of the same parameter st. Therefore, $w_s \cdot w_t - w_{st}$ is nilpotent. Below we make a stronger assumption.

Proposition 12.61. *Suppose for each scalar t, w_t is a characteristic element of parameter t such that*

$$w_s \cdot w_t = w_{st} \tag{12.34}$$

for all s and t. Then $w_1 = \mathtt{H}_O$. If w_{-1} is projective, then $w_{-1} = \mathtt{Tak}$.

Proof. Since $\mathtt{H}_O$ is characteristic of parameter 1, $w_1 - \mathtt{H}_O$ is nilpotent. Setting $s = t = 1$ in (12.34), we obtain that w_1 is idempotent. Hence $w_1 - \mathtt{H}_O$ is both idempotent and nilpotent, and therefore zero.

When w_{-1} is projective, it commutes with $\mathtt{Tak}$, by Lemma 12.30. Setting $s = t = -1$ in (12.34), we obtain that $w_{-1}^2 = w_1 = \mathtt{H}_O$. Thus, w_{-1} and $\mathtt{Tak}$ are commuting involutions, and so they are are simultaneously diagonalizable. On the other hand, since they are both characteristic of parameter -1, their difference is nilpotent. Being both nilpotent and diagonalizable, this difference must be zero. □

12.4.5. Under and over operations. The under and over operations preserve characteristic elements. More precisely:

Lemma 12.62. *Suppose w is a characteristic element of parameter t, with $t \neq 0$. Then:*

- *For any face H, the element $t^{-\mathrm{rk}(H)} w_H$ defined by (12.4) is a characteristic element of parameter t of $\mathcal{A}_H$.*
- *For any flat X, the element w^{X} defined by (12.9) is a characteristic element of parameter t of $\mathcal{A}^{\mathrm{X}}$.*

Proof. The second claim is clear. For the first claim: For any $\mathrm{X} \geq \mathrm{s}(H)$,

$$\lambda_{\mathrm{X}/H}(w_H) = \lambda_{\mathrm{X}}(w) = t^{\mathrm{rk}(\mathrm{X})} = t^{\mathrm{rk}(H)} t^{\mathrm{rk}(\mathrm{X}/H)}.$$ □

12.4.6. Subarrangements. For a subarrangement $\mathcal{A}'$ of $\mathcal{A}$, recall the morphism of algebras $f : \Sigma[\mathcal{A}] \to \Sigma[\mathcal{A}']$ from (9.61).

Lemma 12.63. *Suppose $\mathcal{A}'$ has the same center as $\mathcal{A}$. Then f sends characteristic elements in $\Sigma[\mathcal{A}]$ to characteristic elements in $\Sigma[\mathcal{A}']$.*

PROOF. Let X be a flat of $\mathcal{A}'$. It is also a flat of $\mathcal{A}$. Moreover, since the arrangements have the same center, the ranks of X in $\Pi[\mathcal{A}']$ and in $\Pi[\mathcal{A}]$ coincide. Since X is a flat of $\mathcal{A}'$, for any face F of $\mathcal{A}$ we have $\mathrm{s}(F) \leq \mathrm{X}$ in $\Pi[\mathcal{A}]$ iff $\mathrm{s}(f(F)) \leq \mathrm{X}$ in $\Pi[\mathcal{A}']$. Using (12.1), we deduce that $\lambda_{\mathrm{X}}(w) = \lambda_{\mathrm{X}}(f(w))$. The result follows from (12.28). □

Exercise 12.64. Deduce Lemma 12.63 using Exercise 9.49.

Exercise 12.65. Let $\mathcal{A}'$ be a subarrangement of $\mathcal{A}$. Let r' and r be their respective ranks. Extend Lemma 12.63 by showing that if w is a characteristic element in $\Sigma[\mathcal{A}]$ of parameter $t \neq 0$, then $\frac{1}{t^{r-r'}} f(w)$ is a characteristic element of parameter t in $\Sigma[\mathcal{A}']$. Note this generalizes the first statement in Lemma 12.62. Show also that if $t = 0$ and $r' < r$, then $f(w)$ lies in the radical of $\Sigma[\mathcal{A}']$.

12.4.7. A recursion for the characteristic polynomial. The characteristic polynomial may be calculated recursively by removing one hyperplane at a time. As an application of the theory of characteristic elements, we derive a proof of this formula.

Proposition 12.66. *Let* H *be a hyperplane in* $\mathcal{A}$ *that contains the intersection of the remaining hyperplanes. Then*

$$\chi(\mathcal{A}, t) = \chi(\mathcal{A} \setminus \{\mathrm{H}\}, t) - \chi(\mathcal{A}^{\mathrm{H}}, t). \tag{12.35}$$

PROOF. Let w be a characteristic element of parameter t. We employ (12.29) to calculate both $\chi(\mathcal{A}, t)$ and $\chi(\mathcal{A}^{\mathrm{H}}, t)$ as sums of certain coefficients of w. We obtain

$$\chi(\mathcal{A}, t) + \chi(\mathcal{A}^{\mathrm{H}}, t) = \sum w^C + \sum w^F,$$

the first sum being over all chambers C of $\mathcal{A}$ and the second over all faces F of $\mathcal{A}$ with $\mathrm{s}(F) = \mathrm{H}$.

Let $\mathcal{A}' = \mathcal{A} \setminus \{\mathrm{H}\}$. By hypothesis, $\mathcal{A}'$ has the same center as $\mathcal{A}$. Hence, we may also employ (12.29) to calculate $\chi(\mathcal{A}', t)$ in terms of coefficients of $f(w)$, where f is the morphism in Lemma 12.63. We obtain

$$\chi(\mathcal{A}', t) = \sum f(w)^D = \sum w^G$$

the first sum being over all chambers D of $\mathcal{A}'$ and the second over all faces G of $\mathcal{A}$ with $f(G) = D$ for some such D. These faces G are either chambers of $\mathcal{A}$ or faces with support H. The result follows by comparing the above expressions. □

Exercise 12.67. Let H be a hyperplane in $\mathcal{A}$ that does not contain the intersection of the remaining hyperplanes. Use Exercise 1.45 to deduce that

$$\chi(\mathcal{A} \setminus \{\mathrm{H}\}, t) = \chi(\mathcal{A}^{\mathrm{H}}, t) = \frac{1}{t-1} \chi(\mathcal{A}, t). \tag{12.36}$$

12.4.8. Fulman elements. Assume that $\Bbbk$ has characteristic zero. Let t be a scalar. The *Fulman element* of parameter t is defined to be

$$\mathtt{Ful}_t[\mathcal{A}] := \sum_{\mathrm{X}} \frac{\chi(\mathcal{A}^{\mathrm{X}}, t)}{c^{\mathrm{X}}} \sum_{F:\, \mathrm{s}(F) = \mathrm{X}} \mathtt{H}_F. \tag{12.37}$$

If the arrangement is clear from context, then we may only write $\mathtt{Ful}_t$.

Lemma 12.68. *The element* $\mathtt{Ful}_t$ *is a characteristic element of parameter* t *(in which the coefficients of faces are distributed uniformly). Further, it is invariant under any symmetry of* $\mathcal{A}$.

PROOF. Faces with the same support X have the same coefficient in $\mathtt{Ful}_t$, and further their sum is $\chi(\mathcal{A}^{\mathrm{X}}, t)$. Hence, the first part follows by Lemma 12.55. For the second part, note that if a symmetry takes X to Y, then $\mathcal{A}^{\mathrm{X}}$ and $\mathcal{A}^{\mathrm{Y}}$ are cisomorphic, so they will have the same characteristic polynomial and the same number of chambers. □

Lemma 12.69. *For a good reflection arrangement $\mathcal{A}$, for any face H and flat* X,

$$\mathtt{Ful}_t[\mathcal{A}]_H = t^{\mathrm{rk}(H)}\mathtt{Ful}_t[\mathcal{A}_H] \quad \textit{and} \quad \mathtt{Ful}_t[\mathcal{A}]^{\mathrm{X}} = \mathtt{Ful}_t[\mathcal{A}^{\mathrm{X}}].$$

PROOF. The second claim is clear (and valid for any arrangement). For the first claim, for a face G greater than H and with support X,

$$\begin{aligned}\langle \mathtt{H}_H \cdot \mathtt{Ful}_t[\mathcal{A}], \mathtt{H}_G\rangle &= \langle \mathtt{H}_H \cdot \mathtt{Ful}_t[\mathcal{A}^{\mathrm{X}}], \mathtt{H}_G\rangle \\ &= \frac{1}{c_H^{\mathrm{X}}} \sum_{F:\, \mathrm{s}(HF)=\mathrm{X}} \frac{\chi(\mathcal{A}^{\mathrm{s}(F)}, t)}{c^F} \\ &= \frac{1}{c_H^{\mathrm{X}}} \sum_{\mathrm{Y}:\, \mathrm{s}(H)\vee\mathrm{Y}=\mathrm{X}} \chi(\mathcal{A}^{\mathrm{Y}}, t) \\ &= t^{\mathrm{rk}(H)} \frac{\chi(\mathcal{A}_H^{\mathrm{X}}, t)}{c_H^{\mathrm{X}}}.\end{aligned}$$

Since $\mathcal{A}$ is good, $\mathcal{A}^{\mathrm{X}}$ is a reflection arrangement. The element $\mathtt{H}_H \cdot \mathtt{Ful}_t[\mathcal{A}^{\mathrm{X}}]$ is invariant under all Coxeter symmetries of $\mathcal{A}^{\mathrm{X}}$ which fix H. So to calculate the coefficient of $\mathtt{H}_G$, we can average the coefficients of all faces greater than H and with support X (on which the symmetries act transitively). This is what was done in the second step. In the last step, we used formula (1.55). □

Lemma 12.70. *For a good reflection arrangement $\mathcal{A}$, the uniform section* $\mathtt{u}$ *is an eigensection for* $\mathtt{Ful}_t$, *and in particular* $\mathtt{Ful}_t$ *is diagonalizable.*

PROOF. Since $\mathtt{Ful}_t$ and $\mathtt{u}_\top$ are invariant under the symmetries of $\mathcal{A}$, so is $\mathtt{Ful}_t \cdot \mathtt{u}_\top$. Since $\mathcal{A}$ is a reflection arrangement, its Coxeter symmetries act transitively on chambers, and so $\mathtt{Ful}_t \cdot \mathtt{u}_\top$ must be a multiple of $\mathtt{u}_\top$. So $\mathtt{u}_\top$ is an eigenvector of $\mathtt{Ful}_t$. For the general case, we apply this result to $\mathcal{A}^{\mathrm{X}}$ which is a reflection arrangement (since $\mathcal{A}$ is assumed to be good). □

More generally: Suppose $\mathcal{A}$ is a good reflection arrangement, and w is an element of the Tits algebra of $\mathcal{A}$ such that w^{X} is invariant under the Coxeter symmetries of $\mathcal{A}^{\mathrm{X}}$ for each flat X. Then the uniform section is an eigensection of w.

Theorem 12.71. *For a good reflection arrangement $\mathcal{A}$,*

$$\mathtt{Ful}_t = \sum_{\mathrm{X}} t^{\mathrm{rk}(\mathrm{X})}\, \mathtt{E}_{\mathrm{X}} = \sum_{k=0}^{\mathrm{rk}(\mathcal{A})} t^k\, \mathtt{E}_k, \tag{12.38}$$

with $\mathtt{E}_{\mathrm{X}}$ *as in* (11.51), *and* $\mathtt{E}_k$ *as in* (11.54). *Further, if $t \neq 0$ and t is not a root of unity, then* $\mathtt{E}$ *is the unique Eulerian family for which* (12.38) *holds.*

PROOF. By Proposition 12.14 and Lemma 12.70, $\mathtt{Ful}_t$ can be diagonalized by the Eulerian family associated to the uniform section (given in Theorem 11.53). The first part follows. If $t \neq 0$ and t is not a root of unity, then $\mathtt{Ful}_t$ is separating and uniqueness follows from Theorem 12.17. □

Viewing t as a formal parameter, $\mathtt{Ful}_t$ is a linear combination of faces whose coefficients are polynomials in t. The element $\mathtt{E}_k$ is precisely the coefficient of t^k. In view of Lemma 1.87, this gives an alternative proof of Theorem 11.56.

We now consider the parameter values $t = 0, 1, -1$ of the Fulman element. For $t = 0$, using (1.54a), we see that $\mathtt{Ful}_0[\mathcal{A}]$ is the same as the element given in (11.53). Further, for any good reflection arrangement $\mathcal{A}$, using (11.52),

$$\mathtt{Ful}_0[\mathcal{A}] = \mathtt{E}_\perp, \tag{12.39a}$$

the first Eulerian idempotent of the uniform section. This is consistent with (12.38).

For $t = 1$, using (1.54b), we see that only the minimum flat contributes to the rhs of (12.37). Thus,

$$\mathtt{Ful}_1[\mathcal{A}] = \mathtt{H}_O. \tag{12.39b}$$

In this case, (12.38) is an instance of (11.18).

For $t = -1$, using (1.54c), we obtain

$$\mathtt{Ful}_{-1}[\mathcal{A}] = \mathtt{Tak}[\mathcal{A}], \tag{12.39c}$$

the Takeuchi element. In this case, (12.38) is an instance of (12.24).

12.4.9. Fulman algebra. Let $\mathsf{F}[\mathcal{A}]$ denote the subalgebra of $\Sigma[\mathcal{A}]$ generated by all the Fulman elements. We call $\mathsf{F}[\mathcal{A}]$ the *Fulman algebra*.

Theorem 12.72. *For a good reflection arrangement $\mathcal{A}$, the Fulman algebra is a split-semisimple commutative algebra of dimension $\mathrm{rk}(\mathcal{A}) + 1$. Explicitly,*

$$\mathsf{F}[\mathcal{A}] \xrightarrow{\cong} \Bbbk^{\mathrm{rk}(\mathcal{A})+1}, \qquad \mathtt{Ful}_t \mapsto (1, t, \dots, t^{\mathrm{rk}(\mathcal{A})}).$$

If $t \neq 0$ and t is not a root of unity, then $\mathsf{F}[\mathcal{A}]$ is generated by $\mathtt{Ful}_t$.

PROOF. First note that the subalgebra of $\Sigma[\mathcal{A}]$ generated by the $\mathtt{E}_k$, for $k = 0, \dots, \mathrm{rk}(\mathcal{A})$, is isomorphic to $\Bbbk^{\mathrm{rk}(\mathcal{A})+1}$. Now using Theorem 12.71 yields an injective algebra homomorphism as above. To see that it is surjective, consider the image of $\mathtt{Ful}_t$ for $\mathrm{rk}(\mathcal{A}) + 1$ distinct choices of t, and note that these images are linearly independent, say from the determinant formula of the Vandermonde matrix. The last claim follows similarly. □

Corollary 12.73. *For a good reflection arrangement $\mathcal{A}$, for any scalars s and t,*

$$\mathtt{Ful}_s \cdot \mathtt{Ful}_t = \mathtt{Ful}_{st}. \tag{12.40}$$

PROOF. This follows from the isomorphism in Theorem 12.72. Alternatively, one may argue directly as follows. First note from Lemma 12.54 that $\mathtt{Ful}_s \cdot \mathtt{Ful}_t$ is a characteristic element of parameter st. Further, this element is invariant under the Coxeter symmetries of $\mathcal{A}$. So all chambers must appear with the same coefficient. Since by (12.29), the sum of these coefficients is $\chi(\mathcal{A}, t)$, each coefficient must be $\chi(\mathcal{A}, t)/c(\mathcal{A})$, where $c(\mathcal{A})$ is the number of chambers. Applying this argument to each $\mathcal{A}^{\mathrm{X}}$ establishes (12.40). □

12.5. Type A Eulerian idempotents and Adams elements

Let $\mathcal{A}$ be the braid arrangement on $[p]$ (Sections 6.3–6.6). We work with the uniform section (so it is implicit that the field characteristic is 0). This defines the Eulerian family $\mathtt{E}$ as well as the $\mathtt{Q}$-basis. Recall that the braid arrangement is a good reflection arrangement. This allows us to specialize the results of Section 11.6. We first give the change of basis formulas between the $\mathtt{H}$- and $\mathtt{Q}$-bases, and then cancelation-free formulas for the Eulerian idempotents in the $\mathtt{H}$-basis. Further, we consider an interesting family of elements indexed by integers called Adams elements, and write down an explicit diagonalization for them. This family includes the Takeuchi element (up to a sign), and is also closely related to the Fulman elements. We conclude with a general construction of Eulerian families from Zie elements which includes the above Eulerian family as a special case.

12.5.1. $\mathtt{H}$- and $\mathtt{Q}$-bases.

Lemma 12.74. *For the uniform section, the* $\mathtt{H}$- *and* $\mathtt{Q}$-*bases are related by*

$$\mathtt{H}_F = \sum_{G:\, F \le G} \frac{1}{\deg!(G/F)} \mathtt{Q}_G \quad \textit{and} \quad \mathtt{Q}_F = \sum_{G:\, F \le G} \frac{(-1)^{\mathrm{rk}(G/F)}}{\deg(G/F)} \mathtt{H}_G, \tag{12.41}$$

with degrees and factorials as in Section 6.6.3.

PROOF. These formulas are a specialization of (11.50). The second formula used (6.11). □

For $p = 2$, the change of basis formulas are

$$\mathtt{H}_{1|2} = \mathtt{Q}_{1|2}, \quad \mathtt{H}_{2|1} = \mathtt{Q}_{2|1}, \quad \mathtt{H}_{12} = \mathtt{Q}_{12} + \frac{1}{2}(\mathtt{Q}_{1|2} + \mathtt{Q}_{2|1}), \quad \mathtt{Q}_{12} = \mathtt{H}_{12} - \frac{1}{2}(\mathtt{H}_{1|2} + \mathtt{H}_{2|1}).$$

12.5.2. Eulerian idempotents.

Theorem 12.75. *The Eulerian idempotents for the uniform section are given by*

$$\begin{aligned} \mathtt{E}_{\mathrm{X}} &= \frac{1}{\deg!(\mathrm{X})} \sum_{F:\, \mathrm{s}(F) = \mathrm{X}} \mathtt{Q}_F \\ &= \frac{1}{\deg!(\mathrm{X})} \sum_{F:\, \mathrm{s}(F) = \mathrm{X}} \; \sum_{G:\, F \le G} \frac{(-1)^{\mathrm{rk}(G/F)}}{\deg(G/F)} \mathtt{H}_G, \end{aligned} \tag{12.42}$$

where $\deg!(\mathrm{X})$ *is the factorial of the number of blocks of* X.

In particular, the first Eulerian idempotent is

$$\mathtt{E}_{\perp} = \mathtt{Q}_O = \sum_F \frac{(-1)^{\mathrm{rk}(F)}}{\deg(F)} \mathtt{H}_F. \tag{12.43}$$

PROOF. This is a specialization of Theorem 11.53. □

For $p = 2$, the two Eulerian idempotents are

$$\mathtt{E}_{\top} = \mathtt{E}_{1,2} = \frac{1}{2}(\mathtt{H}_{1|2} + \mathtt{H}_{2|1}) \quad \text{and} \quad \mathtt{E}_{\perp} = \mathtt{E}_{12} = \mathtt{H}_{12} - \frac{1}{2}(\mathtt{H}_{1|2} + \mathtt{H}_{2|1}).$$

For $p = 3$, the five Eulerian idempotents are

$$\begin{aligned}
\mathtt{E}_{\top} = \mathtt{E}_{1,2,3} &= \frac{1}{6}(\mathtt{H}_{1|2|3} + \mathtt{H}_{1|3|2} + \mathtt{H}_{2|3|1} + \mathtt{H}_{2|1|3} + \mathtt{H}_{3|1|2} + \mathtt{H}_{3|2|1}),\\
\mathtt{E}_{1,23} &= \frac{1}{2}(\mathtt{H}_{1|23} + \mathtt{H}_{23|1}) - \frac{1}{4}(\mathtt{H}_{1|2|3} + \mathtt{H}_{1|3|2} + \mathtt{H}_{2|3|1} + \mathtt{H}_{3|2|1}),\\
\mathtt{E}_{2,13} &= \frac{1}{2}(\mathtt{H}_{2|13} + \mathtt{H}_{13|2}) - \frac{1}{4}(\mathtt{H}_{2|1|3} + \mathtt{H}_{2|3|1} + \mathtt{H}_{1|3|2} + \mathtt{H}_{3|2|1}),\\
\mathtt{E}_{3,12} &= \frac{1}{2}(\mathtt{H}_{3|12} + \mathtt{H}_{12|3}) - \frac{1}{4}(\mathtt{H}_{3|1|2} + \mathtt{H}_{3|2|1} + \mathtt{H}_{1|2|3} + \mathtt{H}_{2|1|3}),\\
\mathtt{E}_{\perp} = \mathtt{E}_{123} &= \mathtt{H}_{123} - \frac{1}{2}(\mathtt{H}_{1|23} + \mathtt{H}_{23|1} + \mathtt{H}_{2|13} + \mathtt{H}_{13|2} + \mathtt{H}_{3|12} + \mathtt{H}_{12|3})\\
&\quad + \frac{1}{3}(\mathtt{H}_{1|2|3} + \mathtt{H}_{1|3|2} + \mathtt{H}_{2|3|1} + \mathtt{H}_{2|1|3} + \mathtt{H}_{3|1|2} + \mathtt{H}_{3|2|1}).
\end{aligned}$$

Since the braid arrangement on $[3]$ is cisomorphic to the rank-two arrangement of 3 lines, the above are special cases of the formulas in Proposition 11.65.

Now set

$$\mathtt{E}_k := \sum_{\mathrm{X}:\, \deg(\mathrm{X})=k} \mathtt{E}_{\mathrm{X}} = \frac{1}{k!} \sum_{F:\, \deg(F)=k} \mathtt{Q}_F \quad \text{for } 1 \leq k \leq p. \tag{12.44}$$

Warning. This convention does not match (11.54), where rank of X is employed. However, for type A, usage of degree is standard and appears to be more convenient than rank.

Put

$$\mathtt{T}_k := \sum_{F:\, \deg(F)=k} \mathtt{H}_F. \tag{12.45}$$

Theorem 12.76. *We have*

$$\mathtt{E}_k = \sum_{m:\, m \geq k} s(m,k)\, \frac{\mathtt{T}_m}{m!}, \tag{12.46}$$

where $s(m,k)$ are the Stirling numbers of the first kind.

PROOF. This formula is a specialization of Theorem 11.56. This can be seen from (6.13). For the switch from $k+1$ to k, see the above warning. □

For $p = 3$,

$$\begin{aligned}
\mathtt{E}_1 &= \mathtt{T}_1 - \frac{\mathtt{T}_2}{2!} + 2\frac{\mathtt{T}_3}{3!}\\
\mathtt{E}_2 &= \frac{\mathtt{T}_2}{2!} - 3\frac{\mathtt{T}_3}{3!}\\
\mathtt{E}_3 &= \frac{\mathtt{T}_3}{3!}.
\end{aligned}$$

Lemma 12.77. *We have*

$$\sum_{k=1}^{p} \binom{n}{k} \mathtt{T}_k = \sum_{k=1}^{p} n^k\, \mathtt{E}_k. \tag{12.47}$$

PROOF. Express the binomial coefficients in the lhs as polynomials in n, collect the coefficients of various powers of n, and use (12.46). □

12.5.3. Adams elements. For any integer n, define the *Adams element* of parameter n to be

$$\mathtt{Ads}_n := \sum_F \binom{n}{\deg(F)} \mathtt{H}_F = \sum_{k=1}^{p} \binom{n}{k} \mathtt{T}_k. \tag{12.48}$$

By Lemma 6.12 and definition (12.1), we obtain:

Lemma 12.78. *The eigenvalues of* $\mathtt{Ads}_n$ *are given by* $\lambda_{\mathrm{X}} = n^{\deg(\mathrm{X})}$, *one for each set partition* X.

Further:

Proposition 12.79. *The Adams elements diagonalize as follows.*

$$\mathtt{Ads}_n = \sum_{\mathrm{X}} n^{\deg(\mathrm{X})} \mathtt{E}_{\mathrm{X}} = \sum_{k=1}^{p} n^k \mathtt{E}_k. \tag{12.49}$$

PROOF. This is a restatement of Lemma 12.77. Alternatively, put $w := \mathtt{Ads}_n$. Note that w^{X} is invariant under the action of the symmetric group on the blocks of X. So the uniform section is an eigensection of w. Hence (12.49) holds by Proposition 12.14 in view of Lemma 12.78. □

As a consequence of (12.49):

Lemma 12.80. *For any integers* m *and* n,

$$\mathtt{Ads}_m \cdot \mathtt{Ads}_n = \mathtt{Ads}_{mn}. \tag{12.50}$$

Similarly:

Theorem 12.81. *For the braid arrangement on* $[p]$, *the subalgebra of the Tits algebra generated by the elements* $\mathtt{Ads}_n$ *is a split-semisimple commutative algebra, with primitive idempotents* $\mathtt{E}_k$, *for* $1 \leq k \leq p$.

Let us now consider the parameter value $n = -1$. Using definition (12.48),

$$\mathtt{Ads}_{-1} = \sum_F (-1)^{\deg(F)} \mathtt{H}_F.$$

This is the negative of the Takeuchi element (12.15). It diagonalizes as

$$\mathtt{Ads}_{-1} = \sum_{\mathrm{X}} (-1)^{\deg(\mathrm{X})} \mathtt{E}_{\mathrm{X}} = \sum_{k=1}^{p} (-1)^k \mathtt{E}_k.$$

This can also be seen as an instance of (12.24) (after taking negative of both sides).

Let us now relate the Adams elements to the Fulman elements. A comparison of (6.12) with (12.37) shows that

$$\mathtt{Ads}_n = n\, \mathtt{Ful}_n,$$

where $\mathtt{Ful}_n$ is the Fulman element of parameter n. In particular, $\mathtt{Ads}_{-1} = -\mathtt{Ful}_{-1}$, which is the negative of the Takeuchi element by (12.39c). This is consistent with what we noted above.

The diagonalization (12.49) can also be deduced by multiplying both sides of (12.38) by t and letting $t = n$. Similarly, (12.40) yields (12.50). Also, Theorem 12.81 can be deduced from Theorem 12.72.

Exercise 12.82. Check that $\mathtt{Ads}_n$ is a separating element if $n \neq 0, 1, -1$.

12.5.4. Action on chambers. Let $\mathtt{E}_\perp$ be the first Eulerian idempotent given by (12.43). A formula for its action on the left module of chambers is given below.

Proposition 12.83. *For any chamber C,*

$$\mathtt{E}_\perp \cdot \mathtt{H}_C = \sum_D (-1)^{s-1} \frac{1}{s\binom{p}{s}} \mathtt{H}_D, \tag{12.51}$$

where s is the degree of $\mathrm{Des}(C, D)$, *with the latter as in* (7.1).

PROOF. The crux of the calculation is shown below.

$$\sum_{i=0}^{p-s} \frac{(-1)^i}{i+s} \binom{p-s}{i} = \int_0^1 x^{s-1}(1-x)^{p-s} dx = \frac{1}{s\binom{p}{s}} \quad \text{for } 1 \leq s \leq p.$$

We omit the details. □

Viewing $\Sigma[\mathcal{A}]$ as a subspace of $\widehat{\mathsf{Q}}[\mathcal{A}]$ via (9.44), we have

$$\mathtt{E}_\perp = \sum_{H \leq D} (-1)^{s-1} \frac{1}{s\binom{p}{s}} \mathtt{K}_{H,D}, \tag{12.52}$$

where $s = \deg(H)$. This follows by combining (12.51) and (9.46).

12.5.5. Eulerian idempotents from Zie elements. We now discuss a general construction of a complete system of idempotents of the Tits algebra. The starting data is as follows.

- Suppose for each nonempty subset S of $[p]$, we are given an arbitrary special Zie element $\mathtt{Q}_{(S)}$ of the braid arrangement on S.

For each composition $F = (S_1, \dots, S_k)$ of $[p]$, put

$$\mathtt{Q}_F := \mu_F(\mathtt{Q}_{(S_1)}, \dots, \mathtt{Q}_{(S_k)}). \tag{12.53}$$

The rhs is the external product from Section 6.3.13 defined on the $\mathtt{H}$-basis by

$$\mu_F(\mathtt{H}_{H_1}, \dots, \mathtt{H}_{H_p}) = \mathtt{H}_H,$$

and extended by multilinearity. Here each H_i is a composition of S_i and H is their ordered concatenation. As F varies over all compositions of $[p]$, the $\mathtt{Q}_F$ yield a basis of the Tits algebra. So we can write

$$\mathtt{H}_O = \sum_F \mathtt{u}^F \mathtt{Q}_F, \tag{12.54}$$

for unique scalars $\mathtt{u}^F$. Now define

$$\mathtt{E}_\mathrm{X} := \sum_{F:\, \mathrm{s}(F)=\mathrm{X}} \mathtt{u}^F \mathtt{Q}_F. \tag{12.55}$$

These are the required idempotents. Choosing each $\mathtt{Q}_S$ to be (12.43) recovers the Eulerian idempotents (12.42) (with the $\mathtt{u}^F$ coinciding with the uniform section).

The above is a special case of the construction in Section 11.5.3. Recall from Section 6.3.11 that the arrangement over a flat of the braid arrangement is a cartesian product of smaller braid arrangements. Also, Zie elements are compatible with cartesian product (10.22). Hence, each $\mathtt{Q}_F$ as defined in (12.53) yields a special Zie element in the star of F. This is equivalent to a special Zie family. We see from (11.46) and (11.47) that the resulting homogeneous section $\mathtt{u}$ and Eulerian family $\mathtt{E}$ are precisely as defined above.

Exercise 12.84. Use the Friedrichs criterion (Lemma 10.19) to directly check that the elements (12.53) satisfy (11.29). Deduce that they define an abstract $\mathtt{Q}$-basis in the sense of Section 11.5.5.

Exercise 12.85. For any nonempty subset I of $[p]$, (12.53) applied to compositions F of I defines a $\mathtt{Q}$-basis of the braid arrangement on I. This in turn defines scalars $\mathtt{u}_F^G$ and $\mathtt{a}_F^G$ via (11.33) for any compositions F and G of I with $F \leq G$. When F is the one-block composition of I, we shorten $\mathtt{u}_F^G$ and $\mathtt{a}_F^G$ to $\mathtt{u}^G$ and $\mathtt{a}^G$, respectively. Check that

$$\mathtt{u}_F^G = \prod_i \mathtt{u}^{(G/F)_i} \quad \text{and} \quad \mathtt{a}_F^G = \prod_i \mathtt{a}^{(G/F)_i}.$$

Recall that $(G/F)_i$ is the set composition consisting of those contiguous blocks of G which refine the i-th block of F.

12.6. Type B Eulerian idempotents and Adams elements

Let $\mathcal{A}$ be the arrangement of type B on $[p]$ (Section 6.7). We work with the uniform section (so it is implicit that the field characteristic is 0). This defines the Eulerian family $\mathtt{E}$ as well as the $\mathtt{Q}$-basis. The arrangement of type B is a good reflection arrangement. This allows us to specialize the results of Section 11.6. We first give the change of basis formulas between the $\mathtt{H}$- and $\mathtt{Q}$-bases, and then cancelation-free formulas for the Eulerian idempotents in the $\mathtt{H}$-basis. Further, we consider a family of elements indexed by integers called type B Adams elements, and write down an explicit diagonalization for them. This family includes the Takeuchi element.

The discussion in this section largely proceeds in analogy with Section 12.5, however there are some differences. In contrast to Adams elements, the type B Adams elements indexed by odd integers and even integers work differently. For instance, only those indexed by odd integers are Fulman elements.

12.6.1. $\mathtt{H}$- and $\mathtt{Q}$-bases.

Lemma 12.86. *For the uniform section, the* $\mathtt{H}$- *and* $\mathtt{Q}$-*bases are related by*

$$\mathtt{H}_F = \sum_{G:\, F\leq G} \frac{1}{\mathbf{deg}!(G/F)} \mathtt{Q}_G \quad \textit{and} \quad \mathtt{Q}_F = \sum_{G:\, F\leq G} \frac{(-1)^{\mathrm{rk}(G/F)}}{\mathbf{deg}(G/F)} \mathtt{H}_G, \tag{12.56}$$

with degrees and factorials as in Section 6.7.13.

PROOF. These formulas are a specialization of (11.50). The second used (6.20). ☐

For $p = 1$, the change of basis formulas are

$$\mathtt{H}_{0|1} = \mathtt{Q}_{0|1}, \quad \mathtt{H}_{0|\bar{1}} = \mathtt{Q}_{0|\bar{1}}, \quad \mathtt{H}_{01} = \mathtt{Q}_{01} + \frac{1}{2}(\mathtt{Q}_{0|1} + \mathtt{Q}_{0|\bar{1}}), \quad \mathtt{Q}_{01} = \mathtt{H}_{01} - \frac{1}{2}(\mathtt{H}_{0|1} + \mathtt{H}_{0|\bar{1}}).$$

12.6.2. Eulerian idempotents.

Theorem 12.87. *The Eulerian idempotents for the uniform section are given by*

$$\begin{aligned} \mathtt{E}_{\mathrm{X}} &= \frac{1}{\mathbf{deg}!(\mathrm{X})} \sum_{F:\, \mathrm{s}(F)=\mathrm{X}} \mathtt{Q}_F \\ &= \frac{1}{\mathbf{deg}!(\mathrm{X})} \sum_{F:\, \mathrm{s}(F)=\mathrm{X}} \sum_{G:\, F\leq G} \frac{(-1)^{\mathrm{rk}(G/F)}}{\mathbf{deg}(G/F)} \mathtt{H}_G. \end{aligned} \tag{12.57}$$

In particular, the first Eulerian idempotent is

$$\mathtt{E}_\perp = \mathtt{Q}_O = \sum_F \frac{(-1)^{\mathrm{rk}(F)}}{\mathbf{deg}(F)} \mathtt{H}_F = \sum_F \binom{-1/2}{\mathrm{rk}(F)} \mathtt{H}_F. \tag{12.58}$$

PROOF. This is a special case of Theorem 11.53. The last step used (6.15). □

For instance, for $p = 2$, for the octagon,

$$\mathtt{E}_\perp = \mathtt{Q}_O = \mathtt{H}_O - \frac{1}{2}\sum_P \mathtt{H}_P + \frac{3}{8}\sum_C \mathtt{H}_C.$$

The first sum is over the eight vertices, and the second sum is over the eight edges. This is also the $n = 4$ case of (11.64). In combinatorial notation,

$$\begin{aligned}\mathtt{E}_{012} = \mathtt{H}_{012} &- \frac{1}{2}(\mathtt{H}_{02|1} + \mathtt{H}_{0|12} + \mathtt{H}_{01|2} + \mathtt{H}_{0|\bar{1}2} + \mathtt{H}_{02|\bar{1}} + \mathtt{H}_{0|\bar{1}\bar{2}} + \mathtt{H}_{01|\bar{2}} + \mathtt{H}_{0|1\bar{2}}) \\ &+ \frac{3}{8}(\mathtt{H}_{0|2|1} + \mathtt{H}_{0|1|2} + \mathtt{H}_{0|\bar{1}|2} + \mathtt{H}_{0|2|\bar{1}} + \mathtt{H}_{0|\bar{2}|\bar{1}} + \mathtt{H}_{0|\bar{1}|\bar{2}} + \mathtt{H}_{0|1|\bar{2}} + \mathtt{H}_{0|\bar{2}|1}).\end{aligned}$$

For $p = 3$,

$$\mathtt{E}_\perp = \mathtt{Q}_O = \mathtt{H}_O - \frac{1}{2}\sum_P \mathtt{H}_P + \frac{3}{8}\sum_E \mathtt{H}_E - \frac{5}{16}\sum_C \mathtt{H}_C.$$

The first sum is over all vertices, the second over all edges, and the third over all triangles. This is an instance of (11.67) with $\mathrm{u}^E = 1/8$.

Now set

$$\mathtt{E}_k := \sum_{\mathrm{X}\colon \mathrm{rk}(\mathrm{X})=k} \mathtt{E}_\mathrm{X} \quad \text{and} \quad \mathtt{E}'_k := \sum_{\mathrm{X}\colon \mathrm{rk}(\mathrm{X})=k,\, z(\mathrm{X})=\{0\}} \mathtt{E}_\mathrm{X}. \tag{12.59}$$

The notation $\mathtt{E}_k$ is consistent with (11.54). This is in contrast to type A, see (12.44).

Also put

$$\mathtt{T}_k := \sum_{F\colon \mathrm{rk}(F)=k} \mathtt{H}_F \quad \text{and} \quad \mathtt{T}'_k := \sum_{F\colon \mathrm{rk}(F)=k,\, z(F)=\{0\}} \mathtt{H}_F. \tag{12.60}$$

Theorem 12.88. *We have*

$$\mathtt{E}_k = \sum_{m\colon m\geq k} s^{\pm}(m,k)\, \frac{\mathtt{T}_m}{(2m)!!} \quad \textit{and} \quad \mathtt{E}'_k = \frac{1}{2^k}\sum_{m\colon m\geq k} s(m,k)\, \frac{\mathtt{T}'_m}{m!}, \tag{12.61}$$

where $s(m,k)$ and $s^{\pm}(m,k)$ are the Stirling numbers of types A and B.

PROOF. The first formula is a specialization of Theorem 11.56. This can be seen from (6.23). The combinatorics involved in the second formula is identical to that in (12.46) and can be deduced from it. □

12.6.3. Type B Adams elements. For each integer n, we define the *type B Adams element* of parameter n. The definition splits into two depending on the parity of n as follows.

$$\mathtt{Ads}^{\pm}_{2n+1} := \sum_F \binom{n}{\mathrm{rk}(F)} \mathtt{H}_F \quad \text{and} \quad \mathtt{Ads}^{\pm}_{2n} := \sum_{F\colon z(F)=\{0\}} \binom{n}{\mathrm{rk}(F)} \mathtt{H}_F. \tag{12.62}$$

The first sum is over all faces, while the second sum is over all faces whose zero block is a singleton.

Proposition 12.89. *The type B Adams elements diagonalize as follows.*

$$\mathtt{Ads}^{\pm}_{2n+1} = \sum_{\mathrm{X}} (2n+1)^{\mathrm{rk}(\mathrm{X})}\, \mathtt{E}_{\mathrm{X}} = \sum_{k=0}^{p} (2n+1)^{k}\, \mathtt{E}_{k}. \tag{12.63a}$$

$$\mathtt{Ads}^{\pm}_{2n} = \sum_{\mathrm{X}:\, z(\mathrm{X})=\{0\}} (2n)^{\mathrm{rk}(\mathrm{X})}\, \mathtt{E}_{\mathrm{X}} = \sum_{k=1}^{p} (2n)^{k}\, \mathtt{E}'_{k}. \tag{12.63b}$$

PROOF. One can check that the uniform section is an eigensection for the type B Adams elements. Hence these formulas follow from Lemma 6.22 and Proposition 12.14. ☐

For notational convenience, set $\mathtt{E}'_0 = 0$. Observe that, for any integer n (irrespective of parity),

$$\mathtt{Ads}^{\pm}_{n} = \sum_{k=0}^{p-1} \chi_k(n)(\mathtt{E}_k - \mathtt{E}'_k) + \sum_{k=1}^{p} \chi'_k(n)\mathtt{E}'_k, \tag{12.64}$$

where

$$\chi'_k(n) = n^k \quad \text{and} \quad \chi_k(n) = \begin{cases} n^k & \text{if } n \text{ is odd,} \\ 0 & \text{if } n \text{ is even.} \end{cases}$$

Note that $\chi_k(mn) = \chi_k(m)\chi_k(n)$ and $\chi'_k(mn) = \chi'_k(m)\chi'_k(n)$.

Theorem 12.90. *For the arrangement of type B on [p], the subalgebra of the Tits algebra generated by the elements* $\mathtt{Ads}^{\pm}_{n}$ *is a split-semisimple commutative algebra of dimension* $2p$*, with primitive idempotents* $\mathtt{E}'_k$*, for* $1 \le k \le p$*, and* $\mathtt{E}_k - \mathtt{E}'_k$*, for* $0 \le k \le p-1$*.*

PROOF. By definition, the $\mathtt{E}'_k$ and $\mathtt{E}_k - \mathtt{E}'_k$ form an orthogonal family of idempotents which add up to the unit element. So the algebra generated by them is isomorphic to $\mathbb{k}^{2p}$. Let us denote it by A. By (12.64), the elements $\mathtt{Ads}^{\pm}_{n}$ belong to A. Further, from the determinant formula of the Vandermonde matrix, we deduce that each $\mathtt{E}_k$ and $\mathtt{E}'_k$ is expressible as a linear combination of the $\mathtt{Ads}^{\pm}_{n}$. So the algebra generated by the $\mathtt{Ads}^{\pm}_{n}$ equals A. ☐

In conjunction with (12.64), we deduce that:

Lemma 12.91. *For any integers m and n,*

$$\mathtt{Ads}^{\pm}_{m} \cdot \mathtt{Ads}^{\pm}_{n} = \mathtt{Ads}^{\pm}_{mn}. \tag{12.65}$$

Similarly, (12.63a) yields:

Theorem 12.92. *For the arrangement of type B on [p], the subalgebra of the Tits algebra generated by the elements* $\mathtt{Ads}^{\pm}_{2n+1}$ *is a split-semisimple commutative algebra of dimension* $p+1$*, with primitive idempotents* $\mathtt{E}_k$*, for* $0 \le k \le p$*.*

We note that $\mathtt{Ads}^{\pm}_{-1}$ is the Takeuchi element (12.15). It diagonalizes as

$$\mathtt{Ads}^{\pm}_{-1} = \sum_{\mathrm{X}} (-1)^{\mathrm{rk}(\mathrm{X})}\, \mathtt{E}_{\mathrm{X}} = \sum_{k=0}^{p} (-1)^{k}\, \mathtt{E}_{k}.$$

Similarly, a comparison of (6.22) with (12.37) shows that

$$\mathtt{Ads}^{\pm}_{2n+1} = \mathtt{Ful}_{2n+1}.$$

Thus, $\mathtt{Ads}^{\pm}_{2n+1}$ is the Fulman element of parameter $2n+1$ (which is consistent with the fact that $\mathtt{Ads}^{\pm}_{-1}$ is the Takeuchi element). The implications are given below.

- The diagonalization (12.63a) is an instance of (12.38) for $t = 2n+1$.
- When m and n are both odd integers, (12.65) is an instance of (12.40).
- Theorem 12.92 is an instance of Theorem 12.72.

In contrast, the element $\mathtt{Ads}^{\pm}_{2n}$ is *not* the Fulman element of parameter $2n$. In fact, it is not even a characteristic element of parameter $2n$: the eigenvalues of $\mathtt{Ads}^{\pm}_{2n}$ are $(2n)^{\mathrm{rk(X)}}$ only on flats X with $z(\mathrm{X}) = \{0\}$, and zero on other flats. The companion result to Theorem 12.92, which can be deduced from (12.63b), is given below.

Theorem 12.93. *For the arrangement of type B on* $[p]$*, the subalgebra of the Tits algebra generated by the elements* $\mathtt{Ads}^{\pm}_{2n}$ *is a split-semisimple commutative algebra of dimension* $p+1$*, with primitive idempotents* $\mathtt{E}'_k$*, for* $1 \leq k \leq p$*, and* $1 - \sum_k \mathtt{E}'_k$.

12.6.4. Action on chambers. Let $\mathtt{E}_\perp$ be the first Eulerian idempotent given by (12.58). A formula for its action on the left module of chambers is given below.

Proposition 12.94. *For any chamber* C,

$$\mathtt{E}_\perp \cdot \mathtt{H}_C = \sum_D \binom{p-s-1/2}{p} \mathtt{H}_D, \tag{12.66}$$

where s *is the rank of* $\mathrm{Des}(C,D)$*, with the latter as in* (7.1).

PROOF. This reduces to the following instance of Vandermonde's identity:

$$\sum_{k=s}^{p} \binom{-1/2}{k}\binom{p-s}{k-s} = \binom{p-s-1/2}{p} \quad \text{for } 0 \leq s \leq p.$$

We omit the details. □

Exercise 12.95. View $\Sigma[\mathcal{A}]$ as a subspace of $\widehat{\mathrm{Q}}[\mathcal{A}]$ via (9.44). Use (12.66) and (9.46) to expand $\mathtt{E}_\perp$ in the $\mathtt{K}$-basis of the latter space.

Notes

Diagonalizability and stationary distribution. The diagonalizability of a nonnegative element of the real Tits algebra was shown by Brown and Diaconis [**105**, Theorem 1]. This result was generalized from hyperplane arrangements to left regular bands by Brown [**103**, Theorems 1 and 5]. The two proofs are different. Brown also showed that his method generalized to any field $\Bbbk$ provided the element is separating.

Brown also explicitly constructed a family of primitive orthogonal idempotents for the Tits algebra which diagonalize the given separating element [**103**, Equations (24) and (27)]. They are related to (12.13).

Our treatment of diagonalizability differs from that of Brown. We follow Saliola's method using the notion of eigensections [**370**]: Starting with an element w of the Tits algebra, first construct an eigensection of w, and then apply the recursive construction (11.14) to obtain the family of idempotents which diagonalize w. We applied the method to separating elements; Saliola applies it to nonnegative elements to prove Theorem 12.20, see [**370**, Theorem 4]. An exposition of Brown's diagonalizability argument is given in [**358**, Section IV.3].

Brown and Diaconis give a probabilistic description of the stationary distribution for a separating element w [**105**, Theorem 2, part (b)]. This is equivalent to formula (12.6) which then immediately yields the eigensection (12.12) of the given element w. The Eulerian idempotents associated to this homogeneous section are precisely the ones constructed by Brown. This has to be the case by uniqueness. For rank-three arrangements, Billera, Brown and Diaconis considered the element with uniform weights on the vertices and computed its stationary distribution (12.8), see [**63**, Theorem 1].

Steinberg [**404**] or [**405**, Section 14.5] gives a proof of diagonalizability by establishing Lemma 12.26. Results for stationary distribution and diagonalizability in more general contexts are given by Ayyer, Schilling, Steinberg and Thiéry [**34**, Theorems 4.3 and 4.10].

We have not discussed here the problem of estimating the rate at which the random walk converges to the stationary distribution. The interested reader may look for instance at [**61**, Section 5], [**105**, Theorem 2, part (c)] and [**34**, Section 4.5].

Results related to random walks induced on subarrangements are given in [**29**].

Characteristic polynomial. Together, formulas (12.35) and (12.36) are equivalent to the *deletion-restriction recursion* for the characteristic polynomial [**400**, Lemma 2.2], [**328**, Theorem 2.56]. The proofs in those references rest on (1.51). Our proof via characteristic elements in the Tits algebra is quite different (and limited to the case of real hyperplane arrangements). The recursion goes back to Brylawski [**107**, Theorem 4.2] and Zaslavsky [**440**, Lemma 4A4]. The deletion-contraction recursion for the chromatic polynomial of graphs appears in [**432**, Formula (19.4)], where it is attributed by Whitney to Ronald Martin Foster (albeit misspelling his name).

The polynomial (12.31) was defined by Zaslavsky in [**440**, Section 1C] under the name Möbius polynomial. The term Whitney polynomial is used in [**444**, Section 2] and [**26**, Section 6]. The identity (12.33) relating it to the f-polynomial of the face poset appears in [**440**, Corollary 6.3] and [**26**, Theorem 6.4].

Takeuchi and Fulman elements. The Takeuchi element is related to the antipode of a connected Hopf monoid in species [**9**, Formula (8.27)]. In turn this originates in work of Sweedler and Takeuchi on the antipode of a connected Hopf algebra [**410**, Proof of Lemma 9.2.3], [**411**, Proof of Lemma 14]. Also see [**316**, Lemma 5.2.10] or [**9**, Formula (2.55)]. The existence of the antipode is already pointed out in [**312**, Proposition 8.2]. For a topological analogue see [**430**, Chapter X.2.2]: if a connected CW-complex has an H-space structure, then it is group-like, that is, the multiplication admits a homotopy inverse. The result of Lemma 12.45 is closely related to the fact that the antipode of a Hopf algebra reverses products and coproducts. More results related to the Takeuchi element will be given in future work.

The Fulman element (up to normalization) was considered by Fulman [**191**, Definitions 2 and 3], [**189**, Definition on page 154]. Theorem 12.71 (or the later Theorem 16.51) is equivalent to [**191**, Theorem 3]. He lists out the irreducible good reflection arrangements (Theorem 5.29) and proves this result by verifying it for each case. Our approach gives a conceptual understanding of his calculations.

Eulerian idempotents and Adams elements. The idempotents $\mathtt{E}_k$ and the elements $\mathtt{Ads}_n$ appeared in work of Reutenauer [**359**, Section 3] and Garsia [**196**, Section 7] on the free Lie algebra, of Bayer and Diaconis [**47**, Section 3] on card shuffling (see below), of Patras [**331**, Section II.2] on the polytope group, and of several authors on Hochschild and cyclic homology [**182, 201, 222, 287**]. (In [**287**, Section 1], they are called λ-elements and a different sign convention is used.) In these references, $\mathtt{E}_k$ and $\mathtt{Ads}_n$ are viewed as elements of the symmetric group algebra. See also [**360**, Sections 3.2 and 3.3].

The first Eulerian idempotent $\mathtt{E}_1$ appears in work of Hain [**218**]. A related idempotent (the sum of all higher Eulerian idempotents except the first) goes back to Barr [**44**]; see [**201**, Theorem 1.3]. The map on chambers (12.51) arising from the action of the first Eulerian idempotent is present in early work of Solomon [**388**, Formula (1.2)] and is also considered in [**228**, Theorem (19) and Lemma (21)], [**359**, Corollary 1.6] and [**360**, Corollary 3.16].

We presented the Eulerian idempotents and the Adams elements as elements of the Tits algebra of the braid arrangement. They are in fact elements of the Solomon descent algebra. The precise connection between the two algebras is explained later in Theorem 16.8; see the notes to Chapter 16 for further discussion. The latter is a subalgebra of both the Tits algebra and the symmetric group algebra.

In the context of Hopf monoids in species, the Eulerian idempotents and the Adams elements are studied in [**10**, Section 14]. Formulas (12.42) for the Eulerian idempotents $\mathtt{E}_X$ are [**10**, Formulas (243) and (244)]. The Adams element is defined in [**10**, Equation (256)].

The Adams elements are closely related to the convolution powers of the identity of a connected Hopf algebra. This perspective goes back to Patras [**332**, Section 1] and Gerstenhaber and Schack [**202**, Sections 1–3]; see also [**288**, Section 4.5], [**112**, Section 3.8], and [**10**, Section 14.4]. When the Hopf algebra is commutative, these operators endow the underlying algebra with the structure of a λ-ring for which they serve as the Adams operations (hence the name). See [**333**, Section 5]. The power maps $x \mapsto x^n$ on a topological group induce the Adams operations on its cohomology Hopf algebra. A brief discussion of Adams operations in the context of group representations and K-theory is given in [**50**, Section 5.9] and [**51**, Section 2.5].

In the Hopf algebra literature, the term *Hopf powers* or *Sweedler powers* is employed for the convolution powers of the identity. See [**253**], [**320**], [**252**], [**152**], [**330**], [**7**].

Eulerian and Zie. The construction of a $\mathtt{Q}$-basis starting with Zie elements in Section 12.5.5 is present in work of Schocker [**376**, Section 5]. For related information, see the notes to Chapter 11.

Riffle shuffle. A riffle shuffle is a method commonly employed to shuffle a deck of cards. It is described mathematically by the Gilbert-Shannon-Reeds model: Cut the deck of cards into two heaps according to a binomial distribution, and then riffle them together such that cards drop from the left or right heaps with probability proportional to the number of cards in each heap. The n-shuffle, for any integer $n \geq 2$, can be defined in a similar manner by cutting the deck of cards into n ordered heaps and riffling them together. The 2-shuffle is the same as the riffle shuffle.

The inverse n-shuffle works as follows: Label each card randomly with an integer from 1 to n. Move all the cards labeled 1 to the bottom of the deck, preserving their relative order. Next move all the cards labeled 2 likewise, and proceed in this manner.

These shuffling methods were analyzed by Bayer and Diaconis [**47**]. Mathematically, the inverse n-shuffle is an element of the group algebra of the symmetric group on p letters, p being the size of the deck. It can be explicitly written down using rising sequences or descents, and hence is an element of the Solomon descent algebra of the symmetric group. Thus, it can also be expressed as an element of the Tits algebra. Up to normalization, this is precisely the Adams element $\mathtt{Ads}_n$. This geometric interpretation was discovered by Bidigare, Hanlon and Rockmore, Equation (12.48) is [**61**, Proposition 2.3].

The inverse m-shuffle followed by the inverse n-shuffle is the inverse mn-shuffle. This relation is stated in (12.50). It can be seen directly and is one of the key observations of Bayer and Diaconis [**47**, Lemma 1]. They show that the algebra generated by the inverse riffle shuffle contains all the inverse n-shuffles and is a split-semisimple commutative algebra. This is stated in Theorem 12.81.

An exposition of the above ideas is given in [**296**, Section 2.3.3]. Fine information on riffle shuffling can be found in the comprehensive review by Diaconis [**151**]. For more recent references, see [**24**], [**123**], [**152**, Section 5] and [**286**, Section 8.3].

Type B riffle shuffles. For type B, we work with a signed deck of cards, in which cards may be face up or face down. The analogues of the n-shuffles (and in particular the riffle shuffle) are non-obvious and are defined as follows, following [**296**, Section 2.4.2].

For riffle shuffling a signed deck of cards, first cut the deck into two heaps, making the cut according to a binomial probability distribution, then turn the second heap face up and then riffle them together such that cards drop from the left or right heaps with probability proportional to the number of cards in each heap. More generally, for the signed n-shuffle, cut the deck into n ordered heaps, turn the heaps in even positions face up, and then riffle them.

Let us denote the inverse signed n-shuffle by S_n. It is convenient to split the description into two cases depending on the parity of n.

- S_2: For every card, either flip or do not flip its sign with equal probability. The cards with unchanged signs move to the top in the same relative order and the rest move to the bottom in the reverse relative order. This is the inverse signed riffle shuffle.
- S_{2n}: Do an inverse n-shuffle (of type A) with labels $1, 2, \dots, n$. Then within each of the n blocks with a fixed label do an inverse signed riffle shuffle.
- S_{2n+1}: Do an inverse $(n+1)$-shuffle (of type A) with labels $0, 1, 2, \dots, n$. Then do an inverse signed riffle shuffle on each block except the one labeled 0.

As for type A, the inverse signed n-shuffle can be expressed as an element of the Tits algebra. Up to normalization, this is precisely the type B Adams element $\mathtt{Ads}^{\pm}_n$. The multiplicative identity (12.65) is deducible from the above descriptions.

The signed 2-shuffle, or the signed riffle shuffle, is described by Bayer and Diaconis [**47**, Section 5.3]. The signed 3-shuffle and the algebra it generates is described by Bergeron and Bergeron [**54**, Section 6, pages 127-128]. A unified treatment of the even and odd shuffles is given independently by Fulman [**190**, Section 5] and in [**296**, Section 2.4.2].

Type D riffle shuffle. There are also riffle shuffles of type D, as considered in [**296**, Section 2.5.2]. Here one departs the setting of good reflection arrangements and the constructions become more involved.

For more examples involving shuffles, see [**296**, Chapter 2] and references therein. See Section 2.2 in particular, where the notion of a shuffle algebra is introduced. Examples of shuffle algebras are listed in Theorem 7.

CHAPTER 13

Loewy series and Peirce decompositions

Recall from Section D.5 that for any module over an algebra, one can define its radical series and socle series with the former contained termwise in the latter. These are two extreme examples of Loewy series. For left modules over the Tits algebra, we introduce a third series called the primitive series. It is a Loewy series and hence trapped between the radical series and the socle series. The first nontrivial term (from the bottom) in the primitive series is the primitive part of the module. For the left module of chambers, all three series coincide. Dually, for right modules, we introduce the decomposable series which is also a Loewy series. The first nontrivial term (from the top) in the decomposable series is the decomposable part of the module. For the right module of Zie elements, the radical, decomposable and socle series all coincide.

Recall that decompositions arising from a system of orthogonal idempotents are called Peirce decompositions (left, right, two-sided). Any Eulerian family yields a left Peirce decomposition of a left module over the Tits algebra with components indexed by flats. We provide formulas for the dimensions of the components and relate them to terms in the primitive series. For instance, the component for the minimum flat coincides with the primitive part of the module. For the left module of chambers, the component indexed by a flat identifies with the space of Lie elements in the arrangement over that flat. This can be viewed as an algebraic form of the Zaslavsky formula. Similarly, for the Tits algebra viewed as a left module over itself, the component indexed by a flat identifies with the space of Zie elements in the arrangement over that flat. There are similar results for the right Peirce decompositions of right modules. For the right module of Zie elements, the component indexed by a flat identifies with the space of Lie elements in the arrangement under that flat. Similarly, for the Tits algebra viewed as a right module over itself, the component indexed by a flat identifies with the space of chamber elements in the arrangement under that flat.

Since the Tits algebra is a bimodule over itself, any Eulerian family yields a two-sided Peirce decomposition (obtained by combining the left and right Peirce decompositions). The components are indexed by nested flats. A typical component identifies with the space of Lie elements in the arrangement over the first flat and under the second flat (taken from the nested flat). Further, this identification is compatible with the substitution product of Lie. This can be used to describe the powers of the radical of the Tits algebra and also compute its quiver.

All modules are assumed to be finite dimensional as per Convention 9.1.

Notation 13.1. In this chapter, r denotes the rank of the arrangement $\mathcal{A}$.

13.1. Primitive series and decomposable series

We define the primitive series of any left module over the Tits algebra. The primitive part of a left module is the first nontrivial term (from the bottom) in its primitive series. Dually, we define the decomposable series of a right module. The decomposable part of a right module is the first nontrivial term (from the top) in its decomposable series. The primitive series and decomposable series are both Loewy. Further, in each semisimple summand of the associated graded modules, we determine the multiplicities of the simple modules.

13.1.1. Primitive series of a left module. Let h be a left Σ-module. Recall from Section 9.7 that h has a primitive part $\mathcal{P}(\mathsf{h})$ defined by

$$z \in \mathcal{P}(\mathsf{h}) \iff \mathtt{H}_F \cdot z = 0 \text{ for all } F > O.$$

More generally, for any flat X, define

$$\mathcal{P}_{\mathrm{X}}(\mathsf{h}) := \{z \in \mathsf{h} \mid \mathtt{H}_H \cdot z = 0 \text{ whenever } \mathrm{s}(H) \not\leq \mathrm{X}\}. \tag{13.1}$$

The linear span of $\mathtt{H}_H$ with $\mathrm{s}(H) \not\leq \mathrm{X}$ is an ideal of Σ, and $\mathcal{P}_{\mathrm{X}}(\mathsf{h})$ consists precisely of those elements of h which are annihilated by this ideal.

Observe that for $\mathrm{X} \leq \mathrm{Y}$,

$$\mathcal{P}_{\mathrm{X}}(\mathsf{h}) \subseteq \mathcal{P}_{\mathrm{Y}}(\mathsf{h})$$

with

$$\mathcal{P}_{\bot}(\mathsf{h}) = \mathcal{P}(\mathsf{h}) \quad \text{and} \quad \mathcal{P}_{\top}(\mathsf{h}) = \mathsf{h}.$$

Thus the $\mathcal{P}_{\mathrm{X}}(\mathsf{h})$ define a filtration of h indexed by flats.

Similarly, for any $k \geq 0$, define

$$\mathcal{P}_k(\mathsf{h}) := \{z \in \mathsf{h} \mid \mathtt{H}_H \cdot z = 0 \text{ whenever } \mathrm{rk}(H) \geq k\}. \tag{13.2}$$

The linear span of $\mathtt{H}_H$ with $\mathrm{rk}(H) \geq k$ is an ideal of Σ, and $\mathcal{P}_k(\mathsf{h})$ consists precisely of those elements of h which are annihilated by this ideal.

Observe that

$$0 = \mathcal{P}_0(\mathsf{h}) \subseteq \mathcal{P}_1(\mathsf{h}) \subseteq \mathcal{P}_2(\mathsf{h}) \subseteq \cdots \subseteq \mathcal{P}_{r+1}(\mathsf{h}) = \mathsf{h}.$$

This is the *primitive series* of h. Note that the second term (from the bottom) is indeed the primitive part of h, that is, $\mathcal{P}(\mathsf{h}) = \mathcal{P}_1(\mathsf{h})$.

13.1.2. Decomposable series of a right module. Let h be a right Σ-module. Consider the filtration of h defined in Lemma 9.55. It is indexed by flats, with the X-component given by $\mathcal{D}_{\mathrm{X}}(\mathsf{h})$.

Now, for any $k \geq 0$, define

$$\mathcal{D}_k(\mathsf{h}) := \sum_{F:\,\mathrm{rk}(F) \geq k} \mathsf{h} \cdot \mathtt{H}_F = \sum_{\mathrm{X}:\,\mathrm{rk}(\mathrm{X}) \geq k} \mathcal{D}_{\mathrm{X}}(\mathsf{h}). \tag{13.3}$$

In view of (9.65), it suffices to sum over F with $\mathrm{rk}(F) = k$ in the first sum and over X with $\mathrm{rk}(\mathrm{X}) = k$ in the second sum.

Observe that $\mathcal{D}_1(\mathsf{h}) = \mathcal{D}(\mathsf{h})$, the decomposable part of h defined in (9.68). Further,

$$\mathsf{h} = \mathcal{D}_0(\mathsf{h}) \supseteq \mathcal{D}_1(\mathsf{h}) \supseteq \mathcal{D}_2(\mathsf{h}) \supseteq \cdots \supseteq \mathcal{D}_{r+1}(\mathsf{h}) = 0.$$

This is the *decomposable series* of h.

Observe that:

Lemma 13.2. *The submodule $\mathcal{D}_r(\mathsf{h})$ is the invariant subspace of* h *for the multiplicative character $\chi_\top$ defined in* (9.50). *In other words,*

$$x \in \mathcal{D}_r(\mathsf{h}) \iff x \cdot z = \chi_\top(z)\, x \text{ for all } z \in \Sigma. \tag{13.4}$$

Equivalently, $\mathcal{D}_r(\mathsf{h})$ is the sum of the simple submodules of h *with multiplicative character $\chi_\top$. In particular, $\mathcal{D}_r(\mathsf{h})$ is semisimple.*

Proposition 13.3. *For a left Σ-module* h, *the spaces $\mathcal{P}_k(\mathsf{h})$ and $\mathcal{D}_k(\mathsf{h}^*)$ are orthogonal to each other under the canonical pairing between* h *and* h*.*

The special case $k = 1$ was given in Proposition 9.61.

13.1.3. Primitive and decomposable series are Loewy. Recall from Section D.5.7 that a filtration of a module is a Loewy series if multiplying one term by the radical puts it in the next (smaller) term, or equivalently, if successive quotients are semisimple.

We now show that the primitive series and decomposable series are both Loewy. Further, in each semisimple summand of the associated graded modules, we determine the multiplicities of the simple modules. This refines the eigenvalue-multiplicity theorem (Theorem 9.44).

Proposition 13.4. *Let* h *be a right Σ-module. The decomposable series of* h *is a Loewy series, that is, the quotient*

$$\mathcal{D}_i(\mathsf{h})/\mathcal{D}_{i+1}(\mathsf{h}),$$

for $0 \le i \le r$, is semisimple. Further, the multiplicity of the simple module with multiplicative character χ_{X} in the above quotient is $\eta_{\mathrm{X}}(\mathsf{h})$ if X *has rank i, and zero otherwise.*

The special case $i = r$ is mentioned in Lemma 13.2.

PROOF. This result is contained in the second proof of Theorem 9.44 given in Section 9.6. For convenience, we recall some of the ideas involved. As a consequence of Lemma 9.58,

$$\mathcal{D}_i(\mathsf{h})/\mathcal{D}_{i+1}(\mathsf{h}) \cong \bigoplus_{\mathrm{X}:\, \mathrm{rk}(\mathrm{X})=i} \mathsf{k}_{\mathrm{X}},$$

where k_{X} is the quotient of $\mathcal{D}_{\mathrm{X}}(\mathsf{h})$ by the sum of $\mathcal{D}_{\mathrm{Y}}(\mathsf{h})$ for $\mathrm{Y} > \mathrm{X}$. In k_{X} only the simple module with multiplicative character χ_{X} appears, and so its multiplicity is the dimension of k_{X} which is $\eta_{\mathrm{X}}(\mathsf{h})$. □

Exercise 13.5. Use Lemma 9.18 and (9.63) to check that

$$\mathcal{D}_{\mathrm{X}}(\mathsf{h}) \cdot \mathrm{rad}(\Sigma) \subseteq \sum_{\mathrm{Y}:\, \mathrm{Y}>\mathrm{X}} \mathcal{D}_{\mathrm{Y}}(\mathsf{h}).$$

Conclude that the decomposable series is Loewy. (This is a direct argument from the definitions.)

Proposition 13.6. *Let* h *be a left Σ-module. The primitive series of* h *is a Loewy series, that is, the quotient*

$$\mathcal{P}_{i+1}(\mathsf{h})/\mathcal{P}_i(\mathsf{h}),$$

for $0 \leq i \leq r$, *is semisimple. Further, the multiplicity of the simple module with multiplicative character* χ_X *in the above quotient is* $\eta_X(\mathsf{h})$ *if* X *has rank* i, *and zero otherwise.*

The special case $i = 0$ is mentioned in Lemma 9.60.

PROOF. This result can be deduced from Proposition 13.4 by using the orthogonality relationship between primitive series and decomposable series given in Proposition 13.3. Details follow. For $0 \leq i \leq r$, there is a canonical isomorphism

$$(\mathcal{P}_{i+1}(\mathsf{h})/\mathcal{P}_i(\mathsf{h}))^* \cong \mathcal{D}_i(\mathsf{h}^*)/\mathcal{D}_{i+1}(\mathsf{h}^*).$$

Since any simple module is one-dimensional, its dual is also simple and it has the same multiplicative character. Thus the dual of a semisimple module is semisimple. The above isomorphism now implies that $\mathcal{P}_{i+1}(\mathsf{h})/\mathcal{P}_i(\mathsf{h})$ is semisimple. Further, the multiplicity of a given simple module in this quotient equals its multiplicity in $\mathcal{D}_i(\mathsf{h}^*)/\mathcal{D}_{i+1}(\mathsf{h}^*)$. Finally, recall from (9.59) that $\eta_X(\mathsf{h}) = \eta_X(\mathsf{h}^*)$. □

Exercise 13.7. Use Lemma 9.18 and (13.2) to check that

$$\operatorname{rad}(\Sigma) \cdot \mathcal{P}_{i+1}(\mathsf{h}) \subseteq \mathcal{P}_i(\mathsf{h}).$$

Conclude that the primitive series is Loewy. (This is a direct argument from the definitions.)

13.2. Primitive series and socle series

Since the primitive series is Loewy, by general theory, it is contained termwise in the socle series. We begin with an explicit proof of this fact. We then provide a sufficient condition for the two series to coincide. We refer to this condition as the disjoint-star property. It is satisfied by the left module of chambers.

13.2.1. Socle series of a module. Let h be a left Σ-module. Then, by (D.15), the socle of h, denoted $\operatorname{soc}(\mathsf{h})$, is given by

$$x \in \operatorname{soc}(\mathsf{h}) \iff z \cdot x = 0 \text{ for all } z \in \operatorname{rad}(\Sigma).$$

The socle series of h is

$$0 \subseteq \operatorname{soc}_1(\mathsf{h}) \subseteq \operatorname{soc}_2(\mathsf{h}) \subseteq \cdots \subseteq \operatorname{soc}_{\operatorname{rk}(\mathcal{A})+1}(\mathsf{h}) = \mathsf{h},$$

where $\operatorname{soc}_k(\mathsf{h})$ consists of the elements annihilated by the ideal $\operatorname{rad}(\Sigma)^k$.

Lemma 13.8. *For a left* Σ*-module* h, *the primitive series of* h *is contained termwise in the socle series of* h.

PROOF. Since $\operatorname{rad}(\Sigma)$ consists of linear combinations of noncentral faces only, it follows from (13.2) for $k = 1$ that $\mathcal{P}(\mathsf{h}) \subseteq \operatorname{soc}(\mathsf{h})$. More generally, by Lemma 9.19, $\mathcal{P}_k(\mathsf{h}) \subseteq \operatorname{soc}_k(\mathsf{h})$. □

Lemma 13.10 below gives a sufficient condition for equality to hold.

Exercise 13.9. For a right Σ-module h, using definition (13.3) and Lemma 9.19, check that the radical series of h is contained termwise in the decomposable series of h. For $\mathsf{h} = \Sigma$, this statement is in fact equivalent to Lemma 9.19 since $\mathcal{D}_k(\Sigma)$ is linearly spanned by faces of rank at least k.

13.2.2. Disjoint-star property. We say that a left Σ-module h satisfies the *disjoint-star property* if for any flat X, the sum

$$\sum_{F:\mathrm{s}(F)=\mathrm{X}} \mathtt{H}_F \cdot \mathsf{h}$$

of subspaces inside h is direct. In particular, for any distinct faces F and G with the same support, the spaces $\mathtt{H}_F \cdot \mathsf{h}$ and $\mathtt{H}_G \cdot \mathsf{h}$ intersect trivially, that is, $\mathtt{H}_F \cdot \mathsf{h} \cap \mathtt{H}_G \cdot \mathsf{h} = 0$. (Lemma 9.54 shows that the disjoint-star property is not meaningful for right modules.)

This property can also be defined set-theoretically for left modules over the Tits monoid. We build on the discussion in Section 7.6.2. We say that a left Σ-set h satisfies the *disjoint-star property* if for any distinct faces F and G with the same support, $\mathrm{h}_F \cap \mathrm{h}_G = \emptyset$, that is, the star of F and the star of G are disjoint.

Observe that a left Σ-set h has the disjoint-star property iff its linearization has the disjoint-star property. By Exercise 1.39, the left Σ-sets Γ and Σ satisfy the disjoint-star property, and hence so do their linearizations $\mathsf{\Gamma}$ and $\mathsf{\Sigma}$.

Lemma 13.10. *Let* h *be a left* Σ*-module with the disjoint-star property. Then the primitive series and the socle series of* h *coincide, that is,*

$$\mathcal{P}_k(\mathsf{h}) = \mathrm{soc}_k(\mathsf{h})$$

for all k*. In particular,* $\mathcal{P}(\mathsf{h}) = \mathrm{soc}(\mathsf{h})$*.*

PROOF. In view of Lemma 13.8, we need to show that the socle series is contained in the primitive series. We first show that $\mathrm{soc}(\mathsf{h}) \subseteq \mathcal{P}(\mathsf{h})$. Accordingly, let $x \in \mathrm{soc}(\mathsf{h})$. Let F be any noncentral face. Pick a face G with the same support as F which is distinct from F. Since $\mathtt{H}_F - \mathtt{H}_G \in \mathrm{rad}(\Sigma)$, this element annihilates x. Hence $\mathtt{H}_F \cdot x = \mathtt{H}_G \cdot x$. The lhs belongs to $\mathtt{H}_F \cdot \mathsf{h}$ and the rhs to $\mathtt{H}_G \cdot \mathsf{h}$. Since these spaces intersect trivially by the disjoint-star property, we conclude that $\mathtt{H}_F \cdot x = 0$. Thus $x \in \mathcal{P}(\mathsf{h})$ as required. The general claim $\mathrm{soc}_k(\mathsf{h}) \subseteq \mathcal{P}_k(\mathsf{h})$ follows along the same lines by making use of Lemma 9.20. □

Recall from Section D.5.5 that the socle of a module is the sum of all its simple submodules. Further, the socle is homogeneous if all these submodules are isomorphic.

Lemma 13.11. *Let* h *be a left* Σ*-module with the disjoint-star property. Then:*

(1) *Any simple submodule of* h *is contained in* $\mathcal{P}(\mathsf{h})$*.*
(2) *All simple submodules of* h *are isomorphic and have multiplicative character* $\chi_\perp$*.*
(3) *The socle of* h *is homogeneous.*

PROOF. (1) By definition, any simple submodule of h is contained in $\mathrm{soc}(\mathsf{h})$. The latter equals $\mathcal{P}(\mathsf{h})$ by Lemma 13.10. Alternatively: Suppose k is a simple submodule of h with multiplicative character χ_X as given in (9.48). If $\mathrm{X} = \perp$, then k is contained in $\mathcal{P}(\mathsf{h})$ by the backward implication of (9.67). So suppose $\mathrm{X} \neq \perp$. Let k be spanned by the element $x \neq 0$. Pick distinct faces F and G with support X. Then $\mathtt{H}_F \cdot x = x = \mathtt{H}_G \cdot x$. But $\mathtt{H}_F \cdot \mathsf{h} \cap \mathtt{H}_G \cdot \mathsf{h} = 0$, so $x = 0$. This is a contradiction.

(2) Follows from (1) and the forward implication of (9.67).

(3) Follows from (2). □

Lemma 13.12. *The primitive series and the socle series of the left module of chambers coincide. In particular,*

$$\mathcal{P}(\Gamma) = \mathrm{soc}(\Gamma) = \mathsf{Lie}.$$

Further, the socle of Γ *is homogeneous.*

PROOF. This follows from Lemmas 13.10 and 13.11 (since the left module of chambers satisfies the disjoint star property.) Also the primitive part of Γ is Lie by the Friedrichs criterion (Lemma 10.5). □

Similarly:

Lemma 13.13. *The primitive series and the socle series of* Σ *(viewed as a left module over itself) coincide. In particular,*

$$\mathcal{P}(\Sigma) = \mathrm{soc}(\Sigma) = \mathsf{Zie}.$$

Further, the (left) socle of Σ *is homogeneous.*

Exercise 13.14. Deduce Corollary 9.33 using Lemma 13.12 and (D.16).

13.3. Radical series and primitive series

Since the primitive series is Loewy, by general theory, it contains the radical series termwise. We now give an explicit proof of this fact.

13.3.1. Radical of the Tits algebra and Lie elements.

Recall from Proposition 9.21 that $\mathrm{rad}(\Sigma)^{r+1} = 0$ but $\mathrm{rad}(\Sigma)^r \neq 0$. The r-th power of the radical only contains elements which are linear combinations of chambers. In fact, we show below that these are Lie elements.

Lemma 13.15. *We have* $\mathrm{rad}(\Sigma)^r \subseteq \mathsf{Lie}$.

In fact, we will see later in Proposition 13.66 that equality holds.

PROOF. We proceed in a manner similar to the proof of Lemma 9.19. Consider $x_1 \cdot x_2 \cdot \ldots \cdot x_r \in \mathrm{rad}(\Sigma)^r$, where each x_i is a homogeneous element of $\mathrm{rad}(\Sigma)$. Then

$$\perp \leq \mathbf{s}(x_1) \leq \mathbf{s}(x_1 \cdot x_2) \leq \cdots \leq \mathbf{s}(x_1 \cdot \ldots \cdot x_r).$$

If equality holds in any place, say $\mathbf{s}(x_1 \cdot \ldots \cdot x_{i-1}) = \mathbf{s}(x_1 \cdot \ldots \cdot x_i)$, then $\mathbf{s}(x_i) \leq \mathbf{s}(x_1 \cdot \ldots \cdot x_{i-1})$, and hence $x_1 \cdot \ldots \cdot x_i = 0$ by Lemma 9.18. So suppose

$$\perp < \mathbf{s}(x_1) < \mathbf{s}(x_1 \cdot x_2) < \cdots < \mathbf{s}(x_1 \cdot \ldots \cdot x_r).$$

By rank considerations, $x_1 \cdot x_2 \cdot \ldots \cdot x_r$ is a linear combination of chambers. We now show that it belongs to the primitive part of Γ. Accordingly, let $F > O$. Consider

$$\mathbf{s}(\mathtt{H}_F) \leq \mathbf{s}(\mathtt{H}_F \cdot x_1) \leq \mathbf{s}(\mathtt{H}_F \cdot x_1 \cdot x_2) \leq \cdots \leq \mathbf{s}(\mathtt{H}_F \cdot x_1 \cdot \ldots \cdot x_r).$$

Here equality is forced in at least one place, and by the same argument as above, we conclude that $\mathtt{H}_F \cdot x_1 \cdot \ldots \cdot x_r = 0$. Hence, by the Friedrichs criterion, $x_1 \cdot x_2 \cdot \ldots \cdot x_r \in \mathsf{Lie}$, as required. □

13.3.2. Radical series and primitive series.

Lemma 13.16. *For a left* Σ*-module* h,

$$\mathcal{P}_k(\Sigma) \cdot \mathsf{h} \subseteq \mathcal{P}_k(\mathsf{h}),$$

where the latter is the k*-th term of the primitive series of* h *as in* (13.2).

This is straightforward from the definition. The special case $k = 1$ which deals with the primitive part of h was addressed in Proposition 10.35.

Lemma 13.17. *For* $0 \leq k \leq r$*, and* F *a face of rank at least* $r - k + 1$,

$$\mathtt{H}_F \cdot \mathrm{rad}(\Sigma)^k = 0.$$

The special case $k = r$ was considered in Lemma 13.15.

PROOF. We proceed in a manner similar to the proof of Lemma 13.15. Accordingly, consider $x_1 \cdot x_2 \cdot \ldots \cdot x_k \in \mathrm{rad}(\Sigma)^k$, where each x_i is a homogeneous element of $\mathrm{rad}(\Sigma)$. Let F be a face of rank at least $r - k + 1$. Then

$$\mathrm{s}(\mathtt{H}_F) \leq \mathrm{s}(\mathtt{H}_F \cdot x_1) \leq \mathrm{s}(\mathtt{H}_F \cdot x_1 \cdot x_2) \leq \cdots \leq \mathrm{s}(\mathtt{H}_F \cdot x_1 \cdot \ldots \cdot x_k).$$

By rank considerations, equality is forced in at least one place. Say $\mathrm{s}(\mathtt{H}_F \cdot x_1 \cdot \ldots \cdot x_{i-1}) = \mathrm{s}(\mathtt{H}_F \cdot x_1 \cdot \ldots \cdot x_i)$. Then $\mathrm{s}(x_i) \leq \mathrm{s}(\mathtt{H}_F \cdot x_1 \cdot \ldots \cdot x_{i-1})$, and hence $\mathtt{H}_F \cdot x_1 \cdot \ldots \cdot x_i = 0$ by Lemma 9.18. Hence $\mathtt{H}_F \cdot x_1 \cdot \ldots \cdot x_k = 0$ as required. □

Lemma 13.18. *For a left* Σ*-module* h*, the radical series of* h *is contained termwise in the primitive series of* h*. Explicitly, for* $0 \leq k \leq r$,

$$\mathrm{rad}(\Sigma)^k \cdot \mathsf{h} \subseteq \mathcal{P}_{r-k+1}(\mathsf{h}).$$

In particular,

$$\mathrm{rad}(\Sigma)^k \subseteq \mathcal{P}_{r-k+1}(\Sigma).$$

PROOF. The second part follows from Lemma 13.17. The first part then follows from Lemma 13.16. □

A sufficient condition for equality to hold is given later in Section 13.11.2.

Exercise 13.19. For a right Σ-module h, using (13.3) and Lemma 13.17, check that the decomposable series of h is contained termwise in the socle series of h. (The converse of Lemma 13.17 which is related to the reverse containment is given in Exercise 13.84.)

13.4. Peirce decompositions, and primitive and decomposable series

Let h be a (left or right) module over the Tits algebra Σ. Any Eulerian family yields a (left or right) Peirce decomposition of h indexed by flats, with the dimension of the X-component being $\eta_{\mathrm{X}}(\mathsf{h})$. These numbers were defined in (9.52). They are independent of the choice of the Eulerian family. For a left Σ-module h, the $\bot$-component of the left Peirce decomposition is precisely the primitive part $\mathcal{P}(\mathsf{h})$. More generally, the higher components are related to the primitive series of h. Similarly, for a right Σ-module h, the components of the right Peirce decomposition are related to the decomposable series of h.

13.4.1. Peirce decompositions.

Proposition 13.20. *For any left Σ-module* h *and Eulerian family* $\mathtt{E}$,

$$\mathsf{h} = \bigoplus_{\mathrm{X}} \mathtt{E}_{\mathrm{X}} \cdot \mathsf{h}, \tag{13.5}$$

with

$$\dim(\mathtt{E}_{\mathrm{X}} \cdot \mathsf{h}) = \eta_{\mathrm{X}}(\mathsf{h}), \tag{13.6}$$

where $\eta_{\mathrm{X}}(\mathsf{h})$ *are as in* (9.52).

Similar statement holds for a right Σ*-module.*

PROOF. This is a special case of Proposition D.40. □

We refer to (13.5) as the *left Peirce decomposition* of h. Similarly, for a right Σ-module, we use the term *right Peirce decomposition*.

13.4.2. Primitive series.

Proposition 13.21. *For any left Σ-module* h *and Eulerian family* $\mathtt{E}$,

$$\mathcal{P}(\mathsf{h}) = \mathtt{E}_{\perp} \cdot \mathsf{h}. \tag{13.7}$$

PROOF. This follows from Proposition 10.35 since $\mathtt{E}_{\perp}$ is a special Zie element by Lemma 11.42. □

The fact that $\dim(\mathtt{E}_{\perp} \cdot \mathsf{h}) = \eta_{\perp}(\mathsf{h})$ also follows from Proposition 10.37.

Proposition 13.22. *For any left Σ-module* h *and Eulerian family* $\mathtt{E}$,

$$\mathcal{P}_{\mathrm{X}}(\mathsf{h}) = \bigoplus_{\mathrm{Y}:\, \mathrm{Y} \leq \mathrm{X}} \mathtt{E}_{\mathrm{Y}} \cdot \mathsf{h} \quad \textit{and} \quad \mathcal{P}_k(\mathsf{h}) = \bigoplus_{\mathrm{Y}:\, \mathrm{rk}(\mathrm{Y}) \leq k-1} \mathtt{E}_{\mathrm{Y}} \cdot \mathsf{h},$$

with $\mathcal{P}_{\mathrm{X}}(\mathsf{h})$ *and* $\mathcal{P}_k(\mathsf{h})$ *as in* (13.1) *and* (13.2).

Setting $\mathrm{X} = \perp$ or $k = 1$ recovers the previous result.

PROOF. Consider the first claim. By the Saliola lemma (Lemma 11.12), the rhs is contained in the lhs. Conversely, suppose x belongs to the lhs. Then, by (11.18),

$$x = \sum_{\mathrm{Y}} \mathtt{E}_{\mathrm{Y}} \cdot x = \sum_{\mathrm{Y}:\, \mathrm{Y} \leq \mathrm{X}} \mathtt{E}_{\mathrm{Y}} \cdot x$$

which belongs to the rhs. In the second step, we used that $\mathtt{E}_{\mathrm{Y}}$ only contains faces of support Y or higher, so if $\mathrm{Y} \not\leq \mathrm{X}$, then $\mathtt{E}_{\mathrm{Y}} \cdot x = 0$. The second claim can be proved in the same manner. □

Corollary 13.23. *For any left Σ-module* h,

$$\mathcal{P}_k(\mathsf{h}) = \sum_{\mathrm{X}:\, \mathrm{rk}(\mathrm{X}) \leq k-1} \mathcal{P}_{\mathrm{X}}(\mathsf{h}).$$

Note very carefully that this result makes no reference to any Eulerian family.

13.4.3. Decomposable series.

Proposition 13.24. *For any right* Σ*-module* h *and Eulerian family* E,

$$\mathcal{D}_{\mathrm{X}}(\mathsf{h}) = \bigoplus_{\mathrm{Y}:\, \mathrm{Y} \geq \mathrm{X}} \mathsf{h} \cdot \mathsf{E}_{\mathrm{Y}} \quad \textit{and} \quad \mathcal{D}_{k}(\mathsf{h}) = \bigoplus_{\mathrm{Y}:\, \mathrm{rk}(\mathrm{Y}) \geq k} \mathsf{h} \cdot \mathsf{E}_{\mathrm{Y}},$$

with $\mathcal{D}_{\mathrm{X}}(\mathsf{h})$ *and* $\mathcal{D}_{k}(\mathsf{h})$ *as in* (9.63) *and* (13.3). *In particular,*

$$\mathcal{D}_{r}(\mathsf{h}) = \mathsf{h} \cdot \mathsf{E}_{\top}.$$

PROOF. The first claim follows from the two identities below. For any face F with support X,

$$\mathsf{H}_F \cdot \Big(\sum_{\mathrm{Y}:\, \mathrm{Y} \geq \mathrm{X}} \mathsf{E}_{\mathrm{Y}} \Big) = \mathsf{H}_F \quad \text{and} \quad \Big(\sum_{\mathrm{Y}:\, \mathrm{Y} \geq \mathrm{X}} \mathsf{E}_{\mathrm{Y}} \Big) \cdot \mathsf{H}_F = \sum_{\mathrm{Y}:\, \mathrm{Y} \geq \mathrm{X}} \mathsf{E}_{\mathrm{Y}}.$$

The first identity is (11.17). For the second, note that $\mathsf{E}_{\mathrm{Y}} \cdot \mathsf{H}_F = \mathsf{E}_{\mathrm{Y}}$ since any face in E_{Y} has support at least X.

In view of (13.3), the second claim follows from the first. □

Exercise 13.25. Use Proposition 13.24 to deduce Lemma 9.56.

13.4.4. Modules over the Birkhoff algebra. Take a Π-module h and view it as a Σ-module via the support map. Since h is semisimple, we have $\mathrm{soc}(\mathsf{h}) = \mathsf{h}$ and $\mathrm{rad}(\mathsf{h}) = 0$. Thus, the socle series ascends from 0 to h in one step, while the radical series descends from h to 0 in one step. Now, let us take a look at the primitive series of h. By Proposition 13.22 and (11.23),

$$\mathcal{P}_{\mathrm{X}}(\mathsf{h}) = \bigoplus_{\mathrm{Y}:\, \mathrm{Y} \leq \mathrm{X}} \mathsf{Q}_{\mathrm{Y}} \cdot \mathsf{h} \quad \text{and} \quad \mathcal{P}_{k}(\mathsf{h}) = \bigoplus_{\mathrm{Y}:\, \mathrm{rk}(\mathrm{Y}) \leq k-1} \mathsf{Q}_{\mathrm{Y}} \cdot \mathsf{h}.$$

(The first decomposition is also given in (9.27).) Thus, in contrast to the radical and socle series, the primitive series is nontrivial in general. This gives examples where the inclusions among these series are strict. For a concrete example, take h to be the Birkhoff algebra Π itself.

Similar remarks apply to the decomposable series.

13.5. Left Peirce decomposition of chambers. Lie over flats

Recall that any Eulerian family yields a left Peirce decomposition of a left module over the Tits algebra. Applied to the left module of chambers, this leads to an algebraic form of the Zaslavsky formula. It involves expressing a chamber element as a sum of Lie elements over flats. Similarly, the left Peirce decomposition of the Tits algebra (as a left module over itself) breaks as a sum of Zie elements over flats.

13.5.1. Algebraic form of Zaslavsky formula. Recall from Proposition 13.21 that for an Eulerian family E and a left Σ-module h, the summand $\mathsf{E}_{\perp} \cdot \mathsf{h}$ is the primitive part of h. More generally, the summand $\mathsf{E}_{\mathrm{X}} \cdot \mathsf{h}$ is isomorphic to the primitive part of a certain module over $\Sigma[\mathcal{A}_{\mathrm{X}}]$; this module is constructed by projecting h on some face of support X. We focus on the example of chambers.

For $\mathsf{h} = \Gamma$, the module in question is $\Gamma[\mathcal{A}_{\mathrm{X}}]$, and the decomposition (13.5) can thus be rephrased as

$$\Gamma[\mathcal{A}] \cong \bigoplus_{\mathrm{X}} \mathsf{Lie}[\mathcal{A}_{\mathrm{X}}]. \tag{13.8}$$

(Recall from the Friedrichs criterion (Lemma 10.5) that Lie is the primitive part of Γ.) The isomorphism (13.8) may be viewed as an algebraic form of the Zaslavsky formula (1.48). Also see formulas (11.61) and (11.63) in this regard.

For $\mathsf{h} = \Sigma$, the module in question is $\Sigma[\mathcal{A}_{\mathrm{X}}]$, and the decomposition (13.5) can thus be rephrased as

$$\Sigma[\mathcal{A}] \cong \bigoplus_{\mathrm{X}} \mathsf{Zie}[\mathcal{A}_{\mathrm{X}}]. \tag{13.9}$$

(Recall from the Friedrichs criterion (Lemma 10.19) that Zie is the primitive part of Σ.) The isomorphism (13.9) may be viewed as an algebraic form of (1.49).

The isomorphisms (13.8) and (13.9) are developed in more detail below.

13.5.2. Left Peirce decomposition of chambers. Recall the maps $\beta_{F,\mathrm{X}}$, $\beta_{\mathrm{X},F}$, $\beta_{G,F}$, μ_F and Δ_F from Section 9.8.1.

Lemma 13.26. *Fix a homogeneous section* $\mathtt{u}$ *with associated Eulerian family* $\mathtt{E}$. *For any flat* X, *there is a linear isomorphism*

$$\mathsf{Lie}[\mathcal{A}_{\mathrm{X}}] \xrightarrow{\cong} \mathtt{E}_{\mathrm{X}} \cdot \Gamma[\mathcal{A}], \qquad z \mapsto \sum_{F:\, \mathrm{s}(F)=\mathrm{X}} \mathtt{u}^F \mu_F \beta_{F,\mathrm{X}}(z). \tag{13.10}$$

In particular,

$$\mathsf{Lie}[\mathcal{A}] = \mathtt{E}_{\perp} \cdot \Gamma[\mathcal{A}]. \tag{13.11}$$

The inverse of (13.10) *is given by*

$$\mathtt{E}_{\mathrm{X}} \cdot \Gamma[\mathcal{A}] \to \mathsf{Lie}[\mathcal{A}_{\mathrm{X}}], \qquad z \mapsto \beta_{\mathrm{X},F} \Delta_F(z),$$

where F *is any face of support* X.

We elaborate on the isomorphism (13.10). It says that $\mathtt{E}_{\mathrm{X}} \cdot \Gamma[\mathcal{A}]$ is the image of the composite map

$$\mathsf{Lie}[\mathcal{A}_{\mathrm{X}}] \hookrightarrow \Gamma[\mathcal{A}_{\mathrm{X}}] \to \bigoplus_{F:\, \mathrm{s}(F)=\mathrm{X}} \Gamma[\mathcal{A}_F] \to \Gamma[\mathcal{A}].$$

The second map projected on each F-component is $\beta_{F,\mathrm{X}}$, while the last map restricted to each F-component is μ_F multiplied by the scalar $\mathtt{u}^F$. Informally, the composite map distributes a Lie element of $\mathcal{A}_{\mathrm{X}}$ over the stars of faces F with support X, with the star of F receiving weight $\mathtt{u}^F$.

PROOF. First note that (13.11) is a special case of (13.7) in view of the Friedrichs criterion. We now proceed to the general case. For a face F with support X,

$$\mathsf{Lie}[\mathcal{A}_F] = \mathtt{E}_{\mathrm{X}/F} \cdot \Gamma[\mathcal{A}_F] = \Delta_F(\mathtt{E}_{\mathrm{X}}) \cdot \Delta_F(\Gamma[\mathcal{A}]) = \Delta_F(\mathtt{E}_{\mathrm{X}} \cdot \Gamma[\mathcal{A}]).$$

The first equality holds by (13.11) since $\mathtt{E}_{\mathrm{X}/F}$ is the first Eulerian idempotent of $\mathcal{A}_F$. The second step used (11.21) and the fact that Δ_F maps $\Gamma[\mathcal{A}]$ onto $\Gamma[\mathcal{A}_F]$. The last step used that Δ_F is an algebra homomorphism.

Let V be the image of the map

$$\mathsf{Lie}[\mathcal{A}_{\mathrm{X}}] \to \bigoplus_{F:\, \mathrm{s}(F)=\mathrm{X}} \mathsf{Lie}[\mathcal{A}_F], \qquad z \mapsto \sum_{F:\, \mathrm{s}(F)=\mathrm{X}} \beta_{F,\mathrm{X}}(z). \tag{a}$$

Explicitly, V consists of elements (z_F) such that $\beta_{G,F}(z_F) = z_G$. The map (a) is injective, so V is isomorphic to $\mathsf{Lie}[\mathcal{A}_{\mathrm{X}}]$.

By our first calculation and (9.72), we have a surjective map

$$\text{(b)} \qquad \mathtt{E}_{\mathrm{X}} \cdot \Gamma[\mathcal{A}] \to V, \qquad z \mapsto \sum_{F:\,\mathrm{s}(F)=\mathrm{X}} \Delta_F(z).$$

Moreover, for any $z \in \Gamma[\mathcal{A}]$,

$$\begin{aligned}\sum_{F:\,\mathrm{s}(F)=\mathrm{X}} \mathtt{u}^F \mu_F \Delta_F(\mathtt{E}_{\mathrm{X}} \cdot z) &= \sum_{F:\,\mathrm{s}(F)=\mathrm{X}} \mathtt{u}^F\, \mathtt{H}_F \cdot (\mathtt{E}_{\mathrm{X}} \cdot z)\\ &= \mathtt{u}_{\mathrm{X}} \cdot (\mathtt{E}_{\mathrm{X}} \cdot z)\\ &= (\mathtt{u}_{\mathrm{X}} \cdot \mathtt{E}_{\mathrm{X}}) \cdot z\\ &= \mathtt{E}_{\mathrm{X}} \cdot z.\end{aligned}$$

The first step used (9.75), the second step used (11.2), while the last step used (11.15). Thus, the map (b) is also injective, and its inverse is given by

$$\text{(c)} \qquad V \to \mathtt{E}_{\mathrm{X}} \cdot \Gamma[\mathcal{A}], \qquad (z_F) \mapsto \sum_{F:\,\mathrm{s}(F)=\mathrm{X}} \mathtt{u}^F \mu_F(z_F).$$

Composing (a) with (c) yields the isomorphism (13.10). □

Observe that:

Lemma 13.27. *The isomorphism* (13.10) *arises from the composite map*

$$\mathsf{Lie}[\mathcal{A}_{\mathrm{X}}] \to \Gamma[\mathcal{A}^{\mathrm{X}}] \otimes \mathsf{Lie}[\mathcal{A}_{\mathrm{X}}] \to \Gamma[\mathcal{A}],$$

where the first map sends z to $(\sum_{F:\,\mathrm{s}(F)=\mathrm{X}} \mathtt{u}^F \mathtt{H}_F) \otimes z$, and the second map is the substitution product (10.39).

Proposition 13.28. *For each homogeneous section* $\mathtt{u}$, *there is a linear isomorphism*

$$(13.12) \qquad \bigoplus_{\mathrm{X}} \mathsf{Lie}[\mathcal{A}_{\mathrm{X}}] \xrightarrow{\cong} \Gamma[\mathcal{A}], \qquad (z_{\mathrm{X}}) \mapsto \sum_{\mathrm{X}} \sum_{F:\,\mathrm{s}(F)=\mathrm{X}} \mathtt{u}^F \mu_F \beta_{F,\mathrm{X}}(z_{\mathrm{X}}).$$

The direct sum is over all flats.

Further, for any $\mathrm{X} \le \mathrm{Y}$, *the diagram*

$$(13.13) \qquad \begin{array}{ccc} \mathsf{Lie}[\mathcal{A}^{\mathrm{Y}}_{\mathrm{X}}] \otimes \mathsf{Lie}[\mathcal{A}_{\mathrm{Y}}] & \longrightarrow & \mathsf{Lie}[\mathcal{A}_{\mathrm{X}}] \\ \downarrow & & \downarrow \\ \Gamma[\mathcal{A}^{\mathrm{Y}}] \otimes \mathsf{Lie}[\mathcal{A}_{\mathrm{Y}}] & \longrightarrow & \Gamma[\mathcal{A}] \end{array}$$

commutes. The horizontal maps are the substitution products (10.30) *and* (10.39), *while the vertical maps are the inclusions induced from* (13.12).

PROOF. The first part follows from Lemma 13.26. The second part follows from Lemma 13.27 and (10.41). □

Exercise 13.29. Check that the isomorphism (13.12) restricts to

$$\bigoplus_{\mathrm{X} \neq \top} \mathsf{Lie}[\mathcal{A}_{\mathrm{X}}] \xrightarrow{\cong} \mathrm{rad}(\Gamma[\mathcal{A}]).$$

The direct sum in the lhs is over all non-maximum flats, while the rhs is the radical of $\Gamma[\mathcal{A}]$, see (9.37). This result is an improvement on Lemma 10.4.

13.5.3. Left Peirce decomposition of faces. Recall from Lemma 10.21 that $\mathsf{Zie}[\mathcal{A}_X]$ is a right ideal of $\Sigma[\mathcal{A}_X]$. This yields a right action of $\Sigma[\mathcal{A}]$ on $\mathsf{Zie}[\mathcal{A}_X]$ via the algebra homomorphism $\Sigma[\mathcal{A}] \to \Sigma[\mathcal{A}_X]$ given in (9.76). Using the same argument as with chamber elements, we deduce:

Lemma 13.30. *Fix a homogeneous section* $\mathtt{u}$ *with associated Eulerian family* $\mathtt{E}$. *For any flat* X, *the map*

$$\mathsf{Zie}[\mathcal{A}_X] \xrightarrow{\cong} \mathtt{E}_X \cdot \Sigma[\mathcal{A}], \qquad z \mapsto \sum_{F:\, \mathrm{s}(F)=X} \mathtt{u}^F \mu_F \beta_{F,X}(z)$$

is an isomorphism of right $\Sigma[\mathcal{A}]$*-modules, with inverse*

$$\mathtt{E}_X \cdot \Sigma[\mathcal{A}] \to \mathsf{Zie}[\mathcal{A}_X], \qquad z \mapsto \beta_{X,F}\Delta_F(z),$$

where F *is any face of support* X.

For $X = \bot$, *the map is the identity, that is,*

$$\mathsf{Zie}[\mathcal{A}] = \mathtt{E}_\bot \cdot \Sigma[\mathcal{A}]$$

as right ideals of $\Sigma[\mathcal{A}]$. *Equivalently,* $\mathsf{Zie}[\mathcal{A}]$ *is the right ideal of* $\Sigma[\mathcal{A}]$ *generated by* $\mathtt{E}_\bot$.

Proposition 13.31. *For each homogeneous section* $\mathtt{u}$, *there is a linear isomorphism*

$$\bigoplus_X \mathsf{Zie}[\mathcal{A}_X] \xrightarrow{\cong} \Sigma[\mathcal{A}], \qquad (z_X) \mapsto \sum_X \sum_{F:\, \mathrm{s}(F)=X} \mathtt{u}^F \mu_F \beta_{F,X}(z_X). \tag{13.14}$$

The direct sum is over all flats.

13.5.4. Rank one. Consider the rank-one arrangement with chambers C and $\overline{C}$. There are two flats, namely, $\bot$ and $\top$. The spaces $\mathsf{Lie}[\mathcal{A}_\bot]$ and $\mathsf{Lie}[\mathcal{A}_\top]$ are both one-dimensional, with $\mathtt{H}_C - \mathtt{H}_{\overline{C}}$ and $1 \in \Bbbk$ serving as basis elements, respectively. Recall from Section 11.1.5 that any homogeneous section $\mathtt{u}$ of $\mathcal{A}$ is of the form $\mathtt{u}^O = 1$, $\mathtt{u}^C = p$, $\mathtt{u}^{\overline{C}} = 1-p$ for an arbitrary scalar p. Now applying the isomorphism (13.12) to the two basis elements above yields

$$\mathtt{H}_C - \mathtt{H}_{\overline{C}} \quad \text{and} \quad p\,\mathtt{H}_C + (1-p)\,\mathtt{H}_{\overline{C}},$$

and they indeed form a basis for $\Gamma[\mathcal{A}]$.

Exercise 13.32. Do a similar analysis of the isomorphism (13.14) in rank one.

Exercise 13.33. Analyze the isomorphisms (13.12) and (13.14) for the rank-two arrangement of n lines. For simplicity, start with the case $n = 3$.

13.6. Right Peirce decomposition of Zie. Lie under flats

Recall from Lemma 10.21 that the space of Zie elements is a right ideal of the Tits algebra. So any Eulerian family yields a right Peirce decomposition of the space of Zie elements. The component indexed by a flat identifies with the space of Lie elements in the arrangement under that flat. In other words, the space of Zie elements breaks as a direct sum of spaces of Lie elements in arrangements under flats. Further, this decomposition is compatible with the substitution products of Lie and Zie.

Recall that an Eulerian family is equivalent to a $\mathtt{Q}$-basis of the Tits algebra. The latter is easier to work with in the present setting.

13.6.1. Zie as a sum of Lie under flats. Given a Zie element z (written in the $\mathtt{H}$-basis), one may naively expect that for any flat X, the part of z consisting of faces with support X is a Lie element of $\mathcal{A}^{\mathrm{X}}$. This is not true in general. However, this problem can be fixed using the $\mathtt{Q}$-basis as follows.

Lemma 13.34. *Fix a $\mathtt{Q}$-basis. For any $z \in \Sigma[\mathcal{A}]$, write $z = \sum_F x^F\, \mathtt{Q}_F$. Then z is a Zie element iff for each flat X, the element*

$$\sum_{F:\, \mathrm{s}(F)=\mathrm{X}} x^F\, \mathtt{H}_F$$

is a Lie element of $\mathcal{A}^{\mathrm{X}}$.

PROOF. By (11.29),

$$\mathtt{H}_H \cdot z = \sum_F x^F\, \mathtt{H}_H \cdot \mathtt{Q}_F = \sum_{F:\, FH=F} x^F\, \mathtt{Q}_{HF} = \sum_{G:\, G \geq H} \Big(\sum_{\substack{F:\, FH=F,\\ HF=G}} x^F \Big)\, \mathtt{Q}_G.$$

Rewrite the term in parenthesis using Exercise 1.12 and then apply Lemma 10.19 to get

$$z \in \mathsf{Zie}[\mathcal{A}] \iff \sum_{\substack{F:\, HF=G,\\ \mathrm{s}(F)=\mathrm{s}(G)}} x^F = 0 \quad \text{for all } O < H \leq G.$$

Now group the equations in the rhs according to the support of G, and apply (10.1) to each $\mathcal{A}^{\mathrm{X}}$. □

Exercise 13.35. Deduce Lemma 10.20 using Lemma 13.34 and the fact that $\mathtt{Q}_C = \mathtt{H}_C$ for any chamber C.

Lemma 13.36. *For each $\mathtt{Q}$-basis, there is a linear isomorphism*

$$\Sigma[\mathcal{A}] \xrightarrow{\cong} \Sigma[\mathcal{A}], \qquad \mathtt{H}_F \mapsto \mathtt{Q}_F. \tag{13.15}$$

Further, for any X, the diagram

$$\begin{array}{ccc} \Gamma[\mathcal{A}^{\mathrm{X}}] \otimes \Sigma[\mathcal{A}_{\mathrm{X}}] & \longrightarrow & \Sigma[\mathcal{A}] \\ \downarrow & & \downarrow \\ \Gamma[\mathcal{A}^{\mathrm{X}}] \otimes \Sigma[\mathcal{A}_{\mathrm{X}}] & \longrightarrow & \Sigma[\mathcal{A}] \end{array} \tag{13.16}$$

commutes. Both horizontal maps are the substitution product (10.33). *The vertical maps are induced from* (13.15).

PROOF. The isomorphism is clear. The substitution product (10.33) can be expressed as $\mathtt{H}_F \otimes \mathtt{H}_{K/F} \mapsto \mathtt{H}_K$. By the second claim in (11.37), this is equivalent to $\mathtt{H}_F \otimes \mathtt{Q}_{K/F} \mapsto \mathtt{Q}_K$. The commutativity of the diagram follows. □

Proposition 13.37. *For each $\mathtt{Q}$-basis, there is a linear isomorphism*

$$\bigoplus_{\mathrm{X}} \mathsf{Lie}[\mathcal{A}^{\mathrm{X}}] \xrightarrow{\cong} \mathsf{Zie}[\mathcal{A}], \qquad \Big(\sum_{F:\, \mathrm{s}(F)=\mathrm{X}} x^F\, \mathtt{H}_F \Big) \mapsto \sum_F x^F\, \mathtt{Q}_F. \tag{13.17}$$

The direct sum is over all flats.

Further, for any $\mathrm{X} \leq \mathrm{Y}$*, the diagram*

$$\begin{array}{ccc} \mathsf{Lie}[\mathcal{A}^{\mathrm{X}}] \otimes \mathsf{Lie}[\mathcal{A}^{\mathrm{Y}}_{\mathrm{X}}] & \longrightarrow & \mathsf{Lie}[\mathcal{A}^{\mathrm{Y}}] \\ \downarrow & & \downarrow \\ \mathsf{Lie}[\mathcal{A}^{\mathrm{X}}] \otimes \mathsf{Zie}[\mathcal{A}_{\mathrm{X}}] & \longrightarrow & \mathsf{Zie}[\mathcal{A}] \end{array} \tag{13.18}$$

commutes. The horizontal maps are the substitution products (10.30) *and* (10.34)*, while the vertical maps are the inclusions induced from* (13.17).

PROOF. The isomorphism (13.17) was obtained in Lemma 13.34. In view of (4.22), (10.29) and (10.35), the commutativity of (13.18) follows from that of (13.16). □

In view of the isomorphism (13.17), compare the dimension formulas (10.24) and (10.25).

Exercise 13.38. Check that the isomorphism (13.17) is the sum over X of the composite maps

$$\mathsf{Lie}[\mathcal{A}^{\mathrm{X}}] \to \mathsf{Lie}[\mathcal{A}^{\mathrm{X}}] \otimes \mathsf{Zie}[\mathcal{A}_{\mathrm{X}}] \to \mathsf{Zie}[\mathcal{A}],$$

where the first map sends z to $z \otimes \beta_{\mathrm{X},F}(\mathtt{Q}_F)$, and the second map is the substitution product (10.34). Use this to deduce (13.18) from (10.36).

13.6.2. Rank one. Consider the rank-one arrangement $\mathcal{A}$ with chambers C and $\overline{C}$. Recall from Section 11.4.3 that the Q-basis is

$$\mathtt{Q}_C = \mathtt{H}_C, \quad \mathtt{Q}_{\overline{C}} = \mathtt{H}_{\overline{C}}, \quad \mathtt{Q}_O = \mathtt{H}_O - p\,\mathtt{H}_C - (1-p)\,\mathtt{H}_{\overline{C}},$$

where p is an arbitrary scalar. We calculate

$$x^O\,\mathtt{Q}_O + x^C\,\mathtt{Q}_C + x^{\overline{C}}\,\mathtt{Q}_{\overline{C}} = x^O\,\mathtt{H}_O + (x^C - x^O p)\,\mathtt{H}_C + (x^{\overline{C}} - x^O(1-p))\,\mathtt{H}_{\overline{C}}.$$

Employing (10.26), this is a Zie element iff $x^C + x^{\overline{C}} = 0$. Thus,

$$x^O\,\mathtt{Q}_O + x^C\,\mathtt{Q}_C + x^{\overline{C}}\,\mathtt{Q}_{\overline{C}} \text{ is a Zie element} \iff x^C\,\mathtt{H}_C + x^{\overline{C}}\,\mathtt{H}_{\overline{C}} \text{ is a Lie element.}$$

(Also $x^O\mathtt{H}_O$ is a Lie element of $\mathcal{A}^{\perp}$.) This explicitly verifies Lemma 13.34.

Exercise 13.39. Do a similar verification of Lemma 13.34 for the rank-two arrangement of n lines. (Formulas for the Q-basis elements are given in Section 11.8.)

13.6.3. Right Peirce decomposition of faces. Let us begin with the right Peirce decomposition of the Tits algebra. The component indexed by a flat identifies with the space of chambers in the arrangement under that flat as follows.

Lemma 13.40. *Let* E *be any Eulerian family of* $\mathcal{A}$*. For any flat* Y*, the map*

$$\Gamma[\mathcal{A}^{\mathrm{Y}}] \xrightarrow{\cong} \Sigma[\mathcal{A}] \cdot \mathtt{E}_{\mathrm{Y}}, \qquad z \mapsto z \cdot \mathtt{E}_{\mathrm{Y}}$$

is an isomorphism of left $\Sigma[\mathcal{A}]$*-modules.*

For $\mathrm{Y} = \top$*, the map is the identity, that is,*

$$\Gamma[\mathcal{A}] = \Sigma[\mathcal{A}] \cdot \mathtt{E}_{\top}$$

as left ideals of $\Sigma[\mathcal{A}]$*. Equivalently,* $\Gamma[\mathcal{A}]$ *is the left ideal of* $\Sigma[\mathcal{A}]$ *generated by* $\mathtt{E}_{\top}$.

Here the action of $\Sigma[\mathcal{A}]$ on $\Gamma[\mathcal{A}^{\mathrm{Y}}]$ is via the algebra homomorphism $\Sigma[\mathcal{A}] \to \Sigma[\mathcal{A}^{\mathrm{Y}}]$ given in (9.77). That is, to act by z, we first truncate z to faces contained in Y and then act as usual.

PROOF. Observe using (11.26) that the above map sends $\mathtt{H}_F$ to $\mathtt{Q}_F$ for any face F with support Y. Further, the isomorphism (13.15) breaks as a direct sum over all flats, with the Y-summand being the above map. In particular, the above map is an isomorphism. The fact that it respects the action of $\Sigma[\mathcal{A}]$ can be deduced from Lemma 11.31. The special case $\mathrm{Y} = \top$ can also be seen directly since $\mathtt{E}_\top$ is a linear combination of chambers. □

Exercise 13.41. Deduce that: For any flat X, the elements $\mathtt{Q}_F$, with F of support X, form a linear basis of $\Sigma[\mathcal{A}] \cdot \mathtt{E}_\mathrm{X}$. (See Exercise 11.35 for related ideas.) Similarly, $\oplus_{\mathrm{Y}:\,\mathrm{Y} \geq \mathrm{X}} \Sigma[\mathcal{A}] \cdot \mathtt{E}_\mathrm{Y}$ has a linear basis consisting of $\mathtt{H}_G$, with $\mathrm{s}(G) \geq \mathrm{X}$.

Remark 13.42. For any Eulerian families $\mathtt{E}$ and $\mathtt{E}'$ and flat Y, the idempotents $\mathtt{E}_\mathrm{Y}$ and $\mathtt{E}'_\mathrm{Y}$ are isomorphic. This follows from Theorem D.33. As a result, $\Sigma[\mathcal{A}] \cdot \mathtt{E}_\mathrm{Y}$ and $\Sigma[\mathcal{A}] \cdot \mathtt{E}'_\mathrm{Y}$ are isomorphic as left $\Sigma[\mathcal{A}]$-modules. Lemma 13.40 gives an explicit isomorphism by connecting both to $\Gamma[\mathcal{A}^\mathrm{Y}]$.

13.6.4. Right Peirce decomposition of Zie. Recall from Lemma 10.21 that the space of Zie elements is a right ideal of the Tits algebra. Indeed, its right Peirce decomposition is given by (13.17) as elaborated below.

Lemma 13.43. *Let* $\mathtt{E}$ *be any Eulerian family of* $\mathcal{A}$. *For any flat* Y, *there is a linear isomorphism*

$$\mathsf{Lie}[\mathcal{A}^\mathrm{Y}] \xrightarrow{\cong} \mathsf{Zie}[\mathcal{A}] \cdot \mathtt{E}_\mathrm{Y}, \qquad z \mapsto z \cdot \mathtt{E}_\mathrm{Y}. \tag{13.19}$$

For $\mathrm{Y} = \top$, *the map is the identity, that is,*

$$\mathsf{Lie}[\mathcal{A}] = \mathsf{Zie}[\mathcal{A}] \cdot \mathtt{E}_\top.$$

PROOF. We observe as in the proof of Lemma 13.40 that the map (13.17) induces the above map. Since (13.17) is an isomorphism, so is the above map. □

In the exercises below, $\mathsf{Zie}[\mathcal{A}]$ is viewed as a right ideal of the Tits algebra.

Exercise 13.44. Check that the isomorphism (13.17) restricts to

$$\bigoplus_{\mathrm{X} \neq \bot} \mathsf{Lie}[\mathcal{A}^\mathrm{X}] \xrightarrow{\cong} \mathcal{D}(\mathsf{Zie}[\mathcal{A}]).$$

The direct sum in the lhs is over all non-minimum flats, while the rhs is the decomposable part of $\mathsf{Zie}[\mathcal{A}]$.

Exercise 13.45. Check that $\mathcal{D}(\mathsf{Zie}[\mathcal{A}])$ consists precisely of the Zie elements whose coefficient of the central face is 0. In particular, it is a codimension-one subspace of $\mathsf{Zie}[\mathcal{A}]$.

Exercise 13.46. Check that $\mathcal{D}(\mathsf{Zie}[\mathcal{A}]) = \mathsf{Zie}[\mathcal{A}] \cap \mathrm{rad}(\Sigma[\mathcal{A}])$. (Combine Exercises 10.25 and 13.45.)

Exercise 13.47. Show that

$$\xi_\mathrm{X}(\mathsf{Zie}[\mathcal{A}]) = \sum_{\mathrm{Y}:\,\mathrm{Y} \geq \mathrm{X}} |\mu(\mathcal{A}^\mathrm{Y})| \quad \text{and} \quad \eta_\mathrm{X}(\mathsf{Zie}[\mathcal{A}]) = |\mu(\mathcal{A}^\mathrm{X})|.$$

(To get the first formula, use (9.64) and Proposition 13.24 and Lemma 13.43. To get the second formula, use Proposition 13.20 and Lemma 13.43. The two formulas also imply each other in view of (9.52).)

13.7. Two-sided Peirce decomposition of faces. Lie over & under flats

We now consider the two-sided Peirce decomposition of the Tits algebra (viewed as a bimodule over itself), and identify its components with Lie elements in arrangements over and under flats, with further compatibility with the substitution product of Lie. This is done by combining the left and right Peirce decompositions discussed in the preceding sections. There is also a projective analogue to this story which identifies components of the two-sided Peirce decomposition of the projective Tits algebra with projective Lie elements in arrangements over and under flats.

13.7.1. Two-sided Peirce decomposition of faces. For an Eulerian family $\mathtt{E}$ of $\mathcal{A}$, let $\mathtt{E}_{\mathrm{X}} \cdot \Sigma[\mathcal{A}] \cdot \mathtt{E}_{\mathrm{Y}}$ denote the subspace of $\Sigma[\mathcal{A}]$ consisting of all elements of the form $\mathtt{E}_{\mathrm{X}} \cdot z \cdot \mathtt{E}_{\mathrm{Y}}$ with z an arbitrary element of $\Sigma[\mathcal{A}]$.

Lemma 13.48. *For any Eulerian family* $\mathtt{E}$ *of* $\mathcal{A}$,

$$\Sigma[\mathcal{A}] = \bigoplus_{\mathrm{X} \leq \mathrm{Y}} \mathtt{E}_{\mathrm{X}} \cdot \Sigma[\mathcal{A}] \cdot \mathtt{E}_{\mathrm{Y}}. \tag{13.20}$$

The sum is over both X *and* Y.

PROOF. The point to note is that

$$\mathtt{E}_{\mathrm{X}} \cdot \Sigma[\mathcal{A}] \cdot \mathtt{E}_{\mathrm{Y}} = 0 \quad \text{for} \quad \mathrm{X} \not\leq \mathrm{Y}.$$

This follows from the Saliola lemma (Lemma 11.12). □

We refer to (13.20) as the *two-sided Peirce decomposition* of the Tits algebra. The decomposition depends on the choice of the Eulerian family.

Lemma 13.49. *Let* $\mathtt{E}$ *be an Eulerian family of* $\mathcal{A}$. *Then*

$$\Gamma[\mathcal{A}] = \bigoplus_{\mathrm{X}} \mathtt{E}_{\mathrm{X}} \cdot \Sigma[\mathcal{A}] \cdot \mathtt{E}_{\top} \quad \text{and} \quad \Gamma[\mathcal{A}] = \mathrm{rad}(\Gamma[\mathcal{A}]) \oplus \mathtt{E}_{\top} \cdot \Sigma[\mathcal{A}] \cdot \mathtt{E}_{\top}. \tag{13.21}$$

Similarly,

$$\mathsf{Zie}[\mathcal{A}] = \bigoplus_{\mathrm{X}} \mathtt{E}_{\perp} \cdot \Sigma[\mathcal{A}] \cdot \mathtt{E}_{\mathrm{X}} \quad \text{and} \quad \mathsf{Zie}[\mathcal{A}] = \mathtt{E}_{\perp} \cdot \Sigma[\mathcal{A}] \cdot \mathtt{E}_{\perp} \oplus \mathcal{D}(\mathsf{Zie}[\mathcal{A}]). \tag{13.22}$$

PROOF. Let us first deal with chambers. For the first identity, use Lemma 13.48 and the second part of Lemma 13.40. The second can be seen directly since $\mathrm{rad}(\Gamma[\mathcal{A}])$ consists of chamber elements whose coefficients add up to 0, and $\mathtt{E}_{\top}$ is a chamber element whose coefficients add up to 1.

The story with Zie elements is similar. For the first identity, use Lemma 13.48 and the second part of Lemma 13.30. The second can be seen directly since $\mathcal{D}(\mathsf{Zie}[\mathcal{A}])$ consists of elements whose coefficient of the central face is 0, and $\mathtt{E}_{\perp}$ is a special Zie element, so its coefficient of the central face is 1. Alternatively, it follows from the first and Proposition 13.24 for $k = 1$. □

This explains precisely how the left ideal of chambers and the right ideal of Zie elements relate to the two-sided Peirce decomposition of the Tits algebra.

TABLE 13.1. (Left, right, two-sided) Peirce decompositions.

Left Peirce decompositions	Right Peirce decompositions
$\mathsf{E}_{\mathrm{X}} \cdot \Sigma[\mathcal{A}] = \mathsf{Zie}[\mathcal{A}_{\mathrm{X}}]$	$\Sigma[\mathcal{A}] \cdot \mathsf{E}_{\mathrm{Y}} = \Gamma[\mathcal{A}^{\mathrm{Y}}]$
$\mathsf{E}_{\mathrm{X}} \cdot \Gamma[\mathcal{A}] = \mathsf{Lie}[\mathcal{A}_{\mathrm{X}}]$	$\mathsf{Zie}[\mathcal{A}] \cdot \mathsf{E}_{\mathrm{Y}} = \mathsf{Lie}[\mathcal{A}^{\mathrm{Y}}]$
Two-sided Peirce decomposition	
$$\begin{array}{ccc} & \mathsf{E}_{\mathrm{X}} \cdot \Sigma[\mathcal{A}] \cdot \mathsf{E}_{\mathrm{Y}} & \\ \mathrel{/\!\!/} & & \mathrel{\backslash\!\!\backslash} \\ \mathsf{E}_{\mathrm{X}} \cdot \Gamma[\mathcal{A}^{\mathrm{Y}}] & & \mathsf{Zie}[\mathcal{A}_{\mathrm{X}}] \cdot \mathsf{E}_{\mathrm{Y}} \\ \mathrel{\backslash\!\!\backslash} & & \mathrel{/\!\!/} \\ & \mathsf{Lie}[\mathcal{A}^{\mathrm{Y}}_{\mathrm{X}}] & \end{array}$$	

13.7.2. Combining left and right Peirce decompositions. We studied the left Peirce decomposition of chambers and faces in Section 13.5 and the right Peirce decomposition of Zie and faces in Section 13.6. These results can be combined to understand the two-sided Peirce decomposition of faces in two different ways. This is summarized in Table 13.1. Details follow.

Proposition 13.50. *Let* E *be any Eulerian family of* $\mathcal{A}$. *For any* $\mathrm{X} \leq \mathrm{Y}$, *there is a natural vector space isomorphism*

$$\mathsf{Lie}[\mathcal{A}^{\mathrm{Y}}_{\mathrm{X}}] \xrightarrow{\cong} \mathsf{E}_{\mathrm{X}} \cdot \Sigma[\mathcal{A}] \cdot \mathsf{E}_{\mathrm{Y}}. \tag{13.23}$$

Further, for any $\mathrm{X} \leq \mathrm{Y} \leq \mathrm{Z}$, *the diagram*

$$\begin{array}{ccc} \mathsf{Lie}[\mathcal{A}^{\mathrm{Y}}_{\mathrm{X}}] \otimes \mathsf{Lie}[\mathcal{A}^{\mathrm{Z}}_{\mathrm{Y}}] & \longrightarrow & \mathsf{Lie}[\mathcal{A}^{\mathrm{Z}}_{\mathrm{X}}] \\ \big\downarrow & & \big\downarrow \\ \mathsf{E}_{\mathrm{X}} \cdot \Sigma[\mathcal{A}] \cdot \mathsf{E}_{\mathrm{Y}} \otimes \mathsf{E}_{\mathrm{Y}} \cdot \Sigma[\mathcal{A}] \cdot \mathsf{E}_{\mathrm{Z}} & \longrightarrow & \mathsf{E}_{\mathrm{X}} \cdot \Sigma[\mathcal{A}] \cdot \mathsf{E}_{\mathrm{Z}} \end{array} \tag{13.24}$$

commutes. The top-horizontal map is the substitution product (10.30), *while the bottom-horizontal map is induced from the Tits product.*

PROOF. First method: Apply (13.10) to the arrangement $\mathcal{A}^{\mathrm{Y}}$ and use Lemma 13.40 to obtain

$$\mathsf{Lie}[\mathcal{A}^{\mathrm{Y}}_{\mathrm{X}}] \xrightarrow{\cong} \mathsf{E}^{\mathrm{Y}}_{\mathrm{X}} \cdot \Gamma[\mathcal{A}^{\mathrm{Y}}] \xrightarrow{\cong} \mathsf{E}_{\mathrm{X}} \cdot \Sigma[\mathcal{A}] \cdot \mathsf{E}_{\mathrm{Y}}.$$

(Here $\mathsf{E}^{\mathrm{Y}}_{\mathrm{X}}$ is the Eulerian idempotent of $\mathcal{A}^{\mathrm{Y}}$ for the flat X. It is obtained from E_{X} by truncating to faces contained in Y.) This yields the isomorphism (13.23). Explicitly, it is given by

$$z \mapsto \alpha_{\mathrm{X}}(z) \cdot \mathsf{E}_{\mathrm{Y}},$$

where

$$\alpha_{\mathrm{X}}(z) := \sum_{F:\, \mathrm{s}(F)=\mathrm{X}} \mathtt{u}^{F} \mu_F \beta_{F,\mathrm{X}}(z).$$

Note that $\alpha_{\mathrm{X}}(z) \cdot \mathsf{E}_{\mathrm{Y}} = \mathsf{E}_{\mathrm{X}} \cdot \alpha_{\mathrm{X}}(z) \cdot \mathsf{E}_{\mathrm{Y}}$.

The commutativity of (13.24) follows from the identity

$$\alpha_{\mathrm{X}}(x) \cdot \alpha_{\mathrm{Y}}(y) = \alpha_{\mathrm{X}}(x \circ y),$$

where x is a Lie element of $\mathcal{A}_X^Y$, y is a Lie element of $\mathcal{A}_Y^Z$, and $x \circ y$ denotes the Lie element of $\mathcal{A}_X^Z$ obtained by substitution. This can be deduced from (10.29) and the definition of (4.18).

Second method: Apply (13.19) to the arrangement $\mathcal{A}_X$ and use Lemma 13.30 to obtain

$$\mathsf{Lie}[\mathcal{A}_X^Y] \xrightarrow{\cong} \mathsf{Zie}[\mathcal{A}_X] \cdot \mathtt{E}_{Y/X} \xrightarrow{\cong} \mathtt{E}_X \cdot \Sigma[\mathcal{A}] \cdot \mathtt{E}_Y.$$

(Here $\mathtt{E}_{Y/X}$ is the Eulerian idempotent of $\mathcal{A}_X$ for the flat Y.) One may check that the composite map is the same as before. □

Lemma 13.51. *The following diagrams commute.*

$$\begin{array}{ccc} \Gamma[\mathcal{A}^X] \otimes \mathsf{Lie}[\mathcal{A}_X^Y] & \longrightarrow & \Gamma[\mathcal{A}^Y] \\ \cong\downarrow & & \downarrow\cong \\ \Sigma[\mathcal{A}] \cdot \mathtt{E}_X \otimes \mathtt{E}_X \cdot \Sigma[\mathcal{A}] \cdot \mathtt{E}_Y & \longrightarrow & \Sigma[\mathcal{A}] \cdot \mathtt{E}_Y \end{array} \tag{13.25}$$

The top horizontal map is the substitution product (10.39).

$$\begin{array}{ccc} \mathsf{Lie}[\mathcal{A}_X^Y] \otimes \mathsf{Zie}[\mathcal{A}_Y] & \longrightarrow & \mathsf{Zie}[\mathcal{A}_X] \\ \cong\downarrow & & \downarrow\cong \\ \mathtt{E}_X \cdot \Sigma[\mathcal{A}] \cdot \mathtt{E}_Y \otimes \mathtt{E}_Y \cdot \Sigma[\mathcal{A}] & \longrightarrow & \mathtt{E}_X \cdot \Sigma[\mathcal{A}] \end{array} \tag{13.26}$$

The top horizontal map is the substitution product (10.34).

PROOF. Diagram (13.25) is obtained by summing (13.24) over all X and using (13.13) (and renaming Y and Z respectively to X and Y). Similarly, diagram (13.26) is obtained by summing (13.24) over all Z and using (13.18). □

Exercise 13.52. Each component $\mathtt{E}_X \cdot \Sigma[\mathcal{A}] \cdot \mathtt{E}_Y$, for $X \leq Y$, is nonzero. Further, if $w \in \mathtt{E}_X \cdot \Sigma[\mathcal{A}] \cdot \mathtt{E}_Y$ and $z \in \mathtt{E}_Y \cdot \Sigma[\mathcal{A}] \cdot \mathtt{E}_Z$ are both nonzero, then their product $w \cdot z$ is a nonzero element of $\mathtt{E}_X \cdot \Sigma[\mathcal{A}] \cdot \mathtt{E}_Z$. (Use Exercises 10.46 and 10.48 and Proposition 13.50.)

13.7.3. Cartan invariants. Using Proposition 13.50 and (10.24), we obtain:

Proposition 13.53. *For any Eulerian family* $\mathtt{E}$,

$$\dim(\mathtt{E}_X \cdot \Sigma[\mathcal{A}] \cdot \mathtt{E}_Y) = \begin{cases} |\mu(X,Y)| & \textit{if } X \leq Y, \\ 0 & \textit{otherwise.} \end{cases}$$

The above numbers are by definition the *Cartan invariants* of the Tits algebra.

13.7.4. Summing the components. Consider the vector space

$$\bigoplus_{X \leq Y} \mathsf{Lie}[\mathcal{A}_X^Y]. \tag{13.27}$$

The sum is over both X and Y. This space carries an algebra structure. Elements in the (X, Y)-summand are multiplied with elements in the (Y, Z)-summand by substitution (10.30); the remaining products are all zero.

Theorem 13.54. *There is an algebra isomorphism*

$$\bigoplus_{\mathrm{X}\leq \mathrm{Y}} \mathsf{Lie}[\mathcal{A}_{\mathrm{X}}^{\mathrm{Y}}] \xrightarrow{\cong} \Sigma[\mathcal{A}]$$

obtained by summing the isomorphisms (13.23) *over all* $\mathrm{X} \leq \mathrm{Y}$.

Exercise 13.55. Work out the isomorphism in Theorem 13.54 for the rank-one arrangement by building on the discussion in Section 13.5.4.

Exercise 13.56. Use Theorem 13.54 to deduce that the Tits algebra has trivial center. (This result was also obtained in Corollary 9.31.)

The vector space

$$\bigoplus_{\mathrm{X}} \Gamma[\mathcal{A}^{\mathrm{X}}] \tag{13.28}$$

is a right module over the algebra (13.27). Elements in the X-summand are acted upon by elements in the (X, Y)-summand by substitution (10.39) to yield an element in the Y-summand; the remaining actions are all zero.

Lemma 13.57. *The module* (13.28) *is isomorphic to the right regular representation of the algebra* (13.27). *Further, the diagram*

$$\begin{array}{ccc} \big(\bigoplus_{\mathrm{X}} \Gamma[\mathcal{A}^{\mathrm{X}}]\big) \otimes \big(\bigoplus_{\mathrm{X}\leq \mathrm{Y}} \mathsf{Lie}[\mathcal{A}_{\mathrm{X}}^{\mathrm{Y}}]\big) & \longrightarrow & \bigoplus_{\mathrm{Y}} \Gamma[\mathcal{A}^{\mathrm{Y}}] \\ \big\downarrow & & \big\downarrow \\ \Sigma[\mathcal{A}] \otimes \Sigma[\mathcal{A}] & \longrightarrow & \Sigma[\mathcal{A}] \end{array} \tag{13.29}$$

commutes.

Proof. The isomorphism between (13.28) and (13.27) is obtained from the left Peirce decomposition of chambers. The fact that it is a module map follows from (13.13). This proves the first claim. Diagram (13.29) follows from (13.25). □

The vector space

$$\bigoplus_{\mathrm{Y}} \mathsf{Zie}[\mathcal{A}_{\mathrm{Y}}] \tag{13.30}$$

is a left module over the algebra (13.27). Elements in the (X, Y)-summand act on elements in the Y-summand by substitution (10.34) to yield an element in the X-summand; the remaining actions are all zero.

Lemma 13.58. *The module* (13.30) *is isomorphic to the left regular representation of the algebra* (13.27). *Further, the diagram*

$$\begin{array}{ccc} \big(\bigoplus_{\mathrm{X}\leq \mathrm{Y}} \mathsf{Lie}[\mathcal{A}_{\mathrm{X}}^{\mathrm{Y}}]\big) \otimes \big(\bigoplus_{\mathrm{Y}} \mathsf{Zie}[\mathcal{A}_{\mathrm{Y}}]\big) & \longrightarrow & \bigoplus_{\mathrm{X}} \mathsf{Zie}[\mathcal{A}_{\mathrm{X}}] \\ \big\downarrow & & \big\downarrow \\ \Sigma[\mathcal{A}] \otimes \Sigma[\mathcal{A}] & \longrightarrow & \Sigma[\mathcal{A}] \end{array} \tag{13.31}$$

commutes.

PROOF. The isomorphism between (13.30) and (13.27) is obtained from the right Peirce decomposition of Zie. The fact that it is a module map follows from (13.18). This proves the first claim. Diagram (13.31) follows from (13.26). □

13.7.5. Projective Tits algebra. The preceding results have analogues for the projective Tits algebra, some of which we briefly summarize below. We assume that the field characteristic is not 2.

Consider the subspace of (13.27) consisting of projective Lie elements in each $\mathcal{A}_{\mathrm{X}}^{\mathrm{Y}}$. By Lemma 10.12, this subspace equals

$$\bigoplus_{\mathrm{rk}(\mathrm{Y}/\mathrm{X}) \text{ is even}} \mathsf{Lie}[\mathcal{A}_{\mathrm{X}}^{\mathrm{Y}}]. \tag{13.32}$$

The sum is over all $\mathrm{X} \leq \mathrm{Y}$ such that the difference in their ranks is even. This subspace is clearly a subalgebra of (13.27).

Theorem 13.59. *There is an algebra isomorphism from* (13.32) *to the projective Tits algebra.*

An explicit isomorphism is obtained by fixing a projective Eulerian family and restricting the resulting isomorphism in Theorem 13.54.

PROOF. We need to revisit Lemmas 13.26 and 13.40 which entered into the proof of Proposition 13.50. It is clear from the formulas that the isomorphisms in Lemma 13.26 restrict to the projective setting. Similarly, by considering the isomorphism of the projective Tits algebra which sends $\mathtt{H}_{\{F,\overline{F}\}}$ to $\mathtt{Q}_{\{F,\overline{F}\}}$, we deduce that the isomorphism in Lemma 13.40 restricts to the projective setting. □

The Cartan invariants of the projective Tits algebra are defined in a similar manner to the Tits algebra. They are $|\mu(\mathrm{X},\mathrm{Y})|$ if $\mathrm{rk}(\mathrm{Y}/\mathrm{X})$ is even, and 0 in all other cases.

In contrast to the Tits algebra, the projective Tits algebra does have a nontrivial center in general. In rank one, the algebra is commutative, so the center has dimension 2. For the general case, one can proceed as follows. Consider the graph whose vertices are flats, and there is an edge joining X and Y whenever $\mathrm{rk}(\mathrm{Y}/\mathrm{X}) = 2$. For each connected component c of this graph, put

$$e_c := \sum_{\mathrm{X} \text{ is a vertex of } c} \mathtt{E}_{\mathrm{X}},$$

where $\mathtt{E}$ is any projective Eulerian family. The center of the projective Tits algebra has a basis consisting of the elements e_c, as c varies over all connected components. This can be deduced using Theorem 13.59. We elaborate on this result below.

Proposition 13.60. *For the rank-two arrangement of n lines, the center of the projective Tits algebra has a basis of $n+1$ elements, namely, $\mathtt{E}_{\bot} + \mathtt{E}_{\top}$ and $\mathtt{E}_{\mathrm{X}}$, as X varies over the n lines. Here $\mathtt{E}$ is any projective Eulerian family.*

Exercise 13.61. Use Proposition 11.65 to check that for the rank-two arrangement of n lines, an element $z = \sum_H x^H \mathtt{H}_H$ belongs to the center of the projective Tits algebra iff $2x^C + x^F + x^G = 0$ for any chamber C with F and G as its two vertices.

Proposition 13.62. *For any arrangement of rank not equal to 2, the center of the projective Tits algebra is linearly spanned by $\mathtt{H}_O$ and the Takeuchi element.*

PROOF. For the arrangement of rank zero, the result is clear. For arrangements of odd rank, the graph defined above has exactly 2 connected components, one containing $\perp$ and the other containing $\top$. For arrangements of even rank strictly greater than 2, the graph again has exactly 2 connected components. This requires a small argument which we omit. Now use (12.25). $\square$

Compare this result with the one obtained in Lemma 12.30.

13.8. Generation of Lie elements in rank one

We now show that Lie elements of any arrangement are generated by Lie elements in rank one by iterated substitution. The proof makes use of the description of the left Peirce decomposition of chambers in terms of Lie elements over flats.

Let $\mathcal{A}$ be an arrangement of rank r. The iterated substitution product of Lie (10.32) yields the map

$$\bigoplus_{z} \mathsf{Lie}[\mathcal{A}^{\mathrm{X}_1}] \otimes \mathsf{Lie}[\mathcal{A}^{\mathrm{X}_2}_{\mathrm{X}_1}] \otimes \cdots \otimes \mathsf{Lie}[\mathcal{A}_{\mathrm{X}_{r-1}}] \to \mathsf{Lie}[\mathcal{A}]. \tag{13.33}$$

The sum is over all maximal chains of flats $z = (\perp \lessdot \mathrm{X}_1 \lessdot \cdots \lessdot \mathrm{X}_{r-1} \lessdot \top)$.

Lemma 13.63. *The map* (13.33) *is surjective.*

PROOF. Fix an Eulerian family E. For any flat Y and chain of flats $\mathrm{Y} \lessdot \mathrm{Y}_1 \lessdot \cdots \lessdot \mathrm{Y}_k \lessdot \top$, consider the composite map

$$\mathsf{Lie}[\mathcal{A}^{\mathrm{Y}_1}_{\mathrm{Y}}] \otimes \mathsf{Lie}[\mathcal{A}^{\mathrm{Y}_2}_{\mathrm{Y}_1}] \otimes \cdots \otimes \mathsf{Lie}[\mathcal{A}_{\mathrm{Y}_k}] \to \mathsf{Lie}[\mathcal{A}_{\mathrm{Y}}] \to \mathsf{E}_{\mathrm{Y}} \cdot \Gamma[\mathcal{A}]$$

obtained by iterated substitution followed by (13.10). When $\mathrm{Y} = \top$, this map is

$$\Bbbk = \mathsf{Lie}[\mathcal{A}_\top] \to \Gamma[\mathcal{A}], \qquad 1 \mapsto \mathsf{E}_\top.$$

(Recall that $\mathsf{E}_\top$ is a linear combination of chambers whose coefficients add up to 1.) By summing these maps, we obtain

$$\bigoplus_{\mathrm{Y}} \bigoplus_{z} \mathsf{Lie}[\mathcal{A}^{\mathrm{Y}_1}_{\mathrm{Y}}] \otimes \mathsf{Lie}[\mathcal{A}^{\mathrm{Y}_2}_{\mathrm{Y}_1}] \otimes \cdots \otimes \mathsf{Lie}[\mathcal{A}_{\mathrm{Y}_k}] \to \Gamma[\mathcal{A}]. \tag{13.34}$$

The outside sum is over all flats Y, while the inside sum is over all maximal chains from Y to $\top$.

To prove the lemma, it suffices to show that (13.34) is surjective. We do this by an induction on the rank of $\mathcal{A}$. First note that $\mathsf{E}_\top$ belongs to the image of (13.34). Next, for any hyperplane Z, consider the composite map

$$\Big(\bigoplus_{\mathrm{Y}:\, \mathrm{Y} \leq \mathrm{Z}} \bigoplus_{z'} \mathsf{Lie}[\mathcal{A}^{\mathrm{Y}_1}_{\mathrm{Y}}] \otimes \mathsf{Lie}[\mathcal{A}^{\mathrm{Y}_2}_{\mathrm{Y}_1}] \otimes \cdots \otimes \mathsf{Lie}[\mathcal{A}^{\mathrm{Z}}_{\mathrm{Y}_k}] \Big) \otimes \mathsf{Lie}[\mathcal{A}_{\mathrm{Z}}] \twoheadrightarrow \Gamma[\mathcal{A}^{\mathrm{Z}}] \otimes \mathsf{Lie}[\mathcal{A}_{\mathrm{Z}}] \to \Gamma[\mathcal{A}],$$

where z' runs over maximal chains from Y to Z. The first map on the first tensor factor is (13.34) applied to $\mathcal{A}^{\mathrm{Z}}$, while the second map is (10.39). By (13.13), the composite is indeed the restriction of (13.34). Since the rank of $\mathcal{A}^{\mathrm{Z}}$ is strictly smaller than the rank of $\mathcal{A}$, by induction hypothesis, the first map is surjective. By summing over all hyperplanes Z and applying Lemma 10.51, we see that $\mathrm{rad}(\Gamma[\mathcal{A}])$ is contained in the image of (13.34). This codimension-one subspace along with $\mathsf{E}_\top$ spans $\Gamma[\mathcal{A}]$, so (13.34) is surjective. $\square$

Lemma 13.64. *Let* E *be any Eulerian family of* $\mathcal{A}$. *For any flats* $\mathrm{X} < \mathrm{Y}$, *the map*

$$\bigoplus_{z} \mathtt{E}_{\mathrm{X}} \cdot \Sigma[\mathcal{A}] \cdot \mathtt{E}_{\mathrm{X}_1} \otimes \mathtt{E}_{\mathrm{X}_1} \cdot \Sigma[\mathcal{A}] \cdot \mathtt{E}_{\mathrm{X}_2} \otimes \cdots \otimes \mathtt{E}_{\mathrm{X}_k} \cdot \Sigma[\mathcal{A}] \cdot \mathtt{E}_{\mathrm{Y}} \twoheadrightarrow \mathtt{E}_{\mathrm{X}} \cdot \Sigma[\mathcal{A}] \cdot \mathtt{E}_{\mathrm{Y}}$$

induced from the Tits product is surjective. The sum is over all maximal chains of flats $z = (\mathrm{X} \lessdot \mathrm{X}_1 \lessdot \cdots \lessdot \mathrm{X}_k \lessdot \mathrm{Y})$.

PROOF. Apply Lemma 13.63 to the arrangement $\mathcal{A}_{\mathrm{X}}^{\mathrm{Y}}$ and use Proposition 13.50. □

Exercise 13.65. Use (13.22) and (13.26) in conjunction with Lemma 13.64 to show that: The map

$$\bigoplus_{\mathrm{X} \text{ is rank-one}} \mathsf{Lie}[\mathcal{A}^{\mathrm{X}}] \otimes \mathsf{Zie}[\mathcal{A}_{\mathrm{X}}] \twoheadrightarrow \mathcal{D}(\mathsf{Zie}[\mathcal{A}])$$

is surjective. The sum is over all rank-one flats X. (This result is a companion to Lemma 10.51.)

13.9. Rigidity of the left module of chambers

Recall that the radical of a finite-dimensional algebra is a nilpotent ideal. We show that the largest nonzero power of the radical of the Tits algebra coincides with the space of Lie elements. We apply this result to prove that radical series, socle series and primitive series of the left module of chambers coincide. This means precisely that the left module of chambers is rigid.

We mention that the results of this section are obtained independently and in greater generality in Section 13.11.

13.9.1. Radical of the Tits algebra and Lie elements. We now strengthen the result of Lemma 13.15.

Proposition 13.66. *We have* $\mathrm{rad}(\Sigma)^r = \mathsf{Lie}$.

PROOF. We need to show that the rhs is contained in the lhs. For this, we apply Lemma 13.63. The image of the z-summand in (13.33) is spanned by the element

$$(\mathtt{H}_{F_1} - \mathtt{H}_{G_1}) \cdot \ldots \cdot (\mathtt{H}_{F_r} - \mathtt{H}_{G_r}),$$

where $O \lessdot F_1 \lessdot \cdots \lessdot F_r$ is any maximal chain of faces, with $\mathrm{s}(F_i) = \mathrm{X}_i$ for $1 \le i \le r$ and G_i is the face opposite to F_i in the star of F_{i-1} (with the convention $F_0 = O$). By (9.33) applied to each factor, the above element is in $\mathrm{rad}(\Sigma)^r$. So by surjectivity of the map (13.33), we are done. □

13.9.2. Radical series of the left module of chambers. Consider the radical series of the left module of chambers Γ, namely,

$$0 \subseteq \mathrm{rad}(\Sigma)^r \cdot \Gamma \subseteq \cdots \subseteq \mathrm{rad}(\Sigma)^2 \cdot \Gamma \subseteq \mathrm{rad}(\Sigma) \cdot \Gamma \subseteq \Gamma.$$

The associated graded Σ-module is

$$\bigoplus_{i=0}^{r} \mathrm{rad}(\Sigma)^i \cdot \Gamma / \mathrm{rad}(\Sigma)^{i+1} \cdot \Gamma.$$

Each summand is a semisimple Σ-module. We know from BHR (Theorem 9.46) and related discussion that in the above direct sum, the simple module with multiplicative character χ_{X} as given in (9.48) appears with multiplicity $\eta_{\mathrm{X}}(\Gamma) = |\mu(\mathrm{X}, \top)|$. The finer information of the multiplicity in each graded piece is given by the following result.

Proposition 13.67. *For the semisimple Σ-module*

$$\mathrm{rad}(\Sigma)^i \cdot \Gamma / \mathrm{rad}(\Sigma)^{i+1} \cdot \Gamma,$$

the multiplicity of the simple module with multiplicative character χ_X is $|\mu(X, \top)|$ if X has codimension i, and zero otherwise.

Let us first spell out the two end cases.

- $i = 0$: Recall that $\mathrm{rad}(\Sigma) \cdot \Gamma$ is the radical of Γ. By (9.37), the quotient $\Gamma / \mathrm{rad}(\Sigma) \cdot \Gamma$ is one-dimensional in which all chambers are equivalent to one another. The element $\mathtt{H}_F$ sends $\mathtt{H}_C$ to $\mathtt{H}_{FC}$ but these two elements are equivalent, thus $\mathtt{H}_F$ acts by identity on the quotient. So, the action is by the multiplicative character $\chi_\top$ given in (9.50).
- $i = r$: Since $\mathrm{rad}(\Sigma)^{r+1} \cdot \Gamma = 0$, by Proposition 13.66, the module in question is Lie whose dimension is $|\mu(\perp, \top)|$. By the Friedrichs criterion (Lemma 10.5), all noncentral faces act by zero on Lie. Thus, the action is by the multiplicative character $\chi_\perp$ given in (9.50).

PROOF. Put $J[\mathcal{A}] := \mathrm{rad}(\Sigma[\mathcal{A}])$. For clarity, we make the dependence on $\mathcal{A}$ explicit. Fix an index i and put $\mathsf{k} := J[\mathcal{A}]^i \cdot \Gamma[\mathcal{A}] / J[\mathcal{A}]^{i+1} \cdot \Gamma[\mathcal{A}]$. For each flat X of codimension i, it suffices to locate a submodule k_X of k of dimension $|\mu(X, \top)|$ on which the action is by the multiplicative character χ_X. For that, fix a face F of support X, and consider the space

$$J[\mathcal{A}_F]^i \cdot \Gamma[\mathcal{A}_F] = \mathsf{Lie}[\mathcal{A}_F].$$

(The equality is an instance of Proposition 13.66.) This space has the right dimension, namely, $|\mu(X, \top)|$. View it as a subspace of $J[\mathcal{A}]^i \cdot \Gamma[\mathcal{A}]$ (by viewing each face of $\mathcal{A}_F$ as a face of $\mathcal{A}$). Let k_X denote its image in the quotient k. By Lemma 13.17, $\mathtt{H}_F \cdot J[\mathcal{A}]^{i+1} \cdot \Gamma[\mathcal{A}] = 0$, so no nonzero element of $J[\mathcal{A}]^{i+1} \cdot \Gamma[\mathcal{A}]$ can lie entirely in the star of F. Hence k_X also has dimension $|\mu(X, \top)|$. Further: If H is not contained in X, then by the Friedrichs criterion, one can deduce that $\mathtt{H}_H$ acts by zero on k_X. Now suppose H is contained in X. Then $\mathtt{H}_H$ sends any element in the star of F to an element in the star of HF. Since $\mathtt{H}_F - \mathtt{H}_{HF} \in J[\mathcal{A}]$, the difference between these two elements belongs to $J[\mathcal{A}]^{i+1} \cdot \Gamma[\mathcal{A}]$, hence $\mathtt{H}_H$ acts by the identity on k_X. It follows that the action on k_X is by the multiplicative character χ_X as required. □

13.9.3. Series comparison. We can now compare the three different filtrations of the left module of chambers.

Theorem 13.68. *The left module of chambers Γ is rigid. That is, its radical series, primitive series and socle series all coincide. Explicitly, for $0 \leq k \leq r$,*

$$\mathrm{rad}(\Sigma)^k \cdot \Gamma = \mathcal{P}_{r-k+1}(\Gamma) = \mathrm{soc}_{r-k+1}(\Gamma),$$

and the dimension of each of these spaces is

$$\sum_{Y:\, \mathrm{rk}(Y) \leq r-k} |\mu(Y, \top)|.$$

PROOF. By Lemma 13.12, the primitive series and socle series of Γ coincide. By Lemma 13.18, the radical series is contained termwise in the primitive series. Further, using Proposition 13.22 and formulas (9.56) and (13.6) for the primitive series, and Proposition 13.67 for the radical series, we note that the above dimension formula is valid for both, hence they must coincide. For the primitive series, one may also employ Proposition 13.6 and (9.56). □

In conjunction with Exercise 9.48, we conclude that Γ is rigid but *not* uniserial for $r \geq 2$.

Exercise 13.69. The special case $k = r$ in Lemma 13.18 says that $\mathrm{rad}(\Sigma)^r \subseteq \mathcal{P}(\Sigma)$, or equivalently, $\mathsf{Lie} \subseteq \mathsf{Zie}$ by Proposition 13.66 and the Friedrichs criterion. This inclusion is strict for $r \geq 1$ (since for instance special Zie elements exist). Deduce that for Σ (viewed as a left module over itself), the radical series does not equal the primitive series for $r \geq 1$, and in particular, Σ is *not* rigid for $r \geq 1$.

13.10. Quiver of the Tits algebra

We saw that components of the two-sided Peirce decomposition of faces identify with Lie elements in arrangements over and under flats. We have also seen that Lie elements are generated in rank one. Using these two facts, one can get a handle on the powers of the radical of the Tits algebra and also compute its quiver. In a similar manner, one can also compute the quiver of the projective Tits algebra.

13.10.1. Powers of the radical.

Proposition 13.70. *We have*

$$\mathrm{rad}(\Sigma[\mathcal{A}])^i = \bigoplus_{\mathrm{rk}(\mathrm{Y}/\mathrm{X}) \geq i} \mathsf{E}_{\mathrm{X}} \cdot \Sigma[\mathcal{A}] \cdot \mathsf{E}_{\mathrm{Y}}. \tag{13.35}$$

The sum is over all $\mathrm{X} \leq \mathrm{Y}$ *such that the codimension of* X *in* Y *is greater than* i. *In particular,*

$$\mathrm{rad}(\Sigma[\mathcal{A}]) = \bigoplus_{\mathrm{X}<\mathrm{Y}} \mathsf{E}_{\mathrm{X}} \cdot \Sigma[\mathcal{A}] \cdot \mathsf{E}_{\mathrm{Y}} \tag{13.36}$$

and

$$\mathrm{rad}(\Sigma[\mathcal{A}])^r = \mathsf{E}_{\bot} \cdot \Sigma[\mathcal{A}] \cdot \mathsf{E}_{\top} = \mathsf{Lie}[\mathcal{A}]. \tag{13.37}$$

Note that this result includes Proposition 13.66.

Proof. We first establish the expression for $\mathrm{rad}(\Sigma[\mathcal{A}])$ using (13.20). Since $\mathrm{s}(\mathsf{E}_{\mathrm{X}}) = \mathsf{Q}_{\mathrm{X}}$, it follows that $\mathsf{E}_{\mathrm{X}} \cdot \Sigma[\mathcal{A}] \cdot \mathsf{E}_{\mathrm{Y}}$, for $\mathrm{X} < \mathrm{Y}$, belongs to the kernel of the support map, and hence to the radical. Further, by Proposition 13.50, $\mathsf{E}_{\mathrm{X}} \cdot \Sigma[\mathcal{A}] \cdot \mathsf{E}_{\mathrm{X}}$ is one-dimensional and maps isomorphically to the linear span of Q_{X}. The claim about $\mathrm{rad}(\Sigma[\mathcal{A}])$ follows. We deduce from here that $\mathrm{rad}(\Sigma[\mathcal{A}])^i$ is contained in the sum of $\mathsf{E}_{\mathrm{X}} \cdot \Sigma[\mathcal{A}] \cdot \mathsf{E}_{\mathrm{Y}}$ over all $\mathrm{rk}(\mathrm{Y}/\mathrm{X}) \geq i$. Conversely, by Lemma 13.64, any such term is contained in $\mathrm{rad}(\Sigma[\mathcal{A}])^i$ since it is generated by terms of the form $\mathsf{E}_{\mathrm{X}} \cdot \Sigma[\mathcal{A}] \cdot \mathsf{E}_{\mathrm{Y}}$, with $\mathrm{X} \lessdot \mathrm{Y}$. ☐

Exercise 13.71. Combine (13.21) and (13.22) respectively with (13.36) to deduce the results of Exercises 9.35 and 13.46.

Exercise 13.72. Combine (13.21), (13.22) and (13.37) to deduce Lemma 10.20.

13.10.2. Quiver. Recall from Section D.8.9 that every elementary algebra has an associated quiver.

Theorem 13.73. *The quiver of the Tits algebra is as follows. The vertices are flats, and there is exactly one arrow from* Y *to* X *when* $X \lessdot Y$, *and no arrows otherwise. In other words, the quiver is the Hasse diagram of the poset of flats.*

PROOF. The split-semisimple quotient of the Tits algebra is the Birkhoff algebra. Hence the vertices of its quiver are flats. The arrows can be computed from (13.35). Put $J := \mathrm{rad}(\Sigma[\mathcal{A}])$. Note that

$$J/J^2 \cong \bigoplus_{X \lessdot Y} \mathtt{E}_X \cdot \Sigma[\mathcal{A}] \cdot \mathtt{E}_Y.$$

Thus, $\mathtt{E}_X \cdot (J/J^2) \cdot \mathtt{E}_Y$ is zero unless $X \lessdot Y$. Let us assume this to be the case. Then by Proposition 13.50, $\mathtt{E}_X \cdot (J/J^2) \cdot \mathtt{E}_Y$ is isomorphic to $\mathsf{Lie}[\mathcal{A}^Y_X]$ which is one-dimensional. □

Exercise 13.74. Compare the quivers of the Tits algebra and the algebra of upper triangular matrices to deduce the result of Exercise 9.25. (See Exercise C.15 in this regard.)

13.10.3. Projective Tits algebra. We assume that the field characteristic is not 2.

Theorem 13.75. *The quiver of the projective Tits algebra is as follows. The vertices are flats, and there are* $|\mu(X, Y)|$ *arrows from* Y *to* X *when* $\mathrm{rk}(Y/X) = 2$, *and no arrows otherwise.*

PROOF. The split-semisimple quotient of the projective Tits algebra is the Birkhoff algebra. Hence the vertices of its quiver are flats. Let J denote the radical of the projective Tits algebra. Then

$$J/J^2 \cong \bigoplus_{\mathrm{rk}(Y/X)=2} \mathsf{Lie}[\mathcal{A}^Y_X].$$

(Since Lie elements are generated in rank 1, it follows that projective Lie elements are generated in rank 2.) If $\mathrm{rk}(Y/X) = 2$, then $\mathtt{E}_X \cdot (J/J^2) \cdot \mathtt{E}_Y$ is isomorphic to $\mathsf{Lie}[\mathcal{A}^Y_X]$ which has dimension $|\mu(X, Y)|$. In all other cases, $\mathtt{E}_X \cdot (J/J^2) \cdot \mathtt{E}_Y$ is zero. □

Exercise 13.76. Write down the analogue of Proposition 13.70 for the projective Tits algebra. In particular, check that the nilpotency index of its radical is as stated in Exercise 9.28.

13.11. Applications of Peirce decompositions to Loewy series

The two-sided Peirce decomposition of the powers of the radical of the Tits algebra can be used to study the different Loewy series of modules that we have been considering. We elaborate this technique by reproving some of the earlier results as well as obtaining new ones. In particular, we show that the right module of Zie elements is rigid. This is a companion of the result that the left module of chambers is rigid.

13.11.1. Primitive and decomposable series are Loewy. We begin by giving quick proofs of Propositions 13.4 and 13.6.

Lemma 13.77. *For any left* Σ*-module* h *and Eulerian family* $\mathtt{E}$*,*

$$\mathrm{rad}(\Sigma)\cdot(\mathtt{E}_\mathrm{X}\cdot\mathsf{h})\subseteq\bigoplus_{\mathrm{Y}:\,\mathrm{Y}<\mathrm{X}}\mathtt{E}_\mathrm{Y}\cdot\mathsf{h}.$$

In particular,

$$\mathrm{rad}(\Sigma)\cdot\mathtt{E}_\mathrm{X}\subseteq\bigoplus_{\mathrm{Y}:\,\mathrm{Y}<\mathrm{X}}\mathtt{E}_\mathrm{Y}\cdot\Sigma.$$

Similar statement holds for a right Σ*-module with the sum over all* Y *with* $\mathrm{Y}>\mathrm{X}$.

PROOF. The second fact follows from (13.36). Multiplying by h on the right then yields the first fact. □

SECOND PROOF OF PROPOSITION 13.6. Proposition 13.22 and Lemma 13.77 together imply that $\mathrm{rad}(\Sigma)\cdot\mathcal{P}_{i+1}(\mathsf{h})\subseteq\mathcal{P}_i(\mathsf{h})$. So $\mathrm{rad}(\Sigma)$ acts by zero on the quotient $\mathcal{P}_{i+1}(\mathsf{h})/\mathcal{P}_i(\mathsf{h})$. Further, it can be identified with the direct sum of $\mathtt{E}_\mathrm{X}\cdot\mathsf{h}$ over all X of rank i. Observe that the summand $\mathtt{E}_\mathrm{X}\cdot\mathsf{h}$ only contains the simple module with multiplicative character χ_X. So the multiplicity of this simple module is the dimension of $\mathtt{E}_\mathrm{X}\cdot\mathsf{h}$ which is $\eta_\mathrm{X}(\mathsf{h})$ by (13.6). □

Similarly, one can give a second proof of Proposition 13.4 by employing Proposition 13.24 and Lemma 13.77 (for a right Σ-module).

13.11.2. Sufficient condition for equality of radical and primitive series. Recall from general theory or explicitly from Lemma 13.18 that the radical series is contained termwise in the primitive series. We now give a sufficient condition for equality to hold.

Consider the following condition on a left Σ-module h:

$$\Gamma\cdot\mathsf{h}=\mathsf{h}. \tag{13.38}$$

Recall that Γ is the ideal of chambers. The lhs is the submodule of h spanned by elements of the form $z\cdot h$ with $z\in\Gamma$ and $h\in\mathsf{h}$.

Proposition 13.78. *Let* h *be a left* Σ*-module which satisfies condition* (13.38)*. Then, for any Eulerian family* $\mathtt{E}$*,*

$$\mathrm{rad}(\Sigma)^i\cdot\mathsf{h}=\bigoplus_{\mathrm{X}:\,\mathrm{rk}(\mathrm{X})\le r-i}\mathtt{E}_\mathrm{X}\cdot\mathsf{h}=\mathcal{P}_{r-i+1}(\mathsf{h}).$$

In particular, the radical series and the primitive series of h *coincide.*

PROOF. By applying the decomposition of $\mathrm{rad}(\Sigma)^i$ in (13.35) on h, we deduce that $\mathrm{rad}(\Sigma)^i\cdot\mathsf{h}$ is contained in the direct sum above. This can also be deduced from Lemma 13.18 and Proposition 13.22. This is for any h. For the reverse containment, we use condition (13.38) to obtain

$$\mathtt{E}_\mathrm{X}\cdot\mathsf{h}=\mathtt{E}_\mathrm{X}\cdot\Gamma\cdot\mathsf{h}=\mathtt{E}_\mathrm{X}\cdot\Sigma\cdot\mathtt{E}_\top\cdot\mathsf{h}\subseteq\mathrm{rad}(\Sigma)^i\cdot\mathsf{h}$$

whenever X has codimension greater than i. □

Note that the left module of chambers Γ satisfies condition (13.38). This gives another way to obtain Proposition 13.67, namely, use Proposition 13.78 in conjunction with Proposition 13.6.

Starting with any left Σ-module h, set $\mathsf{k} := \Gamma \cdot \mathsf{h}$. This defines a new left Σ-module which indeed satisfies the condition $\Gamma \cdot \mathsf{k} = \mathsf{k}$. For example, starting with $\mathsf{h} = \Sigma$ yields $\mathsf{k} = \Gamma$.

Something similar can be done for right Σ-modules. This is summarized in the exercise below.

Exercise 13.79. Show that: For the right ideal Zie, the radical series and decomposable series coincide. (Use (13.22), (13.35) and Proposition 13.24.) The same is true for any right Σ-module h which satisfies the condition

$$\mathsf{h} \cdot \mathsf{Zie} = \mathsf{h}.$$

Exercise 13.80. Check that: For a semisimple left Σ-module h, the submodule $\Gamma \cdot \mathsf{h}$ is the invariant subspace of h for the multiplicative character $\chi_{\top}$. Similarly, for a semisimple right Σ-module h, the submodule $\mathsf{h} \cdot \mathsf{Zie}$ is the invariant subspace of h for the multiplicative character $\chi_{\perp}$.

13.11.3. Rigidity of the right module of Zie elements.

We now compare decomposable and socle series.

Lemma 13.81. *The decomposable series and the socle series of Σ (viewed as a right module over itself) coincide. In particular,*

$$\mathcal{D}_r(\Sigma) = \operatorname{soc}(\Sigma) = \Gamma.$$

Further, the (right) socle of Σ is homogeneous.

Proof. We employ Propositions 13.24 and 13.70. We have $\mathcal{D}_{r-k+1}(\Sigma) \subseteq \operatorname{soc}_k(\Sigma)$. For the reverse inclusion: Let $z \in \operatorname{soc}_k(\Sigma)$, that is, z is annihilated on the right by $\operatorname{rad}(\Sigma)^k$. Now write $z = \sum_{X \leq Y} z_{X,Y}$ where $z_{X,Y} \in \mathtt{E}_X \cdot \Sigma \cdot \mathtt{E}_Y$. Let Y be any flat with $\operatorname{rk}(Y) \leq r - k$. We want to show that $z_{X,Y} = 0$. For this, use Exercise 13.52 to pick a nonzero $w \in \mathtt{E}_Y \cdot \Sigma \cdot \mathtt{E}_Z$ with $\operatorname{rk}(Z/Y) = k$. The hypothesis implies that $z_{X,Y} \cdot w = 0$. So again using Exercise 13.52, $z_{X,Y} = 0$ as required.

The homogeneity of the socle can be seen directly or as a special case of Lemma 13.2. □

In a similar manner, by employing (13.22), we obtain:

Lemma 13.82. *The decomposable series and the socle series of* Zie *coincide. In particular,*

$$\mathcal{D}_r(\mathsf{Zie}) = \operatorname{soc}(\mathsf{Zie}) = \mathsf{Lie}.$$

Further, the socle of Zie *is homogeneous.*

Theorem 13.83. *The right module* Zie *is rigid. That is, its radical series, decomposable series and socle series all coincide.*

Proof. Combine Exercise 13.79 and Lemma 13.82. □

Exercise 13.84. Use Lemma 13.81 to deduce the converse of Lemma 13.17 which says: Suppose z is an element of the Tits algebra which is annihilated on the right by $\operatorname{rad}(\Sigma)^k$ for some $0 \leq k \leq r$. Then z can be written as a linear combination of faces each of rank at least $r - k + 1$. The special case $k = 1$ says: if z is annihilated on the right by $\operatorname{rad}(\Sigma)$, then z can be written as a linear combination of chambers.

Exercise 13.85. Give alternative proofs of Lemmas 13.12 and 13.13 by using the proof method of Lemma 13.81.

13.11.4. Series comparison. Here is a summary of how the different series relate for the modules of faces, chambers and Zie elements. We assume that $r \geq 1$.

Module	Series comparison
${}_{\Sigma}\Sigma$	Radical series $\neq$ Primitive series $=$ Socle series
Σ_{Σ}	Radical series $\neq$ Decomposable series $=$ Socle series
${}_{\Sigma}\Gamma$	Radical series $=$ Primitive series $=$ Socle series
Zie_{Σ}	Radical series $=$ Decomposable series $=$ Socle series

We have used subscripts to distinguish left and right modules. We point out that $\mathsf{Lie} \neq \mathsf{Zie}$ and $\mathsf{Lie} \neq \Gamma$ suffice for equality to fail in the first and second row, respectively.

13.11.5. Socles. The Tits algebra has both a left and a right socle. Further, the left socle being a right ideal has a right socle, while the right socle being a left ideal has a left socle. These socles are summarized below.

$$\operatorname{soc}({}_{\Sigma}\Sigma) = \mathsf{Zie}, \quad \operatorname{soc}(\Sigma_{\Sigma}) = \Gamma, \quad \operatorname{soc}(\mathsf{Zie}_{\Sigma}) = \mathsf{Lie} = \operatorname{soc}({}_{\Sigma}\Gamma). \tag{13.39}$$

As a consequence,

$$\begin{aligned} \mathsf{Lie} &= \operatorname{soc}({}_{\Sigma}\Sigma) \cap \operatorname{soc}(\Sigma_{\Sigma}) \\ &= \{z \in \Sigma \mid x \cdot z = 0 = z \cdot x \text{ for all } x \in \operatorname{rad}(\Sigma)\}. \end{aligned}$$

Notes

Primitive series. Modules over the Tits algebra are intimately connected to cocommutative connected bialgebras. (See notes to Chapter 9.) The primitive series of a left module over the Tits algebra corresponds to the coradical filtration of a cocommutative connected bialgebra. The decomposable series of a right module corresponds to the augmentation filtration of a commutative connected bialgebra (the filtration by powers of the augmentation ideal). The fact that the primitive series is Loewy (Proposition 13.6) corresponds to the fact that the associated graded bialgebra to a connected bialgebra is always commutative (a result of Sweedler [**410**, Thm. 11.2.5.a]).

Quiver of the Tits algebra. The Cartan invariants and quiver of the Tits algebra of the braid arrangement were computed by Schocker [**376**, Theorems 6.4 and 8.1] by studying the combinatorics of a specific Q-basis and Eulerian family that he constructed. This is elaborated in Exercise 14.70 and involves elements of classical Lie theory. Schocker in his Section 7 also describes the radical series of the Tits algebra in terms of his chosen Eulerian family.

The Cartan invariants and quiver for an arbitrary arrangement (Proposition 13.53 and Theorem 13.73) were computed by Saliola [**368**, Proposition 6.4 and Corollary 8.4]. Lemma 13.40 is given in the proof of his Proposition 6.2. The two-sided Peirce decomposition (13.20) plays an important role in his work. Lemma 13.64 is contained in [**367**, Lemma 6.5]. Saliola does not mention any connection with Lie theory, but our argument for Lemma 13.64 is essentially the same as his. We also point out that the quiver relations described in [**368**, Proposition 8.5] or [**367**, Lemma 6.6] are what correspond to the Jacobi identities. For more on this, see the notes to Chapter 14.

The action of $\Sigma[\mathcal{A}]$ on $\Gamma[\mathcal{A}^{Y}]$ in Lemma 13.40 is present in earlier work of Brown and Diaconis [**105**, page 1844]. In the semigroup literature, this action is called the (left) Schützenberger representation associated to the flat Y. For more information, see [**303**, Section 5.2], [**304**, Section 4.3.2] and [**405**, Section 5.2].

Starting with any basis for the space of Lie elements of $\mathcal{A}_{X}^{Y}$, one obtains a basis for the space $\mathtt{E}_{X} \cdot \Sigma \cdot \mathtt{E}_{Y}$. More arbitrarily, by Lemma 13.63, one obtains a spanning set for this space by iterated substitution of Lie elements in maximal chains which start at X and end in Y. The special case when Y is the maximal flat is of interest, since $\mathtt{E}_{X} \cdot \Sigma \cdot \mathtt{E}_{\top}$ functions as an eigenspace for the random walk on chambers. In this regard, see [**370**, Section 4] and [**143**, Theorem 3.4].

Proposition 13.50 and Theorem 13.54 are valid for any LRB. We mention that the quiver of the free left regular band was computed by Brown, and this result was generalized to all left regular bands by Saliola [**366**, Theorems 6.2, 8.1 and 13.1]. For related results and generalizations, see [**146**], [**304**, Sections 4.5, 6 and 9.2], [**305**], [**306**] and [**405**, Chapter 17].

CHAPTER 14

Dynkin idempotents

We discuss a construction of special Zie elements of an arrangement $\mathcal{A}$. The starting data is a generic half-space h. The Zie element is defined as an alternating sum of faces contained in h. We call this the Dynkin element associated to h. Its action on chambers gives rise to a basis of the space of Lie elements which we call the Dynkin basis. This yields another proof of the fact that the dimension of the space of Lie elements is the absolute value of the Möbius number of $\mathcal{A}$. We establish certain key properties of Dynkin elements and present two applications to the theory of finite affine hyperplane arrangements. One is a derivation of the Zaslavsky formula for the number of bounded chambers of an affine arrangement. The other is a more surprising result: that the semigroup algebra of such an arrangement possesses a unit element.

We discuss the notion of orientation of an arrangement in terms of maximal chains in its poset of faces. We then prove the Joyal-Klyachko-Stanley (JKS) theorem which identifies, up to orientation, the top-cohomology of the lattice of flats with the space of Lie elements. In effect, it says that the space of Lie elements is freely generated by the orientation space in rank one subject to the Jacobi identities in rank two. (There is also an analogue of the JKS theorem which relates the top-cohomology of the poset of faces with the space of chambers.) The dual of the Dynkin basis is, up to orientation, a basis for the top-homology of the lattice of flats. This is the Björner-Wachs basis. We also discuss another pair of dual bases, the Björner basis for top-homology and the Lyndon basis for top-cohomology. The starting data for this is an ordered coordinate chart. A summary is given below.

Starting data	Top-homology	Lie
Generic half-space	Björner-Wachs basis	Dynkin basis
Ordered coordinate chart	Björner basis	Lyndon basis

We discuss examples in the later sections of the chapter. We begin with rank-two and coordinate arrangements. We then treat the important case of the braid arrangement, which motivates most of our terminology. Contact with the classical Lie bracket, antisymmetry, Jacobi identity, Lie operad, binary trees, the classical JKS theorem, the Dynkin-Specht-Wever theorem and the Lyndon basis is made here. A similar discussion is given for the arrangement of type B.

14.1. Dynkin elements

A Dynkin element is a special Zie element constructed from a generic half-space.

14.1.1. Dynkin element. Generic half-spaces are discussed in Section 1.9.1. Let h be a generic half-space wrt $\mathcal{A}$, and let H denote its bounding hyperplane. Now define

$$\theta_{\mathrm{h}} := \sum_{F:\, F \subseteq \mathrm{h}} (-1)^{\mathrm{rk}(F)}\, \mathtt{H}_F \quad \in \Sigma[\mathcal{A}]. \tag{14.1}$$

The sum is over all faces F of $\mathcal{A}$ which are contained in the fixed half-space h. These are precisely those faces of $\mathcal{A}$ which are not cut by H and which are on the h-side of H. We refer to θ_{h} as the *Dynkin element* associated to the generic half-space h. The central face is contained in h and since its rank is zero, it appears in θ_{h} with coefficient 1.

Note that the Dynkin elements associated to two half-spaces coincide iff the two half-spaces are equivalent. Thus, the number of distinct Dynkin elements is the number of equivalence classes of generic half-spaces which by Lemma 1.54, is also the number of chambers in the adjoint arrangement $\widehat{\mathcal{A}}$.

For any generic h and any non-minimum flat X, $\mathrm{h} \cap \mathrm{X}$ is a generic half-space wrt the arrangement $\mathcal{A}^{\mathrm{X}}$. Note that the Dynkin element $\theta_{\mathrm{h}\cap\mathrm{X}}$ is the truncation of θ_{h} to faces smaller than X, that is,

$$\theta_{\mathrm{h}\cap\mathrm{X}} = (\theta_{\mathrm{h}})^{\mathrm{X}}, \tag{14.2}$$

with notation as in (12.9).

Proposition 14.1. *For any generic half-space* h, *the Dynkin element* θ_{h} *is a special Zie element. In particular, it is an idempotent.*

PROOF. Let $\mathcal{A}'$ denote the arrangement obtained by adding the base of h to $\mathcal{A}$. It is discussed in Section 1.9.1.

To show that θ_{h} is a Zie element, we verify that it satisfies the Ree criterion (10.17b). Accordingly, let L be any non-singleton combinatorial lune of $\mathcal{A}$. Equivalently, the base of L is not the minimum flat. We need to show that

$$\sum_{F:\, F \in \mathrm{Cl}(\mathrm{L}), F \subseteq \mathrm{h}} (-1)^{\mathrm{rk}(F)} = 0.$$

In view of (14.2), by passing to the arrangement under the case of L, we may assume that L is a top-lune. Let L' be the combinatorial top-lune of $\mathcal{A}'$ whose underlying geometric cone is the same as that of L. We claim that the base of h must cut L': If not, then either h or $\overline{\mathrm{h}}$ contains L'. This implies that $\mathrm{b}(\mathrm{h})$ contains $\mathrm{b}(\mathrm{L}')$ which contradicts the fact that $\mathrm{b}(\mathrm{h})$ is a generic hyperplane. As a consequence, $\mathrm{L}' \wedge \mathrm{h}$ contains a chamber and in particular, it cannot be a flat. Hence by Proposition 2.27, $\mathrm{L}' \wedge \mathrm{h}$ must be a topological ball. So its reduced Euler characteristic is zero. This yields:

$$\sum_{F \in \mathrm{Cl}(\mathrm{L}' \wedge \mathrm{h})} (-1)^{\mathrm{rk}(F)} = 0.$$

This is almost what we want. The only difference is that here we have also counted faces contained in $\mathrm{b}(\mathrm{h})$ and their adjoining faces contained in h. (These are among those faces which were created by the addition of h.) However their contribution cancels since the ranks of corresponding faces differ by one. Thus, θ_{h} is a Zie element, as required. Since its coefficient of the central face is 1, it is special.

The above argument is illustrated in the following diagram.

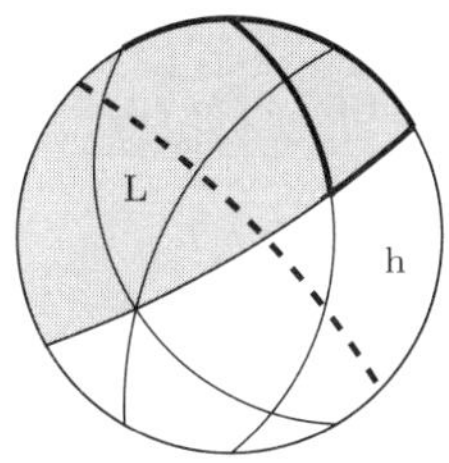

The generic hyperplane is shown dotted, and the half-space h is the region to the right of it. The lune L is the shaded region. It is bounded by two semicircles and is fully visible. The faces in Cl(L) which are contained in h are those either on the thick lines or inside the region defined by the thick lines (consisting of a triangle and a quadrilateral). Topologically, this set is a ball with an edge hanging out. □

Corollary 14.2. *The number of chambers contained in any generic half-space wrt* $\mathcal{A}$ *is given by* $|\mu(\mathcal{A})|$.

PROOF. Apply (10.15) to the special Zie element θ_{h} for the flat $\mathrm{X} = \top$. □

Proposition 14.3. *Let* $z = \sum_{F \subseteq \mathrm{h}} x^F\, \mathtt{H}_F$ *for some generic half-space* h. *Then* z *is a special Zie element iff* $z = \theta_{\mathrm{h}}$.

PROOF. The backward implication was proved in Proposition 14.1. For the forward implication: By induction on rank, we may assume that $x^F = (-1)^{\mathrm{rk}(F)}$ for all faces F which are not chambers. Then $z - \theta_{\mathrm{h}}$ is a Zie element, which is a linear combination of chambers, so it is a Lie element by Lemma 10.20. Since it is entirely contained in an half-space, we conclude from Lemma 10.11 that it must be 0. Thus, $z = \theta_{\mathrm{h}}$ as required. □

Corollary 14.4. *Let* h *be a generic half-space, and* $z = \sum_{F \subseteq \mathrm{h}} x^F\, \mathtt{H}_F$ *be such that*

$$\sum_{F:\, \mathrm{s}(F)=\mathrm{X},\, F\subseteq \mathrm{h}} x^F = \mu(\bot, \mathrm{X})$$

for all flats X. *Then*

$$\theta_{\mathrm{h}} = (\mathtt{H}_O - (\mathtt{H}_O - z)^{r+1})^{r+1},$$

where $r = \mathrm{rk}(\mathcal{A})$.

By Lemma 10.18, the condition on z is equivalent to the condition $\mathrm{s}(z) = \mathtt{Q}_\bot$.

PROOF. Put $z' = (\mathtt{H}_O - (\mathtt{H}_O - z)^{r+1})^{r+1}$. Then from the claim in the first paragraph of the proof of Lemma D.28, z' is an idempotent which lifts $\mathtt{Q}_\bot$. Further, just like z, it only involves faces contained in h. From Lemma 10.24, we first deduce that z' is a special Zie element, and next from Proposition 14.3, we deduce that $z' = \theta_{\mathrm{h}}$. □

Example 14.5. Let $\mathcal{A}$ be the rank-one arrangement with chambers C and $\overline{C}$. The origin is a generic hyperplane. In this case, $\mathcal{A} = \mathcal{A}'$. Thus, there are two generic half-spaces, and $\mathtt{H}_O - \mathtt{H}_C$ and $\mathtt{H}_O - \mathtt{H}_{\overline{C}}$ are the two Dynkin elements. Note that they are special Zie elements. In this case, all preceding results can be checked directly.

14.1.2. Symmetrized Dynkin element. Suppose $\mathcal{A}$ is a reflection arrangement and W is its Coxeter group. Fix a generic half-space h wrt $\mathcal{A}$ as in the preceding discussion. Define

$$d_{\mathrm{h}} := \sum_{\mathrm{h}'} \theta_{\mathrm{h}'}, \tag{14.3}$$

where the sum is over all half-spaces h' of the form $w \cdot \mathrm{h}$ for some $w \in W$. We refer to d_{h} as the *symmetrized Dynkin element* associated to h. It is invariant under the action of W.

Proposition 14.6. *The symmetrized Dynkin element* d_{h} *is a Zie element. Further, it is a quasi-idempotent, that is,*

$$d_{\mathrm{h}}^2 = \alpha_{\mathrm{h}} d_{\mathrm{h}}, \tag{14.4}$$

where α_{h} *is the size of the orbit of* h *under* W.

PROOF. By Proposition 14.1, each $\theta_{\mathrm{h}'}$ is a Zie element. So their sum which is d_{h} is also a Zie element. This proves the first claim. In d_{h}, the coefficient of the central face is the size of the orbit of h. The second claim follows from the first by Lemma 10.22. □

14.2. Dynkin basis for the space of Lie elements

Recall the space of Lie elements $\mathsf{Lie}[\mathcal{A}]$. Any generic half-space h wrt $\mathcal{A}$ gives rise to a basis of $\mathsf{Lie}[\mathcal{A}]$. We call this the Dynkin basis. It is obtained via the action of the Dynkin element θ_{h} on those chambers which are contained in $\overline{\mathrm{h}}$.

14.2.1. Action on chambers and Lie elements. Recall that the Tits algebra $\Sigma[\mathcal{A}]$ acts on the left module of chambers $\Gamma[\mathcal{A}]$.

Proposition 14.7. *The Dynkin element* θ_{h} *is an idempotent operator and projects* $\Gamma[\mathcal{A}]$ *onto* $\mathsf{Lie}[\mathcal{A}]$.

PROOF. This follows from Propositions 10.35 and 14.1 and the Friedrichs criterion which says that $\mathsf{Lie}[\mathcal{A}]$ is the primitive part of $\Gamma[\mathcal{A}]$ (Lemma 10.5). □

The symmetrized Dynkin element d_{h} also sends $\Gamma[\mathcal{A}]$ onto $\mathsf{Lie}[\mathcal{A}]$, and acts on the latter by scalar multiplication by α_{h}, with α_{h} as in (14.4). (We are implicitly assuming characteristic 0.)

We now work towards a formula for the action of the Dynkin element on chambers. For a generic half-space h, and chambers C and D, put

$$\mathrm{A} = \{H \in \Sigma[\mathcal{A}] \mid HC = D\} \quad \text{and} \quad \mathrm{B} = \{H \in \Sigma[\mathcal{A}] \mid H \leq D, H \subseteq \mathrm{h}\}.$$

Both A and B consist of faces of D, with $D \in \mathrm{A}$ and $O \in \mathrm{B}$. Further,

$$\langle \theta_{\mathrm{h}} \cdot \mathtt{H}_C, \mathtt{H}_D \rangle = \sum_{H \in \mathrm{A} \cap \mathrm{B}} (-1)^{\mathrm{rk}(H)}. \tag{14.5}$$

The lhs denotes the coefficient of $\mathtt{H}_D$ in $\theta_{\mathrm{h}} \cdot \mathtt{H}_C$. We would like to understand the rhs.

Simplicial case. For simplicity, let us first assume the arrangement to be simplicial. For a generic half-space h and a chamber D, let $\mathrm{h}(D)$ denote the largest face of D which is contained in h.

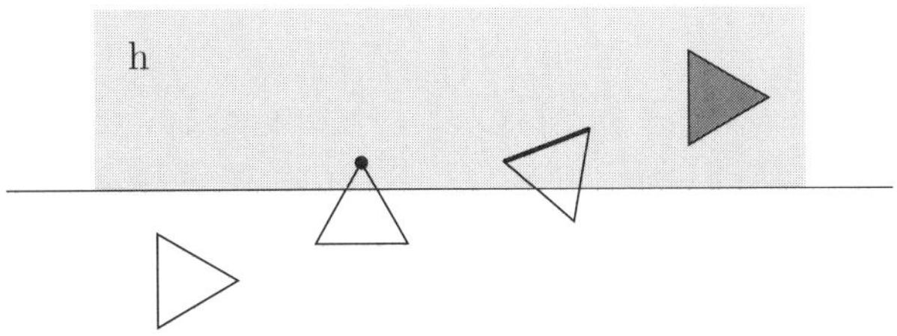

This is illustrated above in rank 3. The half-space h is the shaded region. Since the arrangement is simplicial, each chamber D is a triangle, and there are four possibilities for $\mathrm{h}(D)$ depending on how the vertices of D lie wrt h. Each case is shown separately with the face $\mathrm{h}(D)$ being the central face, a vertex, an edge or the triangle.

Lemma 14.8. *Let $\mathcal{A}$ be a simplicial arrangement. Then*

$$\theta_{\mathrm{h}} \cdot \mathtt{H}_C = \sum_{D\colon \mathrm{Des}(C,D)=\mathrm{h}(D)} (-1)^{\mathrm{rk}(\mathrm{h}(D))}\, \mathtt{H}_D. \tag{14.6}$$

PROOF. Using (7.1) and the definition of $\mathrm{h}(D)$, we obtain

$$\mathrm{A} = \{H \mid \mathrm{Des}(C,D) \le H \le D\} \quad \text{and} \quad \mathrm{B} = \{H \mid H \le \mathrm{h}(D)\}.$$

Combining and substituting in (14.5), we obtain

$$\langle \theta_{\mathrm{h}} \cdot \mathtt{H}_C, \mathtt{H}_D \rangle = \sum_{H\colon \mathrm{Des}(C,D) \le H \le \mathrm{h}(D)} (-1)^{\mathrm{rk}(H)}.$$

The indexing set (which could be empty) is a Boolean poset. So the sum will be zero unless the set is a singleton, that is, $\mathrm{Des}(C,D) = \mathrm{h}(D)$. □

Lemma 14.9. *If* $\mathrm{Des}(C,D) = \mathrm{h}(D)$, *then* $\overline{C} \subseteq \mathrm{h}$.

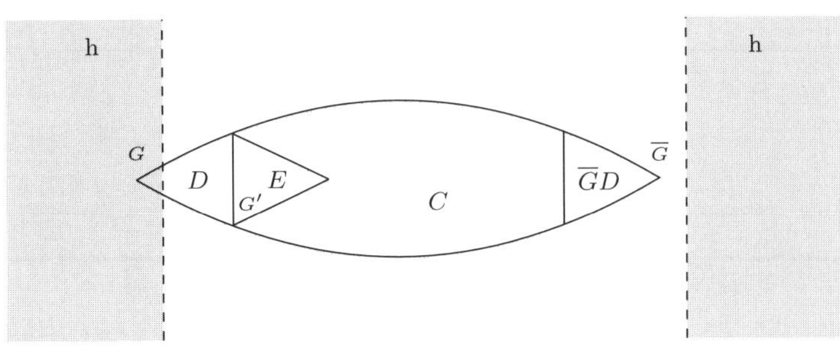

PROOF. For simplicity of notation, put $G := \mathrm{h}(D)$. Let G' be the face of D complementary to G. Then observe that

$$\mathrm{Des}(C,D) = \mathrm{h}(D) \iff C \text{ lies in the gallery interval } [E\colon \overline{G}D],$$

where E is the chamber opposite to D in the star of G'. But this entire gallery interval lies in the interior of $\overline{\mathrm{h}}$. (In the figure, the latter is the region between the two dotted lines.) So if $\mathrm{Des}(C,D) = \mathrm{h}(D)$, then $\overline{C}$ is contained in h. □

Proposition 14.10. *Let $\mathcal{A}$ be a simplicial arrangement. Then*

$$\theta_{\mathrm{h}} \cdot \mathtt{H}_C = \begin{cases} \mathtt{H}_C + (-1)^{\mathrm{rk}(\mathcal{A})}\mathtt{H}_{\overline{C}} + \sum_D (-1)^{\mathrm{rk}(\mathrm{h}(D))}\mathtt{H}_D & \text{if } \overline{C} \subseteq \mathrm{h}, \\ 0 & \text{otherwise.} \end{cases} \tag{14.7}$$

The sum is over chambers D which are cut by the base of h *(so that part of D lies in* h *and part in* $\overline{\mathrm{h}}$*) and which satisfy* $\mathrm{Des}(C, D) = \mathrm{h}(D)$.

PROOF. The second case follows from Lemmas 14.8 and 14.9. So suppose that $\overline{C} \subseteq \mathrm{h}$. Then

$$\mathrm{Des}(C, D) = \mathrm{h}(D) = O \iff D = C \quad \text{and} \quad \mathrm{Des}(C, D) = \mathrm{h}(D) = D \iff D = \overline{C}.$$

This yields the terms $\mathtt{H}_C$ and $(-1)^{\mathrm{rk}(\mathcal{A})}\mathtt{H}_{\overline{C}}$. In the remaining cases, $O < \mathrm{h}(D) < D$ and hence D is cut by the base of h. □

General case. Let us now deal with the general case where $\mathcal{A}$ is not assumed to be simplicial.

Lemma 14.11. *We have*

$$\langle \theta_{\mathrm{h}} \cdot \mathtt{H}_C, \mathtt{H}_C \rangle = \begin{cases} 1 & \text{if } \overline{C} \subseteq \mathrm{h}, \\ 0 & \text{otherwise,} \end{cases} \quad \text{and} \quad \langle \theta_{\mathrm{h}} \cdot \mathtt{H}_C, \mathtt{H}_{\overline{C}} \rangle = \begin{cases} (-1)^{\mathrm{rk}(\mathcal{A})} & \text{if } \overline{C} \subseteq \mathrm{h}, \\ 0 & \text{otherwise.} \end{cases}$$

PROOF. We work with formula (14.5). Suppose $D = C$. Then A consists of all faces of D. If $\overline{C} \subseteq \mathrm{h}$, then B is a singleton consisting of the central face and the sum in the rhs of (14.5) is 1. If not, then the sum is the negative of the reduced Euler characteristic of a ball and hence 0.

Suppose $D = \overline{C}$. Then A is a singleton consisting of D. If $D \subseteq \mathrm{h}$, then $\mathrm{A} \cap \mathrm{B}$ consists of D, and the sum in the rhs of (14.5) is $(-1)^{\mathrm{rk}(\mathcal{A})}$. If not, then $\mathrm{A} \cap \mathrm{B}$ is empty, and the sum is zero. □

Lemma 14.12. *Suppose either ($D \neq \overline{C}$ and $D \subseteq \mathrm{h}$) or ($D \neq C$ and $\overline{D} \subseteq \mathrm{h}$). Then* $\langle \theta_{\mathrm{h}} \cdot \mathtt{H}_C, \mathtt{H}_D \rangle = 0$.

PROOF. Let us consider the first case. Since $D \subseteq \mathrm{h}$, B consists of all faces of D. So the rhs of (14.5) is the same as (7.10), which is zero since $D \neq \overline{C}$.

Now consider the second case. Since $\overline{D} \subseteq \mathrm{h}$, B consists of only the central face. Further, since $D \neq C$, the central face does not belong to A. Thus $\mathrm{A} \cap \mathrm{B}$ is empty, and the sum is zero. □

Lemma 14.13. *If $\overline{C} \not\subseteq \mathrm{h}$, then* $\theta_{\mathrm{h}} \cdot \mathtt{H}_C = 0$.

PROOF. Suppose $\overline{C} \not\subseteq \mathrm{h}$. If $D \subseteq \mathrm{h}$, then $D \neq \overline{C}$, and if $\overline{D} \subseteq \mathrm{h}$, then $D \neq C$. Hence, by Lemma 14.12, $\langle \theta_{\mathrm{h}} \cdot \mathtt{H}_C, \mathtt{H}_D \rangle = 0$ holds if either $D \subseteq \mathrm{h}$ or $\overline{D} \subseteq \mathrm{h}$. So we may assume that D is cut by h. Let $\mathcal{A}'$ be the arrangement obtained by adding H, which is the base of h, to $\mathcal{A}$. Note that $C \cap \mathrm{h}$ and $D \cap \mathrm{h}$ are chambers in $\mathcal{A}'$, and they cannot be opposite since they are both contained in h. Put

$$\mathrm{A}' = \{H \in \Sigma[\mathcal{A}'] \mid H(C \cap \mathrm{h}) = D \cap \mathrm{h}\}.$$

By (7.10),

$$\sum_{H \in \mathrm{A}'} (-1)^{\mathrm{rk}(H)} = 0.$$

Now A' contains $\mathrm{A} \cap \mathrm{B}$ as a subset. In addition, it contains some faces which either lie on the panel $D \cap \mathrm{H}$ or intersect it in a face of one smaller dimension. Since

H is generic, such faces occur in pairs (for instance, $D \cap \mathrm{h}$ and $D \cap \mathrm{H}$ is one such pair). By Proposition 7.15, we deduce that either both faces in a pair belong to A′ or neither. Since their ranks differ by 1, their contribution cancels. So the rhs of (14.5) evaluates to 0. □

Proposition 14.14. *We have*

$$\theta_{\mathrm{h}} \cdot \mathtt{H}_C = \begin{cases} \mathtt{H}_C + (-1)^{\mathrm{rk}(\mathcal{A})}\mathtt{H}_{\overline{C}} + \sum_D a_{C,D}\mathtt{H}_D & \text{if } \overline{C} \subseteq \mathrm{h}, \\ 0 & \text{otherwise.} \end{cases} \tag{14.8}$$

The sum is over chambers D which are cut by the base of h, *and the $a_{C,D}$ are certain integer coefficients.*

PROOF. The second case was proved in Lemma 14.13. The first case follows from Lemmas 14.11 and 14.12. □

Exercise 14.15. Extend Lemma 14.13 as follows. For any face F, if $\overline{F} \not\subseteq \mathrm{h}$, then $\theta_{\mathrm{h}} \cdot \mathtt{H}_F = 0$.

14.2.2. Dynkin basis.

Proposition 14.16. *For any generic half-space* h *wrt $\mathcal{A}$, the set*

$$\{\theta_{\mathrm{h}} \cdot \mathtt{H}_C \mid \overline{C} \subseteq \mathrm{h}\} \tag{14.9}$$

is a basis of $\mathsf{Lie}[\mathcal{A}]$. *In particular, the dimension of* $\mathsf{Lie}[\mathcal{A}]$ *is* $|\mu(\mathcal{A})|$.

PROOF. For $\overline{C} \subseteq \mathrm{h}$, by the first case of formula (14.8), the term $\mathtt{H}_C$ only occurs in $\theta_{\mathrm{h}} \cdot \mathtt{H}_C$, so these elements are linearly independent. Further, by Proposition 14.7, these elements span $\mathsf{Lie}[\mathcal{A}]$. Hence they form a basis. The second statement then follows from Corollary 14.2. □

We call (14.9) the *Dynkin basis* associated to h.

Exercise 14.17. Show that: If the rank of $\mathcal{A}$ is even, then the Dynkin bases associated to h and $\overline{\mathrm{h}}$ coincide, and if the rank of $\mathcal{A}$ is odd, then the two are negatives of each other.

Exercise 14.18. Show that there cannot exist a nonzero Lie element which involves only chambers cut by a generic hyperplane.

Exercise 14.19. Prove Lemma 10.11 as follows. Pick a half-space h such that $\overline{\mathrm{h}}$ contains the given chamber D. Check using (14.8) that (10.5) is true for the Lie elements $\theta_{\mathrm{h}} \cdot \mathtt{H}_C$.

Exercise 14.20. Let $\mathcal{A}$ be a simplicial arrangement. For a generic half-space h, and chambers C and D, check that: $\mathrm{Des}(C, \overline{D})$ is the face of $\overline{D}$ complementary to $\overline{\mathrm{Des}(C, D)}$, while $\mathrm{h}(\overline{D})$ is the face of $\overline{D}$ complementary to $\overline{\mathrm{h}(D)}$. Check using (14.7) that (10.5) is true for the Lie elements $\theta_{\mathrm{h}} \cdot \mathtt{H}_C$.

14.3. Applications to affine hyperplane arrangements

We employ the theory of Dynkin elements to deduce the Zaslavsky formula for the number of bounded chambers of an affine arrangement. We also prove that the semigroup algebra of such an arrangement does possess a unit element.

14.3.1. Affine hyperplane arrangements. An *affine hyperplane arrangement* is a finite set of affine hyperplanes in a finite-dimensional real vector space V. An *affine hyperplane* is a translate of a (linear) hyperplane. An affine arrangement is *central* if the intersection of all hyperplanes is nonempty.

Faces and flats are defined as in Chapter 1. They are nonempty subsets of V. The set of faces is a semigroup under the Tits product. It does not possess a unit element unless the arrangement is central. The set of flats is a join-semilattice.

An affine arrangement is essential if the minimal faces are points (geometrically). There is no loss in generality in restricting attention to essential arrangements. For instance, the arrangement of three parallel lines in the plane is not essential. It can be made essential by cutting with a transverse line. Faces of the original arrangement are in correspondence with faces of the essentialization. The rank of a face is the dimension of the corresponding face in the essentialization. The minimal faces are of rank 0. The same considerations apply to flats.

Let $\mathcal{A}$ be an affine arrangement in V. Build a vector space V' in which V sits as a nonlinear hyperplane. The linear span in V' of an affine hyperplane in V is then a linear hyperplane in V'. The *coning* of $\mathcal{A}$ is the linear arrangement $\mathcal{A}'$ in V' consisting of the linear spans of the affine hyperplanes in $\mathcal{A}$.

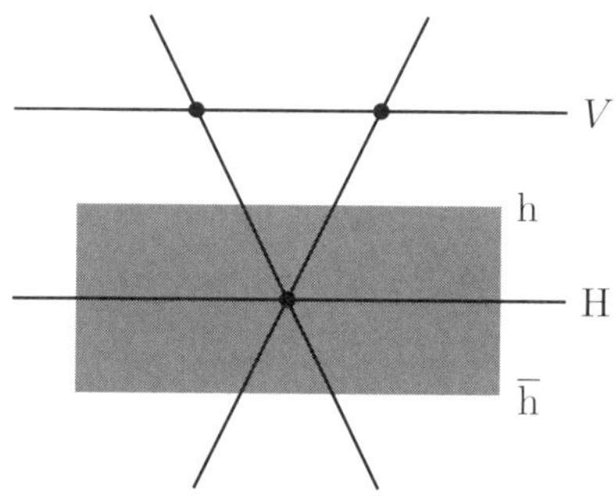

Let H be the linear hyperplane in V' which is parallel to V. Note that H is generic wrt $\mathcal{A}'$. Let h be the half-space in V' with base H which contains V. Note also that faces of $\mathcal{A}$ are in correspondence with those faces of $\mathcal{A}'$ which are not contained in the half-space opposite to h. Moreover, the Tits semigroup $\Sigma[\mathcal{A}]$ identifies in this manner with the subsemigroup of $\Sigma[\mathcal{A}']$ consisting of the latter faces.

14.3.2. Zaslavsky formula for the number of bounded chambers.

Theorem 14.21. *For any essential affine arrangement $\mathcal{A}$, the number of bounded chambers equals*

$$|\sum_{\mathrm{X}} \mu(\mathrm{X}, \top)|,$$

where X *runs over all flats of $\mathcal{A}$.*

Proof. Let $\mathcal{A}'$, H and h be as above. Observe that the chambers of $\mathcal{A}'$ contained in h correspond to the bounded chambers of $\mathcal{A}$. By Corollary 14.2, their cardinality is $|\mu(\mathcal{A}')|$. Now by (C.5b),

$$\mu(\mathcal{A}') = -\sum_{\mathrm{X}} \mu(\mathrm{X}, \top),$$

where X varies over all non-minimum flats of $\mathcal{A}'$. But the non-minimum flats of $\mathcal{A}'$ correspond to flats of $\mathcal{A}$. The result follows. □

SECOND PROOF. A direct proof along the lines of the argument in Theorem 1.81 is sketched below. It uses the fact that the cell complex of all bounded faces of an essential affine arrangement is contractible [**82**, Theorem 4.5.7, part (ii)]. Augment the poset of flats of $\mathcal{A}$ by a minimum element $\bot$. Let a^{X} denote the number of bounded faces of support X. By convention $a^{\bot} = 1$. Put

$$f(\mathrm{X}) := (-1)^{\mathrm{rk}(\mathrm{X})} a^{\mathrm{X}}.$$

Then

$$g(\mathrm{Y}) := \sum_{\mathrm{X}:\, \mathrm{X} \leq \mathrm{Y}} f(\mathrm{X}) = 0,$$

being the reduced Euler characteristic of a contractible cell complex. Hence by Möbius inversion (C.12), $f(\top) = \mu(\bot, \top)$, which yields the desired formula. □

Theorem 14.21 continues to hold for nonessential arrangements with the count now yielding the number of *essentially bounded chambers* (those that correspond to bounded chambers of the essentialization). By passing to the arrangement under a flat, we deduce the following result.

Corollary 14.22. *Let* Y *be a flat of an affine arrangement. The number of essentially bounded faces of support* Y *equals*

$$\Big| \sum_{\mathrm{X}:\, \mathrm{X} \leq \mathrm{Y}} \mu(\mathrm{X}, \mathrm{Y}) \Big|.$$

14.3.3. The unit of the semigroup algebra.

Theorem 14.23. *The semigroup algebra of an affine arrangement $\mathcal{A}$ possesses a unit element. The unit is the element*

$$v = \sum_{F} (-1)^{\mathrm{rk}(F)} \mathtt{H}_F,$$

with F running over the set of essentially bounded faces of $\mathcal{A}$.

PROOF. Let $\mathcal{A}'$, H and h be as above. We carry out the calculation in the subalgebra of $\Sigma[\mathcal{A}']$ linearly spanned by faces of $\mathcal{A}'$ not contained in $\overline{\mathrm{h}}$. A face of rank i in $\mathcal{A}$ corresponds to a face of rank $i+1$ in $\mathcal{A}'$. The essentially bounded faces of $\mathcal{A}$ correspond to the faces of $\mathcal{A}'$ contained in h, except for the central face of $\mathcal{A}'$. Therefore, v corresponds precisely to the element $\mathtt{H}_O - \theta_{\mathrm{h}}$ of $\Sigma[\mathcal{A}']$. To show this is the required unit, we need to check that

$$\mathtt{H}_F \cdot \theta_{\mathrm{h}} = 0 = \theta_{\mathrm{h}} \cdot \mathtt{H}_F$$

for all faces F of $\mathcal{A}'$ not contained $\overline{\mathrm{h}}$. The first equality is clear since θ_{h} is a Zie element of $\mathcal{A}'$ (Proposition 14.1), and F is not the central face of $\mathcal{A}'$. For the second equality, apply Exercise 14.15. □

The set of flats $\Pi[\mathcal{A}]$ is a join-semilattice, hence a commutative semigroup. There is a top element $\top$ but no bottom element (and no unit for the semigroup) unless the arrangement is central. In the semigroup algebra, we may define a $\mathtt{Q}$-basis as in (9.1). Just as before, the $\mathtt{Q}_{\mathrm{X}}$ turn out to be orthogonal idempotents. Hence the semigroup algebra $\mathsf{\Pi}[\mathcal{A}]$ possesses a unit, namely, the sum of all the $\mathtt{Q}_{\mathrm{X}}$.

Exercise 14.24. The support map is surjective, hence it preserves units: $\mathrm{s}(v) = \sum_{\mathrm{X}} \mathtt{Q}_{\mathrm{X}}$.

(i) Deduce this fact from (10.16).
(ii) Deduce the result in Corollary 14.22 from this fact.

Exercise 14.25. In the semigroup algebra of an affine arrangement, starting with a homogeneous section, define the $\mathtt{E}_{\mathrm{X}}$ by formula (11.14). Check that the Saliola lemma (Lemma 11.12) continues to hold. Deduce that the sum of all the $\mathtt{E}_{\mathrm{X}}$ is a right identity for the semigroup algebra, and hence equals the unit v in Theorem 14.23.

14.4. Orientation space

We discuss the notion of orientation for any arrangement.

14.4.1. Orientation space. For any arrangement $\mathcal{A}$, let $\mathsf{E}^{\mathsf{o}}[\mathcal{A}]$ denote the space spanned by maximal chains in the poset of faces $\Sigma[\mathcal{A}]$ subject to the relations: If two maximal chains differ in exactly one position, then they are negatives of each other. We call $\mathsf{E}^{\mathsf{o}}[\mathcal{A}]$ the *orientation space* of $\mathcal{A}$. We denote the image of a maximal chain f in the orientation space by $[f]$. An *orientation* of $\mathcal{A}$ is an element of $\mathsf{E}^{\mathsf{o}}[\mathcal{A}]$ of the form $[f]$ for some maximal chain f.

Example 14.26. Let $\mathcal{A}$ be the rank-one arrangement with chambers C and $\overline{C}$. There are two maximal chains, namely, $O \lessdot C$ and $O \lessdot \overline{C}$. Since they differ in exactly one position, we write

$$[O \lessdot C] = -[O \lessdot \overline{C}].$$

So $\mathsf{E}^{\mathsf{o}}[\mathcal{A}]$ is one-dimensional. It has two orientations, namely, $[O \lessdot C]$ which we call the right orientation, and $[O \lessdot \overline{C}]$ which we call the left orientation.

Example 14.27. Let $\mathcal{A}$ be the rank-two arrangement of n lines. A maximal chain has the form $O \lessdot P \lessdot C$. There are $4n$ maximal chains. The relations can be expressed as

$$[O \lessdot P \lessdot C] = -[O \lessdot Q \lessdot C],$$

where P and Q are the two vertices of C, and

$$[O \lessdot P \lessdot C] = -[O \lessdot P \lessdot D],$$

where C and D are the two chambers greater than P. Again we note that $\mathsf{E}^{\mathsf{o}}[\mathcal{A}]$ is one-dimensional. There are two orientations, which we can think of as clockwise and anticlockwise. This is illustrated below for $n = 3$.

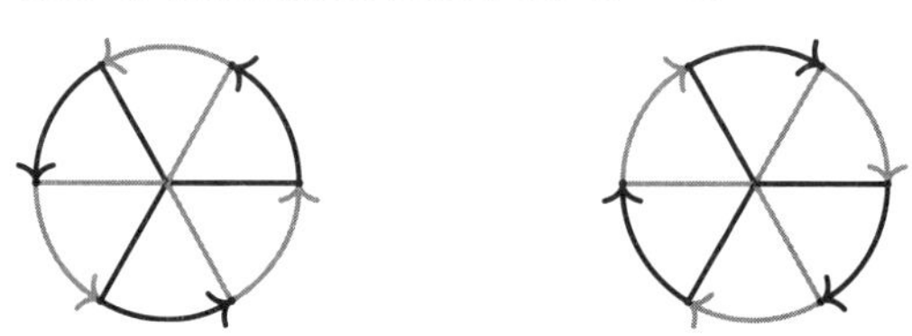

The six maximal chains which give the anticlockwise orientation are shown on the left, while the six which give the clockwise orientation are shown on the right.

14.4.2. Connection with orientation of real vector spaces. Recall the familiar notion of orientation of a finite-dimensional real vector space W. An orientation is an equivalence class of ordered bases of W, where two bases are equivalent if the determinant of the linear transformation that takes one to the other is positive. Thus, W has two orientations, say σ and τ. Let $\mathrm{Det}(W)$ denote the one-dimensional space spanned by σ and τ subject to the relation $\sigma = -\tau$.

Lemma 14.28. *The orientation space of any arrangement is one-dimensional. In fact,*

$$\mathsf{E}^{\mathsf{o}}[\mathcal{A}] \xrightarrow{\cong} \mathrm{Det}(\top/\bot), \tag{14.10}$$

where $\top/\bot$ is the quotient of the maximum flat (ambient space) by the minimum flat (center) of $\mathcal{A}$.

PROOF. Let $O \lessdot F_1 \lessdot \cdots \lessdot F_n$ be a maximal chain of faces. For each F_i, choose any vector in the ambient space in the interior of F_i. Let v_{F_i} denote its image in the quotient $\top/\bot$. Then the ordered basis $(v_{F_1}, \ldots, v_{F_n})$ represents an orientation of $\top/\bot$. This orientation does not depend on the specific choice of the vectors. Moreover, if two maximal chains differ in exactly one position, then they give rise to opposite orientations. This yields the map (14.10). It is surjective by construction. Further, by Lemma 1.34 and Proposition B.10, one can pass from one maximal chain to another by a sequence of maximal chains in which two consecutive chains differ in exactly one position. Hence (14.10) must be an isomorphism. □

Thus, any arrangement has two orientations. We will use the letter σ to denote an orientation; the opposite orientation will be $-\sigma$.

Exercise 14.29. Check that a half-flat of $\mathcal{A}$ is the same as a triple $(\mathrm{X}, \sigma, \mathrm{Y})$, where $\mathrm{X} \lessdot \mathrm{Y}$ are flats and σ is an orientation of the rank-one arrangement $\mathcal{A}_{\mathrm{X}}^{\mathrm{Y}}$. The opposite half-flat then corresponds to $(\mathrm{X}, -\sigma, \mathrm{Y})$.

14.4.3. Concatenation of orientations. There is a canonical isomorphism

$$\mathsf{E}^{\mathsf{o}}[\mathcal{A}] \otimes \mathsf{E}^{\mathsf{o}}[\mathcal{A}] \xrightarrow{\cong} \Bbbk, \qquad \sigma \otimes \sigma \mapsto 1, \tag{14.11}$$

where σ is either of the two orientations of $\mathcal{A}$. Changing σ to $-\sigma$ incurs two minus signs, so the map is well-defined.

For any flat X, there is an isomorphism

$$\mathsf{E}^{\mathsf{o}}[\mathcal{A}^{\mathrm{X}}] \otimes \mathsf{E}^{\mathsf{o}}[\mathcal{A}_{\mathrm{X}}] \xrightarrow{\cong} \mathsf{E}^{\mathsf{o}}[\mathcal{A}], \qquad \sigma_1 \otimes \sigma_2 \mapsto \tau, \tag{14.12}$$

where τ is obtained by "concatenating" σ_1 and σ_2: Suppose c_1 is a maximal chain of faces in $\mathcal{A}^{\mathrm{X}}$ which represents σ_1. Let c_1' denote the corresponding chain in $\mathcal{A}$. It ends at a face with support X. Call that face F. Similarly, let c_2 be a maximal chain of faces in $\mathcal{A}_{\mathrm{X}}$ which represents σ_2. By using the canonical identification $\Sigma[\mathcal{A}_{\mathrm{X}}] \xrightarrow{\cong} \Sigma[\mathcal{A}]_F$, we obtain a chain c_2' in $\mathcal{A}$ which starts at F. The concatenation of c_1' and c_2' is a maximal chain in $\mathcal{A}$. Its class is the required τ.

In terms of the identification (14.10), the map (14.12) can be described as follows. Let V denote the ambient space of $\mathcal{A}$, and U denote the ambient space of $\mathcal{A}^{\mathrm{X}}$. An orientation of U/O and on V/U together determines an orientation of V/O.

Iterating this procedure yields an isomorphism

$$\mathsf{E}^{\mathsf{o}}[\mathcal{A}^{\mathrm{X}_1}] \otimes \mathsf{E}^{\mathsf{o}}[\mathcal{A}_{\mathrm{X}_1}^{\mathrm{X}_2}] \otimes \cdots \otimes \mathsf{E}^{\mathsf{o}}[\mathcal{A}_{\mathrm{X}_k}] \xrightarrow{\cong} \mathsf{E}^{\mathsf{o}}[\mathcal{A}] \tag{14.13}$$

for any strict chain of flats $(\bot < \mathrm{X}_1 < \cdots < \mathrm{X}_k < \top)$.

14.4.4. Signature space. We mention here another interesting construction of a one-dimensional space from an arrangement. Start with $\Gamma[\mathcal{A}]$, the vector space with basis indexed by chambers in $\mathcal{A}$. Take its quotient by the subspace spanned by elements of the form

$$\mathtt{H}_C - (-1)^{\mathrm{dist}(C,D)} \mathtt{H}_D,$$

as C and D vary over chambers. Since chambers are gallery connected and the distance function v_{-1} given in (8.17) is log-antisymmetric, the quotient space is one-dimensional. We call it the *signature space* of $\mathcal{A}$ and denote it by $\mathsf{E}^-[\mathcal{A}]$.

14.5. Joyal-Klyachko-Stanley. Presentation of Lie

We review poset (co)homology with emphasis on the poset of flats of an arrangement. The Joyal-Klyachko-Stanley (JKS) theorem identifies, up to orientation, the top-cohomology of the lattice of flats with the space of Lie elements through the unbracketing operation. We discuss the Björner-Wachs basis for the top-homology of the lattice of flats, and show that its dual basis for top-cohomology identifies with the Dynkin basis for the space of Lie elements. Both bases are defined relative to a generic half-space. We also explain how the JKS theorem is equivalent to a presentation of the space of Lie elements with the relations being anti-symmetry and Jacobi identity.

We give two proofs of the JKS theorem, each with its own advantage. The first proof makes critical use of the duality between the Björner-Wachs and Dynkin bases. The existence of the substitution product of Lie is a consequence of this proof. The second proof is basis-free, but it relies on Lemma 13.63 which says that the space of Lie elements is generated in rank one. The existence of the Björner-Wachs basis is a consequence of this proof.

14.5.1. Order (co)homology of a poset. Fix a field $\Bbbk$. Let P be a graded poset of rank $r \geq 1$ with minimum element $\bot$ and maximum element $\top$. The strict chains in P starting at $\bot$ and ending at $\top$ (or equivalently, the strict chains in $P \setminus \{\bot, \top\}$) form a simplicial complex $\Delta(P)$. This is the *order complex* of P. Each chain defines a simplex. The order (co)homology of P is the reduced simplicial (co)homology of $\Delta(P)$ over the field $\Bbbk$. Let us make this more explicit.

The chain complex for the order homology of P is as follows. For $-1 \leq k \leq r-2$, the chain group $\mathcal{C}_k(P)$ is the vector space over $\Bbbk$ with basis consisting of strict chains $\bot < x_1 < \cdots < x_{k+1} < \top$. The remaining chain groups are 0. Note that $\mathcal{C}_{r-2}(P)$ has a basis of maximal chains, while $\mathcal{C}_{-1}(P)$ is one-dimensional and spanned by the chain $\bot < \top$. The boundary operator $\partial_k : \mathcal{C}_k(P) \to \mathcal{C}_{k-1}(P)$ is given by

$$\partial_k(\bot < x_1 < \cdots < x_{k+1} < \top) = \sum_{i=1}^{k+1} (-1)^i (\bot < x_1 < \cdots < \hat{x}_i < \cdots < x_{k+1} < \top),$$

where by standard convention, $\hat{x}_i$ means that x_i has been deleted from the chain.

The cochain complex is obtained by dualizing the chain complex. We denote the cochain groups by $\mathcal{C}^k(P)$ and the coboundary operators by $\delta_k : \mathcal{C}_k(P) \to \mathcal{C}_{k+1}(P)$.

Explicitly,

$$\begin{aligned}(14.14)\quad &\delta_k(\bot < x_1 < \cdots < x_{k+1} < \top)^* \\ &= \sum_{i=1}^{k+2} (-1)^i \sum_{x_{i-1}<x<x_i} (\bot < x_1 < \cdots < x_{i-1} < x < x_i < \cdots < x_{k+1} < \top)^*,\end{aligned}$$

with the convention that $x_0 = \bot$ and $x_{k+2} = \top$. The superscript $*$ stands for the dual basis.

We write $\mathcal{H}_k(P)$ and $\mathcal{H}^k(P)$ for the homology and cohomology groups in position k. They are duals of each other, with the duality induced by the duality between the chain and cochain complexes. The top-dimensional homology and cohomology groups of P are $\mathcal{H}_{r-2}(P)$ and $\mathcal{H}^{r-2}(P)$.

Proposition 14.30. *If P is a geometric lattice, then P has (co)homology only in the top dimension, where it is $\Bbbk^{|\mu(\bot,\top)|}$. In particular, for any arrangement $\mathcal{A}$, the lattice of flats $\Pi[\mathcal{A}]$ has (co)homology only in the top dimension, where it is $\Bbbk^{|\mu(\mathcal{A})|}$.*

Geometric lattice is a standard notion from lattice theory (Section B.3.2). The main point for us is that the (opposite of the) lattice of flats is geometric (Lemma 1.30). We list here some well-known general facts (without proof) which lead to the above result. If P is geometric, then the simplicial complex $\Delta(P)$ is pure and shellable [**74**]. So it is homotopy equivalent to a wedge of $(r-2)$-dimensional spheres, hence it has (co)homology only in dimension $r-2$. Further, its rank, up to sign, is the Euler characteristic of the (co)chain complex, so it must be $|\mu(\bot,\top)|$.

For any geometric lattice P of rank r, define its *Whitney (co)homology* by

$$(14.15)\quad \mathcal{WH}_k(P) := \bigoplus_{\substack{x:\\ r-\mathrm{rk}(x)=k}} \mathcal{H}_{k-2}(x,\top) \quad\text{and}\quad \mathcal{WH}^k(P) := \bigoplus_{\substack{x:\\ r-\mathrm{rk}(x)=k}} \mathcal{H}^{k-2}(x,\top).$$

The index k varies between 0 and r. The summand for x computes the top order (co)homology of the interval $[x,\top]$ and has dimension $|\mu(x,\top)|$. Observe that

$$\mathcal{WH}_r(P) = \mathcal{H}_{r-2}(P) \quad\text{and}\quad \mathcal{WH}^r(P) = \mathcal{H}^{r-2}(P).$$

By convention, $\mathcal{WH}_0(P) = \mathcal{WH}^0(P) = \Bbbk$. Thus,

$$\dim \mathcal{WH}_k(P) = (-1)^k \mathrm{wy}(P, r-k),$$

where wy denotes the Whitney numbers of the first kind. (The definition given in (1.56) extends to any geometric lattice.)

We let $\mathcal{WH}_*(P)$ denote the direct sum of all Whitney homology groups of P. Similarly, $\mathcal{WH}^*(P)$ denotes the direct sum of all Whitney cohomology groups.

14.5.2. Björner-Wachs basis for homology. Fix an arrangement $\mathcal{A}$. For any strict chain of faces $f = (F_1 < \cdots < F_k)$, define its support by

$$\mathrm{s}(f) := (\mathrm{s}(F_1) < \cdots < \mathrm{s}(F_k)).$$

This is a strict chain of flats. For any chamber C, define

$$(14.16)\qquad \mathtt{BW}_C := \Big(\sum_f (\sigma : f)\, \mathrm{s}(f)\Big) \otimes \sigma,$$

where σ is any orientation of $\mathcal{A}$, the sum is over maximal chains of faces f which end in C, and

$$(\sigma : f) = \begin{cases} 1 & \text{if } \sigma = [f], \\ -1 & \text{if } \sigma = -[f]. \end{cases}$$

Note that if σ is changed to $-\sigma$, then the signs inside the sum also negate, so $\mathtt{BW}_C$ is well-defined. It may also be expressed as

$$\mathtt{BW}_C = \sum_f \mathrm{s}(f) \otimes [f], \tag{14.17}$$

with f varying as in (14.16).

Lemma 14.31. *Consider the element of $\mathcal{C}_{r-2}(\Pi[\mathcal{A}])$ given by the term inside the parenthesis in* (14.16). *It is a cycle, that is, ∂_{r-2} applied to it is* 0*; so it is an element of $\mathcal{H}_{r-2}(\Pi[\mathcal{A}])$.*

PROOF. Applying ∂_{r-2} to the element yields a sum over strict chains g which end in C. Any such g contains a unique pair $H < K$ of faces whose ranks differ by 2. There are exactly two faces that lie between H and K, so g can be extended to a maximal chain in exactly two ways, and these have opposite orientations. Thus, the coefficient of $\mathrm{s}(g)$ is zero, as claimed.

Topologically, what has happened is the following. A strict chain of faces ending in C can be identified with a face in the barycentric subdivision of the boundary of C. The latter is a simplicial sphere. So a cycle can be constructed by taking a linear combination with ± 1 coefficients of the top-simplices, which in this case are maximal chains ending at C. □

Proposition 14.32. *For any generic half-space* h *wrt $\mathcal{A}$, the set*

$$\{\mathtt{BW}_C \mid \overline{C} \subseteq \mathrm{h}\} \tag{14.18}$$

is a basis of $\mathcal{H}_{r-2}(\Pi[\mathcal{A}]) \otimes \mathsf{E}^{\mathrm{o}}[\mathcal{A}]$, where $r = \mathrm{rk}(\mathcal{A})$.

PROOF. See [**84**, Theorem 4.2]. □

We call (14.18) the *Björner-Wachs basis* associated to h. For simplicity, we abbreviate it to BW-basis. We will make critical use of the BW-basis in the proof of the JKS theorem below. Later in Section 14.5.7, we will give a second basis-free proof of the JKS theorem and deduce Proposition 14.32 as a consequence.

Exercise 14.33. For a maximal chain of faces f, let $\overline{f}$ denote the maximal chain obtained by replacing each face occuring in f by its opposite. Show that $[f] = (-1)^{\mathrm{rk}(\mathcal{A})}[\overline{f}]$. Use this to deduce that

$$\mathtt{BW}_C = (-1)^{\mathrm{rk}(\mathcal{A})} \mathtt{BW}_{\overline{C}}$$

for any chamber C.

14.5.3. Joyal-Klyachko-Stanley. Put $r := \mathrm{rk}(\mathcal{A})$. For a maximal chain of faces f, let $\mathrm{last}(f)$ denote the last face in the chain f (which is necessarily a chamber).

We now define a linear map

$$\mathcal{C}^{r-2}(\Pi[\mathcal{A}]) \otimes \mathsf{E}^{\mathrm{o}}[\mathcal{A}] \to \Gamma[\mathcal{A}]. \tag{14.19}$$

We provide a number of definitions and then indicate why they coincide. In some cases, the map is only specified on a spanning set with the understanding that it is extended linearly.

The map (14.19) is given by

$$\varphi \otimes \sigma \mapsto \sum_{f} \varphi(\mathrm{s}(f))(\sigma : f)\, \mathtt{H}_{\mathrm{last}(f)}, \tag{14.20a}$$

where φ is a cochain, σ is an orientation, and the sum is over all maximal chains of faces f.

The map (14.19) is given by

$$z^* \otimes \sigma \mapsto \sum_{f:\, \mathrm{s}(f)=z} (\sigma : f)\, \mathtt{H}_{\mathrm{last}(f)}, \tag{14.20b}$$

where z is any maximal chain of flats and σ is an orientation.

The map (14.19) is given by

$$z^* \otimes \sigma \mapsto \sum_{D} \pm \mathtt{H}_D, \tag{14.20c}$$

where z is any maximal chain of flats and σ is an orientation. The sum is over those chambers D for which there exists a maximal chain of faces f with $\mathrm{last}(f) = D$ and $\mathrm{s}(f) = z$. The coefficient of $\mathtt{H}_D$ is $+1$ if $[f] = \sigma$ and -1 if $[f] = -\sigma$.

The map (14.19) is given by

$$\alpha \mapsto \sum_{D} \langle \alpha, \mathtt{BW}_D \rangle\, \mathtt{H}_D, \tag{14.20d}$$

where $\langle\ ,\ \rangle$ is the canonical pairing between $\mathcal{C}^{r-2}(\Pi[\mathcal{A}])$ and $\mathcal{C}_{r-2}(\Pi[\mathcal{A}])$ tensored with (14.11).

The map (14.19) is given by

$$\mathrm{s}(f)^* \otimes [f] \mapsto (\mathtt{H}_{F_1} - \mathtt{H}_{G_1}) \cdot \ldots \cdot (\mathtt{H}_{F_r} - \mathtt{H}_{G_r}), \tag{14.20e}$$

where $f = (O \lessdot F_1 \lessdot \cdots \lessdot F_r)$ is a maximal chain of faces, and for $1 \le i \le r$, G_i is the face opposite to F_i in the star of F_{i-1} (with the convention $F_0 = O$).

Lemma 14.34. *Definitions* (14.20a)–(14.20e) *all coincide.*

PROOF. Let us begin from (14.20a). Evaluating it on $z^* \otimes \sigma$ yields (14.20b). By Proposition 1.17, distinct faces of a chamber have distinct supports. Hence there can be at most one chamber D such that $\mathrm{last}(f) = D$ and $\mathrm{s}(f) = z$. This yields (14.20c). Using definition (14.16), this then verifies (14.20d) on the basis elements $\alpha = z^* \otimes \sigma$. For (14.20e): Observe that there are exactly 2^r chains of faces k whose support is $\mathrm{s}(f)$, and they are given by

$$k = (O \lessdot K_1 \lessdot K_1K_2 \lessdot \cdots \lessdot K_1 \ldots K_r),$$

where each K_i is either F_i or G_i. (Also see Exercise 1.42.) Thus the term $\mathtt{H}_D$ appears in the rhs of (14.20e) iff there exists a maximal chain of faces k ending at D whose support is $\mathrm{s}(f)$. Further, the coefficient of $\mathtt{H}_D$ is $+1$ if $[k] = [f]$ and -1 if $[k] = -[f]$. (This follows from a small orientation argument.) This is the same as (14.20c). ∎

In view of (9.33), each $\mathtt{H}_{F_i} - \mathtt{H}_{G_i}$ belongs to the radical of the Tits algebra, so their product is in the r-th power of the radical which is contained in $\mathsf{Lie}[\mathcal{A}]$ by Lemma 13.15. Thus, the rhs of (14.20e) is a Lie element. As a consequence, (14.19) induces a map

$$\mathcal{C}^{r-2}(\Pi[\mathcal{A}]) \otimes \mathsf{E}^{\mathsf{o}}[\mathcal{A}] \to \mathsf{Lie}[\mathcal{A}]. \tag{14.21}$$

We refer to the image of $z^* \otimes \sigma$ under (14.20c) as the *unbracketing* of z wrt σ. It determines a Lie element. For convenience, we may sometimes just say 'unbracketing of a maximal chain' keeping the orientation implicit. An illustration is given below.

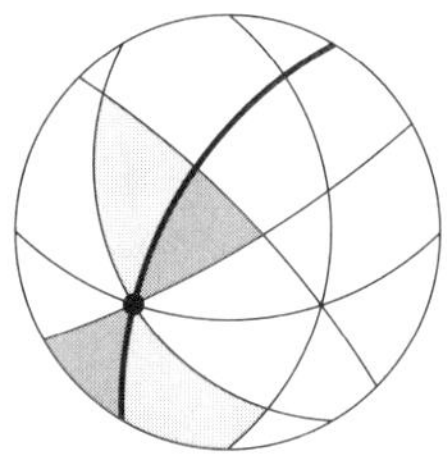

Let X_1 denote the support of the thick vertex, and X_2 denote the thick line. Unbracketing the maximal chain $z = (\bot < \mathrm{X}_1 < \mathrm{X}_2 < \top)$ yields a sum of 8 chambers with coefficients ± 1. Four of them are seen in the picture, while the remaining four are on the backside. The chambers in light shade have one coefficient, while those in dark shade have the opposite coefficient.

Exercise 14.35. Let $O \lessdot F_1 \lessdot \cdots \lessdot F_r$ be a maximal chain of faces. Suppose the chamber F_r is a simplex. For $1 \leq i \leq r$, let P_i be the vertex of F_i which is not in F_{i-1}, while G_i be the face opposite to F_i in the star of F_{i-1} (with the convention $F_0 = O$). Show that

$$(\mathtt{H}_{F_1} - \mathtt{H}_{G_1}) \cdot \ldots \cdot (\mathtt{H}_{F_r} - \mathtt{H}_{G_r}) = (\mathtt{H}_{P_1} - \mathtt{H}_{\overline{P_1}}) \cdot \ldots \cdot (\mathtt{H}_{P_r} - \mathtt{H}_{\overline{P_r}}).$$

In fact, for any i, the product of the first i terms in the lhs equals the product of the first i terms in the rhs.

Exercise 14.36. Prove (10.5) by checking that a Lie element arising by unbracketing a maximal chain of flats has this property. (The surjectivity of (14.21) is shown below.)

A coboundary relation is an element of $\mathcal{C}^{r-2}(\Pi[\mathcal{A}])$ of the form $\delta_{r-3}(z^*)$ for some strict chain of flats z. (The coboundary map is given in (14.14).) Note that the top cohomology group $\mathcal{H}^{r-2}(\Pi[\mathcal{A}])$ is the quotient of $\mathcal{C}^{r-2}(\Pi[\mathcal{A}])$ by the subspace spanned by the coboundary relations.

Lemma 14.37. *The map* (14.19) *sends any coboundary relation (tensored with an orientation) to zero.*

PROOF. This follows from (14.20d) and Lemma 14.31 in view of the fact that any coboundary relation evaluated on a cycle is zero. □

Thus (14.21) induces a map

$$\mathcal{H}^{r-2}(\Pi[\mathcal{A}]) \otimes \mathsf{E}^{\mathsf{o}}[\mathcal{A}] \to \mathsf{Lie}[\mathcal{A}]. \tag{14.22}$$

By tensoring both sides by $\mathsf{E}^{\mathrm{o}}[\mathcal{A}]$ and using (14.11), it can be expressed in the equivalent form:

$$\mathcal{H}^{r-2}(\Pi[\mathcal{A}]) \to \mathsf{Lie}[\mathcal{A}] \otimes \mathsf{E}^{\mathrm{o}}[\mathcal{A}]. \tag{14.23}$$

Theorem 14.38. *The maps* (14.22) *and* (14.23) *are natural isomorphisms. Naturality is wrt cisomorphisms of arrangements.*

Naturality means that for any cisomorphism between arrangements $\mathcal{A}$ and $\mathcal{A}'$, the diagram

$$\begin{array}{ccc} \mathcal{H}^{r-2}(\Pi[\mathcal{A}]) \otimes \mathsf{E}^{\mathrm{o}}[\mathcal{A}] & \longrightarrow & \mathsf{Lie}[\mathcal{A}] \\ \downarrow & & \downarrow \\ \mathcal{H}^{r-2}(\Pi[\mathcal{A}']) \otimes \mathsf{E}^{\mathrm{o}}[\mathcal{A}'] & \longrightarrow & \mathsf{Lie}[\mathcal{A}'] \end{array}$$

commutes. This is true because a cisomorphism preserves flats, gallery distances, opposite faces, and the Tits product (Section 1.11), and the entities involved in the map are expressible in terms of these notions.

Proof. Let us write ψ for the map (14.22). Both sides of this map have the same dimension, namely, $|\mu(\mathcal{A})|$, so it suffices to show that ψ is injective. Fix a generic half-space h. Let

$$\{\mathtt{BW}^*_C \mid \overline{C} \subseteq \mathrm{h}\} \tag{14.24}$$

be the basis of $\mathcal{H}^{r-2}(\Pi[\mathcal{A}]) \otimes \mathsf{E}^{\mathrm{o}}[\mathcal{A}]$ dual to the BW-basis (14.18) associated to h. Let us compute $\psi(\mathtt{BW}^*_C)$. Pick any cochain α which represents $\mathtt{BW}^*_C$. Then $\psi(\mathtt{BW}^*_C)$ is the same as the map (14.19) applied to α. Now for any chamber D inside $\overline{\mathrm{h}}$,

$$\langle \alpha, \mathtt{BW}_D \rangle = \langle \mathtt{BW}^*_C, \mathtt{BW}_D \rangle = \begin{cases} 1 & \text{if } C = D, \\ 0 & \text{otherwise.} \end{cases}$$

(Note very carefully that up to orientation the first pairing is between cochains and chains, while the second is between cohomology and homology.) Formula (14.20d) implies that $\psi(\mathtt{BW}^*_C)$ involves $\mathtt{H}_C$ but no other chamber inside $\overline{\mathrm{h}}$. So these elements, as C varies inside $\overline{\mathrm{h}}$, are linearly independent, and ψ is injective. □

We call this the *Joyal-Klyachko-Stanley theorem*, or JKS for short. We refer to (14.22) as the JKS isomorphism.

Corollary 14.39. *The Björner-Wachs basis* (14.18) *and the Dynkin basis* (14.9) *are duals via the JKS isomorphism* (14.22). *More precisely, the map* (14.22) *sends the dual* BW*-basis* (14.24) *to the Dynkin basis.*

Proof. For $C \subseteq \overline{\mathrm{h}}$, let A_C denote the inverse of $\theta_{\mathrm{h}} \cdot \mathtt{H}_C$ under (14.22). Recall from (14.8) that $\theta_{\mathrm{h}} \cdot \mathtt{H}_C$ involves $\mathtt{H}_C$ but no other chamber inside $\overline{\mathrm{h}}$. Hence by (14.20d), for any $C, D \subseteq \overline{\mathrm{h}}$,

$$\langle A_C, \mathtt{BW}_D \rangle = \begin{cases} 1 & \text{if } C = D, \\ 0 & \text{otherwise.} \end{cases}$$

This shows that $A_C = \mathtt{BW}^*_C$, as required. □

14.5.4. Whitney cohomology. The JKS theorem allows us to describe the Whitney cohomology (14.15) of the poset of flats in terms of Lie elements:

Theorem 14.40. *For each $0 \leq k \leq \mathrm{rk}(\mathcal{A})$, there is a natural isomorphism*

$$\mathcal{WH}^k(\Pi[\mathcal{A}]) \otimes \mathsf{E}^{\mathsf{o}}[\mathcal{A}] \xrightarrow{\cong} \bigoplus_{\mathrm{X}:\,\mathrm{rk}(\mathcal{A}_{\mathrm{X}})=k} \mathsf{E}^{\mathsf{o}}[\mathcal{A}^{\mathrm{X}}] \otimes \mathsf{Lie}[\mathcal{A}_{\mathrm{X}}].$$

These induce a natural isomorphism

$$\mathcal{WH}^*(\Pi[\mathcal{A}]) \otimes \mathsf{E}^{\mathsf{o}}[\mathcal{A}] \xrightarrow{\cong} \bigoplus_{\mathrm{X}} \mathsf{E}^{\mathsf{o}}[\mathcal{A}^{\mathrm{X}}] \otimes \mathsf{Lie}[\mathcal{A}_{\mathrm{X}}]. \tag{14.25}$$

Proof. This follows by using the JKS isomorphism (14.23) for each $\mathcal{A}_{\mathrm{X}}$ in conjunction with (14.12). □

We deduce from the Zaslavsky formula (1.48) that

$$\dim \mathcal{WH}^*(\Pi[\mathcal{A}]) = c(\mathcal{A}), \tag{14.26}$$

the number of chambers in $\mathcal{A}$.

14.5.5. Presentation of Lie. Lie elements of any arrangement are "generated" by Lie elements of rank-one arrangements with the "relations" being Jacobi identities in rank-two arrangements (Section 10.2). We refer to this as the presentation of $\mathsf{Lie}[\mathcal{A}]$. The existence of this presentation is equivalent to the JKS theorem. This is explained below.

For the rank-one arrangement $\mathcal{A}$ with chambers C and $\overline{C}$,

$$\mathsf{E}^{\mathsf{o}}[\mathcal{A}] \xrightarrow{\cong} \mathsf{Lie}[\mathcal{A}], \qquad [O \lessdot C] \mapsto \mathtt{H}_C - \mathtt{H}_{\overline{C}}.$$

(Both spaces are 1-dimensional.) This isomorphism is an instance of (14.22).

Now suppose $\mathcal{A}$ is the rank-two arrangement of n lines. Then (14.21) along with the identification (14.12) can be rewritten as

$$\bigoplus_{i=1}^{n} \mathsf{E}^{\mathsf{o}}[\mathcal{A}^{\mathrm{X}_i}] \otimes \mathsf{E}^{\mathsf{o}}[\mathcal{A}_{\mathrm{X}_i}] \to \mathsf{Lie}[\mathcal{A}],$$

where the X_i are the n lines (one-dimensional flats) of $\mathcal{A}$. This map is surjective. The lhs is n-dimensional while the rhs is $(n-1)$-dimensional. The kernel is spanned by the element

$$\sum_{i=1}^{n} \tau^i \otimes \tau_i \tag{14.27}$$

where τ^i and τ_i are orientations of $\mathcal{A}^{\mathrm{X}_i}$ and $\mathcal{A}_{\mathrm{X}_i}$ such that their concatenation is (say) the anticlockwise orientation of $\mathcal{A}$. This element corresponds to the Jacobi identity.

Now let $\mathcal{A}$ be arbitrary. The map (14.21) can be rewritten as

$$\bigoplus_{z} \mathsf{E}^{\mathsf{o}}[\mathcal{A}^{\mathrm{X}_1}] \otimes \mathsf{E}^{\mathsf{o}}[\mathcal{A}^{\mathrm{X}_2}_{\mathrm{X}_1}] \otimes \cdots \otimes \mathsf{E}^{\mathsf{o}}[\mathcal{A}_{\mathrm{X}_{r-1}}] \to \mathsf{Lie}[\mathcal{A}], \tag{14.28}$$

where the sum is over all maximal chains of flats $z = (\bot \lessdot \mathrm{X}_1 \lessdot \cdots \lessdot \mathrm{X}_{r-1} \lessdot \top)$. (In this rewriting, the second tensor factor $\mathsf{E}^{\mathsf{o}}[\mathcal{A}]$ in the lhs of (14.21) is identified with the summands in the lhs above via (14.13).) Theorem 14.38 says that the kernel of (14.28) is the subspace generated by (14.27). (The latter corresponds to the coboundary relations.) To summarize:

Theorem 14.41. *The space* $\mathsf{Lie}[\mathcal{A}]$ *is freely generated by the orientation space in rank one subject to the Jacobi identities in rank two.*

14.5.6. Concatenation of chains. It is convenient to write $\mathcal{H}^{\mathrm{top}}(\Pi[\mathcal{A}])$ for the top-dimensional cohomology of the lattice of flats of $\mathcal{A}$. For any flat X, there is a map

$$\mathcal{H}^{\mathrm{top}}(\Pi[\mathcal{A}^{\mathrm{X}}]) \otimes \mathcal{H}^{\mathrm{top}}(\Pi[\mathcal{A}_{\mathrm{X}}]) \to \mathcal{H}^{\mathrm{top}}(\Pi[\mathcal{A}]) \tag{14.29}$$

obtained by concatenating: A maximal chain of flats in $\mathcal{A}^{\mathrm{X}}$ can be identified with a chain of flats in $\mathcal{A}$ ending at X, while a maximal chain of flats in $\mathcal{A}_{\mathrm{X}}$ can be identified with a chain of flats in $\mathcal{A}$ starting at X. So, concatenating the two yields a maximal chain of flats in $\mathcal{A}$. The map (14.29) is obtained by passing to the homology classes.

Exercise 14.42. Combining (14.29) with (14.12) and using the JKS isomorphism (14.22), we obtain a map

$$\mathsf{Lie}[\mathcal{A}^{\mathrm{X}}] \otimes \mathsf{Lie}[\mathcal{A}_{\mathrm{X}}] \to \mathsf{Lie}[\mathcal{A}].$$

Check that this coincides with the substitution product of Lie defined in (10.28). (Use (14.20e) to verify the commutativity of (10.29).)

14.5.7. Second proof of JKS. We now give a second proof of the JKS theorem. The starting point is the substitution product of Lie defined in (10.28). By iterated substitution, we obtain the map (13.33). Observe that this map is the same as the map (14.28). By Lemma 13.63, this map is surjective. Going back, this says that the JKS map (14.22) is surjective. Since both sides of the JKS map have the same dimension, we deduce that the JKS map is an isomorphism. Note that this also proves Theorem 14.41.

The above proof of the JKS theorem is basis-free. In fact, following up on this approach, one can use the argument in Corollary 14.39 to deduce both Proposition 14.32 and Corollary 14.39.

14.5.8. JKS and cartesian product. JKS is compatible with taking cartesian product of arrangements.

Lemma 14.43. *We have isomorphisms*

$$\mathsf{E}^{\mathrm{o}}[\mathcal{A}] \otimes \mathsf{E}^{\mathrm{o}}[\mathcal{A}'] \xrightarrow{\cong} \mathsf{E}^{\mathrm{o}}[\mathcal{A} \times \mathcal{A}']$$
$$\mathcal{H}^{\mathrm{top}}(\Pi[\mathcal{A}]) \otimes \mathcal{H}^{\mathrm{top}}(\Pi[\mathcal{A}']) \xrightarrow{\cong} \mathcal{H}^{\mathrm{top}}(\Pi[\mathcal{A} \times \mathcal{A}'])$$
$$\mathsf{Lie}[\mathcal{A}] \otimes \mathsf{Lie}[\mathcal{A}'] \xrightarrow{\cong} \mathsf{Lie}[\mathcal{A} \times \mathcal{A}'].$$

PROOF. Recall that $\mathcal{A}$ and $\mathcal{A}'$ arise, respectively, as the arrangements under and over a certain flat of $\mathcal{A} \times \mathcal{A}'$. The above maps then arise as special cases of (14.12), (14.29) and (10.28). The first map is clearly an isomorphism. The isomorphism for Lie is given in (10.7). This forces the middle map to be an isomorphism as well. □

Thus, the JKS isomorphism for $\mathcal{A} \times \mathcal{A}'$ can be identified with the tensor product of the JKS isomorphisms for $\mathcal{A}$ and $\mathcal{A}'$.

14.5.9. Order cohomology of flats and Tits algebra. In Section 13.7, we saw a connection between Lie elements and the Tits algebra (Proposition 13.50 and Theorem 13.54). By using the JKS isomorphism (14.22), these results can also be stated as a connection between order cohomology of the lattice of flats and the Tits algebra:

Proposition 14.44. *Let* $\mathtt{E}$ *be any Eulerian family of* $\mathcal{A}$*. For flats* $\mathrm{X} \leq \mathrm{Y}$*, there is an isomorphism*

$$\mathcal{H}^{\mathrm{top}}(\Pi[\mathcal{A}_{\mathrm{X}}^{\mathrm{Y}}]) \otimes \mathsf{E}^{\mathrm{o}}[\mathcal{A}_{\mathrm{X}}^{\mathrm{Y}}] \xrightarrow{\cong} \mathtt{E}_{\mathrm{X}} \cdot \Sigma[\mathcal{A}] \cdot \mathtt{E}_{\mathrm{Y}}. \tag{14.30}$$

The map (14.30) involves unbracketing a maximal chain from X to Y using a specified orientation of $\mathcal{A}_{\mathrm{X}}^{\mathrm{Y}}$ and viewing the resulting Lie element of $\mathcal{A}_{\mathrm{X}}^{\mathrm{Y}}$ as an element of $\mathtt{E}_{\mathrm{X}} \cdot \Sigma[\mathcal{A}] \cdot \mathtt{E}_{\mathrm{Y}}$.

Theorem 14.45. *There is an algebra isomorphism*

$$\bigoplus_{\mathrm{X} \leq \mathrm{Y}} \mathcal{H}^{\mathrm{top}}(\Pi[\mathcal{A}_{\mathrm{X}}^{\mathrm{Y}}]) \otimes \mathsf{E}^{\mathrm{o}}[\mathcal{A}_{\mathrm{X}}^{\mathrm{Y}}] \xrightarrow{\cong} \Sigma[\mathcal{A}] \tag{14.31}$$

obtained by summing (14.30) *over all* $\mathrm{X} \leq \mathrm{Y}$*. The product on order cohomology is induced from concatenation of chains* (14.29).

14.5.10. Order cohomology of poset of faces. The poset of faces $\Sigma[\mathcal{A}]$ has a minimum element, but many maximal elements. So the definition of order cohomology does not directly apply. One way to rectify this is as follows.

For each chamber D, consider the interval $[O, D]$. This poset has a minimum and a maximum. Its order complex Δ is the barycentric subdivision of the boundary of D. Thus, the order cohomology of $[O, D]$ is $\Bbbk$ in the top dimension, and 0 otherwise. Now let

$$\mathcal{C}^{\bullet}(\Sigma[\mathcal{A}]) := \bigoplus_{D} \mathcal{C}^{\bullet}([O, D]), \tag{14.32}$$

where the summands in the rhs are the cochain complexes that compute order cohomology. This complex has cohomology only in top dimension, where it is $\Bbbk^{c(\mathcal{A})}$, with $c(\mathcal{A})$ being the number of chambers. More explicitly, consider the map

$$\mathcal{C}^{r-2}(\Sigma[\mathcal{A}]) \otimes \mathsf{E}^{\mathrm{o}}[\mathcal{A}] \to \Gamma[\mathcal{A}], \qquad f^* \otimes \sigma \mapsto (\sigma : f)\, \mathtt{H}_{\mathrm{last}(f)}. \tag{14.33}$$

The coboundary relations are spanned by $f^* + g^*$, where f and g are maximal chains from O to say D which differ in exactly one positition. Hence, $(\sigma : f) = (\sigma : g)$, and $f^* + g^*$ maps to zero. As a consequence:

Theorem 14.46. *There is a natural isomorphism*

$$\mathcal{H}^{r-2}(\Sigma[\mathcal{A}]) \otimes \mathsf{E}^{\mathrm{o}}[\mathcal{A}] \xrightarrow{\cong} \Gamma[\mathcal{A}].$$

Naturality is wrt cisomorphisms of arrangements.

This is the analogue of the JKS theorem for the poset of faces. Moreover, the two results relate to one another as follows. The map

$$\mathcal{C}^{\bullet}(\Pi[\mathcal{A}]) \to \mathcal{C}^{\bullet}(\Sigma[\mathcal{A}]), \qquad z^* \mapsto \sum_{f:\, \mathrm{s}(f) = z} f^*$$

is a cochain map. It induces the commutative diagram

$$\begin{array}{ccc} \mathcal{H}^{r-2}(\Pi[\mathcal{A}]) \otimes \mathsf{E}^{\mathsf{o}}[\mathcal{A}] & \xrightarrow{\cong} & \mathsf{Lie}[\mathcal{A}] \\ \downarrow & & \downarrow \\ \mathcal{H}^{r-2}(\Sigma[\mathcal{A}]) \otimes \mathsf{E}^{\mathsf{o}}[\mathcal{A}] & \xrightarrow[\cong]{} & \Gamma[\mathcal{A}]. \end{array} \tag{14.34}$$

This follows by comparing (14.21) and (14.33).

14.6. Björner and Lyndon bases

We discuss another pair of dual bases for the top (co)homology of the lattice of flats. They are constructed out of ordered coordinate charts.

14.6.1. Ordered coordinate charts and labeled chains. An *ordered coordinate chart* is a sequence $(\mathrm{H}_1, \dots, \mathrm{H}_r)$ of hyperplanes taken from the arrangement $\mathcal{A}$ such that the intersection of all the H_i is the minimum flat, and $r = \mathrm{rk}(\mathcal{A})$. A *hyperplane-labeled maximal chain* consists of a diagram

$$\bot \overset{\mathrm{H}_1}{—} \mathrm{X}_1 \overset{\mathrm{H}_2}{—} \dots \overset{\mathrm{H}_{r-1}}{—} \mathrm{X}_{r-1} \overset{\mathrm{H}_r}{—} \top$$

such that $(\bot \lessdot \mathrm{X}_1 \lessdot \dots \lessdot \mathrm{X}_{r-1} \lessdot \top)$ is a maximal chain of flats, and for each i, H_i contains X_{i-1} but not X_i. (For uniformity, we may write $\bot = \mathrm{X}_0$ and $\top = \mathrm{X}_r$.) For convenience, we shorten 'hyperplane-labeled maximal chain' to 'labeled chain'.

Lemma 14.47. *Ordered coordinate charts and labeled chains are equivalent notions.*

PROOF. The labels of a labeled chain form an ordered coordinate chart. Conversely, given an ordered coordinate chart, the flats can be constructed by

$$\mathrm{X}_{r-1} = \mathrm{H}_r, \quad \mathrm{X}_{r-2} = \mathrm{H}_r \wedge \mathrm{H}_{r-1}, \quad \dots, \quad \mathrm{X}_1 = \mathrm{H}_r \wedge \dots \wedge \mathrm{H}_2. \qquad \square$$

We will use the letters g and h to denote labeled chains (as for charts). For a labeled chain g, we let $\mathrm{cf}(g)$ denote the maximal chain of flats underlying g.

A *choice function* is a function γ which assigns to every non-maximum flat X a hyperplane H containing X. We write $\gamma(\mathrm{X}) = \mathrm{H}$. For instance, any linear order ℓ on the set of hyperplanes gives rise to a choice function γ: For any flat X, define $\gamma(\mathrm{X})$ to be the first hyperplane in ℓ among those that contain X.

Let γ be a choice function. A labeled chain is *γ-compatible* if $\gamma(\mathrm{X}_i) = \mathrm{H}_{i+1}$ for each i. Let $\mathrm{B}(\gamma)$ denote the set of such γ-compatible labeled chains.

Lemma 14.48. *Suppose γ is a choice function. For any flat X, let γ_{X} be the induced choice function on $\mathcal{A}_{\mathrm{X}}$. Then there is a bijection*

$$\bigsqcup_{\mathrm{X}:\, \mathrm{rk}(\mathrm{X})=1, \mathrm{X} \not\leq \mathrm{H}} \mathrm{B}(\gamma_{\mathrm{X}}) \longrightarrow \mathrm{B}(\gamma),$$

where $\gamma(\bot) = \mathrm{H}$.

PROOF. For any rank-one flat X not contained in H, appending $\bot \overset{\mathrm{H}_1}{—}$ (on the left) to any γ_{X}-compatible labeled chain in $\mathcal{A}_{\mathrm{X}}$ yields a γ-compatible labeled chain in $\mathcal{A}$ (from which X can be recovered uniquely). $\square$

Lemma 14.49. *For a choice function γ, the number of γ-compatible labeled chains is $|\mu(\mathcal{A})|$.*

PROOF. Let $g(\mathcal{A})$ denote the cardinality of $\mathrm{B}(\gamma)$. Then by Lemma 14.48,

$$g(\mathcal{A}) = \sum_{\mathrm{X}:\, \mathrm{rk}(\mathrm{X})=1, \mathrm{X} \nleq \mathrm{H}} g(\mathcal{A}_{\mathrm{X}}),$$

where $\gamma(\perp) = \mathrm{H}$. By induction, the Weisner formula (1.46b) and (1.47), we conclude that $g(\mathcal{A}) = |\mu(\mathcal{A})|$. □

14.6.2. Björner basis. For an ordered coordinate chart $g = (\mathrm{H}_1, \dots, \mathrm{H}_r)$, let

(14.35) $\mathtt{B}_g :=$

$$\sum_{\sigma \in \mathrm{S}_r} \mathrm{sgn}(\sigma)\, (\perp = \mathrm{H}_{\sigma(1)} \wedge \ldots \wedge \mathrm{H}_{\sigma(r)} \lessdot \cdots \lessdot \mathrm{H}_{\sigma(r-1)} \wedge \mathrm{H}_{\sigma(r)} \lessdot \mathrm{H}_{\sigma(r)} \lessdot \top).$$

Here S_r is the permutation group on r letters, and $\mathrm{sgn}(\sigma)$ denotes the sign of the permutation σ.

All maximal chains of flats that occur in $\mathtt{B}_g$ are distinct. The summand for $\sigma = \mathrm{id}$ is precisely $\mathrm{cf}(g)$ with g viewed as a labeled chain. Each $\mathtt{B}_g$ is, to start with, an element of the chain group $\mathcal{C}_{r-2}(\Pi[\mathcal{A}])$, but in fact, one can check that it is a cycle, that is, an element of the homology group $\mathcal{H}_{r-2}(\Pi[\mathcal{A}])$.

Lemma 14.50. *Suppose g and h are any γ-compatible labeled chains. Then* $\mathrm{cf}(g)$ *appears in* $\mathtt{B}_h$ *iff $g = h$, in which case it appears once and with coefficient* $+1$.

PROOF. Suppose $\mathrm{cf}(g)$ appears in $\mathtt{B}_h$. Since g and h are γ-compatible, the first hyperplane in both equals $\gamma(\perp)$. The rank-one flat in $\mathrm{cf}(g)$ and $\mathrm{cf}(h)$ are not contained in $\gamma(\perp)$, and since $\mathrm{cf}(g)$ appears in $\mathtt{B}_h$, we deduce that these rank-one flats coincide, and equal say X_1. Then the second hyperplane in both g and h equals $\gamma(\mathrm{X}_1)$. We continue in this manner to deduce that $g = h$. □

Proposition 14.51. *Let γ be any choice function. The set of elements $\mathtt{B}_g$, as g varies over all γ-compatible labeled chains, is a basis of $\mathcal{H}_{r-2}(\Pi[\mathcal{A}])$.*

PROOF. By Lemma 14.50, the $\mathtt{B}_g$, as g varies over all γ-compatible labeled chains, are linearly independent. By Lemma 14.49, their cardinality equals the dimension of $\mathcal{H}_{r-2}(\Pi[\mathcal{A}])$. So they form a basis. □

We refer to this as the *Björner basis* of $\mathcal{H}_{r-2}(\Pi[\mathcal{A}])$. It depends on a choice function γ.

14.6.3. Lyndon basis. For any labeled chain g, consider the dual element

$$\mathrm{cf}(g)^* \in \mathcal{C}^{r-2}(\Pi[\mathcal{A}]).$$

We write $\mathtt{L}_g$ for its image in the quotient $\mathcal{H}^{r-2}(\Pi[\mathcal{A}])$.

Proposition 14.52. *Let γ be any choice function. The set of elements $\mathtt{L}_g$, as g varies over all γ-compatible labeled chains, is a basis of $\mathcal{H}^{r-2}(\Pi[\mathcal{A}])$. Further, for γ-compatible labeled chains g and h,*

$$\langle \mathtt{L}_g, \mathtt{B}_h \rangle = \begin{cases} 1 & \text{if } g = h, \\ 0 & \text{otherwise,} \end{cases} \tag{14.36}$$

under the canonical pairing between cohomology and homology.

PROOF. Under the canonical pairing between $\mathcal{C}^{r-2}(\Pi[\mathcal{A}])$ and $\mathcal{C}_{r-2}(\Pi[\mathcal{A}])$,

$$\langle \mathrm{cf}(g)^*, \mathsf{B}_h \rangle = \begin{cases} 1 & \text{if } g = h, \\ 0 & \text{otherwise.} \end{cases}$$

This is a restatement of Lemma 14.50. This proves (14.36) and the result follows. □

We refer to this as the *Lyndon basis* of $\mathcal{H}^{r-2}(\Pi[\mathcal{A}])$. It is dual to the Björner basis. Using the JKS isomorphism (14.23), this yields a basis of $\mathsf{Lie}[\mathcal{A}] \otimes \mathsf{E}^{\mathrm{o}}[\mathcal{A}]$ which we continue to refer as the Lyndon basis.

Proposition 14.53. *Let* H *be any hyperplane. Then there is an isomorphism*

$$\bigoplus_{\mathrm{X}:\,\mathrm{rk}(\mathrm{X})=1, \mathrm{X} \nleq \mathrm{H}} \mathcal{H}^{\mathrm{top}}(\Pi[\mathcal{A}_{\mathrm{X}}]) \to \mathcal{H}^{\mathrm{top}}(\Pi[\mathcal{A}]).$$

The map on the X-summand is obtained by extending a maximal chain in $[\mathrm{X}, \top]$ to a maximal chain in $[\bot, \top]$. This is a special case of (14.29) after noting that $\mathcal{H}^{\mathrm{top}}(\Pi[\mathcal{A}^{\mathrm{X}}]) = \Bbbk$ when X has rank 1.

PROOF. Pick any choice function γ with $\gamma(\bot) = \mathrm{H}$. Then from the bijection in Lemma 14.48, we see that the union over X of the Lyndon basis of each summand in the lhs maps to the Lyndon basis of the rhs. □

Equivalently:

Proposition 14.54. *Let* H *be any hyperplane. Then the map induced from* (10.28)

$$\bigoplus_{\mathrm{X}:\,\mathrm{rk}(\mathrm{X})=1, \mathrm{X} \nleq \mathrm{H}} \mathsf{Lie}[\mathcal{A}^{\mathrm{X}}] \otimes \mathsf{Lie}[\mathcal{A}_{\mathrm{X}}] \xrightarrow{\cong} \mathsf{Lie}[\mathcal{A}]$$

is an isomorphism.

The previous two results were proved using the Lyndon basis but neither statement makes any reference to it. In fact, these results provide an inductive procedure for the construction of new bases. For instance, picking a basis for $\mathsf{Lie}[\mathcal{A}_{\mathrm{X}}]$ for each rank-one flat X not contained in a fixed hyperplane leads to a basis for $\mathsf{Lie}[\mathcal{A}]$.

Dualizing the map in Proposition 14.53 (and taking inverse) yields

$$\bigoplus_{\mathrm{X}:\,\mathrm{rk}(\mathrm{X})=1, \mathrm{X} \nleq \mathrm{H}} \mathcal{H}_{\mathrm{top}}(\Pi[\mathcal{A}_{\mathrm{X}}]) \xrightarrow{\cong} \mathcal{H}_{\mathrm{top}}(\Pi[\mathcal{A}]). \tag{14.37}$$

On the X-summand, this map is explicitly given as follows. For any maximal chain $(\mathrm{X} \lessdot \mathrm{Y}_1 \lessdot \cdots \lessdot \mathrm{Y}_{r-2} \lessdot \top)$ in $[\mathrm{X}, \top]$, define a linear combination of maximal chains by intersecting with H at all possible positions and inserting signs:

$$\begin{aligned} &(\bot \lessdot \mathrm{X} \lessdot \mathrm{Y}_1 \lessdot \cdots \lessdot \mathrm{Y}_{r-2} \lessdot \top) \\ &\quad + \sum_{i=1}^{r-2} (-1)^i (\bot \lessdot \mathrm{Y}_1 \wedge \mathrm{H} \lessdot \cdots \lessdot \mathrm{Y}_i \wedge \mathrm{H} \lessdot \mathrm{Y}_i \lessdot \cdots \lessdot \mathrm{Y}_{r-2} \lessdot \top) \\ &\qquad\qquad + (-1)^{r-1} (\bot \lessdot \mathrm{Y}_1 \wedge \mathrm{H} \lessdot \cdots \lessdot \mathrm{Y}_{r-2} \wedge \mathrm{H} \lessdot \mathrm{H} \lessdot \top). \end{aligned} \tag{14.38}$$

Linearizing yields a map $\mathcal{C}_{\mathrm{top}}(\Pi[\mathcal{A}_{\mathrm{X}}]) \to \mathcal{C}_{\mathrm{top}}(\Pi[\mathcal{A}])$ which preserves cycles and is the required map.

PROOF. The map (14.37) preserves the Björner basis since the dual map in Proposition 14.53 preserves the Lyndon basis. Thus, to show that (14.38) is the correct formula, we only need to check that it preserves the Björner basis. This is a straightforward observation. □

14.7. Coordinate arrangement

We specialize $\mathcal{A}$ to the coordinate arrangement of rank n (Section 6.1). It consists of the hyperplanes $x_i = 0$ for $1 \leq i \leq n$. By convention, let $x_i > 0$ be the $+$ side and $x_i < 0$ be the $-$ side. Faces correspond to n-tuples consisting of $+$, $-$ and 0 signs, while chambers correspond to n-tuples consisting of $+$ and $-$ signs.

14.7.1. Lie elements. The space $\mathsf{Lie}[\mathcal{A}]$ is one-dimensional and is spanned by the element

$$\sum_{D} (-1)^{m_+(D)} \mathtt{H}_D, \tag{14.39}$$

where the sum is over all chambers D, and $m_+(D)$ is the number of $+$ signs in the sign sequence of D. For $n = 1, 2$, the spanning Lie elements are

$$\mathtt{H}_- - \mathtt{H}_+ \quad \text{and} \quad \mathtt{H}_{--} - \mathtt{H}_{+-} - \mathtt{H}_{-+} + \mathtt{H}_{++}.$$

To see this: Recall that a vertex-based top-lune is a pair of adjacent chambers. Hence, by the Ree criterion, the sum of the coefficients of two adjacent chambers in a Lie element is 0. Since adjacent chambers are related by one sign change, the result follows.

Another way to see this is to use (10.7) and the fact that $\mathcal{A}$ is the n-fold cartesian product of the rank-one arrangement. Thus, the spanning element of $\mathsf{Lie}[\mathcal{A}]$ is the n-fold tensor product of $\mathtt{H}_- - \mathtt{H}_+$, which is the spanning Lie element in rank one.

Also recall that $\mu(\mathcal{A}) = (-1)^n$. This is consistent with (10.24).

Exercise 14.55. Make the isomorphism (13.12) explicit for the coordinate arrangement. The rank-one case $n = 1$ was explained in Section 13.5.4.

14.7.2. Substitution product. Let X be a flat of $\mathcal{A}$. Let z_1 be the Lie element (14.39) for $\mathcal{A}^{\mathrm{X}}$, and z_2 the same element for $\mathcal{A}_{\mathrm{X}}$. Their substitution product (10.28) then coincides with the Lie element (14.39) for $\mathcal{A}$. To see this, we employ (10.29) to calculate with chambers. The claim boils down to the fact that if D_1 and D_2 are chambers of $\mathcal{A}^{\mathrm{X}}$ and $\mathcal{A}_{\mathrm{X}}$ respectively, and D is their substitution product, then $m_+(D) = m_+(D_1) + m_+(D_2)$. This may be readily seen from the description in Section 6.1.7.

14.7.3. JKS. Specializing Theorem 14.38, we obtain:

Theorem 14.56. *There is an isomorphism of $\mathbb{Z}_2^n$-modules*

$$\mathcal{H}^{r-2}(\Pi[\mathcal{A}]) \otimes \mathsf{E}^{\mathrm{o}}[\mathcal{A}] \xrightarrow{\cong} \mathsf{Lie}[\mathcal{A}].$$

All spaces are one-dimensional. On $\mathsf{Lie}[\mathcal{A}]$ and $\mathsf{E}^{\mathrm{o}}[\mathcal{A}]$, the generator of each $\mathbb{Z}_2$ factor multiplies by -1, while on $\mathcal{H}^{r-2}(\Pi[\mathcal{A}])$, it acts by the identity.

One can also see this result directly. Since $\Pi[\mathcal{A}]$ is the Boolean poset, its order complex is a topological sphere. So $\mathcal{H}^{r-2}(\Pi[\mathcal{A}])$ is one-dimensional and any maximal chain of flats is a basis element. Further, unbracketing of any such maximal chain produces a Lie element with ± 1 coefficients of the chambers. Also note that any coboundary relation is the sum of two maximal chains which differ in exactly one position, and unbracketing them produces Lie elements which are negatives of each other.

Another way is to start with the $n = 1$ case for which the result is a triviality, take n-fold cartesian product, and use the compatibilities given in Section 14.5.8.

14.7.4. Dynkin elements. The hyperplane H defined by $x_1 + \cdots + x_n = 0$ is generic wrt the coordinate arrangement of rank n. Let h be the half-space $x_1 + \cdots + x_n \geq 0$. It contains only one chamber C, namely, the one whose sign sequence has all signs $+$. Barring this chamber and its opposite, the remaining chambers are all cut by H. The Dynkin element can be written as

$$\theta_{\mathrm{h}} = \sum_{F\colon F \leq C} (-1)^{m(F)}\, \mathtt{H}_F,$$

where $m(F)$ is the number of $+$ signs in the sign sequence of F. Observe that

$$\theta_{\mathrm{h}} \cdot \mathtt{H}_{\overline{C}} = \sum_{D} (-1)^{m(D)}\, \mathtt{H}_D,$$

where the rhs is the spanning Lie element given in (14.39). This formula is an instance of (14.7).

Exercise 14.57. Use the Ree criterion to check that θ_{h} is a special Zie element. (This was the method used to prove Proposition 14.1, but the point is that the argument is much simpler in the present case.) Alternatively, use the Friedrichs criterion to prove this fact.

Exercise 14.58. Show that $\mathsf{Zie}[\mathcal{A}]$ has dimension 2^n and that the set of all Dynkin elements form a basis for it.

Exercise 14.59. Check that for any choice function γ, there is exactly one γ-compatible labeled chain.

14.8. Rank-two arrangements

Let $\mathcal{A}$ be the rank-two arrangement of n lines, with $n \geq 2$. The spherical model is the $2n$-gon. A line passing through the origin is generic wrt $\mathcal{A}$ if it cuts two opposite sides of the $2n$-gon. For definiteness, we demand that the lines bisect the two sides that they cut. We discuss the Dynkin basis, Lyndon basis and JKS theorem for this arrangement. Lie and Zie elements were discussed in Sections 10.2.2 and 10.5.2.

14.8.1. Hexagon. Let us first consider the case $n = 3$ which is the hexagon.

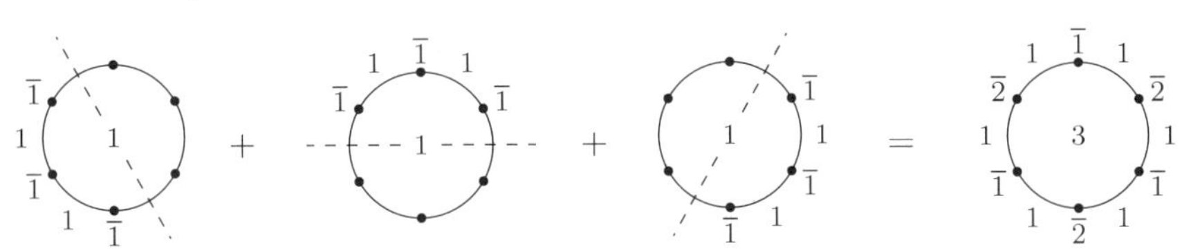

(By convention, $\overline{1}$ denotes -1.) The summands in the lhs represent Dynkin elements associated to three half-spaces as shown. These three half-spaces form an orbit under the Coxeter symmetries of the hexagon. (Reflection in a dotted line is not an allowed symmetry.) Hence their sum is a symmetrized Dynkin element d which is shown in the rhs. It is clear that there is another orbit consisting of three half-spaces (which pick the other side of the dotted line). This gives rise to another symmetrized Dynkin element $\overline{d}$. As indicated by the notation, d and $\overline{d}$ are related by the opposition map.

14.8.2. Octagon. Let us now consider the case $n = 4$ which is the octagon. Here there are eight generic half-spaces, and unlike the previous case, they form a single orbit. So there is only one symmetrized Dynkin element as shown below.

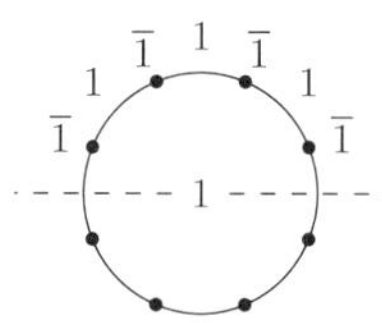

$+$ other terms $=$

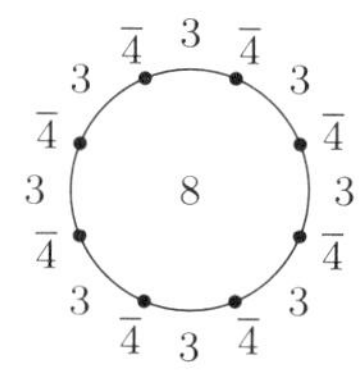

14.8.3. General case. There are two cases depending on whether n is even or odd. The analysis is similar to that of the octagon and hexagon, respectively.

If $n = 2k$, then all generic half-spaces lie in the same orbit. So there is exactly one symmetrized Dynkin element. Explicitly, it is given by

$$d = \sum_F a^F \mathtt{H}_F,$$

with

$$a^F := \begin{cases} 4k & \text{if } F = O, \\ -2k & \text{if } F \text{ is a vertex}, \\ 2k-1 & \text{if } F \text{ is an edge}. \end{cases} \tag{14.40}$$

For $k = 2$, this gives the formula for the octagon.

If $n = 2k+1$, then there are two orbits of generic half-spaces. So there are two symmetrized Dynkin elements. They are opposites of each other. Explicitly, the coefficients are given by

$$a^F := \begin{cases} 2k+1 & \text{if } F = O, \\ -k \text{ or } -(k+1) & \text{if } F \text{ is a vertex}, \\ k & \text{if } F \text{ is an edge}. \end{cases} \tag{14.41}$$

There are two types of vertices. Vertices of one type have coefficient k and of the other type have coefficient $k+1$. For $k = 1$, this gives the formula for the hexagon.

Exercise 14.60. For the arrangement $\mathcal{A}$ of n lines, check that the set of all Dynkin elements form a basis of $\mathsf{Zie}[\mathcal{A}]$. (Recall from Section 10.5.2 that the dimension of the latter is $2n$.)

Exercise 14.61. Consider the rank-two arrangement of 3 lines, and let h be a generic half-space. Let P, Q and R be the vertices and C and D be the chambers contained in h. Check that the possible choices for z in Corollary 14.4 are

$$z_{\alpha,\beta} = \mathtt{H}_O - \mathtt{H}_P - \mathtt{H}_Q - \mathtt{H}_R + \alpha\,\mathtt{H}_C + \beta\,\mathtt{H}_D$$

with $\alpha+\beta=2$. Check by explicit calculation that $(\mathtt{H}_O-(\mathtt{H}_O-z_{\alpha,\beta})^3)^3 = z_{1,1} = \theta_{\mathrm{h}}$.

14.8.4. Dynkin basis. In a $2n$-gon, fix any two opposite chambers, say D and $\overline{D}$. Let h be one of the two half-spaces whose base bisects D and $\overline{D}$. Then the set

$$\{\mathtt{H}_C+\mathtt{H}_{\overline{C}}-\mathtt{H}_D-\mathtt{H}_{\overline{D}} \mid \overline{C}\subseteq \mathrm{h}\}$$

is a basis for $\mathsf{Lie}[\mathcal{A}]$. It has $n-1$ elements. This is precisely the Dynkin basis associated to h and also to $\overline{\mathrm{h}}$ (consistent with Exercise 14.17). For instance, for $n=4$, a choice for the Dynkin basis is shown below.

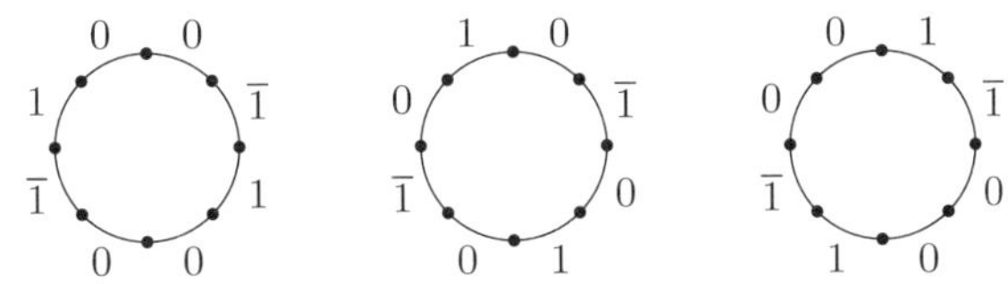

We now describe the action of the symmetrized Dynkin elements on chambers. For n even,

$$d\cdot \mathtt{H}_C=(n-1)\,\mathtt{H}_C+(n-1)\,\mathtt{H}_{\overline{C}}-\sum_{D\neq C,\overline{C}}\mathtt{H}_D. \tag{14.42}$$

For n odd, there are two symmetrized Dynkin elements. The action of one of them is given by

$$d\cdot \mathtt{H}_C=\frac{n-1}{2}\,\mathtt{H}_C+\frac{n-1}{2}\,\mathtt{H}_{\overline{C}}-\sum_{D\in \mathrm{A}}\mathtt{H}_D, \tag{14.43}$$

where A consists of edges which are either at odd distance from C while moving clockwise till $\overline{C}$, or at even distance from C while moving anticlockwise till $\overline{C}$. The action of $\overline{d}$ is similar with odd and even swapping places.

14.8.5. JKS and Jacobi identity. Any maximal chain of flats has the form $\bot\lessdot \mathrm{H}\lessdot \top$. So maximal chains correspond to hyperplanes. The top-cohomology $\mathcal{H}^{\mathrm{top}}(\Pi[\mathcal{A}])$ is spanned by all maximal chains subject to the coboundary relation

$$\sum_{\mathrm{H}}(\bot\lessdot \mathrm{H}\lessdot \top)^*.$$

The sum is over all hyperplanes. In particular, $\mathcal{H}^{\mathrm{top}}(\Pi[\mathcal{A}])$ has dimension $n-1$.

Unbracketing a maximal chain $\bot\lessdot \mathrm{H}\lessdot \top$ yields a Lie element of the form $\mathtt{H}_C+\mathtt{H}_{\overline{C}}-\mathtt{H}_D-\mathtt{H}_{\overline{D}}$, where C and D are adjacent, and their common panel has support H. These elements, as H varies, span $\mathsf{Lie}[\mathcal{A}]$ with the coboundary relation mapping to the Jacobi identity. This is illustrated below for $n=3$.

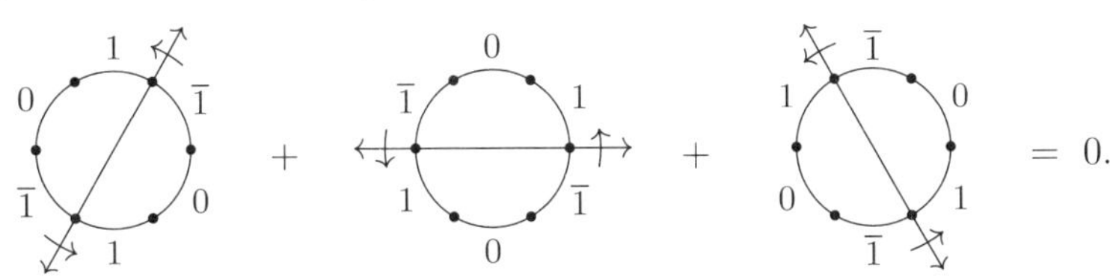

In the figure, the orientation chosen for unbracketing is the anticlockwise direction. (An illustration of how terms in the Jacobi identity arise from the substitution product of Lie is given in Example 10.44.)

14.8.6. Lyndon basis. Since $\perp$ is the only non-maximum flat which is not a hyperplane, a choice function is the same as choosing a hyperplane. Accordingly, fix a hyperplane H. The Lyndon basis is given by maximal chains

$$(\perp \lessdot \mathrm{H}' \lessdot \top)^*,$$

where H' varies over all hyperplanes distinct from H. There are $n-1$ of them. Fix a half-space h with base H. Then by unbracketing each of these maximal chains, we see that

$$\{\mathtt{H}_C + \mathtt{H}_{\overline{C}} - \mathtt{H}_D - \mathtt{H}_{\overline{D}} \mid C, D \subseteq \mathrm{h}, \text{ and } C \text{ is adjacent to } D \text{ and to the left of } D\}$$

is a basis for $\mathsf{Lie}[\mathcal{A}]$.

14.9. Classical (type A) Lie elements

We specialize $\mathcal{A}$ to the braid arrangement on $[n]$ (Sections 6.3–6.6). For convenience, we shorten all notations of the form $p[\mathcal{A}]$ to $p[n]$. For instance, we write $\Pi[n]$ for $\Pi[\mathcal{A}]$. In case we work with the braid arrangement on a finite set I, we will write $p[I]$.

Classical Lie elements are defined using bracket expressions. Such expressions are closely linked to binary trees which is turn are related to maximal chains of faces in the braid arrangement which in turn is the combinatorics involved in the JKS isomorphism. Unbracketing a bracket expression allows us to view a Lie element as a linear combination of linear orders.

Any coordinate hyperplane is generic wrt the braid arrangement. The corresponding (symmetrized) Dynkin elements act on chambers via the left- and right-bracketing operators. In particular, we obtain the Dynkin-Specht-Wever theorem which says that the image of the left-bracketing operator on chambers is the space of Lie elements. Along with the Dynkin basis, we also discuss the Björner-Wachs basis and the Lyndon basis.

We assume that the field characteristic is 0.

14.9.1. Classical Lie elements. Let $\mathsf{Lie}[n]$ denote the space of Lie elements. It is the primitive part of the space of linear orders $\Gamma[n]$. Using Theorem 14.41, one can deduce the following more classical way of defining Lie elements. More explanation is given in the discussion below on the JKS theorem.

Recall the external product on linear orders from Section 6.3.13 given by ordered concatenation. We extend it by multilinearity on the H-basis. Now for any $I = S \sqcup T$, and linear orders ℓ_1 on S and ℓ_2 on T, put

$$[\mathtt{H}_{\ell_1}, \mathtt{H}_{\ell_2}] := \mu_{(S,T)}(\mathtt{H}_{\ell_1}, \mathtt{H}_{\ell_2}) - \mu_{(T,S)}(\mathtt{H}_{\ell_2}, \mathtt{H}_{\ell_1}) = \mathtt{H}_{\ell_1 \cdot \ell_2} - \mathtt{H}_{\ell_2 \cdot \ell_1}, \tag{14.44}$$

where $\ell_1 \cdot \ell_2$ is the linear order in which elements of S precede the elements of T. We refer to this as the unbracketing operation. Lie elements are those elements of $\Gamma[n]$ which are generated by these unbracketing operations from linear orders on singletons in the set $[n]$.

For example, $\mathsf{Lie}[2]$ is one-dimensional and spanned by

$$[\mathtt{H}_1, \mathtt{H}_2] = -[\mathtt{H}_2, \mathtt{H}_1] = \mathtt{H}_{1|2} - \mathtt{H}_{2|1}.$$

The space $\mathsf{Lie}[3]$ is 2-dimensional and spanned by elements of the form

$$[[\mathtt{H}_1, \mathtt{H}_3], \mathtt{H}_2] = [\mathtt{H}_{1|3} - \mathtt{H}_{3|1}, \mathtt{H}_2] = \mathtt{H}_{1|3|2} - \mathtt{H}_{3|1|2} - \mathtt{H}_{2|1|3} + \mathtt{H}_{2|3|1}.$$

An element of $\mathsf{Lie}[4]$ is shown below.

$$\begin{aligned}[[\mathtt{H}_1, \mathtt{H}_3], [\mathtt{H}_2, \mathtt{H}_4]] &= [\mathtt{H}_{1|3} - \mathtt{H}_{3|1}, \mathtt{H}_{2|4} - \mathtt{H}_{4|2}] \\ &= \mathtt{H}_{1|3|2|4} - \mathtt{H}_{2|4|1|3} - \mathtt{H}_{3|1|2|4} + \mathtt{H}_{2|4|3|1} - \mathtt{H}_{1|3|4|2} + \mathtt{H}_{4|2|1|3} + \mathtt{H}_{3|1|4|2} - \mathtt{H}_{4|2|3|1}.\end{aligned}$$

These unbracketing operations satisfy antisymmetry and the classical Jacobi identity. Further, when two bracket expressions define the same Lie element, one can pass from one to the other by using these two relations.

Given $[n] = S \sqcup T$, where S and T are nonempty sets, and ℓ_1 and ℓ_2 are linear orders on S and T, respectively, let $\ell_1 \sqcup\!\sqcup\, \ell_2$ denote the sum of all shuffles of ℓ_1 and ℓ_2. Define $\mathsf{Lie}^{\perp}[n]$ to be the span of elements of the form $\ell_1 \sqcup\!\sqcup\, \ell_2$. This is the decomposable part of $\Gamma^*[n]$.

For example, $\mathsf{Lie}^{\perp}[2]$ is one-dimensional and spanned by

$$\mathtt{M}_{1|2} + \mathtt{M}_{2|1}$$

(which is the shuffle of 1 and 2). The space $\mathsf{Lie}^{\perp}[3]$ is 4-dimensional:

$$\mathtt{M}_{1|2|3} + \mathtt{M}_{1|3|2} + \mathtt{M}_{3|1|2},\ \mathtt{M}_{2|1|3} + \mathtt{M}_{2|3|1} + \mathtt{M}_{3|2|1}, \\ \mathtt{M}_{1|3|2} + \mathtt{M}_{1|2|3} + \mathtt{M}_{2|1|3},\ \text{and } \mathtt{M}_{3|1|2} + \mathtt{M}_{3|2|1} + \mathtt{M}_{2|3|1}$$

form a basis. (They are shuffles of $1|2$ and 3, $2|1$ and 3, $1|3$ and 2, and $3|1$ and 2.)

Lemma 14.62. *The subspaces* $\mathsf{Lie}[n]$ *and* $\mathsf{Lie}^{\perp}[n]$ *are orthogonal to each other under the canonical pairing between* $\Gamma[\mathcal{A}]$ *and* $\Gamma[\mathcal{A}]^*$.

PROOF. This is a specialization of Lemma 10.10. ☐

For $n = 2$, the orthogonality is illustrated below.

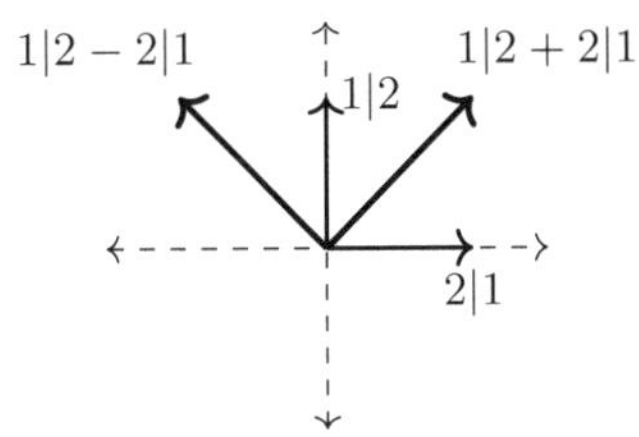

The vector in the north-west direction belongs to $\mathsf{Lie}[2]$, while the one in the north-east direction belongs to $\mathsf{Lie}^{\perp}[2]$.

14.9.2. JKS. Recall the symmetric group S_n on n letters. It is the Coxeter group of the braid arrangement on $[n]$. Specializing Theorem 14.38, we obtain:

Theorem 14.63. *There is an isomorphism of* S_n*-modules*

$$\mathcal{H}^{r-2}(\Pi[n]) \otimes \mathsf{E}^{\mathrm{o}}[n] \xrightarrow{\cong} \mathsf{Lie}[n]. \tag{14.45}$$

Let us describe it explicitly. For clarity, we work with the braid arrangement on a finite set I. Setting $I = [n]$ would yield the above map. A maximal chain of faces f corresponds to a binary tree with $|I|$ leaves each with a distinct label from I, and whose internal nodes are linearly ordered in such a manner that each node appears before both of its children. (In particular, the linear order begins with the root.) An illustration of the bijection for $I = \{a, b, c, d, e\}$ is shown below.

$$(abcde \lessdot ab|cde \lessdot ab|cd|e \lessdot a|b|cd|e \lessdot a|b|c|d|e) \longleftrightarrow$$

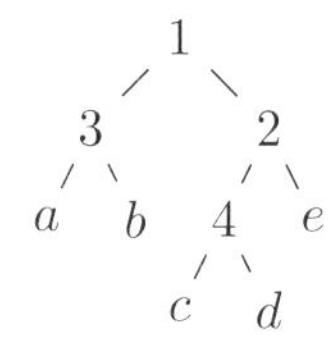

Now a binary tree gives rise to a bracket expression, and hence yields a Lie element. Continuing the example, with f denoting the maximal chain of faces, the map is

$$\mathrm{s}(f)^* \otimes [f] \mapsto [[\mathtt{H}_a, \mathtt{H}_b], [[\mathtt{H}_c, \mathtt{H}_d], \mathtt{H}_e]].$$

Unbracketing the rhs results in a linear combination of linear orders on $\{a, b, c, d, e\}$, and this is what (14.20e) gives.

14.9.3. Classical Jacobi identity. The braid arrangement on $I = \{a, b, c\}$ is shown below.

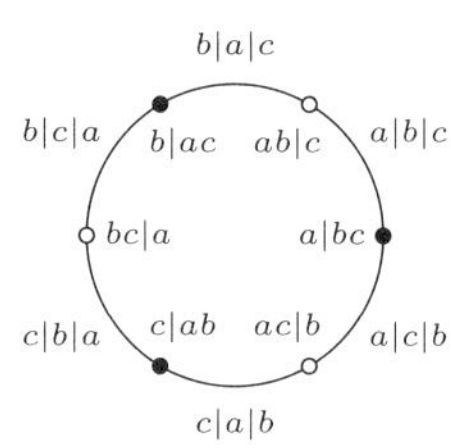

The Lie element obtained by unbracketing the maximal chain of flats

$$(\{abc\} \lessdot \{c, ab\} \lessdot \{a, b, c\})$$

with the anticlockwise orientation is given by

$$[\mathtt{H}_c, [\mathtt{H}_a, \mathtt{H}_b]] = \mathtt{H}_{c|a|b} - \mathtt{H}_{c|b|a} - \mathtt{H}_{a|b|c} + \mathtt{H}_{b|a|c}.$$

Similarly, one can unbracket the remaining two maximal chains of flats. The sum of the three Lie elements thus obtained is zero. This is the classical Jacobi identity and is illustrated below.

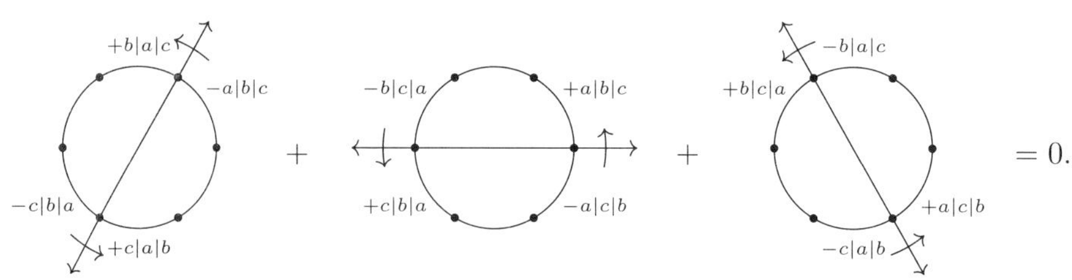

$[c, [a, b]]$ $\qquad$ $[a, [b, c]]$ $\qquad$ $[b, [c, a]]$

Unbracketing with the clockwise orientation yields the following form of the identity.

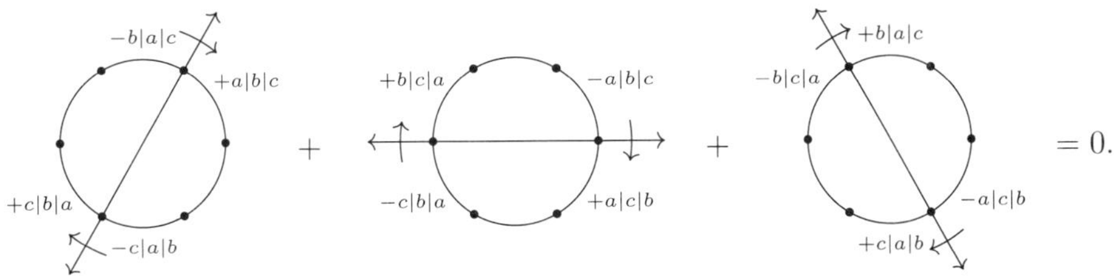

$[[a, b], c]$ $[[b, c], a]$ $[[c, a], b]$

14.9.4. Orientation and signature spaces. We explain the relation between the orientation space $\mathsf{E}^{\mathrm{o}}[n]$ and the signature space $\mathsf{E}^{-}[n]$.

Let $\mathsf{mc}[n]$ denote the space spanned by maximal chains of faces. Consider the map

$$\Gamma[n] \to \mathsf{mc}[n], \qquad \mathtt{H}_{l_1|l_2|\dots|l_n} \mapsto (O, F_1, F_2, \dots, F_{n-1}),$$

where

$$F_1 = l_1|l_2 \dots l_n, \quad F_2 = l_1|l_2|l_3 \dots l_n, \quad \dots, \quad F_{n-1} = l_1|l_2|\dots|l_n.$$

In other words, F_i consists of the first i singletons followed by the remaining elements. One may check that by passing to quotients, this map induces an isomorphism of S_n-modules

$$\mathsf{E}^{-}[n] \xrightarrow{\cong} \mathsf{E}^{\mathrm{o}}[n].$$

The key observation is that maximal chains of faces arising from adjacent chambers have opposite orientations.

The image of the element $\mathtt{H}_{l_1|\dots|l_n}$ in $\mathsf{E}^{-}[n]$ may be denoted $l_1 \wedge \dots \wedge l_n$, with the usual understanding that switching two adjacent letters incurs a minus sign. Thus, $\mathsf{E}^{-}[n]$ (and hence $\mathsf{E}^{\mathrm{o}}[n]$) coincides with the sign representation of S_n.

14.9.5. Whitney cohomology. We now relate Whitney cohomology of the lattice of flats to Lie elements. For the braid arrangement $\mathcal{A}$ on $[n]$, recall that $\mathcal{A}^{\mathrm{X}}$ is also a braid arrangement on the number of blocks of X, while $\mathcal{A}_{\mathrm{X}}$ is the cartesian product of braid arrangements, one for each block of X. Since $\mathsf{E}^{\mathrm{o}}[n]$ is the sign representation of S_n, the rhs of (14.25) can be identified with $\bigwedge \mathsf{Lie}[n]$, the multilinear part of the exterior algebra on the free Lie algebra on $[n]$.

Explicitly, for $n = 2$, this space is two-dimensional with basis

$$\mathtt{H}_1 \wedge \mathtt{H}_2 \qquad \text{and} \qquad [\mathtt{H}_1, \mathtt{H}_2];$$

thus, it is a direct sum of two sign representations. For $n = 3$, this space is 6-dimensional with basis

$$[[\mathtt{H}_1, \mathtt{H}_2], \mathtt{H}_3], \quad [[\mathtt{H}_1, \mathtt{H}_3], \mathtt{H}_2], \quad \mathtt{H}_1 \wedge [\mathtt{H}_2, \mathtt{H}_3], \quad \mathtt{H}_2 \wedge [\mathtt{H}_1, \mathtt{H}_3], \quad \mathtt{H}_3 \wedge [\mathtt{H}_1, \mathtt{H}_2], \quad \mathtt{H}_1 \wedge \mathtt{H}_2 \wedge \mathtt{H}_3.$$

The relations are antisymmetry of the wedge, as well as antisymmetry and Jacobi identity of the bracket.

Thus, specializing Theorem 14.40, we obtain:

Theorem 14.64. *There is an isomorphism of* S_n*-modules*

$$\mathcal{WH}^*(\Pi[n]) \otimes \mathsf{E}^{\mathrm{o}}[n] \longrightarrow \bigwedge \mathsf{Lie}[n].$$

14.9.6. Classical Lie operad. Let $\mathcal{A}$ be the braid arrangement on the set I. For a partition X of I, the substitution product (10.28) is as follows. An element of $\mathsf{Lie}[\mathcal{A}^{\mathrm{X}}]$ is a bracket expression on the blocks of X, while an element of $\mathsf{Lie}[\mathcal{A}_{\mathrm{X}}]$ is a family of bracket expressions, one on each block of X. The two together specify a bracket expression on I, which is an element of $\mathsf{Lie}[\mathcal{A}]$. For example, for the partition $\mathrm{X} = \{abd, fg, ce\}$,

$$([[\mathtt{H}_{fg}, \mathtt{H}_{ce}], \mathtt{H}_{abd}], \{[\mathtt{H}_a, [\mathtt{H}_d, \mathtt{H}_b]], [\mathtt{H}_g, \mathtt{H}_f], [\mathtt{H}_c, \mathtt{H}_e]\}) \mapsto [[[\mathtt{H}_g, \mathtt{H}_f], [\mathtt{H}_c, \mathtt{H}_e]], [\mathtt{H}_a, [\mathtt{H}_d, \mathtt{H}_b]]].$$

Bracket expressions on finite sets with these structure maps constitute the *classical Lie operad.*

Using the structure maps of the classical associative operad in Section 6.5.10, one can check that diagram (10.29) indeed commutes.

14.9.7. Linear orders as symmetrized Lie elements. Proposition 13.28 for the uniform section of the braid arrangement on I specializes as follows.

Proposition 14.65. *Fix a finite set I. The linear map*

$$\bigoplus_{\mathrm{X} \vdash I} \bigotimes_{S \in \mathrm{X}} \mathsf{Lie}[S] \to \Gamma[I], \qquad \bigotimes_{S \in \mathrm{X}} z_S \mapsto \frac{1}{\deg!(\mathrm{X})} \sum_{\substack{F=(S_1,\dots,S_k)\vDash I,\\ \mathrm{s}(F)=\mathrm{X}}} \mu_F(z_{S_1}, z_{S_2}, \dots, z_{S_k})$$

is an isomorphism.

In the expression for the map, the sum is over compositions F of I, and μ_F refers to the external product on linear orders given by ordered concatenation. An illustration on the partition $\mathrm{X} = \{ace, bd\}$ is given below.

$$[\mathtt{H}_c, [\mathtt{H}_a, \mathtt{H}_e]] \otimes [\mathtt{H}_d, \mathtt{H}_b] \mapsto$$
$$\frac{1}{2}\big(\mu_{ace,bd}([\mathtt{H}_c, [\mathtt{H}_a, \mathtt{H}_e]], [\mathtt{H}_d, \mathtt{H}_b]) + \mu_{bd,ace}([\mathtt{H}_d, \mathtt{H}_b], [\mathtt{H}_c, [\mathtt{H}_a, \mathtt{H}_e]])\big).$$

14.9.8. Dynkin elements. Recall that the braid arrangement is not essential. Let H_0 denote the hyperplane $x_1 + \cdots + x_n = 0$. The arrangement under this flat of the braid arrangement is the essential braid arrangement. Its faces are indexed by set compositions.

For $1 \leq i \leq n$, let H_i denote the intersection of the coordinate hyperplane $x_i = 0$ with H_0. This is a generic hyperplane wrt the essential braid arrangement. Let h_i denote the half-space of H_i corresponding to $x_i \geq 0$. Observe that:

Lemma 14.66. *A set composition F is contained in h_i iff i belongs to the last block of F.*

Let θ_i denote the Dynkin element associated to h_i. Thus,

$$\theta_i = \sum_F (-1)^{\mathrm{rk}(F)}\, \mathtt{H}_F, \tag{14.46}$$

where the sum is over all set compositions F whose last block contains i. The half-spaces $\{\mathrm{h}_i\}_{1 \leq i \leq n}$ form an orbit under the action of the symmetric group. So the symmetrized Dynkin elements of all the θ_i are the same and given by their sum

$$d_n := \theta_1 + \cdots + \theta_n.$$

Explicitly,

$$d_n = \sum_F (-1)^{\mathrm{rk}(F)}\, w^F \mathtt{H}_F, \tag{14.47}$$

where w^F is the size of the last block of F. The sign in front is the negative of the parity of the number of blocks of F. For example,

$$d_2 = 2\,\mathtt{H}_{12} - \mathtt{H}_{1|2} - \mathtt{H}_{2|1}$$

and

$$d_3 = 3\,\mathtt{H}_{123} - 2\,\mathtt{H}_{1|23} - 2\,\mathtt{H}_{2|13} - 2\,\mathtt{H}_{3|12} - \mathtt{H}_{12|3} - \mathtt{H}_{13|2} - \mathtt{H}_{23|1} \\ + \mathtt{H}_{1|2|3} + \mathtt{H}_{2|1|3} + \mathtt{H}_{1|3|2} + \mathtt{H}_{2|3|1} + \mathtt{H}_{3|1|2} + \mathtt{H}_{3|2|1}.$$

The preceding discussion also applies to the braid arrangement on I. There is a Dynkin element θ_i one for each $i \in I$. Their sum is a symmetrized Dynkin element which we denote by d_I. It is an element of $\Sigma[I]$.

Let $\overline{\mathrm{h}_i}$ denote the half-space opposite to h_i. It corresponds to the points $x_i \leq 0$. Let $\overline{\theta_i}$ be the Dynkin element associated to $\overline{\mathrm{h}_i}$. Explicitly, $\overline{\theta_i}$ is given as in (14.46), where the sum is over all set compositions F whose first block contains i. The symmetrized Dynkin element can be expressed as

$$\overline{d_n} = \sum_F (-1)^{\mathrm{rk}(F)}\, w^{\overline{F}} \mathtt{H}_F,$$

where $w^{\overline{F}}$ is the size of the first block of F.

The symmetrized Dynkin elements d_n and $\overline{d_n}$ are Zie elements of $\Sigma[n]$. This is an instance of Proposition 14.6. Further, by (14.4),

$$d_n^2 = n d_n.$$

Note that d_n/n is a special Zie element.

Exercise 14.67. Show that, up to equivalence, the rank-three braid arrangement $\mathcal{A}$ has 32 generic half-spaces. (This is consistent with (6.13) in view of Lemma 1.54.) Only 8 of these are accounted for by the h_i and their opposites. Write down the Dynkin elements for the other 24 half-spaces. Check that the dimension of $\mathsf{Zie}[\mathcal{A}]$ is 26. Hence the set of Dynkin elements is not linearly independent. Give one explicit linear dependency relation.

Exercise 14.68. A family $\mathcal{F}$ of nonempty proper subsets of $[n]$ is a *positive sum system* if there exist real numbers $a_1, \dots, a_n$ summing to zero such that for any nonempty proper subset S, the sum $\sum_{i \in S} a_i$ is positive if $S \in \mathcal{F}$ and negative if $S \notin \mathcal{F}$.

(i) Let $\mathcal{F}$ be a positive sum system. Show that there exists a generic half-space h of the braid arrangement on $[n]$ for which the vertices that lie in h correspond to the subsets in $\mathcal{F}$.
(ii) Deduce that $\mathcal{F}$ consists of $2^{n-1} - 1$ sets.
(iii) View $\mathcal{F}$ as a poset under inclusion. Show that chains in $\mathcal{F}$ correspond to faces in h, and maximal chains to chambers. Thus, the order complex of $\mathcal{F}$ is the subcomplex of faces that lie in h.
(iv) Use Corollary 14.2 and (6.9) to deduce that the number of maximal chains in $\mathcal{F}$ is $(n-1)!$.

14.9.9. Complete system from Zie elements. We discuss a special case of the construction given in Section 12.5.5. Fix a finite set I. For each composition $F = (S_1, \ldots, S_k)$ of I, put

$$\mathtt{Q}_F := \mu_F\Big(\frac{d_{S_1}}{|S_1|}, \ldots, \frac{d_{S_k}}{|S_k|}\Big). \tag{14.48}$$

The Dynkin elements are multiplied using the external product. This defines a $\mathtt{Q}$-basis. It has a corresponding Eulerian family $\mathtt{E}$ and a homogeneous section $\mathtt{u}$. The first Eulerian idempotent is

$$\mathtt{E}_{\{I\}} = \mathtt{Q}_{(I)} = \frac{d_I}{|I|}.$$

The homogeneous section is determined by (12.54). It is given by the following formula. For each composition $\alpha = (a_1, \ldots, a_k)$, put

$$\mathtt{u}^\alpha := \prod_{i=1}^{k} \frac{a_i}{a_1 + \cdots + a_i}. \tag{14.49}$$

For any set composition F, let

$$\mathtt{u}^F := \mathtt{u}^{\mathrm{t}(F)}. \tag{14.50}$$

To see that this is the correct formula, we need to understand the cancelations in the rhs of (12.54). For any noncentral face G, we want to show that the coefficient of $\mathtt{H}_G$ is zero. The contributions come from faces $F \leq G$. Consider those F whose first block equals the first block of G. To any such F, there is a corresponding F' obtained by merging its first two blocks. The key observation is that the contributions of F and F' cancel. Thus, we obtain:

Theorem 14.69. *Let $\mathtt{Q}_F$ be as in (14.48) and $\mathtt{u}^F$ as in (14.50). As X varies over all compositions of I, the elements*

$$\mathtt{E}_{\mathrm{X}} := \sum_{F:\mathrm{s}(F)=\mathrm{X}} \mathtt{u}^F \mathtt{Q}_F$$

form a complete system of primitive orthogonal idempotents of the Tits algebra of the braid arrangement on I.

Exercise 14.70. For any ordered set S, consider $z_S := \theta_{\max S}$, the Dynkin element in (14.46). The sum is over all compositions of S whose last block contains the maximum element of S. Now fix a positive integer n. For each composition $F = (S_1, \ldots, S_k)$ of $[n]$, put

$$\mathtt{Q}_F := \mu_F(z_{S_1}, \ldots, z_{S_k}),$$

where for each S_i, we use the order induced from $[n]$. This yields a $\mathtt{Q}$-basis for the Tits algebra of the braid arrangement on $[n]$. Show that the corresponding homogeneous section $\mathtt{u}$ determined by (12.54) is set-theoretic and given by

$$\mathtt{u}^F = \begin{cases} 1 & \text{if } \max S_1 < \cdots < \max S_k \text{ with } F = (S_1, \ldots, S_k), \\ 0 & \text{otherwise.} \end{cases}$$

This leads to a complete system as in Theorem 14.69. For instance, for $n = 2$,

$$\mathtt{Q}_{1|2} = z_1 z_2 = \mathtt{H}_1\mathtt{H}_2 = \mathtt{H}_{1|2}, \quad \mathtt{Q}_{2|1} = z_2 z_1 = \mathtt{H}_2\mathtt{H}_1 = \mathtt{H}_{2|1}, \quad \mathtt{Q}_{12} = z_{12} = \mathtt{H}_{12} - \mathtt{H}_{1|2},$$

with corresponding $\mathtt{u}$ given by $\mathtt{u}^{12} = 1$, $\mathtt{u}^{1|2} = 1$ and $\mathtt{u}^{2|1} = 0$, and Eulerian family $\mathtt{E}$ given by $\mathtt{E}_{1,2} = \mathtt{H}_{1|2}$ and $\mathtt{E}_{12} = \mathtt{H}_{12} - \mathtt{H}_{1|2}$.

14.9.10. Dynkin-Specht-Wever theorem. Let ℓ and ℓ' be linear orders on some finite set. We say that ℓ' is *peakless* wrt ℓ if in the linear order of ℓ, the entries of ℓ' decrease (till they reach the first entry of ℓ) and then increase.

For example, for

$$\ell = 3|1|5|2|4 \qquad \text{and} \qquad \ell' = 4|1|3|5|2,$$

ℓ' is peakless wrt ℓ since $4 > 1 > 3 < 5 < 2$ in the order $3 < 1 < 5 < 2 < 4$.

Lemma 14.71. *For a linear order $\ell = i_1|i_2|\ldots|i_n$,*

$$[\ldots[\mathtt{H}_{i_1}, \mathtt{H}_{i_2}], \ldots, \mathtt{H}_{i_n}] = \sum_{\substack{I=S\sqcup T\\ i_1\in T}} (-1)^{|S|}\mathtt{H}_{\overline{\ell|_S}\cdot\ell|_T} = \sum (-1)^k \mathtt{H}_{\ell'}. \tag{14.51}$$

The second sum is over all ℓ' which are peakless wrt ℓ, and k is the number of elements that precede i_1 in ℓ'.

PROOF. Expanding $\big[\ldots[\mathtt{H}_{i_1}, \mathtt{H}_{i_2}], \ldots, \mathtt{H}_{i_n}\big]$ yields a sum of elements of the form $\pm\mathtt{H}_{\ell'}$, where each linear order ℓ' arises from a sequence of 2^{n-1} choices as follows. First we choose whether i_2 precedes or follows i_1. Then we choose whether i_3 precedes or follows both of i_1 and i_2. We continue in this manner until we finally choose whether i_n precedes or follows all the other elements. Each time we choose to precede, a minus sign is picked up. Let S consist of those elements i_j that were chosen to precede the elements $i_1, \ldots, i_{j-1}$, and T of the remaining elements of I. Then i_1 belongs to T. In the linear order ℓ', the elements of S appear first, and reversed from the manner in which they appear in ℓ; the elements of T appear last and in the same manner as in ℓ. Thus, $\ell' = \overline{\ell|_S} \cdot \ell|_T$ and the element that appears in the expansion is $(-1)^{|S|}\mathtt{H}_{\overline{\ell|_S}\cdot\ell|_T}$. □

Recall that $\mathrm{h}_k(D)$ denotes the largest face of D which is contained in the half-space h_k (corresponding to $x_k \geq 0$). Explicitly, $\mathrm{h}_k(D)$ is the largest face of D for which k is in the last block. For example,

$$\mathrm{h}_2(4|5|2|1|3) = 4|5|213.$$

For any chambers C and D,

$$\mathrm{Des}(C, D) = \mathrm{h}_k(D) \iff \text{The first entry of } C \text{ is } k, \text{ and } D \text{ is peakless wrt } C.$$

This is straightforward to check.

Lemma 14.72. *For $k \in I$ and $i_1|i_2|\ldots|i_n$ a linear order on I,*

$$\theta_k \cdot \mathtt{H}_{i_1|i_2|\ldots|i_n} = \begin{cases} [\ldots[\mathtt{H}_{i_1}, \mathtt{H}_{i_2}], \ldots, \mathtt{H}_{i_n}] & \text{if } i_1 = k, \\ 0 & \text{otherwise.} \end{cases} \tag{14.52}$$

PROOF. We use formula (14.6) in conjunction with the previous observation for $C = i_1|i_2|\ldots|i_n$: If $i_1 \neq k$, then $\mathrm{Des}(C, D) = \mathrm{h}_k(D)$ has no solutions, so the lhs is zero. If $i_1 = k$, then the lhs is a signed sum over all D which are peakless wrt C. By (14.51), this sum is precisely $[\ldots[\mathtt{H}_{i_1}, \mathtt{H}_{i_2}], \ldots, \mathtt{H}_{i_n}]$. □

Since d_I is the sum of the θ_k, we obtain:

Lemma 14.73. *For any linear order $i_1|i_2|\ldots|i_n$ on I,*

$$d_I \cdot \mathtt{H}_{i_1|i_2|\ldots|i_n} = [\ldots[\mathtt{H}_{i_1}, \mathtt{H}_{i_2}], \ldots, \mathtt{H}_{i_n}]. \tag{14.53}$$

In words, d_I is the left bracketing operator *on $\Gamma[I]$.*

If we use $\overline{d_I}$ instead of d_I, then we get the right-bracketing operator.

Theorem 14.74. *The image of the left bracketing operator on* $\mathsf{\Gamma}[I]$ *is* $\mathsf{Lie}[I]$*. In addition,* $\mathsf{Lie}[I]$ *is the eigenspace of eigenvalue* $|I|$ *of this linear operator.*

This is the *Dynkin-Specht-Wever theorem.*

Exercise 14.75. Establish (14.52) directly by induction on n.

Exercise 14.76. Check that formula (14.53) is consistent with (14.43) when I has three elements.

14.9.11. Dynkin basis. Recall from Proposition 14.16 that to any generic half-space is associated a Dynkin basis of the space of Lie elements. Let us understand this explicitly for the half-space h_k. A linear order $i_1|i_2|\dots|i_n$ on I is contained in $\overline{\mathrm{h}_k}$ iff $i_1 = k$. To get the Dynkin basis, we need to apply θ_k on each of these linear orders. Formula (14.52) now implies:

Proposition 14.77. *For a fixed* $k \in I$*, the elements*

$$[\dots[\mathtt{H}_{i_1}, \mathtt{H}_{i_2}], \dots, \mathtt{H}_{i_n}],$$

as $i_1|i_2|\dots|i_n$ *varies over all linear orders on* I *with* $i_1 = k$*, form a basis of* $\mathsf{Lie}[I]$*. This is the Dynkin basis associated to the half-space* h_k*.*

14.9.12. Dynkin element inside top-nested faces and pairs of chambers. Recall from Section 9.4.7 that $\Sigma[I]$ can be viewed as a subspace of $\widehat{\mathsf{Q}}[I]$, which in turn can be viewed as a subspace of $\mathsf{\Gamma}[I]^* \otimes \mathsf{\Gamma}[I]$. For the latter, we put

$$\mathtt{K}_{C,D} := \mathtt{M}_C \otimes \mathtt{H}_D.$$

The formulas for the Dynkin element viewed inside these larger spaces are as follows.

Lemma 14.78. *We have*

$$d_I = \sum (-1)^{\mathrm{rk}(H)}\, \mathtt{K}_{H,D} \quad \textit{and} \quad d_I = \sum (-1)^{\mathrm{rk}(\mathrm{Des}(C,D))}\, \mathtt{K}_{C,D}. \tag{14.54}$$

The first sum is over all pairs $H \le D$ *such that all blocks of* H*, except possibly the last, are singletons. The second sum is over all* C *and* D *such that* D *is peakless wrt* C*. In this situation,* $\mathrm{rk}(\mathrm{Des}(C,D))$ *is the number of elements in* D *which precede* i_1*, where* i_1 *is the first element of* C*.*

PROOF. The calculation for the first formula is as follows. Using (9.43),

$$\sum (-1)^{\mathrm{rk}(H)} \mathtt{K}_{H,D} = \sum_{K \le D} (-1)^{\mathrm{rk}(K)} \sum_{H} \mathtt{H}_{K,D} = \sum_{K \le D} (-1)^{\mathrm{rk}(K)} w^K \mathtt{H}_{K,D}.$$

In the middle expression, the inside sum is over all H between K and D such that all blocks of H, except possibly the last, are singletons. The key observation is that there are w^K such choices for H, where w^K is the size of the last block of K.

The second formula follows from the first by using (9.45). The key point is that the condition on the blocks of H translates to the peakless condition on D wrt C. Observe that the second formula, assuming (14.51), gives another proof of (14.53). □

14.9.13. q-Dynkin element. For any scalar q, define the *q-Dynkin element* by

$$d_{I,q} := \sum_F (-1)^{\mathrm{rk}(F)}\, q^{|I|-w^F} (1+q+\cdots+q^{w^F-1})\, \mathtt{H}_F. \tag{14.55}$$

The sum is over all compositions F of I, and w^F is the size of the last block of F.

For example,

$$d_{2,q} = (1+q)\,\mathtt{H}_{12} - q\,\mathtt{H}_{1|2} - q\,\mathtt{H}_{2|1}$$

and

$$d_{3,q} = (1+q+q^2)\,\mathtt{H}_{123} - q(1+q)\,(\mathtt{H}_{1|23}+\mathtt{H}_{2|13}+\mathtt{H}_{3|12}) - q^2\,(\mathtt{H}_{12|3}+\mathtt{H}_{13|2}+\mathtt{H}_{23|1}) + q^2\,(\mathtt{H}_{1|2|3}+\mathtt{H}_{2|1|3}+\mathtt{H}_{1|3|2}+\mathtt{H}_{2|3|1}+\mathtt{H}_{3|1|2}+\mathtt{H}_{3|2|1}).$$

Setting $q=1$ recovers d_I, while $q=-1$ yields

$$d_{I,-1} = (-1)^{|I|-1} \sum_{F:\, w^F \text{ is odd}} (-1)^{\mathrm{rk}(F)} \mathtt{H}_F. \tag{14.56}$$

Also note that $d_{I,0} = \mathtt{H}_{(I)}$.

The q-Dynkin element viewed inside the spaces of nested faces and pairs of chambers is given by

$$d_{I,q} = \sum (-q)^{\mathrm{rk}(H)}\, \mathtt{K}_{H,D} \quad \text{and} \quad d_{I,q} = \sum (-q)^{\mathrm{rk}(\mathrm{Des}(C,D))}\, \mathtt{K}_{C,D}. \tag{14.57}$$

The conditions on H, C and D are as in (14.54). Observe that for $q=-1$, the signs go away.

For any $I = S \sqcup T$, and linear orders ℓ_1 on S and ℓ_2 on T, put

$$[\mathtt{H}_{\ell_1}, \mathtt{H}_{\ell_2}]_q := \mathtt{H}_{\ell_1 \cdot \ell_2} - q\,\mathtt{H}_{\ell_2 \cdot \ell_1}.$$

This is the *q-commutator*. Formula (14.51) generalizes, with (-1) replaced by $(-q)$, and (14.53) generalizes to

$$d_{I,q} \cdot \mathtt{H}_{i_1|i_2|\dots|i_n} = [\dots[\mathtt{H}_{i_1}, \mathtt{H}_{i_2}]_q, \dots, \mathtt{H}_{i_n}]_q. \tag{14.58}$$

Thus $d_{I,q}$ is the *left q-bracketing operator* on $\Gamma[I]$.

14.9.14. Björner-Wachs basis. For a chamber $C = i_1|\dots|i_n$, let

$$f_C := (i_1\dots i_n \lessdot i_1\dots i_{n-1}|i_n \lessdot i_1\dots i_{n-2}|i_{n-1}|i_n \lessdot \cdots \lessdot i_1|\dots|i_n).$$

This is a maximal chain of faces ending at C (obtained by putting bars from right to left). Under the correspondence between maximal chains and binary trees, the maximal chain f_C goes to the 'left-comb' binary tree in which the right child of each internal node is a leaf. The internal nodes are ordered in the only way possible.

Note from (14.17) that

$$\mathtt{BW}_C = \Big(\sum_f \pm\, \mathrm{s}(f)\Big) \otimes [f_C], \tag{14.59}$$

where the sum is over all maximal chains of faces f ending at C. The sign of $\mathrm{s}(f)$ is $+1$ if $[f] = [f_C]$ and -1 if $[f] = -[f_C]$. In particular, $\mathtt{BW}_C$ contains the term $\mathrm{s}(f_C) \otimes [f_C]$ (with a plus sign).

Recall the generic half-space h_k corresponding to $x_k \geq 0$. A linear order $i_1|i_2|\dots|i_n$ on $[n]$ is contained in $\overline{\mathrm{h}_k}$ iff $i_1 = k$. Thus, the BW-basis associated to the half-space h_k consists of the elements $\mathtt{BW}_C$, as C varies over all linear orders of $[n]$ starting with k.

Proposition 14.79. *The dual of the* BW*-basis associated to the half-space* h_k *is given by*

$$\mathtt{BW}^*_C = \mathrm{s}(f_C)^* \otimes [f_C], \tag{14.60}$$

where C *varies over all linear orders of* $[n]$ *starting with* k.

In the rhs, when we write $\mathrm{s}(f_C)^*$, we mean the cohomology class of $\mathrm{s}(f_C)^*$.

PROOF. Let P denote the vertex (two-block set composition) whose first block consists of the singleton k. Note that the BW-basis is indexed by chambers in the star of P. If C and D are two such chambers, then it is easy to check that $\mathrm{s}(f_C)$ appears in $\mathtt{BW}_D$ iff $C = D$. The claim follows.

Alternatively: In view of Corollary 14.39, it suffices to check that the JKS isomorphism sends $\mathtt{BW}^*_C$ to $\theta_k \cdot \mathtt{H}_C$. Since f_C corresponds to the left-comb binary tree, the JKS isomorphism sends it to the Lie element obtained by left-bracketing. By (14.52), this is precisely $\theta_k \cdot \mathtt{H}_C$. □

Let us illustrate for $n = 3$ for the half-space h_1. The BW-basis consists of the two elements

$$\mathtt{BW}_{1|2|3} = \big((\{123\} \lessdot \{12,3\} \lessdot \{1,2,3\}) - (\{123\} \lessdot \{1,23\} \lessdot \{1,2,3\})\big) \otimes [f_{1|2|3}]$$

and

$$\mathtt{BW}_{1|3|2} = \big((\{123\} \lessdot \{13,2\} \lessdot \{1,2,3\}) - (\{123\} \lessdot \{1,23\} \lessdot \{1,2,3\})\big) \otimes [f_{1|3|2}].$$

Note that $\{12,3\}$ and $\{1,23\}$ are the supports of the two vertices of $1|2|3$. These gave the two terms in $\mathtt{BW}_{1|2|3}$.

The dual BW-basis consists of

$$\mathtt{BW}^*_{1|2|3} = (\{123\} \lessdot \{12,3\} \lessdot \{1,2,3\})^* \otimes [f_{1|2|3}]$$

and

$$\mathtt{BW}^*_{1|3|2} = (\{123\} \lessdot \{13,2\} \lessdot \{1,2,3\})^* \otimes [f_{1|3|2}].$$

These are simpler and have only one term each.

14.9.15. Lyndon basis. Fix a chamber C in the braid arrangement on I. To each chamber D whose first element coincides with the first element of C, we associate a maximal chain of faces $f_{C,D}$ ending in D as follows. Let us use combinatorial notation and write ℓ for the linear order of C. Similarly, let w denote the linear order of D. By assumption, the first element of w is the same as the first element of ℓ. One can uniquely write w as a concatenation uv, where the first element of v is the second element of ℓ. Then the chain $f_{C,D}$ starts as $I \lessdot S|T$, where S is the underlying set of u and T is the underlying set of v. To get the higher terms of $f_{C,D}$, we recursively apply this procedure first on u wrt the restricted linear order $\ell|_S$ and then on v wrt $\ell|_T$.

For example, for $C = a|b|c|d$, there are six chambers D which have the same first element as C. The resulting maximal chains of faces $f_{C,D}$ are shown below.

$(abcd \lessdot a\|bcd \lessdot a\|b\|cd \lessdot a\|b\|c\|d)$	$[\mathtt{H}_a, [\mathtt{H}_b, [\mathtt{H}_c, \mathtt{H}_d]]]$
$(abcd \lessdot a\|bdc \lessdot a\|bd\|c \lessdot a\|b\|d\|c)$	$[\mathtt{H}_a, [[\mathtt{H}_b, \mathtt{H}_d], \mathtt{H}_c]]$
$(abcd \lessdot ac\|bd \lessdot a\|c\|bd \lessdot a\|c\|b\|d)$	$[[\mathtt{H}_a, \mathtt{H}_c], [\mathtt{H}_b, \mathtt{H}_d]]$
$(abcd \lessdot ad\|bc \lessdot a\|d\|bc \lessdot a\|d\|b\|c)$	$[[\mathtt{H}_a, \mathtt{H}_d], [\mathtt{H}_b, \mathtt{H}_c]]$
$(abcd \lessdot acd\|b \lessdot a\|cd\|b \lessdot a\|c\|d\|b)$	$[[\mathtt{H}_a, [\mathtt{H}_c, \mathtt{H}_d]], \mathtt{H}_b]$
$(abcd \lessdot adc\|b \lessdot ad\|c\|b \lessdot a\|d\|c\|b)$	$[[[\mathtt{H}_a, \mathtt{H}_d], \mathtt{H}_c], \mathtt{H}_b]$

The corresponding Lie elements obtained from the map (14.45) are shown on the right. One can check that this is a basis of $\mathsf{Lie}[I]$ for $I = \{a, b, c, d\}$. The general result is:

Proposition 14.80. *Fix a chamber C. The Lie elements corresponding to $f_{C,D}$, as D varies over all chambers with the same first element as C, form a basis for* $\mathsf{Lie}[I]$.

This basis is a manifestation of the Lyndon basis as we now proceed to explain.

The chamber C determines a linear order on I. Put the lexicographic order on the set of hyperplanes. (Recall that a hyperplane is a pair (i, j) with $i, j \in I$ and $i < j$.) This induces a choice function γ_C: For each partition X of I, $\gamma_C(\mathrm{X})$ is the smallest hyperplane (i, j) such that both i and j appear in the same block of X. The main observation is:

Lemma 14.81. *The set of chambers D with the same first element as C is in bijection with the set of γ_C-compatible labeled chains.*

Proof. We explain the map in the forward direction. Suppose D is a chamber with the same first element as C. Then $\mathrm{s}(f_{C,D})$ is a maximal chain of flats. Denote it temporarily by $(\bot \lessdot \mathrm{X}_1 \lessdot \cdots \lessdot \mathrm{X}_{r-1} \lessdot \top)$. Label the edge joining X_k and X_{k+1} by the hyperplane (i, j), where i and j are the two smallest elements in the block of X_k that is split to obtain X_{k+1}. This is the desired labeled chain. It is straightforward to verify that this is a bijection. □

Recall from Proposition 14.52 that a choice function gives rise to a Lyndon basis. It follows from the above proof that:

Lemma 14.82. *The Lyndon basis associated to the choice function γ_C consists of the classes of $\mathrm{s}(f_{C,D})^*$, with D varying over all chambers with the same first element as C.*

As a consequence, the images of $\mathrm{s}(f_{C,D})^* \otimes [f_{C,D}]$ under the map (14.45) is a basis of $\mathsf{Lie}[I]$, and Proposition 14.80 follows.

14.10. Type B Lie elements

We specialize $\mathcal{A}$ to the arrangement of type B (Section 6.7). Recall the notation $\mathbf{I} = I \sqcup \bar{I} \sqcup \{0\}$ and $[\mathbf{n}] = [n] \sqcup \overline{[n]} \sqcup \{0\}$. These sets have a canonical involution. We use the terms involution-exclusive and involution-inclusive subsets in this context. We will shorten all notations of the form $g[\mathcal{A}]$ to $g[\mathbf{I}]$ or $g[\mathbf{n}]$ as may be the case.

We write $\mathsf{Lie}[\mathbf{I}]$ for the space of Lie elements. For clarity, we call these type B Lie elements. They sit inside $\Gamma[\mathbf{I}]$ which is spanned by type B linear orders. This inclusion can be understood through the combinatorics of brackets, the interesting part is that now there are *two* commutators to consider. The bracket expressions are related to type B binary trees which is turn are related to maximal chains of faces in the type B arrangement. This is the combinatorics involved in the JKS isomorphism.

The hyperplane orthogonal to the vector $(1, 2, \dots, 2^{n-1})$ is generic wrt the arrangement of type B on $[n]$. We give an explicit formula for the action of the corresponding symmetrized Dynkin element on chambers. The coefficients are products of some subset of odd numbers between 1 and $2n - 1$. The largest coefficient is $(2n - 1)!!$, which is the absolute value of the Möbius number of the arrangement.

We assume that the field characteristic is 0.

14.10.1. Type *B* Lie elements. To construct type B Lie elements, we need to consider two different commutators. They are as follows.

For any disjoint involution-exclusive nonempty subsets S and T, and linear orders ℓ_1 on S and ℓ_2 on T, put

$$[\mathtt{H}_{\ell_1}, \mathtt{H}_{\ell_2}] := \mathtt{H}_{\ell_1 \cdot \ell_2} - \mathtt{H}_{\ell_2 \cdot \ell_1}, \tag{14.61}$$

where $\ell_1 \cdot \ell_2$ is the linear order in which elements of S precede the elements of T. This is the same operation as in (14.44).

For any involution-inclusive subset $\mathbf{S}$ and involution-exclusive nonempty subset T disjoint from each other, and type B linear order ℓ_1 on $\mathbf{S}$ and linear order ℓ_2 on T, put

$$[\mathtt{H}_{\ell_1}, \mathtt{H}_{\ell_2}] := \mathtt{H}_{\ell_1 \cdot \ell_2} - \mathtt{H}_{\ell_1 \cdot \overline{\ell_2}}, \tag{14.62}$$

where $\overline{\ell_2}$ is the linear order on $\overline{T}$ obtained by reversing ℓ_2 and switching the parity of each letter. Note that we are using a thicker bracket here to distinguish it from the previous bracket.

Type B Lie elements are those elements of $\Gamma[\mathbf{I}]$ which are generated by these unbracketing operations from singletons $\{a\}$, where a is any element of $\mathbf{I}$, including 0. For example, $\mathsf{Lie}[a, 0, \overline{a}]$ is one-dimensional and spanned by

$$[\mathtt{H}_0, \mathtt{H}_a] = -[\mathtt{H}_0, \mathtt{H}_{\overline{a}}] = \mathtt{H}_{0|a} - \mathtt{H}_{0|\overline{a}}. \tag{14.63}$$

The first equality is the antisymmetry relation of type B.

Typical elements of $\mathsf{Lie}[a, b, 0, \overline{a}, \overline{b}]$ are

$$[[\mathtt{H}_0, \mathtt{H}_a], \mathtt{H}_{\overline{b}}] = [\mathtt{H}_{0|a} - \mathtt{H}_{0|\overline{a}}, \mathtt{H}_{\overline{b}}] = \mathtt{H}_{0|a|\overline{b}} - \mathtt{H}_{0|\overline{a}|\overline{b}} - \mathtt{H}_{0|a|b} + \mathtt{H}_{0|\overline{a}|b},$$

$$[\mathtt{H}_0, [\mathtt{H}_a, \mathtt{H}_{\overline{b}}]] = [\mathtt{H}_0, \mathtt{H}_{a|\overline{b}} - \mathtt{H}_{\overline{b}|a}] = \mathtt{H}_{0|a|\overline{b}} - \mathtt{H}_{0|\overline{b}|a} - \mathtt{H}_{0|b|\overline{a}} + \mathtt{H}_{0|\overline{a}|b}.$$

One may check that

$$[[\mathtt{H}_0, \mathtt{H}_a], \mathtt{H}_b] + [\mathtt{H}_0, [\mathtt{H}_b, \mathtt{H}_a]] + [[\mathtt{H}_0, \mathtt{H}_{\overline{b}}], \mathtt{H}_a] + [\mathtt{H}_0, [\mathtt{H}_a, \mathtt{H}_{\overline{b}}]] = 0. \tag{14.64}$$

Expanding out yields a sum of 16 terms which cancel in pairs. This is the type B Jacobi identity.

An element of $\mathsf{Lie}[a, b, c, 0, \overline{a}, \overline{b}, \overline{c}]$ is shown below.

$$\begin{aligned}
[[\mathtt{H}_0, \mathtt{H}_{\overline{a}}], [\mathtt{H}_b, \mathtt{H}_{\overline{c}}]] &= [\mathtt{H}_{0|\overline{a}} - \mathtt{H}_{0|a}, \mathtt{H}_{b|\overline{c}} - \mathtt{H}_{\overline{c}|b}] \\
&= \mathtt{H}_{0|\overline{a}|b|\overline{c}} - \mathtt{H}_{0|\overline{a}|\overline{c}|b} - \mathtt{H}_{0|a|b|\overline{c}} + \mathtt{H}_{0|a|\overline{c}|b} - \mathtt{H}_{0|\overline{a}|c|\overline{b}} + \mathtt{H}_{0|\overline{a}|\overline{b}|c} + \mathtt{H}_{0|a|c|\overline{b}} - \mathtt{H}_{0|a|\overline{b}|c}.
\end{aligned}$$

To get antisymmetry and Jacobi identity in general, we replace each letter by a Lie monomial. A bar on top of a Lie monomial means that we take bar of each letter and also reverse the bracket expression. For example,

$$\overline{[\mathtt{H}_a, \mathtt{H}_b]} := [\mathtt{H}_{\overline{b}}, \mathtt{H}_{\overline{a}}] \quad \text{and} \quad \overline{[\mathtt{H}_a, [\mathtt{H}_{\overline{b}}, \mathtt{H}_c]]} := [[\mathtt{H}_{\overline{c}}, \mathtt{H}_b], \mathtt{H}_{\overline{a}}].$$

This is consistent with the way the bar operation was defined on linear orders in (14.62). Thus,

$$[\mathtt{H}_0, [\mathtt{H}_a, \mathtt{H}_b]] = -[\mathtt{H}_0, [\mathtt{H}_{\overline{b}}, \mathtt{H}_{\overline{a}}]] \quad \text{and} \quad [\mathtt{H}_0, [\mathtt{H}_a, [\mathtt{H}_{\overline{b}}, \mathtt{H}_c]]] = -[\mathtt{H}_0, [[\mathtt{H}_{\overline{c}}, \mathtt{H}_b], \mathtt{H}_{\overline{a}}]]$$

are instances of antisymmetry, while

$$[[\mathtt{H}_0, \mathtt{H}_a], [\mathtt{H}_b, \mathtt{H}_c]] + [\mathtt{H}_0, [[\mathtt{H}_b, \mathtt{H}_c], \mathtt{H}_a]] + [[\mathtt{H}_0, \overline{[\mathtt{H}_b, \mathtt{H}_c]}], \mathtt{H}_a] + [\mathtt{H}_0, [\mathtt{H}_a, \overline{[\mathtt{H}_b, \mathtt{H}_c]}]] = 0$$

is an instance of the Jacobi identity.

14.10.2. Type B binary trees. A type B binary tree on $\mathbf{I}$ is a binary tree with $|I|+1$ leaves with the leftmost leaf labeled 0 and the remaining leaves labeled from $\mathbf{I}$ with exactly one of i and $\overline{i}$ occuring for each $i \in I$. For instance,

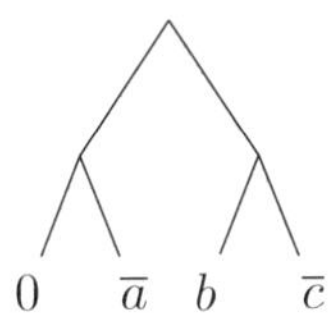

and

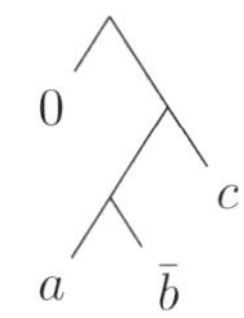

are type B binary trees on $\mathbf{I} = \{a, b, c, 0, \overline{a}, \overline{b}, \overline{c}\}$. A type B binary tree gives rise to a bracket expression, and hence yields a type B Lie element. For instance, the above type B binary trees yield the Lie elements

$$[[\mathtt{H}_0, \mathtt{H}_{\overline{a}}], [\mathtt{H}_b, \mathtt{H}_{\overline{c}}]] \quad \text{and} \quad [\mathtt{H}_0, [[\mathtt{H}_a, \mathtt{H}_{\overline{b}}], \mathtt{H}_c]].$$

Unbracketing them results in a linear combination of type B linear orders.

14.10.3. JKS. Let $\mathrm{S}_n^{\pm}$ denote the signed symmetric group on n letters. This is the Coxeter group of the arrangement of type B on $[n]$. Specializing Theorem 14.38, we obtain:

Theorem 14.83. *There is an isomorphism of* $\mathrm{S}_n^{\pm}$*-modules*

$$\mathcal{H}^{r-2}(\Pi[\mathbf{n}]) \otimes \mathsf{E}^{\mathsf{o}}[\mathbf{n}] \to \mathsf{Lie}[\mathbf{n}].$$

Let us describe this map explicitly. For clarity, we work with the arrangement of type B on a finite set I. A maximal chain of faces f corresponds to a type B binary tree on $\mathbf{I}$ whose internal nodes are linearly ordered in such a manner that each node appears before both of its children. An illustration of the bijection for $\mathbf{I} = \{a, b, c, 0, \overline{a}, \overline{b}, \overline{c}\}$ is shown below.

$$(0abc \lessdot 0a|b\overline{c} \lessdot 0|\overline{a}|b\overline{c} \lessdot 0|\overline{a}|b|\overline{c}) \longleftrightarrow$$

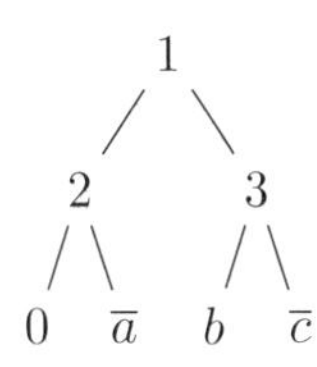

A type B binary tree gives rise to a bracket expression, and hence yields a type B Lie element. Continuing the above example, with f denoting the maximal chain of faces, the map is

$$\mathrm{s}(f)^* \otimes [f] \mapsto [[\mathtt{H}_0, \mathtt{H}_{\overline{a}}], [\mathtt{H}_b, \mathtt{H}_{\overline{c}}]].$$

14.10.4. Type B Jacobi identity. The arrangement of type B on $I = \{a, b\}$ is shown below.

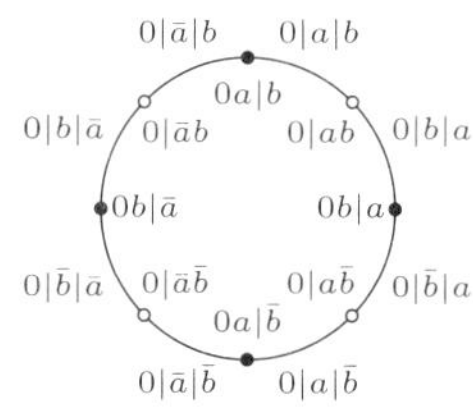

The Lie element obtained by unbracketing the maximal chain of flats

$$(\{\overline{b}\overline{a}0ab\} \lessdot \{\overline{b}, \overline{a}0a, b\} \lessdot \{\overline{b}, \overline{a}, 0, a, b\})$$

with the anticlockwise orientation is given by

$$[[\mathtt{H}_0, \mathtt{H}_{\overline{a}}], \mathtt{H}_b] = \mathtt{H}_{0|\overline{a}|b} - \mathtt{H}_{0|a|b} - \mathtt{H}_{0|\overline{a}|\overline{b}} + \mathtt{H}_{0|a|\overline{b}}.$$

The Lie element obtained by unbracketing the maximal chain

$$(\{\overline{b}\overline{a}0ab\} \lessdot \{\overline{b}\overline{a}, 0, ab\} \lessdot \{\overline{b}, \overline{a}, 0, a, b\})$$

with the anticlockwise orientation is given by

$$[\mathtt{H}_0, [\mathtt{H}_a, \mathtt{H}_b]] = \mathtt{H}_{0|a|b} - \mathtt{H}_{0|b|a} - \mathtt{H}_{0|\overline{b}|\overline{a}} + \mathtt{H}_{0|\overline{a}|\overline{b}}.$$

Similarly, one can unbracket the remaining two maximal chains of flats. The sum of the four Lie elements thus obtained is zero. This is the type B Jacobi identity and is illustrated below.

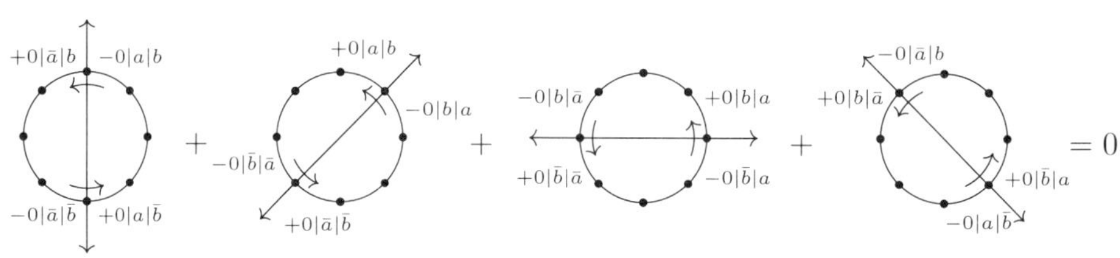

14.10.5. Orientation and signature spaces. We explain the relation between the orientation space $\mathsf{E}^{\mathsf{o}}[\mathbf{n}]$ and the signature space $\mathsf{E}^{-}[\mathbf{n}]$.

Let $\mathsf{mc}[\mathbf{n}]$ denote the space spanned by maximal chains of faces. Consider the map

$$\Gamma[\mathbf{n}] \to \mathsf{mc}[\mathbf{n}], \qquad \mathtt{H}_{0|l_1|l_2|\dots|l_n} \mapsto (O, F_1, F_2, \dots, F_n),$$

where

$$F_1 = 0|l_1 \dots l_n, \quad F_2 = 0|l_1|l_2 \dots l_n, \quad \dots, \quad F_n = 0|l_1|l_2|\dots|l_n$$

with F_i having i nonzero blocks and $z(F_i) = \{0\}$. By passing to quotients, this map induces an isomorphism of $\mathrm{S}_n^{\pm}$-modules

$$\mathsf{E}^{-}[\mathbf{n}] \xrightarrow{\cong} \mathsf{E}^{\mathsf{o}}[\mathbf{n}].$$

The key observation is that maximal chains of faces arising from adjacent chambers have opposite orientations.

The image of the element $\mathtt{H}_{0|l_1|\dots|l_n}$ in $\mathsf{E}^-[\mathbf{n}]$ may be denoted $0 \wedge l_1 \wedge \cdots \wedge l_n$, with the understanding that switching l_i and l_{i+1}, or changing l_i to $\overline{l_i}$ incurs a minus sign. We call this the sign representation of $\mathrm{S}_n^{\pm}$.

14.10.6. Whitney cohomology. Let us now relate Whitney cohomology of the poset of flats to Lie elements. Let $\mathcal{A}$ denote the type B arrangement of rank n. Set

$$\bigwedge \mathsf{Lie}[\mathbf{n}] := \bigoplus_{\mathrm{X}} \mathsf{E}^{\mathrm{o}}[\mathcal{A}^{\mathrm{X}}] \otimes \mathsf{Lie}[\mathcal{A}_{\mathrm{X}}].$$

To describe this space more explicitly, we first recall three facts.

- $\mathcal{A}^{\mathrm{X}}$ is a type B arrangement (whose letters are the blocks of X).
- $\mathcal{A}_{\mathrm{X}}$ is the cartesian product of a type B arrangement (whose letters are the elements of the zero block of X) and several braid arrangements (one for each pair of nonzero blocks of X).
- $\mathsf{E}^{\mathrm{o}}[\mathbf{n}]$ is the sign representation of $\mathrm{S}_n^{\pm}$.

Now we describe $\bigwedge \mathsf{Lie}[\mathbf{n}]$. It is linearly spanned by elements of the form $x_0 \otimes (x_1 \wedge \cdots \wedge x_k)$, where x_0 is a type B Lie element while the rest are type A Lie elements, and each letter from 1 to n appears (with either parity) in exactly one of the x_i. For relations: We have linearity in each variable along with the usual antisymmetry and Jacobi identity for Lie elements (of type B in x_0 and of type A in the rest). In addition, interchanging x_i and x_{i+1} or switching the parity of all letters in x_i (for $x \geq 1$) incurs a minus sign.

For instance, for $n = 1$, this space is two-dimensional with basis

$$[\mathtt{H}_0, \mathtt{H}_1] = -[\mathtt{H}_0, \mathtt{H}_{\overline{1}}] \quad \text{and} \quad \mathtt{H}_0 \otimes \mathtt{H}_1 = -(\mathtt{H}_0 \otimes \mathtt{H}_{\overline{1}}).$$

For $n = 2$, this space is 8-dimensional with basis

$$[\mathtt{H}_0, [\mathtt{H}_1, \mathtt{H}_2]],\ [\mathtt{H}_0, [\mathtt{H}_1, \mathtt{H}_{\overline{2}}]],\ [[\mathtt{H}_0, \mathtt{H}_1], \mathtt{H}_2],\ [\mathtt{H}_0, \mathtt{H}_1] \otimes \mathtt{H}_2,\ [\mathtt{H}_0, \mathtt{H}_2] \otimes \mathtt{H}_{\overline{1}},$$
$$\mathtt{H}_0 \otimes [\mathtt{H}_1, \mathtt{H}_2],\ \mathtt{H}_0 \otimes [\mathtt{H}_1, \mathtt{H}_{\overline{2}}],\ \mathtt{H}_0 \otimes (\mathtt{H}_{\overline{1}} \wedge \mathtt{H}_2).$$

There are many ways in which a particular basis element can be written. For instance,

$$[\mathtt{H}_0, \mathtt{H}_2] \otimes \mathtt{H}_{\overline{1}} = -[\mathtt{H}_0, \mathtt{H}_{\overline{2}}] \otimes \mathtt{H}_{\overline{1}} = [\mathtt{H}_0, \mathtt{H}_{\overline{2}}] \otimes \mathtt{H}_1,$$
$$\mathtt{H}_0 \otimes (\mathtt{H}_{\overline{1}} \wedge \mathtt{H}_2) = -(\mathtt{H}_0 \otimes (\mathtt{H}_1 \wedge \mathtt{H}_2)) = \mathtt{H}_0 \otimes (\mathtt{H}_2 \wedge \mathtt{H}_1).$$

In general, $\bigwedge \mathsf{Lie}[\mathbf{n}]$ has dimension $(2n)!!$, which is the number of chambers in $\mathcal{A}$. Specializing Theorem 14.40, we obtain:

Theorem 14.84. *There is an isomorphism of* $\mathrm{S}_n^{\pm}$*-modules*

$$\mathcal{WH}^*(\Pi[\mathbf{n}]) \otimes \mathsf{E}^{\mathrm{o}}[\mathbf{n}] \longrightarrow \bigwedge \mathsf{Lie}[\mathbf{n}].$$

14.10.7. Substitution product. For a type B partition X on the set I, the substitution product (10.28) is as follows. An element of $\mathsf{Lie}[\mathcal{A}^{\mathrm{X}}]$ is a type B Lie element on the blocks of X, while an element of $\mathsf{Lie}[\mathcal{A}_{\mathrm{X}}]$ consists of a type B Lie element on the zero block of X and type A Lie elements, one on each nonzero block of X. (Recall that the nonzero blocks occur in pairs $(J, \overline{J})$. It is understood here that the Lie element on $\overline{J}$ is the bar of the Lie element on J.) This data together specifies a type B Lie element on $\mathbf{I}$. For example, for the type B partition $X = \{\overline{ce}, f\overline{g}, \overline{db\overline{a}0abd}, \overline{f}g, ce\}$,

$$([[\mathtt{H}_{0abd}, \mathtt{H}_{f\overline{g}}], \mathtt{H}_{ce}], \{[[\mathtt{H}_0, \mathtt{H}_{\overline{a}}], [\mathtt{H}_{\overline{b}}, \mathtt{H}_d]], [\mathtt{H}_e, \mathtt{H}_c], [\mathtt{H}_{\overline{c}}, \mathtt{H}_{\overline{e}}], [\mathtt{H}_f, \mathtt{H}_{\overline{g}}], [\mathtt{H}_g, \mathtt{H}_{\overline{f}}]\}) \\ \mapsto [[[[\mathtt{H}_0, \mathtt{H}_{\overline{a}}], [\mathtt{H}_{\overline{b}}, \mathtt{H}_d]], [\mathtt{H}_f, \mathtt{H}_{\overline{g}}]], [\mathtt{H}_e, \mathtt{H}_c]].$$

This may also be expressed as

$$([[\mathtt{H}_{0abd}, \mathtt{H}_{f\overline{g}}], \mathtt{H}_{ce}], \{[[\mathtt{H}_0, \mathtt{H}_{\overline{a}}], [\mathtt{H}_{\overline{b}}, \mathtt{H}_d]], [\mathtt{H}_e, \mathtt{H}_c], [\mathtt{H}_f, \mathtt{H}_{\overline{g}}]\}) \\ \mapsto [[[[\mathtt{H}_0, \mathtt{H}_{\overline{a}}], [\mathtt{H}_{\overline{b}}, \mathtt{H}_d]], [\mathtt{H}_f, \mathtt{H}_{\overline{g}}]], [\mathtt{H}_e, \mathtt{H}_c]].$$

Using the substitution product of chambers in Section 6.7.12, one can check that diagram (10.29) indeed commutes.

Exercise 14.85. Make Proposition 13.28 explicit for the uniform section of the arrangement of type B on I. This is the type B analogue of Proposition 14.65.

14.10.8. Another viewpoint on Lie. Let $\mathcal{L}$ denote the Lie subalgebra of the free Lie algebra on $[n] \sqcup \overline{[n]}$ which is invariant under the canonical involution $i \mapsto \overline{i}$. (In this action, the bracket expression is kept the same, and *not* reversed.) Let $\mathcal{E}[n]$ denote the multilinear part of the universal enveloping algebra of $\mathcal{L}$. Explicitly, $\mathcal{E}[1]$ is one-dimensional and spanned by $\mathtt{H}_1 + \mathtt{H}_{\overline{1}}$. Similarly, $\mathcal{E}[2]$ is 3-dimensional with basis

$$[\mathtt{H}_1, \mathtt{H}_2] + [\mathtt{H}_{\overline{1}}, \mathtt{H}_{\overline{2}}], \quad [\mathtt{H}_1, \mathtt{H}_{\overline{2}}] + [\mathtt{H}_{\overline{1}}, \mathtt{H}_2], \quad (\mathtt{H}_1 + \mathtt{H}_{\overline{1}}) \cdot (\mathtt{H}_2 + \mathtt{H}_{\overline{2}}).$$

Here $\cdot$ represents the product in the universal enveloping algebra. Thus, we have the relation

$$(\mathtt{H}_1 + \mathtt{H}_{\overline{1}}) \cdot (\mathtt{H}_2 + \mathtt{H}_{\overline{2}}) - (\mathtt{H}_2 + \mathtt{H}_{\overline{2}}) \cdot (\mathtt{H}_1 + \mathtt{H}_{\overline{1}}) \\ = [\mathtt{H}_1 + \mathtt{H}_{\overline{1}}, \mathtt{H}_2 + \mathtt{H}_{\overline{2}}] = ([\mathtt{H}_1, \mathtt{H}_2] + [\mathtt{H}_{\overline{1}}, \mathtt{H}_{\overline{2}}]) + ([\mathtt{H}_1, \mathtt{H}_{\overline{2}}] + [\mathtt{H}_{\overline{1}}, \mathtt{H}_2]).$$

In general, $\mathcal{E}[n]$ has a basis consisting of elements which are products of symmetrized Lie monomials. An example of such an element for $n = 5$ is

$$([\mathtt{H}_3, \mathtt{H}_{\overline{5}}] + [\mathtt{H}_{\overline{3}}, \mathtt{H}_5]) \cdot ([\mathtt{H}_1, \mathtt{H}_2] + [\mathtt{H}_{\overline{1}}, \mathtt{H}_{\overline{2}}]) \cdot (\mathtt{H}_{\overline{4}} + \mathtt{H}_4).$$

Theorem 14.86. *There is an isomorphism of* $\mathrm{S}_n^{\pm}$*-modules*

$$\mathcal{E}[n] \otimes \mathsf{E}^{\mathsf{o}}[n] \xrightarrow{\cong} \mathsf{Lie}[\mathbf{n}] \otimes \mathsf{E}^{\mathsf{o}}[\mathbf{n}].$$

Note very carefully that we have put $\mathsf{E}^{\mathsf{o}}[n]$ in the lhs. This is the sign representation of the usual symmetric group S_n. The signed symmetric group $\mathrm{S}_n^{\pm}$ acts on it by ignoring parity.

We illustrate the isomorphism on the above example.

$$([\mathtt{H}_3, \mathtt{H}_{\overline{5}}] + [\mathtt{H}_{\overline{3}}, \mathtt{H}_5]) \cdot ([\mathtt{H}_1, \mathtt{H}_2] + [\mathtt{H}_{\overline{1}}, \mathtt{H}_{\overline{2}}]) \cdot (\mathtt{H}_{\overline{4}} + \mathtt{H}_4) \otimes (3 \wedge 5 \wedge 1 \wedge 2 \wedge 4) \mapsto \\ [[[\mathtt{H}_0, [\mathtt{H}_3, \mathtt{H}_{\overline{5}}]], [\mathtt{H}_1, \mathtt{H}_2]], \mathtt{H}_{\overline{4}}] \otimes (3 \wedge \overline{5} \wedge 1 \wedge 2 \wedge \overline{4}).$$

This map is well-defined by type B antisymmetry. The universal enveloping algebra relation corresponds to the type B Jacobi identity.

Combining with Theorem 14.83, we obtain:

Theorem 14.87. *There is an isomorphism of* $\mathrm{S}_n^{\pm}$*-modules*

$$\mathcal{H}^{r-2}(\Pi[\mathbf{n}]) \otimes \mathsf{E}^{\mathsf{o}}[n] \xrightarrow{\cong} \mathcal{E}[\mathbf{n}].$$

14.10.9. Dynkin elements. Let $\mathcal{A}$ be the arrangement of type B on I. In this subsection and the next, we work with an alternative combinatorial formulation of faces and chambers which is as follows. A chamber C is the same as a linear order ℓ_C on $I \cup \{0\}$ in which 0 appears first, along with a parity function

$$\epsilon_C : I \to \{1, -1\}.$$

This is the same as a type B linear order. More generally, a face F is the same as a set composition ℓ_F on $I \cup \{0\}$ in which 0 appears in the first block (also called the zero block), along with a parity function $\epsilon_F : I' \to \{1, -1\}$, where I' is the union of the nonzero blocks of ℓ_F. This is the same as a type B set composition.

Now fix a chamber C. Consider the function

$$p_C : I \to \mathbb{R}, \qquad p_C(a) = \epsilon_C(a)\, 2^{k-2},$$

where k is the position of $a \in I$ in the linear order ℓ_C. Note that p_C is a point in the interior of the chamber C. Let h_C denote the half-space containing C whose base is orthogonal to (the vector determined by) p_C. The half-space h_C is then generic wrt $\mathcal{A}$. The faces contained in h_C can be described as follows.

Lemma 14.88. *We have $F \subseteq \mathrm{h}_C$ iff F and C satisfy the following compatibility property. The largest element (wrt the linear order ℓ_C), say a, in the last nonzero block of ℓ_F satisfies $\epsilon_F(a) = \epsilon_C(a)$, similarly the largest element, say b, in the last two nonzero blocks of F satisfies $\epsilon_F(b) = \epsilon_C(b)$, and so on till we touch the zero block of F.*

PROOF. Note that $F \subseteq \mathrm{h}_C$ iff every point in F has a nonnegative inner product with the vector p_C. The latter condition can be analysed by repeatedly using the simple inequality $2^k > 1 + \cdots + 2^{k-1}$. The outline of the argument is given below.

Forward implication. Suppose F and C are not compatible. Then there is an element, say a, which is largest (wrt the linear order ℓ_C) in some final segment of nonzero blocks of ℓ_F, and $\epsilon_F(a) \neq \epsilon_C(a)$. Consider the point q whose value on a is $\epsilon_F(a)$ if a belongs to this final segment, and 0 otherwise. Then q belongs to the face F but its inner product with p_C is strictly negative.

Backward implication. Suppose F and C are compatible. The set of blocks of ℓ_F can be uniquely partitioned into contiguous sets of blocks such that the largest element (wrt the linear order ℓ_C) in any partition occurs in the last block (of that partition). Now let q be any point in F. Then the inner product of q with p_C can be split as a sum over these partitions, and each partition makes a nonnegative contribution. □

The Dynkin element θ_{h_C}, which we abbreviate to θ_C, is

$$\theta_C = \sum_{F \subseteq \mathrm{h}_C} (-1)^{\mathrm{rk}(F)} \mathtt{H}_F.$$

The sum is over all faces F which are compatible with the chamber C in the sense of Lemma 14.88. Since the central face has no nonzero blocks, it is compatible with every chamber. This is consistent with the fact that the central face is contained in every half-space.

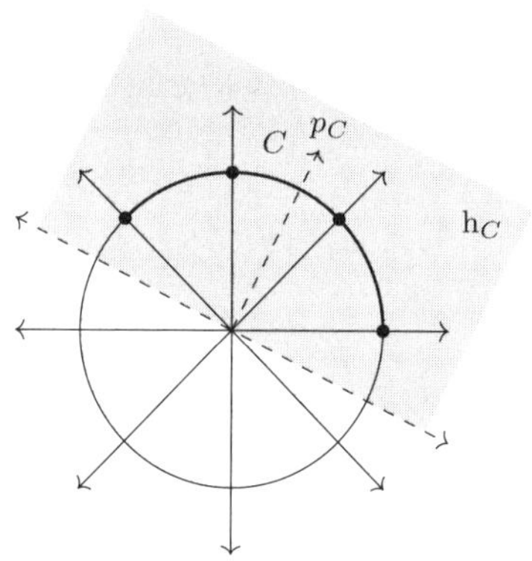

The figure illustrates the situation when I has two elements. Denoting the fixed chamber by $C = 0|a|b$,

$$\theta_C = \mathtt{H}_{0ab} - \mathtt{H}_{0|\bar{a}b} - \mathtt{H}_{0a|b} - \mathtt{H}_{0|ab} - \mathtt{H}_{0b|a} + \mathtt{H}_{0|\bar{a}|b} + \mathtt{H}_{0|a|b} + \mathtt{H}_{0|b|a}.$$

It is the linear combination given by the central face minus 4 contiguous vertices plus the 3 edges in-between.

If a Coxeter symmetry takes chamber C to chamber D, then it takes the point p_C to the point p_D. It follows that the set of half-spaces $\{\mathrm{h}_C\}$ form a single orbit under the action of the signed symmetric group. Thus the symmetrized Dynkin element is given by

$$d_I = \sum_C \theta_C.$$

14.10.10. Action of symmetrized Dynkin element. Fix n to be the cardinality of I, that is, $n = |I|$. Let T be any subset of $S := \{s_0, s_1, \dots, s_{n-1}\}$. Put

$$\alpha_T = \prod_{i=1}^{n-1} T(i), \tag{14.65}$$

where

$$T(i) = \begin{cases} 2n - 2i + 1 & \text{if } s_{i-1}, s_i \in T, \text{ or } s_{i-1}, s_i \notin T, \\ 1 & \text{otherwise.} \end{cases}$$

This is a product of some subset of odd numbers between 3 and $2n - 1$ (both inclusive). Note that $\alpha_T = \alpha_{S \setminus T}$. One extreme case is when T is either $\emptyset$ or S in which case $\alpha_T = (2n-1)!!$. The other extreme case is when T and $S \setminus T$ consist of alternate elements of S, that is, either $s_0, s_2, \dots \in T$ and $s_1, s_3, \dots \in S \setminus T$, or vice-versa in which case $\alpha_T = 1$.

For example, for $n = 2$,

$$\alpha_\emptyset = \alpha_{s_0,s_1} = 3, \qquad \alpha_{s_0} = \alpha_{s_1} = 1,$$

and for $n = 3$,

$$\alpha_\emptyset = \alpha_{s_0,s_1,s_2} = 15,\ \alpha_{s_0} = \alpha_{s_1,s_2} = 3,\ \alpha_{s_1} = \alpha_{s_0,s_2} = 1,\ \alpha_{s_2} = \alpha_{s_0,s_1} = 5.$$

Lemma 14.89. *We have*

$$\sum_{T:\, T \subseteq S} \alpha_T = (2n)!!,$$

which is the order of the signed symmetric group on n letters.

This will follow from the considerations below. For $n = 2$ and $n = 3$, the sum of the α_T is 8 and 48, respectively, which can be readily checked from the above calculations.

The *type* of any face F, denoted $\mathrm{t}(F)$, is a subset of S defined as follows. $s_i \in \mathrm{t}(F)$ if the total size of any initial segment of blocks of F is $i+1$. For example, for $n = 4$,

$$\mathrm{t}(0a|bc|d) = \{s_1, s_3\}, \qquad \mathrm{t}(0|ab|c|d) = \{s_0, s_2, s_3\}.$$

In other words, the locations of the bars of F are encoded in its type. (This description of the type map is equivalent to the one given in Section 6.7.8.)

Lemma 14.90. *Let D and E be chambers. For any $0 \le i \le |I|$, let $\ell_{D,i}$ denote the restriction of ℓ_D to the last $|I| - i$ elements. Let a denote the largest element in $\ell_{D,i}$ wrt the linear order ℓ_E. Then*

$$s_i \in \mathrm{t}(\mathrm{h}_E(D)) \iff \epsilon_E(a) = \epsilon_D(a).$$

Proof. Let P be the vertex of D of type s_i. Then ℓ_P has two blocks, and the nonzero block is precisely the underlying set of $\ell_{D,i}$. Thus, by Lemma 14.88, $P \subseteq \mathrm{h}_E$ iff $\epsilon_E(a) = \epsilon_P(a)$. Now the first condition is the same as the condition $s_i \in \mathrm{t}(\mathrm{h}_E(D))$. Also, since P is a face of D, $\epsilon_P(a)$ is the same as $\epsilon_D(a)$. □

Theorem 14.91. *We have*

$$d_I \cdot \mathtt{H}_C = \sum_D (-1)^{\mathrm{rk}(\mathrm{Des}(C,D))} \alpha_{\mathrm{t}(\mathrm{Des}(C,D))} \mathtt{H}_D,$$

where $\mathrm{t}(\mathrm{Des}(C,D))$ *is the type of the face* $\mathrm{Des}(C,D)$, *and* α_T *is as in* (14.65).

Proof. The arrangement of type B is simplicial. Thus, by Lemma 14.8,

$$d_I \cdot \mathtt{H}_C = \sum_D (-1)^{\mathrm{rk}(\mathrm{Des}(C,D))} \alpha_{C,D} \mathtt{H}_D,$$

where $\alpha_{C,D}$ is the number of chambers E for which $\mathrm{h}_E(D) = \mathrm{Des}(C,D)$. For notational simplicity, put $T = \mathrm{t}(\mathrm{Des}(C,D))$. Also write $D = 0|p_1|\dots|p_n$. We now give an algorithm that lists all E for which the type of $\mathrm{h}_E(D)$ is T. It works by inserting the elements $p_n, \dots, p_1$ in that order (each with same or reversed parity). It begins by inserting p_n into the empty list. Now suppose that $p_n, \dots, p_{i+1}$ have been inserted. While inserting p_i, there are four cases. We employ Lemma 14.90 in each case.

- Suppose $s_{i-1} \in T, s_i \notin T$. Then p_i must be inserted to the right end of the list with same parity.
- Suppose $s_{i-1} \notin T, s_i \in T$. Then p_i must be inserted to the right end of the list with parity reversed.
- Suppose $s_{i-1}, s_i \in T$. Then p_i can be inserted either to the right end of the list with same parity, or in any of the remaining $n - i$ positions with any parity.
- Suppose $s_{i-1}, s_i \notin T$. Then p_i can be inserted either to the right end of the list with parity reversed, or in any of the remaining $n - i$ positions with any parity.

Thus, for inserting p_i, there is a unique choice in the first two cases, and $2n-2i+1$ choices in the next two cases. This shows that $\alpha_{C,D}$ equals α_T as required. □

When I has 2 elements, the α_T values are 1 and 3, and the above formula specializes to (14.42) for $n = 4$. (Recall that in this case, the arrangement of type B is cisomorphic to the rank-two arrangement of 4 lines.)

Proposition 14.92. *For any $T \subseteq S$, α_T equals the number of chambers E such that $\mathrm{t}(\mathrm{h}_E(D)) = T$, where D is any fixed chamber. Dually, α_T equals the number of chambers D such that $\mathrm{t}(\mathrm{h}_E(D)) = T$, where E is any fixed chamber.*

PROOF. The first part is contained in the proof of Theorem 14.91. Consider the set of all pairs (D, E) such that $\mathrm{t}(\mathrm{h}_E(D)) = T$. The signed symmetric group acts freely on this set. So we can count the number of orbits either by fixing D or by fixing E. Since doing it the first way gives α_T, so must doing it the second way. □

By either interpretation, we deduce that the sum of the α_T equals the number of chambers. This proves Lemma 14.89. We may also rederive (6.18) as a consequence.

Corollary 14.93. *The absolute value of the Möbius number of the arrangement of type B on $[n]$ is $(2n-1)!!$ (which also equals α_S or $\alpha_\emptyset$).*

PROOF. Put $T = S$ in the second interpretation of α_T given by Proposition 14.92. A chamber D satisfies $\mathrm{t}(\mathrm{h}_E(D)) = S$ precisely when it is contained in the half-space h_E (which is generic). Now apply Proposition 14.16. □

Notes

Generic halfspaces. Corollary 14.2 is given in [**210**, Theorem 3.2]. Positive sum systems are studied by Björner in [**79**]. Exercise 14.68 provides a geometric approach to his Theorem 5.3. The exercise and Lemma 1.54 imply that positive sum systems correspond to chambers in the restricted all-subset arrangement (Section 6.3.12). See also the notes to Chapter 6.

Affine arrangements and bounded faces. Theorem 14.21 is due to Zaslavsky [**440**, Theorem C and Corollary 2.2]. It is also proved in [**82**, Theorem 4.6.5]. For information on the topology of the complex of bounded faces, see [**400**, Chapter 1, Exercise 7] and [**158**].

Margolis, Saliola and Steinberg characterize when a left regular band algebra has a unit [**304**, Theorem 4.15]. (The left regular band is not assumed to be a monoid.) Their criterion implies that the semigroup algebra of an affine arrangement has a unit. Theorem 14.23 proves this independently and in addition gives an explicit formula for the unit. The result of Exercise 14.25 is contained in [**304**, Theorem 4.2].

Poset homology. Proposition 14.30 is due to Folkman [**185**, Theorem 4.1]. The sketch for our proof is taken from [**76**, Theorem 2.1]. Whitney homology for geometric lattices was introduced by Baclawski [**36**]. Its connection with the formulation given here is explained in [**78**, Exercise 7.53]. For more information on poset topology, see the survey article by Wachs [**423**].

JKS. The results in Section 14.5.9 relating order homology of the lattice of flats and the Tits algebra are due to Saliola [**368**, Propositions 10.1 and 10.5]. The map (14.30) is equivalent to his map φ defined in [**368**, Section 8.3.1]. In the argument given in [**367**, Section 6], he proves Lemma 13.64 which in closely linked to Lemma 13.63. (See the notes to Chapter 13.) The latter was used in the second basis-free proof of JKS.

Recall that Lie elements can be defined for any LRB. Generalizing the presentation of $\mathsf{Lie}[\mathcal{A}]$ given by Theorem 14.41 to LRBs is however nontrivial. Important progress on this problem has been made by Margolis, Saliola and Steinberg [**303**, Theorems 4.1 and 4.2, and Corollary 4.3] (without any mention of Lie theory).

Björner basis and Björner-Wachs basis. The Björner basis for homology of the lattice of flats was constructed by Björner [**76**]. He implicitly uses a choice function to first construct neat base-families and then the basis elements using (14.35). Proposition 14.51 corresponds to [**76**, Theorem 4.2]. Lemma 14.49 corresponds to [**76**, Proposition 3.6]. He also defines the map (14.38) and shows that it preserves the basis elements [**76**, Lemmas 2.2 and 4.1]. A later survey with emphasis on the matroid aspects is given in [**78**].

The Björner-Wachs basis was introduced by Björner and Wachs [**84**] using the technique of generic hyperplanes. They give specific choices of a generic hyperplane for types A, B and D. We have made use of these in our discussion on types A and B. Lemmas 14.66 and 14.88 are given in [**84**, Propositions 6.2 and 7.2]. A small distinction is that the latter describe only the chambers contained in the generic half-space (as opposed to all faces). The counting argument using insertions given in the proof of Theorem 14.91 generalizes the argument given in [**84**, Corollary 7.4]. More precisely, Björner and Wachs are counting α_S in a dual manner, as stated in the second part of Proposition 14.92.

We defined the Björner-Wachs basis for top-homology tensored with the orientation space, rather than just for top-homology. This avoids the orientation-fixing issue. We made the choice $\overline{C} \subseteq \mathrm{h}$ in (14.18) so that the BW-basis became exactly dual to the Dynkin basis.

Type *A*. The left-bracketing operator and Theorem 14.74 (in the context of Lie polynomials) appeared in work of Dynkin [**166**]; also see the papers of Specht [**390**] and Wever [**429**], and the later paper of Ree [**354**, Theorem 2.3]. The Dynkin basis in Proposition 14.77 is called the (left) comb basis by Barcelo-Sundaram [**43**, Theorem 1.3] and Wachs [**422**]. A discussion of Dynkin idempotents (without mention of generic hyperplanes) is given in [**10**, Sections 14.5 and 14.6]. The elements $\overline{\theta_1}$ and $\overline{d_n}$ are considered by Schocker [**376**, Lemma 4.5 and Section 9.1]. The Q-basis and the complete system in Exercise 14.70 also occur in his work [**376**, Corollary 6.3]. He works with minimum instead of maximum elements. For related information, see the notes to Chapter 11. For information about q-Dynkin elements, see the notes to Chapter 16. Proposition 14.80 is a classical result. A direct proof is given by Garsia [**196**, Sections 3 and 4, Theorem 4.4].

The character of $\mathcal{H}^{r-2}(\Pi[n])$ was computed by Stanley [**395**, Lemma 7.1]. Via Hopf-Lefschetz Theorem [**37**, Theorem 1.1], it coincides with the Möbius number of the subposet of $\Pi[n]$ fixed by the permutation whose character we are computing. The formula for this Möbius number follows from a result of Hanlon [**220**, Theorem 4]. Stanley [**395**, Theorem 7.3] or [**423**, Theorem 4.4.7] further relates this character to the character of a representation of S_n induced from a certain 1-dimensional representation of the cyclic group of order n. (A different proof of Stanley's result was given by Sundaram [**409**, Example 1.6].) By a result of Klyachko [**258**], the latter is the same as the character of $\mathsf{Lie}[n]$. (The character of the degree n-component of the free Lie algebra on n generators was computed much earlier by Brandt [**100**, Theorem III].) Putting the two together implies Theorem 14.63. (A more general character result is given by Robinson and Whitehouse [**362**, Theorem 3.1].) This result was first stated in the present form by Joyal [**247**, Theorem 4, Chapter 4]. His proof used species and he also computed the character using this technique [**247**, Proposition 4, Chapter 4]. The connection of this result to Koszul duality was explained by Fresse [**187**, Prolog and Epilog]. The first combinatorial proof was given by Barcelo [**40**, Theorem 10.1] by identifying the dual of the Björner basis with the Lyndon basis using the bijection of Lemma 14.81. The bijection is explained at the beginning of [**40**, Section 10]: Chambers D with the same first element as C correspond to Lyndon permutations, and γ_C-compatible labeled chains correspond to NBC bases. The description of the JKS isomorphism (14.45) by relating chains and Lie elements to binary trees (similar to what we do) is given by Wachs [**422**, Theorem 5.4]. She proves Theorem 14.63 by checking that the cohomology relations corresponds to Jacobi dentities. The same argument is sketched in [**423**, Sections 1.5 and 1.6]. The bases (14.59) and (14.60) are also considered by Wachs [**422**, Proposition 2.3] under the name splitting basis. The fact that the former is a specialization of the BW-basis for an appropriate choice of half-space is pointed out in [**84**, Theorem 6.3]. The fact that the dual splitting basis goes to the comb basis of $\mathsf{Lie}[n]$ under the JKS isomorphism is given in [**422**, Theorems 4.3 and 5.5]. Wachs mentions that the first of these results was obtained jointly with S. Sundaram. She also explains the Björner-Lyndon duality in [**422**, Proposition 2.2 and Theorem 4.1].

Theorem 14.64 can be deduced from a character computation of Lehrer and Solomon [**282**, Corollary 4.6]. This result explicitly appeared in work of Barcelo and Bergeron [**41**, Theorem 3.1]. (They phrase it in terms of the Orlik-Solomon algebra which is isomorphic to Whitney homology.) They prove it by extending the duality between the Björner and Lyndon bases on $\mathsf{Lie}[n]$ (established by Barcelo [**40**]) to $\bigwedge \mathsf{Lie}[n]$. The same isomorphism is explained in a basis-free manner (similar to what we do) by Wachs [**422**, Theorem 7.2].

The description of the classical Lie operad given in Section 14.9.6 can be found for instance in [**9**, Example B.5]. There are also other ways to formulate this operad, see for instance [**307**, Definition 1.28] and [**290**, Section 13.2]. Proposition 14.65 is the operadic version of the classical result that the tensor algebra is the symmetrization of the free Lie algebra [**360**, Theorem 3.7].

Type *B*. Bergeron [**56**] generalized his work with Barcelo to type B. He uses the term 'pair of hedge-rows' for a type B binary tree, and denotes the space $\mathcal{E}[\mathbf{n}]$ by $\mathscr{L}_{(n,\emptyset)}(n)$. His Theorem 4.1 is equivalent to our Theorem 14.84. (He works with the Orlik-Solomon algebra rather than Whitney homology, and with $\mathcal{E}[\mathbf{n}]$ rather than $\mathsf{Lie}[\mathbf{n}]$.) The cohomology space $\mathcal{H}^{r-2}(\Pi[\mathbf{n}])$ is studied by Gottlieb and Wachs [**204**, Theorem 5.6 and Corollary 5.7]. They associate to a maximal chain of flats a type B binary tree, and give a presentation of the cohomology space using such trees. They introduce elementary relations of types 1, 2, 3 and 4. They give a similar presentation of $\mathcal{E}[\mathbf{n}]$ in their Theorem 6.1. Their Corollary 7.4 is our Theorem 14.87. In their Corollary 8.6, they generalize this result to components of Whitney homology. This corresponds to Bergeron's Theorem 4.1 and to our Theorem 14.84. We mention that Gottlieb and Wachs obtain their results in the more general setting of Dowling lattices. Bergeron, Gottlieb and Wachs do not consider the object $\mathsf{Lie}[\mathbf{n}]$. (The precise relation between $\mathcal{E}[\mathbf{n}]$ and $\mathsf{Lie}[\mathbf{n}]$ is given in Theorem 14.86.) For results related to character computations, see [**221, 281, 56**].

The dual BW-basis is not expressible using a single maximal chain of flats in contrast to type A (14.60). This is because there are chambers in type B contained in a generic half-space h which have no contact with any of the chambers cut by b(h). Even the octagon presents such an example. This is consistent with the observation in [**204**, Theorem 9.2], whic gives a nice basis for cohomology but it is not the dual BW-basis. It picks maximal chains (not linear combinations) to represent cohomology classes just like the Lyndon basis.

CHAPTER 15

Incidence algebras

Incidence algebras are reviewed in Section C.1. We begin with the flat-incidence algebra. It is the incidence algebra of the poset of flats. It is related to the Birkhoff algebra in the sense that the zeta function and the Möbius function intervene in the change of basis formulas between the H- and Q-bases.

Next, we introduce the lune-incidence algebra, which is a certain reduced incidence algebra of the poset of faces. It can also be viewed as the incidence algebra of the category of lunes. In particular, it has a basis indexed by lunes. In this algebra, one can define noncommutative zeta functions and noncommutative Möbius functions, which are inverse to each other. This algebra connects to the Tits algebra. Recall that homogeneous sections and special Zie families can be used to construct and characterize Eulerian families. The punchline is that homogeneous sections correspond to noncommutative zeta functions, while special Zie families correspond to noncommutative Möbius functions. In effect, noncommutative zeta and Möbius functions intervene in the change of basis formulas between the H- and Q-bases of the Tits algebra.

There are three important subspaces of the lune-incidence algebra that we consider, namely, the Lie-incidence algebra, the space of additive functions and the space of Weisner functions. All three subspaces have the same dimension given by the number of faces in the arrangement. The Lie-incidence algebra is a subalgebra of the lune-incidence algebra. Its elements can be identified with Lie elements in arrangements over and under various flats. In fact, it is isomorphic to the Tits algebra. The space of additive functions contains the space of noncommutative zeta functions. It is a right module over the Lie-incidence algebra with action induced from the product in the lune-incidence algebra. In fact, it is isomorphic to the right regular representation of the Lie-incidence algebra. Similarly, the space of Weisner functions contains the space of noncommutative Möbius functions. It is a left module over the Lie-incidence algebra isomorphic to the left regular representation.

There are similarities between the flat-incidence algebra, lune-incidence algebra and Lie-incidence algebra. All three algebras are elementary, with the Birkhoff algebra as their split-semisimple quotient. Their quivers have flats as their vertices, and arrows can go only from a flat to a smaller flat which it covers.

An encompassing picture for these observations involves a generalization of the classical notion of operad.

15.1. Flat-incidence algebra

The flat-incidence algebra is the incidence algebra of the poset of flats. We discuss this briefly.

15.1.1. Flat-incidence algebra. Recall that a nested flat is a pair of flats (X, Y) with $\mathrm{X} \leq \mathrm{Y}$. Let $\mathrm{I}_{\mathrm{flat}}[\mathcal{A}]$ denote the incidence algebra of the poset of flats. We call it the *flat-incidence algebra*. It consists of functions f on nested flats, with the product of f and g given by

$$(fg)(\mathrm{X}, \mathrm{Z}) = \sum_{\mathrm{Y}:\, \mathrm{X} \leq \mathrm{Y} \leq \mathrm{Z}} f(\mathrm{X}, \mathrm{Y}) g(\mathrm{Y}, \mathrm{Z}). \tag{15.1}$$

This is a specialization of (C.1).

Incidence algebras are compatible with cartesian product of posets. As a consequence: For arrangements $\mathcal{A}$ and $\mathcal{A}'$,

$$\mathrm{I}_{\mathrm{flat}}[\mathcal{A} \times \mathcal{A}'] = \mathrm{I}_{\mathrm{flat}}[\mathcal{A}] \otimes \mathrm{I}_{\mathrm{flat}}[\mathcal{A}']. \tag{15.2}$$

15.1.2. A substitution product. For any arrangement $\mathcal{A}$, let $\mathsf{E}[\mathcal{A}] = \Bbbk$. For any flat X, we have the map

$$\mathsf{E}[\mathcal{A}^{\mathrm{X}}] \otimes \mathsf{E}[\mathcal{A}_{\mathrm{X}}] \to \mathsf{E}[\mathcal{A}], \qquad \Bbbk \otimes \Bbbk \xrightarrow{\cong} \Bbbk. \tag{15.3}$$

We call this the substitution product of E.

Consider the vector space

$$\bigoplus_{\mathrm{X} \leq \mathrm{Y}} \mathsf{E}[\mathcal{A}_{\mathrm{X}}^{\mathrm{Y}}].$$

The sum is over both X and Y. This space carries an algebra structure. Elements in the (X, Y)-summand are multiplied with elements in the (Y, Z)-summand by substitution; the remaining products are all zero.

There is an isomorphism of algebras

$$\mathrm{I}_{\mathrm{flat}}[\mathcal{A}] \xrightarrow{\cong} \bigoplus_{\mathrm{X} \leq \mathrm{Y}} \mathsf{E}[\mathcal{A}_{\mathrm{X}}^{\mathrm{Y}}]. \tag{15.4}$$

Given $f \in \mathrm{I}_{\mathrm{flat}}[\mathcal{A}]$, its image under (15.4) in the (X, Y)-summand is the scalar $f(\mathrm{X}, \mathrm{Y})$.

15.1.3. Flat-incidence module. Let $\mathrm{M}_{\mathrm{flat}}[\mathcal{A}]$ denote the incidence module of the poset of flats. We call it the *flat-incidence module*. It consists of functions on flats. The module structure is given by

$$(fg)(\mathrm{X}) = \sum_{\mathrm{Y}:\, \mathrm{X} \leq \mathrm{Y}} f(\mathrm{X}, \mathrm{Y}) g(\mathrm{Y}). \tag{15.5}$$

This is a specialization of (C.10).

In view of (15.4), we note that there is an isomorphism of modules

$$\mathrm{M}_{\mathrm{flat}}[\mathcal{A}] \xrightarrow{\cong} \bigoplus_{\mathrm{X}} \mathsf{E}[\mathcal{A}_{\mathrm{X}}]. \tag{15.6}$$

15.1.4. Radical of the flat-incidence algebra. Structure results on incidence algebras are given in Section C.2. We now specialize some of them to the flat-incidence algebra.

Proposition 15.1. *The flat-incidence algebra is elementary. Its split-semisimple quotient is the Birkhoff algebra, with the quotient map*

$$\mathrm{I}_{\mathrm{flat}}[\mathcal{A}] \twoheadrightarrow \Pi[\mathcal{A}], \qquad f \mapsto \sum_{\mathrm{X}} f(\mathrm{X}, \mathrm{X})\, \mathsf{Q}_{\mathrm{X}}. \tag{15.7}$$

In particular, the radical of the flat-incidence algebra consists of functions which are zero on nested flats of the form (X, X).

PROOF. This is a special case of Proposition C.10. □

Theorem 15.2. *The quiver of the flat-incidence algebra is as follows. The vertices are flats, and there is exactly one arrow from* Y *to* X *when* $\mathrm{X} \lessdot \mathrm{Y}$, *and no arrows otherwise. In other words, the quiver is the Hasse diagram of the poset of flats.*

PROOF. This is a special case of Theorem C.14. □

15.2. Lune-incidence algebra

We introduce the lune-incidence algebra. It is a certain reduced incidence algebra of the poset of faces. It has a basis indexed by lunes. It can also be viewed as the incidence algebra of the category of lunes, and is thus linked to the substitution product of chambers. It is related to the flat-incidence algebra via the base-case map. We also introduce the lune-incidence module. The lune-incidence algebra is elementary, and we describe its radical series and quiver.

Familiarity with the notions and results from Chapters 3 and 4 is a prerequisite for this section.

15.2.1. Face-incidence algebra. Recall that a nested face is a pair of faces (F, H) with $F \leq H$. Let $\mathrm{I}_{\mathrm{face}}[\mathcal{A}]$ denote the incidence algebra of the poset of faces. We call it the *face-incidence algebra*. It consists of functions f on nested faces, with the product of f and g given by

$$(fg)(F, H) = \sum_{G:\, F \leq G \leq H} f(F, G) g(G, H). \tag{15.8}$$

15.2.2. Lune-incidence algebra. Recall the equivalence relation (3.13) on the set of nested faces. For convenience, it is reproduced below.

$$(A, F) \sim (B, G) \iff AB = A,\ BA = B,\ AG = F \text{ and } BF = G. \tag{15.9}$$

Let $\mathrm{I}_{\mathrm{lune}}[\mathcal{A}]$ denote the subspace of $\mathrm{I}_{\mathrm{face}}[\mathcal{A}]$ consisting of those functions f on nested faces such that $f(A, F) = f(B, G)$ whenever $(A, F) \sim (B, G)$.

Lemma 15.3. *The equivalence relation* (15.9) *is order-compatible, or equivalently,* $\mathrm{I}_{\mathrm{lune}}[\mathcal{A}]$ *is a subalgebra of* $\mathrm{I}_{\mathrm{face}}[\mathcal{A}]$. *It has a basis indexed by lunes.*

PROOF. The poset isomorphism in Lemma 1.37 restricts to an isomorphism between intervals $[A, F]$ and $[B, G]$ when $(A, F) \sim (B, G)$. Proposition C.16 then ensures that the relation is order-compatible. Since equivalence classes correspond to lunes (Proposition 3.13), the remaining assertion holds. □

We refer to $\mathrm{I}_{\mathrm{lune}}[\mathcal{A}]$ as the *lune-incidence algebra*. It is an example of a reduced incidence algebra of the poset of faces.

The relation (15.9) is strictly stronger than poset isomorphism. For example, $(F, F) \sim (G, G)$ only if F and G have the same support, while $[F, F]$ and $[G, G]$ are always isomorphic as posets, being singletons.

15.2.3. Substitution product of chambers. Consider the vector space

$$\bigoplus_{\mathrm{X}\leq\mathrm{Y}} \Gamma[\mathcal{A}_{\mathrm{X}}^{\mathrm{Y}}].$$

The sum is over both X and Y. This space carries an algebra structure. Elements in the (X, Y)-summand are multiplied with elements in the (Y, Z)-summand by substitution (10.27); the remaining products are all zero.

Lemma 15.4. *There is an isomorphism of algebras*

$$\mathrm{I}_{\mathrm{lune}}[\mathcal{A}] \xrightarrow{\cong} \bigoplus_{\mathrm{X}\leq\mathrm{Y}} \Gamma[\mathcal{A}_{\mathrm{X}}^{\mathrm{Y}}]. \tag{15.10}$$

Given $f \in \mathrm{I}_{\mathrm{lune}}[\mathcal{A}]$, its image under (15.10) in the (X, Y)-summand is the element

$$\sum_{F:\, F\geq A,\, \mathrm{s}(F)=\mathrm{Y}} f(A, F)\, \mathtt{H}_{F/A}, \tag{15.11}$$

where A is a fixed face of support X.

Exercise 15.5. Use Lemma 15.4 to deduce that the lune-incidence algebra has trivial center.

15.2.4. Base-case map. We relate the lune-incidence algebra to the flat-incidence algebra. The former is the reduced incidence algebra of the poset of faces under the relation (15.9). We may view the latter as the reduced incidence algebra of the poset of flats under the equality relation.

Proposition 15.6. *There is an algebra homomorphism*

$$\mathrm{bc} : \mathrm{I}_{\mathrm{lune}}[\mathcal{A}] \to \mathrm{I}_{\mathrm{flat}}[\mathcal{A}] \tag{15.12}$$

given by

$$\mathrm{bc}(f)(\mathrm{X}, \mathrm{Y}) = \sum_{F:\, F\geq A,\, \mathrm{s}(F)=\mathrm{Y}} f(A, F),$$

where A is a fixed face of support X.

PROOF. Proposition 3.14 implies that the support map $\mathrm{s} : \Sigma \to \Pi$ creates relations, in the sense of Section C.3. Indeed, it says that both conditions (C.19) and (C.22) hold. The result then follows from Proposition C.18. □

SECOND PROOF. A comparison of (15.4) and (15.10) shows that the map bc is induced from maps $\Gamma[\mathcal{A}] \to \mathsf{E}[\mathcal{A}]$, as $\mathcal{A}$ varies, which send all basis elements to 1. The fact that it is an algebra homomorphism is then encapsulated in the commutative diagram

$$\begin{array}{ccc} \Gamma[\mathcal{A}^{\mathrm{X}}] \otimes \Gamma[\mathcal{A}_{\mathrm{X}}] & \longrightarrow & \Gamma[\mathcal{A}] \\ \downarrow & & \downarrow \\ \mathsf{E}[\mathcal{A}^{\mathrm{X}}] \otimes \mathsf{E}[\mathcal{A}_{\mathrm{X}}] & \longrightarrow & \mathsf{E}[\mathcal{A}]. \end{array} \tag{15.13}$$

□

We call bc the *base-case map*. Its relationship with the base-case map in (4.4) is explained in (15.15) below.

15.2.5. Category of lunes. Incidence algebras can be defined for any locally finite category (Section C.1.13). Recall the category of lunes from Section 4.4. It is a finite category.

Proposition 15.7. *The lune-incidence algebra is the incidence algebra of the category of lunes. Explicitly, it consists of functions f on lunes, with the product of f and g given by*

$$(fg)(\mathrm{N}) = \sum_{\mathrm{L} \circ \mathrm{M} = \mathrm{N}} f(\mathrm{L}) g(\mathrm{M}). \tag{15.14}$$

The sum is over both L *and* M. *The unit element is the function which is* 1 *on flats, and* 0 *on lunes which are not flats.*

In the language of Proposition 15.7, the base-case map (15.12) is given by

$$\mathrm{bc}(f)(\mathrm{X}, \mathrm{Y}) = \sum_{\mathrm{L}:\, \mathrm{b}(\mathrm{L}) = \mathrm{X},\, \mathrm{c}(\mathrm{L}) = \mathrm{Y}} f(\mathrm{L}). \tag{15.15}$$

The sum is over all lunes L whose base is X and case is Y, or equivalently, over all lunes L whose base-case is (X, Y).

Exercise 15.8. Use formula (15.15) along with (15.14) to deduce that the base-case map is an algebra homomorphism. (This is formally equivalent to the second proof of Proposition 15.6.)

Incidence algebras are compatible with cartesian product of categories. Also recall from Section 4.4.4 that the category of lunes is compatible with cartesian product of arrangements. As a consequence of Proposition 15.7: For arrangements $\mathcal{A}$ and $\mathcal{A}'$,

$$\mathrm{I}_{\mathrm{lune}}[\mathcal{A} \times \mathcal{A}'] = \mathrm{I}_{\mathrm{lune}}[\mathcal{A}] \otimes \mathrm{I}_{\mathrm{lune}}[\mathcal{A}']. \tag{15.16}$$

15.2.6. Lune-incidence module. Recall the equivalence relation (3.7) on the set of top-nested faces. Define the vector space $\mathrm{M}_{\mathrm{lune}}[\mathcal{A}]$ as follows. It consists of functions g on top-nested faces such that $g(A, C) = g(B, D)$ whenever $(A, C) \sim (B, D)$. This is a left module over the lune-incidence algebra with action given by

$$(fg)(F, C) = \sum_{G:\, F \leq G \leq C} f(F, G) g(G, C). \tag{15.17}$$

We call it the *lune-incidence module*. It has a basis indexed by top-lunes. Along the lines of Proposition 15.7, we obtain:

Proposition 15.9. *The lune-incidence module consists of functions g on top-lunes, with the action of f on g given by*

$$(fg)(\mathrm{N}) = \sum_{\mathrm{L} \circ \mathrm{M} = \mathrm{N}} f(\mathrm{L}) g(\mathrm{M}). \tag{15.18}$$

Here N *is a top-lune. (Note that this forces* M *to be a top-lune.)*

Observe that the lune-incidence module can be viewed as a left ideal of the lune-incidence algebra consisting of functions which are 0 on lunes which are not top-lunes. Further, the isomorphism of algebras (15.10) induces an isomorphism of left ideals

$$\mathrm{M}_{\mathrm{lune}}[\mathcal{A}] \xrightarrow{\cong} \bigoplus_{\mathrm{X}} \Gamma[\mathcal{A}_{\mathrm{X}}]. \tag{15.19}$$

15.2.7. Radical of the lune-incidence algebra. We work with the description given in Proposition 15.7.

Proposition 15.10. *The lune-incidence algebra is elementary. Its split-semisimple quotient is the Birkhoff algebra, with the quotient map*

$$\mathrm{I}_{\mathrm{lune}}[\mathcal{A}] \twoheadrightarrow \Pi[\mathcal{A}], \qquad f \mapsto \sum_{\mathrm{X}} f(\mathrm{X})\, \mathtt{Q}_{\mathrm{X}}. \tag{15.20}$$

In particular, the radical of the lune-incidence algebra consists of functions on lunes which vanish on flats.

Proof. Let J denote the set of functions on lunes which are zero on flats. Observe that J is a nilpotent ideal and all nilpotent elements belong to J. Hence J is the radical of the lune-incidence algebra. All claims follow. □

Proposition 15.11. *For each flat* X, *let* e_{X} *denote the function on lunes which is* 1 *on* X *and* 0 *otherwise. The* e_{X}, *as* X *varies, is a complete system of primitive orthogonal idempotents of* $\mathrm{I}_{\mathrm{lune}}[\mathcal{A}]$.

Proof. We only have to note that the map $\mathtt{Q}_{\mathrm{X}} \mapsto e_{\mathrm{X}}$ is an algebra section of (15.20). □

Recall from (3.2) that sk(L) denotes the slack of the lune L.

Proposition 15.12. *The* i*-th power of the radical of the lune-incidence algebra consists of functions* f *on lunes such that* $f(\mathrm{L}) = 0$ *whenever* $\mathrm{sk}(\mathrm{L}) < i$.

Proof. This can be deduced from Exercise 4.43 and the description of the radical given in Proposition 15.10. □

As a consequence:

Proposition 15.13. *The nilpotency index of the radical of the lune-incidence algebra is* $r+1$, *where* r *is the rank of the arrangement. The* r*-th power of the radical consists of functions which are zero on all lunes which are not chambers.*

The powers of the radical can also be described using the identification (15.10). The i-th power equals

$$\bigoplus_{\mathrm{rk(Y/X)} \geq i} \Gamma[\mathcal{A}^{\mathrm{Y}}_{\mathrm{X}}].$$

The sum is over all $\mathrm{X} \leq \mathrm{Y}$ such that the codimension of X in Y is greater than i. In particular, the radical is $\bigoplus_{\mathrm{X}<\mathrm{Y}} \Gamma[\mathcal{A}^{\mathrm{Y}}_{\mathrm{X}}]$ and the r-th power is $\Gamma[\mathcal{A}]$.

Theorem 15.14. *The quiver of the lune-incidence algebra is as follows. The vertices are flats, and there are exactly two arrows from* Y *to* X *when* $\mathrm{X} \lessdot \mathrm{Y}$, *and no arrows otherwise.*

Proof. The split-semisimple quotient of the lune-incidence algebra is the Birkhoff algebra. Hence the vertices of its quiver are flats. The arrows can be computed from Proposition 15.12. Note that $J/J^2 \cong \bigoplus_{\mathrm{X}\lessdot\mathrm{Y}} e_{\mathrm{X}} \mathrm{I}_{\mathrm{lune}}\, e_{\mathrm{Y}}$, where J is the radical of the lune-incidence algebra, and the e_{X} are as in Proposition 15.11. Thus, $e_{\mathrm{X}}(J/J^2)e_{\mathrm{Y}}$ is zero unless $\mathrm{X} \lessdot \mathrm{Y}$, and in this case, its dimension is 2. This is because, for $\mathrm{X} \lessdot \mathrm{Y}$, there are two lunes with base X and case Y. They correspond to the two chambers in the rank-one arrangement $\mathcal{A}^{\mathrm{Y}}_{\mathrm{X}}$. □

15.2.8. Projective lune-incidence algebra. Every nested face (A, F) has an *opposite nested face* given by $(A, A\overline{F})$. Note that (A, F) is its own opposite iff $A = F$. Let $\equiv$ denote the equivalence relation on nested faces in which opposite nested faces are equivalent. A function f on nested faces is projective if $f(A, F) = f(A, A\overline{F})$ for all nested faces (A, F).

Let $\sim'$ denote the equivalence relation on nested faces generated by $\sim$ and $\equiv$. Explicitly, in view of (3.14),

$$(A, F) \sim' (B, G) \iff \text{Either } (A, F) \sim (B, G) \text{ or } (A, F) \sim (B, B\overline{G}). \tag{15.21}$$

The relation $(A, F) \equiv (A, A\overline{F})$ is covered by the second alternative. The equivalence classes under $\sim'$ are projective lunes.

Lemma 15.15. *The equivalence relation* (15.21) *is order-compatible. In other words, the space of all projective functions in the lune-incidence algebra form a subalgebra. It has a basis indexed by projective lunes.*

PROOF. We extend the proof of Lemma 15.3 to include the case $(A, F) \equiv (A, A\overline{F})$ as follows. The intervals $[A, F]$ and $[A, A\overline{F}]$ are isomorphic under the map $K \mapsto A\overline{K}$. Further, $(A, K) \equiv (A, A\overline{K})$, and $(K, F) \sim (A\overline{K}, A\overline{K}F) \equiv (A\overline{K}, A\overline{F})$. Note very carefully that the last step employs both $\sim$ and $\equiv$. □

We call this the *projective lune-incidence algebra*.

A function f on lunes is projective if $f(\mathrm{L}) = f(\overline{\mathrm{L}})$. In the language of Proposition 15.7, the projective lune-incidence algebra is the subalgebra of the lune-incidence algebra consisting of projective functions. The fact that it is a subalgebra can also be deduced from Lemma 4.39.

Exercise 15.16. Show that the equivalence relation $\equiv$ on nested faces is *not* order-compatible in general. In fact, order-compatibility fails in rank-two arrangements.

The subspace of the lune-incidence module consisting of projective functions (either on top-nested faces or on top-lunes depending on viewpoint) is a module over the projective lune-incidence algebra. We call this the *projective lune-incidence module*.

15.2.9. Rank one. Let $\mathcal{A}$ be the arrangement of rank one. It has four lunes, namely, the flats $\bot$ and $\top$ and chambers C and $\overline{C}$. Except $\bot$, the rest are top-lunes. The lune-incidence algebra of $\mathcal{A}$ consists of functions f on lunes, with product of f and g given by

$$\begin{aligned} (fg)(\bot) &= f(\bot)g(\bot), & (fg)(C) &= f(\bot)g(C) + f(C)g(\top), \\ (fg)(\top) &= f(\top)g(\top), & (fg)(\overline{C}) &= f(\bot)g(\overline{C}) + f(\overline{C})g(\top). \end{aligned} \tag{15.22}$$

It is the incidence algebra of the category with two objects and two parallel (non-identity) arrows (Example 4.28). Its radical is the ideal consisting of functions f on lunes for which $f(\bot) = f(\top) = 0$.

The lune-incidence module is the left ideal consisting of functions f on lunes for which $f(\bot) = 0$.

For the projective case, we impose the additional condition $f(C) = f(\overline{C})$.

15.3. Noncommutative zeta and Möbius functions

Noncommutative zeta functions are elements of the lune-incidence algebra characterized by lune-additivity. Similarly, noncommutative Möbius functions are elements of this algebra characterized by the noncommutative Weisner formula. Interestingly, these functions are not unique, but they are inverse to each other. They can be seen to correspond to homogeneous sections and special Zie families, respectively.

15.3.1. Noncommutative zeta functions. Recall the lune-incidence algebra $\mathrm{I}_{\text{lune}}[\mathcal{A}]$. A *noncommutative zeta function* is an element $\boldsymbol{\zeta} \in \mathrm{I}_{\text{lune}}[\mathcal{A}]$ such that $\boldsymbol{\zeta}(A, A) = 1$ for all A and

$$\boldsymbol{\zeta}(H, G) = \sum_{\substack{F:\, F \geq A,\, HF = G,\\ \mathrm{s}(F)=\mathrm{s}(G)}} \boldsymbol{\zeta}(A, F) \tag{15.23}$$

for all $A \leq H \leq G$. Note that when $A = H$, it holds automatically, as the only F that contributes to the sum is $F = G$.

Lemma 15.17. *All noncommutative zeta functions $\boldsymbol{\zeta} \in \mathrm{I}_{\text{lune}}[\mathcal{A}]$ have the same base-case: the zeta function $\zeta \in \mathrm{I}_{\text{flat}}[\mathcal{A}]$.*

PROOF. Fix $\mathrm{X} \leq \mathrm{Y}$, and a face A of support X. Let G be any face greater than A with support Y. Then

$$\mathrm{bc}(\boldsymbol{\zeta})(\mathrm{X}, \mathrm{Y}) = \sum_{F:\, F \geq A,\, \mathrm{s}(F)=\mathrm{Y}} \boldsymbol{\zeta}(A, F) = \boldsymbol{\zeta}(G, G) = 1.$$

For the second step, we put $H = G$ in (15.23). □

Lemma 15.18. *A noncommutative zeta function is equivalent to a homogeneous section of the support map.*

PROOF. Given $\boldsymbol{\zeta}$, put $\mathtt{u}^F := \boldsymbol{\zeta}(O, F)$. Putting $A = O$ and $H = G$ in (15.23), we see that (11.3) holds. Thus, $(\mathtt{u}^F)$ is a homogeneous section. Conversely, given a homogeneous section $\mathtt{u}$, put $\boldsymbol{\zeta}(H, G) := \mathtt{u}^G_H$. Then $\boldsymbol{\zeta}$ defines a noncommutative zeta function: (11.10) says that it belongs to the lune-incidence algebra, while (11.11) is the same as (15.23). □

Exercise 15.19. A noncommutative zeta function $\boldsymbol{\zeta}$ is

- *set-theoretic* if the scalars $\boldsymbol{\zeta}(O, F)$ are either 0 or 1, and among all faces F of a given support, exactly one $\boldsymbol{\zeta}(O, F)$ is 1,
- *projective* if $\boldsymbol{\zeta}(O, F) = \boldsymbol{\zeta}(O, \overline{F})$ for all faces F, and
- *uniform* if $\boldsymbol{\zeta}(O, F) = \boldsymbol{\zeta}(O, G)$ whenever F and G have the same support.

Check that: Under the identification in Lemma 15.18, the above correspond to set-theoretic, projective and uniform sections, respectively.

Exercise 15.20. Check that a noncommutative zeta function $\boldsymbol{\zeta}$ is projective if $\boldsymbol{\zeta}(A, F) = \boldsymbol{\zeta}(A, A\overline{F})$ for all $A \leq F$. (This is equivalent to the second assertion in Exercise 11.8.)

In the language of Proposition 15.7:

Lemma 15.21. *A noncommutative zeta function is the same as a function* $\boldsymbol{\zeta}$ *on lunes such that* $\boldsymbol{\zeta}(\mathrm{Z}) = 1$ *for any flat* Z, *and for any lune* M *and flat* X *with* $\mathrm{X} \leq \mathrm{b}(\mathrm{M})$,

$$\boldsymbol{\zeta}(\mathrm{M}) = \sum_{\substack{\mathrm{L}:\, \mathrm{b}(\mathrm{L})=\mathrm{X},\, \mathrm{L}\leq \mathrm{M} \\ \mathrm{c}(\mathrm{L})=\mathrm{c}(\mathrm{M})}} \boldsymbol{\zeta}(\mathrm{L}). \tag{15.24}$$

The sum is over all lunes L *which are contained in* M, *have base* X *and the same case as* M.

A comparison with (3.24) shows that $\boldsymbol{\zeta}$ is a function on lunes which is 1 on flats and additive over any lune decomposition of a lune (into smaller lunes). We refer to either (15.23) or (15.24) as *lune-additivity*.

Lemma 15.22. *Let* $\boldsymbol{\zeta}$ *be a noncommutative zeta function. Then:*

For any combinatorial lune M,

$$\boldsymbol{\zeta}(\mathrm{M}) = \sum_{F\in \mathrm{M}} \boldsymbol{\zeta}(F). \tag{15.25}$$

The sum is over all top-dimensional faces F *of* M.

For any flats $\mathrm{X} \leq \mathrm{Y}$,

$$\sum_{\mathrm{L}:\, \mathrm{b}(\mathrm{L})=\mathrm{X},\, \mathrm{c}(\mathrm{L})=\mathrm{Y}} \boldsymbol{\zeta}(\mathrm{L}) = \boldsymbol{\zeta}(\mathrm{Y}). \tag{15.26}$$

PROOF. Setting $\mathrm{X} = \perp$ in (15.24) yields (15.25). (Recall that faces are the only lunes whose base is the minimum flat.)

Similarly, setting $\mathrm{M} = \mathrm{Y}$ in (15.24) yields (15.26). □

Observe that condition (15.25) implies and hence is equivalent to (15.24). This is also the content of Lemma 15.18.

15.3.2. Noncommutative Möbius functions.

A *noncommutative Möbius function* is an element $\boldsymbol{\mu} \in \mathrm{I}_{\mathrm{lune}}[\mathcal{A}]$ such that $\boldsymbol{\mu}(A, A) = 1$ for all A and

$$\sum_{F:\, F\geq A,\, HF=G} \boldsymbol{\mu}(A, F) = 0 \tag{15.27}$$

for all $A < H \leq G$.

Lemma 15.23. *All noncommutative Möbius functions* $\boldsymbol{\mu} \in \mathrm{I}_{\mathrm{lune}}[\mathcal{A}]$ *have the same base-case: the Möbius function* $\mu \in \mathrm{I}_{\mathrm{flat}}[\mathcal{A}]$.

PROOF. We check that $\mathrm{bc}(\boldsymbol{\mu})$ satisfies the Weisner formula (C.7a), that is, for $\mathrm{Z} < \mathrm{Y} \leq \mathrm{W}$,

$$\sum_{\mathrm{X}:\, \mathrm{X}\geq \mathrm{Z},\, \mathrm{Y}\vee \mathrm{X}=\mathrm{W}} \mathrm{bc}(\boldsymbol{\mu})(\mathrm{Z}, \mathrm{X}) = 0.$$

Fix a face A of support Z, and face H greater than A of support Y. Then

$$\begin{aligned}\sum_{\mathrm{X}:\, \mathrm{X}\geq \mathrm{Z},\, \mathrm{Y}\vee \mathrm{X}=\mathrm{W}} \mathrm{bc}(\boldsymbol{\mu})(\mathrm{Z}, \mathrm{X}) &= \sum_{F:\, F\geq A,\, \mathrm{Y}\vee \mathrm{s}(F)=\mathrm{W}} \boldsymbol{\mu}(A, F) \\ &= \sum_{F:\, F\geq A,\, \mathrm{s}(HF)=\mathrm{W}} \boldsymbol{\mu}(A, F) \\ &= \sum_{G:\, \mathrm{s}(G)=\mathrm{W}} \ \sum_{F:\, F\geq A,\, HF=G} \boldsymbol{\mu}(A, F) \ = \ 0. \end{aligned}$$

□

Lemma 15.24. *A noncommutative Möbius function is equivalent to a special Zie family.*

PROOF. Given $\boldsymbol{\mu}$, for each F, the element $x_F := \sum_{G:\, F \leq G} \boldsymbol{\mu}(F, G)\, \mathtt{H}_{G/F}$ is a Zie element of $\mathcal{A}_F$. (The Friedrichs criterion reduces to the noncommutative Weisner formula.) The special Zie family P can now be defined as follows. For each flat X, pick a face F with support X, and let $\mathsf{P}_{\mathrm{X}} := \beta_{\mathrm{X},F}(x_F)$. Since $\boldsymbol{\mu}$ belongs to the lune-incidence algebra, this does not depend on the choice of F. These steps can be reversed. □

Exercise 15.25. By definition, a noncommutative Möbius function is projective if $\boldsymbol{\mu}(A, F) = \boldsymbol{\mu}(A, A\overline{F})$ for all $A \leq F$. Check that: Under the identification in Lemma 15.24, a projective noncommutative Möbius function corresponds to a projective special Zie family.

In the language of Proposition 15.7:

Lemma 15.26. *A noncommutative Möbius function is the same as a function* $\boldsymbol{\mu}$ *on lunes such that* $\boldsymbol{\mu}(\mathrm{Z}) = 1$ *for any flat* Z, *and for any lune* M *and flat* X *with* $\mathrm{X} < \mathrm{b}(\mathrm{M})$,

$$\sum_{\mathrm{L}:\, \mathrm{b}(\mathrm{L}) = \mathrm{X},\, \mathrm{L} \preceq \mathrm{M}} \boldsymbol{\mu}(\mathrm{L}) = 0. \tag{15.28}$$

The sum is over all lunes L *whose base is* X *and whose interior is contained in the interior of* M, *see* (4.7).

We refer to either (15.27) or (15.28) as the *noncommutative Weisner formula.*

Lemma 15.27. *Let* $\boldsymbol{\mu}$ *be a noncommutative Möbius function. Then:*

For any lune M *which is not a face,*

$$\sum_{F \in \mathrm{M}^{\circ}} \boldsymbol{\mu}(F) = 0. \tag{15.29}$$

The sum is over all interior faces F *of* M.

For any flats $\mathrm{X} < \mathrm{Y}$,

$$\sum_{\mathrm{L}:\, \mathrm{b}(\mathrm{L}) = \mathrm{X},\, \mathrm{c}(\mathrm{L}) \leq \mathrm{Y}} \boldsymbol{\mu}(\mathrm{L}) = 0. \tag{15.30}$$

PROOF. Setting $\mathrm{X} = \perp$ in (15.28) yields (15.29). (Recall that faces are the only lunes whose base is the minimum flat.)

Similarly, setting $\mathrm{M} = \mathrm{Y}$ in (15.28) and using (4.8) yields (15.30). □

15.3.3. Zeta and Möbius as inverses.

The main result is the following.

Theorem 15.28. *In the lune-incidence algebra, the inverse of a noncommutative zeta function is a noncommutative Möbius function, and vice-versa.*

This is a part of a more general result given in Theorem 15.44. For convenience, we indicate here a way to obtain this result directly.

SKETCH OF A DIRECT PROOF. Let us start afresh forgetting all that we discussed about the Tits algebra. Let Σ denote the linearization of the set of faces, and let

$\mathtt{H}$ and $\mathtt{Q}$ be two bases of Σ indexed by faces. Consider the following two binary operations on Σ.

$$\mathtt{H}_F \cdot_1 \mathtt{H}_G = \mathtt{H}_{FG} \quad \text{and} \quad \mathtt{H}_F \cdot_2 \mathtt{Q}_G = \begin{cases} \mathtt{Q}_{FG} & \text{if } GF = G, \\ 0 & \text{if } GF > G. \end{cases}$$

The theorem follows from the following two lemmas.

Lemma 15.29. *Suppose* $\boldsymbol{\mu} \in \mathrm{I}_{\mathrm{lune}}[\mathcal{A}]$ *such that* $\boldsymbol{\mu}(A, A) = 1$ *for all* A, *and* $\mathtt{H}$ *and* $\mathtt{Q}$ *are related by*

$$\mathtt{Q}_F = \sum_{G:\, F \leq G} \boldsymbol{\mu}(F, G)\, \mathtt{H}_G.$$

Then $\boldsymbol{\mu}$ *is a noncommutative Möbius function iff the above two operations coincide.*

To prove this, start with the second operation, express $\mathtt{Q}_G$ in the $\mathtt{H}$-basis, and compare this with the first operation. We omit the details.

Lemma 15.30. *Suppose* $\boldsymbol{\zeta} \in \mathrm{I}_{\mathrm{lune}}[\mathcal{A}]$ *such that* $\boldsymbol{\zeta}(A, A) = 1$ *for all* A, *and* $\mathtt{H}$ *and* $\mathtt{Q}$ *are related by*

$$\mathtt{H}_F = \sum_{K:\, F \leq K} \boldsymbol{\zeta}(F, K)\, \mathtt{Q}_K.$$

Then $\boldsymbol{\zeta}$ *is a noncommutative zeta function iff the above two operations coincide.*

This is similar. We start with the first operation, express $\mathtt{H}_G$ in the $\mathtt{Q}$-basis, and compare this with the second operation. ☐

Let $\boldsymbol{\zeta}$ and $\boldsymbol{\zeta}'$ be noncommutative zeta functions for arrangements $\mathcal{A}$ and $\mathcal{A}'$, respectively. Under the identification (15.16), we have that

(15.31) $\qquad \boldsymbol{\zeta} \otimes \boldsymbol{\zeta}'$ is a noncommutative zeta function for $\mathcal{A} \times \mathcal{A}'$.

(One may deduce this from Lemma 15.18, for example.) Taking inverses we obtain that if $\boldsymbol{\mu}$ and $\boldsymbol{\mu}'$ are noncommutative Möbius functions for $\mathcal{A}$ and $\mathcal{A}'$, then

(15.32) $\qquad \boldsymbol{\mu} \otimes \boldsymbol{\mu}'$ is a noncommutative Möbius function for $\mathcal{A} \times \mathcal{A}'$.

Exercise 15.31. Use Lemma 15.18 or (15.25) to deduce that all noncommutative zeta functions form an affine space whose dimension equals the number of faces minus the number of flats. Use Lemma 15.24 and formulas (1.49) and (10.25) to do the same for all noncommutative Möbius functions. (By general principles, the dimensions must be equal. So alternatively, one can use this fact and induction to deduce formula (10.25).)

Exercise 15.32. Using Lemma D.25, item (2), deduce that: Projective noncommutative zeta and Möbius functions are in bijection with each other under taking inverses in the projective lune-incidence algebra. Both sets form affine spaces whose dimension equals the number of projective faces minus the number of flats (assuming that the field characteristic is not 2).

15.3.4. Examples. Some examples of noncommutative zeta and Möbius functions are given below.

Example 15.33. Consider the rank-one arrangement with chambers C and $\overline{C}$. Now fix an arbitrary scalar p.

Any noncommutative zeta function $\boldsymbol{\zeta}$ is of the form

$$\boldsymbol{\zeta}(O,O) = \boldsymbol{\zeta}(C,C) = \boldsymbol{\zeta}(\overline{C},\overline{C}) = 1, \quad \boldsymbol{\zeta}(O,C) = p, \quad \boldsymbol{\zeta}(O,\overline{C}) = 1-p.$$

Lune-additivity says

$$\boldsymbol{\zeta}(O,C) + \boldsymbol{\zeta}(O,\overline{C}) = 1.$$

To find the inverse, we solve the equations

$$\boldsymbol{\zeta}(O,O)\boldsymbol{\mu}(O,C) + \boldsymbol{\zeta}(O,C)\boldsymbol{\mu}(C,C) = 0, \quad \boldsymbol{\zeta}(O,O)\boldsymbol{\mu}(O,\overline{C}) + \boldsymbol{\zeta}(O,\overline{C})\boldsymbol{\mu}(\overline{C},\overline{C}) = 0.$$

Thus, the corresponding noncommutative Möbius function $\boldsymbol{\mu}$ is given by

$$\boldsymbol{\mu}(O,O) = \boldsymbol{\mu}(C,C) = \boldsymbol{\mu}(\overline{C},\overline{C}) = 1, \quad \boldsymbol{\mu}(O,C) = -p, \quad \boldsymbol{\mu}(O,\overline{C}) = p-1.$$

The noncommutative Weisner formula says

$$\boldsymbol{\mu}(O,O) + \boldsymbol{\mu}(O,C) + \boldsymbol{\mu}(O,\overline{C}) = 0.$$

As functions on lunes, we write

$$\boldsymbol{\zeta}(\bot) = \boldsymbol{\zeta}(\top) = 1, \quad \boldsymbol{\zeta}(C) = p, \quad \boldsymbol{\zeta}(\overline{C}) = 1-p,$$
$$\boldsymbol{\mu}(\bot) = \boldsymbol{\mu}(\top) = 1, \quad \boldsymbol{\mu}(C) = -p, \quad \boldsymbol{\mu}(\overline{C}) = p-1.$$

There is a unique projective noncommutative zeta function and a unique projective noncommutative Möbius function; they correspond to $p = 1/2$.

Example 15.34. Consider the coordinate arrangement (Section 6.1). For any lune L, let $m_+(\mathrm{L})$ denote the number of $+$ in the tuple of L, and $m_-(\mathrm{L})$ denote the number of $-$ in the tuple of L. Thus, $\mathrm{sk}(\mathrm{L}) = m_+(\mathrm{L}) + m_-(\mathrm{L})$.

For an arbitrary scalar p, we have a noncommutative zeta function and a noncommutative Möbius function defined by

$$\boldsymbol{\zeta}(\mathrm{L}) = p^{m_+(\mathrm{L})}(1-p)^{m_-(\mathrm{L})} \quad \text{and} \quad \boldsymbol{\mu}(\mathrm{L}) = (-p)^{m_+(\mathrm{L})}(p-1)^{m_-(\mathrm{L})}. \tag{15.33}$$

This follows from Example 15.33 and (15.31)–(15.32). One may also check directly that (15.24) and (15.28) hold. (Use Exercise 6.2 for the second check.) Note that $\boldsymbol{\mu}(\mathrm{L}) = (-1)^{\mathrm{sk}(\mathrm{L})}\boldsymbol{\zeta}(\mathrm{L})$.

We now consider special values of p which yield uniform and set-theoretic $\boldsymbol{\zeta}$ as defined in Exercise 15.19. The choice $p = 1/2$ yields

$$\boldsymbol{\zeta}(\mathrm{L}) = \left(\frac{1}{2}\right)^{\mathrm{sk}(\mathrm{L})} \quad \text{and} \quad \boldsymbol{\mu}(\mathrm{L}) = \left(\frac{-1}{2}\right)^{\mathrm{sk}(\mathrm{L})}. \tag{15.34}$$

This is the uniform $\boldsymbol{\zeta}$. The choice $p = 1$ yields

$$\boldsymbol{\zeta}(\mathrm{L}) = \begin{cases} 1 & \text{if } m_-(\mathrm{L}) = 0, \\ 0 & \text{otherwise,} \end{cases} \quad \text{and} \quad \boldsymbol{\mu}(\mathrm{L}) = \begin{cases} (-1)^{\mathrm{sk}(\mathrm{L})} & \text{if } m_-(\mathrm{L}) = 0, \\ 0 & \text{otherwise.} \end{cases} \tag{15.35}$$

The first alternative says that the tuple of L involves only 0, $+$ and $\pm$. This choice of $\boldsymbol{\zeta}$ is set-theoretic. Similar situation arises for the choice $p = 0$.

We can also define $\boldsymbol{\zeta}$ and $\boldsymbol{\mu}$ more generally by fixing scalars $p_1, \dots, p_n$, with p_i corresponding to the i-th coordinate. When all the p_i are equal, we recover (15.33).

Example 15.35. For a good reflection arrangement, a canonical choice is

$$\boldsymbol{\zeta}(F,G) = \frac{1}{c_F^G} \qquad \text{and} \qquad \boldsymbol{\mu}(F,G) = \frac{\mu(\mathcal{A}_F^G)}{c_F^G}.$$

This is the uniform $\boldsymbol{\zeta}$. This choice of $\boldsymbol{\zeta}$ and $\boldsymbol{\mu}$ is equivalent to (11.49). Both functions relate to their commutative analogues by the same scaling factor c_F^G. One may also check directly that $\boldsymbol{\mu}(F,G)$ satisfies the noncommutative Weisner formula.

For the coordinate arrangement, this choice specializes to (15.34).

Example 15.36. Another nice example arises from a separating element of the Tits algebra (Section 12.2.2). The noncommutative zeta and Möbius functions are given by formulas (12.12) and (12.14).

Exercise 15.37. Write down the noncommutative zeta and Möbius functions of Example 15.35 for the braid arrangement. Consider (15.23) and (15.27) with $A = O$ for these functions. Recover the count of shuffles as a multinomial coefficient in Exercise 6.16 from the former. Deduce the following identity from the latter:

$$\sum_{p=\max\{p_1,\dots,p_k\}}^{p_1+\dots+p_k} \frac{(-1)^p}{p} \binom{p}{p_1,\dots,p_k}_{\mathrm{qs}} = 0. \tag{15.36}$$

Exercise 15.38. Take a look at Section 14.9.9. Write down the noncommutative zeta and Möbius functions for the braid arrangement corresponding to the homogeneous section (14.49) and $\mathtt{Q}$-basis (14.48). Do the same for the example in Exercise 14.70. (Use Exercise 12.85.)

15.3.5. Noncommutative Hall formula. Given a noncommutative zeta function $\boldsymbol{\zeta}$, its inverse $\boldsymbol{\mu}$ is given by

$$\boldsymbol{\mu}(A,F) = \sum_{k\geq 0}(-1)^k \sum_{A=G_0<G_1<\dots<G_k=F} \boldsymbol{\zeta}(G_0,G_1)\dots\boldsymbol{\zeta}(G_{k-1},G_k). \tag{15.37}$$

This is a special case of (C.4) and is equivalent to (11.32).

In terms of lunes, the inverse is given by

$$\boldsymbol{\mu}(\mathrm{L}) = \sum_{k\geq 0}(-1)^k \sum_{\mathrm{L}_1\circ\dots\circ\mathrm{L}_k=\mathrm{L}} \boldsymbol{\zeta}(\mathrm{L}_1)\dots\boldsymbol{\zeta}(\mathrm{L}_k). \tag{15.38}$$

The sum is over all sequences of composable lunes (in the category of lunes) which are not flats (identity morphisms). The summand corresponding to $k = 0$ is 0 unless L is a flat, in which case it is 1. We refer to (15.38) as the *noncommutative Hall formula*.

15.3.6. Left regular bands. Theorem 15.28 generalizes to any left regular band (LRB). Recall that every LRB carries the structure of a poset via (E.2) which then has an incidence algebra. The relation (15.9) is an equivalence relation which is order-compatible, so it yields a subalgebra of the incidence algebra. Lune-additivity (15.23) and noncommutative Weisner formula (15.27) make sense, and we use them to define noncommutative zeta functions and noncommutative Möbius functions. They are inverses of each other in this subalgebra; the proof sketched above goes through.

For the Tits monoid, this subalgebra is the lune-incidence algebra, and we recover Theorem 15.28. For the Birkhoff monoid, the relation (15.9) is trivial, and this subalgebra is the flat-incidence algebra. In this case, there is a unique zeta function and a unique Möbius function, and they are inverses of each other. Formula (15.23) reduces to the tautology $1 = 1$, while (15.27) reduces to the Weisner formula (1.46a).

15.4. Noncommutative Möbius inversion. Group-likes and primitives

We introduce noncommutative Möbius inversion for the lune-incidence module. We then define the notion of group-likes and primitives in the lune-incidence module, and show that they correspond to each other under noncommutative Möbius inversion.

15.4.1. Noncommutative Möbius inversion. Recall the lune-incidence module $\mathrm{M}_{\mathrm{lune}}[\mathcal{A}]$. Let $\boldsymbol{\zeta}$ be a noncommutative zeta function, and $\boldsymbol{\mu}$ be its inverse. Then for functions f and g in $\mathrm{M}_{\mathrm{lune}}[\mathcal{A}]$,

$$g(F,C) = \sum_{G:\, F \leq G \leq C} \boldsymbol{\zeta}(F,G) f(G,C) \iff f(F,C) = \sum_{G:\, F \leq G \leq C} \boldsymbol{\mu}(F,G) g(G,C). \tag{15.39}$$

This is equivalent to $g = \boldsymbol{\zeta} f \iff f = \boldsymbol{\mu} g$. We call this *noncommutative Möbius inversion*. In this situation, we say that g is the *exponential* of f, and f is the *logarithm* of g.

Given lunes L and N with $\mathrm{L} \leq \mathrm{N}$ and $\mathrm{b(L)} = \mathrm{b(N)}$, let $\mathrm{L} \backslash \mathrm{N}$ be the unique lune M such that $\mathrm{L} \circ \mathrm{M} = \mathrm{N}$, as in Lemma 4.32. If N is a top-lune, then so is $\mathrm{L} \backslash \mathrm{N}$. Noncommutative Möbius inversion can be reformulated as follows:

$$g(\mathrm{N}) = \sum_{\substack{\mathrm{L}:\, \mathrm{L} \leq \mathrm{N}, \\ \mathrm{b(L)} = \mathrm{b(N)}}} \boldsymbol{\zeta}(\mathrm{L}) f(\mathrm{L} \backslash \mathrm{N}) \iff f(\mathrm{N}) = \sum_{\substack{\mathrm{L}:\, \mathrm{L} \leq \mathrm{N}, \\ \mathrm{b(L)} = \mathrm{b(N)}}} \boldsymbol{\mu}(\mathrm{L}) f(\mathrm{L} \backslash \mathrm{N}). \tag{15.40}$$

Exercise 15.39. Consider the coordinate arrangement and the functions given in (15.33).

(i) Show that if $p \neq 0, 1$, then (15.40) is equivalent to

$$g(\mathrm{N}) = \sum_{\mathrm{M}:\, \mathrm{N} \leq \mathrm{M}} f(\mathrm{M}) \iff f(\mathrm{N}) = \sum_{\mathrm{M}:\, \mathrm{N} \leq \mathrm{M}} (-1)^{\mathrm{sk(N)} - \mathrm{sk(M)}} g(\mathrm{M}).$$

This is Möbius inversion (C.11) in the poset of top-lunes. (Use Exercise 6.3, part (ii), and rescale the functions.)

(ii) Show that if $p = 0, 1$, then (15.40) is equivalent to Möbius inversion in a Boolean poset.

15.4.2. Group-likes and primitives. We say $g \in \mathrm{M}_{\mathrm{lune}}[\mathcal{A}]$ is *group-like* if it satisfies

$$g(H,D) = \sum_{C:\, C \geq A,\, HC = D} g(A,C) \tag{15.41}$$

for all $A \leq H \leq D$.

Lemma 15.40. *A group-like g is equivalent to a choice of scalars $g(O,C)$, one for each chamber C.*

PROOF. Similar to the proof of Lemma 15.18. □

Similarly, we say $f \in \mathrm{M}_{\mathrm{lune}}[\mathcal{A}]$ is *primitive* if it satisfies

$$\sum_{C\colon C \geq A, HC = D} f(A, C) = 0 \tag{15.42}$$

for all $A < H \leq D$.

Lemma 15.41. *A primitive f is equivalent to a choice of a Lie element of $\mathcal{A}_{\mathrm{X}}$, one for each flat* X.

PROOF. Similar to the proof of Lemma 15.24. For each face A, the Lie element in $\mathcal{A}_A$ is given by $\sum_{C\colon C \geq A} f(A, C)\mathtt{H}_{C/A}$. □

Theorem 15.42. *Let ζ be a noncommutative zeta function, and $\boldsymbol{\mu}$ be its inverse. Then under the correspondence* (15.39), *f is primitive iff g is group-like.*

PROOF. Suppose f is primitive. We check that g is group-like. For $A \leq H \leq D$,

$$\begin{aligned}
\sum_{C\colon C \geq A, HC = D} g(A, C) &= \sum_{C\colon C \geq A, HC = D} \ \sum_{K\colon A \leq K \leq C} \zeta(A, K) f(K, C) \\
&= \sum_{K\colon A \leq K, HK \leq D} \zeta(A, K) \sum_{C\colon C \geq K, HC = D} f(K, C) \\
&= \sum_{\substack{K\colon A \leq K, HK \leq D, \\ KH = K}} \zeta(A, K) f(K, KD) \\
&= \sum_{\substack{K\colon A \leq K, HK \leq D, \\ KH = K}} \zeta(A, K) f(HK, D) \\
&= \sum_{G\colon H \leq G \leq D} \Big(\sum_{\substack{K\colon A \leq K, HK = G, \\ \mathrm{s}(K) = \mathrm{s}(G)}} \zeta(A, K) \Big) f(G, D) \\
&= \sum_{G\colon H \leq G \leq D} \zeta(H, G) f(G, D) \\
&= g(H, D).
\end{aligned}$$

The first and last steps used the definition (15.39). The second step interchanged the order of the sums. Since f is primitive, by (15.42) the inner sum is zero unless $KH = K$. (Otherwise $KH > K$ and $HC = D$ is the same as $KHC = KD$.) This was used in the third step. The fourth step used that $(K, KD) \sim (HK, D)$. In the fifth step, we introduced a new variable G for HK. The sixth step used lune-additivity (15.23).

Conversely, suppose g is group-like. We check below that f is primitive. For $A < H \leq D$,

$$\begin{aligned}
\sum_{C:\, C \geq A, HC=D} f(A,C) &= \sum_{C:\, C\geq A, HC=D} \ \sum_{K:\, A \leq K \leq C} \boldsymbol{\mu}(A,K) g(K,C) \\
&= \sum_{K:\, A\leq K, HK \leq D} \boldsymbol{\mu}(A,K) \sum_{C:\, C\geq K, HC=D} g(K,C) \\
&= \sum_{K:\, A\leq K, HK\leq D} \boldsymbol{\mu}(A,K) g(KH,KD) \\
&= \sum_{K:\, A\leq K, HK\leq D} \boldsymbol{\mu}(A,K) g(HK,D) \\
&= \sum_{G:\, H\leq G\leq D} \Big(\sum_{K:\, A\leq K, HK=G} \boldsymbol{\mu}(A,K) \Big) f(G,D) \\
&= 0.
\end{aligned}$$

The first step used the definition (15.39). The second step interchanged the order of the sums. The third step used that g is group-like (15.41) and the fact that $HC = D$ is the same as $KHC = KD$. The fourth step used that $(KH, KD) \sim (HK, D)$. In the fifth step, we introduced a new variable G for HK. The last step used the noncommutative Weisner formula (15.27). □

An application is given below.

SECOND PROOF OF THEOREM 11.57. Fix a noncommutative zeta function $\boldsymbol{\zeta}$, and let $\boldsymbol{\mu}$ be its inverse. Put $g(H,D) := v_H^D$ for all $H > O$. We want to extend g to a function on all top-nested faces such that g is group-like in the sense of (15.41). The conditions (11.57) and (11.58) say that the given data is consistent with g being group-like. Define $f(H,D)$ for all $H > O$ using (15.39). We extend f to a function on all top-nested faces by picking a Lie element $\sum_C x^C \mathtt{H}_C$ and setting $f(O,C) := x^C$. By Lemma 15.41, f is primitive in the sense of (15.42). Now define $g(O,D)$ using (15.39). By Theorem 15.42, g is group-like. □

Exercise 15.43. Show that:

- g is group-like iff for any top-lune M and flat X with $\mathrm{X} \leq \mathrm{b(M)}$,

$$g(\mathrm{M}) = \sum_{\mathrm{L}:\, \mathrm{b(L)=X},\, \mathrm{L} \leq \mathrm{M}} g(\mathrm{L}). \tag{15.43}$$

- f is primitive iff for any top-lune M and flat X with $\mathrm{X} < \mathrm{b(M)}$,

$$\sum_{\mathrm{L}:\, \mathrm{b(L)=X},\, \mathrm{L}\leq\mathrm{M}} f(\mathrm{L}) = 0. \tag{15.44}$$

(Both sums are over top-lunes L.) Then reprove Theorem 15.42 in this language.

15.5. Characterizations of Eulerian families

Recall that in Chapter 11, we characterized Eulerian families of the Tits algebra in terms of homogeneous sections and special Zie families. Noncommutative zeta and Möbius functions can now be added to the story. Similar considerations apply to the projective Tits algebra.

15.5.1. Tits algebra.

Theorem 15.44. *The following pieces of data are equivalent.*

(1) *A noncommutative zeta function* $\boldsymbol{\zeta}$ *of* $\mathcal{A}$.
(2) *A homogeneous section* $\mathtt{u}$ *of* $\mathcal{A}$.
(3) *An Eulerian family* $\mathtt{E}$ *of* $\mathcal{A}$.
(4) *A complete system of primitive orthogonal idempotents of* $\Sigma[\mathcal{A}]$.
(5) *An algebra section of the support map* $\mathrm{s} : \Sigma[\mathcal{A}] \to \Pi[\mathcal{A}]$.
(6) *A* $\mathtt{Q}$*-basis of the Tits algebra of* $\mathcal{A}$.
(7) *A special Zie family* P *of* $\mathcal{A}$.
(8) *A noncommutative Möbius function* $\boldsymbol{\mu}$ *of* $\mathcal{A}$.

Proof. Combine Lemmas 15.18 and 15.24 with Theorems 11.20 and 11.40 and Proposition 11.43. □

For convenience, the interactions between the different pieces of data are summarized in Table 15.1.

Table 15.1. Interactions in a nutshell.

(1) → (2)	$\mathtt{u}^F := \boldsymbol{\zeta}(O, F)$
(2) → (1)	$\boldsymbol{\zeta}(H, G) := \mathtt{u}_H^G$ with $\mathtt{u}_H := \Delta_H(\mathtt{u})$
(2) → (3)	$\mathtt{E}_{\mathrm{X}} := \mathtt{u}_{\mathrm{X}} - \sum_{\mathrm{Y}:\mathrm{Y}>\mathrm{X}} \mathtt{u}_{\mathrm{X}} \cdot \mathtt{E}_{\mathrm{Y}}$ with $\mathtt{u}_{\mathrm{X}} := \sum_{F:\, \mathrm{s}(F)=\mathrm{X}} \mathtt{u}^F \mathtt{H}_F$
(3) → (2)	$\mathtt{u}_{\mathrm{X}}$ is the base term of $\mathtt{E}_{\mathrm{X}}$
(3) → (4)	An Eulerian family $\{\mathtt{E}_{\mathrm{X}}\}$ is a complete system of $\Sigma[\mathcal{A}]$
(3) → (5)	$\mathrm{s}(\mathtt{E}_{\mathrm{X}}) = \mathtt{Q}_{\mathrm{X}}$ and $\mathtt{Q}_{\mathrm{X}} \mapsto \mathtt{E}_{\mathrm{X}}$ is an algebra section of s
(4) ↔ (5)	General fact about elementary algebras
(3) → (6)	$\mathtt{Q}_F := \mathtt{H}_F \cdot \mathtt{E}_{\mathrm{s}(F)}$
(6) → (2), (3)	$\mathtt{H}_O = \sum_F \mathtt{u}^F \mathtt{Q}_F$ and $\mathtt{E}_{\mathrm{X}} := \sum_{F:\, \mathrm{s}(F)=\mathrm{X}} \mathtt{u}^F \mathtt{Q}_F$
(6) → (7)	$\mathsf{P}_{\mathrm{X}} := \beta_{\mathrm{X},F}(\Delta_F(\mathtt{Q}_F))$
(7) → (6)	$\mathtt{Q}_F := \mu_F(\beta_{F,\mathrm{X}}(\mathsf{P}_{\mathrm{X}}))$
(7) → (8)	$\sum_{G:\, F \le G} \boldsymbol{\mu}(F, G)\, \mathtt{H}_{G/F} = \beta_{F,\mathrm{X}}(\mathsf{P}_{\mathrm{X}})$
(8) → (7)	$\mathsf{P}_{\mathrm{X}} := \beta_{\mathrm{X},F}(\sum_{G:\, F \le G} \boldsymbol{\mu}(F, G)\, \mathtt{H}_{G/F})$

Formulas (11.33) relating the $\mathtt{H}$- and $\mathtt{Q}$-bases can be rewritten as

$$\mathtt{H}_F = \sum_{K:\, F \le K} \boldsymbol{\zeta}(F, K)\, \mathtt{Q}_K \quad \text{and} \quad \mathtt{Q}_F = \sum_{G:\, F \le G} \boldsymbol{\mu}(F, G)\, \mathtt{H}_G. \tag{15.45}$$

In particular, $\boldsymbol{\zeta}$ and $\boldsymbol{\mu}$ are inverses of each other in the lune-incidence algebra. This gives a proof of Theorem 15.28.

The $\mathtt{Q}$-basis expansion (11.28) for the Eulerian idempotents can be written as

$$\mathtt{E}_{\mathrm{X}} = \sum_{F:\, \mathrm{s}(F)=\mathrm{X}} \boldsymbol{\zeta}(O, F)\, \mathtt{Q}_F. \tag{15.46}$$

Exercise 15.45. Show that: In rank at least one,

$$\mathtt{Q}_O \cdot \Big(\sum_C \zeta(O,C)\,\mathtt{H}_C\Big) = 0. \tag{15.47}$$

This is equivalent to the result of Exercise 11.14.

Example 15.46. Consider the coordinate arrangement with ζ and μ as in (15.33). The corresponding homogeneous section $\mathtt{u}$ is given by

$$\mathtt{u}^F = p^{m_+(F)}(1-p)^{m_-(F)},$$

where $m_+(F)$ is the number of $+$ in the tuple of F, while $m_-(F)$ is the number of $-$ in the tuple of F.

The corresponding $\mathtt{Q}$-basis is given by

$$\mathtt{Q}_F = \sum_{G\colon F\leq G} (-p)^{m_+(G/F)}(p-1)^{m_-(G/F)}\,\mathtt{H}_G.$$

Here $m_+(G/F)$ is the number of $+$ in the tuple of G restricted to coordinates where F has 0 entries. The definition of $m_-(G/F)$ is similar with $-$ instead of $+$. The above formula can be deduced from (15.45) and Exercise 6.1.

The corresponding Eulerian idempotents are given by

$$\mathtt{E}_{\mathrm{X}} = (-1)^{\mathrm{rk(X)}} \sum (-1)^{\mathrm{rk}(G)} p^{m_+(G)}(1-p)^{m_-(G)}\,\mathtt{H}_G.$$

The sum is over those faces G which have either $+$ or $-$ in all entries where the flat X has $\pm$. For instance, when $\mathrm{X} = \top$, the sum is over all chambers, and when $\mathrm{X} = \bot$, the sum is over all faces. The above formula can be deduced using (15.46).

For the rank-one arrangement, the above formulas for the $\mathtt{Q}$-basis and Eulerian idempotents specialize to those in Section 11.4.3. For the uniform case $p = 1/2$, they are special cases of formulas (11.50) and (11.51): put $c_F^G = 2^{\mathrm{rk}(G/F)}$ and $\mu(\mathcal{A}_F^G) = (-1)^{\mathrm{rk}(G/F)}$.

15.5.2. Projective Tits algebra. The analogue of Theorem 15.44 for the projective Tits algebra is given below.

Theorem 15.47. *Assume that the field characterstic is not* 2. *The following pieces of data are equivalent.*

(1) *A projective noncommutative zeta function of* $\mathcal{A}$.
(2) *A projective section of* $\mathcal{A}$.
(3) *A projective Eulerian family of* $\mathcal{A}$.
(4) *A complete system of the projective Tits algebra.*
(5) *An algebra section from the Birkhoff algebra to the projective Tits algebra.*
(6) *A* $\mathtt{Q}$*-basis of the projective Tits algebra of* $\mathcal{A}$.
(7) *A projective special Zie family of* $\mathcal{A}$.
(8) *A projective noncommutative Möbius function of* $\mathcal{A}$.

PROOF. Combine Exercises 15.19 and 15.25 with Propositions 11.16, 11.45 and 11.46 and Exercise 11.24. □

15.5.3. Left regular bands. Theorem 15.44 generalizes to any left regular band (extending the considerations in Section 15.3.6). In particular, by Example E.2, it applies to any lattice (including the Birkhoff monoid). In this case, each piece of data in Theorem 15.44 is unique: there is a unique zeta function, a unique complete system of primitive orthogonal idempotents, a unique Möbius function and so on. Also note that $\mathtt{Q}_\perp$ is the unique special Zie element. Compare (15.45) with (D.22).

15.6. Lie-incidence algebra

We introduce the Lie-incidence algebra. It is constructed as a subalgebra of the lune-incidence algebra. The conditions defining this subalgebra are similar to the conditions defining Lie elements as a subspace of chambers. This connection is made precise. The dimension of the Lie-incidence algebra equals the number of faces. In fact, the Lie-incidence algebra is isomorphic to the Tits algebra; the isomorphism depends on the choice of an Eulerian family. We also introduce the Lie-incidence module. It is isomorphic to the left module of chambers.

15.6.1. Subspace of the lune-incidence algebra. Recall the lune-incidence algebra $\mathrm{I}_{\mathrm{lune}}[\mathcal{A}]$. Let $\mathrm{I}_{\mathrm{Lie}}[\mathcal{A}]$ denote the subspace of $\mathrm{I}_{\mathrm{lune}}[\mathcal{A}]$ consisting of functions f on nested faces which satisfy

$$\sum_{\substack{F:\, F\geq A, HF=G,\\ \mathrm{s}(F)=\mathrm{s}(G)}} f(A,F) = 0 \tag{15.48}$$

for all $A < H \leq G$. This is a linear system of equations.

In the language of Proposition 15.7:

Proposition 15.48. *The subspace* $\mathrm{I}_{\mathrm{Lie}}[\mathcal{A}]$ *consists of functions* f *on lunes such that for any lune* M *and flat* X *with* $\mathrm{X} < \mathrm{b}(\mathrm{M})$,

$$\sum_{\substack{\mathrm{L}:\, \mathrm{b}(\mathrm{L})=\mathrm{X},\, \mathrm{L}\leq\mathrm{M}\\ \mathrm{c}(\mathrm{L})=\mathrm{c}(\mathrm{M})}} f(\mathrm{L}) = 0. \tag{15.49}$$

In view of (3.24), the sum in (15.49) is over all lunes L which appear in the lune decomposition of M over the flat X. Since X is strictly smaller than b(M), this decomposition is nontrivial. The simplest case is when $\mathrm{X} \lessdot \mathrm{b}(\mathrm{M})$ and M is a flat. For instance, X could be a hyperplane and M the maximum flat. This yields the following.

For $f \in \mathrm{I}_{\mathrm{Lie}}[\mathcal{A}]$,

$$f(\mathrm{L}) + f(\overline{\mathrm{L}}) = 0 \tag{15.50}$$

for any half-flat L. (Recall that a half-flat is a lune L for which $\mathrm{b}(\mathrm{L}) \lessdot \mathrm{c}(\mathrm{L})$.) In particular, for $f \in \mathrm{I}_{\mathrm{Lie}}[\mathcal{A}]$,

$$f(\mathrm{h}) + f(\overline{\mathrm{h}}) = 0$$

for any half-space h.

Also note that the equations (15.49) do not impose any contraints on the value of f on flats.

15.6.2. Lie-incidence algebra. We now proceed to show that $\mathrm{I}_{\mathrm{Lie}}[\mathcal{A}]$ is a subalgebra of the lune-incidence algebra $\mathrm{I}_{\mathrm{lune}}[\mathcal{A}]$. We begin with some preliminary lemmas.

Lemma 15.49. *Let* f *and* g *be any functions on lunes. Then, for any lune* N',

$$\sum_{\substack{\mathrm{L},\mathrm{M}:\\ \mathrm{L}\circ\mathrm{M}\preceq\mathrm{N}'}} f(\mathrm{L})g(\mathrm{M}) = \sum_{\substack{\mathrm{L}',\mathrm{M}':\\ \mathrm{L}'\circ\mathrm{M}'=\mathrm{N}'}} \sum_{\substack{\mathrm{L},\mathrm{M}:\\ \mathrm{L}\preceq\mathrm{L}',\, \mathrm{M}\preceq\mathrm{M}'\\ \mathrm{c}(\mathrm{L})=\mathrm{b}(\mathrm{M})}} f(\mathrm{L})g(\mathrm{M}). \tag{15.51}$$

In addition, for any flats $X \le Z$,

$$\sum_{\substack{L,M:\\ L\circ M\preceq N',\\ b(L)=X,\, c(M)=Z}} f(L)g(M) = \sum_{\substack{L',M':\\ L'\circ M'=N'}} \sum_{\substack{L,M:\\ L\preceq L',\, M\preceq M'\\ c(L)=b(M)\\ b(L)=X,\, c(M)=Z}} f(L)g(M). \tag{15.52}$$

PROOF. This follows from Corollary 4.34. $\square$

Lemma 15.50. *Let* f *and* g *be functions on lunes such that for any* $L' \circ M' = N'$ *with* $X < b(N')$ *and* $Z = c(N')$,

$$\sum_{\substack{L,M:\\ L\preceq L',\, M\preceq M'\\ c(L)=b(M)\\ b(L)=X,\, c(M)=Z}} f(L)g(M) = 0. \tag{15.53}$$

Then $fg \in I_{\mathrm{Lie}}[\mathcal{A}]$.

PROOF. Let us analyze the condition $fg \in I_{\mathrm{Lie}}[\mathcal{A}]$ using (15.49). It says that for any lune N' and flat X with $X < b(N')$,

$$\sum_{N:\, N\le N',\, b(N)=X,\, c(N)=Z} (fg)(N) = 0,$$

where $Z := c(N')$. Equivalently,

$$\sum_{\substack{L,M:\\ L\circ M\le N',\, b(L)=X,\, c(M)=Z}} f(L)g(M) = 0.$$

Since M and N′ have the same case, $L \circ M \le N'$ is the same as $L \circ M \preceq N'$. Thus, $fg \in I_{\mathrm{Lie}}[\mathcal{A}]$ is equivalent to the lhs of (15.52) being zero. Condition (15.53) says that the inner sum in the rhs of (15.52) is zero. The result follows. $\square$

Proposition 15.51. *The subspace* $I_{\mathrm{Lie}}[\mathcal{A}]$ *is a subalgebra of the lune-incidence algebra* $I_{\mathrm{lune}}[\mathcal{A}]$. *That is, functions on lunes satisfying* (15.49) *are closed under the product* (15.14).

PROOF. Let $f, g \in I_{\mathrm{Lie}}[\mathcal{A}]$. We want to show that $fg \in I_{\mathrm{Lie}}[\mathcal{A}]$. For this, we employ Lemma 15.50 and show that (15.53) holds. Accordingly, let $L' \circ M' = N'$ with $X < b(N')$ and $Z = c(N')$. Depending on whether $c(L) = b(M)$ is greater than $b(N')$ or not, we split the sum in the lhs of (15.53) into two smaller sums.

<u>$c(L) = b(M) \ge b(N')$</u>. The first smaller sum can be manipulated as follows.

$$\sum_{\substack{M:\\ M\preceq M'\\ c(M)=Z}} g(M)\Big(\sum_{\substack{L:\\ L\preceq L'\\ c(L)=b(M)\ge b(L')\\ b(L)=X}} f(L)\Big) = \sum_{\substack{M:\\ M\preceq M'\\ b(M)=c(L')\\ c(M)=Z}} g(M)\Big(\sum_{\substack{L:\\ L\preceq L'\\ c(L)=c(L')\\ b(L)=X}} f(L)\Big) = 0$$

since f satisfies (15.49). (The conditions $L \preceq L'$ and $c(L) \ge b(L')$ imply that $c(L) = c(L')$. Also, since L and L′ have the same case, $L \preceq L'$ is the same as $L \le L'$.) In this case, one can also deduce from Corollary 3.17 that $M = M'$ (since they have the same base and same case).

This case is illustrated below. The lune N′ is the fully visible vertex-based top-lune shown in dark shade, and $X = \bot$. The lune L′ is the semicircle shown as a thick line, while $\mathrm{M}' = \mathrm{M}$ is the hemisphere containing N' whose bounding hyperplane includes L'.

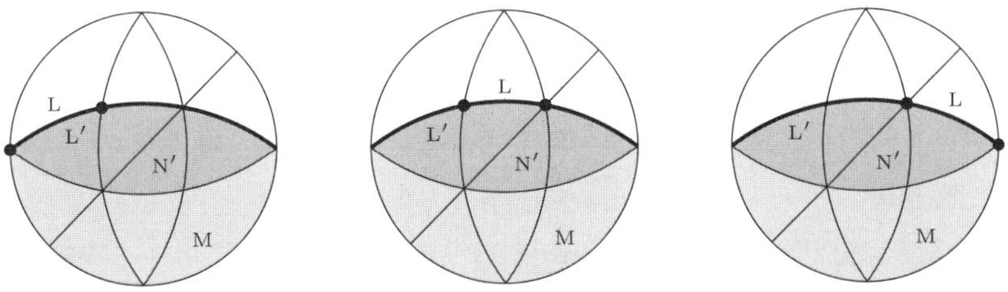

The lune L can be either of the three edges of L' (demarcated by the two black vertices) as indicated in the three pictures. The sum of $f(\mathrm{L})$, as L varies over the three edges, is zero.

$\underline{\mathrm{c(L)} = \mathrm{b(M)} \not\geq \mathrm{b(N')}}$. The second smaller sum can be manipulated as follows.

$$\sum_{\substack{\mathrm{L}:\\ \mathrm{L}\preceq\mathrm{L}'\\ \mathrm{b(L)}=\mathrm{X}}} f(\mathrm{L})\Big(\sum_{\substack{\mathrm{M}:\\ \mathrm{M}\preceq\mathrm{M}'\\ \mathrm{c(L)}=\mathrm{b(M)}\not\geq\mathrm{b(L')}\\ \mathrm{c(M)}=\mathrm{Z}}} g(\mathrm{M})\Big) = \sum_{\substack{\mathrm{L}:\\ \mathrm{L}\preceq\mathrm{L}'\\ \mathrm{c(L)}\not\geq\mathrm{b(L')}\\ \mathrm{b(L)}=\mathrm{X}}} f(\mathrm{L})\Big(\sum_{\substack{\mathrm{M}:\\ \mathrm{M}\preceq\mathrm{M}'\\ \mathrm{b(M)}=\mathrm{c(L)}\\ \mathrm{c(M)}=\mathrm{Z}}} g(\mathrm{M})\Big) = 0$$

since g satisfies (15.49). (The conditions $\mathrm{L} \preceq \mathrm{L}'$ and $\mathrm{c(L)} \not\geq \mathrm{b(L')}$ imply that $\mathrm{c(L)} < \mathrm{c(L')}$. Thus, $\mathrm{b(M)} < \mathrm{b(M')}$. Also, since M and M' have the same case, $\mathrm{M} \preceq \mathrm{M}'$ is the same as $\mathrm{M} \leq \mathrm{M}'$.)

This case is illustrated below. The lune N' is the fully visible vertex-based top-lune shown shaded, and $\mathrm{X} = \bot$. The lune L is the black vertex, L' is the semicircle shown as a thick line, while M' is the hemisphere containing N' whose bounding hyperplane includes L'.

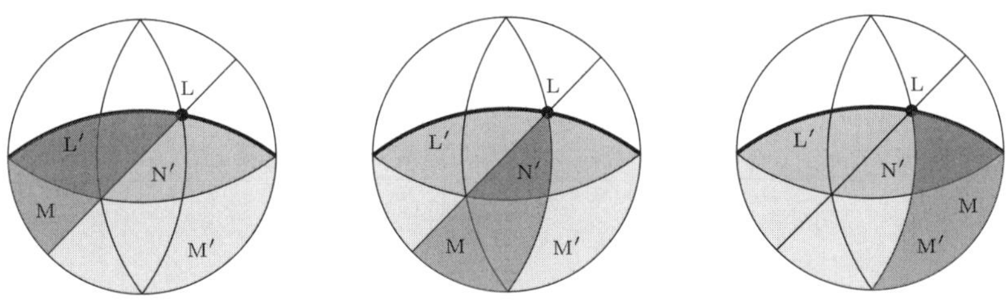

The lune M can be either of the three vertex-based top-lunes as indicated in the three pictures. The sum of $g(\mathrm{M})$, as M varies over the three lunes, is zero. □

We refer to $\mathrm{I}_{\mathrm{Lie}}[\mathcal{A}]$ as the *Lie-incidence algebra*. The motivation for this terminology is given below.

15.6.3. Substitution product of Lie. Recall the algebra (13.27) obtained by summing Lie elements in arrangements over and under flats with product induced from the substitution product of Lie.

Lemma 15.52. *There is an isomorphism of algebras*

$$\mathrm{I}_{\mathrm{Lie}}[\mathcal{A}] \xrightarrow{\cong} \bigoplus_{\mathrm{X}\leq\mathrm{Y}} \mathsf{Lie}[\mathcal{A}^{\mathrm{Y}}_{\mathrm{X}}] \tag{15.54}$$

such that the diagram of algebras

$$
\begin{array}{ccc}
\mathrm{I}_{\mathrm{lune}}[\mathcal{A}] & \xrightarrow{\cong} & \bigoplus_{\mathrm{X} \leq \mathrm{Y}} \Gamma[\mathcal{A}_{\mathrm{X}}^{\mathrm{Y}}] \\
\uparrow & & \uparrow \\
\mathrm{I}_{\mathrm{Lie}}[\mathcal{A}] & \xrightarrow[\cong]{} & \bigoplus_{\mathrm{X} \leq \mathrm{Y}} \mathsf{Lie}[\mathcal{A}_{\mathrm{X}}^{\mathrm{Y}}].
\end{array} \tag{15.55}
$$

commutes.

The top horizontal map is the algebra isomorphism (15.10). The vertical maps are the canonical inclusions. The inclusion on the right is an algebra homomorphism by (10.29).

PROOF. We only need to check the existence of the bottom horizontal map. By (10.1), we see that condition (15.48) translates to the fact that (15.11) is a Lie element in $\mathcal{A}_{\mathrm{X}}^{\mathrm{Y}}$. Alternatively, we see that condition (15.49) relates to Lie using the Ree criterion (Lemma 10.7) applied to each $\mathcal{A}_{\mathrm{X}}^{\mathrm{Y}}$. □

An upshot of this discussion is that Propositions 10.42 and 15.51 imply each other. The two proofs are essentially the same; the two cases considered in the proof of Proposition 10.42 correspond to the two cases in the proof of Proposition 15.51.

Proposition 15.53. *The dimension of the Lie-incidence algebra* $\mathrm{I}_{\mathrm{Lie}}[\mathcal{A}]$ *equals the number of faces of* $\mathcal{A}$.

PROOF. This follows from (15.54), (10.24) and (1.49). □

For arrangements $\mathcal{A}$ and $\mathcal{A}'$,

$$\mathrm{I}_{\mathrm{Lie}}[\mathcal{A} \times \mathcal{A}'] = \mathrm{I}_{\mathrm{Lie}}[\mathcal{A}] \otimes \mathrm{I}_{\mathrm{Lie}}[\mathcal{A}']. \tag{15.56}$$

This is a restriction of the identification (15.16). One way to deduce this is to use (10.7) and (15.54).

15.6.4. Lie-incidence module. Recall the lune-incidence module $\mathrm{M}_{\mathrm{lune}}[\mathcal{A}]$. Let $\mathrm{M}_{\mathrm{Lie}}[\mathcal{A}]$ denote the subspace of $\mathrm{M}_{\mathrm{lune}}[\mathcal{A}]$ consisting of functions g on top-nested faces which satisfy

$$\sum_{C:\, C \geq A, HC = D} g(A, C) = 0 \tag{15.57}$$

for all $A < H \leq D$.

In the language of Proposition 15.9:

Proposition 15.54. *The subspace* $\mathrm{M}_{\mathrm{Lie}}[\mathcal{A}]$ *consists of functions* g *on top-lunes such that for any top-lune* M *and flat* X *with* $\mathrm{X} < \mathrm{b}(\mathrm{M})$,

$$\sum_{\mathrm{L}:\, \mathrm{b}(\mathrm{L}) = \mathrm{X},\, \mathrm{L} \leq \mathrm{M}} g(\mathrm{L}) = 0. \tag{15.58}$$

The subspace $\mathrm{M}_{\mathrm{Lie}}[\mathcal{A}]$ is a left module over the Lie-incidence algebra with the action given by restricting (15.17). We call it the *Lie-incidence module*. It can also be viewed as a left ideal. The isomorphism of algebras (15.54) induces an isomorphism of left ideals

$$\mathrm{M}_{\mathrm{Lie}}[\mathcal{A}] \xrightarrow{\cong} \bigoplus_{\mathrm{X}} \mathsf{Lie}[\mathcal{A}_{\mathrm{X}}]. \tag{15.59}$$

This map is also a restriction of (15.19).

Proposition 15.55. *The dimension of the Lie-incidence module* $\mathrm{M}_{\mathrm{Lie}}[\mathcal{A}]$ *equals the number of chambers of* $\mathcal{A}$.

PROOF. This follows from (15.59), (10.24) and (1.48). □

15.6.5. Rank one. Let $\mathcal{A}$ be the arrangement of rank one. It has four lunes, namely, the flats $\bot$ and $\top$ and chambers C and $\overline{C}$. The Lie-incidence algebra of $\mathcal{A}$ has dimension three: Elements are functions f on lunes, with $f(\bot)$ and $f(\top)$ arbitrary, and $f(C)+f(\overline{C})=0$. The latter condition is an instance of (15.50). The product of f and g is given by (15.22). (This is the product in the lune-incidence algebra of $\mathcal{A}$.)

The Lie-incidence module of $\mathcal{A}$ has dimension two: It is the left ideal of the Lie-incidence algebra consisting of functions f which satisfy in addition $f(\bot)=0$.

15.6.6. Connection with Tits algebra. Recall from Theorem 13.54 that there is an intimate connection between the Tits algebra and Lie elements in arrangements under and over flats. Combining this result with Lemma 15.52, we obtain:

Theorem 15.56. *There is an algebra isomorphism*

$$\mathrm{I}_{\mathrm{Lie}}[\mathcal{A}] \xrightarrow{\cong} \Sigma[\mathcal{A}]$$

from the Lie-incidence algebra to the Tits algebra. It induces an isomorphism

$$\mathrm{M}_{\mathrm{Lie}}[\mathcal{A}] \xrightarrow{\cong} \Gamma[\mathcal{A}]$$

from the Lie-incidence module to the left module of chambers.

This is consistent with the results obtained in Propositions 15.53 and 15.55. As a consequence, results about powers of the radical and quiver of the Tits algebra obtained in Section 13.10 apply to the Lie-incidence algebra.

15.6.7. Projective Lie-incidence algebra. The *projective Lie-incidence algebra* is the subalgebra of the Lie-incidence algebra consisting of projective functions. It is isomorphic to

$$\bigoplus_{\mathrm{rk}(\mathrm{Y}/\mathrm{X}) \text{ is even}} \mathsf{Lie}[\mathcal{A}_{\mathrm{X}}^{\mathrm{Y}}]$$

(assuming that the field characteristic is not 2). The sum is over all $\mathrm{X} \leq \mathrm{Y}$ such that the difference in their ranks is even. This follows from (15.54) and Lemma 10.12. In conjunction with Theorem 13.59, we obtain:

Theorem 15.57. *The projective Lie-incidence algebra is isomorphic to the projective Tits algebra.*

An explicit isomorphism is obtained by fixing a projective Eulerian family and restricting the resulting isomorphism in Theorem 15.56.

15.7. Additive and Weisner functions on lunes

We consider additive and Weisner functions on lunes. These define linear subspaces inside the lune-incidence algebra which respectively contain the affine subspaces of noncommutative zeta and Möbius functions. Further, they are respectively right and left modules over the Lie-incidence algebra, with actions induced from the product of the lune-incidence algebra. In fact, these modules are isomorphic to the right and left regular representations of the Lie-incidence algebra, and in turn to the right and left regular representations of the Tits algebra.

15.7.1. Additive functions on lunes. A function on lunes is *additive* if it satisfies condition (15.24). Let $\mathrm{Zet}[\mathcal{A}]$ denote the set of all additive functions on lunes. It is a linear subspace of the lune-incidence algebra $\mathrm{I}_{\mathrm{lune}}[\mathcal{A}]$.

Note that a noncommutative zeta function is the same as an additive function on lunes which is 1 on all flats. Let $\mathrm{Zet}_1[\mathcal{A}]$ denote the set of all noncommutative zeta functions. It is an affine subspace of $\mathrm{Zet}[\mathcal{A}]$. Let $\mathrm{Zet}_0[\mathcal{A}]$ denote the linear subspace which is parallel to it. Explicitly, it consists of additive functions which are 0 on all flats.

Lemma 15.58. *There is a linear isomorphism*

$$\mathrm{Zet}[\mathcal{A}] \xrightarrow{\cong} \bigoplus_{\mathrm{X}} \Gamma[\mathcal{A}^{\mathrm{X}}], \qquad f \mapsto \Big(\sum_{F:\, \mathrm{s}(F)=\mathrm{X}} f(F)\, \mathtt{H}_F \Big). \tag{15.60}$$

The direct sum is over all flats. In particular, the dimension of $\mathrm{Zet}[\mathcal{A}]$ *equals the number of faces.*

PROOF. The isomorphism extends the identification in Lemma 15.18 by removing the condition on the value of the function on flats. □

Consider arrangements $\mathcal{A}$ and $\mathcal{A}'$ and the identification (15.16). As in (15.31), we have

$$\mathrm{Zet}[\mathcal{A} \times \mathcal{A}'] = \mathrm{Zet}[\mathcal{A}] \otimes \mathrm{Zet}[\mathcal{A}'] \quad \text{and} \quad \mathrm{Zet}_1[\mathcal{A} \times \mathcal{A}'] \supseteq \mathrm{Zet}_1[\mathcal{A}] \otimes \mathrm{Zet}_1[\mathcal{A}'].$$

The inclusion is strict in general.

15.7.2. Weisner functions on lunes. A function on lunes is *Weisner* if it satisfies condition (15.28). Let $\mathrm{Mob}[\mathcal{A}]$ denote the set of all Weisner functions on lunes. It is a linear subspace of $\mathrm{I}_{\mathrm{lune}}[\mathcal{A}]$.

Note that a noncommutative Möbius function is the same as a Weisner function on lunes which is 1 on all flats. Let $\mathrm{Mob}_1[\mathcal{A}]$ denote the set of all noncommutative Möbius functions. It is an affine subspace of $\mathrm{Mob}[\mathcal{A}]$. Let $\mathrm{Mob}_0[\mathcal{A}]$ denote the linear subspace which is parallel to it. Explicitly, it consists of Weisner functions which are 0 on all flats.

Lemma 15.59. *A Weisner function is equivalent to a Zie family. More precisely: There is a linear isomorphism*

$$\mathrm{Mob}[\mathcal{A}] \xrightarrow{\cong} \bigoplus_{\mathrm{X}} \mathsf{Zie}[\mathcal{A}_{\mathrm{X}}], \qquad f \mapsto \Big(\sum_{\mathrm{L}:\, \mathrm{b}(\mathrm{L})=\mathrm{X}} f(\mathrm{L})\, \mathtt{H}_{\mathrm{L}} \Big). \tag{15.61}$$

The direct sum is over all flats. In particular, the dimension of $\mathrm{Mob}[\mathcal{A}]$ *equals the number of faces.*

PROOF. The isomorphism extends the identification in Lemma 15.24 by removing the condition on the value of the function on flats. As a result, we obtain a Zie family instead of a special Zie family. The dimension claim follows from (1.49) and (10.25). Alternatively, use the isomorphism (13.9). □

Consider arrangements $\mathcal{A}$ and $\mathcal{A}'$ and the identification (15.16). As in (15.32), we have

$$\mathrm{Mob}[\mathcal{A} \times \mathcal{A}'] = \mathrm{Mob}[\mathcal{A}] \otimes \mathrm{Mob}[\mathcal{A}'] \quad \text{and} \quad \mathrm{Mob}_1[\mathcal{A} \times \mathcal{A}'] \supseteq \mathrm{Mob}_1[\mathcal{A}] \otimes \mathrm{Mob}_1[\mathcal{A}'].$$

The inclusion is strict in general.

15.7.3. Product of Weisner and additive functions. Interestingly, the product of a Weisner function and an additive function taken in the lune-incidence algebra lands inside the Lie-incidence algebra.

Proposition 15.60. *Let $f \in \mathrm{Mob}[\mathcal{A}]$ and $g \in \mathrm{Zet}[\mathcal{A}]$. Then $fg \in \mathrm{I}_{\mathrm{Lie}}[\mathcal{A}]$.*

Thus we have the commutative diagram

$$\begin{array}{ccc} \mathrm{Mob}[\mathcal{A}] \otimes \mathrm{Zet}[\mathcal{A}] & \longrightarrow & \mathrm{I}_{\mathrm{Lie}}[\mathcal{A}] \\ \downarrow & & \downarrow \\ \mathrm{I}_{\mathrm{lune}}[\mathcal{A}] \otimes \mathrm{I}_{\mathrm{lune}}[\mathcal{A}] & \longrightarrow & \mathrm{I}_{\mathrm{lune}}[\mathcal{A}] \end{array}$$

where the vertical maps are inclusions.

PROOF. We employ Lemma 15.50 and show that (15.53) holds. Accordingly, let $\mathrm{L}' \circ \mathrm{M}' = \mathrm{N}'$ with $\mathrm{X} < \mathrm{b}(\mathrm{N}')$ and $\mathrm{Z} = \mathrm{c}(\mathrm{N}')$. We calculate:

$$\sum_{\substack{\mathrm{L},\mathrm{M}:\\ \mathrm{L}\preceq\mathrm{L}',\,\mathrm{M}\preceq\mathrm{M}'\\ \mathrm{c}(\mathrm{L})=\mathrm{b}(\mathrm{M})\\ \mathrm{b}(\mathrm{L})=\mathrm{X},\,\mathrm{c}(\mathrm{M})=\mathrm{Z}}} f(\mathrm{L})g(\mathrm{M}) = \sum_{\substack{\mathrm{L}:\\ \mathrm{L}\preceq\mathrm{L}'\\ \mathrm{b}(\mathrm{L})=\mathrm{X}}} f(\mathrm{L})\Big(\sum_{\substack{\mathrm{M}:\\ \mathrm{M}\preceq\mathrm{M}'\\ \mathrm{c}(\mathrm{L})=\mathrm{b}(\mathrm{M})\\ \mathrm{c}(\mathrm{M})=\mathrm{Z}}} g(\mathrm{M})\Big) = \Big(\sum_{\substack{\mathrm{L}:\\ \mathrm{L}\preceq\mathrm{L}'\\ \mathrm{b}(\mathrm{L})=\mathrm{X}}} f(\mathrm{L})\Big) g(\mathrm{M}') = 0.$$

The second equality used $g \in \mathrm{Zet}[\mathcal{A}]$, while the third used $f \in \mathrm{Mob}[\mathcal{A}]$. (Since M and M' have the same case, $\mathrm{M} \preceq \mathrm{M}'$ is the same as $\mathrm{M} \leq \mathrm{M}'$.) □

15.7.4. Additive functions as a right module. We now show that the space of additive functions is a right module over the Lie-incidence algebra. We start with a preliminary lemma. It is a companion of Lemma 15.50 and has a similar proof.

Lemma 15.61. *Let f and g be functions on lunes such that for any $\mathrm{L}' \circ \mathrm{M}' = \mathrm{N}'$ with $\mathrm{X} \leq \mathrm{b}(\mathrm{N}')$ and $\mathrm{Z} = \mathrm{c}(\mathrm{N}')$,*

$$\sum_{\substack{\mathrm{L},\mathrm{M}:\\ \mathrm{L}\preceq\mathrm{L}',\,\mathrm{M}\preceq\mathrm{M}'\\ \mathrm{c}(\mathrm{L})=\mathrm{b}(\mathrm{M})\\ \mathrm{b}(\mathrm{L})=\mathrm{X},\,\mathrm{c}(\mathrm{M})=\mathrm{Z}}} f(\mathrm{L})g(\mathrm{M}) = f(\mathrm{L}')g(\mathrm{M}'). \tag{15.62}$$

Then $fg \in \mathrm{Zet}[\mathcal{A}]$.

PROOF. We analyze the condition $fg \in \mathrm{Zet}[\mathcal{A}]$ using (15.24). It says that for any lune N' and flat X with $\mathrm{X} \leq \mathrm{b}(\mathrm{N}')$,

$$\sum_{\mathrm{N}:\, \mathrm{N}\leq\mathrm{N}',\, \mathrm{b}(\mathrm{N})=\mathrm{X},\, \mathrm{c}(\mathrm{N})=\mathrm{Z}} (fg)(\mathrm{N}) = (fg)(\mathrm{N}'),$$

where $\mathrm{Z} := \mathrm{c}(\mathrm{N}')$. Equivalently,

$$\sum_{\substack{\mathrm{L},\mathrm{M}:\\ \mathrm{L}\circ\mathrm{M}\leq\mathrm{N}',\,\mathrm{b}(\mathrm{L})=\mathrm{X},\,\mathrm{c}(\mathrm{M})=\mathrm{Z}}} f(\mathrm{L})g(\mathrm{M}) = \sum_{\substack{\mathrm{L}',\mathrm{M}':\\ \mathrm{L}'\circ\mathrm{M}'=\mathrm{N}'}} f(\mathrm{L}')g(\mathrm{M}').$$

Since M and N′ have the same case, $\mathrm{L}\circ\mathrm{M}\leq\mathrm{N}'$ is the same as $\mathrm{L}\circ\mathrm{M}\preceq\mathrm{N}'$. Now rewrite the above lhs as a double sum using (15.52). Condition (15.62) arises by equating the summands on both sides for each L′ and M′. □

Proposition 15.62. *Let $f \in \mathrm{Zet}[\mathcal{A}]$ and $g \in \mathrm{I}_{\mathrm{Lie}}[\mathcal{A}]$. Then $fg \in \mathrm{Zet}[\mathcal{A}]$.*

Thus the space of additive functions is a right module over the Lie-incidence algebra. The module structure is such that the diagram

$$\begin{array}{ccc} \mathrm{Zet}[\mathcal{A}] \otimes \mathrm{I}_{\mathrm{Lie}}[\mathcal{A}] & \longrightarrow & \mathrm{Zet}[\mathcal{A}] \\ \downarrow & & \downarrow \\ \mathrm{I}_{\mathrm{lune}}[\mathcal{A}] \otimes \mathrm{I}_{\mathrm{lune}}[\mathcal{A}] & \longrightarrow & \mathrm{I}_{\mathrm{lune}}[\mathcal{A}] \end{array}$$

commutes.

Proof. The analysis is similar to the proof of Proposition 15.51. For this, we employ Lemma 15.61 and show that (15.62) holds. Accordingly, let $\mathrm{L}'\circ\mathrm{M}' = \mathrm{N}'$ with $\mathrm{X} \leq \mathrm{b}(\mathrm{N}')$ and $\mathrm{Z} = \mathrm{c}(\mathrm{N}')$. Depending on whether $\mathrm{c}(\mathrm{L}) = \mathrm{b}(\mathrm{M})$ is greater than $\mathrm{b}(\mathrm{N}')$ or not, we split the sum in the lhs of (15.62) into two smaller sums. As we will see, the first smaller sum equals the rhs, while the second smaller sum is zero.

<u>$\mathrm{c}(\mathrm{L}) = \mathrm{b}(\mathrm{M}) \geq \mathrm{b}(\mathrm{N}')$.</u> The first smaller sum can be manipulated as follows.

$$\sum_{\substack{\mathrm{M}:\\ \mathrm{M}\preceq\mathrm{M}'\\ \mathrm{c}(\mathrm{M})=\mathrm{Z}}} g(\mathrm{M})\Big(\sum_{\substack{\mathrm{L}:\\ \mathrm{L}\preceq\mathrm{L}'\\ \mathrm{c}(\mathrm{L})=\mathrm{b}(\mathrm{M})\geq\mathrm{b}(\mathrm{L}')\\ \mathrm{b}(\mathrm{L})=\mathrm{X}}} f(\mathrm{L})\Big) = \sum_{\substack{\mathrm{M}:\\ \mathrm{M}\preceq\mathrm{M}'\\ \mathrm{b}(\mathrm{M})=\mathrm{c}(\mathrm{L}')\\ \mathrm{c}(\mathrm{M})=\mathrm{Z}}} g(\mathrm{M})\Big(\sum_{\substack{\mathrm{L}:\\ \mathrm{L}\preceq\mathrm{L}'\\ \mathrm{c}(\mathrm{L})=\mathrm{c}(\mathrm{L}')\\ \mathrm{b}(\mathrm{L})=\mathrm{X}}} f(\mathrm{L})\Big)$$

$$= \sum_{\substack{\mathrm{M}:\\ \mathrm{M}\preceq\mathrm{M}'\\ \mathrm{b}(\mathrm{M})=\mathrm{c}(\mathrm{L}')\\ \mathrm{c}(\mathrm{M})=\mathrm{Z}}} g(\mathrm{M})f(\mathrm{L}') = g(\mathrm{M}')f(\mathrm{L}').$$

The conditions $\mathrm{L}\preceq\mathrm{L}'$ and $\mathrm{c}(\mathrm{L}) \geq \mathrm{b}(\mathrm{L}')$ imply that $\mathrm{c}(\mathrm{L}) = \mathrm{c}(\mathrm{L}')$. This was used in the first step. Since L and L′ have the same case, $\mathrm{L}\preceq\mathrm{L}'$ is the same as $\mathrm{L}\leq\mathrm{L}'$. Since f satisfies (15.24), we get the second step. Finally, from Corollary 3.17, we deduce that $\mathrm{M} = \mathrm{M}'$ (since they have the same base and same case). This was used in the last step.

<u>$\mathrm{c}(\mathrm{L}) = \mathrm{b}(\mathrm{M}) \not\geq \mathrm{b}(\mathrm{N}')$.</u> The second smaller sum can be manipulated as follows.

$$\sum_{\substack{\mathrm{L}:\\ \mathrm{L}\preceq\mathrm{L}'\\ \mathrm{b}(\mathrm{L})=\mathrm{X}}} f(\mathrm{L})\Big(\sum_{\substack{\mathrm{M}:\\ \mathrm{M}\preceq\mathrm{M}'\\ \mathrm{c}(\mathrm{L})=\mathrm{b}(\mathrm{M})\not\geq\mathrm{b}(\mathrm{L}')\\ \mathrm{c}(\mathrm{M})=\mathrm{Z}}} g(\mathrm{M})\Big) = \sum_{\substack{\mathrm{L}:\\ \mathrm{L}\preceq\mathrm{L}'\\ \mathrm{c}(\mathrm{L})\not\geq\mathrm{b}(\mathrm{L}')\\ \mathrm{b}(\mathrm{L})=\mathrm{X}}} f(\mathrm{L})\Big(\sum_{\substack{\mathrm{M}:\\ \mathrm{M}\preceq\mathrm{M}'\\ \mathrm{b}(\mathrm{M})=\mathrm{c}(\mathrm{L})\\ \mathrm{c}(\mathrm{M})=\mathrm{Z}}} g(\mathrm{M})\Big) = 0$$

since g satisfies (15.49). (The conditions $\mathrm{L}\preceq\mathrm{L}'$ and $\mathrm{c}(\mathrm{L}) \not\geq \mathrm{b}(\mathrm{L}')$ imply that $\mathrm{c}(\mathrm{L}) < \mathrm{c}(\mathrm{L}')$. Thus, $\mathrm{b}(\mathrm{M}) < \mathrm{b}(\mathrm{M}')$. Also, since M and M' have the same case, $\mathrm{M}\preceq\mathrm{M}'$ is the same as $\mathrm{M}\leq\mathrm{M}'$.) □

Proposition 15.63. *The space of additive functions is isomorphic to the Lie-incidence algebra, viewed as a right module over itself. More precisely, for any noncommutative zeta function* ζ, *the map*

$$\mathrm{I}_{\mathrm{Lie}}[\mathcal{A}] \xrightarrow{\cong} \mathrm{Zet}[\mathcal{A}], \qquad f \mapsto \zeta f$$

is an isomorphism of right $\mathrm{I}_{\mathrm{Lie}}[\mathcal{A}]$*-modules.*

More generally, instead of ζ, one may take any additive function which is invertible in the lune-incidence algebra. These are additive functions which are nonzero on each flat.

PROOF. By Theorem 15.28, we have $\zeta^{-1} \in \mathrm{Mob}[\mathcal{A}]$. So, in view of Proposition 15.60, there is a map

$$\mathrm{Zet}[\mathcal{A}] \to \mathrm{I}_{\mathrm{Lie}}[\mathcal{A}], \qquad g \mapsto \zeta^{-1} g,$$

and it is clearly inverse to the map stated in the proposition. □

Recall from (15.54) that $\mathrm{I}_{\mathrm{Lie}}[\mathcal{A}]$ is isomorphic to the algebra $\bigoplus_{\mathrm{X}\leq\mathrm{Y}} \mathsf{Lie}[\mathcal{A}^{\mathrm{Y}}_{\mathrm{X}}]$. Further, recall the right module (13.28) over the latter algebra obtained by summing chamber elements in arrangements under flats. Observe that:

Lemma 15.64. *The map* (15.60) *is an isomorphism of right modules in the sense that the diagram*

$$\begin{array}{ccc} \mathrm{Zet}[\mathcal{A}] \otimes \mathrm{I}_{\mathrm{Lie}}[\mathcal{A}] & \longrightarrow & \mathrm{Zet}[\mathcal{A}] \\ \cong\downarrow & & \downarrow\cong \\ \big(\bigoplus_{\mathrm{X}} \Gamma[\mathcal{A}^{\mathrm{X}}]\big) \otimes \big(\bigoplus_{\mathrm{X}\leq\mathrm{Y}} \mathsf{Lie}[\mathcal{A}^{\mathrm{Y}}_{\mathrm{X}}]\big) & \longrightarrow & \bigoplus_{\mathrm{Y}} \Gamma[\mathcal{A}^{\mathrm{Y}}] \end{array} \tag{15.63}$$

commutes.

An upshot of this discussion is that Propositions 10.50 and 15.62 imply each other. We further point out that the results obtained in Propositions 13.28 and 15.63 are equivalent.

15.7.5. Weisner functions as a left module.

We now show that the space of Weisner functions is a left module over the Lie-incidence algebra. We again start with the appropriate companion of Lemma 15.50.

Lemma 15.65. *Let* f *and* g *be functions on lunes such that for any* $\mathrm{L}' \circ \mathrm{M}' = \mathrm{N}'$ *with* $\mathrm{X} < \mathrm{b}(\mathrm{N}')$,

$$\sum_{\substack{\mathrm{L},\mathrm{M}:\\ \mathrm{L}\preceq\mathrm{L}',\,\mathrm{M}\preceq\mathrm{M}'\\ \mathrm{b}(\mathrm{L})=\mathrm{X},\,\mathrm{c}(\mathrm{L})=\mathrm{b}(\mathrm{M})}} f(\mathrm{L})g(\mathrm{M}) = 0. \tag{15.64}$$

Then $fg \in \mathrm{Mob}[\mathcal{A}]$.

PROOF. We analyze the condition $fg \in \mathrm{Mob}[\mathcal{A}]$ using (15.28). It says that for any lune N' and flat X with $\mathrm{X} < \mathrm{b}(\mathrm{N}')$,

$$\sum_{\mathrm{N}:\,\mathrm{N}\leq\mathrm{N}',\,\mathrm{b}(\mathrm{N})=\mathrm{X}} (fg)(\mathrm{N}) = 0.$$

Equivalently,

$$\sum_{\substack{\mathrm{L,M}:\\ \mathrm{L}\circ\mathrm{M}\preceq\mathrm{N}',\ \mathrm{b(L)=X}}} f(\mathrm{L})g(\mathrm{M}) = 0.$$

Thus, $fg \in \mathrm{Mob}[\mathcal{A}]$ is equivalent to the lhs of (15.52) (summed over all Z) being zero. Condition (15.64) says that the inner sum in the rhs of (15.52) (summed over all Z) is zero. The result follows. □

Proposition 15.66. *Let* $f \in \mathrm{I}_{\mathrm{Lie}}[\mathcal{A}]$ *and* $g \in \mathrm{Mob}[\mathcal{A}]$. *Then* $fg \in \mathrm{Mob}[\mathcal{A}]$.

Thus the space of Weisner functions is a left module over the Lie-incidence algebra. The module structure is such that the diagram

$$\begin{array}{ccc} \mathrm{I}_{\mathrm{Lie}}[\mathcal{A}] \otimes \mathrm{Mob}[\mathcal{A}] & \longrightarrow & \mathrm{Mob}[\mathcal{A}] \\ \big\downarrow & & \big\downarrow \\ \mathrm{I}_{\mathrm{lune}}[\mathcal{A}] \otimes \mathrm{I}_{\mathrm{lune}}[\mathcal{A}] & \longrightarrow & \mathrm{I}_{\mathrm{lune}}[\mathcal{A}] \end{array}$$

commutes.

PROOF. The analysis is similar to the proof of Proposition 15.51. For this, we employ Lemma 15.65 and show that (15.64) holds. Accordingly, let $\mathrm{L}' \circ \mathrm{M}' = \mathrm{N}'$ with $\mathrm{X} < \mathrm{b(N')}$. Depending on whether $\mathrm{c(L) = b(M)}$ is greater than $\mathrm{b(N')}$ or not, we split the sum in the lhs of (15.64) into two smaller sums.

$\underline{\mathrm{c(L) = b(M) \geq b(N')}}$. The first smaller sum can be manipulated as follows.

$$\sum_{\substack{\mathrm{M}:\\ \mathrm{M}\preceq\mathrm{M}'}} g(\mathrm{M})\Big(\sum_{\substack{\mathrm{L}:\\ \mathrm{L}\preceq\mathrm{L}'\\ \mathrm{c(L)=b(M)\geq b(L')}\\ \mathrm{b(L)=X}}} f(\mathrm{L})\Big) = \sum_{\substack{\mathrm{M}:\\ \mathrm{M}\preceq\mathrm{M}'\\ \mathrm{b(M)=c(L')}}} g(\mathrm{M})\Big(\sum_{\substack{\mathrm{L}:\\ \mathrm{L}\preceq\mathrm{L}'\\ \mathrm{c(L)=c(L')}\\ \mathrm{b(L)=X}}} f(\mathrm{L})\Big) = 0$$

since f satisfies (15.49). (The conditions $\mathrm{L} \preceq \mathrm{L}'$ and $\mathrm{c(L)} \geq \mathrm{b(L')}$ imply that $\mathrm{c(L) = c(L')}$. Also, since L and L′ have the same case, $\mathrm{L} \preceq \mathrm{L}'$ is the same as $\mathrm{L} \leq \mathrm{L}'$.) In this case, one may also deduce that $\mathrm{M} = \mathrm{M}'$.

$\underline{\mathrm{c(L) = b(M) \not\geq b(N')}}$. The second smaller sum can be manipulated as follows.

$$\sum_{\substack{\mathrm{L}:\\ \mathrm{L}\preceq\mathrm{L}'\\ \mathrm{b(L)=X}}} f(\mathrm{L})\Big(\sum_{\substack{\mathrm{M}:\\ \mathrm{M}\preceq\mathrm{M}'\\ \mathrm{c(L)=b(M)\not\geq b(L')}}} g(\mathrm{M})\Big) = \sum_{\substack{\mathrm{L}:\\ \mathrm{L}\preceq\mathrm{L}'\\ \mathrm{c(L)\not\geq b(L')}\\ \mathrm{b(L)=X}}} f(\mathrm{L})\Big(\sum_{\substack{\mathrm{M}:\\ \mathrm{M}\preceq\mathrm{M}'\\ \mathrm{b(M)=c(L)}}} g(\mathrm{M})\Big) = 0$$

since g satisfies (15.28). (The conditions $\mathrm{L} \preceq \mathrm{L}'$ and $\mathrm{c(L)} \not\geq \mathrm{b(L')}$ imply that $\mathrm{c(L) < c(L')}$. Thus, $\mathrm{b(M) < b(M')}$.) □

Proposition 15.67. *The space of Weisner functions is isomorphic to the Lie-incidence algebra, viewed as a left module over itself. More precisely, for any noncommutative Möbius function* $\boldsymbol{\mu}$, *the map*

$$\mathrm{I}_{\mathrm{Lie}}[\mathcal{A}] \xrightarrow{\cong} \mathrm{Mob}[\mathcal{A}], \qquad f \mapsto f\boldsymbol{\mu}$$

is an isomorphism of left $\mathrm{I}_{\mathrm{Lie}}[\mathcal{A}]$*-modules.*

More generally, instead of $\boldsymbol{\mu}$, one may take any Weisner function which is invertible in the lune-incidence algebra. These are Weisner functions which are nonzero on each flat.

PROOF. By Theorem 15.28, we have $\boldsymbol{\mu}^{-1} \in \mathrm{Zet}[\mathcal{A}]$. So, in view of Proposition 15.60, there is a map

$$\mathrm{Mob}[\mathcal{A}] \to \mathrm{I}_{\mathrm{Lie}}[\mathcal{A}], \qquad g \mapsto g\boldsymbol{\mu}^{-1},$$

and it is clearly inverse to the map stated in the proposition. □

Recall from (15.54) that $\mathrm{I}_{\mathrm{Lie}}[\mathcal{A}]$ is isomorphic to the algebra $\bigoplus_{\mathrm{X}\leq\mathrm{Y}} \mathsf{Lie}[\mathcal{A}_{\mathrm{X}}^{\mathrm{Y}}]$. Further, recall the left module (13.30) over the latter algebra obtained by summing Zie elements in arrangements over flats. Observe that:

Lemma 15.68. *The map* (15.61) *is an isomorphism of left modules in the sense that the diagram*

$$\begin{array}{ccc} \mathrm{I}_{\mathrm{Lie}}[\mathcal{A}] \otimes \mathrm{Mob}[\mathcal{A}] & \longrightarrow & \mathrm{Mob}[\mathcal{A}] \\ \cong\downarrow & & \downarrow\cong \\ \big(\bigoplus_{\mathrm{X}\leq\mathrm{Y}} \mathsf{Lie}[\mathcal{A}_{\mathrm{X}}^{\mathrm{Y}}]\big) \otimes \big(\bigoplus_{\mathrm{Y}} \mathsf{Zie}[\mathcal{A}_{\mathrm{Y}}]\big) & \longrightarrow & \bigoplus_{\mathrm{X}} \mathsf{Zie}[\mathcal{A}_{\mathrm{X}}] \end{array} \tag{15.65}$$

commutes.

An upshot of this discussion is that Propositions 10.49 and 15.66 imply each other. We further point out that the results obtained in Propositions 13.37 and 15.67 are equivalent. This can be deduced from the second formula in (15.45) which gives the precise connection between a Q-basis and a noncommutative Möbius function $\boldsymbol{\mu}$.

15.7.6. Connection with the Tits algebra.

Recall from Theorem 15.56 that the Lie-incidence algebra is isomorphic to the Tits algebra. In addition, the right module of additive functions is isomorphic to the Tits algebra as a right module over itself, while the left module of Weisner functions is isomorphic to the Tits algebra as a left module over itself. This leads to the following commutative diagrams.

$$\begin{array}{ccc} \mathrm{Zet}[\mathcal{A}] \otimes \mathrm{I}_{\mathrm{Lie}}[\mathcal{A}] & \longrightarrow & \mathrm{Zet}[\mathcal{A}] \\ \cong\downarrow & & \downarrow\cong \\ \Sigma[\mathcal{A}] \otimes \Sigma[\mathcal{A}] & \longrightarrow & \Sigma[\mathcal{A}] \end{array} \qquad \begin{array}{ccc} \mathrm{I}_{\mathrm{Lie}}[\mathcal{A}] \otimes \mathrm{Mob}[\mathcal{A}] & \longrightarrow & \mathrm{Mob}[\mathcal{A}] \\ \cong\downarrow & & \downarrow\cong \\ \Sigma[\mathcal{A}] \otimes \Sigma[\mathcal{A}] & \longrightarrow & \Sigma[\mathcal{A}] \end{array}$$

Explicitly, the left diagram is obtained by combining (13.29) and (15.63), while the right diagram is obtained by combining (13.31) and (15.65).

15.8. Subalgebras of the lune-incidence algebra

Define two subspaces S_l and S_r of the lune-incidence algebra. The former consists of functions f on lunes such that for any flats X and Y with $\mathrm{X} \leq \mathrm{Y}$,

$$f(\mathrm{Y}) = \sum_{\mathrm{L}:\, \mathrm{b}(\mathrm{L})=\mathrm{X},\, \mathrm{c}(\mathrm{L})\leq\mathrm{Y}} f(\mathrm{L}). \tag{15.66}$$

The latter consists of functions f on lunes such that for any flats X and Y with $\mathrm{X} < \mathrm{Y}$,

$$\sum_{\mathrm{L}:\, \mathrm{b}(\mathrm{L})=\mathrm{X},\, \mathrm{c}(\mathrm{L})=\mathrm{Y}} f(\mathrm{L}) = 0. \tag{15.67}$$

Consider also the subspace $S := S_l \cap S_r$. Note that if f satisfies both (15.66) and (15.67), it is constant on flats. Conversely, this propery together with either one of (15.66) or (15.67) implies the other.

Lemma 15.69. *The subspaces S_l and S_r (and hence S) are subalgebras of the lune-incidence algebra.*

PROOF. We check that S_l is a subalgebra. The check for S_r is similar. The unit of the lune-incidence algebra satisfies (15.66): There is exactly one flat which appears in the rhs, namely, X, so both sides evaluate to 1. Now suppose f and g satisfy (15.66). Take $\mathrm{X} \le \mathrm{Z}$. Then

$$(fg)(\mathrm{Z}) = f(\mathrm{Z})g(\mathrm{Z})$$

and

$$\sum_{\substack{\mathrm{N}:\\ \mathrm{b}(\mathrm{N})=\mathrm{X}\\ \mathrm{c}(\mathrm{N})\le \mathrm{Z}}} (fg)(\mathrm{N}) = \sum_{\substack{\mathrm{N}:\\ \mathrm{b}(\mathrm{N})=\mathrm{X}\\ \mathrm{c}(\mathrm{N})\le \mathrm{Z}}} \sum_{\substack{\mathrm{L},\mathrm{M}:\\ \mathrm{L}\circ\mathrm{M}=\mathrm{N}}} f(\mathrm{L})g(\mathrm{M}) = \sum_{\substack{\mathrm{L},\mathrm{M}:\\ \mathrm{b}(\mathrm{L})=\mathrm{X},\, \mathrm{c}(\mathrm{M})\le \mathrm{Z}\\ \mathrm{c}(\mathrm{L})=\mathrm{b}(\mathrm{M})}} f(\mathrm{L})g(\mathrm{M})$$

$$= \sum_{\substack{\mathrm{Y}:\\ \mathrm{X}\le\mathrm{Y}\le\mathrm{Z}}} \Big(\sum_{\substack{\mathrm{L}:\\ \mathrm{b}(\mathrm{L})=\mathrm{X}\\ \mathrm{c}(\mathrm{L})=\mathrm{Y}}} f(\mathrm{L})\Big)\Big(\sum_{\substack{\mathrm{M}:\\ \mathrm{b}(\mathrm{M})=\mathrm{Y}\\ \mathrm{c}(\mathrm{M})\le\mathrm{Z}}} g(\mathrm{M})\Big) = \Big(\sum_{\substack{\mathrm{L}:\\ \mathrm{b}(\mathrm{L})=\mathrm{X}\\ \mathrm{c}(\mathrm{L})\le\mathrm{Z}}} f(\mathrm{L})\Big) g(\mathrm{Z}) = f(\mathrm{Z})g(\mathrm{Z}).$$

Thus fg also satisfies (15.66) as required. □

Exercise 15.70. For any subalgebra A of the lune-incidence algebra, the radical of A consists of those functions in A which are zero on flats. (Clearly, this is the largest nilpotent ideal of A.) Deduce that the radicals of S_l, S_r and S coincide.

Lemma 15.71. *The subspaces* $\mathrm{Zet}_0[\mathcal{A}]$ *and* $\mathrm{Mob}_0[\mathcal{A}]$ *are contained in the radical of S.*

PROOF. We make two important observations.

- $\mathrm{Zet}_0[\mathcal{A}]$ is contained in S_r. (Apply (15.26) and use $\boldsymbol{\zeta}(\mathrm{Y}) = 0$.)
- $\mathrm{Mob}_0[\mathcal{A}]$ is contained in S_l. (Apply (15.30) and use $\boldsymbol{\mu}(\mathrm{Y}) = 0$.)

Both $\mathrm{Zet}_0[\mathcal{A}]$ and $\mathrm{Mob}_0[\mathcal{A}]$ consist of functions that vanish on flats, and in particular, are constant on flats. Thus, both $\mathrm{Zet}_0[\mathcal{A}]$ and $\mathrm{Mob}_0[\mathcal{A}]$ are contained in S. Vanishing on flats gives the stronger conclusion that they are contained in the radical of S. See Exercise 15.70. □

15.9. Commutative, associative and Lie operads

We give a brief sketch (without complete definitions or proofs) on how many ideas in this monograph relate to the notion of operads. Details will be provided in a future work.

15.9.1. Operads. We have often encountered maps of the form

$$\mathsf{p}[\mathcal{A}^{\mathrm{X}}] \otimes \mathsf{p}[\mathcal{A}_{\mathrm{X}}] \to \mathsf{p}[\mathcal{A}].$$

This can be abstracted into the notion of an operad. More formally, an operad is a monoid under the monoidal structure defined by

$$(\mathsf{p} \circ \mathsf{q})[\mathcal{A}] := \bigoplus_{\mathrm{X}} \mathsf{p}[\mathcal{A}^{\mathrm{X}}] \otimes \mathsf{q}[\mathcal{A}_{\mathrm{X}}], \tag{15.68}$$

with X running over all flats in $\mathcal{A}$. For morphisms, one may take either cisomorphisms or gisomorphisms.

The above definition of an operad generalizes the classical definition of an operad. The context for the latter is the family of braid arrangements. However, we mention that for morphisms, more intricate choices than cisomorphisms or gisomorphisms are required to make full contact with the classical theory.

We mention that there is another monoidal structure closely related to (15.68) in which we sum only over those flats X which are factors of $\mathcal{A}$. This monoidal structure is useful for studying the cartesian aspects of the theory. The compatibility of various objects such as flats, chambers, and so on with the cartersian product has been witnessed in many places such as Sections 1.8, 3.7 and 10.1.7.

15.9.2. Commutative, associative and Lie operads. The substitution products (15.3), (10.27) and (10.28) correspond to the commutative operad, associative operad and Lie operad, respectively. Let us denote them by Com, As and Lie. Diagrams (10.29) and (15.13) say that there are morphisms of operads

$$\mathsf{Lie} \to \mathsf{As} \to \mathsf{Com}.$$

Substitution products for the classical associative operad and the classical Lie operad are given in Sections 6.5.10 and 14.9.6, respectively.

15.9.3. Poisson operad. By linking Com and Lie by a distributive law, one can also extend the classical Poisson operad to all arrangements. The definition of the distributive law involves summing over modular complements in the lattice of flats. Exercise 10.47 plays a role in this analysis.

15.9.4. Operad incidence algebras. Given an operad p, one can define the p-incidence algebra as

$$\mathrm{I}_{\mathsf{p}}[\mathcal{A}] := \bigoplus_{\mathrm{X} \leq \mathrm{Y}} \mathsf{p}[\mathcal{A}^{\mathrm{Y}}_{\mathrm{X}}], \tag{15.69}$$

and the p-incidence module over it as

$$\mathrm{M}_{\mathsf{p}}[\mathcal{A}] := \bigoplus_{\mathrm{X}} \mathsf{p}[\mathcal{A}_{\mathrm{X}}]. \tag{15.70}$$

These constructions applied to Com yield the flat-incidence algebra (15.4) and the flat-incidence module (15.6), applied to As yield the lune-incidence algebra (15.10) and the lune-incidence module (15.19), and applied to Lie yield the Lie-incidence algebra (15.54) and the Lie-incidence module (15.59).

The p-incidence algebra can be viewed as the incidence algebra of a linear category associated to p whose objects are flats and the space of morphisms from Y to X is $\mathsf{p}[\mathcal{A}^{\mathrm{Y}}_{\mathrm{X}}]$. For the associative operad, this category is the linearization of the category of lunes. (See Proposition 15.7.) For the commutative operad, it is the linearization of the opposite of the category of the poset of flats.

15.9.5. Elementary algebras. An operad p is connected if $\mathsf{p}[\mathcal{A}] = \Bbbk$ when $\mathcal{A}$ is any rank-zero arrangement. For a connected operad p, the p-incidence algebra is elementary, and its split-semisimple quotient is the Birkhoff algebra. The operads Com, As and Lie are connected. This explains Propositions 15.1, 15.10 and 9.22. For the latter, we go through Theorem 15.56. We mention that Theorem 15.56 is a manifestation of the classical Cartier-Milnor-Moore theorem.

15.9.6. Binary quadratic and Koszulness. The operads Com, As and Lie are binary quadratic. The relevant results are Propositions 4.41 and 4.42 and Theorem 14.41. Further, Com and Lie are quadratic duals of each other, while As is self-dual. The operads Com, As and Lie are also Koszul. The Koszulness of Com and Lie can be deduced from Proposition 14.30. The passage from p to I_{p} preserves quadratic duality as well as Koszulness. As a consequence, the flat-incidence algebra, the lune-incidence algebra and the Tits algebra are Koszul. The lune-incidence algebra is self-dual, while the flat-incidence algebra and Tits algebra are duals of each other. This also explains why the flat-incidence algebra and the Tits algebra have the same quiver.

15.9.7. External product. The map (10.33) says that Σ is a left module over the associative operad. In operadic language, one says that Σ is an associative monoid. Specializing to the braid arrangement yields the external product of set compositions discussed in Section 6.3.13. Similarly, the external product of linear orders arises from the associative monoid Γ which is the associative operad viewed as a left module over itself.

It also makes sense to consider right modules over an operad. The map (10.38) says that $\widehat{\Lambda}$ is a right module over the associative operad.

15.9.8. Zie as a free Lie monoid. Let us also briefly look at Zie elements. The map (10.34) says that Zie is a left module over the Lie operad. In operadic language, one says that Zie is a Lie monoid. Further, Proposition 13.37 implies that it is free as such.

Notes

For the classical theory of operads, see [**307, 283, 290**]. For basic information on monoidal categories, see [**9**, Part I] and references therein.

The notion of operad proposed in (15.68) is similar to the one present in the work of Rains [**349**, page 794].

The Koszulness of the flat-incidence algebra is a special case of the result which says that a poset P is Cohen-Macaulay iff the incidence algebra of P is Koszul. This is present in work of Cibils [**118**], Polo [**345**] and Woodcock [**436**]. The fact that the Tits algebra is Koszul and dual to the flat-incidence algebra is a result of Saliola [**368**, Propositions 9.4 and 9.6]. A generalization is given in [**304**, Theorem 7.5].

CHAPTER 16

Invariant Birkhoff algebra and invariant Tits algebra

We studied the Birkhoff algebra and Tits algebra of an arrangement in detail. Recall that a reflection arrangement is acted upon by its Coxeter group. Hence, in this situation, it makes sense to consider the invariant part of these algebras. We call these the invariant Birkhoff algebra and invariant Tits algebra. The former is a split-semisimple commutative algebra whose primitive idempotents are indexed by flat-types. The latter is elementary and has a basis indexed by face-types. Its radical is the invariant part of the radical of the Tits algebra, and the quotient by this is the invariant Birkhoff algebra. Thus, the simple modules over the invariant Tits algebra are one-dimensional and indexed by flat-types. Complete systems of primitive orthogonal idempotents for the invariant Tits algebra are the same as invariant Eulerian families. They correspond to invariant sections and to invariant special Zie families, and similarly to invariant noncommutative zeta and Möbius functions. The latter belong to the invariant lune-incidence algebra. For a good reflection arrangement, for the uniform section, there are cancelation-free formulas for the Eulerian idempotents. The two-sided Peirce decomposition of the invariant Tits algebra can be used to shed light on its quiver. This necessitates the study of invariant Lie and Zie elements.

Recall that there is an injective map from the Tits algebra to the space indexed by pairs of chambers. Taking invariants induces an injective map from the invariant Tits algebra to the Coxeter group algebra. The image of this map is a subalgebra of the Coxeter group algebra which is known as the Solomon descent algebra. This induces an anti-isomorphism of algebras between the invariant Tits algebra and the Solomon descent algebra. This result makes it possible to study the Solomon descent algebra using the invariant Tits algebra.

The structure constants of the invariant Tits algebra are of great theoretical significance. They intervene in the invariant formulation of lune-additivity and the noncommutative Weisner formula. They are also intimately connected to enumeration of face-types.

We illustrate some of the above ideas for the braid arrangement. This makes contact with the Garsia-Reutenauer idempotents and the Bayer-Diaconis-Garsia-Loday formula. We also briefly discuss the arrangement of type B.

Notation 16.1. For a reflection arrangement $\mathcal{A}$, for any face-types $T \le U$, we let $\mathcal{A}_T^U$ refer to any arrangement $\mathcal{A}_H^G$, with $\mathrm{t}(H) = T$ and $\mathrm{t}(G) = U$. Similarly, we let c_T^U denote the number of chambers in $\mathcal{A}_T^U$. In particular, c^T is the number of chambers in $\mathcal{A}^T$.

Convention 16.2. Throughout this chapter, $\mathcal{A}$ is a reflection arrangement with Coxeter group W. We assume that the field characteristic does not divide the order of W whenever $\mathcal{A}$ has rank at least one, that is, when W is not the trivial group.

16.1. Invariant Birkhoff algebra

Recall the Birkhoff algebra $\Pi[\mathcal{A}]$. It has a H-basis and a Q-basis. It is a split-semisimple commutative algebra whose primitive idempotents are the Q-basis elements. The Coxeter group W acts on this algebra via

$$w \cdot \mathtt{H}_{\mathrm{X}} := \mathtt{H}_{w(\mathrm{X})}.$$

Let $\Pi[\mathcal{A}]^W$ denote the subalgebra of $\Pi[\mathcal{A}]$ consisting of the W-invariant elements. We call this the *invariant Birkhoff algebra*.

For each flat-type λ, put

$$\mathtt{H}_\lambda := |\lambda| \sum_{\mathrm{X}:\, \mathrm{t}(\mathrm{X})=\lambda} \mathtt{H}_{\mathrm{X}} \qquad \text{and} \qquad \mathtt{Q}_\lambda := \sum_{\mathrm{X}:\, \mathrm{t}(\mathrm{X})=\lambda} \mathtt{Q}_{\mathrm{X}}, \tag{16.1}$$

with $|\lambda|$ as defined in Section 5.5.1. As λ varies over all flat-types, these define the H- and Q-bases of the invariant Birkhoff algebra.

Theorem 16.3. *The invariant Birkhoff algebra is a split-semisimple commutative algebra. Its dimension equals the number of flat-types in $\mathcal{A}$. The unique complete system of primitive orthogonal idempotents is given by the* Q*-basis.*

PROOF. This can be deduced from Theorem 9.2 using Lemma D.14. □

One may check using (9.1) and (16.1) that

$$\mathtt{H}_\mu = \sum_{\lambda:\, \mu \leq \lambda} R_{\lambda\mu} \mathtt{Q}_\lambda, \tag{16.2}$$

where $R_{\lambda\mu}$ is $|\mu|$ times the number of flats of type μ contained in a given flat of type λ.

16.2. Invariant Tits algebra

The structure theory of the invariant Tits algebra proceeds in analogy with that of the Tits algebra (and can mostly be deduced from it). The invariant Tits algebra is elementary and its split-semisimple quotient is the invariant Birkhoff algebra. This story is continued in Section 16.8.

16.2.1. Invariant Tits algebra. The Coxeter group W acts on the Tits algebra $\Sigma[\mathcal{A}]$ via

$$w \cdot \mathtt{H}_F := \mathtt{H}_{wF}.$$

Let $\Sigma[\mathcal{A}]^W$ denote the subalgebra of $\Sigma[\mathcal{A}]$ consisting of the W-invariant elements. We call this the *invariant Tits algebra*.

For each face-type T, put

$$\mathtt{H}_T := \sum_{F:\, \mathrm{t}(F)=T} \mathtt{H}_F. \tag{16.3}$$

As T varies over all face-types, this defines the H-basis of the invariant Tits algebra. The K-basis is then defined by

$$\mathtt{H}_T = \sum_{U:\, U \leq T} \mathtt{K}_U \qquad \text{or equivalently} \qquad \mathtt{K}_T = \sum_{U:\, U \leq T} (-1)^{|T \setminus U|}\, \mathtt{H}_U. \tag{16.4}$$

Recall the space of top-nested faces from Section 9.4.7. It contains the Tits algebra and hence the invariant Tits algebra as a subspace. On the H- and K-bases,

$$\mathtt{H}_T = \sum_{H \le D,\, \mathrm{t}(H)=T} \mathtt{H}_{H,D} \quad \text{and} \quad \mathtt{K}_T = \sum_{H \le D,\, \mathrm{t}(H)=T} \mathtt{K}_{H,D}. \tag{16.5}$$

The sum is over all pairs $H \le D$ such that the type of H is T. This follows from (9.43) and (9.44).

16.2.2. Invariant support map. The support map from the Tits algebra to the Birkhoff algebra restricts to a linear map

$$\mathrm{s} : \Sigma[\mathcal{A}]^W \twoheadrightarrow \Pi[\mathcal{A}]^W. \tag{16.6}$$

We call this the *invariant support map*. On the H-basis,

$$\mathrm{s}(\mathtt{H}_T) = \mathtt{H}_{\mathrm{s}(T)}. \tag{16.7}$$

16.2.3. Radical. We now turn to the radical of the invariant Tits algebra.

Proposition 16.4. *The invariant Tits algebra is elementary. Its split-semisimple quotient is the invariant Birkhoff algebra, with the invariant support map as the quotient map. Its radical is the kernel of the invariant support map and equals* $\mathrm{rad}(\Sigma[\mathcal{A}])^W$ *which is the invariant part of the radical of the Tits algebra.*

PROOF. This can be deduced from Proposition 9.22 using Lemma D.46. Also note Convention 16.2. □

16.2.4. Multiplicative characters. Applying Theorem D.35 to the invariant Tits algebra yields:

Theorem 16.5. *The simple modules over* $\Sigma[\mathcal{A}]^W$ *are one-dimensional and indexed by flat-types. Let* χ_λ *denote the multiplicative character corresponding to the flat-type* λ. *It is specified by*

$$\mathrm{s}(z) = \sum_{\lambda} \chi_\lambda(z)\, \mathtt{Q}_\lambda, \tag{16.8}$$

where s *is the invariant support map* (16.6).

16.3. Solomon descent algebra

In any pointed arrangement, there is a descent map from the Coxeter group W to the set of all face-types. Inside the group algebra W, there is a subalgebra linearly spanned by sums of group elements that have the same descent. This is the Solomon descent algebra. It is anti-isomorphic to the invariant Tits algebra.

16.3.1. Descents and shuffles. Fix a pointed arrangement $(\mathcal{A}, C)$. A chamber obtained by projecting C on a face of type T is called a T-gate wrt C.

For $\sigma \in W$, let $\mathrm{Des}(\sigma)$ denote the type of the face $\mathrm{Des}(C, \sigma C)$, with the latter defined by (7.1). This is the standard notion of descent in Coxeter theory.

The descent maps on pairs of chambers and on group elements are related by the commutative diagram

$$\begin{array}{ccc} \Gamma[\mathcal{A}] \times \Gamma[\mathcal{A}] & \xrightarrow{\text{Des}} & \Sigma[\mathcal{A}] \\ {\scriptstyle d}\downarrow & & \downarrow{\scriptstyle \mathrm{t}} \\ W & \xrightarrow[\text{Des}]{} & \Sigma[\mathcal{A}]^W, \end{array} \tag{16.9}$$

with d being the W-valued gallery distance function.

For any face-type T, define the set of T-shuffles by

$$\mathrm{Sh}_T := \{\sigma \in W \mid \mathrm{Des}(\sigma) \leq T\}.$$

Since $\sigma = d(C, \sigma C)$, it follows from (7.1) that all T-shuffles are of the form $d(C, D)$, as D varies over all T-gates. In other words,

$$\mathrm{Sh}_T = \{\sigma \in W \mid \sigma(C) = D \text{ for some } T\text{-gate } D\}. \tag{16.10}$$

Lemma 16.6. *For any face-type T, every element $w \in W$ can be uniquely written as $w = vu$ such that $u \in W_T$ and $v \in \mathrm{Sh}_T$, and further $l(w) = l(u) + l(v)$.*

Here recall that W_T is the parabolic subgroup of W which leaves F invariant, with F being the face of C of type T. Also, $l(w)$ denotes the length of w as in (5.6).

PROOF. This can be deduced from the gate property. □

16.3.2. Solomon descent algebra. Let W denote the group algebra of W over a field $\Bbbk$. We use the letter $\mathtt{K}$ to denote its canonical basis.

Theorem 16.7. *The subspace of W linearly spanned by the elements*

$$y_T := \sum_{w\colon \mathrm{Des}(w) \leq T} \mathtt{K}_w = \sum_{w \in \mathrm{Sh}_T} \mathtt{K}_w, \tag{16.11}$$

as T varies over subsets of S, is a subalgebra of W.

This subalgebra is known as the *Solomon descent algebra*. The elements y_T define a basis for this algebra. It also has another basis consisting of the elements

$$z_T := \sum_{w\colon \mathrm{Des}(w) = T} \mathtt{K}_w. \tag{16.12}$$

We now relate the Solomon descent algebra to the invariant Tits algebra. In particular, this will also yield a proof of Theorem 16.7.

Recall that the action of the Tits algebra $\Sigma[\mathcal{A}]$ on $\Gamma[\mathcal{A}]$ yields the injective algebra homomorphism (9.42). The Coxeter group W acts on this map. The space of W-invariants of $\Sigma[\mathcal{A}]$ is the invariant Tits algebra. The W-invariants of $\Gamma[\mathcal{A}]^* \otimes \Gamma[\mathcal{A}]$ can be identified with the opposite of the group algebra via

$$\mathsf{W}^{\mathrm{op}} \xrightarrow{\cong} (\Gamma[\mathcal{A}]^* \otimes \Gamma[\mathcal{A}])^W, \qquad \mathtt{K}_w \mapsto \sum_{(D,C)\colon d(D,C) = w} \mathtt{M}_D \otimes \mathtt{H}_C. \tag{16.13}$$

The main assertion here is that this is an algebra homomorphism. This is a consequence of (5.7) and (9.41). Thus, we obtain the following commutative diagram of algebras.

$$\begin{array}{ccc} \Sigma[\mathcal{A}] & \longrightarrow & \Gamma[\mathcal{A}]^* \otimes \Gamma[\mathcal{A}] \\ \uparrow & & \uparrow \\ \Sigma[\mathcal{A}]^W & \longrightarrow & \mathsf{W}^{\mathrm{op}} \end{array}$$

It follows from (7.1) and (16.9) that the bottom-horizontal map sends $\mathtt{H}_T$ to y_T. Equivalently, it sends $\mathtt{K}_T$ to z_T. Hence the image of the bottom-horizontal map is precisely the Solomon descent algebra. As a consequence:

Theorem 16.8. *The Solomon descent algebra is isomorphic to the opposite of the invariant Tits algebra.*

16.4. Enumeration of face-types

In a reflection arrangement, the problem of counting faces of a given type T is related to T-shuffles and dually to chambers in a T-based top-lune. (A T-based lune is a lune whose base supports a face of type T.) More generally, one can do a q-enumeration of faces of type T by using gallery distances between chambers. The Poincaré polynomial is obtained when we do a q-enumeration of chambers (which are faces of type S).

16.4.1. Counting of face-types. For any pointed arrangement $\boldsymbol{\alpha} = (\mathcal{A}, C)$, let d_T denote the number of faces of type T in $\boldsymbol{\alpha}$. In particular, d_S is the number of chambers. More generally, for $T \leq U$, let $d_{U/T}$ denote the number of faces of type U/T in $\boldsymbol{\alpha}_T$.

Lemma 16.9. *For $T \leq U \leq V$,*

$$d_{V/T} = d_{V/U}\, d_{U/T}. \tag{16.14}$$

In particular, $d_V = d_{V/U} d_U$ (by letting $T = \emptyset$).

PROOF. It suffices to show that $d_V = d_{V/U} d_U$. Let Σ_U and Σ_V denote the set of faces of types U and V, respectively. They have cardinalities d_U and d_V, respectively. Every face of type V has a unique subface of type U. This defines a function $\Sigma_V \to \Sigma_U$ which is clearly surjective. The fibers of this map are of the same cardinality, namely, $d_{V/U}$: This is the number of faces of type V which contain a given face of type U. The result follows. □

Recall from Section 16.3.1 that a T-gate wrt C is a chamber obtained by projecting C on a face of type T. The number of T-gates is d_T since each face of type T contributes to a unique T-gate. It follows from (16.10) that:

Lemma 16.10. *The number of T-shuffles is d_T.*

Equivalently, d_T is the number of pairs (C, D), where the first coordinate is fixed to be C, while the second coordinate varies over T-gates wrt C. There is also a 'dual' way of obtaining d_T by fixing the second coordinate and varying the first. It works as follows.

Lemma 16.11. *The number of chambers in any T-based top-lune is d_T. This is the same as c/c_T. Here c_T is the number of chambers in $\boldsymbol{\alpha}_T$, and in particular, c is the number of chambers in $\boldsymbol{\alpha}$.*

PROOF. Fix a chamber E. Move each T-gate to E by a (unique) Coxeter symmetry. Under such a symmetry, the reference chamber C moves into the T-based top-lune formed at E, and all chambers in this top-lune get occupied exactly once as the T-gate and resulting symmetry vary. This proves the first claim. The second claim follows from Lemma 5.21. □

Lemma 16.12. *If T and T' are face-types with the same support, then $d_T = d_{T'}$.*

PROOF. Since T and T' have the same support, there is no distinction between T-based top-lunes and T'-based top-lunes, so $d_T = d_{T'}$ by Lemma 16.11. □

Exercise 16.13. Deduce (16.14) from Proposition 3.21 and Lemma 16.11.

16.4.2. q-numbers. Fix a scalar q. Define the i-th q-number to be

$$(i)_q := 1 + q + \cdots + q^{i-1}.$$

The *q-factorial* is

$$(n)_q! := (n)_q(n-1)_q \dots (1)_q,$$

and the *q-binomial coefficient* is

$$\binom{n}{i}_q = \frac{(n)_q!}{(i)_q!(n-i)_q!}.$$

For $q = 1$, this is the usual binomial coefficient.

16.4.3. q-counting of face-types. Fix a scalar q. Recall the distance function on chambers v_q defined in (8.15). For any face-type T in $\boldsymbol{\alpha} = (\mathcal{A}, C)$, define a polynomial in q by

$$d_T(q) := \sum_{F:\, \mathrm{t}(F)=T} (v_q)_{C,FC} = \sum_{F:\, \mathrm{t}(F)=T} q^{\mathrm{dist}(C,FC)}. \tag{16.15}$$

The sum is over all faces of type T. Setting $q = 1$ recovers d_T. More generally, for $T \leq U$, define $d_{U/T}(q)$ by applying (16.15) to the face-type U/T of $\boldsymbol{\alpha}_T$.

Lemma 16.14. *For $T \leq U \leq V$,*

$$d_{V/T}(q) = d_{V/U}(q)\, d_{U/T}(q). \tag{16.16}$$

PROOF. We build on the proof of (16.14). For any $F \leq G$, by (8.2e),

$$(v_q)_{C,GC} = (v_q)_{C,FC}(v_q)_{FC,GC}.$$

Now sum over all F of type U, and all G of type V with $F \leq G$. This yields (16.16) for $T = \emptyset$, from which the general case also follows. □

Lemmas 16.10 and 16.11 generalize as follows. We have

$$d_T(q) = \sum_{\sigma \in \mathrm{Sh}_T} q^{l(\sigma)}, \tag{16.17}$$

where $l(\sigma)$ is the length of σ as in (5.6). The sum is over all T-shuffles. (The chambers FC in (16.15) are precisely the T-gates.) Dually,

$$d_T(q) := \sum_{C':\, HC'=D} q^{\mathrm{dist}(C',D)}, \tag{16.18}$$

where D is any chamber and H is the face of D of type T. Thus, we sum over chambers C' in a T-based top-lune, and each summand contributes a power of q.

Lemma 16.15. *We have*

$$\sum_T (-1)^{|T|} d_T(q) = (-1)^{|S|} q^{\operatorname{dist}(C,\overline{C})}. \tag{16.19}$$

PROOF. The lhs above can be manipulated as follows.

$$\begin{aligned}\sum_T (-1)^{|T|} \sum_{F:\,\mathrm{t}(F)=T} q^{\operatorname{dist}(C,FC)} &= \sum_F (-1)^{\operatorname{rk}(F)} q^{\operatorname{dist}(C,FC)} \\ &= \sum_D q^{\operatorname{dist}(C,D)} \Big(\sum_{F:\,FC=D} (-1)^{\operatorname{rk}(F)} \Big)\end{aligned}$$

Now use (7.10). Alternatively, put $x^C := q^{\operatorname{dist}(C,D)}$ in (7.14) and use (16.18). □

Exercise 16.16. For $q = 1$, (16.19) reduces to the Witt identity (7.15). This follows from Lemma 16.11. Now give an alternative proof of (7.15) by proving the $q = 1$ case of (16.19) using (1.41).

16.4.4. Poincaré polynomial. We refer to $d_S(q)$ as the *Poincaré polynomial* of W. Note that

$$d_S(q) = \sum_{w \in W} q^{l(w)},$$

where $l(w)$ is the length of w. This identity is a special case of (16.17) or can also be directly seen from the definition (16.15):

$$d_S(q) = \sum_{F:\,\mathrm{t}(F)=S} q^{\operatorname{dist}(C,FC)} = \sum_D q^{\operatorname{dist}(C,D)} = \sum_{w\in W} q^{\operatorname{dist}(C,wC)} = \sum_{w \in W} q^{l(w)},$$

with C as the reference chamber. Specializing (16.16): For any face-type T,

$$d_S(q) = d_{S/T}(q)\, d_T(q). \tag{16.20}$$

Thus the Poincaré polynomial always factorizes. An important result in this direction is stated below.

Theorem 16.17. *Let (W, S) be a Coxeter system, and $n := |S|$. Then there exist positive integers $e_1, \dots, e_n$ such that*

$$d_S(q) = \prod_{i=1}^n (e_i + 1)_q. \tag{16.21}$$

PROOF. See [**238**, Theorem on page 73] or [**80**, Theorem 7.1.5]. □

By setting $q = 1$, respectively, looking at the exponent of the highest power of q, we obtain

$$|W| = \prod_{i=1}^n (e_i + 1) \quad \text{and} \quad \operatorname{dist}(C, \overline{C}) = \sum_{i=1}^n e_i.$$

The integers $e_1, \dots, e_n$ appearing in (16.21) are called the exponents of (W, S).

For a rank-one arrangement, $d_S(q) = 1 + q$ and in particular, $d_S(-1) = 0$. Hence in any reflection arrangement $d_{S/T}(-1) = 0$ whenever T is the type of a panel. Substituting this in (16.20), one can deduce that

$$d_S(-1) = \sum_{w\in W} (-1)^{l(w)} = \begin{cases} 1 & \text{if } \boldsymbol{\alpha} \text{ has rank } 0, \\ 0 & \text{otherwise.} \end{cases} \tag{16.22}$$

This result can also be deduced from (16.21) by noting that 1 is always an exponent, and $(2)_q = 1 + q = 0$ for $q = -1$.

Exercise 16.18. Consider the augmentation map $\mathsf{W} \to \Bbbk$ which sends each group element w to 1. Under this map, the element $\sum_{w \in W} q^{l(w)} w$ maps to $d_S(q)$. Use Lemma 8.28 to deduce that $d_S(q)$ is invertible when q is not a root of unity. This fact can also be deduced from (16.21).

16.5. Structure constants of the invariant Tits algebra

Unlike the Tits algebra, the invariant Tits algebra is not the linearization of a monoid. So its structure constants in the $\mathtt{H}$-basis can take integer values other than 0 and 1. We provide geometric interpretations for these constants, discuss a number of identities involving them, and also relate them to face-type enumeration.

16.5.1. Structure constants. Fix a pointed arrangement $\boldsymbol{\alpha} = (\mathcal{A}, C)$. For face-types T and T', write

$$\mathtt{H}_T \cdot \mathtt{H}_{T'} := \sum_{U:\, U \geq T} a^{TUT'} \mathtt{H}_U \tag{16.23}$$

for suitable scalars $a^{TUT'}$. These are the structure constants of the invariant Tits algebra in the $\mathtt{H}$-basis.

Lemma 16.19. *Let T, T' and U be face-types with $T \leq U$. Then $a^{TUT'}$ is the number of faces F of type T' such that $HF = G$, where (H, G) is a fixed nested face of type (T, U).*

PROOF. Let G be any face of type U. Then by definition, $a^{TUT'}$ is the number of ways in which G can be written in the form HF where H has type T and F has type T'. This forces H to be the unique face of G of type T. So we are left with counting the number of faces F of type T' such that $HF = G$. □

Just like d_T, the $a^{TUT'}$ can also be counted in a 'dual' manner using a top-lune based at T' as follows.

Lemma 16.20. *Let T, T' and U be face-types with $T \leq U$. Let L be a T'-based top-lune. Fix a face F' of type T' in the base of L. Then $a^{TUT'}$ is the number of chambers C' in L such that $H'F' = G'$, where H' and G' are the faces of C' of type T and U, respectively.*

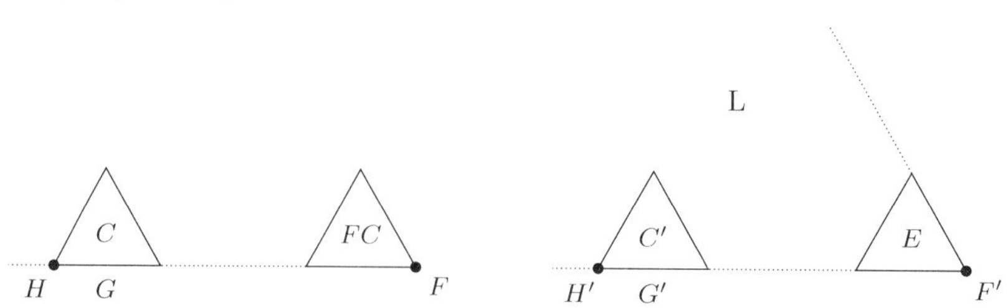

PROOF. Let E be the unique chamber in L with F' as a face. Let H and G be the faces of C of types T and U. Consider the T'-gates of the form FC, as F varies over faces of type T' such that $HF = G$. They are $a^{TUT'}$ in number. Move each such T'-gate to E by a (unique) Coxeter symmetry. Under such a symmetry C moves into

L to a chamber say C'. As the T'-gate and resulting symmetry vary, C' occupies precisely those chambers in L with the property stated in the proposition. □

Let us apply the above discussion to $\boldsymbol{\alpha}_Z$, where Z is any face-type of $\boldsymbol{\alpha}$. For each face-type T greater than Z, put

$$\mathtt{H}_{T/Z} := \sum_{F:\,\mathrm{t}(F)=T,\, F\geq A} \mathtt{H}_{F/A}, \tag{16.24}$$

where A is the face of C of type Z. This defines the $\mathtt{H}$-basis of the invariant Tits algebra of $\boldsymbol{\alpha}_Z$. Let $a_Z^{TUT'}$ denote its structure constants, that is,

$$\mathtt{H}_{T/Z} \cdot \mathtt{H}_{T'/Z} := \sum_{U:\,U\geq T} a_Z^{TUT'} \mathtt{H}_{U/Z}. \tag{16.25}$$

Lemma 16.21. *For any face-types T, T', V and Y with $T \leq V \leq Y$,*

$$a^{VYT'} = \sum_{U:\,U\geq T} a_T^{VYU}\, a^{TUT'}. \tag{16.26}$$

PROOF. We explain this identity by means of a picture.

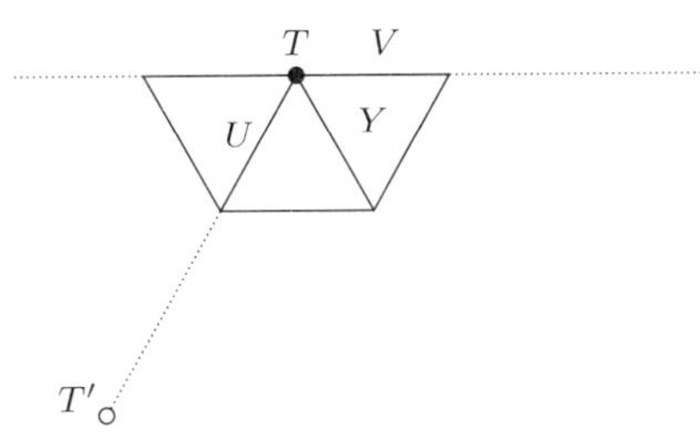

Fix faces F and G of types V and Y, respectively, with $F \leq G$. In the picture, F is the edge labeled V, while G is the triangle labeled Y. By Lemma 16.19, the lhs of (16.26) counts the number of faces of type T' in the region strictly below the horizontal dotted line. We are given that T is a face-type smaller than V. Let H be the face of F of type T. The lhs can be expressed as a sum indexed by U, where U is the type of the face obtained by projecting faces of type T' on H. In the picture, T is a vertex-type shown in black, T' is also a vertex-type shown in white, U is a variable quantity depending on which face of type T' is used. For the particular choice in the picture, U is an edge-type. For each U, we need to multiply two independent counts. The first is the number of faces of type T' in the interior of the half-line determined by T and U which is $a^{TUT'}$. The second is the number of faces of type U in the interior of the lune in the star of H determined by F and G (consisting of the three triangles) which is a_T^{VYU}. □

Lemma 16.22. *For any face-types T and T', we have $a^{TST'} = a^{T'ST}$.*

PROOF. Consider the set of pairs (F, F'), where F has type T, F' has type T', and FF' is a chamber. Then $a^{TST'}$ is the cardinality of this set divided by the number of chambers. The identity follows by noting that FF' is a chamber iff $F'F$ is a chamber. □

Lemma 16.23. *For any $T \leq U$, we have $\sum_{T'} (-1)^{|T'|} a^{TUT'} = (-1)^{|U|}$. The sum is over all face-types T'.*

PROOF. This follows from (7.12a) and Lemma 16.19. □

Exercise 16.24. Recall from Section 5.2.4 the equivalence relation on nested face-types. Show that: If (T, U) and (T', U') are equivalent, then $a^{TUV} = a^{T'U'V}$ for any face-type V.

16.5.2. q-analogue. The structure constants of the invariant Tits algebra can be deformed as follows. Fix a scalar q. For any face-types T, T' and U with $T \leq U$, define

$$a^{TUT'}(q) := \sum_{F} q^{\operatorname{dist}(H,F)} = \sum_{F} (\upsilon_q)_{H,F}. \tag{16.27}$$

Here υ_q is the distance function on faces given by (8.16). The sum is over all faces F of type T' such that $HF = G$, where (H, G) is a fixed nested face of type (T, U). Note from Lemma 16.19 that setting $q = 1$ recovers the structure constants.

Lemma 16.25. *Let E be any chamber and F' be its face of type T'. Then*

$$a^{TUT'}(q) = \sum_{H'} q^{\operatorname{dist}(H',F')}. \tag{16.28}$$

The sum is over all faces H' of type T such that $F'H' \leq E$ and $H'F'$ has type U.

This is the q-analogue of Lemma 16.20 but formulated somewhat differently. See Lemma 3.39 in this regard.

Lemma 16.26. *For any face-types T, T', V and Y with $T \leq V \leq Y$,*

$$a^{VYT'}(q) = \sum_{U:\, U \geq T} a_T^{VYU}(q)\, a^{TUT'}(q). \tag{16.29}$$

PROOF. This is a q-analogue of (16.26). The same argument can be repeated, the additional ingredient is to use (8.9e) for $\upsilon = \upsilon_q$. ☐

Lemma 16.27. *For any face-types T and T', we have $a^{TST'}(q) = a^{T'ST}(q)$.*

PROOF. The argument of Lemma 16.22 can be repeated, with the additional ingredient being $(\upsilon_q)_{F,F'} = (\upsilon_q)_{F',F}$. ☐

For any face-type T, define

$$\operatorname{dist}(T, \overline{T}) := \operatorname{dist}(F, \overline{F}), \tag{16.30}$$

where F is any face of type T.

Lemma 16.28. *For any face-types $T \leq U$,*

$$\sum_{T'} (-1)^{|T'|} a^{TUT'}(q) = (-1)^{|U|} q^{\operatorname{dist}(T,\overline{T})}. \tag{16.31}$$

PROOF. This follows from (7.13a) and (16.27). ☐

Lemma 16.29. *For any face-types U and V,*

$$\sum_{T:\, T \leq V} (-1)^{|T|} a^{TVU}(q) = \begin{cases} (-1)^{|V|} q^{\operatorname{dist}(U,\overline{U})} & \textit{if } V \geq \overline{U}, \\ 0 & \textit{otherwise.} \end{cases} \tag{16.32}$$

PROOF. Let E be any chamber, and F be the face of E of type U. Then by using Lemma 16.25, the lhs of (16.32) can be written as

$$\sum_{\substack{G:\, FG\leq E,\\ \mathrm{t}(G)=V}} q^{\mathrm{dist}(G,F)} \sum_{H:\, HF=G} (-1)^{\mathrm{rk}(H)}.$$

By (7.11a), this equals

$$\sum_{\substack{G:\, FG\leq E,\, \overline{F}\leq G,\\ \mathrm{t}(G)=V}} q^{\mathrm{dist}(G,F)}(-1)^{\mathrm{rk}(G)}.$$

The conditions $FG \leq E$ and $\overline{F} \leq G$ are equivalent to $\overline{F} \leq G \leq \overline{F}E$. Now if $V \not\geq \overline{U}$, then no G works and the sum is 0, while if $V \geq \overline{U}$, then the face of $\overline{F}E$ of type V is the unique G which works. The result follows. □

16.5.3. Structure constants and face-type enumeration. Recall that d_T is the number of faces of type T in $\boldsymbol{\alpha}$, and for $T \leq U$, $d_{U/T}$ is the number of faces of type U/T in $\boldsymbol{\alpha}_T$.

Since any S-based lune is the maximum flat, Lemmas 16.19 and 16.20 yield

$$d_{T'} = a^{SST'} = a^{T'SS}. \tag{16.33}$$

Lemma 16.30. *For any face-types T and T',*

$$d_{T'} = \sum_{U:\, U\geq T} d_{U/T}\, a^{TUT'}. \tag{16.34}$$

PROOF. This is a special case of (16.26): Put $V = Y := S$. The same argument in more direct terms goes as follows. Let $\Sigma_{T'}$ denote the set of all faces of type T'. Define an equivalence relation on $\Sigma_{T'}$: $F \sim F'$ iff $HF = HF'$, where H is the face of C of type T. If $U = \mathrm{t}(HF)$, then we say that the equivalence class of F is of type U. The size of an equivalence class only depends on its type. Further, there are $d_{U/T}$ equivalence classes of type U, and each such equivalence class has size $a^{TUT'}$. The result follows. □

Note that setting $T := S$ in (16.34) forces $U = S$, and the identity specializes to $d_{T'} = a^{SST'}$.

SECOND PROOF. We give another argument based on the lune interpretation of d_T. Fix a T'-based top-lune L and another type T. By Lemma 16.11, $d_{T'}$ is the number of chambers in L. In what follows we employ the notations of Lemma 3.39. By applying it to each face H of type T, we see that L can be written as a disjoint union of smaller top-lunes (in T-stars). Now group these smaller top-lunes L′ according to the type of HF. In any one particular group say of type U, the number of chambers in each L′ is $d_{U/T}$ and the number of these is $a^{TUT'}$ by Lemma 16.20. This conclude the argument.

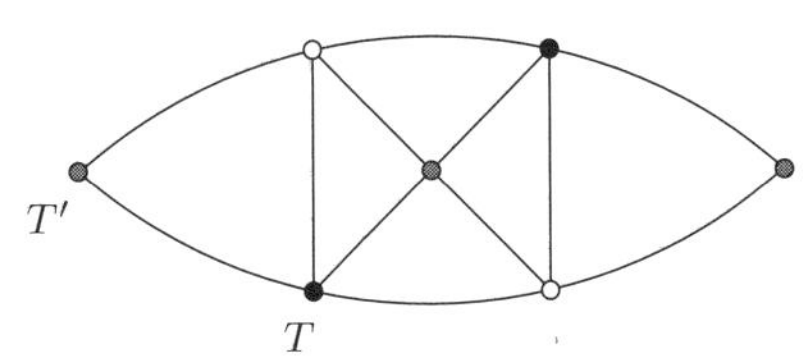

The picture shows L, with T' being the gray vertex-type, and T being the black vertex-type. (This lune occurs in the rank-three braid arrangement.) The top-lune L breaks into two parts of three triangles each. Note however that the two parts fall into different groups, the U for one is the edge-type black and gray, while the U for the other is the edge-type black and white. So the identity reads $6 = 3 \cdot 1 + 3 \cdot 1$. The remaining face-types do not contribute. (One can consider other scenarios in the same picture by taking T to be some other face-type.) $\square$

We briefly discuss q-analogues of the above results. We have

$$d_{T'}(q) = a^{SST'}(q) = a^{T'SS}(q).$$

Observe that setting $T = U := S$ in (16.31) and then using the above identity recovers (16.19).

Lemma 16.31. *For any face-types T and T',*

$$d_{T'}(q) = \sum_{U:\, U \geq T} d_{U/T}(q)\, a^{TUT'}(q). \tag{16.35}$$

Proof. This is a special case of (16.29): Put $V = Y := S$. Alternatively, it can be deduced from the proof of (16.34), the additional ingredient being (8.2f):

$$(v_q)_{C,FC} = (v_q)_{C,HFC}(v_q)_{HFC,FC}. \qquad \square$$

Note that setting $T := S$ in (16.35) forces $U = S$, and the identity specializes to $d_{T'}(q) = a^{SST'}(q)$.

16.6. Invariant Lie and Zie elements

Recall Lie and Zie elements from Chapter 10. We briefly consider the invariant part of these spaces under the action of the Coxeter group W. Results on Peirce decompositions from Sections 13.5 and 13.6 and the JKS theorem from Section 14.5 will be used in the discussion.

16.6.1. Invariant Lie elements. Recall the space of Lie elements $\mathsf{Lie}[\mathcal{A}]$. Let $\mathsf{Lie}[\mathcal{A}]^W$ denote the subspace of $\mathsf{Lie}[\mathcal{A}]$ consisting of the W-invariant Lie elements.

Proposition 16.32. *We have*

$$\mathsf{Lie}[\mathcal{A}]^W = \begin{cases} \Bbbk & \text{if } \mathrm{rk}(\mathcal{A}) = 0, \\ 0 & \text{otherwise.} \end{cases} \tag{16.36}$$

Proof. If $\mathcal{A}$ has rank zero, then $\mathsf{Lie}[\mathcal{A}] = \Bbbk$ by Lemma 10.1 and W is the trivial group; so this case is clear. Now assume that $\mathcal{A}$ has rank at least 1. Fix a maximal chain of flats z and an orientation σ of $\mathcal{A}$. Let α be the Lie element of $\mathcal{A}$ obtained by unbracketing z wrt σ. Let $s \in W$ be the reflection in the hyperplane which appears in the chain z. Then $s(\alpha) = -\alpha$. This is because the action of s preserves z but reverses σ. Therefore

$$\sum_{w \in W} w(\alpha) = \sum_{w \in W} w(s(\alpha)) = -\Big(\sum_{w \in W} w(\alpha)\Big) = 0.$$

By the JKS theorem, elements α as above obtained by unbracketing maximal chains linearly span $\mathsf{Lie}[\mathcal{A}]$. Hence, elements of the form $\sum_{w \in W} w(\alpha)$ linearly span $\mathsf{Lie}[\mathcal{A}]^W$. (The assumption on the field characteristic made in Convention 16.2 is used here.) But any such sum is zero. The result follows. $\square$

More generally:

Proposition 16.33. *Fix a flat-type* λ. *Then*

$$\Big(\bigoplus_{X:\,t(X)=\lambda} \mathsf{Lie}[\mathcal{A}_X]\Big)^W = \begin{cases} \Bbbk & \text{if } \lambda \text{ is the maximum flat-type,} \\ 0 & \text{otherwise.} \end{cases} \tag{16.37}$$

The sum is over all flats X *of type* λ.

Setting λ to be the minimum flat-type recovers (16.36).

PROOF. The argument given for Proposition 16.32 generalizes. Alternatively, one can make use of the left Peirce decomposition of chambers given in (13.12). This isomorphism commutes with the action of W if the homogeneous section u is invariant under W. (The latter notion is elaborated in Section 16.8.1.) Hence, we obtain

$$\Big(\bigoplus_{X} \mathsf{Lie}[\mathcal{A}_X]\Big)^W = \bigoplus_{\lambda}\Big(\bigoplus_{X:\,t(X)=\lambda} \mathsf{Lie}[\mathcal{A}_X]\Big)^W \cong \Gamma[\mathcal{A}]^W.$$

Here λ runs over all flat-types. The space $\Gamma[\mathcal{A}]^W$ is one-dimensional spanned by the sum of all chambers. It corresponds to the summand of the maximum flat-type. As a consequence, the summands of the remaining flat-types are all 0 as required. □

16.6.2. Invariant Zie elements. Recall the space of Zie elements $\mathsf{Zie}[\mathcal{A}]$. Let $\mathsf{Zie}[\mathcal{A}]^W$ denote the subspace of $\mathsf{Zie}[\mathcal{A}]$ consisting of the W-invariant Zie elements.

Recall the right Peirce decomposition of Zie given in (13.17). This isomorphism commutes with the action of W if the involved Q-basis is invariant under W. (The latter notion is elaborated in Section 16.8.3.) Hence, we obtain

$$\mathsf{Zie}[\mathcal{A}]^W \cong \Big(\bigoplus_{X} \mathsf{Lie}[\mathcal{A}^X]\Big)^W = \bigoplus_{\mu}\Big(\bigoplus_{X:\,t(X)=\mu} \mathsf{Lie}[\mathcal{A}^X]\Big)^W. \tag{16.38}$$

Consider the special Zie element $\mathtt{Q}_O$ (which is assumed to be invariant). Its linear span is a one-dimensional subspace of $\mathsf{Zie}[\mathcal{A}]^W$. It corresponds to the summand of the minimum flat-type.

16.7. Invariant lune-incidence algebra

Recall the lune-incidence algebra from Section 15.2. It is acted upon by the Coxeter group W, and thus we obtain the invariant lune-incidence algebra. This algebra can also be viewed as a reduced incidence algebra of the poset of face-types. It has a basis indexed by lune-types.

Recall noncommutative zeta and Möbius functions from Section 15.3. They are elements of the lune-incidence algebra characterized by lune-additivity and the noncommutative Weisner formula. For those noncommutative zeta and Möbius functions which belong to the invariant lune-incidence algebra, we reformulate lune-additivity and the noncommutative Weisner formula using face-types.

16.7.1. Invariant face-incidence algebra. Recall the incidence algebra of the poset of faces. It is called the face-incidence algebra and denoted $\mathrm{I}_{\mathrm{face}}[\mathcal{A}]$. It consists of functions on nested faces with product given by (15.8). We denote its W-invariant subalgebra by $\mathrm{I}_{\mathrm{face}}[\mathcal{A}]^W$. It consists of functions f on nested faces which are *invariant*, that is, $f(H,G) = f(wH, wG)$ for $w \in W$. An alternative description is given below.

Let $\mathrm{I}_{\mathrm{facetype}}[\mathcal{A}]$ denote the incidence algebra of the poset of face-types. It consists of functions f on nested face-types, with the product of f and g given by

$$(fg)(T,V) = \sum_{U:\, T \leq U \leq V} f(T,U)g(U,V). \tag{16.39}$$

There is an isomorphism of algebras

$$\mathrm{I}_{\mathrm{facetype}}[\mathcal{A}] = \mathrm{I}_{\mathrm{face}}[\mathcal{A}]^W.$$

Compare (16.39) with (15.8). This fact can also seen as a formal consequence of Exercise 5.1. See the discussion in Section C.4.6 for more details.

We refer to $\mathrm{I}_{\mathrm{facetype}}[\mathcal{A}]$ as the *invariant face-incidence algebra.*

16.7.2. Invariant lune-incidence algebra. Recall the lune-incidence algebra $\mathrm{I}_{\mathrm{lune}}[\mathcal{A}]$. We denote its W-invariant subalgebra by $\mathrm{I}_{\mathrm{lune}}[\mathcal{A}]^W$. An alternative description is given below.

Recall from Section 5.2.4 the equivalence relation on nested face-types. Let $\mathrm{I}_{\mathrm{lunetype}}[\mathcal{A}]$ denote the subspace of $\mathrm{I}_{\mathrm{facetype}}[\mathcal{A}]$ consisting of those functions f on nested face-types such that $f(T,U) = f(T',U')$ whenever $(T,U) \sim (T',U')$.

Lemma 16.34. *The equivalence relation on nested face-types is order-compatible, or equivalently,* $\mathrm{I}_{\mathrm{lunetype}}[\mathcal{A}]$ *is a subalgebra of* $\mathrm{I}_{\mathrm{facetype}}[\mathcal{A}]$. *It has a basis indexed by lune-types.*

PROOF. This can be seen as a special case of Proposition C.16, by following the proof of Lemma 15.3. □

Thus, $\mathrm{I}_{\mathrm{lunetype}}[\mathcal{A}]$ is an example of a reduced incidence algebra of the poset of face-types. The following is a commutative diagram of algebras.

$$\begin{array}{ccc} \mathrm{I}_{\mathrm{face}}[\mathcal{A}] & \longleftarrow & \mathrm{I}_{\mathrm{face}}[\mathcal{A}]^W = \mathrm{I}_{\mathrm{facetype}}[\mathcal{A}] \\ \uparrow & & \uparrow \\ \mathrm{I}_{\mathrm{lune}}[\mathcal{A}] & \longleftarrow & \mathrm{I}_{\mathrm{lune}}[\mathcal{A}]^W = \mathrm{I}_{\mathrm{lunetype}}[\mathcal{A}] \end{array} \tag{16.40}$$

All maps are inclusions. Compare with (5.4).

We refer to $\mathrm{I}_{\mathrm{lunetype}}[\mathcal{A}]$ as the *invariant lune-incidence algebra.*

16.7.3. Invariant noncommutative zeta and Möbius functions. A noncommutative zeta function $\boldsymbol{\zeta}$ is invariant if $\boldsymbol{\zeta}(H,G) = \boldsymbol{\zeta}(wH,wG)$ for $w \in W$, and a noncommutative Möbius function $\boldsymbol{\mu}$ is invariant if $\boldsymbol{\mu}(H,G) = \boldsymbol{\mu}(wH,wG)$ for $w \in W$. In this situation, the expressions $\boldsymbol{\zeta}(T,U)$ and $\boldsymbol{\mu}(T,U)$ are meaningful.

Lemma 16.35. *For an invariant noncommutative zeta function* $\boldsymbol{\zeta}$, *lune-additivity* (15.23) *can be rewritten as*

$$\boldsymbol{\zeta}(T,U) = \sum_{\substack{V:\, V \geq Z, \\ \mathrm{s}(V/Z) = \mathrm{s}(U/Z)}} a_Z^{TUV}\, \boldsymbol{\zeta}(Z,V) \tag{16.41}$$

for all $Z \leq T \leq U$*. Similarly, for an invariant noncommutative Möbius function* $\boldsymbol{\mu}$*, the noncommutative Weisner formula* (15.27) *can be rewritten as*

$$\sum_{V:\,V \geq Z} a_Z^{TUV}\, \boldsymbol{\mu}(Z,V) = 0 \tag{16.42}$$

for all $Z < T \leq U$*.*

Here a_Z^{TUV} are the structure constants of the invariant Tits algebra studied in Section 16.5. The definition is given in (16.25).

PROOF. This can deduced using Lemma 16.19. □

Remark 16.36. The condition $\mathrm{s}(U/Z) = \mathrm{s}(V/Z)$ is local to $\boldsymbol{\alpha}_Z$. It implies the condition $\mathrm{s}(U) = \mathrm{s}(V)$, but is in general stronger. For instance, consider the rank-three braid arrangement. Recall that face-types are compositions of 4. Take $Z = (2,2)$, $U = (1,1,2)$ and $V = (2,1,1)$. Then $\mathrm{s}(U) = \mathrm{s}(V)$ but $\mathrm{s}(U/Z) \neq \mathrm{s}(V/Z)$. Despite this, it does not matter which of the two conditions we write in (16.41). This is because $a_Z^{TUV} = 0$ in the case that $\mathrm{s}(U) = \mathrm{s}(V)$ but $\mathrm{s}(U/Z) \neq \mathrm{s}(V/Z)$.

Theorem 16.37. *In the invariant lune-incidence algebra, the inverse of an invariant noncommutative zeta function is an invariant noncommutative Möbius function, and vice-versa.*

PROOF. This follows from Theorem 15.28. □

16.7.4. Invariant lune-incidence module. Recall the lune-incidence module $\mathrm{M}_{\mathrm{lune}}[\mathcal{A}]$. We denote its W-invariant part by $\mathrm{M}_{\mathrm{lune}}[\mathcal{A}]^W$. It is a left module over $\mathrm{I}_{\mathrm{lune}}[\mathcal{A}]^W$ with the action induced from (15.17). An alternative description of this module is given below.

Define the vector space $\mathrm{M}_{\mathrm{lunetype}}[\mathcal{A}]$ as follows. It consists of functions f on face-types such that $f(T) = f(U)$ whenever T and U have the same support. This is a left module over the invariant lune-incidence algebra with action given by

$$(fg)(T) = \sum_{U:\,T \leq U} f(T,U) g(U). \tag{16.43}$$

There is an isomorphism of modules

$$\mathrm{M}_{\mathrm{lunetype}}[\mathcal{A}] = \mathrm{M}_{\mathrm{lune}}[\mathcal{A}]^W.$$

We refer to $\mathrm{M}_{\mathrm{lunetype}}[\mathcal{A}]$ as the *invariant lune-incidence module.*

16.7.5. Noncommutative Möbius inversion. Let $\boldsymbol{\zeta}$ be an invariant noncommutative zeta function, and $\boldsymbol{\mu}$ be its inverse. Then for $f, g \in \mathrm{M}_{\mathrm{lunetype}}[\mathcal{A}]$,

$$g(T) = \sum_{U:\,T \leq U} \boldsymbol{\zeta}(T,U) f(U) \iff f(T) = \sum_{U:\,T \leq U} \boldsymbol{\mu}(T,U) g(U). \tag{16.44}$$

This is equivalent to $g = \boldsymbol{\zeta} f \iff f = \boldsymbol{\mu} g$. In this situation, we say that g is the *exponential* of f, and f is the *logarithm* of g.

16.7.6. Group-likes and primitives. Recall that d_T denotes the number of faces of type T. We say $g \in \mathrm{M}_{\mathrm{lunetype}}[\mathcal{A}]$ is *group-like* if $g(T) = \alpha\, d_T$ for some scalar α which is independent of T. Similarly, we say $f \in \mathrm{M}_{\mathrm{lunetype}}[\mathcal{A}]$ is *primitive* if $f(T) = 0$ for $T < S$.

Group-likes and primitives relate to each other via the exponential and logarithm. To show this, we need a couple of preliminary results.

Lemma 16.38. *For any invariant noncommutative zeta function $\boldsymbol{\zeta}$, and face-types $T \le U$,*

$$\boldsymbol{\zeta}(U,S) = d_{U/T}\, \boldsymbol{\zeta}(T,S) \qquad \text{or equivalently} \qquad d_T\, \boldsymbol{\zeta}(U,S) = d_U\, \boldsymbol{\zeta}(T,S).$$

PROOF. Applying (16.41) to $T \le U \le S$, we see that the sum in the rhs contains only one term. Thus, we get

$$\boldsymbol{\zeta}(U,S) = a_T^{USS}\, \boldsymbol{\zeta}(T,S).$$

But $a_T^{USS} = d_{U/T}$ by (16.33) applied to $\boldsymbol{\alpha}_T$. This proves the first identity. To get the other identity, use (16.14). □

Lemma 16.39. *For any invariant noncommutative Möbius function $\boldsymbol{\mu}$, and face-type $T < S$,*

$$\sum_{U:\, T \le U} \boldsymbol{\mu}(T,U) d_U = 0.$$

PROOF. Apply (16.42) to $T < S \le S$ to obtain

$$\sum_{U:\, T \le U} a_T^{SSU}\, \boldsymbol{\mu}(T,U) = 0.$$

But $a_T^{SSU} = d_{U/T}$ by (16.33) applied to $\boldsymbol{\alpha}_T$. Now multiply throughout by d_T and use (16.14). □

Theorem 16.40. *Let $\boldsymbol{\zeta}$ be an invariant noncommutative zeta function, and $\boldsymbol{\mu}$ be its inverse. Then under the correspondence* (16.44), *f is primitive iff g is group-like.*

PROOF. Suppose f is primitive. Then for any T,

$$g(T) = \sum_{U:\, T \le U} \boldsymbol{\zeta}(T,U) f(U) = \boldsymbol{\zeta}(T,S) f(S).$$

By Lemma 16.38, we conclude that g is group-like.

Conversely, suppose g is group-like. Then for $T < S$,

$$f(T) = \sum_{U:\, T \le U} \boldsymbol{\mu}(T,U) g(U) = \alpha \sum_{U:\, T \le U} \boldsymbol{\mu}(T,U) d_U = 0$$

by Lemma 16.39. Thus, f is primitive. □

16.8. Invariant Eulerian idempotents

Recall from Section 15.5 that complete systems of primitive orthogonal idempotents for the Tits algebra can be constructed and characterized in various ways. They involve the notions of homogeneous section, Eulerian family, $\mathtt{Q}$-basis and Zie family from Chapter 11. A similar characterization can be given for the invariant Tits algebra. The Takeuchi and Fulman elements belong to the invariant Tits algebra, and we revisit their diagonalization.

16.8.1. Invariant sections. Recall the notion of homogeneous section from Section 11.1. A homogeneous section $\mathtt{u}$ is *invariant* if it is fixed by the action of the Coxeter group W. That is, $\mathtt{u}^F = \mathtt{u}^G$ whenever F and G are faces of the same type. In this case, it makes sense to write $\mathtt{u}^T$ where T is a face-type. Further, we write $\mathtt{u}_T^U$ for $\mathtt{u}_H^G$, with $H \leq G$ and $\mathrm{t}(H,G) = (T,U)$. In particular, $\mathtt{u}_\emptyset^U = \mathtt{u}^U$. By (11.5),

$$\mathtt{u}_T^U = \sum_{\substack{F:\, HF=G,\\ \mathrm{s}(F)=\mathrm{s}(G)}} \mathtt{u}^{\mathrm{t}(F)}, \tag{16.45}$$

where $H \leq G$ are fixed with $\mathrm{t}(H,G) = (T,U)$.

An alternative way to think about invariant sections is given below. Recall the numbers $|\lambda|$ from Section 5.5.1.

Lemma 16.41. *An invariant section* $\mathtt{u}$ *is a family of scalars* $(\mathtt{u}^T)$ *such that for each flat-type* λ,

$$|\lambda| \sum_{T:\, \mathrm{s}(T)=\lambda} \mathtt{u}^T = 1. \tag{16.46}$$

In particular, $|W|\,\mathtt{u}^S = 1$ *(arising from the maximum flat-type).*

In order to state the next result in complete generality, we temporarily set aside Convention 16.2.

Lemma 16.42. *Assume that* $\mathcal{A}$ *has rank at least one. Let* $\Bbbk$ *be any field. An invariant section of* $\mathcal{A}$ *exists iff the characteristic of* $\Bbbk$ *does not divide* $|W|$.

PROOF. We employ Lemma 16.41. For the linear system (16.46) to have a solution, the characteristic of $\Bbbk$ must not divide $|W|$. This is also a sufficient condition since $|\lambda|$ divides $|W|$ by Exercise 5.16. □

Resuming Convention 16.2, we see that invariant sections exist. Moreover, the dimension of the affine space of all invariant sections is equal to the number of face-types minus the number of flat-types.

16.8.2. Invariant Eulerian families. Recall the notion of an Eulerian family from Section 11.2. An Eulerian family $\mathtt{E}$ is *invariant* if $w \cdot \mathtt{E}_\mathrm{X} = \mathtt{E}_{w\mathrm{X}}$ for $w \in W$.

Now suppose $\mathtt{E}$ is an invariant Eulerian family. For each flat-type λ, define

$$\mathtt{E}_\lambda := \sum_{\mathrm{X}:\, \mathrm{t}(\mathrm{X})=\lambda} \mathtt{E}_\mathrm{X}. \tag{16.47}$$

Note that the summands $\mathtt{E}_\mathrm{X}$ can be recovered from $\mathtt{E}_\lambda$. We deduce from (11.23) and (16.1) that

$$\mathrm{s}(\mathtt{E}_\lambda) = \mathtt{Q}_\lambda. \tag{16.48}$$

Similarly, the map (11.24) restricts to

$$\Pi[\mathcal{A}]^W \hookrightarrow \Sigma[\mathcal{A}]^W, \qquad \mathtt{Q}_\lambda \mapsto \mathtt{E}_\lambda. \tag{16.49}$$

This is an algebra section of the invariant support map. Equivalently, by Theorem D.32 (with A being the invariant Tits algebra and $\bar{A}$ being the invariant Birkhoff algebra), we obtain:

Theorem 16.43. *The elements* $\mathtt{E}_\lambda$*, as* λ *varies over all flat-types of* $\mathcal{A}$*, yield a complete system of primitive orthogonal idempotents of* $\Sigma[\mathcal{A}]^W$.

By Theorem D.31, an arbitrary algebra section is obtained by conjugating (16.49) by an element of $\mathtt{H}_O + \mathrm{rad}(\Sigma[\mathcal{A}])^W$. Now:

Lemma 16.44. *Conjugation of any invariant Eulerian family by an invertible element of the invariant Tits algebra produces another invariant Eulerian family.*

PROOF. Let z be an invertible element and $\mathtt{E}$ be an invariant Eulerian family. By Lemma 11.22, the conjugate $z \cdot \mathtt{E} \cdot z^{-1}$ is an Eulerian family. Further, for $w \in W$,

$$w(z \cdot \mathtt{E}_{\mathrm{X}} \cdot z^{-1}) = w(z) \cdot w(\mathtt{E}_{\mathrm{X}}) \cdot w(z^{-1}) = z \cdot \mathtt{E}_{w\mathrm{X}} \cdot z^{-1},$$

and $z \cdot \mathtt{E} \cdot z^{-1}$ is invariant. □

We conclude that every algebra section of the invariant support map arises from a unique invariant Eulerian family.

16.8.3. $\mathtt{Q}$-bases. Fix an invariant Eulerian family $\mathtt{E}$. Let $\mathtt{Q}$ be the associated basis of the Tits algebra (11.26). By invariance, $w \cdot \mathtt{Q}_F = \mathtt{Q}_{wF}$. For each face-type T, put

$$\mathtt{Q}_T := \sum_{F:\, \mathrm{t}(F)=T} \mathtt{Q}_F. \tag{16.50}$$

As T varies over all face-types, this defines the $\mathtt{Q}$-basis of the invariant Tits algebra. Observe from (11.28) and (16.47) that

$$\mathtt{E}_\lambda = \sum_{T:\, \mathrm{s}(T)=\lambda} \mathtt{u}^T \mathtt{Q}_T. \tag{16.51}$$

We now build on the discussion in Section 11.4.5. Let $(\mathtt{a}_T^U)$ denote the inverse of $(\mathtt{u}_T^U)$ in the poset of face-types. Equivalently, we invert $(\mathtt{u}_F^G)$ in the poset of faces, and put $\mathtt{a}_T^U = \mathtt{a}_F^G$, with $\mathrm{t}(F,G) = (T,U)$. This does not depend on the specific choice of F and G.

Lemma 16.45. *The* $\mathtt{H}$- *and* $\mathtt{Q}$-*bases of the invariant Tits algebra are related by*

$$\mathtt{H}_T = \sum_{U:\, T \le U} \mathtt{u}_T^U \, \mathtt{Q}_U \quad \textit{and} \quad \mathtt{Q}_T = \sum_{U:\, T \le U} \mathtt{a}_T^U \, \mathtt{H}_U. \tag{16.52}$$

In particular,

$$\mathtt{H}_\emptyset = \sum_T \mathtt{u}^T \, \mathtt{Q}_T. \tag{16.53}$$

PROOF. This can be deduced from (11.33). □

As a companion to (16.7), on the $\mathtt{Q}$-basis

$$\mathrm{s}(\mathtt{Q}_T) = |\mathrm{s}(T)| \, \mathtt{Q}_{\mathrm{s}(T)}. \tag{16.54}$$

This formula is consistent with (16.46), (16.48) and (16.51).

Exercise 16.46. Establish the following identities.

$$\mathtt{H}_T \cdot \mathtt{E}_{\mathrm{s}(T)} = \mathtt{Q}_T \cdot \mathtt{E}_{\mathrm{s}(T)} = \mathtt{Q}_T, \qquad \mathtt{E}_{\mathrm{s}(T)} \cdot \mathtt{Q}_T = |\mathrm{s}(T)| \, \mathtt{E}_{\mathrm{s}(T)},$$

$$\mathtt{H}_T \cdot \mathtt{Q}_U = \mathtt{Q}_T \cdot \mathtt{Q}_U = |\lambda| \, \mathtt{Q}_T \text{ if } \mathrm{s}(T) = \mathrm{s}(U) = \lambda.$$

(Use methods similar to the ones in Exercise 11.34.)

Exercise 16.47. Use (11.29) and Lemma 16.19 to show that: For face-types T and T',

$$\mathtt{H}_T \cdot \mathtt{Q}_{T'} = \sum_{U:\, U \geq T,\, \mathrm{s}(U)=\mathrm{s}(T')} a^{TUT'} \mathtt{Q}_U, \tag{16.55}$$

with $a^{TUT'}$ as in (16.23). What happens when $T = T'$? Compare and contrast (16.55) with (16.23).

Recall from Section 11.5.5 that Q-bases of the Tits algebra can also be defined directly without any reference to Eulerian families. One can take a similar approach to Q-bases of the invariant Tits algebra. It works as follows. We say that a Q-basis of the Tits algebra is *invariant* if $w \cdot \mathtt{Q}_F = \mathtt{Q}_{wF}$. Further, we say that a basis of the invariant Tits algebra is a Q-basis if it is obtained from an invariant Q-basis of the Tits algebra via (16.50).

16.8.4. Invariant Zie families. Recall the notion of a Zie family from Section 11.5. A Zie family P is *invariant* if $w \cdot \mathsf{P}_{\mathrm{X}} = \mathsf{P}_{w\mathrm{X}}$ for $w \in W$. In this case, $\mathsf{P}_\perp$ is an invariant Zie element.

16.8.5. Characterizations of invariant Eulerian families.

Theorem 16.48. *For a reflection arrangement $\mathcal{A}$, the following pieces of data are equivalent.*

1. *An invariant noncommutative zeta function $\boldsymbol{\zeta}$ of $\mathcal{A}$.*
2. *An invariant section* u *of $\mathcal{A}$.*
3. *An invariant Eulerian family* E *of $\mathcal{A}$.*
4. *A complete system of primitive orthogonal idempotents of $\Sigma[\mathcal{A}]^W$.*
5. *An algebra section $\Pi[\mathcal{A}]^W \to \Sigma[\mathcal{A}]^W$ of the invariant support map.*
6. *A* Q*-basis of the invariant Tits algebra of $\mathcal{A}$.*
7. *An invariant special Zie family* P *of $\mathcal{A}$.*
8. *An invariant noncommutative Möbius function $\boldsymbol{\mu}$ of $\mathcal{A}$.*

This is the analogue of Theorem 15.44. It largely follows from the fact that in Theorem 15.44, the steps involved in the passage from one piece of data to another respect the action of W. The equivalence between invariant Eulerian families, complete systems and algebra sections is explained in Section 16.8.2.

Formulas (16.52) relating the H- and Q-bases can be rewritten as

$$\mathtt{H}_T = \sum_{U:\, U \geq T} \boldsymbol{\zeta}(T,U)\, \mathtt{Q}_U \quad \text{and} \quad \mathtt{Q}_T = \sum_{U:\, U \geq T} \boldsymbol{\mu}(T,U)\, \mathtt{H}_U. \tag{16.56}$$

Compare with (15.45).

16.8.6. Good reflection arrangements. Recall good reflection arrangements from Section 5.7. The uniform section of a good reflection arrangement is invariant. We revisit Section 11.6 from this perspective.

Lemma 16.49. *For the uniform section of a good reflection arrangement $\mathcal{A}$, the* H- *and* Q*-bases of the invariant Tits algebra are related by*

$$\mathtt{H}_T = \sum_{U:\, T \leq U} \frac{1}{c_T^U} \mathtt{Q}_U \quad \textit{and} \quad \mathtt{Q}_T = \sum_{U:\, T \leq U} \frac{\mu(\mathcal{A}_T^U)}{c_T^U} \mathtt{H}_U. \tag{16.57}$$

PROOF. This can be deduced from (11.50). □

Theorem 16.50. *The invariant Eulerian idempotents for the uniform section of a good reflection arrangement $\mathcal{A}$ are given by*

$$\mathtt{E}_\lambda = \sum_{\mathrm{s}(T)=\lambda} \frac{1}{c^T}\mathtt{Q}_T = \frac{1}{c^\lambda} \sum_{\mathrm{s}(T)=\lambda} \mathtt{Q}_T = \frac{1}{c^\lambda} \sum_{\mathrm{s}(T)=\lambda} \sum_{U:\, T\leq U} \frac{\mu(\mathcal{A}_T^U)}{c_T^U}\, \mathtt{H}_U, \tag{16.58}$$

where $c^\lambda := c^T$ for any face-type T of support λ.

PROOF. This follows from (16.51) and the second formula in (16.57). ☐

Now, for $k \geq 0$, observe that

$$\mathtt{E}_k = \sum_{\lambda:\, \mathrm{rk}(\lambda)=k} \mathtt{E}_\lambda = \sum_{T:\, |T|=k} \frac{1}{c^T}\, \mathtt{Q}_T, \tag{16.59}$$

with $\mathtt{E}_k$ as in (11.54).

16.8.7. Takeuchi and Fulman elements. The Takeuchi element defined in (12.15) belongs to the invariant Tits algebra, and can be rewritten as

$$\mathtt{Tak} = \sum_T (-1)^{|T|}\, \mathtt{H}_T.$$

From (12.23), we obtain

$$\mathtt{Tak} = \sum_\lambda (-1)^{\mathrm{rk}(\lambda)}\, \mathtt{E}_\lambda,$$

for any Eulerian family $\mathtt{E}$ arising from an invariant section which is also projective.

The Fulman element of parameter t defined in (12.37) belongs to the invariant Tits algebra, and can be rewritten as

$$\mathtt{Ful}_t = \sum_T \frac{\chi(\mathcal{A}^T, t)}{c^T}\, \mathtt{H}_T.$$

Theorem 12.71 may be restated as follows.

Theorem 16.51. *For a good reflection arrangement,*

$$\mathtt{Ful}_t = \sum_\lambda t^{\mathrm{rk}(\lambda)}\, \mathtt{E}_\lambda = \sum_{k\geq 0} t^k\, \mathtt{E}_k, \tag{16.60}$$

with $\mathtt{E}_\lambda$ as in (16.58) *and $\mathtt{E}_k$ as in* (16.59).

16.9. Peirce decompositions

We briefly discuss the left, right and two-sided Peirce decompositions of the invariant Tits algebra, their connection with invariant Zie, chamber and Lie elements, and some consequences for its quiver.

In this section, $\mathtt{E}$ is an arbitrary but fixed invariant Eulerian family of $\mathcal{A}$. In view of Theorem 16.48, this is the same as fixing an invariant section, a $\mathtt{Q}$-basis, and so on.

16.9.1. Left Peirce decomposition of chambers. Recall that there is a left action of the (invariant) Tits algebra on $\Gamma[\mathcal{A}]$. By Theorem 16.43,

$$\Gamma[\mathcal{A}] = \bigoplus_{\lambda} \mathrm{E}_\lambda \cdot \Gamma[\mathcal{A}]. \tag{16.61}$$

Further, by (16.47),

$$\dim \mathrm{E}_\lambda \cdot \Gamma[\mathcal{A}] = \sum_{\mathrm{X}:\, \mathrm{t}(\mathrm{X})=\lambda} \dim \mathrm{E}_\mathrm{X} \cdot \Gamma[\mathcal{A}] = \sum_{\mathrm{X}:\, \mathrm{t}(\mathrm{X})=\lambda} |\mu(\mathcal{A}_\mathrm{X})|.$$

The second equality used (13.10) and (10.24). Equivalently, by Theorem 5.18:

Theorem 16.52. *The dimension of* $\mathrm{E}_\lambda \cdot \Gamma[\mathcal{A}]$ *is the number of elements of* W *whose cycle-type is* λ.

16.9.2. Right Peirce decomposition of Zie. Recall that there is a right action of the (invariant) Tits algebra on $\mathsf{Zie}[\mathcal{A}]$. By Theorem 16.43,

$$\mathsf{Zie}[\mathcal{A}] = \bigoplus_{\lambda} \mathsf{Zie}[\mathcal{A}] \cdot \mathrm{E}_\lambda. \tag{16.62}$$

Further, by (16.47),

$$\dim \mathsf{Zie}[\mathcal{A}] \cdot \mathrm{E}_\lambda = \sum_{\mathrm{X}:\, \mathrm{t}(\mathrm{X})=\lambda} \dim \mathsf{Zie}[\mathcal{A}] \cdot \mathrm{E}_\mathrm{X} = \sum_{\mathrm{X}:\, \mathrm{t}(\mathrm{X})=\lambda} |\mu(\mathcal{A}^\mathrm{X})|.$$

The second equality used (13.19) and (10.24).

16.9.3. Peirce decompositions of the invariant Tits algebra. We have

$$\Sigma[\mathcal{A}]^W = \bigoplus_{\lambda} \mathrm{E}_\lambda \cdot \Sigma[\mathcal{A}]^W \quad \text{and} \quad \Sigma[\mathcal{A}]^W = \bigoplus_{\mu} \Sigma[\mathcal{A}]^W \cdot \mathrm{E}_\mu.$$

These are the left and right Peirce decompositions, respectively. The connection of these Peirce decompositions with those of the Tits algebra is as follows.

$$\mathrm{E}_\lambda \cdot \Sigma[\mathcal{A}]^W = \Big(\bigoplus_{\mathrm{X}:\, \mathrm{t}(\mathrm{X})=\lambda} \mathrm{E}_\mathrm{X} \cdot \Sigma[\mathcal{A}] \Big)^W \tag{16.63}$$

and

$$\Sigma[\mathcal{A}]^W \cdot \mathrm{E}_\mu = \Big(\bigoplus_{\mathrm{Y}:\, \mathrm{t}(\mathrm{Y})=\mu} \Sigma[\mathcal{A}] \cdot \mathrm{E}_\mathrm{Y} \Big)^W. \tag{16.64}$$

Similarly, we have the two-sided Peirce decomposition

$$\Sigma[\mathcal{A}]^W = \bigoplus_{\lambda \leq \mu} \mathrm{E}_\lambda \cdot \Sigma[\mathcal{A}]^W \cdot \mathrm{E}_\mu. \tag{16.65}$$

The sum is over both λ and μ. It is the invariant analogue of (13.20). Again the point to note is that

$$\mathrm{E}_\lambda \cdot \Sigma[\mathcal{A}]^W \cdot \mathrm{E}_\mu = 0 \quad \text{for} \quad \lambda \not\leq \mu.$$

Further,

$$\mathrm{E}_\lambda \cdot \Sigma[\mathcal{A}]^W \cdot \mathrm{E}_\mu = \Big(\bigoplus_{\substack{\mathrm{X} \leq \mathrm{Y} \\ \mathrm{t}(\mathrm{X})=\lambda,\, \mathrm{t}(\mathrm{Y})=\mu}} \mathrm{E}_\mathrm{X} \cdot \Sigma[\mathcal{A}] \cdot \mathrm{E}_\mathrm{Y} \Big)^W. \tag{16.66}$$

In conjunction with (13.36) and Proposition 16.4, we deduce that

$$\operatorname{rad}(\Sigma[\mathcal{A}]^W) = \bigoplus_{\lambda<\mu} \mathsf{E}_\lambda \cdot \Sigma[\mathcal{A}]^W \cdot \mathsf{E}_\mu. \tag{16.67}$$

16.9.4. Invariant Lie elements. Since E is invariant, the isomorphism in Theorem 13.54 respects the action of W. Hence, by taking invariants, we obtain

$$\Sigma[\mathcal{A}]^W \cong \Big(\bigoplus_{X \leq Y} \mathsf{Lie}[\mathcal{A}_X^Y] \Big)^W. \tag{16.68}$$

The sum is over both X and Y.

The components of the two-sided Peirce decomposition of the invariant Tits algebra given in (16.65) can also be understood in terms of Lie elements. We have

$$\mathsf{E}_\lambda \cdot \Sigma[\mathcal{A}]^W \cdot \mathsf{E}_\mu \cong \Big(\bigoplus_{\substack{X \leq Y \\ t(X)=\lambda,\, t(Y)=\mu}} \mathsf{Lie}[\mathcal{A}_X^Y] \Big)^W. \tag{16.69}$$

This follows by combining (13.23) and (16.66).

In a similar manner, by using Proposition 13.31 and Lemma 13.40, we obtain

$$\Sigma[\mathcal{A}]^W \cong \Big(\bigoplus_{X} \mathsf{Zie}[\mathcal{A}_X] \Big)^W \quad \text{and} \quad \Sigma[\mathcal{A}]^W \cong \Big(\bigoplus_{Y} \Gamma[\mathcal{A}^Y] \Big)^W.$$

Further, the components of the left and right Peirce decompositions are

$$\mathsf{E}_\lambda \cdot \Sigma[\mathcal{A}]^W \cong \Big(\bigoplus_{X:\, t(X)=\lambda} \mathsf{Zie}[\mathcal{A}_X] \Big)^W \quad \text{and} \quad \Sigma[\mathcal{A}]^W \cdot \mathsf{E}_\mu \cong \Big(\bigoplus_{Y:\, t(Y)=\mu} \Gamma[\mathcal{A}^Y] \Big)^W.$$

We made use of (16.63) and (16.64). The second isomorphism yields the following.

Theorem 16.53. *For any flat-type μ, the dimension of $\Sigma[\mathcal{A}]^W \cdot \mathsf{E}_\mu$ equals the number of face-types whose support is μ.*

More information is given in the exercise below. It is the invariant analogue of Exercise 11.35.

Exercise 16.54. Fix a flat-type λ. Put $\mathsf{h}_\lambda := \Sigma^W \cdot \mathsf{E}_\lambda$. Show that:

- h_λ has a linear basis consisting of the elements Q_T, as T varies over face-types with support λ.
- The radical of h_λ is linearly spanned by elements of the form $\mathsf{Q}_T - \mathsf{Q}_U$, where T and U both have support λ.
- The quotient of h_λ by its radical is one-dimensional with multiplicative character χ_λ. (The latter is defined in (16.8).)

16.9.5. Quiver. Recall the quiver of the Tits algebra from Theorem 13.73. It was computed using the connection of the Tits algebra with Lie elements. In a similar manner, one can try to compute the quiver of the invariant Tits algebra by employing (16.68). However, analyzing this is nontrivial and we only give the following partial result.

Proposition 16.55. *Let Q denote the quiver of the invariant Tits algebra. The vertices of Q are flat-types. If there is an arrow from μ to λ, then $\lambda < \mu$. In particular, Q is acyclic. Further, the maximum flat-type is an isolated vertex, that is, there is no arrow to or from the maximum flat-type.*

PROOF. The split-semisimple quotient of the invariant Tits algebra is the invariant Birkhoff algebra. Hence the vertices of its quiver are flat-types. The second claim follows from (16.67). The last claim follows similarly from (16.37) and (16.69). □

The Cartan invariants of the invariant Tits algebra are the dimensions of the spaces (16.69). It would be nice to have formulas for them. Theorem 16.53 is a step in that direction.

16.10. Bilinear forms

We discuss some bilinear forms on face-types and related objects. The involved constants are closely related to the structure constants of the invariant Tits algebra. In this discussion, a pointed arrangement $(\mathcal{A}, C)$ is fixed.

16.10.1. Bilinear form on nested face-types. Let (T, U) and (T', U') be nested face-types. Let $a^{TUT'U'}$ denote the number of nested faces (F', G') of type (T', U') such that

$$FF' = G \quad \text{and} \quad F'F = G'$$

for a fixed nested face (F, G) of type (T, U). There is an inbuilt symmetry, namely,

$$a^{TUT'U'} = a^{T'U'TU}.$$

The latter is the number of nested faces (F, G) of type (T, U) such that the above equations hold for a fixed nested face (F', G') of type (T', U'). Thus,

$$\langle (T, U), (T', U') \rangle := a^{TUT'U'}$$

defines a symmetric bilinear form on the linear space with basis of nested face-types.

More generally, for any scalar q, with notation as above, define

$$a^{TUT'U'}(q) := \sum q^{\operatorname{dist}(G,G')}. \tag{16.70}$$

The sum is either over (F, G) or over (F', G') depending on the point of view. This can also be expressed as

$$a^{TUT'U'}(q) = \sum q^{l(w)}. \tag{16.71}$$

The sum is over those $w \in W$ such that

$$F_0'(wF_0) = G_0', \ (wF_0)F_0' = wG_0 \text{ and } (wG_0)C = wC,$$

where F_0, G_0, F_0', G_0' are faces of C of types T, U, T', U'.

This defines a q-analogue of the above bilinear form. It continues to be symmetric. The deformed structure constants (16.27) of the invariant Tits algebra are contained in this bilinear form:

$$a^{TUT'T'}(q) = \begin{cases} a^{TUT'}(q) & \text{if } \mathrm{s}(T') = \mathrm{s}(U), \\ 0 & \text{otherwise,} \end{cases} \tag{16.72}$$

and

$$\sum_{U'} a^{TUT'U'}(q) = a^{TUT'}(q) \text{ or equivalently } \sum_{U} a^{TUT'U'}(q) = a^{T'U'T}(q). \tag{16.73}$$

Setting $U = S$ in the first sum, and noting that $a^{TST'U'}(q)$ is 0 if $U' \neq S$, we obtain

$$a^{TST'S}(q) = a^{TST'}(q).$$

The symmetry of the bilinear form gives another proof of $a^{TST'}(q) = a^{T'ST}(q)$.

16.10.2. Bilinear form on face-types. For any face-types T and T', define

$$a^{TT'}(q) := \sum_{U,U'} a^{TUT'U'}(q).$$

This can also be expressed as

$$a^{TT'}(q) = \sum q^{l(w)}.$$

The sum is over those $w \in W$ such that $(wF_0)C = wC$ and $F_0'(wC) = C$, where F_0 and F_0' are faces of C of types T and T'.

We have

$$a^{TT'}(q) = a^{T'T}(q).$$

Hence

$$\langle T, T' \rangle := a^{TT'}(q)$$

defines a symmetric bilinear form on the linear space with basis of face-types.

Also note that

$$a^{ST}(q) = a^{SST}(q) = d_T(q).$$

16.10.3. Bilinear form on double-nested face-types. A double-nested face is a triple (F, G, H) of faces with $F \leq G \leq H$. Similarly, a double-nested face-type is a triple (T, U, V) of face-types with $T \leq U \leq V$.

Let (T, U, V) and (T', U', V') be double-nested face-types. Let $a^{TUVT'U'V'}$ denote the number of triples (F', G', H') of type (T', U', V') such that

$$FF' = G, \ F'F = G', \ FH' = H, \ F'H = H',$$

where (F, G, H) is a fixed double-nested face of type (T, U, V). This defines a bilinear form on double-nested face-types. Note that $\operatorname{dist}(F, F') = \operatorname{dist}(G, G') = \operatorname{dist}(H, H')$.

More generally, for any scalar q, with notation as above, define

$$a^{TUVT'U'V'}(q) := \sum_{(F',G',H')} q^{\operatorname{dist}(G,G')}.$$

This defines a q-analogue of the above bilinear form.

For $T \leq U \leq V$ and $T' \leq V'$, set

$$a^{TUVT'V'}(q) := \sum_{U'} a^{TUVT'U'V'}(q).$$

Lemma 16.56. *For face-types $T \leq Y$ and $T' \leq V'$,*

$$a^{T'V'Y}(q) = \sum_{U \leq V} a_T^{UVY}(q)\, a^{TUVT'V'}(q). \tag{16.74}$$

The sum is over both U and V.

PROOF. Let us first consider the case $q = 1$. This is similar to the proof of (16.26). The relevant picture is given below.

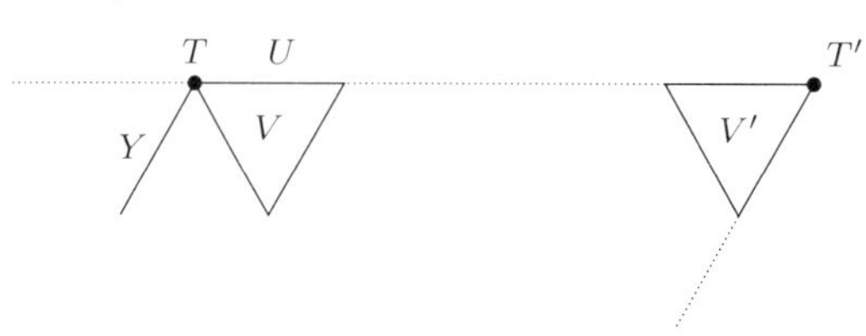

The lhs counts the number of faces of type Y in the interior of the lune marked by dotted lines. It can be expressed as a sum indexed by U and V, where (U, V) is the type of the nested face obtained by projecting the nested face of type (T', V') defining the lune on subfaces of type T. For each such U and V, we need to multiply two independent counts, which can be seen to be a_T^{UVY} and $a^{TUVT'V'}(q)$.

For the case of general q, the additional ingredient is to use (8.9e) for $v = v_q$. □

16.11. Garsia-Reutenauer idempotents (Type A)

Let $\mathcal{A}$ be the braid arrangement on $[p]$ (Sections 6.3–6.6). We discuss the Garsia-Reutenauer idempotents and the more general Krob-Leclerc-Thibon idempotents. The former are the Eulerian idempotents of the invariant Tits algebra associated to the uniform section. We discuss the Bayer-Diaconis-Garsia-Loday formula for them. We also write down the Bergeron formula for the symmetrized Dynkin element viewed as an element of the invariant Tits algebra.

16.11.1. Invariant Birkhoff algebra. Recall that flat-types correspond to integer partitions of p. Thus they index the basis elements of the invariant Birkhoff algebra. Let $\lambda = (l_1, \dots, l_h)$ and $\mu = (m_1, \dots, m_k)$ be two partitions. Let $R_{\lambda\mu}$ denote the change of basis coefficient between the $\mathtt{H}$- and $\mathtt{Q}$-bases as in (16.2). Explicitly, $R_{\lambda\mu}$ is the number of compositions $(S_1, \dots, S_k)$ of the set $[h]$ such that

$$m_j = \sum_{i \in S_j} l_i, \qquad 1 \leq j \leq k. \tag{16.75}$$

16.11.2. Invariant Tits algebra. Recall that face-types correspond to integer compositions of p. Thus they index the basis elements of the invariant Tits algebra. Specializing (16.57):

Lemma 16.57. *For the uniform section, the* $\mathtt{H}$- *and* $\mathtt{Q}$-*bases are related by*

$$\mathtt{H}_\alpha = \sum_{\beta:\, \alpha \leq \beta} \frac{1}{\deg!(\beta/\alpha)} \mathtt{Q}_\beta \quad \text{and} \quad \mathtt{Q}_\alpha = \sum_{\beta:\, \alpha \leq \beta} \frac{(-1)^{\operatorname{rk}(\beta/\alpha)}}{\deg(\beta/\alpha)} \mathtt{H}_\beta, \tag{16.76}$$

with degrees and factorials as in Section 6.6.3.

In particular,

$$\mathtt{Q}_{(p)} = \sum_{\beta} \frac{(-1)^{\operatorname{rk}(\beta)}}{\deg(\beta)} \mathtt{H}_\beta. \tag{16.77}$$

The sum is over all compositions β of p.

Compare (16.76) with (12.41). For $p = 2$, the change of basis formulas are

$$\mathtt{H}_{(1,1)} = \mathtt{Q}_{(1,1)}, \quad \mathtt{H}_{(2)} = \mathtt{Q}_{(2)} + \frac{1}{2}\mathtt{Q}_{(1,1)}, \quad \mathtt{Q}_{(2)} = \mathtt{H}_{(2)} - \frac{1}{2}\mathtt{H}_{(1,1)}.$$

Lemma 16.19 along with the explicit description of the Tits product (6.1) yields the following description of the structure constants of the invariant Tits algebra.

Lemma 16.58. *Suppose α and β are compositions of p with k parts and l parts, respectively, and γ is a composition of p which refines α. Then $a^{\alpha\gamma\beta}$ is the number of ways to fill a $k \times l$ matrix with nonnegative integers such that the row-sums give α and the column-sums give β, and reading the matrix left to right and then top to bottom and deleting the zero entries yields γ.*

16.11.3. Garsia-Reutenauer idempotents. For each partition λ of p, define

$$\mathtt{E}_\lambda := \frac{1}{k!} \sum_{\alpha:\, \mathrm{s}(\alpha)=\lambda} \mathtt{Q}_\alpha = \sum_{\mathrm{X}:\, \mathrm{t}(\mathrm{X})=\lambda} \mathtt{E}_\mathrm{X}, \tag{16.78}$$

where k is the number of parts of λ, and $\mathtt{E}_\mathrm{X}$ is as in (12.42). We call these the *Garsia-Reutenauer idempotents.*

Theorem 16.59. *The elements $\mathtt{E}_\lambda$, as λ varies over all partitions of p, yield a complete system of primitive orthogonal idempotents of the invariant Tits algebra.*

PROOF. This is a special case of Theorem 16.43. □

Recall the idempotents $\mathtt{E}_k$ from (12.44). For $1 \le k \le p$,

$$\mathtt{E}_k = \sum_{\lambda:\, \deg(\lambda)=k} \mathtt{E}_\lambda = \frac{1}{k!} \sum_{\alpha:\, \deg(\alpha)=k} \mathtt{Q}_\alpha, \tag{16.79}$$

In particular, $\mathtt{E}_1 = \mathtt{Q}_{(p)}$. The $\mathtt{E}_k$ form a system of orthogonal idempotents of the invariant Tits algebra, but it is not complete when $p > 1$.

16.11.4. Bayer-Diaconis-Garsia-Loday formula. For α a composition of p, let

$$\mathtt{H}_\alpha := \sum_{F:\, \mathrm{t}(F)=\alpha} \mathtt{H}_F.$$

The sum is over all set compositions whose underlying composition is α. Set

$$\mathtt{K}_\beta := \sum_{\alpha:\, \alpha\le\beta} (-1)^{\deg(\beta)-\deg(\alpha)} \mathtt{H}_\alpha \qquad \text{or equivalently} \qquad \mathtt{H}_\alpha = \sum_{\beta:\, \beta\le\alpha} \mathtt{K}_\beta. \tag{16.80}$$

Put

$$\mathtt{T}_k = \sum_{\alpha:\, \deg(\alpha)=k} \mathtt{H}_\alpha \qquad \text{and} \qquad \mathtt{U}_k = \sum_{\beta:\, \deg(\beta)=k} \mathtt{K}_\beta.$$

(This agrees with the $\mathtt{T}_k$ defined in (12.45).) The Adams element defined in (12.48) is then given by

$$\mathtt{Ads}_n = \sum_{k=1}^{p} \binom{n}{k} \mathtt{T}_k = \sum_{r=1}^{p} \binom{n+p-r}{p} \mathtt{U}_r. \tag{16.81}$$

This boils down to the combinatorial identity: For any composition β of p,

$$\sum_{\alpha:\, \alpha\ge\beta} \binom{n}{\deg(\alpha)} = \binom{n+p-\deg(\beta)}{p}.$$

Lemma 16.60. *We have*

$$\mathtt{E}_k = \frac{1}{p!} \sum_{r=1}^{p} e_{p-k}(1-r, 2-r, \ldots, p-r)\, \mathtt{U}_r, \tag{16.82}$$

where e_k is the k-th elementary symmetric function.

PROOF. We know from (12.49) that $\mathtt{E}_k$ is the coefficient of n^k in $\mathtt{Ads}_n$. Write the binomials in the last expression for this element in (16.81) as polynomials in n:

$$\begin{aligned}\binom{n+p-r}{p} &= \frac{1}{p!}(n+(p-r))\dots(n+(1-r)) \\ &= \frac{1}{p!}\sum_{i=0}^{p-1} n^{p-i} e_i(1-r, 2-r, \dots, p-r).\end{aligned}$$

The sum goes only till $p-1$ because $(1-r)\dots(p-r)=0$. Now put $k=p-i$ and extract the coefficient of n^k. □

We call (16.82) the *Bayer-Diaconis-Garsia-Loday formula*.

Exercise 16.61. Use (12.52) and (16.5) to express $\mathtt{E}_\perp$ in the $\mathtt{K}$-basis of the invariant Tits algebra. Check that it agrees with the case $k=1$ of (16.82).

16.11.5. Bergeron formula. Recall the symmetrized Dynkin element d_p given in (14.47) and its opposite $\overline{d_p}$. As elements of the invariant Tits algebra, in the $\mathtt{H}$-basis, they are given by

$$d_p = \sum_\alpha (-1)^{\mathrm{rk}(\alpha)} w^\alpha \mathtt{H}_\alpha \quad \text{and} \quad \overline{d_p} = \sum_\alpha (-1)^{\mathrm{rk}(\alpha)} w^{\overline{\alpha}} \mathtt{H}_\alpha, \tag{16.83}$$

where w^α is the last part of α, while $w^{\overline{\alpha}}$ is the first part of α. In the $\mathtt{K}$-basis, they are given by

$$d_p = \sum_{k=0}^{p-1} (-1)^k \mathtt{K}_{(1,\dots,1,p-k)} \quad \text{and} \quad \overline{d_p} = \sum_{k=0}^{p-1} (-1)^k \mathtt{K}_{(p-k,1,\dots,1)}. \tag{16.84}$$

We call this the *Bergeron formula*. In the first formula, the last part of the composition is $p-k$ and the remaining k parts are all 1. Further, as an element of the symmetric group algebra in the $\mathtt{K}$-basis, d_p is a sum over all peakless permutations with signs in front. (A permutation is *peakless* if it is a sequence of descents followed by a sequence of ascents.) These claims directly follow from (14.54), (16.5) and (16.13). More generally, from (14.57),

$$d_{p,q} = \sum_{k=0}^{p-1} (-q)^k\, \mathtt{K}_{(1,\dots,1,p-k)}. \tag{16.85}$$

As an element of the group algebra of the symmetric group, $d_{p,-1}$ is a sum over all peakless permutations. (The signs go away.)

16.11.6. Krob-Leclerc-Thibon idempotents. We now discuss the invariant analogue of the construction given in Section 12.5.5. The starting data is as follows.

- Suppose for each $1 \le j \le p$, we are given an arbitrary invariant special Zie element $\mathtt{Q}_{(j)}$ of the braid arrangement on $[j]$.

For each composition $\alpha = (a_1, \dots, a_k)$ of p, define

$$\mathtt{Q}_\alpha := \mu_\alpha(\mathtt{Q}_{(a_1)}, \dots, \mathtt{Q}_{(a_k)}). \tag{16.86}$$

The rhs is the external product on integer compositions from Section 6.3.13 (defined on the $\mathtt{H}$-basis by ordered concatenation and extended by multilinearity). As α

varies over all compositions of p, the $\mathtt{Q}_\alpha$ yield a basis of the invariant Tits algebra. So we can write

$$\mathtt{H}_{(p)} = \sum_\alpha \mathtt{u}^\alpha \mathtt{Q}_\alpha,$$

for unique scalars $\mathtt{u}^\alpha$. Now define

$$\mathtt{E}_\lambda := \sum_{\alpha:\, \mathrm{s}(\alpha)=\lambda} \mathtt{u}^\alpha \mathtt{Q}_\alpha. \tag{16.87}$$

These are the required idempotents. We call them the *Krob-Leclerc-Thibon idempotents*. Choosing each $\mathtt{Q}_{(j)}$ to be as in (16.77) recovers the Garsia-Reutenauer idempotents (16.78) (with the $\mathtt{u}^\alpha$ coinciding with the uniform section).

Exercise 16.62. Apply instead the construction in Section 12.5.5 to $\mathtt{Q}_{(S)}$ obtained from the Zie elements $\mathtt{Q}_{(j)}$. Verify that the resulting section, Eulerian family and $\mathtt{Q}$-basis are invariant. Employ the discussion in Section 16.8 to match these objects with the ones obtained through the above construction.

Exercise 16.63. Put $\mathtt{Q}_{(j)} := d_j/j$, with d_j being the symmetrized Dynkin element given in (14.47). It is an invariant special Zie element of the braid arrangement on $[j]$. Use (16.83) to express (16.86) in the $\mathtt{H}$-basis. Use the discussion in Section 14.9.9 to deduce that the resulting invariant section $\mathtt{u}$ is given by (14.49). Write down formulas for the idempotents (16.87) in the $\mathtt{H}$-basis. Write down formulas for $\boldsymbol{\zeta}(\alpha,\beta)$ and $\boldsymbol{\mu}(\alpha,\beta)$, where $\boldsymbol{\zeta}$ and $\boldsymbol{\mu}$ are the corresponding invariant noncommutative zeta and Möbius function. (See Exercise 15.38.)

16.11.7. Two-sided Peirce decomposition. Specializing (16.65), we have

$$\Sigma[\mathcal{A}]^{\mathrm{S}_p} = \bigoplus_{\lambda \le \mu} \mathtt{E}_\lambda \cdot \Sigma[\mathcal{A}]^{\mathrm{S}_p} \cdot \mathtt{E}_\mu. \tag{16.88}$$

The sum is over both λ and μ. The partial order on partitions is as in Section 6.3.9. Here the $\mathtt{E}_\lambda$ could either be the Garsia-Reutenauer idempotents or the more general Krob-Leclerc-Thibon idempotents.

16.11.8. Invariant Lie elements. Specializing (16.36), we obtain:

$$\mathsf{Lie}[p]^{\mathrm{S}_p} = \begin{cases} \Bbbk & \text{if } p = 1, \\ 0 & \text{otherwise.} \end{cases} \tag{16.89}$$

Thus, there are no invariant Lie elements for $p \geq 2$.

16.11.9. Poincaré polynomial. For the braid arrangement on $[p]$, the exponents of the Coxeter system are $e_i = i$ for $1 \le i \le p-1$. Hence, by (16.21), the Poincaré polynomial is

$$d_S(q) = (p)_q!, \tag{16.90}$$

which is the q-factorial. One can then deduce from (16.20) that for the composition $T = (i, p-i)$,

$$d_T(q) = \binom{p}{i}_q,$$

which is the q-binomial coefficient. In general, we get a q-multinomial coefficient.

16.12. Bergeron idempotents (Type B)

We let $\mathcal{A}$ be the arrangement of type B on $[p]$ (Section 6.7). We discuss the Bergeron idempotents. These are the Eulerian idempotents of the invariant Tits algebra associated to the uniform section.

16.12.1. Invariant Tits algebra. Specializing (16.57):

Lemma 16.64. *For the uniform section, the* $\mathtt{H}$- *and* $\mathtt{Q}$-*bases of the invariant Tits algebra are related by*

$$\mathtt{H}_\alpha = \sum_{\beta:\, \alpha\leq\beta} \frac{1}{\mathbf{deg}!(\beta/\alpha)} \mathtt{Q}_\beta \quad \textit{and} \quad \mathtt{Q}_\alpha = \sum_{\beta:\, \alpha\leq\beta} \frac{(-1)^{\mathrm{rk}(\beta/\alpha)}}{\mathbf{deg}(\beta/\alpha)} \mathtt{H}_\beta, \tag{16.91}$$

with degrees and factorials as in Section 6.7.13.

Compare with (12.56). For $p=1$, the change of basis formulas are

$$\mathtt{H}_{(0,1)} = \mathtt{Q}_{(0,1)}, \quad \mathtt{H}_{(1)} = \mathtt{Q}_{(1)} + \frac{1}{2}\mathtt{Q}_{(0,1)}, \quad \mathtt{Q}_{(1)} = \mathtt{H}_{(1)} - \frac{1}{2}\mathtt{H}_{(0,1)}.$$

16.12.2. Bergeron idempotents. For each type B partition λ of p, define

$$\mathtt{E}_\lambda := \frac{1}{(2k)!!} \sum_{\alpha:\, \mathrm{s}(\alpha)=\lambda} \mathtt{Q}_\alpha = \sum_{\mathrm{X}:\, \mathrm{t}(\mathrm{X})=\lambda} \mathtt{E}_\mathrm{X}, \tag{16.92}$$

where k is the number of nonzero parts of λ, and $\mathtt{E}_\mathrm{X}$ is as in (12.57). We call these the *Bergeron idempotents*. For $p=2$, we have

$$\mathtt{E}_{(2)} = \mathtt{Q}_{(2)} = \mathtt{H}_{(2)} - \frac{1}{2}\mathtt{H}_{(1,1)} - \frac{1}{2}\mathtt{H}_{(0,2)} + \frac{3}{8}\mathtt{H}_{(0,1,1)}.$$

Theorem 16.65. *The elements* $\mathtt{E}_\lambda$, *as* λ *varies over all type* B *partitions of* p, *yield a complete system of primitive orthogonal idempotents of the invariant Tits algebra.*

PROOF. This is a special case of Theorem 16.43. □

Recall the idempotents $\mathtt{E}_k$ from the first formula in (12.59). We have

$$\mathtt{E}_k = \sum_\lambda \mathtt{E}_\lambda = \frac{1}{(2k)!!} \sum_\alpha \mathtt{Q}_\alpha. \tag{16.93}$$

The sum is over all λ (or α) with k nonzero parts. Similarly, from the first formula in (12.60),

$$\mathtt{T}_k = \sum_\alpha \mathtt{H}_\alpha, \tag{16.94}$$

where the sum is over all α with k nonzero parts. According to (12.60), (12.62) and (12.63a), the type B Adams element of odd parameter is

$$\mathtt{Ads}^{\pm}_{2n+1} = \sum_{k=0}^{p} \binom{n}{k} \mathtt{T}_k = \sum_{k=0}^{p} (2n+1)^k\, \mathtt{E}_k. \tag{16.95}$$

16.12.3. Invariant Lie elements. Specializing (16.36), we obtain:

$$\mathsf{Lie}[\mathbf{p}]^{\mathrm{S}^{\pm}_p} = \begin{cases} \Bbbk & \text{if } p=0, \\ 0 & \text{otherwise.} \end{cases} \tag{16.96}$$

Thus, there are no invariant type B Lie elements for $p\geq 1$.

Notes

Solomon descent algebra. The Solomon descent algebra was first defined by Solomon [**389**, Theorem 1]. His paper has an interesting appendix by Tits [**416**]. Numerous proofs of Theorem 16.7 have appeared since. An early reference is [**30**]. Theorem 16.8 which connects the Solomon descent algebra to the invariant Tits algebra is due to Bidigare [**62**, Theorem 3.81]. A geometric version of his proof was given by Brown [**103**, Section 9.7]. Further related ideas can be found in [**9**, Section 10.8]. The H- and K-bases considered in (16.4) are two standard bases of the invariant Tits algebra. In Brown's notation, H is σ and K is τ [**103**, Section 9.5]. In Solomon's original notation, $y_K = \mathtt{K}_K$, and $x_K = \mathtt{H}_{S\setminus K}$, where K is a subset of the generating set S of the Coxeter group [**389**, Equation (1.3)].

Formulas (16.1), (16.5) and (16.7) are given in [**8**, Definitions 5.7.2, 5.7.4 and 5.7.12]. Theorem 16.3 is given in [**8**, Lemmas 2.6.5 and 5.7.1]. The assertion about the radical of the invariant Tits algebra in Proposition 16.4 (phrased in a different language) was obtained by Solomon [**389**, Theorem 3]. It is also given in [**8**, Lemma 2.6.6]. Both Theorem 16.3 and Proposition 16.4 are also explained in [**367**, Proposition 4.1 and its proof]. Information on the nilpotency index of the radical (which we have not discussed) can be found in [**94**, Section 6] and [**369**, Theorem 6.5]. For results on eigenvalues and multiplicities of descent algebra elements, see [**350**] and references therein.

Complete systems. Primitive orthogonal idempotents for the Solomon descent algebra were first constructed in a unified manner by Bergeron, Bergeron, Howlett and Taylor [**55**]. They start with data similar to an invariant section u and then define the Q-basis by formula (16.52). This is their formula (11); they denote the Q-basis by $\{e_K\}$. They denote $\mathtt{u}^T$ by μ_K^{Π}, where Π is the set of simple roots and $K = \Pi \setminus T$. More generally, they denote $\mathtt{u}_T^U$ by μ_K^J, where $J = \Pi\setminus T$ and $K = \Pi\setminus U$. Their formula (10) (after language translation) is our (16.45). Formulas (16.51) and (16.53) and Theorem 16.43 are stated in the second paragraph on their page 25. Formula (16.55) which multiplies a H-basis element with a Q-basis element is given in their Theorem 7.8. Formula (16.41) after translating invariant noncommutative zeta functions to an invariant section is the same as their Lemma 7.5. Theorem 16.52 is given in their Theorem 7.15. Theorem 16.43 is also given by Saliola [**365**, Proposition 2.4], [**367**, Theorem 5.2] and our proof follows his argument.

The quiver of the Solomon descent algebra is a topic of active research. The partial result given in Proposition 16.55 is due to Saliola [**367**, Propositions 7.1, 7.3 and 7.5] or [**369**, Propositions 5.2 and 5.3]. For related work, see the paper by Pfeiffer [**342**].

Face-type enumeration and structure constants. We discussed many identities involving structure constants of the invariant Tits algebra and enumeration of face-types. These identities, developed from first principles here, are related to and can be derived from the existence and other properties of certain Coxeter bialgebras or bosonic Fock spaces. This connection will be explained in a future work.

The results of Lemmas 16.10 and 16.11 are contained in [**8**, Lemmas 5.3.1 and 5.3.2]. Theorem 16.17 was proved by Chevalley [**117**] for Weyl groups and by Solomon [**386**, Corollary 2.3] for Coxeter groups. Formula (16.19) is due to Solomon [**386**, Formula (7)]. It is also given in [**238**, Proposition on page 21] and [**80**, Formula (7.3)]. More information on the polynomial $d_S(q)$ can be found in [**80**, Section 7.1]. Lemma 16.19 is stated in [**8**, Lemma 2.6.3]. Equation (16.26) is stated in [**55**, Proposition 2.6]. The proof given there is by an algebraic manipulation, so it has a different flavor than ours.

Type A. The Solomon descent algebra of compositions arises out of Theorem 16.7 for the symmetric group. Early references are [**180**, Theorem 7], [**203**, Corollary 12] and [**318**]. This algebra was studied in depth by Garsia and Reutenauer [**198**]. The construction of the idempotents $\mathtt{E}_\lambda$ in (16.78) is given in their Theorems 3.1 and 3.2. The change of basis formulas (16.76) are given in their formulas (3.28) and (3.24). In this reference, the letters B and I are used instead of $\mathtt{H}$ and $\mathtt{Q}$. The identities in Exercise 16.46 generalize those given in their Theorem 4.2. The formula for the product of a $\mathtt{H}$-basis and a $\mathtt{Q}$-basis element is given in their Theorem 4.1. The $\mathtt{Q}$-basis is also given in Malvenuto's thesis [**298**, Section 4.3, pages 43-44]. (The notation P^* is used for the $\mathtt{Q}$-basis.) The first formula in (16.76) is [**298**, Formula (4.24)]. It also appears in her paper with Reutenauer [**299**, Formula (2.11)]. The idempotents $\mathtt{E}_\lambda$ are also discussed in Reutenauer's book [**360**, Section 9.2, Theorem 9.27]. References for the idempotents $\mathtt{E}_k$ in (16.79) have already been given in the notes to Chapter 12. The Bayer-Diaconis-Garsia-Loday formula (16.82) for these idempotents is given in [**47**, Corollary 3] and (up to a sign convention) in [**288**, third formula in Proposition 4.5.6]. An equivalent formula is given in [**196**, Theorem 7.4]. Also see [**287**, Proposition 2.8, part d] and [**198**, Remark 4.2]. The matrix description of the structure constants given in Lemma 16.58 is explicitly given in [**198**, Proposition 1.1] and [**31**] with related ideas in [**197**]. See also [**88**, Section 4].

The Solomon descent algebras, as the symmetric groups vary, form the graded components of the Hopf algebra of noncommuting symmetric functions introduced in [**199**]. The $\mathtt{Q}$-basis in this context is up to normalization the same as the power-sum basis of the second kind of noncommutative symmetric functions in [**199**, Definition 3.4], where it is denoted by the letter Φ. The normalization is $\Phi^\alpha = a_1 \dots a_k \, \mathtt{Q}_\alpha$, for $\alpha = (a_1, \dots, a_k)$. The $\mathtt{H}$-basis corresponds to complete noncommutative symmetric functions. The change of basis formulas (16.76) are given in [**199**, Proposition 4.9]. The idempotents $\mathtt{E}_\lambda$ of Garsia and Reutenauer are discussed in [**265**, Section 3.3]. The more general construction (16.87) of the Krob-Leclerc-Thibon idempotents is given in [**265**, Section 3.4]. In this reference, the letter F is instead of $\mathtt{Q}$. For related ideas, see [**334**, Section 5]. The $\mathtt{Q}$-basis in Exercise 16.63 was considered by Blessenohl and Laue [**90**, (27)], and later by Schocker [**374**, Section 5.2] or [**376**, Section 9]. The formula for the product of a $\mathtt{H}$-basis and a $\mathtt{Q}$-basis element is given in [**90**, Formula (16)].

Consider the right Peirce decomposition in Exercise 16.54. For the braid arrangement, the first such decomposition was obtained in [**198**, Theorem 4.3]. The decomposition using the $\mathtt{Q}$-basis in Exercise 16.63 is given in [**90**, Lemma 1.3] and [**376**, Theorem 9.3]. For a unified treatment using the Krob-Leclerc-Thibon construction, see [**374**, Theorem 5.2] and [**375**, Proposition 10.1].

The first formula in (16.84) can be found in [**196**, Lemma 1.1], where it is attributed to Nantel Bergeron. The q-Dynkin element was introduced via formula (16.85) by Krob, Leclerc and Thibon [**265**, Equation (66)]. The authors use the letter R instead of $\mathtt{K}$ and call it the ribbon Schur basis, and denote the q-Dynkin element by $\Theta_n(q)$. (For $q = 1$ this becomes the Dynkin element, which they denote by Ψ_n and call the "power sums of the first kind".) They show that the corresponding left q-bracketing operator (14.58) is diagonalizable and describe its spectral decomposition [**265**, Sections 5.3 and 5.4]. In particular, they show in their Proposition 5.6 that the space of Lie elements is an eigenspace. Apart from Section 5, see also Example 3.6 in the same paper. These results are also described in the survey paper [**414**, Sections 2.4 and 4.3], see in particular [**414**, Equations (55), (56), (63) and (64)]. The image of the left q-bracketing operator as a S_n-module is studied in [**109**]. There is a lot of interest in the case $q = -1$, see for instance [**58**] and references therein.

The factorization of the polynomial $d_S(q)$ in the case of the symmetric group (16.90) is in Dickson's book [**153**, page 114] and credited to an old paper of Rodrigues [**363**]. This polynomial is studied in [**163**, Section 4] and [**223**]. The description of the coefficients $R_{\lambda\mu}$ in Section 16.11.1 is given in [**399**, Proposition 7.7.1]; also see [**8**, Fact 5.7.2].

Let $\Bbbk$ be a field of arbitrary characteristic. The space $\bigoplus_n \mathsf{Lie}[n]^{\mathrm{S}_n}$ carries a structure of a *restricted Lie algebra*. It is in fact the free restricted Lie algebra on one generator. This is a special case of a result of Fresse [**186**, Theorem 1.2.5]. In characteristic 0, this algebra is one dimensional by (16.89). In positive characteristic, it is infinite dimensional. See [**9**, Example 15.38] for additional comments.

The quiver of the Solomon descent algebra of compositions was first computed by Schocker [**374**, Theorem 5.1] using results of Blessenohl and Laue [**90, 91**]. As a directed graph, this quiver had appeared earlier in the work of Garsia and Reutenauer [**198**, Section 5]. In their Theorem 5.6, they specify precisely which of the components in (16.88) are nonzero. This directed graph also appears in later work of Bauer [**45**]. The quiver has also since been computed by Saliola [**367**, Theorem 8.1] and by Bishop and Pfeiffer [**73**, Proposition 22]. The Cartan invariants are described in [**198**, Theorem 5.4], [**265**, Theorem 3.24], [**91**, Corollary 2.1] and [**374**, Theorem 5.3]. (One may check that they are consistent with the result obtained in Theorem 16.53.) The dimension of the space of invariant Zie elements is given by the Witt formula [**291**, Corollary 5.3.5] and [**360**, Corollary 4.14]. The nilpotency index of the radical is discussed in [**31**, Corollary 3.5] and [**198**, Theorem 5.7]. For information on the radical series, see [**90**, Theorem 2.5], [**376**, Theorem 9.10] and [**367**, Theorem 8.2]. We have not discussed these results in the text.

Type *B*. Bergeron gave an explicit construction of primitive orthogonal idempotents for the Solomon descent algebra of type B [**57**, Theorem 2.11]. The dictionary with his paper is as follows. He indexes the Solomon descent algebra of type B_n by compositions p of $m \leq n$. We have preferred to do it by type B compositions of n, where the first part may be 0, since these are the type B face-types. (To transform from his indexing, add a first part equal to $n - m \geq 0$ to his composition of $m \leq n$.) Bergeron uses the letters B and I instead of $\mathtt{H}$ and $\mathtt{Q}$. He defines the $\mathtt{Q}$-basis using the second formula in (16.91). This is his equation (2.1). His $I_\emptyset$ is $\mathtt{Q}_{(n)}$ and his formula on page 106 is the second formula in (16.91) specialized to $\alpha = (n)$. He also singles out $I_{(n)}$ which is $\mathtt{Q}_{(0,n)}$. Formula (16.92) is given on page 113. His Proposition 2.5 confirms that the dimension of Lie is the absolute value of the Möbius number.

The idempotents $\mathtt{E}_k$ in the form (16.93) were introduced by Bergeron and Bergeron [**54**, Formula (2.4)]. Formula (12.61), with $\mathtt{T}_k$ as in (16.94), and formula (16.95) are immediate from [**54**, Formula (2.9)]. These authors view the rhs of (16.95) as a generating function of $\mathtt{E}_k$ in the variable $2n+1$, and the lhs of (16.95) as a simplification. In effect, they diagonalize the odd type B Adams elements. The following quotation is taken from [**54**, Remark 2.2].

> Computer algebra manipulations have shown that an analogous formula (16.95) holds for the exceptional Coxeter group H_3, but does not hold for the groups D_n, $n > 3$, and F_4.

We now have a conceptual explanation of why this is happening. The good cases are precisely those of good reflection arrangements.

A matrix description similar to Lemma 16.58 can be given for the structure constants of the Solomon descent algebras of type B [**53**, Section 2] and type D [**60**]. The quiver of the Solomon descent algebra of type B was first computed by Saliola [**367**, Theorem 9.1]. For further work on this subject, see [**72**]. We have not discussed these results in the text.

Other work. We conclude by mentioning some references (by no means exhaustive) on topics which we have not touched upon. For connections of the descent algebra to character theory of the symmetric group, see [**89, 92, 178, 246, 419**], for variants and generalizations, see [**46, 95, 229, 233, 301, 305, 308, 319, 344**]. For the related topic of peak algebras, see [**5, 11, 58, 59, 266, 322, 338, 375**]. For cyclic descents, see [**12, 113, 114, 155, 337**]. For enumeration of descents, see [**339**] and references therein.

Appendices

APPENDIX A

Regular cell complexes

We review regular cell complexes, with emphasis on minimal galleries, gated sets and the gate property. All cell complexes are assumed to be finite, by which we mean that they have finitely many cells.

A.1. Cell complexes

A.1.1. Regular cell complexes. A cell complex X can be defined inductively: We start with the 0-skeleton consisting of a finite set of points. Then we add one-cells by identifying the two boundary points of each one-cell with points in the 0-skeleton, and so on. A cell complex is also called a CW complex.

A cell complex is *regular* if the identification of the boundary of each n-cell to the $(n-1)$-skeleton is done in a nice manner: the attachment must be injective, and its image must be a union of closed $(n-1)$-cells. The set of cells is ordered by inclusion. This poset is graded: an n-cell has rank $n+1$. (We assume that there is a unique cell of rank 0.) By convention, we also allow the *empty cell complex*. It has no cells in any rank, including 0.

A fundamental property of a regular cell complex is that its topology is determined by the cell poset: The geometric realization of the order complex of the cell poset (the barycentric subdivision of the cell complex) is homeomorphic to the underlying space of the complex.

A regular cell complex is *pure* if all maximal cells have the same rank. In this case, the maximal rank is the *rank* of the regular cell complex. We employ the following terminology for a pure regular cell complex. Any cell of X is called a *face*, a face of rank one is called a *vertex*, a face of rank two is called an *edge*, a maximal face is called a *chamber*, and a face of corank one is called a *panel*. We refer to the face of rank 0 as the *central face*.

A *subcomplex* of a regular cell complex is a subset of its cells which is closed under inclusion. Subcomplexes are cell complexes in their own right.

A.1.2. Examples. Some examples of cell complexes to keep in mind are as follows.

- The faces of a convex polytope are the cells of a regular cell complex.
- The geometric realization of a simplicial complex is a regular cell complex.
- The faces of a hyperplane arrangement are the cells of a regular cell complex.

A.1.3. Euler characteristic. Define the *reduced Euler characteristic* of a cell complex X to be

$$\chi(X) := -c_{-1} + c_0 - c_1 + c_2 - \dots, \tag{A.1}$$

where c_i is the number of i-cells of X.

If X is the empty cell complex, then $\chi(X) = 0$ with each $c_i = 0$. In all remaining cases, $c_{-1} = 1$ (since X has a unique central face).

The reduced Euler characteristic of a cell complex, regular or not, only depends on its underlying topology, since it is also the alternating sum of the ranks of its homology groups. Some well-known examples are recalled below. The reduced Euler characteristic

- of a ball is 0,
- of the sphere of dimension n is $(-1)^n$, and, more generally,
- of the wedge of k spheres each of dimension n is $(-1)^n k$.

A.2. Minimal galleries and gate property

We now focus on pure regular cell complexes. This setting allows us to define gallery distance between chambers and notions such as convexity and gated sets.

A.2.1. Minimal galleries. Let X be a pure regular cell complex. We say two chambers are *adjacent* if they are distinct and share a panel. A *gallery* is a sequence of chambers such that consecutive chambers are adjacent. We say that X is *gallery connected* if for any two chambers C and D, there is a gallery from C to D. The *length* of a gallery is the number of chambers in the gallery minus 1. Thus, a chamber is a gallery of length 0, an ordered pair of adjacent chambers is a gallery of length 1, and so on.

Assume from now on that X is gallery connected. For any chambers C and D, define the *gallery distance* $\operatorname{dist}(C, D)$ to be the minimum length of a gallery connecting C and D. Any gallery which achieves this minimum is a *minimal gallery* from C to D.

For chambers C, D and E, let $C \text{ -- } E \text{ -- } D$ mean that there is a minimal gallery from C to D passing through E. Note that

$$C \text{ -- } E \text{ -- } D \iff \operatorname{dist}(C, D) = \operatorname{dist}(C, E) + \operatorname{dist}(E, D). \tag{A.2}$$

Also note that

$$C \text{ -- } E \text{ -- } D \iff D \text{ -- } E \text{ -- } C.$$

More generally, the notation $C \text{ -- } E_1 \text{ -- } \dots \text{ -- } E_n \text{ -- } D$ means that there is a minimal gallery from C to D passing through the intermediate chambers E_i in the specified order. This is equivalent to additivity of the distance.

There are two useful operations on these gallery notations, namely, deletion and refinement. We illustrate them by examples.

$$C \text{ -- } E \text{ -- } E' \text{ -- } D \implies C \text{ -- } E \text{ -- } D$$

(a minimal gallery going through E and E' goes in particular through E), and

$$C \text{ -- } E \text{ -- } D \text{ and } E \text{ -- } E' \text{ -- } D \implies C \text{ -- } E \text{ -- } E' \text{ -- } D$$

(by additivity of the distance).

A.2.2. Convexity. A nonempty set of chambers is *convex* if for any C and D in this set, every minimal gallery from C to D is contained in this set.

For a nonempty set of chambers A, its *convex closure* is the smallest convex set which contains A. We now describe an inductive procedure to construct this set. Put $\mathrm{A}_1 = \mathrm{A}$. Given A_n, let A_{n+1} consist of those chambers E such that $C \text{ -- } E \text{ -- } D$ for some chambers C and D in A_n. Thus $\mathrm{A}_1 \subseteq \mathrm{A}_2 \subseteq \mathrm{A}_3 \subseteq \dots$. Since the set of chambers is finite, there is an index i, such that $\mathrm{A}_n = \mathrm{A}_i$ for all n greater than or equal to i. Then A_i is the convex closure of A.

Note that the intersection of two convex sets is convex provided it is nonempty.

A.2.3. Gallery intervals. For chambers C and D, define

$$[C:D] := \{E \mid C \text{--} E \text{--} D\}.$$

We refer to such a set as a *gallery interval*. In general, a gallery interval can be expressed in the form $[C : D]$ in many different ways aside from the triviality $[C:D] = [D:C]$.

Note that a nonempty set of chambers A is convex iff for any C and D in A, the gallery interval $[C:D]$ is contained in A.

A.2.4. Gated sets. Let A be a set of chambers, and let C be any chamber. Then A is *gated* wrt C if the following holds: There exists a chamber $D \in \mathrm{A}$ such that

$$\mathrm{dist}(C, E) = \mathrm{dist}(C, D) + \mathrm{dist}(D, E)$$

for all chambers $E \in \mathrm{A}$. In this situation, we say that D is the gate of A wrt C (since D is the closest entry point into A from C). It follows that the gate D is unique. We say A is a *gated set* if it is gated wrt every chamber.

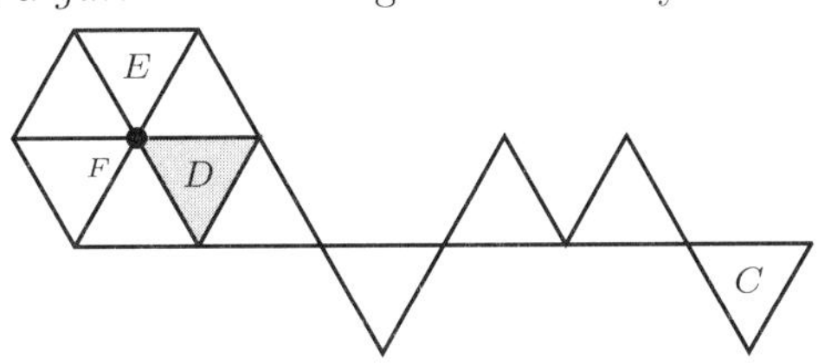

The figure shows a portion of a rank-three simplicial complex. Let A be the set of all chambers which contain a face F. Then the shaded chamber D is the gate of A wrt the chamber C.

Proposition A.1. *A gated set is necessarily convex.*

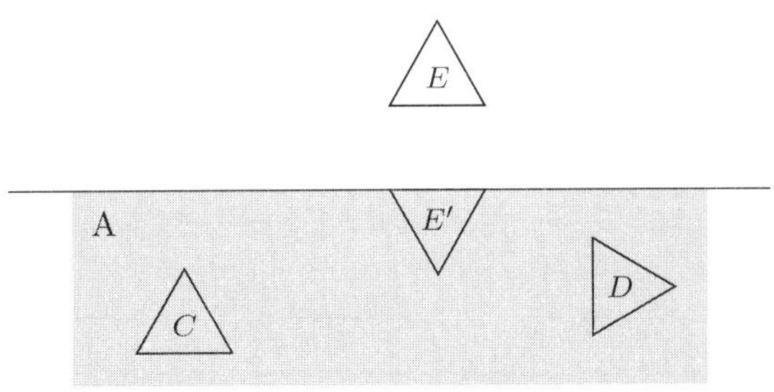

PROOF. Let A be a gated set. To show convexity, take any minimal gallery $C \text{--} E \text{--} D$ with C and D in A. We want to show that E also belongs to A. Let E' denote the gate of A wrt E. Then

$$\mathrm{dist}(C, E) = \mathrm{dist}(C, E') + \mathrm{dist}(E', E) \text{ and } \mathrm{dist}(D, E) = \mathrm{dist}(D, E') + \mathrm{dist}(E', E).$$

Adding and using (A.2), we obtain

$$\mathrm{dist}(C, D) = (\mathrm{dist}(C, E') + \mathrm{dist}(E', D)) + 2\,\mathrm{dist}(E', E).$$

But the sum in parenthesis is greater than $\mathrm{dist}(C, D)$. Hence, $\mathrm{dist}(E', E) = 0$ which is the same as $E = E'$. □

Two sets of chambers A and B are *gated* wrt each other if there exist chambers C in A and D in B such that C is the gate of A wrt D, and D is the gate of B wrt C. In this situation, C uniquely determines D and vice-versa, but the choice of C and D may not be unique. We call (C, D) a *gate pair*.

Proposition A.2. *If* A *and* B *are gated sets, then they are gated wrt each other.*

PROOF. Define a map $p : \mathrm{A} \to \mathrm{B}$ which sends C to the gate of B wrt C. The map $q : \mathrm{B} \to \mathrm{A}$ is similarly defined. Then $pqp = p$ and $qpq = q$, and there is an induced bijection between $q(\mathrm{A})$ and $p(\mathrm{B})$. Any pair of chambers corresponding under this bijection serve as a gate pair. □

The converse is false.

Exercise A.3. Suppose A and B are two sets of chambers with $\mathrm{A} \cap \mathrm{B} \neq \emptyset$. Then A and B are gated wrt each other. For any $C \in \mathrm{A} \cap \mathrm{B}$, (C, C) is a gate pair. In fact, these are all the gate pairs.

A.2.5. Top-stars and gate property. We let Σ denote the set of faces, and Γ denote the set of chambers. Also let Σ_F denote the set of faces which contain a given face F. This is the *star* of F. Similarly, let Γ_F denote the set of chambers which contain F. This is the *top-star* of F.

Definition A.4. A pure regular cell complex has the *gate property* if top-stars of all its faces are gated sets.

A cell complex may or may not have the gate property. For instance, a polygon with an odd number of sides is a rank-two simplicial complex without the gate property.

Proposition A.5. *Suppose the gate property holds. Then all top-stars are convex. In other words, if D and E are any two chambers which contain a given face F, then any minimal gallery from D to E lies entirely in the top-star of F.*

This is a special case of Proposition A.1.

Notes

The notion of CW complex is due to Whitehead [**431**]. Regular cell complexes are discussed in detail by Cooke and Finney [**125**], or Lundell and Weingram [**292**]. Short introductions can be found in [**77**], [**82**, Appendix 4.7] or [**2**, Appendix A.2]. The gate property originated in the work of Tits [**415**, Section 3.19.6]. The abstract notion of gated sets was introduced by Dress and Scharlau [**371, 161**]. Results related to Proposition A.2 and its converse are given in [**227**, Theorem 1.9], [**240**, Theorem 1.8] and [**161**, Theorem, page 116].

APPENDIX B

Posets

We review generalities on partial orders and important classes of lattices such as semimodular and geometric. We also review adjunctions between posets, closure operators and convex geometries.

B.1. Poset terminology

For any poset P, we denote its partial order by $\leq$. When $x \leq y$, we say x is smaller than y, and when $x < y$, we say x is strictly smaller than y. The symbol $x \lessdot y$ means that y covers x, that is, y is strictly greater than x, and there is no element strictly between x and y. We let $[x, y]$ denote the interval consisting of all elements which lie between x and y, that is,

$$[x, y] = \{z \mid x \leq z \leq y\}.$$

We write $\bot$ for the bottom (minimum) element, $\top$ for the top (maximum) element, $\wedge$ for meet and $\vee$ for join, whenever they exist.

A subset I of P is an *upper set* if $x \in I$ and $x \leq y$, then $y \in I$. Similarly, a subset I of P is a *lower set* if $y \in I$ and $x \leq y$, then $x \in I$.

Let P be a subposet of Q (a subset of Q with the induced partial order). We say P is *convex* in Q if

$$x \leq y \leq z \text{ in } Q \text{ and } x, z \in P \text{ imply that } y \in P. \tag{B.1}$$

When both $\bot$ and $\top$ exist, we say P is *bounded*. We say P is a *meet-semilattice* if the meet of any two elements exists, a *join-semilattice* if the join of any two elements exists, and a *lattice* if it is both a meet-semilattice and a join-semilattice. A lattice P is *complete* if arbitrary meets and joins exist. Any nonempty finite lattice is complete, and in particular, bounded.

A poset P is *of finite height* if it has no infinite strict chains, that is, P does not have any infinite totally ordered subposet. A poset P is *locally finite* if each interval of P has finite height. A locally finite poset is *connected* if for any two elements, there is a finite sequence of relations linking them. A poset with bottom or top element is automatically connected.

Every poset has a dual or opposite poset obtained by reversing the partial order. Similarly, for any property of a poset, there is a dual property. For instance, the dual property of minimum is maximum, and of meet is join. If P has a particular property, then its dual poset has the dual property. For instance, if P is a meet-semilattice, then its dual is a join-semilattice.

Let P and Q be posets. A map $f : P \to Q$ is order-preserving if $x \leq y$ implies $f(x) \leq f(y)$ for all $x, y \in P$. In addition: It is strictly order-preserving if $x < y$ implies $f(x) < f(y)$ for all $x, y \in P$. It preserves cover relations if $x \lessdot y$ implies $f(x) \lessdot f(y)$ for all $x, y \in P$.

B.2. Graded posets

Let $\mathbb{N}$ denote the set of nonnegative integers. It is a poset under the standard ordering. It is locally finite and connected but not of finite height.

A locally finite poset P is *graded* if there exists a function $\operatorname{rk} : P \to \mathbb{N}$ such that

$$x \lessdot y \implies \operatorname{rk}(x) + 1 = \operatorname{rk}(y)$$

for all $x, y \in P$. We say that rk is a rank function of P. Once this is fixed, we refer to $\operatorname{rk}(x)$ as the rank of x. In this situation, it follows that

$$x < y \implies \operatorname{rk}(x) < \operatorname{rk}(y).$$

The poset $\mathbb{N}$ is graded and the identity is a rank function.

A useful way to show that a poset P is graded is to find another graded poset Q and an order-preserving map $f : P \to Q$ which preserves cover relations. Composing f with any rank function of Q then yields a rank function of P.

Lemma B.1. *Let P be a graded poset and $x \leq y$. Then all maximal chains from x to y have the same length. Conversely, if P is finite, bounded, and all maximal chains from $\bot$ to $\top$ have the same length, then P is graded.*

PROOF. If P is graded, then for $x \leq y$, $\operatorname{rk}(y) - \operatorname{rk}(x)$ is the length of any maximal chain from x to y. In particular, all such chains have the same length. For the converse, first note that for $x \leq y$, all maximal chains from x to y have the same length. Then let $\operatorname{rk}(z)$ be the length of any maximal chain in $[\bot, z]$. □

It follows from (the proof of) Lemma B.1 that if P is graded and connected, then any two rank functions on P differ by translation by a fixed number.

Let P be a finite, connected, graded poset. The *rank* of P, denoted $\operatorname{rk}(P)$, is the difference between the maximum and minimum values of any rank function of P. If P is empty, its rank is defined to be -1.

For a graded poset with bottom element $\bot$, it is customary to work with the rank function for which $\operatorname{rk}(\bot) = 0$. Then $\operatorname{rk}(z)$ is the length of any maximal chain in $[\bot, z]$. If in addition P is finite, $\operatorname{rk}(P)$ is the maximum of ranks of all elements of P. For a finite graded poset with top element $\top$, it is customary to work with the rank function for which $\operatorname{rk}(\top) = \operatorname{rk}(P)$. When P is bounded, the choices $\operatorname{rk}(\bot) = 0$ and $\operatorname{rk}(\top) = \operatorname{rk}(P)$ are consistent.

A locally finite poset is *locally graded* if every interval $[x, y]$ is graded. By Lemma B.1, this happens precisely when all maximal chains from x to y have the same length, for all $x \leq y$. A graded poset is clearly locally graded. Conversely, a locally graded and bounded poset is graded.

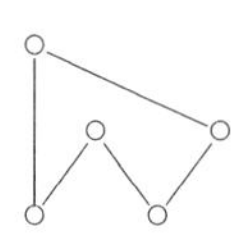

The poset shown above is locally graded but not graded. It has neither a top element nor a bottom element, so it is not bounded.

B.3. Semimodularity and join-distributivity

We briefly review semimodular and geometric lattices and join-distributive join-semilattices.

B.3.1. Semimodularity. A finite lattice is *lower semimodular* if the following property holds.

If $x \neq y$ and $x \vee y$ covers x and y, then x and y cover $x \wedge y$.

The dual property is *upper semimodular*.

Proposition B.2. *A finite (lower or upper) semimodular lattice is graded.*

PROOF. This result is given for instance in [**205**, Section IV.2, Theorem 1]. It can also be seen as a consequence of Lemma B.8. □

In fact, a finite lattice is lower semimodular iff it is graded and the rank function satisfies

$$\mathrm{rk}(x) + \mathrm{rk}(y) \leq \mathrm{rk}(x \wedge y) + \mathrm{rk}(x \vee y).$$

For upper semimodularity, the inequality goes the other way, see [**401**, Proposition 3.3.2].

B.3.2. Geometric lattices. Let P be a finite lattice. An element y in P is a *complement* of an element x if $x \vee y = \top$ and $x \wedge y = \bot$. If every element admits a complement, P is called *complemented*. If every interval is complemented, P is called *relatively complemented*.

Proposition B.3. *A finite lattice is relatively complemented iff it contains no* 3-*element intervals.*

PROOF. See [**75**, Theorem 2]. □

An element that covers $\bot$ is said to be a *point* or an *atom* of P. If every element is the join of some points, P is called *atomic* (and sometimes *atomistic*). If P is graded, the points are the elements of rank 1.

Proposition B.4. *Let P be a finite lattice. If P is relatively complemented, then it is atomic. The converse holds if P is upper semimodular.*

PROOF. See [**70**, Theorem 15, Section I.9, and Theorem 6, Section IV.5]. □

A lattice is called *geometric* if it is finite, upper semimodular, and relatively complemented (or equivalently, atomic). An interval in a geometric lattice is again geometric.

B.3.3. Join-distributivity. Let P be a finite join-semilattice. We say P is *join-distributive* if given an element x and distinct elements $x_1, \dots, x_k$ all covering x, the interval $[x, y]$ with $y = x_1 \vee \dots \vee x_k$ is a Boolean poset. In this case $[x, y]$ is of rank k and the x_i are the join-irreducibles therein.

Given x, let $z(x)$ be the join of *all* elements which cover x. Since intervals $[x, y]$ as above are subintervals of $[x, z(x)]$, we have that P is join-distributive iff each $[x, z(x)]$ is Boolean.

Dually, one defines the meet-distributive property for a finite meet-semilattice.

Proposition B.5. *Every interval in a join-distributive join-semilattice is upper semimodular. Dually, every interval in a meet-distributive meet-semilattice is lower semimodular.*

PROOF. See [**171**, Theorem 1.1] or [**401**, Exercise 3.47(a)]. □

B.4. Strongly connected posets

We discuss strongly connected posets. Such a poset is locally graded and the category associated to it has a nice presentation, the generators are cover relations, and the relations are squares.

B.4.1. Strongly connected posets. A finite poset P is *strongly connected* if given $x < y$ in P, one can transform any maximal chain from x to y to any other by successively changing one element of the chain to a different element. Clearly a strongly connected poset is locally graded.

A finite poset P is *coatom connected* if in any interval $[z, y]$ of P, whenever y covers two elements x and x', there is a sequence of elements $x = x_0, x_1, \ldots, x_k = x'$ all covered by y and such that for every $1 \leq i \leq k$, there exists an element which both x_{i-1} and x_i cover.

This is illustrated below for $k = 2$.

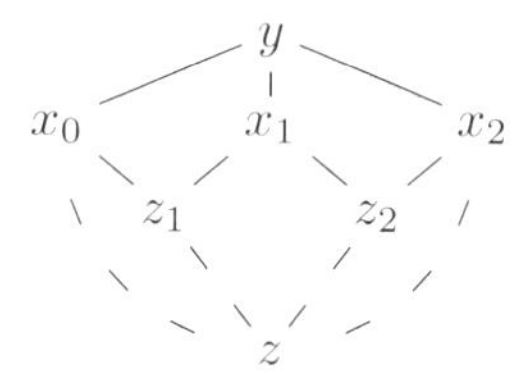

Similarly, a poset is *atom connected* if its opposite poset is coatom connected.

Lemma B.6. *For a finite poset P, we have*

$$P \text{ is coatom connected} \iff P \text{ is strongly connected} \iff P \text{ is atom connected.}$$

PROOF. The second equivalence follows from the first by applying it to the opposite of P. We now establish the first equivalence. Suppose P is coatom connected. We apply induction on the number of elements in P. The base case is clear. For the induction step, consider the interval $[z, y]$. Take two maximal chains from z to y. In the preceding figure, these are the chains shown on the outside. We need to pass from one to the other. Use coatom connectedness to fill in the diagram as shown. Let z_i denote the element covered by x_{i-1} and x_i. For each z_i, pick a maximal chain from z to z_i. By the induction hypothesis, within each interval $[z, x_i]$, one can pass from any maximal chain to any other. Further, one can use the diamonds at the top to pass from a maximal chain containing x_i to a maximal chain containing x_{i+1}. It follows that P is strongly connected. These steps can be reversed, so the reverse implication holds as well. □

Lemma B.7. *The face lattice of a convex polytope is strongly connected.*

PROOF. Any interval in the face lattice of a convex polytope is again the face lattice of a convex polytope [**448**, Theorem 2.7, item (ii)]. Hence, atom connectedness reduces to the fact that the graph of a convex polytope is connected: Given vertices P and P', there exists a sequence of vertices $P = P_0, P_1, \ldots, P_k = P'$ such that there is an edge joining P_{i-1} and P_i for every $1 \leq i \leq k$. This is a special case of Balinski's theorem; see [**215**, Section 11.3] or [**448**, Theorem 3.14]. □

Lemma B.8. *A finite lower or upper semimodular lattice is strongly connected.*

PROOF. A lower (upper) semimodular lattice satisfies the coatom (atom) connectedness property in one step. □

B.4.2. Category associated to a poset. To a finite poset P, one can associate the category whose objects are elements of P, with a unique morphism $x \to y$ whenever $x \leq y$ in P.

Proposition B.9. *The category associated to P has a presentation given by generators $x \xrightarrow{\Delta} y$, with $x \leq y$, and relations*

$$\begin{array}{ccc} & x & \\ {\scriptstyle \Delta}\nearrow & & \searrow{\scriptstyle \Delta} \\ z & \xrightarrow[\Delta]{} & y \end{array} \qquad\qquad (x \xrightarrow{\Delta} x) = \mathrm{id},$$

whenever $z \leq x \leq y$.

When P is strongly connected, there is a nice alternative presentation for this category. The number of generators is minimized, and the resulting relations are commutative squares instead of commutative triangles as follows.

Proposition B.10. *For a strongly connected poset P, the associated category has a presentation given by generators $\Delta : x \to y$, where y covers x, and relations*

$$\begin{array}{ccc} x' & \xrightarrow{\Delta} & y \\ {\scriptstyle \Delta}\uparrow & & \uparrow{\scriptstyle \Delta} \\ z & \xrightarrow[\Delta]{} & x \end{array}$$

whenever y covers both x and x', and they in turn cover z.

PROOF. Let D denote the category with the above presentation. For any $x \leq y$, there is a morphism in D from x to y arising from a maximal chain from x to y. To finish the argument, we need to show that this morphism is unique. For this, observe that two maximal chains which differ in one position are equal by the above relation. Hence uniqueness follows by strong connectedness. □

Note that for an arbitrary finite poset P, the cover relations will generate a category whose morphisms from x to y are maximal chains from x to y. In order to obtain the category associated to P, we need to impose the relations that any two maximal chains from x to y define the same morphism. The role of strong connectedness is that it makes it possible to work with a nice small set of relations.

B.5. Adjunctions between posets

We briefly review adjunctions betweeen posets, closure operators and convex geometries. We also formulate a notion of (super)tightness for join-preserving maps between lattices.

B.5.1. Galois connections. Let P and Q be two finite posets. Suppose

$$\lambda : P \to Q \quad \text{and} \quad \rho : Q \to P$$

are order-preserving maps. We say (λ, ρ) is a *Galois connection* or an *adjunction* if

$$\lambda(x) \leq y \iff x \leq \rho(y) \tag{B.2}$$

for all $x \in P$ and $y \in Q$. We also say that ρ is the right adjoint of λ, and λ is the left adjoint of ρ.

Adjoints (left or right) may not exist, but they are unique whenever they exist.

Remark B.11. Recall from Section B.4.2 that each poset has an associated category. Further, an order-preserving map of posets yields a functor between the associated categories. Moreover, a Galois connection between posets yields an adjunction between the associated categories.

Suppose (λ, ρ) is an adjunction. Then

$$\lambda(\rho(y)) \leq y \quad \text{and} \quad x \leq \rho(\lambda(x)). \tag{B.3}$$

The first one follows by putting $x = \rho(y)$ in (B.2). For the second, put $y = \lambda(x)$. It is of interest to know when equalities hold. There are inverse bijections

$$\{x \in P \mid x = \rho(\lambda(x))\} \rightleftarrows \{y \in Q \mid \lambda(\rho(y)) = y\} \tag{B.4}$$

obtained by restricting λ and ρ. These are inverse isomorphisms of posets, viewing the two sides as subposets of P and Q, respectively.

Adjunctions can be composed: If (λ, ρ) is an adjunction between P and Q, and (μ, δ) is an adjunction between Q and R, then $(\mu\lambda, \rho\delta)$ is an adjunction between P and R.

B.5.2. Closure operators. A *closure operator* on a finite poset P is a map $c : P \to P$ such that for every $x, y \in P$, we have:

- $x \leq c(x)$;
- if $x \leq c(y)$, then $c(x) \leq c(y)$.

We refer to $c(x)$ as the closure of x. An element $z \in P$ is *closed* if $c(z) = z$, that is, the closure of z is itself. Let P_c denote the subposet of all closed elements. If P is a lattice, then so is P_c. The meet in P_c coincides with the meet in P, while the join in P_c is the closure of the join in P.

A *coclosure operator* on a poset is a closure operator on the opposite poset.

In any adjunction (λ, ρ) between P and Q, the composite $\rho\lambda$ is a closure operator on P while $\lambda\rho$ is a coclosure operator on Q, and λ and ρ restrict to inverse poset isomorphisms between their closed sets, see (B.4).

B.5.3. Meets and joins. Assume from now that P and Q are finite lattices. Given λ, the right adjoint ρ exists iff λ preserves finite joins. Dually, given ρ, the left adjoint λ exists iff ρ preserves finite meets. We elaborate on the first statement in one direction. Suppose λ preserves finite joins. In particular, it preserves the minimum element which is the join over the empty set. Then define ρ by

$$\rho(y) := \max\{x \in P \mid \lambda(x) \leq y\}.$$

The minimum element of P is always in the above set. Further, since λ preserves joins, the join of all elements in the set is also in the set, so the set has a maximum. One can check that ρ defined in this manner is the right adjoint of λ.

B.5.4. Tight join-preserving maps. We say an order-preserving map is *join-preserving* if it preserves finite joins.

Suppose $\lambda : P \to Q$ is a join-preserving map. Let ρ denote its right adjoint. Fix $x \leq y$ in Q. Consider the interval $[x, y]$ in Q and the interval $[\rho(x), \rho(y)]$ in P. Let

$$\rho_{x,y} : [x, y] \to [\rho(x), \rho(y)]$$

denote the restriction of ρ. Let

$$\lambda_{x,y} : [\rho(x), \rho(y)] \to [x, y], \qquad z \mapsto \lambda(z) \vee x.$$

Taking join with x ensures that the image of $\lambda_{x,y}$ lands in the interval $[x, y]$.

Lemma B.12. *Given an adjunction (λ, ρ) between P and Q, for any $x \le y$ in Q, $(\lambda_{x,y}, \rho_{x,y})$ is an adjunction between $[\rho(x), \rho(y)]$ and $[x, y]$.*

Given an adjunction (λ, ρ) between P and Q, and an adjunction (μ, δ) between Q and R, for any $x \le y$ in R, the diagram

$$\begin{array}{ccccc} & & [\delta(x), \delta(y)] & & \\ & \overset{\lambda_{\delta(x),\delta(y)}}{\nearrow} & & \overset{\mu_{x,y}}{\searrow} & \\ [\rho\delta(x), \rho\delta(y)] & & \xrightarrow[(\mu\lambda)_{x,y}]{} & & [x, y] \end{array}$$

commutes.

Proof. The first part is a straightforward check. The second part follows from the first by composing adjunctions, and using uniqueness of adjoints. □

Definition B.13. A join-preserving map $\lambda : P \to Q$ is *tight* if

$$\lambda(x) = \top \text{ implies } x = \top.$$

We say λ is *supertight* if $\lambda_{x,y}$ is tight for all $x \le y$ in Q. That is, for any $\rho(x) \le z \le \rho(y)$,

$$\lambda(z) \vee x = y \text{ implies } z = \rho(y).$$

Lemma B.14. *The composite of tight (join-preserving) maps is tight, and the composite of supertight maps is supertight.*

Proof. Suppose $\lambda : P \to Q$ and $\mu : Q \to R$ are tight maps. Then

$$(\mu\lambda)(x) = \top \implies \mu(\lambda(x)) = \top \implies \lambda(x) = \top \implies x = \top.$$

We first used tightness of μ and then of λ. This shows that $\mu\lambda$ is tight.

Suppose λ and μ are supertight. Let ρ be the right adjoint of λ, and δ be the right adjoint of μ. Then $\rho\delta$ is the right adjoint of $\mu\lambda$. Let $x \le y$ in R. Then by Lemma B.12,

$$(\mu\lambda)_{x,y} = \mu_{x,y}\lambda_{\delta(x),\delta(y)}.$$

The maps in the rhs are tight by hypothesis. Since composition preserves tight maps, we deduce that the map in the lhs is tight, and hence $\mu\lambda$ is supertight. □

Exercise B.15. Give an example of a join-preserving map which is not tight.

Lemma B.16. *Suppose $\lambda : P \to Q$ is a supertight join-preserving map of lattices. Then for any $x, y \in Q$, and $z \in P$,*

$$\lambda(z) \vee x = y \iff z \vee \rho(x) = \rho(y) \text{ and } \lambda\rho(y) \vee x = y, \tag{B.5}$$

where ρ is the right adjoint of λ.

As the proof below shows, we only require $\lambda_{x,y}$ to be tight for (B.5) to hold. Also, tightness plays a role only in the forward implication.

Proof. We may assume $x \le y$, else neither side of (B.5) holds. Forward implication. The hypothesis implies $\lambda(z) \le y$. Hence, by the adjunction property (B.2), $z \le \rho(y)$. Since $\lambda\rho(x) \le x$,

$$\lambda(z \vee \rho(x)) \vee x = \lambda(z) \vee \lambda\rho(x) \vee x = \lambda(z) \vee x = y.$$

Since $\rho(x) \leq z \vee \rho(x) \leq \rho(y)$, by supertightness, $z \vee \rho(x) = \rho(y)$. Also $\lambda(z) \vee x \leq \lambda\rho(y) \vee x \leq y$ which implies that $\lambda(\rho(y)) \vee x = y$.

Backward implication. Apply λ to $z \vee \rho(x) = \rho(y)$ to obtain $\lambda(z) \vee \lambda\rho(x) = \lambda\rho(y)$. Now take join with x on both sides, and use $\lambda\rho(x) \leq x$ and $\lambda(\rho(y)) \vee x = y$ to deduce $\lambda(z) \vee x = y$. □

B.5.5. Convex geometries. Fix a finite set X. A *convex geometry* with ground set X is a closure operator c on the Boolean poset on X which satisfies the *antiexchange axiom*:

- if $x \in c(A \cup \{y\})$, $x \neq y$ and $x, y \notin A$, then $y \notin c(A \cup \{x\})$,

for all closed sets A and $x, y \in X$.

In this context, closed sets are also called convex sets. Since the Boolean poset is a lattice, so is the subposet of convex sets. The meet is given by intersection, while the join is given by taking closure of the union.

Given a convex set A, an element $x \in A$ is called an *extreme point* of A if $x \notin c(A \setminus \{x\})$.

Proposition B.17. *Let c be a closure operator on a Boolean poset. Then c is a convex geometry iff the lattice of closed sets of c is meet-distributive.*

PROOF. See [**172**, Theorem 4.1]. □

Proposition B.18. *Let c be a convex geometry. Then an interval $[A, B]$ in the lattice of convex sets of c is a Boolean poset iff all elements in B which are not in A are extreme points of B.*

PROOF. See [**172**, Theorem 4.2]. □

Notes

Posets and lattices originated in work of Boole [**96**], Peirce [**336**] and Schröder [**378**]. The first book on the subject is that of Birkhoff [**69**]. The notion of modularity goes back to Dedekind [**140, 141**]. Semimodular lattices were considered by Birkhoff [**66**, Sections 8 and 9]. Geometric lattices go back to Birkhoff [**67**], Whitney [**433**] and MacLane [**294**]. Meet-distributive lattices were first considered by Dilworth [**156**]. Another early reference is [**33**]. The concept of a closure operator goes back to Moore [**317**]. The concept of a Galois connection between posets originated in works of Birkhoff [**69**] and Ore [**323**]. This predated the appearance of adjunctions in category theory. Some other early references on these topics are [**425, 181**]. Convex geometries were discovered independently by Edelman [**169**] and Jamison [**245**]. Additional references are [**172**] and [**86**]. A convex geometry is equivalent to the notion of an antimatroid [**244**]. Extensive treatments of lattice theory are given in [**70**] and [**205**]. The book [**408**] concentrates on semimodular lattices. Posets and lattices are the subject of [**401**, Chapter 3]. For closure operators and Galois connections, see [**136**, Chapter 7] or [**93**]. Strongly connected posets appear prominently in the theory of abstract polytopes [**311**, Sections 1 and 2]. See also [**184**] and [**234**, Section 6]. Proposition B.10 is closely related to [**304**, Lemma 7.4].

APPENDIX C

Incidence algebras of posets

We review incidence algebras of posets, Möbius functions, reduced incidence algebras and deformations arising from poset cocycles.

C.1. Incidence algebras and Möbius functions

We review the incidence algebra of a locally finite poset, zeta and Möbius functions, the Weisner formula, and Möbius inversion on the incidence module. We also discuss the Rota formula relating the Möbius functions of posets linked by a Galois connection.

C.1.1. Incidence algebra of a poset. Let P be a locally finite poset. A 1-*chain* in P is a pair $(x, y) \in P^2$ with $x \leq y$.

Fix a field $\Bbbk$. An *incidence function* on P is a $\Bbbk$-valued function on the set of 1-chains in P

$$f : \{(x, y) \in P^2 \mid x \leq y\} \to \Bbbk.$$

Let $\mathrm{I}(P)$ denote the vector space of all incidence functions, with pointwise addition and scalar multiplication. For $f, g \in \mathrm{I}(P)$, define a new function $fg \in \mathrm{I}(P)$ by

$$(fg)(x, z) = \sum_{y:\, x \leq y \leq z} f(x, y) g(y, z). \tag{C.1}$$

This turns $\mathrm{I}(P)$ into an algebra. It is called the *incidence algebra* of P. The unit element δ is given by

$$\delta(x, y) = \begin{cases} 1 & \text{if } x = y, \\ 0 & \text{otherwise.} \end{cases} \tag{C.2}$$

In other words, $\delta f = f \delta = f$ for any $f \in \mathrm{I}(P)$.

If P is a convex subposet of Q, there is a morphism of algebras

$$\mathrm{I}(Q) \to \mathrm{I}(P) \tag{C.3}$$

given by restriction. Since every interval in P is an interval in Q, this map is surjective.

Proposition C.1. *Let $f \in \mathrm{I}(P)$ be such that $f(x, x) = 1$ for all $x \in P$. Then f is invertible in $\mathrm{I}(P)$ and*

$$f^{-1}(x, y) = \sum_{k \geq 0} (-1)^k \sum_{x = x_0 < x_1 < \cdots < x_k = y} f(x_0, x_1) \ldots f(x_{k-1}, x_k). \tag{C.4}$$

The summand corresponding to $k = 0$ is 0 unless $x = y$, in which case it is 1.

PROOF. If r is the maximum length of a strict chain from x to y, then $(\delta - f)^k(x,y) = 0$ for all $k > r$. The result follows by expanding

$$f^{-1} = (\delta - (\delta - f))^{-1} = \sum_{k\geq 0} (\delta - f)^k. \qquad \square$$

More generally:

Proposition C.2. *An incidence function f is invertible in $\mathrm{I}(P)$ iff $f(x,x) \neq 0$ for all $x \in P$.*

PROOF. As a special case of (C.3), for fixed $x \in P$, the map $f \mapsto f(x,x)$ is an algebra morphism $\mathrm{I}(P) \to \Bbbk$. Thus the condition for invertibility is necessary. Conversely, we may decompose $f = f_0 f_+$ by defining

$$f_0(x,y) = \begin{cases} f(x,x) & \text{if } x = y, \\ 0 & \text{if not} \end{cases} \quad \text{and} \quad f_+(x,y) = f(x,x)^{-1} f(x,y).$$

Then f_0 is invertible with inverse given by

$$f_0^{-1}(x,y) = \begin{cases} f(x,x)^{-1} & \text{if } x = y, \\ 0 & \text{if not,} \end{cases}$$

and f_+ is invertible by Proposition C.1. $\square$

C.1.2. Zeta and Möbius functions of a poset. The *zeta function* $\zeta \in \mathrm{I}(P)$ is defined by

$$\zeta(x,y) = 1$$

for all $x \leq y$. It is invertible. Its inverse is the *Möbius function* $\mu \in \mathrm{I}(P)$. This may also be defined recursively as follows.

For any element x,

$$\mu(x,x) := 1 \tag{C.5a}$$

and for $x < y$,

$$\mu(x,y) := - \sum_{z:\, x\leq z<y} \mu(x,z) = - \sum_{z:\, x<z\leq y} \mu(z,y),$$

or equivalently,

$$\sum_{z:\, x\leq z\leq y} \mu(x,z) = \sum_{z:\, x\leq z\leq y} \mu(z,y) = 0. \tag{C.5b}$$

For more clarity, we may sometimes write μ_P instead of μ.

Proposition C.3 (Philip Hall formula). *For any $x \leq y$ in P,*

$$\mu(x,y) = \sum_{k\geq 0} (-1)^k c_k(x,y), \tag{C.6}$$

where $c_k(x,y)$ is the number of strict chains of length k from x to y.

PROOF. This is a special case of (C.4). $\square$

C.1.3. Weisner formula.

Proposition C.4. *Let P be a finite lattice with $\bot$ and $\top$ as the minimum and maximum elements. Suppose $y > \bot$. Then for any element z,*

$$\sum_{x:\, y \vee x = z} \mu(\bot, x) = 0. \tag{C.7a}$$

Suppose $y < \top$. Then for any element z,

$$\sum_{x:\, y \wedge x = z} \mu(x, \top) = 0. \tag{C.7b}$$

PROOF. We prove (C.7a) by induction on the size of the interval $[\bot, z]$. The base case is clear. We may assume $y \leq z$, otherwise the lhs is clearly 0.

$$\sum_{x:\, y \vee x = z} \mu(\bot, x) = \sum_{x:\, x \leq z} \mu(\bot, x) - \sum_{x:\, y \vee x < z} \mu(\bot, x).$$

The first term is zero by definition (C.5b), while the second term is zero by the induction hypothesis applied to the intervals $[\bot, z']$ with $z' < z$. This completes the induction step.

The second identity (C.7b) follows by duality. □

We refer to either of (C.7a) or (C.7b) as the *Weisner formula*.

Proposition C.5. *For any finite lattice P, consider the linear system*

$$\sum_{x:\, y \vee x = z} c_x = 0, \qquad \bot < y \leq z.$$

The solution space is one-dimensional and spanned by $(c_x = \mu(\bot, x))$.

PROOF. It is clear from (C.7a) that any scalar multiple of $(c_x = \mu(\bot, x))$ solves the above linear system. To see that these are the only solutions, restrict to the subsystem of equations satisfying $y = z$:

$$\sum_{x:\, x \leq z} c_x = 0, \qquad \bot < z.$$

Starting with an arbitrary value for $c_\bot$, we now see that $c_x = c_\bot \mu(\bot, x)$ is forced. □

C.1.4. Möbius function of a semimodular lattice.

Proposition C.6. *Let P be a finite upper semimodular lattice of rank r. Then*

$$(-1)^r \mu(\bot, \top) = |\mu(\bot, \top)|. \tag{C.8}$$

In other words, the sign of the Möbius function is the same as the parity of the rank. Moreover, if P is geometric, the Möbius function is nonzero.

PROOF. We induct on the rank, the base case $r = 0$ being clear. For the induction step, fix a point y. Then $x \vee y = \top$ implies that either $x = \top$ or $\mathrm{rk}(x) = r - 1$. By the Weisner formula (1.46a) applied to $z = \top$, the sum of $\mu(\bot, x)$ over all such x is 0. So

$$\mu(\bot, \top) = -\sum_{x} \mu(\bot, x),$$

where the sum is over certain elements x of rank $r - 1$. Applying the induction hypothesis to each of the intervals $[\bot, x]$ we obtain that the sign of each summand is the parity of $r-1$. Thus the sign of $\mu(\bot,\top)$ is the parity of r. When P is geometric, y possesses a complement. Thus the above sum is over a nonempty set. A similar inductive argument implies that $\mu(\bot,\top)$ is in this case nonzero. □

The 3-element chain is a modular lattice with $\mu(\bot,\top)=0$.

C.1.5. Möbius function of a join-distributive join-semilattice.

Lemma C.7. *Let P be a finite join-semilattice which is join-distributive. Then for $x \le y$,*

$$\mu(x,y) = \begin{cases} (-1)^{\mathrm{rk}(y)-\mathrm{rk}(x)} & \text{if } [x,y] \text{ is a Boolean poset,} \\ 0 & \text{otherwise.} \end{cases}$$

The first alternative is equivalent to saying that y is the join of the elements covering x in the interval $[x,y]$.

PROOF. This is equivalent to [**171**, Theorem 1.3]. □

C.1.6. Eulerian posets. A finite graded poset P is called *Eulerian* if for any $x \le y$ in P,

$$\mu(x,y) = (-1)^{\mathrm{rk}(y)-\mathrm{rk}(x)}. \tag{C.9}$$

The Boolean poset is Eulerian. More generally, the poset of faces a convex polytope is Eulerian.

C.1.7. Incidence module. Let $\mathrm{M}(P)$ denote the vector space of $\Bbbk$-valued functions on P. The incidence algebra $\mathrm{I}(P)$ acts on $\mathrm{M}(P)$ on the left: For $f \in \mathrm{I}(P)$ and $g \in \mathrm{M}(P)$, define $fg \in \mathrm{M}(P)$ by

$$(fg)(x) = \sum_{y:\, x \le y} f(x,y)g(y). \tag{C.10}$$

Thus, $\mathrm{M}(P)$ is a left module over $\mathrm{I}(P)$. We call it the *incidence module* of P.

For functions f and g on P,

$$g(x) = \sum_{y:\, x\le y} f(y) \iff f(x) = \sum_{y:\, x \le y} \mu(x,y)g(y). \tag{C.11}$$

This is equivalent to $g = \zeta f \iff f = \mu g$. The passage from the first equation to the second is called *Möbius inversion*. In this situation, we say that g is the *exponential* of f, and f is the *logarithm* of g.

Similar to (C.10), there is also a right action of $\mathrm{I}(P)$ on $\mathrm{M}(P)$. Using it, we deduce

$$g(y) = \sum_{x:\, x\le y} f(x) \iff f(y) = \sum_{x:\, x \le y} g(x)\mu(x,y). \tag{C.12}$$

C.1.8. Example: algebra of upper triangular matrices. Consider the poset $P=[n]$ under the usual order $1 < 2 < \cdots < n$. Then $\mathrm{I}(P)$ can be identified with the algebra of upper triangular matrices of size n: An incidence function f corresponds to the upper triangular matrix whose ij-th entry is $f(i,j)$. Proposition C.2 says that an upper triangular matrix is invertible iff its diagonal entries are nonzero. The proof of Proposition C.1 proceeds by writing a unipotent matrix as a sum of the identity matrix and a nilpotent matrix.

The zeta function is the upper triangular matrix whose nonzero entries are all 1. Its inverse, which is the Möbius function, is the matrix whose diagonal entries are 1, entries just above the diagonal are -1, and remaining entries are 0. That is,

$$\mu(i,j) = \begin{cases} 1 & \text{if } i = j, \\ -1 & \text{if } i+1 = j, \\ 0 & \text{otherwise.} \end{cases}$$

The incidence module $\mathrm{M}(P)$ can be identified with $\Bbbk^n$: A function g corresponds to the vector whose i-th coordinate is $g(i)$. The left and right action of $\mathrm{I}(P)$ on $\mathrm{M}(P)$ corresponds to the left and right action of matrices on column and row vectors, respectively.

C.1.9. Incidence bimodule. Let P, Q and R be posets and let $\varphi : P \to R$ and $\psi : Q \to R$ be order-preserving maps. Let $\mathrm{I}(\varphi, \psi)$ denote the space of functions defined on the set

$$\{(x,w) \in P \times Q \mid \varphi(x) \leq \psi(w)\} \tag{C.13}$$

and with values in $\Bbbk$. Then $\mathrm{I}(\varphi, \psi)$ carries a structure of $\mathrm{I}(P)$-$\mathrm{I}(Q)$-bimodule, with actions defined by

$$(fm)(x,w) = \sum_{\substack{y \in P:\, x \leq y \\ \varphi(y) \leq \psi(w)}} f(x,y)m(y,w)$$

and

$$(mg)(x,w) = \sum_{\substack{v \in Q:\, v \leq w \\ \varphi(x) \leq \psi(v)}} m(x,v)g(v,w)$$

for $f \in \mathrm{I}(P)$, $g \in \mathrm{I}(Q)$, and $m \in \mathrm{I}(\varphi, \psi)$. We call this an *incidence bimodule*.

A special case arises when Q and R are the one-element poset. We obtain a left $\mathrm{I}(P)$-module, and this is precisely the incidence module $\mathrm{M}(P)$ of P.

C.1.10. Rota formula. Let (λ, ρ) be a Galois connection between posets P and Q as in Section B.5.1. Consider the incidence bimodules $\mathrm{I}(\lambda, \mathrm{id}_Q)$ and $\mathrm{I}(\mathrm{id}_P, \rho)$. They both consist of scalar-valued functions defined on the set

$$X_{\lambda,\rho} = \{(x,w) \in P \times Q \mid \lambda(x) \leq w\} = \{(x,w) \in P \times Q \mid x \leq \rho(w)\},$$

and thus coincide. Let $I_{\lambda,\rho}$ denote this common bimodule.

Consider now the functions δ_λ, δ_ρ and $\zeta_{\lambda,\rho} \in I_{\lambda,\rho}$ defined by

$$\delta_\lambda(x,w) = \begin{cases} 1 & \text{if } \lambda(x) = w, \\ 0 & \text{if not,} \end{cases} \qquad \delta_\rho(x,w) = \begin{cases} 1 & \text{if } x = \rho(w), \\ 0 & \text{if not,} \end{cases}$$

and

$$\zeta_{\lambda,\rho}(x,w) = 1 \text{ for all } (x,w) \in X_{\lambda,\rho}.$$

One readily finds that

$$\delta_\lambda \zeta_Q = \zeta_{\lambda,\rho} = \zeta_P \delta_\rho.$$

We acted on δ_λ from the right with the zeta function of Q, and on δ_ρ from the left with the zeta function of P. Acting on both sides of $\zeta_{\lambda,\rho}$ with the Möbius functions, it follows that

$$\mu_P \delta_\lambda = \delta_\rho \mu_Q.$$

Evaluating on $(x, w) \in X_{\lambda,\rho}$ we obtain

$$\sum_{y\in P:\, \lambda(y)=w} \mu_P(x,y) = \sum_{v\in Q:\, \rho(v)=x} \mu_Q(v,w). \tag{C.14}$$

This is the *Rota formula*.

C.1.11. Example: mapping cylinder. Let $\varphi : P \to Q$ be an order-preserving map. The *mapping cylinder* M_φ is the disjoint union of the posets P and Q together with the relations

$$x \leq w \text{ if } \varphi(x) \leq w,$$

for $x \in P$ and $w \in Q$.

Let us determine the Möbius function of the mapping cylinder. Both P and Q are convex subposets of M_φ. Therefore,

$$\mu_{M_\varphi}(x,y) = \mu_P(x,y) \quad \text{and} \quad \mu_{M_\varphi}(v,w) = \mu_Q(v,w)$$

when $x, y \in P$ and $v, w \in Q$. In addition, the inclusion $\rho : Q \to M_\varphi$ and the map $\lambda : M_\varphi \to Q$ given by

$$\lambda(x) = \varphi(x) \quad \text{and} \quad \lambda(w) = w,$$

$(x \in P,\ w \in Q)$ define a Galois connection. Applying (C.14) to $x \in P$ and $w \in Q$ with $\varphi(x) \leq w$, we obtain that

$$\mu_{M_\varphi}(x,w) = -\sum_{\substack{x'\in P:\, x\leq x' \\ \varphi(x')=w}} \mu_P(x,x').$$

The incidence algebra of M_φ consists of matrices of the following form:

$$\begin{bmatrix} f & m \\ 0 & g \end{bmatrix}$$

where $f \in \mathrm{I}(P)$, $g \in \mathrm{I}(Q)$, and $m \in \mathrm{I}(\varphi, \mathrm{id}_Q)$. Elements multiply using the algebra structure of each of the incidence algebras and the bimodule structure of the incidence bimodule:

$$\begin{bmatrix} f & m \\ 0 & g \end{bmatrix} \begin{bmatrix} f' & m' \\ 0 & g' \end{bmatrix} = \begin{bmatrix} ff' & fm' + mg' \\ 0 & gg' \end{bmatrix}.$$

The zeta function of the mapping cylinder is

$$\zeta_{M_\varphi} = \begin{bmatrix} \zeta_P & \zeta_\varphi \\ 0 & \zeta_Q \end{bmatrix},$$

where $\zeta_\varphi(x,w) = 1$ whenever $\varphi(x) \leq w$.

Exercise C.8. Rederive the expression for the Möbius function of M_φ employing the previous description of the incidence algebra in terms of matrices.

C.1.12. Example: homotopy colimit. Let I be a poset, $\{P_i\}_{i\in I}$ a family of posets, and

$$\varphi_{i,j} : P_i \to P_j$$

a family of order-preserving maps, one for each pair (i, j) with $i \leq j$ in I, and such that

$$\varphi_{i,i} = \mathrm{id}_{P_i}$$

and

$$\begin{array}{ccc} P_i & \xrightarrow{\varphi_{i,k}} & P_k \\ & {\scriptstyle \varphi_{i,j}} \searrow \quad \nearrow {\scriptstyle \varphi_{j,k}} & \\ & P_{i,j} & \end{array}$$

commutes whenever $i \leq j \leq k$ in I.

The *homotopy colimit* associated to this data is the disjoint union of the posets P_i, $i \in I$, together with the relations

$$x \leq y \text{ if } x \in P_i,\ y \in P_j,\ i \leq j \text{ in } I, \text{ and } \varphi_{i,j}(x) \leq y \text{ in } P_j.$$

The conditions above guarantee that this is indeed a partial order.

Exercise C.9. Use Hall formula (C.6) to determine the Möbius function of the homotopy colimit: given x and y as above,

$$\mu(x,y) = \mu_I(i,j) \sum_{\substack{x' \in P_i:\, x \leq x' \\ \varphi_{i,j}(x') = y}} \mu_{P_i}(x,x').$$

The incidence algebra of the homotopy colimit consists of matrices $(f_{i,j})_{i,j \in I}$ where $f_{i,i} \in \mathrm{I}(P_i)$, $f_{i,j} \in \mathrm{I}(\varphi_{i,j}, \mathrm{id}_{P_j})$ when $i \leq j$ in I, and $f_{i,j} = 0$ otherwise. Elements multiply using the algebra structure of each of the incidence algebras and the bimodule structure of the incidence bimodules.

The mapping cylinder is the special case of the homotopy colimit construction in which $I = \{1 < 2\}$.

C.1.13. Incidence algebra of a locally finite category. A *locally finite category* is a category in which each arrow can be written as a composition of non-identity arrows in only finitely many ways.

Let C be a locally finite category. The *incidence algebra* of C, denoted $\mathrm{I}(\mathsf{C})$, consists of functions f on morphisms in C, with the product of f and g given by

$$(fg)(\gamma) = \sum_{\alpha \circ \beta = \gamma} f(\alpha) g(\beta).$$

By local finiteness, the sum in the rhs is finite. The unit element is the function which is 1 on identity morphisms, and 0 otherwise.

For a locally finite poset P, observe that the incidence algebra of P is the opposite of the incidence algebra of the category associated to P.

Note that

$$\mathrm{I}(\mathsf{C} \times \mathsf{C}') = \mathrm{I}(\mathsf{C}) \otimes \mathrm{I}(\mathsf{C}')$$

for locally finite categories C and C'. This generalizes the fact that

$$\mathrm{I}(P \times P') = \mathrm{I}(P) \otimes \mathrm{I}(P')$$

for locally finite posets P and P'.

C.2. Radical of an incidence algebra

We now give some structure results on incidence algebras. They involve the notion of an elementary algebra reviewed in Section D.8.

Let P be a graded finite lattice with minimum element $\perp$ and maximum element $\top$. Its linearization $\Bbbk P$ is a split-semisimple commutative algebra with primitive idempotents $\mathtt{Q}_x$. This is discussed in Section D.9.

Proposition C.10. *The incidence algebra* $\mathrm{I}(P)$ *of a finite lattice* P *is elementary. Its split-semisimple quotient is* $\Bbbk P$*, with the quotient map*

$$\mathrm{I}(P) \twoheadrightarrow \Bbbk P, \qquad f \mapsto \sum_{x \in P} f(x,x)\, \mathtt{Q}_x. \tag{C.15}$$

In particular, the radical of $\mathrm{I}(P)$ *consists of incidence functions which are zero on* 1*-chains of the form* (x,x)*.*

PROOF. Let J denote the set of incidence functions which are zero on 1-chains of the form (x,x). Observe that J is a nilpotent ideal and all nilpotent elements belong to J. Hence J is the radical of $\mathrm{I}(P)$. All claims follow. □

Proposition C.11. *For each* $x \in P$*, let* e_x *denote the incidence function which is* 1 *on* (x,x) *and* 0 *otherwise. The* e_x*, as* x *varies, is a complete system of primitive orthogonal idempotents of* $\mathrm{I}(P)$*.*

PROOF. We only have to note that the map $\mathtt{Q}_x \mapsto e_x$ is an algebra section of (C.15). □

Proposition C.12. *The* i*-th power of the radical of* $\mathrm{I}(P)$ *consists of incidence functions* f *such that* $f(x,y) = 0$ *whenever* $\mathrm{rk}(y) - \mathrm{rk}(x) < i$*.*

PROOF. This is straightforward. □

As a consequence:

Proposition C.13. *The nilpotency index of the radical of* $\mathrm{I}(P)$ *is* $r+1$*, where* r *is the rank of* P*. The* r*-th power of the radical consists of incidence functions which are zero on all* 1*-chains except* $(\bot, \top)$*.*

Theorem C.14. *The quiver of the incidence algebra* $\mathrm{I}(P)$ *is as follows. The vertices are elements of* P*, and there is exactly one arrow from* y *to* x *when* $x \lessdot y$*, and no arrows otherwise. In other words, the quiver is the Hasse diagram of* P*.*

PROOF. Since the split-semisimple quotient of $\mathrm{I}(P)$ is $\Bbbk P$, the vertices of its quiver are elements of P. The arrows can be computed from Proposition C.12. Note that $J/J^2 \cong \bigoplus_{x \lessdot y} e_x \mathrm{I}(P)\, e_y$, where J is the radical of $\mathrm{I}(P)$, and the e_x are as in Proposition C.11. Thus, $e_x(J/J^2)e_y$ is zero unless $x \lessdot y$, and in this case, its dimension is 1. □

Exercise C.15. Describe the quiver of the algebra of upper triangular matrices. (Use the discussion in Section C.1.8.)

C.3. Reduced incidence algebras

We review reduced incidence algebras. These are subalgebras that arise from order-compatible equivalence relations on the 1-chains of a poset.

C.3.1. Reduced incidence algebras. Let P be a poset and $\sim$ be an equivalence relation on the set of 1-chains in P. Let $\mathrm{I}_\sim(P)$ consist of those incidence functions f such that $f(x,y) = f(x',y')$ whenever $(x,y) \sim (x',y')$. Then $\mathrm{I}_\sim(P)$ is a subspace of $\mathrm{I}(P)$, but in general not a subalgebra. If it is, then we say that the equivalence relation $\sim$ is *order-compatible*, and refer to the subalgebra as the *reduced incidence algebra* associated to $\sim$. When $\sim$ is simply equality, we have $\mathrm{I}_\sim(P) = \mathrm{I}(P)$.

The zeta function of P belongs to any reduced incidence algebra $\mathrm{I}_\sim(P)$, being constant on 1-chains. The Möbius function does too, by Lemma D.25.

Proposition C.16. *Let P be a poset and $\sim$ be an equivalence relation on the 1-chains in P. Suppose that whenever $(x, z) \sim (x', z')$, there exists a bijection $\psi : [x, z] \to [x', z']$ such that*

$$(x, y) \sim (x', \psi(y)) \quad \text{and} \quad (y, z) \sim (\psi(y), z') \tag{C.16}$$

for all $y \in [x, z]$. Then $\sim$ is order-compatible.

The proof is straightforward.

Exercise C.17. Let $\sim$ be an equivalence relation on the 1-chains of a poset P. Employ the zeta function to show that if $\sim$ is order-compatible, then for any two equivalent 1-chains $(x, z) \sim (x', z')$, the intervals $[x, z]$ and $[x', z']$ are equinumerous. Show more generally that $[x, z]$ and $[x', z']$ contain the same number of chains of length k, for all $k \geq 0$.

C.3.2. Classical Möbius function. Let P be the poset of divisors. Elements of P are positive integers with $a \leq b$ if a divides b. Let $(a, b) \sim (c, d)$ if $b/a = d/c$. This is an equivalence relation on the 1-chains in P. It is order-compatible. Indeed, multiplication by c/a specifies a bijection between $[a, b]$ and $[c, d]$ whenever $(a, b) \sim (c, d)$, and (C.16) holds.

Observe that $\mathrm{I}_{\sim}(P)$ can be described as functions f on P with the product of f and g given by

$$(fg)(n) = \sum_{a:\, a \text{ divides } n} f(a)g(n/a).$$

The sum is over all divisors a of n. One may thus identify $\mathrm{I}_{\sim}(P)$ with the algebra of *Dirichlet series*

$$\hat{f}(s) = \sum_{n \geq 1} \frac{f(n)}{n^s}.$$

The zeta function of P corresponds to the Riemann zeta function

$$\zeta(s) = \sum_{n \geq 1} \frac{1}{n^s}.$$

Each interval in P is a lattice: least common multiples and greatest common divisors are joins and meets, respectively. Let p be a prime divisor of n. Employing (C.7a) with $\bot = 1$, $x = p$ and $z = n$, we deduce that

$$\mu(n) = \begin{cases} -\mu(n/p) & \text{if } p \text{ does not divide } n/p, \\ 0 & \text{otherwise.} \end{cases}$$

It follows that

$$\mu(n) = \begin{cases} (-1)^r & \text{if } n \text{ has } r \text{ distinct prime factors, all of multiplicity 1,} \\ 0 & \text{if } n \text{ has repeated prime factors.} \end{cases} \tag{C.17}$$

This is the classical Möbius function.

C.3.3. Functoriality. Let P and Q be posets, each with an order-compatible relation $\sim$. Let $\varphi : P \to Q$ be an order-preserving map. We say that φ *creates relations* if it satisfies the following condition. Given equivalent 1-chains $(u, v) \sim (u', v')$ in Q, and x, x' in P with $\varphi(x) = u$ and $\varphi(x') = u'$, there exists a bijection

$$\psi : \{y \in P \mid x \leq y,\ \varphi y = v\} \to \{y' \in P \mid x' \leq y',\ \varphi y' = v'\} \tag{C.18}$$

such that for every y in its domain

$$(x, y) \sim (x', \psi y). \tag{C.19}$$

The bijection ψ depends on x, x', v and v' and need not be order-preserving.

Assume now that $\varphi : P \to Q$ is surjective and creates relations. Consider the map

$$\varphi_* : \mathrm{I}_\sim(P) \to \mathrm{I}_\sim(Q)$$

defined as follows. Given an incidence function f and a 1-chain (u, v) on Q, pick x in P such that $\varphi(x) = u$. Then set

$$\varphi_*(f)(u, v) = \sum_{\substack{y:\, x \le y,\\ \varphi y = v}} f(x, y). \tag{C.20}$$

Proposition C.18. *In this situation, φ_* is a well-defined morphism of algebras.*

PROOF. Suppose $\varphi x' = u = \varphi x$. Consider the special case of (C.18) in which $u = u'$ and $v = v'$. There is a bijection ψ such that for any y contributing a summand to the rhs of (C.20) we have $(x, y) \sim (x', \psi y)$. Changing variables by means of ψ and noting that f is constant on classes, we see that $\varphi_*(f)(u, v)$ does not depend on the choice of x.

Pick $\varphi(x) = u$. We calculate

$$\begin{aligned}
(\varphi_*(f)\varphi_*(g))(u, w) &= \sum_{v:\, u \le v \le w} \varphi_*(f)(u, v)\varphi_*(g)(v, w) \\
&= \sum_{v:\, u \le v \le w} \sum_{\substack{y:\, x \le y,\\ \varphi y = v}} f(x, y) \sum_{\substack{z:\, y \le z,\\ \varphi z = w}} g(y, z) = \sum_{\substack{y,z:\, x \le y \le z\\ \varphi(z) = w}} f(x, y)g(y, z) \\
&= \sum_{\substack{z:\, x \le z\\ \varphi(z) = w}} (fg)(x, z) = \varphi_*(fg)(u, w).
\end{aligned}$$

Thus, φ_* preserves multiplications. A similar calculation shows that φ_* preserves units.

Finally, we check that $\varphi_*(f)$ is constant on equivalence classes. Given equivalent 1-chains $(u, v) \sim (u', v')$ in Q, pick $\varphi(x) = u$, $\varphi(x') = u'$, and ψ as in (C.18). Then for each y contributing a summand to $\varphi_*(f)(u, v)$ in (C.20), $y' = \psi y$ contributes to $\varphi_*(f)(u', v')$, and $(x, y) \sim (x', y')$ in P. Since f is constant on classes, we have $\varphi_*(f)(u, v) = \varphi_*(f)(u', v')$ as needed. $\square$

Consider now the special case in which the relation on the 1-chains of Q is equality. In this situation, the following condition guarantees that φ creates relations. Given x and x' in P with $\varphi x = \varphi x'$, there exists a bijection

$$\psi : \{y \in P \mid x \le y\} \to \{y' \in P \mid x' \le y'\} \tag{C.21}$$

such that (C.19) holds, and in addition the diagram

$$\begin{array}{ccc}
\{y \in P \mid x \le y\} & \xrightarrow{\ \psi\ } & \{y' \in P \mid x' \le y'\} \\
& {\scriptstyle\varphi} \searrow \quad \swarrow {\scriptstyle\varphi} & \\
& Q &
\end{array} \tag{C.22}$$

commutes. Indeed, the commutativity guarantees that ψ restricts to a bijection as in (C.18).

C.4. Poset cocycles and deformations of incidence algebras

Functoriality of the incidence algebra construction naturally leads to the consideration of poset 2-cocycles and the associated deformations. We study these notions and illustrate the discussion with examples for posets of triangular type. We also consider group actions on posets and their effect on incidence algebras.

C.4.1. Poset cocycles. Let P be a poset and $\mathbb{A}$ a commutative monoid, written multiplicatively. We do not require that $\mathbb{A}$ be a group.

A 1-*cochain* is a map

$$\alpha : \{(x, y) \in P^2 \mid x \leq y\} \to \mathbb{A}.$$

A 1-*cocycle* is a 1-cochain α such that

$$\alpha(x, z) = \alpha(x, y)\alpha(y, z) \tag{C.23}$$

for all $x \leq y \leq z$ in P.

A 2-*cochain* is a map

$$\gamma : \{(x, y, z) \in P^3 \mid x \leq y \leq z\} \to \mathbb{A}.$$

A 2-*cocycle* is a 2-cochain γ such that

$$\gamma(w, x, z)\gamma(x, y, z) = \gamma(w, y, z)\gamma(w, x, y) \tag{C.24}$$

for all $w \leq x \leq y \leq z$ in P.

The constant cochain (either 1 or 2-cochain) whose only value is the identity of $\mathbb{A}$ is a cocycle, the *trivial* one.

Two 2-cochains γ and γ' are *cohomologous* along a 1-cochain α if

$$\gamma(x, y, z)\alpha(x, z) = \gamma'(x, y, z)\alpha(y, z)\alpha(x, y) \tag{C.25}$$

for all $x \leq y \leq z$ in P. If γ' is trivial, so that

$$\gamma(x, y, z)\alpha(x, z) = \alpha(y, z)\alpha(x, y), \tag{C.26}$$

we say that γ *cobounds* α. The trivial 2-cocycle cobounds any 1-cocycle.

A 1-cochain α is *normalized* if

$$\alpha(x, x) = 1 \tag{C.27}$$

for all x in P. A 2-cochain γ is *normalized* if

$$\gamma(x, x, y) = 1 = \gamma(x, y, y) \tag{C.28}$$

for all $x \leq y$ in P.

Cochains (either 1 or 2-cochains) form a commutative monoid under pointwise multiplication. The unit element is the trivial cochain. Cocycles and normalized cocycles constitute submonoids.

Suppose $\mathbb{A}$ is a commutative group. Then all of the above are groups. Given a 1-cochain α, its *coboundary* is the 2-cochain $\partial(\alpha)$ defined by

$$\partial(\alpha)(x, y, z) = \alpha(y, z)\alpha(x, z)^{-1}\alpha(x, y). \tag{C.29}$$

It is always a 2-cocycle. Two 2-cochains are cohomologous along α iff $\gamma = \partial(\alpha)\gamma'$. A 2-cochain γ cobounds α iff $\gamma = \partial(\alpha)$. A 1-cochain α is a 1-cocycle iff $\partial(\alpha)$ is the trivial 2-cocycle. The coboundary map ∂ is a morphism of groups that preserves normalized cochains.

Exercise C.19. Let $P = 2^{[n]}$ be the collection of subsets of $[n]$ ordered by inclusion (a *Boolean* poset). Let $\mathbb{A} = \mathbb{N}$ be the additive monoid of nonnegative integers.

(i) Given $R \subseteq S \subseteq T$ in P, define

$$\gamma(R, S, T) = |\{(s,t) \in (S \setminus R) \times (T \setminus S) \mid s > t\}|.$$

Verify that γ is a 2-cocycle on P with values in $\mathbb{A}$.

(ii) Given $S \subseteq T$ in P, define

$$\alpha(S, T) = |\{(s,t) \in S \times (T \setminus S) \mid s > t\}|.$$

Verify that γ cobounds α.

If view these cochains as having values in the additive group $\mathbb{Z}$ of integers, then $\partial(\alpha) = \gamma$ is in fact a coboundary.

Remark C.20. Suppose $\mathbb{A}$ is an abelian group. With such coefficients, poset cohomology can be calculated from the classifying space of P. When P has a bottom or a top element, this space is contractible. In particular, in this case, any cocycle is a coboundary.

C.4.2. Cocycle deformations of the incidence algebra. Fix a field $\Bbbk$. Let P be a finite poset. Let γ be a normalized 2-cocycle on P with values in the multiplicative monoid $\Bbbk$. On the space of $\Bbbk$-valued incidence functions on P, define a new multiplication as follows:

$$(fg)(x,z) = \sum_{y:\, x \leq y \leq z} \gamma(x,y,z)\, f(x,y) g(y,z). \tag{C.30}$$

This operation is associative and unital, with the unit being the function δ from (C.2). These properties follow from (C.24) and (C.28). We use $\mathrm{I}(P;\gamma)$ to denote the resulting algebra.

If γ is the trivial 2-cocycle, then $\mathrm{I}(P;\gamma)$ recovers the incidence algebra $\mathrm{I}(P)$ from Section C.1.1.

Suppose now that γ and γ' are cohomologous along α, where γ and γ' are 2-cocycles and α is a 1-cochain, all normalized. Given an incidence function f, let f_α denote the incidence function defined by

$$f_\alpha(x,y) = \alpha(x,y) f(x,y)$$

for all $x \leq y$ in P.

Lemma C.21. *The map*

$$\mathrm{I}(P;\gamma) \to \mathrm{I}(P;\gamma'), \quad f \mapsto f_\alpha,$$

is a morphism of algebras. If α takes values in the multiplicative group $\Bbbk^\times$, this is an isomorphism.

PROOF. It follows readily from (C.25) that the map preserves multiplications. It preserves units because α is normalized. When α takes invertible values, the inverse isomorphism sends g to $g_{\alpha^{-1}}$. □

Let P and γ be as above. Let $\sim$ be an equivalence relation on the set of 1-chains in P. As in Section C.3, we consider the space of incidence functions that are constant on equivalence classes. If this subspace is a subalgebra of $\mathrm{I}(P;\gamma)$, we say that the relation $\sim$ is *γ-compatible*, and refer to the subalgebra as the *reduced incidence algebra* associated to $\sim$ and γ.

Lemma C.22. *Suppose that whenever* $(x, z) \sim (x', z')$, *there exists a bijection* $\psi : [x, z] \to [x', z']$ *such that* (C.16) *holds, and in addition*

$$\gamma(x, y, z) = \gamma(x', \psi(y), z') \tag{C.31}$$

for all $y \in [x, z]$. *Then* $\sim$ *is* γ*-compatible.*

C.4.3. Transfer of cocycles. Let $\varphi : P \to Q$ be an order-preserving map between posets. Let γ be a 2-cochain on P with values on the multiplicative monoid $\Bbbk$. We discuss certain conditions under which γ can be transferred to Q.

Given $x \leq z$ in P and $v \in Q$ such that $\varphi(x) \leq v \leq \varphi(z)$, consider the set

$$\Gamma_{x,z}(v) = \{y \in [x, z] \mid \varphi(y) = v\}.$$

The conditions are as follows. For any $x \leq z$ in P,

$$\Gamma_{x,z}(\varphi x) = \{x\} \quad \text{and} \quad \Gamma_{x,z}(\varphi z) = \{z\}. \tag{C.32a}$$

For any $x \leq z$ in P and $v \in Q$ as above, the scalar

$$\sum_{y \in \Gamma_{x,z}(v)} \gamma(x, y, z) \tag{C.32b}$$

depends only on v, $u = \varphi(x)$ and $w = \varphi(z)$ (but is otherwise independent of x and z). This scalar may be zero and the set $\Gamma_{x,z}(v)$ may be empty.

When conditions (C.32a) and (C.32b) hold, we say that γ *can be transferred along* φ. We construct the transfer γ_* next.

Take $u \leq v \leq w$ in Q. First of all, if there are no $x \leq z$ in P with $\varphi(x) = u$ and $\varphi(z) = w$, we set

$$\gamma_*(u, v, w) = \begin{cases} 1 & \text{if } u = v \text{ or } v = w, \\ 0 & \text{otherwise.} \end{cases} \tag{C.33}$$

These values of γ_* will play no essential role. On the other hand, if there is at least one such pair $x \leq y$, we set

$$\gamma_*(u, v, w) = \sum_{y \in \Gamma_{x,z}(v)} \gamma(x, y, z). \tag{C.34}$$

Condition (C.32b) states that this is well-defined. We obtain a 2-cochain γ_* on Q.

It follows from (C.33) and (C.34) that if $u < v < w$ and there is no $x < y < z$ such that $\varphi(x) = u$, $\varphi(y) = v$ and $\varphi(z) = w$, then $\gamma_*(u, v, w) = 0$.

Lemma C.23. *Let* γ *be a normalized* 2*-cocycle on* P *that can be transferred along* φ. *Then the transfer* γ_* *is a normalized* 2*-cocycle on* Q.

PROOF. When γ_* is defined by (C.34), the fact that it is normalized follows from condition (C.32a) (and the fact that so is γ). Otherwise, the property follows from (C.33).

Take $t \leq u \leq v \leq w$ in Q. In order to derive the cocycle condition

$$\gamma_*(t, u, w)\gamma_*(u, v, w) = \gamma_*(t, v, w)\gamma_*(t, u, v),$$

we may assume that $t < u < v < w$, since γ_* is normalized. We claim that either there exist $a < b < c < d$ in P such that

$$\varphi(a) = t, \quad \varphi(b) = u, \quad \varphi(c) = v \quad \text{and} \quad \varphi(d) = w,$$

or else both sides are 0. Indeed, suppose the lhs is different from 0. From $\gamma_*(t, u, w) \neq 0$ we deduce the existence of $a < b < d$ in P such that $\varphi(a) = t$,

$\varphi(b) = u$, and $\varphi(d) = w$, as noted above. Then, from $\gamma_*(u, v, w) \neq 0$ and (C.32b), we deduce the existence of c such that $b < c < d$ and $\varphi(c) = v$. A similar analysis leads to the same conclusion when the rhs is nonzero.

It remains to show that when such elements exist, the two sides are equal. Fix $a < d$ with $\varphi(a) = t$ and $\varphi(d) = w$. Consider the set

$$\Gamma_{a,d}(u, v) = \{(b', c') \in P^2 \mid a \leq b' \leq c' \leq d,\ \varphi(b') = u,\ \varphi(c') = v\}.$$

There are obvious bijections

$$\bigsqcup_{b' \in \Gamma_{a,d}(u)} \Gamma_{b',d}(v) \cong \Gamma_{a,d}(u, v) \cong \bigsqcup_{c' \in \Gamma_{a,d}(v)} \Gamma_{a,c'}(u).$$

In view of (C.34),

$$\begin{aligned}
\gamma_*(t, u, w)\gamma_*(u, v, w) &= \sum_{b' \in \Gamma_{a,d}(u)} \gamma(a, b', d) \sum_{c' \in \Gamma_{b',d}(v)} \gamma(b', c', d) \\
= \sum_{(b',c') \in \Gamma_{a,d}(u,v)} &\gamma(a, b', d)\gamma(b', c', d) = \sum_{(b',c') \in \Gamma_{a,d}(u,v)} \gamma(a, c', d)\gamma(a, b', c') \\
&= \sum_{c' \in \Gamma_{a,d}(v)} \gamma(a, c', d) \sum_{b' \in \Gamma_{a,c'}(u)} \gamma(a, b', c') = \gamma_*(t, v, w)\gamma_*(t, u, v). \quad \square
\end{aligned}$$

In the exercises below, we carry out two consecutive cocycle transfers. They show in particular that γ_* need not be trivial, even when γ is.

Exercise C.24 (Schubert symbol). Let q be a prime power, $\Bbbk$ a field with q elements, and V an n-dimensional vector space over $\Bbbk$. Fix a complete flag of subspaces

$$0 = V_0 \subset V_1 \subset \cdots \subset V_n = V.$$

(The symbol $\subset$ denotes proper inclusions.) Given a subspace W of V, let

$$\varphi(W) = \{i \in [n] \mid W \cap V_{i-1} \subset W \cap V_i\}.$$

Equivalently, $i \notin \varphi(W) \iff W \cap V_{i-1} = W \cap V_i$. Let P be the poset of subspaces of V, ordered by inclusion, and Q the Boolean poset $2^{[n]}$.

(i) Show that $\varphi : P \to Q$ is order-preserving.

(ii) Show that the trivial cocycle on P (with values on the multiplicative monoid $\Bbbk$) can be transferred along φ and that

$$\gamma_*(R, S, T) = q^{|\{(s,t) \in (S \setminus R) \times (T \setminus S) \mid s > t\}|}. \tag{C.35}$$

Note that γ_* also arises from the cocycle in Exercise C.19 by composing with the morphism of monoids $\mathbb{N} \to \Bbbk$, $n \mapsto q^n$. This shows that γ_* is defined for all scalars q, not just prime powers.

The following exercise employs the q-binomials and factorials defined in Section 16.4.2.

Exercise C.25. Let P be the Boolean poset $2^{[n]}$ and Q the chain $\{0 < 1 < \cdots < n\}$. Consider the order-preserving map $\varphi : P \to Q$ given by

$$\varphi(S) = |S|.$$

Fix a scalar $q \in \Bbbk$ and consider the 2-cocycle γ on P given by (C.35).

(i) Show that γ can be transferred along φ and that

$$\gamma_*(i,j,k) = \binom{k-i}{j-i}_q, \tag{C.36}$$

a q-binomial coefficient.

(ii) Show that if q is not a root of unity, or if $q = 1$ and $\Bbbk$ is of characteristic 0, this cocycle is a coboundary: $\gamma_* = \partial(\alpha)$ with

$$\alpha(i,j) = \frac{1}{(j-i)_q!}.$$

If $q = 1$, γ is trivial, while $\gamma_*(i,j,k) = \binom{k-i}{j-i}$ is the classical binomial coefficient.

Cocycle transfer is functorial. We employ the preceding examples to illustrate this assertion.

Let P be the poset of subspaces of a finite vector space V, as in Exercise C.24, and Q the chain of length $\dim V$, as in Exercise C.25. Composing the maps in those exercises we obtain an order-preserving map $P \to Q$. It sends a subspace to its dimension. The trivial cocycle on P can be transferred and the result is the cocycle (C.36) on Q.

We return to the general discussion: $\varphi : P \to Q$ is order-preserving, γ is a normalized 2-cocycle on P that can be transferred along φ, and γ_* is the resulting 2-cocycle on Q. Consider now the map

$$\varphi^* : \mathrm{I}(Q;\gamma_*) \to \mathrm{I}(P;\gamma)$$

defined by

$$\varphi^*(f)(x,z) = f(\varphi x, \varphi z), \tag{C.37}$$

and the equivalence relation on 1-chains in P defined by

$$(x,z) \sim (x',z') \text{ if } \varphi(x) = \varphi(x') \text{ and } \varphi(z) = \varphi(z'). \tag{C.38}$$

Proposition C.26. *In this situation, φ^* is a morphism of algebras. Moreover, the relation (C.38) is γ-compatible and the image of φ^* is the corresponding reduced incidence algebra of P.*

PROOF. First note that if $x < z$, then $\varphi(x) < \varphi(z)$. This is equivalent to (C.32a). This shows that φ^* preserves the incidence unit δ. We now compare multiplications. Choose $x \le z$ in P and let $\varphi(x) = u$, $\varphi(z) = w$. We have

$$\begin{aligned}
(\varphi^*(f)\varphi^*(g))(x,z) &= \sum_{y:\, x\le y\le z} \gamma(x,y,z) f(\varphi x,\varphi y) g(\varphi y,\varphi z) \\
= \sum_{v:\, u\le v\le w}\ \sum_{y\in\Gamma_{x,z}(v)} \gamma(x,y,z) f(u,v) g(v,w) &= \sum_{v:\, u\le v\le w} \gamma_*(u,v,w)\, f(u,v) g(v,w) \\
&= (fg)(u,w) = \varphi^*(fg)(x,z).
\end{aligned}$$

Any function in the image of φ^* is constant on equivalence classes. Conversely, let g be such a function. If (u,w) is a 1-chain in Q and there is a 1-chain (x,z) in P such that $\varphi(x) = u$ and $\varphi(z) = w$, we set

$$f(u,w) = g(x,z).$$

By the assumption on g, $f(u,w)$ is well-defined. If there is no 1-chain (x,z) as above, we choose an arbitrary value for $f(u,w)$. The incidence function f on Q then satisfies $\varphi^*(f) = g$. It follows that the image of φ^* is the reduced incidence algebra of P associated to the relation (C.38) (and that this relation is γ-compatible). $\square$

We conclude the section with a couple of remarks on the transfer conditions (C.32a) and (C.32b). Assume $\varphi : P \to Q$ is an order-preserving map and γ is a normalized 2-cocycle on P.

Condition (C.32a) is equivalent to any of the following:

(C.39) The map φ is strictly increasing: if $x < z$, then $\varphi(x) < \varphi(z)$.

(C.40) The fibers of φ are antichains of P.

The following is a stronger condition than (C.32b): for any $u \leq w$ in Q, $x \leq z$ and $x' \leq z'$ in P with $\varphi(x) = \varphi(x') = u$ and $\varphi(z) = \varphi(z') = w$, there exists a bijection $\psi : [x,z] \to [x',z']$ such that (C.31) holds and in addition

(C.41)
$$\begin{array}{ccc} [x,z] & \xrightarrow{\psi} & [x',z'] \\ & {\scriptstyle\varphi}\searrow \quad \swarrow{\scriptstyle\varphi} & \\ & [u,w] & \end{array}$$

commutes. The bijection ψ need not be order-preserving.

Consider the relation (C.38). Note that (C.41) is equivalent to (C.16) for this relation. Lemma C.22 says that under the preceding conditions, this relation is γ-compatible. This also follows from Proposition C.26.

C.4.4. Posets of triangular type. Let P be a graded poset with bottom element $\perp$. Let rk be the rank function such that $\mathrm{rk}(\perp) = 0$. Consider the equivalence relation on 1-chains in P defined by

(C.42) $$(x,z) \sim (x',z') \text{ if } \mathrm{rk}(x) = \mathrm{rk}(x') \text{ and } \mathrm{rk}(z) = \mathrm{rk}(z').$$

This is the special case of (C.38) arising from the order-preserving map $\mathrm{rk} : P \to \mathbb{N}$. We say P is of *of triangular type* if $(x,z) \sim (x',z')$ implies that the number of maximal chains in $[x,z]$ equals the number of maximal chains in $[x',z']$.

Let γ be the trivial cocycle on P with values on the multiplicative group $\mathbb{Q}$.

Proposition C.27. *If P is of triangular type, γ can be transferred along* $\mathrm{rk} : P \to \mathbb{N}$. *Moreover, $\gamma_* = \partial(\alpha)$, with the 1-cochain α on $\mathbb{N}$ given by*

$$\alpha(i,j) = \frac{1}{m(i,j)},$$

where $m(i,j)$ is the number of maximal chains in $[x,y]$, x and y being any elements of P such that $x \leq y$, $\mathrm{rk}(x) = i$ and $\mathrm{rk}(y) = j$.

PROOF. Note that m and α are well-defined since P is of triangular type. Condition (C.32a) holds since rk is strictly increasing, which is (C.39). We next note that, given $i \leq j \leq k$ in $\mathbb{N}$ and $x \leq z$ in P with $\mathrm{rk}(x) = i$ and $\mathrm{rk}(z) = k$,

$$|\{y \in [x,z] : \mathrm{rk}(y) = j\}| m(j,k) m(i,j) = m(i,k).$$

Indeed, the rhs counts maximal chains from x to z, while the lhs counts these chains according to the element of rank j along the chain. This shows that (C.32b) holds, so γ can be transferred, and also that $\gamma_* = \partial(\alpha)$, according to (C.26). $\square$

Example C.28. The Boolean poset $2^{[n]}$ is of triangular type: the rank of a subset is its cardinality, and given $S \subseteq T$, the number of maximal chains from S to T is $(j-i)!$, where $i = |S|$, and $j = |T|$. The cocycle afforded by Proposition C.27 is the case $q = 1$ of the cocycle (C.36).

The poset of subspaces of a finite vector space, as in Exercise C.24, is of triangular type. The rank of a subspace is its dimension. The number of maximal chains from X to Y is $(j-i)_q!$, where $i = \dim X$ and $j = \dim Y$. The cocycle afforded by Proposition C.27 is (C.36).

C.4.5. Covering maps. Let P and Q be posets. An order-preserving map $\varphi : P \to Q$ is a *covering* if the following condition is satisfied. For any $y \in P$ and $u, v \in Q$ with

$$\varphi(y) = v \quad \text{and} \quad u \leq v,$$

there exists a unique $x \in P$ such that

$$\varphi(x) = u \quad \text{and} \quad x \leq y.$$

In this situation, we say that x is the lift of u to y.

Lemma C.29. *A covering map φ restricts to an (order-preserving) bijection*

$$[x, y] \to [\varphi x, \varphi y] \tag{C.43}$$

between intervals.

PROOF. Given w in $[\varphi x, \varphi y]$, let z be its lift to y and let x' be the lift of $\varphi(x)$ to z. Then $x' \leq z \leq y$ and $\varphi(x') = \varphi(x)$. Then both x and x' are lifts of $\varphi(x)$ to y, and hence must coincide. Then $x \leq z \leq y$ and $\varphi(z) = w$, proving that the map (C.43) is surjective. Injectivity follows from the uniqueness of lifts. □

Proposition C.30. *The trivial cocycle can be transferred along a covering map. Moreover, the transferred cocycle is trivial.*

PROOF. The injectivity of the map (C.43) gives us condition (C.32a), in the equivalent form (C.39). The lemma also implies that φ satisfies condition (C.41). Since (C.31) holds trivially, we also have (C.32b). In addition, the sets $\Gamma_{x,z}(v)$ are singletons, so the transferred cocycle is trivial. □

From Propositions C.26 and C.30 we deduce that the canonical map $\varphi^* : \mathrm{I}(Q) \to \mathrm{I}(P)$ is a morphism of algebras. Moreover, its image is the reduced incidence algebra of P arising from (C.38).

If in addition φ is surjective, then any interval $[u, v]$ in Q is the image of an interval $[x, y]$ in P (choose y such that $\varphi(y) = v$ and let x be the lift of u to y). It follows that φ^* is an injective morphism of algebras.

Exercise C.31. Show that the equivalence relation (C.38) satisfies condition (C.16) and deduce (again) that it is order-compatible.

C.4.6. Group actions on posets. Let G be a group acting on a poset P by order-preserving maps. Then G acts on the set of 1-chains in P and on the incidence algebra of P by

$$g \cdot (x, y) = (g \cdot x, g \cdot y) \quad \text{and} \quad (g \cdot f)(x, y) = f(g^{-1} \cdot x, g^{-1} \cdot y).$$

The equivalence relation on 1-chains in resulting from the G-action satisfies condition (C.16). We may then consider the corresponding reduced incidence algebra. It consists of functions which are constant on G-orbits of intervals and is therefore the invariant subalgebra $\mathrm{I}(P)^G$ of the incidence algebra $\mathrm{I}(P)$.

Assume now that the G-action on P satisfies in addition that

$$\text{if } x \leq g \cdot x \text{ for some } g \in G, \text{ then } g \cdot x = x. \tag{C.44}$$

Equivalently, each G-orbit is an antichain of P. We say in this case that the action is *regular*.

Let $\overline{x}$ denote the orbit of an element x. When the action is regular, there is a partial order on G-orbits defined by

$$\overline{x} \leq \overline{y} \text{ if there exists } g \in G \text{ such that } g \cdot x \leq y.$$

This is well-defined since the action of each element is order-preserving. Condition (C.44) guarantees antisymmetry of this order. We let P_G denote the resulting poset of G-orbits. The canonical map

$$\pi : P \to P_G, \quad x \mapsto \overline{x}$$

is order-preserving.

Proposition C.32. *Let G act on P by order-preserving maps. The action is regular in either of the following cases.*

(i) *The poset P is graded and the G-action is rank-preserving.*
(ii) *G is finite, or more generally if each element of G has finite order.*
(iii) *P is of finite height.*

Proof. In case (i), elements of the same rank constitute an antichain. In case (ii), note that $x \leq g \cdot x$ implies $g \cdot x \leq g^2 \cdot x$, and then if n is the order of g,

$$x \leq g \cdot x \leq \cdots \leq g^n \cdot x = x.$$

A similar argument applies in case (iii). □

Consider now the following conditions.

$$\text{If } x \leq y \text{ and } g \cdot x \leq y \text{ for some } x, y \in P \text{ and } g \in G, \text{ then there is } h \in G \text{ such that } h \cdot x = g \cdot x \text{ and } h \cdot y = y. \tag{C.45}$$

$$\text{If } x \leq y \text{ and } g \cdot x \leq y \text{ for some } x, y \in P \text{ and } g \in G, \text{ then } g \cdot x = x. \tag{C.46}$$

Proposition C.33. *Let G act on P by order-preserving maps. Then:*

(a) *(C.46) $\Rightarrow$ (C.45) $\Rightarrow$ (C.44).*
(b) *(C.45) $\Rightarrow$ the trivial cocycle on P can be transferred along π.*
(c) *(C.46) $\iff$ π is a covering.*

Proof. When (C.46) holds, we may choose $h = 1$ in (C.45). When (C.45) holds, choosing $y = g \cdot x$ yields $h \in G$ such that $h \cdot x = g \cdot x$ and $hg \cdot x = g \cdot x$. Canceling h yields $x = g \cdot x$, which shows (C.44) holds. This proves (a).

As already mentioned, (C.44) states that the fibers of π are antichains. This is condition (C.40) for π, which is equivalent to (C.32a). Now suppose (C.45) holds. We show below that (C.41) follows. This implies (C.32b) and completes the proof of (b).

Take intervals $[x,z]$ and $[x',z']$ as stipulated by (C.41). Then $x' = f\cdot x$ and $z' = k\cdot z$ for some $f,k\in G$. Let $g = k^{-1}f$. Then

$$g\cdot x = k^{-1}\cdot x' \leq k^{-1}\cdot z' = z.$$

Condition (C.45) affords $h\in G$ such that $h\cdot x = g\cdot x$ and $h\cdot z = z$. Then

$$kh\cdot x = kg\cdot x = f\cdot x = x' \quad \text{and} \quad kh\cdot z = k\cdot z = z'.$$

The action of kh then defines a bijection $[x,z]\to[x',z']$ commuting with π. Condition (C.31) holds trivially. Thus, (C.41) holds.

Finally, if $\overline{x}\leq\overline{y}$, there is an element in the orbit of x that is below y. Condition (C.46) guarantees this element is unique, so π is a covering. This proves (c). □

Suppose (C.45) holds. Let γ_G be the cocycle on P_G transferred from the trivial one along π. From Propositions C.26 and C.33 we deduce that

$$\pi^* : \mathrm{I}(P_G,\gamma_G)\to\mathrm{I}(P)$$

is an injective morphism of algebras which identifies $\mathrm{I}(P_G,\gamma_G)$ with the reduced incidence algebra of P arising from the following equivalence relation on 1-chains in:

$$(x,y)\sim(x',y') \text{ if there are } g,h\in G \text{ such that } g\cdot x = x' \text{ and } h\cdot y = y'. \tag{C.47}$$

We claim that this algebra coincides with the invariant subalgebra $\mathrm{I}(P)^G$. As explained above, the latter is the reduced incidence algebra arising from the G-action on intervals. Clearly, two intervals in the same G-orbit are equivalent under (C.47). The proof of (b) in Proposition C.33 shows the converse. Thus, the two relations on intervals, and the corresponding reduced incidence algebras, coincide.

In summary, condition (C.45) affords a canonical isomorphism of algebras

$$\mathrm{I}(P_G,\gamma_G)\cong\mathrm{I}(P)^G.$$

Under the stronger assumption that (C.46) holds, the cocycle γ_G is trivial (Proposition C.30) and we obtain an isomorphism of algebras

$$\mathrm{I}(P_G)\cong\mathrm{I}(P)^G.$$

Condition (C.46) is rare: if P has a top element, or if P is a join-semilattice (joins exist), then only the trivial action satisfies it.

Exercise C.34. Find an action of a finite group on a finite poset by order-preserving maps for which (C.45) does not hold. (By Proposition C.32, (C.44) does hold.)

Exercise C.35. Let I be a finite set, P be the Boolean poset 2^I and G the symmetric group S_I. G acts on P canonically. Let $n = |I|$. Show that:

(i) The canonical G-action on P satisfies (C.45) but not (C.46).

(ii) We have

$$P_G = \{0<1<\cdots<n\}$$

and the canonical map $\pi : P\to P_G$ sends an element of P to its rank.

(iii) The map $\mathrm{I}(P_G)\to\mathrm{I}(P)$ sending f to $\tilde{f}$ defined by

$$\tilde{f}(S,T) = (t-s)!\,f(s,t),$$

where $s=|S|$ and $t=|T|$, is a morphism of algebras. Its image is $\mathrm{I}(P)^G$.

Notes

Incidence algebras and Möbius functions. Möbius functions go back to work of Möbius [**314**]. His context is the reduced incidence algebra of the poset of divisors given in Section C.3.2. He explains formula (C.4) on pages 106-107. On page 108, he considers the zeta function. He then turns to its inverse, states the recursive formula (C.5b), and then determines it by a case analysis on pages 109-111. This is the classical Möbius function (C.17). A more detailed historical account is given in [**328**, Section 2.2]. The generalization to arbitrary posets appeared much later in the combined work of Weisner [**427**], Ward [**424**] and Hall [**219**]. Proposition C.4 (and its proof) is due to Weisner [**427**, Theorems 9 and 10]. Proposition C.3 is due to Hall [**219**, Section 2.2].

A main source for incidence algebras and Möbius functions is the classical paper by Rota [**364**]. (Proposition C.6 is his Theorem 4.) Other important references are those of Aigner [**13**, Chapter IV], Greene [**209**] and Stanley [**401**, Chapter 3]. For a survey on Eulerian posets, see [**396**]. Reduced incidence algebras are studied in [**159**, Section 4]. A characterization of order-compatible equivalence relations is given in [**264**].

Möbius inversion is closely related to the exp-log correspondence in Lie theory. An illustration is provided in Section 6.6.5. More details will be provided in a future work.

Ring-theoretical properties of incidence algebras are studied in [**391**]. Representations of incidence algebras and their cocycle deformations are studied in [**239**].

Galois connections. The set (C.13) is the object set of the comma-category $\varphi \downarrow \psi$. An adjunction gives rise to an isomorphism of comma-categories; this specializes to the observation in Section C.1.10 that $\mathrm{I}(\lambda, \mathrm{id}_Q) = \mathrm{I}(\mathrm{id}_P, \rho)$. The Rota formula (C.14) goes back to [**364**, Theorem 1]; see also [**209**, 5.4]. The approach in the text follows [**6**, Section 2].

Incidence algebras of categories. Incidence algebras of categories appeared in work of Mitchell [**313**, Section 7] and Gabriel [**194**, Section II.1] (though not under that name). In connection with Möbius inversion, they are considered by Content, Lemay, and Leroux [**124**], Haigh [**217**], Schwab [**380, 381**] and more recently by Leinster [**284, 285**]. These papers focus on the invertibility of the standard zeta function (which maps every morphism to 1), in contrast to the noncommutative zeta functions introduced in this work. Functoriality with respect to functors which are the identity on objects is given in [**124**, Proposition 5.6], see also [**285**, Proposition 2.1]. Functoriality with respect to coverings is a special case of [**285**, Theorem 5.1]. Both are special cases of a more general type of functoriality considered in [**4**, Sections 5.3 and 9.1].

Covering maps and cocycle deformations. Covering maps between categories go back to [**192**, Appendix I.1]. They are also called *discrete fibrations*. The notion of covering map between posets is a special case. Lemma C.29 is a special case of a more general assertion for categories: a discrete fibration is always a *Grothendieck fibration*.

The connection between poset 2-cocycles and reduced incidence algebras presented in Section C.4.3 is somewhat implicit in [**159**]. Closely related ideas are developed in [**164**, Chapter I, Section 4], where cocycles are prominent; in particular, Lemma C.23 generalizes [**164**, Proposition 1.48]. Posets of triangular type are the subject of [**159**, Section 9]. Interesting examples and results on lattices of triangular type are given there and in [**437**]. *Binomial posets* in the sense of [**401**, Definition 3.18.2] are particular posets of triangular type. Their theory is developed in [**159**, Sections 7–9] and [**401**, Section 3.18], but without an explicit connection to poset cocycles. Proposition C.27 corresponds to (3.89) in [**401**]. The posets in Example C.28 are binomial. Additional examples of binomial posets are given in [**401**, Example 3.18.3].

Group actions on posets. Regular group actions on simplicial complexes and posets appear in [**101**, Chapter III, Definition 1.2] and [**134**, Section 1]. Algebraic and enumerative aspects of group actions on posets are studied in [**395**].

APPENDIX D

Algebras and modules

We review some general aspects of associative algebras and their modules. We are interested mainly in split-semisimple commutative algebras and elementary algebras. We analyze in detail the split-semisimple commutative algebras which are obtained by linearizing lattices.

Convention D.1. In this appendix, $\mathbb{k}$ is an arbitrary field, A is a $\mathbb{k}$-algebra (not necessarily commutative), and M is a (left or right) module over A. We assume that A and M are finite-dimensional (though not everything we do requires this assumption.) Dimension refers to dimension as a vector space over $\mathbb{k}$.

D.1. Modules

We review some basic concepts related to modules over an algebra and set up the required notation.

D.1.1. Modules and representations. Let M be a left module over A. That is, M is a vector space over $\mathbb{k}$ equipped with a bilinear map

$$A \times M \to M, \qquad (a, m) \mapsto am,$$

such that $a(bm) = (ab)m$ and $1m = m$ for all $a, b \in A$ and $m \in M$.

Any $w \in A$ gives rise to a linear operator

$$\Psi_M(w) : M \to M, \qquad m \mapsto wm$$

defined by left multiplication by w. This gives rise to an algebra homomorphism

$$\Psi_M : A \to \operatorname{End}_{\mathbb{k}}(M).$$

The latter is the *algebra of endomorphisms* of M, where the product is composition:

$$(fg)(m) = f(g(m)).$$

We say that Ψ_M is the *representation* of A associated to the module M.

Similarly, a right A-module M is defined by a bilinear map

$$M \times A \to M, \qquad (m, a) \mapsto ma,$$

In this case, we let $\Psi_M(w)$ denote right multiplication by w. The resulting map Ψ_M is an algebra antimorphism.

Standard terms of linear algebra apply to the operator $\Psi_M(w)$. For instance, we can consider the eigenvalues of $\Psi_M(w)$ and their multiplicities. (By multiplicity, we always mean algebraic multiplicity.) Similarly, we say $\Psi_M(w)$ is *diagonalizable* if M can be expressed as a direct sum of subspaces such that $\Psi_M(w)$ acts on each subspace by multiplication by a scalar. The scalars are the eigenvalues of $\Psi_M(w)$ and the subspaces the eigenspaces.

For a left module M, let wM denote the image of the linear operator $\Psi_M(w)$. In other words, wM consists of all elements of the form wm, as m varies over elements of M. For a right module M, we denote the image by Mw.

D.1.2. Faithful and simple modules. A left A-module M is *faithful* if the representation Ψ_M is injective. The annihilator $\operatorname{ann}(M)$ of a left A-module M is the kernel of Ψ_M:

$$\operatorname{ann}(M) := \{a \in A \mid am = 0 \text{ for all } m \in M\}.$$

Thus, M is faithful iff $\operatorname{ann}(M) = 0$. Similar considerations apply to right A-modules.

A (left or right) module over A is *simple* if it is nonzero and has no proper submodules. Any one-dimensional A-module M is simple.

D.1.3. Characters. Let A be a $\Bbbk$-algebra. The *character* of a (left or right) A-module M is the linear functional

$$\chi_M : A \to \Bbbk, \qquad \chi_M(w) = \operatorname{tr}(\Psi_M(w)), \tag{D.1}$$

where $\operatorname{tr}(\Psi_M(w))$ denotes trace of the linear operator $\Psi_M(w)$. A linear functional on A is called a *character* of A if it is the character of some A-module M.

A *multiplicative character* of A is an algebra homomorphism $\chi : A \to \Bbbk$.

Lemma D.2. *If M is a one-dimensional A-module M, then χ_M is multiplicative. Conversely, given a multiplicative character χ, there exists a one-dimensional module M, unique up to isomorphism, such that $\chi_M = \chi$.*

PROOF. If M is one-dimensional, then $\operatorname{tr} : \operatorname{End}_{\Bbbk}(M) \to \Bbbk$ is an isomorphism of algebras, so χ_M is an algebra morphism. For the converse, $M \cong \Bbbk$ with $am = \chi(a)m$. □

For any A-module M, we have $\chi_M(1) = (\dim_{\Bbbk} M) \cdot 1$, where on the right 1 denotes the unit element of the ground field $\Bbbk$. It follows that if $\Bbbk$ is of characteristic 0, then

$$\chi_M(1) = 1 \iff \dim_{\Bbbk} M = 1 \iff \chi_M \text{ is a multiplicative character.}$$

D.1.4. Dual module. Let M be a left A-module. Write M^* for the linear dual of M. Then M^* is a right A-module with the dual action: For $a \in A$ and $f \in M^*$,

$$(fa)(m) := f(am). \tag{D.2}$$

Similarly, the dual of a right A-module is a left A-module.

Lemma D.3. *For a left A-module M and $w \in A$, we have* $\dim wM = \dim M^*w$.

PROOF. The space wM is the image of the operator $\Psi_M(w)$, while M^*w is the image of $\Psi_{M^*}(w)$. By (D.2), these operators are dual. Therefore, the image of one is orthogonal to the kernel of the other, and their images have the same dimension. □

For a commutative algebra A, there is no distinction between left and right modules. In this case, we say a module M is *self-dual* if there exists an isomorphism $M \cong M^*$ of A-modules.

D.1.5. Frobenius algebras. Fix an algebra A. A bilinear form $\langle\ ,\ \rangle : A \times A \to \Bbbk$ is *associative* if

$$\langle a, bc \rangle = \langle ab, c \rangle \tag{D.3}$$

holds for all $a, b, c \in A$. Any linear functional $f : A \to \Bbbk$ gives rise to an associative bilinear form via $\langle a, b \rangle := f(ab)$, and moreover one can recover the functional from the form by $f(a) := \langle a, 1 \rangle = \langle 1, a \rangle$. Thus, there is a correspondence between linear functionals and associative bilinear forms on A.

A *Frobenius algebra* is an algebra A equipped with a nondegenerate associative bilinear form. The corresponding linear functional is called the *Frobenius functional*. A Frobenius algebra yields an isomorphism $A \to A^*$ of left A-modules, and another isomorphism $A \to A^*$ of right A-modules. Thus, Frobenius structures on A correspond to self-duality isomorphisms of A (either as left or as right A-modules).

D.2. Idempotents and nilpotents

We review idempotent and nilpotent elements in an algebra, and discuss complete systems of primitive orthogonal idempotents.

D.2.1. Idempotents. Let A be an algebra. An element $e \in A$ is an *idempotent* if $e^2 = e$. Idempotents e and f are *orthogonal* if $ef = fe = 0$. In this case, $e + f$ is also an idempotent. Note that for any idempotent e, $1 - e$ is also an idempotent and it is orthogonal to e. A nonzero idempotent e is *primitive* if it cannot be written as a sum of two orthogonal nonzero idempotents.

Lemma D.4. *Every nonzero idempotent of A can be expressed as a sum of mutually orthogonal primitive idempotents.*

PROOF. Let e be the given idempotent. If e is primitive, then we are done. If not, then write $e = f + g$, with both f and g nonzero orthogonal idempotents. If f (or g) is not primitive, then write it as a sum of two orthogonal nonzero idempotents. Continue this procedure. If at some stage we have $e = e_1 + \cdots + e_k$, then $eA = e_1 A \oplus \cdots \oplus e_k A$, with each $e_i A \neq 0$. So by finite-dimensionality of A, this procedure must terminate. □

Applying Lemma D.4 to the unit element 1, we deduce that there exists a family of mutually orthogonal primitive idempotents which sum up to 1. Any such family is called a *complete system* of primitive orthogonal idempotents of A. Complete refers to the fact that the idempotents sum up to 1. Thus:

Proposition D.5. *Every algebra has a complete system of primitive orthogonal idempotents.*

Let $e \in A$ be an idempotent. Let M be a left A-module. The linear operator $\Psi_M(e)$ is diagonalizable. Its eigenvalues are 1 and 0 with eigenspaces eM and $(1 - e)M$, respectively. Its trace is the dimension of eM. Thus,

$$\chi_M(e) = \operatorname{tr}(\Psi_M(e)) = \dim eM. \tag{D.4}$$

D.2.2. Nilpotents. An element $a \in A$ is *nilpotent* if there exists an integer $k \geq 1$ such that $a^k = 0$.

For any nilpotent element $a \in A$ and left A-module M,

$$\chi_M(a) = \operatorname{tr}(\Psi_M(a)) = 0. \tag{D.5}$$

This is because the trace of any nilpotent matrix is 0.

Note that there is only one element in A which is both idempotent and nilpotent, namely, 0.

D.3. Split-semisimple commutative algebras

We discuss split-semisimple commutative algebras. They have a unique complete system of primitive idempotents. The simple modules over such an algebra are one-dimensional, and any module breaks as a direct sum of simple submodules. Further, all modules are self-dual.

D.3.1. Split-semisimple commutative algebras. Let A be a commutative $\Bbbk$-algebra. It is *split-semisimple* if it is isomorphic as an algebra to a product of copies of $\Bbbk$, that is, $A \cong \Bbbk^n$ for some n. For $1 \leq i \leq n$, let e_i denote the element of A which corresponds to $(0, \dots, 1, \dots, 0) \in \Bbbk^n$ which is 1 in the i-th coordinate and zero elsewhere. Observe that $f \in A$ is an idempotent iff f is a sum of some of the e_i. In particular, the e_i are the only primitive idempotents of A. These elements constitute a complete system of primitive orthogonal idempotents of A, and this system is unique. The only algebra automorphisms of A are those obtained by permuting the e_i.

A split-semisimple commutative algebra does not contain any nonzero nilpotent elements. So an algebra such as $\Bbbk[x]/(x^n)$ for $n > 1$ cannot be split-semisimple.

D.3.2. Modules. Suppose A is a split-semisimple commutative algebra, and M is an A-module. Then each $e_i M$ is a submodule of M, and further

$$M = \bigoplus_{i=1}^{n} e_i M. \tag{D.6}$$

An element $z \in A$ acts on $e_i M$ by scalar multiplication by the coefficient of e_i in z. Note that each $e_i A$ is one-dimensional.

For each $1 \leq i \leq n$, put

$$\eta_i(M) := \chi_M(e_i) = \dim e_i M. \tag{D.7}$$

The second equality can be seen directly, or as an instance of (D.4).

Some important consequences of the above discussion are summarized below.

Theorem D.6. *A split-semisimple commutative algebra A of dimension n has n distinct simple modules (up to isomorphism). They are one-dimensional. For $1 \leq i \leq n$, the i-th simple module is given by $e_i A$, or equivalently, by the multiplicative character*

$$\chi_i : A \to \Bbbk, \qquad z \mapsto \langle z, e_i \rangle,$$

where $\langle z, e_i \rangle$ denotes the coefficient of e_i in z.

By definition of χ_i, for any $w \in A$, we have $w = \sum_i \chi_i(w)\, e_i$. Thus,

$$\chi_M(w) = \sum_{i=1}^{n} \chi_i(w)\, \eta_i(M). \tag{D.8}$$

Theorem D.7. *Let A be a split-semisimple commutative algebra. Each A-module M is a direct sum of simple modules with the multiplicity of the i-th simple module being $\eta_i(M)$. In particular, M is faithful iff $\eta_i(M) > 0$ for each i.*

Theorem D.8. *Let A be a split-semisimple commutative algebra. For any element $w \in A$, the linear operator $\Psi_M(w)$ is diagonalizable. Writing $w = \sum_i \lambda_i e_i$, the operator $\Psi_M(w)$ has eigenvalues λ_i and the eigenspace of λ_i is $e_i M$. In particular, the multiplicity of λ_i is $\eta_i(M)$.*

It is possible that $e_i M$ is 0 for some i in which case the eigenvalue λ_i does not occur. It may also happen that the λ_i are not distinct. In that case, the eigenspaces are obtained by lumping together the corresponding $e_i M$. For instance, if $w = e_i$, then the eigenvalues are 1 and 0. The eigenspace for 1 is $e_i M$ and the eigenspace for 0 is the sum of the remaining $e_j M$.

Proposition D.9. *The characters of a split-semisimple commutative algebra A of dimension n correspond to families $(\eta_i)_{1 \le i \le n}$ of nonnegative integers, with the multiplicative ones corresponding to those families in which exactly one η_i is 1 and the rest are 0.*

The character χ and the family (η_i) relate by $\chi(e_i) = \eta_i$. Note that the character determines the module M (up to isomorphism), with η_i being the number of times the i-th simple module occurs in M.

D.3.3. Self-duality of modules. Suppose A is a split-semisimple commutative algebra, and M is an A-module. Consider the dual module M^*. It has two decompositions, one obtained by virtue of it being an A-module, and the other obtained by dualizing (D.6). However, there is a canonical identification

$$(e_i M)^* \xrightarrow{\cong} e_i M^*,$$

and so the two decompositions coincide. It follows that for disjoint subsets S and T whose union is $[n]$, the subspaces

$$\bigoplus_{i \in S} e_i M \qquad \text{and} \qquad \bigoplus_{j \in T} e_j M^*$$

are orthogonal complements of each other under the canonical pairing between M and M^*.

Lemma D.10. *Any module over a split-semisimple commutative algebra is self-dual.*

Proof. For each i, choose any linear isomorphism between $e_i M$ and $(e_i M)^*$. □

The isomorphism for self-duality is not canonical. We concentrate on the case when M is A itself.

Lemma D.11. *Let A be a split-semisimple commutative algebra. Then an A-module isomorphism $A \cong A^*$ corresponds to a family (β_i) of nonzero scalars indexed by the primitive idempotents e_i of A.*

Proof. An isomorphism $A \cong A^*$ of A-modules is obtained by choosing linear isomorphisms $e_i A \cong (e_i A)^*$ for each i. But each $e_i A$ is one-dimensional, so the corresponding linear isomorphism is given by multiplication by a nonzero scalar β_i. □

Equivalently:

Lemma D.12. *Let A be a split-semisimple commutative algebra. A linear functional $f : A \to \Bbbk$ corresponds to a sequence of scalars β_i via $f(e_i) = \beta_i$. Further, f is a Frobenius functional iff the β_i are nonzero.*

D.3.4. Subalgebras of split-semisimple commutative algebras. We now state two results related to subalgebras of split-semisimple commutative algebras. The proofs are straightforward.

Lemma D.13. *Any subalgebra of a split-semisimple commutative algebra is again split-semisimple commutative.*

Lemma D.14. *Let G be a finite group which acts on a split-semisimple commutative algebra A. Then: G acts on the set of primitive idempotents $\{e_1, \ldots, e_n\}$ of A. The invariant subalgebra A^G is split-semisimple commutative with primitive idempotents $\sum_{i \in I} e_i$, as I varies over all orbits of the G-action on $\{e_1, \ldots, e_n\}$.*

D.4. Diagonalizability and Jordan-Chevalley decomposition

An element of an algebra is diagonalizable if it can be expressed as a linear combination of mutually orthogonal idempotents. A diagonalizable element can be characterized using the factors of its minimum polynomial or the split-semisimplicity of the subalgebra it generates. We also discuss the Jordan-Chevalley decomposition of an element into its diagonalizable and nilpotent parts.

D.4.1. Minimum polynomial. Let A be an algebra over $\Bbbk$. For $w \in A$, let $\Bbbk[w]$ denote the subalgebra of A generated by w. Explicitly, it consists of all elements of A of the form

$$\alpha_0 + \alpha_1 w + \cdots + \alpha_k w^k$$

for scalars $\alpha_i \in \Bbbk$. There is a surjective algebra homomorphism

$$\Bbbk[x] \twoheadrightarrow \Bbbk[w], \qquad x \mapsto w,$$

where $\Bbbk[x]$ denotes the algebra of polynomials in the variable x. Since $\Bbbk[w]$ is finite-dimensional, the kernel is nonzero. Further, since all ideals in $\Bbbk[x]$ are principal, the kernel is generated by a unique monic polynomial. It is called the *minimum polynomial* of w. By construction, the quotient of $\Bbbk[x]$ by the ideal generated by the minimum polynomial of w is isomorphic to $\Bbbk[w]$.

The minimum polynomial of an idempotent $e \in A$ is x if $e = 0$, $x - 1$ if $e = 1$, and $x(x-1)$ otherwise. The minimum polynomial of any nilpotent element has the form x^k.

D.4.2. Diagonalizable elements. We say that $w \in A$ is *diagonalizable* if it can be expressed as a linear combination of mutually orthogonal idempotents. Note:

- For the endomorphism algebra $A = \operatorname{End}_{\Bbbk} V$, this specializes to the usual notion of diagonalizability.
- Every element of a split-semisimple commutative algebra is diagonalizable.

A diagonalizable element can be put in the form

$$w = \lambda_1 e_1 + \cdots + \lambda_n e_n, \tag{D.9}$$

where $e_1, \ldots, e_n$ are mutually orthogonal nonzero idempotents of A which add up to 1, and the λ_i are distinct scalars: If the e_i do not add up to 1, then $1 - \sum_i e_i$ is an idempotent which is orthogonal to all the e_i, so we can write $w = \lambda_1 e_1 + \cdots + \lambda_n e_n + 0(1 - \sum_i e_i)$. We can further ensure that the scalars are distinct by lumping together the idempotents with the same coefficient.

Also note that diagonalizable elements are precisely those which are in the linear span of some complete system of primitive orthogonal idempotents. This follows from Lemma D.4.

Any idempotent $e \in A$ is diagonalizable. In the form (D.9), $e = 0 \cdot 1$ if $e = 0$, $e = 1 \cdot 1$ if $e = 1$ and $e = e + 0(1 - e)$ otherwise.

Theorem D.15. *For $w \in A$, the following are equivalent.*

(1) *w is diagonalizable.*
(2) *$\Bbbk[w]$ is a split-semisimple commutative algebra.*
(3) *The minimum polynomial of w factorizes into distinct linear factors.*

PROOF. Suppose (1) holds, that is, w is of the form (D.9). Then $\Bbbk[w]$ is contained in the subalgebra of A generated by the e_i, for $i = 1, \dots, n$. The latter is isomorphic to $\Bbbk^n$. This yields an injective algebra homomorphism

$$\text{(a)} \qquad \Bbbk[w] \to \Bbbk^n, \qquad w \mapsto (\lambda_1, \dots, \lambda_n).$$

Further, the formula for the determinant of the Vandermonde matrix implies that the images of $1, w, \dots, w^{n-1}$ are linearly independent, so the map (a) is surjective as well, and hence an isomorphism. This proves (2).

Suppose (2) holds. That is, we are given an isomorphism of the form (a). We deduce that the λ_i are distinct. It follows that the kernel of the surjective algebra homomorphism

$$\Bbbk[x] \twoheadrightarrow \Bbbk^n, \qquad x \mapsto (\lambda_1, \dots, \lambda_n)$$

is the ideal generated by the polynomial $(x - \lambda_1) \dots (x - \lambda_n)$. So this is the minimum polynomial of w proving (3).

The above steps can be reversed, so the three statements are equivalent. □

Note that a nonzero nilpotent element in A is *not* diagonalizable. This is because a diagonalizable element generates a split-semisimple commutative algebra and the latter has no nonzero nilpotent elements. Alternatively, the minimum polynomial of a nonzero nilpotent element has 0 as a repeated root.

Theorem D.16. *Suppose $w \in A$ is diagonalizable. Then the λ_i and e_i in (D.9) are uniquely determined. The minimum polynomial of w is $(x - \lambda_1) \dots (x - \lambda_n)$. The algebra $\Bbbk[w]$ equals the subalgebra of A generated by the e_i and is isomorphic to $\Bbbk^n$. In particular, the e_i belong to $\Bbbk[w]$. In fact,*

$$\text{(D.10)} \qquad e_i = \prod_{j \neq i} \frac{w - \lambda_j}{\lambda_i - \lambda_j}.$$

PROOF. The claims related to the minimum polynomial of w and the algebra $\Bbbk[w]$ follow from the proof of Theorem D.15. For uniqueness of λ_i and e_i: Suppose $w = \lambda'_1 e'_1 + \cdots + \lambda'_m e'_m$ is another expression of the form (D.9). Then $\Bbbk[w] \cong \Bbbk^m$. Now use that $\Bbbk^n \cong \Bbbk^m$ implies $m = n$, and the only algebra isomorphisms are those obtained by permuting the coordinates. The uniqueness of the λ_i can also be seen from the minimum polynomial of w. Formula (D.10) can be checked directly. □

A discussion related to formula (D.10) (in a more general setting) is given later in the proof of Theorem D.20.

Proposition D.17. *Let $\varphi : A \to B$ be an algebra homomorphism. If $w \in A$ is diagonalizable, then so is $\varphi(w)$. Converse holds if φ is injective.*

PROOF. Any algebra homomorphism preserves idempotents and orthogonality, so the first claim follows. Now suppose φ is injective. Let $p(x)$ be any polynomial. Then $p(w) = 0$ iff $p(\varphi(w)) = 0$. Thus, the minimum polynomials of w and $\varphi(w)$ coincide, and the second claim follows by Theorem D.15, item (3). □

Applying this result to the representation $\Psi_M : A \to \operatorname{End}_{\Bbbk}(M)$, we obtain:

Corollary D.18. *Let M be a left A-module, and $w \in A$. If w is diagonalizable, then so is $\Psi_M(w)$. Explicitly, writing w in the form* (D.9), *the eigenvalues of $\Psi_M(w)$ are $\lambda_1, \dots, \lambda_n$, and the eigenspace corresponding to λ_i is e_iM. Converse holds if M is faithful.*

Faithful modules always exist. For instance, A is a faithful module over itself. It follows that an element of A is diagonalizable iff its action on all left A-modules is diagonalizable.

D.4.3. Jordan-Chevalley decomposition. If the minimum polynomial of an element of A factorizes into linear factors, then it can be uniquely written as a sum of commuting diagonalizable and nilpotent elements. To show this, we begin with a preliminary lemma.

Lemma D.19. *Let u and v be two diagonalizable (nilpotent) elements in A such that $uv = vu$. Then $u + v$ is also diagonalizable (nilpotent).*

PROOF. We explain the diagonalizable part. Let the e_i (respectively f_j) be the unique idempotents of u (respectively v). Since e_i and f_j are polynomials in u and v, and u and v commute, we obtain $e_if_j = f_je_i$. From here, we deduce that the e_if_j as i and j vary form a family of mutually orthogonal idempotents which sum to 1, and they simultaneously diagonalize u and v. In particular, $u + v$ is diagonalizable. □

Theorem D.20. *Let $w \in A$ be such that its minimum polynomial factorizes into (not necessarily distinct) linear factors. Then there exist unique $w_d, w_n \in A$ such that*

$$w = w_d + w_n, \tag{D.11}$$

and w_d is diagonalizable, w_n is nilpotent, and w_d and w_n commute. Further, w_d and w_n can be expressed as polynomials in w.

We refer to (D.11) as the *Jordan-Chevalley decomposition* of w.

PROOF. Let the minimum polynomial of w be $m(x) := (x - \lambda_1)^{m_1} \dots (x - \lambda_k)^{m_k}$ with $\lambda_1, \dots, \lambda_k$ distinct and $m_i \geq 1$. Put

$$p_i(x) := \prod_{j \neq i} (x - \lambda_j)^{m_j}.$$

Since the gcd of the $p_i(x)$ in the polynomial algebra $\Bbbk[x]$ is 1, there exist polynomials $q_i(x)$ such that $\sum_{i=1}^k q_i(x)p_i(x) = 1$. Put

$$e_i := q_i(w)p_i(w).$$

Note that for $i \neq j$, $m(x)$ divides $p_i(x)p_j(x)$. Thus, the e_i are pairwise orthogonal elements and $\sum_{i=1}^k e_i = 1$. It is then automatic that each e_i is an idempotent.

Further, $m(x)$ does not divide $q_j(x)p_j(x)$ for any j. (If it did, then $x - \lambda_j$ would divide $\sum_{i=1}^k q_i(x)p_i(x)$ giving a contradiction.) Thus, each e_j is nonzero. Put

$$w_d := \sum_{i=1}^{k} \lambda_i e_i \quad \text{and} \quad w_n := w - w_d.$$

It is clear that w_d is diagonalizable, w_d and w_n are polynomials in w, and hence commute. It remains to show that w_n is nilpotent. For this, write $w_n = \sum_{i=1}^k (w - \lambda_i)e_i$. Then

$$w_n^N = \sum_{i=1}^{k} (w - \lambda_i)^N e_i = 0$$

when N is greater than each exponent m_i. This is because e_i contains $p_i(w)$ as a factor, and the minimum polynomial vanishes at w.

For uniqueness: Suppose $w = w'_d + w'_n$ is another decomposition. Since w'_d and w'_n commute with each other, they commute with w, and hence with w_d and w_n (since these are polynomials in w). Lemma D.19 implies that $w_d - w'_d = w'_n - w_n$ is both diagonalizable and nilpotent. Hence it must be zero as required. □

It follows from the above proof that the minimum polynomials of w and w_d have the same roots. More precisely, if the minimum polynomial of w is $(x - \lambda_1)^{m_1} \dots (x - \lambda_k)^{m_k}$, then the minimum polynomial of w_d is $(x - \lambda_1) \dots (x - \lambda_k)$.

D.5. Radical, socle and semisimplicity

Nilpotent ideals play an important role in the structure theory of algebras. The radical of an algebra is its largest nilpotent ideal. An algebra is semisimple if its radical is zero. The notions of radical and semisimplicity also make sense for any module. There is also a related notion of the socle of a module. More generally, one can consider the radical series and socle series of a module. Closely related notions are those of Loewy series, composition series, rigid modules and uniserial modules. We briefly review these notions (mostly without proofs).

D.5.1. Radical of an algebra. Let A be an algebra. A (two-sided) ideal N of A is *nilpotent* if there exists an integer $k \geq 1$ such that $N^k = 0$. The smallest k for which this happens is the *nilpotency index* of N. In other words: N has nilpotency index k iff the product of any k elements in N is zero, and there exist $k-1$ elements whose product is nonzero.

The sum of all nilpotent ideals of A is again a nilpotent ideal. This ideal is defined to be the *radical* of A. In other words, the radical of A is the largest nilpotent ideal of A. We denote it by $\mathrm{rad}(A)$. It is contained in the set of all nilpotent elements of A.

Notation D.21. For any ideal I of A, we have the quotient map $A \twoheadrightarrow A/I$. In such a situation, for $z \in A$, we will write $\bar{z}$ for its image in A/I.

D.5.2. Semisimple algebras. We say A is *semisimple* if $\mathrm{rad}(A) = 0$. The Wedderburn structure theorem says that an algebra is semisimple iff it is isomorphic to a product of matrix algebras over division $\Bbbk$-algebras. If the division $\Bbbk$-algebras involved are all $\Bbbk$, then we say that the semisimple algebra is split. In other words, A is *split-semisimple* iff it is isomorphic to a product of matrix algebras over $\Bbbk$.

A semisimple algebra is commutative iff it is isomorphic to a product of fields which are finite extensions of $\Bbbk$. Similarly, a split-semisimple algebra is commutative iff it is isomorphic to a product of copies of $\Bbbk$. The latter notion was elaborated in Section D.3.

Note very carefully that a semisimple algebra does not have any nonzero nilpotent ideals but it can have nonzero nilpotent elements. However, if the semisimple algebra is commutative, then it cannot have nonzero nilpotent elements.

Proposition D.22. *Suppose N is a nilpotent ideal of A such that the quotient A/N is a semisimple commutative algebra. Then $N = \operatorname{rad}(A)$ and it consists precisely of the nilpotent elements of A.*

PROOF. Since N is nilpotent, it is contained in $\operatorname{rad}(A)$, which in turn is contained in the set of all nilpotent elements. Suppose $z \in A$ is nilpotent. Then, so is its image $\bar{z} \in A/N$. However, since A/N is a semisimple commutative algebra, it has no nonzero nilpotent elements. Hence, $\bar{z} = 0$, and $z \in N$. Thus N consists precisely of the nilpotent elements of A, and equals $\operatorname{rad}(A)$. $\square$

D.5.3. Semisimple modules. Let M be an A-module. We say M is *semisimple* if any of the following equivalent conditions hold.

- Every submodule of M is a direct summand of M (that is, has a complementary submodule).
- M is the direct sum of a family of simple modules.
- M is the sum of a family of simple modules.

D.5.4. Radical of a module. For a left A-module M, the *radical* of M is the intersection of all maximal submodules of M. We denote it by $\operatorname{rad}(M)$. It is also given by

$$\operatorname{rad}(M) = \operatorname{rad}(A)M. \tag{D.12}$$

Also,

$$\operatorname{rad}(M) = 0 \iff M \text{ is semisimple.} \tag{D.13}$$

More generally: The *radical series* of a left A-module M is defined to be the filtration

$$0 = J^k M \subseteq \cdots \subseteq J^2 M \subseteq JM \subseteq M, \tag{D.14}$$

where $J = \operatorname{rad}(A)$ with nilpotency index k. Note that the second term in the radical series of M (from the top) is indeed the radical of M.

Similar considerations apply to right modules.

D.5.5. Socle of a module. For a left A-module M, the *socle* of M is the sum of all simple submodules of M. We denote it by $\operatorname{soc}(M)$. The socle is *homogeneous* if all the simple submodules of M are isomorphic. The socle consists precisely of those elements of M which are annihilated by the radical of A. That is,

$$m \in \operatorname{soc}(M) \iff zm = 0 \text{ for all } z \in \operatorname{rad}(A). \tag{D.15}$$

Also,

$$\operatorname{soc}(M) = M \iff M \text{ is semisimple.} \tag{D.16}$$

More generally: For any $i \geq 1$, let $\mathrm{soc}_i(M)$ consist of those elements of M which are annihilated by the ideal $\mathrm{rad}(A)^i$. Observe that

$$0 \subseteq \mathrm{soc}_1(M) \subseteq \mathrm{soc}_2(M) \subseteq \cdots \subseteq \mathrm{soc}_k(M) = M, \tag{D.17}$$

where k is the nilpotency index of $\mathrm{rad}(A)$. This is the *socle series* of M. Note that the second term in the socle series of M (from the bottom) is indeed the socle of M, that is, $\mathrm{soc}(M) = \mathrm{soc}_1(M)$.

Similar considerations apply to right modules.

The radical series and socle series are dual in the following sense. For a left A-module M, the subspaces $\mathrm{rad}(A)^i M$ and $\mathrm{soc}_i(M^*)$ are orthogonal under the canonical pairing between M and M^*.

D.5.6. Rigid modules. For any module, the radical series is contained termwise in the socle series: Suppose J is the radical of M with nilpotency index k. Then $J^{k-i}M$ is annihilated by J^i, which says that $J^{k-i}M$ is contained in $\mathrm{soc}_i(M)$.

A module is *rigid* when its radical and socle series coincide.

D.5.7. Loewy series. A *Loewy series* of a module M is a filtration $0 = M_0 \subseteq M_1 \subseteq \cdots \subseteq M_k = M$ in which each quotient M_j/M_{j-1} is semisimple. A filtration is a Loewy series precisely when multiplying one term by the radical puts it in the next (smaller) term.

The radical and socle series are two extreme examples of Loewy series. They are also called the descending and ascending Loewy series, respectively.

D.5.8. Composition series. A *composition series* of a module M is a filtration $0 = M_0 \lessdot M_1 \lessdot \cdots \lessdot M_k = M$ in which each M_{j-1} is a maximal proper submodule of M_j (as indicated by the notation $\lessdot$). Each quotient M_j/M_{j-1} is called a *composition factor* of M. It is a simple module. In particular, any composition series is a Loewy series.

D.5.9. Uniserial modules. A module is *uniserial* if any of the following equivalent conditions hold.

- The lattice of submodules is totally ordered by inclusion (and so constitutes a finite chain).
- The module possesses a unique composition series.
- The radical series is a composition series.
- The socle series is a composition series.

Note that uniserial modules are rigid.

D.6. Invertible elements and zero divisors

Let A be an algebra. An element $u \in A$ is *invertible* if there exists an element $v \in A$ such that $uv = vu = 1$. The element v, if it exists, is unique and is called the inverse of u. It is denoted by u^{-1}. The set $A^\times$ of invertible elements of A is a group under multiplication.

If $a \in A$ is a nilpotent element, then $1+a$ is invertible with the inverse given by $1 - a + a^2 - \dots$. Thus, for any nilpotent ideal N of A, the set $1 + N$ is a subgroup of $A^\times$.

Two elements a and b in A are *conjugate* if there exists $u \in A^\times$ such that $a = ubu^{-1}$.

Proposition D.23. *Let N be a nilpotent ideal of A. Then for $u \in A$,*

$$u \in A^\times \iff \bar{u} \in (A/N)^\times,$$

where $\bar{u}$ is the image of u under the canonical quotient map.

PROOF. Since any algebra homomorphism preserves invertibility, $u \in A^\times$ implies $\bar{u} \in (A/N)^\times$. Conversely, suppose $\bar{u} \in (A/N)^\times$. Pick $v \in A$ such that $\bar{v}$ is the inverse of $\bar{u}$. Then $\overline{uv} = \bar{u}\bar{v} = 1$, so uv belongs to $1 + N$ and hence is invertible. This shows that u has a right inverse. Similarly, it has a left inverse. By general principles, the left and right inverses coincide and $u \in A^\times$. □

An element $a \in A$ is called a *zero divisor* if there exists a nonzero element $b \in A$ such that either $ab = 0$ or $ba = 0$. Clearly, a zero divisor cannot be invertible.

Lemma D.24. *Any element of an algebra A is either invertible or a zero divisor (but not both).*

PROOF. Let $w \in A$. Consider the minimum polynomial of w. If its constant term is 0, then w is a zero divisor, else it is invertible. □

Lemma D.25. *Let A be an algebra. Then:*

(1) *If $w \in A$ is invertible, then w^{-1} can be expressed as a polynomial in w.*
(2) *If an element w of a subalgebra B of A is invertible in A, then w^{-1} belongs to B.*
(3) *If $w \in A$ acts invertibly on a faithful module, then w is invertible.*

PROOF. For (1): If $w \in A$ is invertible, then its minimum polynomial has a nonzero constant term, so it can be used to express w^{-1} as a polynomial in w. (2) follows from (1). For (3), view A as a subalgebra of the endomorphism algebra of the module and apply (2). □

D.7. Lifting idempotents

We discuss the idempotent lifting problem for the canonical map from an algebra to its quotient by a nilpotent ideal. In particular, this nilpotent ideal could be the radical of the algebra.

D.7.1. Isomorphic idempotents. Let A be an algebra. Let $e, f \in A$ be two idempotents. We say e and f are *isomorphic* if there exist $a \in eAf$ and $b \in fAe$ such that $e = ab$ and $f = ba$. In this case, we write $e \cong f$.

For idempotents e and f:

- If $e = ab$ and $f = ba$ for some a and b, then $e \cong f$. (To ensure that $a \in eAf$ and $b \in fAe$, replace a by aba and b by bab.)
- If e and f are conjugate, then $e \cong f$.
- If A is commutative, then $e \cong f$ iff $e = f$.
- $e \cong f$ iff $Ae \cong Af$ as left A-modules iff $eA \cong fA$ as right A-modules.

Lemma D.26. *Let $1 = e_1 + \cdots + e_n = f_1 + \cdots + f_n$ be two decompositions of the unit element into orthogonal idempotents such that $e_i \cong f_i$ for each i. Then there exist elements $a_i \in e_iAf_i$ such that $u = a_1 + \cdots + a_n \in A^\times$ satisfies $uf_iu^{-1} = e_i$ for each i.*

PROOF. Since $e_i \cong f_i$, we can pick a_i, b_i such that $a_i \in e_iAf_i$, $b_i \in f_iAe_i$, $e_i = a_ib_i$ and $f_i = b_ia_i$. Put $u = a_1 + \cdots + a_n$ and $v = b_1 + \cdots + b_n$. We have $a_ib_j = 0$ for $i \neq j$ (since $a_i \in e_iAf_i$, $b_j \in f_jAe_j$, and $f_if_j = 0$). It follows that $uv = a_1b_1 + \cdots + a_nb_n = e_1 + \cdots + e_n = 1$. Similarly, $vu = 1$. Thus, $u \in A^\times$. A similar calculation yields $uf_iv = e_i$. □

Lemma D.27. *Let N be a nilpotent ideal of A. Let e and f be idempotents in A, and $\bar{e}$ and $\bar{f}$ be their images in A/N. Then $e \cong f$ iff $\bar{e} \cong \bar{f}$.*

PROOF. The forward implication is clear. For the backward implication: Pick $a \in eAf$ and $b \in fAe$ such that $\bar{e} = \bar{a}\bar{b}$ and $\bar{f} = \bar{b}\bar{a}$. The point is that $e = ab$ and $f = ba$ may not hold. To fix this, put $z = e - ab$. Then $z \in eAe$ and $z \in N$, so z is nilpotent. Thus, $e = ab(e + z + z^2 + \dots)$. Put $a' = a$ and $b' = b(e + z + z^2 + \dots)$. One may indeed check that $e = a'b'$ and $f = b'a'$. □

D.7.2. Lifting idempotents. Let I be an ideal of A. We say an idempotent $e \in A/I$ can be lifted to A if there exists an idempotent $\hat{e} \in A$ which maps to e under the canonical projection $A \twoheadrightarrow A/I$.

Lemma D.28. *Let N be a nilpotent ideal of A. Then any idempotent of A/N can be lifted to A. Further, any lift of a primitive idempotent is primitive.*

PROOF. Suppose e is an idempotent of A/N. Since N is nilpotent, $N^k = 0$ for some $k \geq 1$. Choose any $a \in A$ which maps to e. One may check that $\hat{e} := (1-(1-a)^k)^k$ is an idempotent of A which lifts e.

Now suppose e is a primitive idempotent of A/N. We claim that any lift $\hat{e}$ is primitive: Let $\hat{e} = f + g$ be a decomposition into orthogonal idempotents. Then $e = \bar{f} + \bar{g}$. Since e is primitive, either $\bar{f} = 0$ or $\bar{g} = 0$. For definiteness, say $\bar{f} = 0$. But then $f \in N$, making f both idempotent and nilpotent, and hence $f = 0$. Thus, $\hat{e}$ is primitive. □

Lemma D.29. *Let N be a nilpotent ideal of A. Then any family $e_1, \dots, e_n$ of mutually orthogonal idempotents of A/N can be lifted to a family $\hat{e}_1, \dots, \hat{e}_n$ of mutually orthogonal idempotents of A. Further, if the e_i sum to 1, then so do the $\hat{e}_i$.*

PROOF. There is an inductive procedure to lift idempotents one at a time such that the idempotent constructed at a given step is orthogonal to the ones previously constructed. At step one, we use Lemma D.28 to obtain $\hat{e}_1$. Now assume that we have constructed the lifts $\hat{e}_1, \dots, \hat{e}_i$. Put $e := \hat{e}_1 + \cdots + \hat{e}_i$. Now use Lemma D.28 to first construct an idempotent f lifting e_{i+1}, and then set

$$\hat{e}_{i+1} := (1 - e)(1 - fe)^{-1}f(1 - fe).$$

This new element also lifts e_{i+1} and further, it is orthogonal to e, and hence to all the previous $\hat{e}_j$. This completes the inductive step.

Now suppose the e_i sum to 1. Then $1 - \sum_i \hat{e}_i$ is both idempotent and nilpotent, the latter because it belongs to N. So this element must be zero, as required. □

Lemma D.30. *Let N be a nilpotent ideal of A. Let $e \in A$ be an idempotent such that $\bar{e} = f + g$, where f and g are mutually orthogonal idempotents of A/N. Then there exist mutually orthogonal lifts $\hat{f}$ and $\hat{g}$ of f and g such that $e = \hat{f} + \hat{g}$. In particular: If $e \in A$ is primitive, then so is its image $\bar{e} \in A/N$.*

PROOF. Use Lemma D.29 to lift the family $1-\bar{e}, f, g$ to $1-e, \hat{f}, \hat{g}$. Since the family downstairs sums to 1, so does the lifted family. Hence, $e = \hat{f} + \hat{g}$ as required. □

D.8. Elementary algebras

We discuss elementary algebras. An algebra is elementary if the quotient by its radical is a split-semisimple commutative algebra. In an elementary algebra, a complete system of primitive idempotents exists but it is not necessarily unique. (This is related to the idempotent lifting problem.) Every element has a Jordan-Chevalley decomposition into a diagonalizable part and a nilpotent part. The simple modules over an elementary algebra are one-dimensional, but an arbitrary module may not break as a direct sum of simple submodules. Invertible elements and zero divisors of the algebra can be characterized using its multiplicative characters. We also briefly mention quivers.

D.8.1. Elementary algebras. Let A be an algebra. Recall that $\operatorname{rad}(A)$ is the largest nilpotent ideal of A. We say A is *elementary* if the quotient $A/\operatorname{rad}(A)$ is a split-semisimple commutative algebra.

It is convenient to denote the quotient $A/\operatorname{rad}(A)$ by $\bar{A}$. We assume that $\bar{A}$ has dimension n and denote its primitive idempotents by $e_1, \dots, e_n$. Also following Notation D.21, for $z \in A$, we write $\bar{z}$ for its image in $\bar{A}$.

D.8.2. Complete systems and algebra sections. For an elementary algebra A, an *algebra section* is an algebra homomorphism $\bar{A} \hookrightarrow A$ which is a section of the canonical projection $A \twoheadrightarrow \bar{A}$.

Theorem D.31. *Let A be elementary. Then there exists an algebra section $\bar{A} \hookrightarrow A$. Further, any two algebra sections are conjugates of each other by an element of $1 + \operatorname{rad}(A)$.*

PROOF. Applying Lemma D.29 to $e_1, \dots, e_n$ yields a family $\hat{e}_1, \dots, \hat{e}_n$ of mutually orthogonal idempotents of A which sum to 1. The map $e_i \mapsto \hat{e}_i$ defines an algebra section $\bar{A} \hookrightarrow A$. This proves the first part. For the second part: Suppose φ and ψ are two algebra sections. Then by Lemma D.27, $\varphi(e_i) \cong \psi(e_i)$ for each i. Now apply Lemma D.26 to get $u \in A^\times$ and $a_i \in \varphi(e_i)A\psi(e_i)$ such that φ and ψ conjugates of each other by u. In fact, since $e_i\bar{A}e_i$ is one-dimensional, by multiplying each a_i by a suitable nonzero scalar, one may assume that $\bar{a_i} = e_i$. It then follows that $\bar{u} = 1$ and $u \in 1 + \operatorname{rad}(A)$. □

Theorem D.32. *Let A be elementary. Then there is a correspondence between complete systems of primitive orthogonal idempotents of A and algebra sections $\bar{A} \hookrightarrow A$.*

PROOF. An algebra section $\varphi : \bar{A} \hookrightarrow A$ yields the family $\varphi(e_1), \dots, \varphi(e_n)$. These are mutually orthogonal idempotents and they sum to 1 (since φ is an algebra homomorphism). Further, they are primitive by Lemma D.28. Conversely, given such a family in A, its image in $\bar{A}$ must be $e_1, \dots, e_n$ (since the canonical projection is an algebra homomorphism and preserves primitives by the last part of Lemma D.30). So we can write the original family as $f_1, \dots, f_n$ with $\bar{f_i} = e_i$. The corresponding algebra section is $e_i \mapsto f_i$. □

Theorem D.33. *Let A be elementary. Then for idempotents $e, f \in A$,*

$$e \cong f \iff 1 - e \cong 1 - f \iff e - f \in \operatorname{rad}(A)$$
$$\iff \bar{e} = \bar{f} \iff f = u^{-1}eu \text{ for some } u \in 1 + \operatorname{rad}(A).$$

In particular, if e and f lift the same idempotent of $\bar{A}$, then they are conjugate by an element of $1 + \operatorname{rad}(A)$.

Proof. By Lemma D.27 and the fact that $\bar{A}$ is commutative, $e \cong f$ iff $\bar{e} = \bar{f}$. The intermediate equivalences also follow. It only remains to understand the forward implication of the last equivalence. Suppose $\bar{e} = \bar{f}$. Refine $1 = e + (1 - e)$ (resp. $1 = f + (1 - f)$) into a complete system. By Theorems D.31 and D.32, these two complete systems are conjugate by say $u \in 1 + \operatorname{rad}(A)$. Since $\bar{e} = \bar{f}$, we deduce that e and f are also conjugate by u. □

Theorem D.34. *Let A be elementary. Then $w \in A$ is diagonalizable iff w is in the image of some algebra section $\bar{A} \hookrightarrow A$. In particular, every idempotent of A is in the image of some algebra section.*

Proof. This follows from Theorem D.32 since diagonalizable elements are precisely those which are in the linear span of some complete system of primitive orthogonal idempotents. □

D.8.3. Simple modules.

Theorem D.35. *Let A be elementary. Then A has n distinct simple modules (up to isomorphism). They are one-dimensional. For $1 \leq i \leq n$, the i-th simple module is defined by the multiplicative character*

$$\chi_i : A \to \mathbb{k}, \qquad z \mapsto \langle \bar{z}, e_i \rangle.$$

In fact, there is a correspondence between simple modules over A and over $\bar{A}$.

Proof. Let M be a simple A-module. Then JM is a submodule of M, where $J = \operatorname{rad}(A)$. By simplicity of M, this submodule is either M or 0. The nilpotency of J forces $JM = 0$. So the action of A factors through the quotient map $A \twoheadrightarrow \bar{A}$, and M is a simple $\bar{A}$-module. Conversely, any simple $\bar{A}$-module is a simple A-module. This establishes the correspondence between simple modules over A and over $\bar{A}$. Now apply Theorem D.6. □

By definition of χ_i,

$$\bar{z} = \sum_i \chi_i(z)\, e_i. \tag{D.18}$$

It follows that $z \in \operatorname{rad}(A)$ iff $\chi_i(z) = 0$ for all i.

D.8.4. Modules. Let M be a (left or right) A-module. As a consequence of (D.5), the character of M factors through the quotient map $A \to \bar{A}$ yielding the commutative diagram

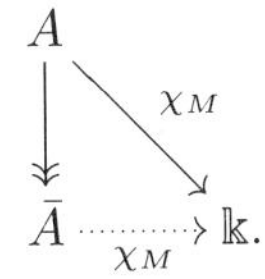

We continue to denote the induced functional on $\bar{A}$ by χ_M. For $1 \le i \le n$, put

$$\eta_i(M) := \chi_M(e_i). \tag{D.19}$$

Observe that for any $w \in A$,

$$\chi_M(w) = \sum_{i=1}^{n} \chi_i(w)\, \eta_i(M). \tag{D.20}$$

Let $0 = M_0 \lessdot M_1 \lessdot \cdots \lessdot M_k = M$ be any composition series of M. Each composition factor M_j/M_{j-1} is a simple module and hence one-dimensional by Theorem D.35. The associated graded module of the filtration, namely,

$$\bar{M} := \bigoplus_{j=1}^{k} M_j/M_{j-1}$$

is both an A-module and an $\bar{A}$-module. Thus, for $w \in A$, the operators $\Psi_{\bar{M}}(w)$ and $\Psi_{\bar{M}}(\bar{w})$ coincide. Further, we claim that the eigenvalues (and hence trace) of the operator $\Psi_M(w)$ coincide with those of the operator $\Psi_{\bar{M}}(\bar{w})$. To see this, pick a basis of M by first picking a nonzero element from M_1, followed by an element of M_2 which is not in M_1, and so on. This basis does not depend on w. It induces a basis of $\bar{M}$. In these bases, $\Psi_{\bar{M}}(\bar{w})$ is a diagonal matrix, while $\Psi_M(w)$ is an upper triangular matrix whose diagonal part agrees with $\Psi_{\bar{M}}(\bar{w})$. This proves the claim. In particular, the induced functional χ_M on $\bar{A}$ equals the character $\chi_{\bar{M}}$ of $\bar{M}$.

Example D.36. Let A be the algebra of upper triangular matrices of size n. It is elementary. The radical is the ideal of strictly upper triangular matrices. Elements of the quotient $\bar{A}$ can be identified with diagonal matrices. Since A is an example of an incidence algebra (Section C.1.8), these facts can also be deduced from Proposition C.10.

Let M be the left A-module of column vectors. For $0 \le i \le n$, let M_i denote the submodule consisting of vectors whose last $n-i$ entries are zero. This defines a composition series of M. Let $\bar{M}$ denote its associated graded module. The action of any upper triangular matrix on $\bar{M}$ is via its diagonal part.

Some consequences of the above discussion are stated below.

Theorem D.37. *Let A be elementary and M be an A-module. Then in any composition series of M, the number of times the simple module associated to χ_i appears as a composition factor is $\eta_i(M)$.*

Proof. We have $\eta_i(M) = \chi_M(e_i) = \chi_{\bar{M}}(e_i)$. Now use Theorem D.7. ☐

The fact that the multiplicity of a simple module does not depend on the choice of the composition series is the *Jordan-Hölder theorem*.

Theorem D.38. *Let A be elementary and M be an A-module. Then all elements of A are simultaneously triangularizable on M. For $w \in A$, the eigenvalues of the linear operator $\Psi_M(w)$ are $\chi_i(w)$, and the multiplicity of $\chi_i(w)$ is $\eta_i(M)$.*

Proof. For the second part, we can use Theorem D.8 since $\Psi_M(w)$ and $\Psi_{\bar{M}}(\bar{w})$ have the same eigenvalues. ☐

It is interesting that all eigenvalues of $\Psi_M(w)$ belong to the ground field $\Bbbk$. The number $\eta_i(M)$ which is the multiplicity of $\chi_i(w)$ only depends on i and not on

w. We call it the *generic multiplicity* associated to the index i. It is possible that $\eta_i(M)$ is 0 for some i in which case the eigenvalue $\chi_i(w)$ does not occur. It may also happen that $i \neq j$ but $\chi_i(w) = \chi_j(w) = \lambda$ (say). In this case, the multiplicity of λ will be the sum of $\eta_i(M)$ over those i for which $\chi_i(w) = \lambda$.

Note very carefully that Theorem D.38 makes no claim about the diagonalizability of $\Psi_M(w)$.

Proposition D.39. *For an elementary algebra A, there is a correspondence between (multiplicative) characters of A and (multiplicative) characters of $\bar{A}$. Thus, a character of A corresponds to a family $(\eta_i)_{1\le i\le n}$ of nonnegative integers. It is multiplicative if exactly one η_i is 1 and the rest are 0.*

PROOF. For the second part, we can use Proposition D.9. □

D.8.5. Peirce decomposition. Decompositions arising from a system of orthogonal idempotents are called *Peirce decompositions* (left, right, two-sided).

Proposition D.40. *Let A be elementary and M be a left A-module. Let $\hat{e}_1, \dots, \hat{e}_n$ be a complete system of primitive orthogonal idempotents of A such that $\hat{e}_i$ lifts e_i. Then*

$$M = \bigoplus_i \hat{e}_i M \quad \text{and} \quad \dim \hat{e}_i M = \eta_i(M).$$

Similar statement holds for a right A-module.

PROOF. The decomposition of M is clear. For the formula:

$$\dim \hat{e}_i M = \chi_M(\hat{e}_i) = \chi_M(e_i) = \eta_i(M).$$

We used (D.4) and (D.19). Alternatively: Since $\hat{e}_i$ is an idempotent, the operator $\Psi_M(\hat{e}_i)$ has eigenvalues 0 and 1, and $\dim \hat{e}_i M$ is the multiplicity of the eigenvalue 1. Now apply Theorem D.38 with $w := \hat{e}_i$. Then, by (D.18), $\chi_i(w) = 1$ while all the remaining $\chi_j(w)$ are zero. So the multiplicity is $\eta_i(M)$ as required. □

Note very carefully that Proposition D.40 does *not* claim that the $\hat{e}_i M$ are submodules of M. However, the $\hat{e}_i M$ do serve as eigenspaces for the operator $\Psi_M(w)$ for any $w \in A$ of the form $\lambda_1 \hat{e}_1 + \dots + \lambda_n \hat{e}_n$.

D.8.6. Faithful modules.

Proposition D.41. *Let A be elementary and M be an A-module. Then the following are equivalent.*

(1) $\operatorname{ann}(M) \subseteq \operatorname{rad}(A)$.
(2) *No nonzero diagonalizable element of A annihilates M.*
(3) $\eta_i(M) > 0$ *for all i.*

PROOF. (1) implies (2). This is because a nonzero diagonalizable element cannot be nilpotent.

(2) implies (3). Suppose $\eta_j(M) = 0$ for some j. Then by Proposition D.40, any idempotent $\hat{e}_j$ lifting e_j will annihilate M, which is a contradiction.

(3) implies (1). Let $z \in \operatorname{ann}(M)$, that is, $\Psi_M(z) = 0$. Then by Theorem D.38, all eigenvalues $\chi_i(z)$ are forced to be zero since they appear with nonzero multiplicity. This implies that $z \in \operatorname{rad}(A)$. □

Corollary D.42. *A module M over an elementary algebra A is faithful iff $\eta_i(M) > 0$ for all i, and no nonzero nilpotent element annihilates M.*

A split-semisimple commutative algebra has no nonzero nilpotent elements. So, in this case, the above criterion specializes to the one given in Theorem D.7.

D.8.7. Invertible elements and zero divisors. Invertible elements and zero divisors of an elementary algebra can be characterized as follows.

Proposition D.43. *Let A be elementary. Then, for $u \in A$,*

$$u \in A^\times \iff \bar{u} \in \bar{A}^\times \iff \chi_i(u) \neq 0 \text{ for all } i.$$

Similarly, for $u \in A$,

$$u \text{ is a zero divisor} \iff \chi_i(u) = 0 \text{ for some } i.$$

PROOF. We first prove the claim about invertible elements. Since $\mathrm{rad}(A)$ is a nilpotent ideal, the first equivalence follows from Proposition D.23. Recall from (D.18) that $\bar{u} = \sum_i \chi_i(u) e_i$. This element is invertible in $\bar{A}$ precisely when the coefficients of the e_i are all nonzero.

The claim about zero divisors now follows from Lemma D.24. □

D.8.8. Jordan-Chevalley decomposition. For A elementary and $w \in A$, put

$$d(x) := (x - \lambda_1) \dots (x - \lambda_k),$$

where $\lambda_1, \dots, \lambda_k$ are the distinct scalars occuring in the list $\chi_1(w), \dots, \chi_n(w)$.

Proposition D.44. *Let A be elementary. Then the minimum polynomial of any $w \in A$ factorizes into linear factors of the form $(x - \lambda_j)$. In particular, every $w \in A$ has a Jordan-Chevalley decomposition. Further, w is diagonalizable iff $d(w) = 0$.*

PROOF. Let $\bar{w}$ denote the image of w in $\bar{A}$. Then

$$\bar{w} = \sum_{i=1}^{n} \chi_i(w) e_i = \sum_{i=1}^{k} \lambda_i f_i,$$

where f_i are pairwise orthogonal idempotents of $\bar{A}$ which sum to 1. It follows that the image of $d(w)$ in $\bar{A}$ is zero. Hence $d(w)$ is nilpotent. Say $d(w)^N = 0$. Then, the minimum polynomial of w divides $d(x)^N$ and hence factorizes into linear factors of the form $(x - \lambda_j)$. This proves the first statement. The rest follow from Theorem D.20 and Theorem D.15, item (3). □

D.8.9. Quivers. Quivers are an important tool in the study of elementary algebras. A *quiver* is a finite directed graph (V, E), where V is the set of vertices, and E is the set of arrows between vertices. To every quiver Q is associated an algebra $\Bbbk Q$ called the *path algebra*. It has a basis consisting of directed paths. When the end of one path equals the beginning of another path, the two paths can be concatenated. This is how the product in the path algebra is defined. When paths cannot be concatenated, the product is zero. The path algebra is finite-dimensional iff Q is acyclic. Further, the path algebra of an acyclic quiver is elementary.

Now let A be an elementary algebra with radical $J = \mathrm{rad}(A)$. Let $\bar{A}$ be its split-semisimple quotient with primitive idempotents $e_1, \dots, e_n$. Observe that J/J^2 is a $\bar{A}$-bimodule. The quiver Q of A is defined as follows. It has vertices $1, \dots, n$, with the number of arrows from i to j equal to the dimension of $e_j(J/J^2)e_i$. A point of this construction is that there exists a surjective algebra morphism $\Bbbk Q \twoheadrightarrow A$ from the path algebra of Q to A. For details, see [**32**, Section III.1].

D.8.10. Subalgebras of elementary algebras.

Lemma D.45. *Let A be an elementary algebra and B be any subalgebra. Then*

- *B is an elementary algebra,*
- $\mathrm{rad}(B) = B \cap \mathrm{rad}(A)$, *and*
- *$B/\mathrm{rad}(B)$ is a subalgebra of $A/\mathrm{rad}(A)$.*

In addition, if B and $\mathrm{rad}(A)$ linearly span A, then $B/\mathrm{rad}(B) = A/\mathrm{rad}(A)$.

PROOF. Let $J = B \cap \mathrm{rad}(A)$. Then J is a nilpotent ideal of B. Since $\mathrm{rad}(A)$ consists of the nilpotent elements of A, J consists of the nilpotent elements of B. Therefore, J is the largest nilpotent ideal of B, that is, $J = \mathrm{rad}(B)$. It follows that $\mathrm{rad}(B)$ is the kernel of the algebra map $B \to A/\mathrm{rad}(A)$. This yields an injective algebra map $B/\mathrm{rad}(B) \to A/\mathrm{rad}(A)$. Using Lemma D.13, we deduce that B is elementary. If B and $\mathrm{rad}(A)$ linearly span A, then the map $B \to A/\mathrm{rad}(A)$ is also surjective, so $B/\mathrm{rad}(B) = A/\mathrm{rad}(A)$. □

Lemma D.46. *Let G be a finite group which acts on an elementary algebra A. Then the invariant subalgebra A^G is elementary and $\mathrm{rad}(A^G) = \mathrm{rad}(A)^G$. In addition, if the field characteristic does not divide the order of G, then $A^G/\mathrm{rad}(A^G) = (A/\mathrm{rad}(A))^G$.*

PROOF. Since $\mathrm{rad}(A)$ is the largest nilpotent ideal of A, the action of G preserves $\mathrm{rad}(A)$. Also, clearly $\mathrm{rad}(A)^G = A^G \cap \mathrm{rad}(A)$. The first claim now follows from Lemma D.45. If the field characteristic does not divide the order of G, then taking G-invariants is exact. In this situation, in the diagram

$$\begin{array}{ccccccccc} 0 & \longrightarrow & \mathrm{rad}(A) & \longrightarrow & A & \longrightarrow & A/\mathrm{rad}(A) & \longrightarrow & 0 \\ & & \uparrow & & \uparrow & & \uparrow & & \\ 0 & \longrightarrow & \mathrm{rad}(A)^G & \longrightarrow & A^G & \longrightarrow & (A/\mathrm{rad}(A))^G & \longrightarrow & 0, \end{array}$$

the top-row is exact, and hence so is the bottom row. The second claim follows. □

D.9. Algebra of a finite lattice

The linearization of a finite lattice is an algebra with product given by the join operation. This algebra is a split-semisimple commutative algebra; the primitive idempotents can be explicitly written down in terms of the canonical basis using the Möbius function of the lattice. As a consequence, this algebra is self-dual (as a module over itself), and the isomorphism inducing self-duality is controlled by a family of nonzero scalars indexed by elements of the lattice. These scalars can be chosen in a consistent manner for lattices related to one another by join-preserving maps satisfying a mild hypothesis related to tightness.

D.9.1. Algebra of a lattice. Let P be a finite lattice with minimum element $\bot$ and maximum element $\top$. Let $\Bbbk P$ denote the linearization of P over the field $\Bbbk$. This is a commutative $\Bbbk$-algebra with product induced from the join operation in P. Letting $\mathtt{H}$ denote the canonical basis,

$$\mathtt{H}_x \cdot \mathtt{H}_y := \mathtt{H}_{x \vee y}. \tag{D.21}$$

We use the symbol $\cdot$ to denote the product. We will make use of this symbol in other similar situations as well. We will use the letters $x, y, z, \dots$ to denote elements of the lattice. The algebra $\Bbbk P$ is often called the *Möbius algebra* of the lattice P.

D.9.2. Q-basis and split-semisimplicity. Define the Q-basis of $\Bbbk P$ by

$$\mathtt{H}_x = \sum_{y\colon y \geq x} \mathtt{Q}_y \qquad \text{or equivalently} \qquad \mathtt{Q}_x = \sum_{y\colon y \geq x} \mu(x,y)\, \mathtt{H}_y. \tag{D.22}$$

Here μ refers to the Möbius function of the lattice P. In particular,

$$\mathtt{H}_\perp = \sum_y \mathtt{Q}_y. \tag{D.23}$$

Theorem D.47. *The algebra of a finite lattice is a split-semisimple commutative algebra. The unique complete system of primitive orthogonal idempotents is given by the* Q*-basis. In other words,*

$$\mathtt{Q}_x \cdot \mathtt{Q}_y = \begin{cases} \mathtt{Q}_x & \textit{if } x = y, \\ 0 & \textit{otherwise.} \end{cases} \tag{D.24}$$

PROOF. An easy way to establish (D.24) is to assume it and deduce (D.21) from it. The required calculation is shown below.

$$\mathtt{H}_x \cdot \mathtt{H}_y = \Big(\sum_{z\colon z \geq x} \mathtt{Q}_z \Big) \cdot \Big(\sum_{w\colon w \geq y} \mathtt{Q}_w \Big) = \sum_{u\colon u \geq x \vee y} \mathtt{Q}_u = \mathtt{H}_{x \vee y}. \qquad \square$$

Also from (D.22) and (D.24), we obtain

$$\mathtt{H}_y \cdot \mathtt{Q}_x = \begin{cases} \mathtt{Q}_x & \text{if } x \geq y, \\ 0 & \text{otherwise.} \end{cases} \tag{D.25}$$

In particular,

$$\mathtt{H}_y \cdot \mathtt{Q}_\perp = 0 \text{ for } y > \perp.$$

Let us make this explicit.

$$\begin{aligned} \mathtt{H}_y \cdot \mathtt{Q}_\perp &= \mathtt{H}_y \cdot \big(\sum_w \mu(\perp, w)\, \mathtt{H}_w \big) \\ &= \sum_w \mu(\perp, w)\, \mathtt{H}_{y \vee w} \\ &= \sum_{w'} \big(\sum_{w\colon y \vee w = w'} \mu(\perp, w) \big)\, \mathtt{H}_{w'}. \end{aligned}$$

So each term in the parenthesis must be 0. This is exactly the Weisner formula (C.7a) (and we have proved it again).

Exercise D.48. Starting with (D.21), use the Weisner formula (C.7a) to first prove (D.25) and then deduce (D.24) from it.

D.9.3. Linear functionals, exponential and logarithm. Suppose $f : \Bbbk P \to \Bbbk$ is a linear map. Then define (set-theoretic) maps $\xi, \eta : P \to \Bbbk$ as follows. For each $x \in P$, let

$$\xi_x = f(\mathtt{H}_x) \qquad \text{and} \qquad \eta_x = f(\mathtt{Q}_x). \tag{D.26}$$

We deduce from (D.22) that ξ and η are the exponential and logarithm of each other in the lattice P, that is,

$$\xi_x = \sum_{y\colon y \geq x} \eta_y \quad \text{and} \quad \eta_x = \sum_{y\colon y \geq x} \mu(x,y)\, \xi_y. \tag{D.27}$$

Further, linearizing ξ in the $\mathtt{H}$-basis or η in the $\mathtt{Q}$-basis recovers f. Thus among f, ξ and η, knowing any one determines the remaining two.

Some interesting choices for ξ and η are given below.

Example D.49. For $x \in P$, put

$$\xi_x = \begin{cases} 1 & \text{if } x = \top, \\ 0 & \text{otherwise} \end{cases} \qquad \text{and} \qquad \eta_x = \mu_P(x, \top). \tag{D.28}$$

In general, η will take both positive and negative values.

Example D.50. Let M be a module over $\Bbbk P$. For each element $x \in P$, define

$$\xi_x(M) := \dim(\mathtt{H}_x M) \quad \text{and} \quad \eta_x(M) := \dim(\mathtt{Q}_x M). \tag{D.29}$$

These scalars are always nonnegative integers, since they are dimensions of spaces.

Recall from (D.4) that for any idempotent operator, the dimension of its image is its trace. Since $\mathtt{H}_x$ and $\mathtt{Q}_x$ are idempotents, the linear functional f associated to $\xi_x(M)$ (or to $\eta_x(M)$) is the character χ_M of M. (The character of a module is defined in (D.1).)

D.9.4. Simple modules and diagonalizability.

Theorem D.51. *The algebra $\Bbbk P$ has $|P|$ distinct simple modules (up to isomorphism). They are one-dimensional. The simple module corresponding to $x \in P$ is defined by the multiplicative character*

$$\chi_x : \Bbbk P \to \Bbbk, \qquad \sum_y b^y \, \mathtt{Q}_y \mapsto b^x.$$

On the $\mathtt{H}$-basis, the multiplicative character is given by

$$\chi_x : \Bbbk P \to \Bbbk, \qquad \sum_y a^y \, \mathtt{H}_y \mapsto \sum_{y:\, y \le x} a^y.$$

PROOF. The claim about simple modules and the character formula on the $\mathtt{Q}$-basis follows from Theorems D.6 and D.47. The formula on the $\mathtt{H}$-basis can then be deduced as follows.

$$\sum_y a^y \, \mathtt{H}_y \mapsto \sum_y a^y \sum_{z:\, z \ge y} \mathtt{Q}_z \mapsto \sum_z \Big(\sum_{y:\, y \le z} a^y \Big) \mathtt{Q}_z \mapsto \sum_{y:\, y \le x} a^y. \qquad \square$$

For a module M over $\Bbbk P$, let $\eta_x(M)$ be as in (D.29). Then: For any $\alpha \in \Bbbk P$,

$$\chi_M(\alpha) = \sum_{x \in P} \chi_x(\alpha) \, \eta_x(M). \tag{D.30}$$

This is a special case of (D.8).

Theorem D.52. *Any module M over the algebra $\Bbbk P$ is a direct sum of simple modules with $\eta_x(M)$ being the multiplicity of the simple module corresponding to $x \in P$. In particular, M is faithful iff $\eta_x(M) > 0$ for each $x \in P$.*

PROOF. This follows from Theorems D.7 and D.47. $\square$

Theorem D.53. *Let M be a module over $\Bbbk P$. For $\alpha = \sum_x a^x \, \mathtt{H}_x$, the linear operator $\Psi_M(\alpha)$ is diagonalizable. It has an eigenvalue*

$$\lambda_x(\alpha) = \chi_x(\alpha) = \sum_{y:\, y \le x} a^y \tag{D.31}$$

for each $x \in P$, with multiplicity $\eta_x(M)$.

PROOF. This follows from Theorems D.8 and D.47 and the $\mathtt{H}$-basis formula in Theorem D.51. □

D.9.5. Primitive part of a module. Let M be a module over $\Bbbk P$. The *primitive part* of M is the submodule defined by

$$\mathcal{P}(M) := \{m \in M \mid \mathtt{H}_x \cdot m = 0 \text{ for all } x > \bot\}.$$

More generally, for $x \in P$, let

$$\mathcal{P}_x(M) := \{m \in M \mid \mathtt{H}_y \cdot m = 0 \text{ for all } y \not\leq x\}.$$

Equivalently,

$$\mathcal{P}_x(M) = \bigcap_{y:\, y \not\leq x} \ker(\Psi_M(\mathtt{H}_y) : M \to M).$$

Observe that for $x \leq y$,

$$\mathcal{P}_x(M) \subseteq \mathcal{P}_y(M)$$

with

$$\mathcal{P}_\bot(M) = \mathcal{P}(M) \quad \text{and} \quad \mathcal{P}_\top(M) = M.$$

Thus the $\mathcal{P}_x(M)$ define a filtration of M indexed by elements of P.

Lemma D.54. *We have* $\mathcal{P}(M) = \mathtt{Q}_\bot \cdot M$. *More generally,*

$$\mathcal{P}_x(M) = \bigoplus_{y:\, y \leq x} \mathtt{Q}_y \cdot M \quad \textit{and} \quad \dim \mathcal{P}_x(M) = \sum_{y:\, y \leq x} \eta_y(M).$$

PROOF. We directly prove the decomposition. By (D.25), the rhs is contained in the lhs. Conversely, suppose m belongs to the lhs. Then by (D.23),

$$m = \sum_y \mathtt{Q}_y \cdot m = \sum_{y:\, y \leq x} \mathtt{Q}_y \cdot m$$

which belongs to the rhs. In the second step, we used (D.22): $\mathtt{Q}_y$ expressed in the $\mathtt{H}$-basis only contains terms indexed by elements greater than y, so if $y \not\leq x$, then $\mathtt{Q}_y \cdot m = 0$. □

In particular: The primitive part $\mathcal{P}(\Bbbk P)$ (with $\Bbbk P$ viewed as a module over itself) is one-dimensional and spanned by $\mathtt{Q}_\bot$. This statement is equivalent to Proposition C.5: A simple calculation shows that $\sum_x c_x \mathtt{H}_x \in \mathcal{P}(\Bbbk P)$ iff (c_x) solves the linear system given in Proposition C.5.

D.9.6. Decomposable part of a module. Let M be a module over $\Bbbk P$. The *decomposable part* of M is the submodule defined by

$$\mathcal{D}(M) := \sum_{x:\, x > \bot} \mathtt{H}_x \cdot M.$$

More generally, for $x \in P$, let

$$\mathcal{D}_x(M) := \sum_{y:\, y \not\leq x} \mathtt{H}_y \cdot M.$$

Observe that for $x \leq y$,

$$\mathcal{D}_y(M) \subseteq \mathcal{D}_x(M)$$

with

$$\mathcal{D}_\top(M) = 0 \quad \text{and} \quad \mathcal{D}_\bot(M) = \mathcal{D}(M).$$

Lemma D.55. *We have*

$$\mathcal{D}_x(M) = \bigoplus_{y:\, y \not\leq x} \mathtt{Q}_y \cdot M \quad \textit{and} \quad \dim \mathcal{D}_x(M) = \sum_{y:\, y \not\leq x} \eta_y(M).$$

This can be proved along the same lines as Lemma D.54. Using these two results, we deduce:

Proposition D.56. *Let M be a module over $\Bbbk P$. Then the subspaces $\mathcal{P}_k(M)$ and $\mathcal{D}_k(M^*)$ ($\mathcal{D}_k(M)$ and $\mathcal{P}_k(M^*)$) are orthogonal complements of each other under the canonical pairing between M and M^*.*

D.9.7. Associative bilinear forms and self-duality. Recall from Lemma D.12 that Frobenius structures on a split-semisimple algebra of dimension n are classified by a sequence of n nonzero scalars. We elaborate on this for the algebra $\Bbbk P$ first in the language of bilinear forms and then in the language of self-duality.

Let f, ξ and η be related by (D.26) and (D.27). This gives rise to an associative bilinear form on $\Bbbk P$ by

$$\Bbbk P \times \Bbbk P \to \Bbbk, \qquad \langle \alpha, \beta \rangle = f(\alpha \cdot \beta), \tag{D.32}$$

or equivalently,

$$\langle \mathtt{H}_x, \mathtt{H}_y \rangle = \xi_{x \vee y} \qquad \text{or} \qquad \langle \mathtt{Q}_x, \mathtt{Q}_y \rangle = \begin{cases} \eta_x & \text{if } x = y, \\ 0 & \text{otherwise.} \end{cases} \tag{D.33}$$

This yields:

Lemma D.57. *The determinant of the bilinear form* (D.33) *is $\prod_{x \in P} \eta_x$. In particular, the bilinear form is nondegenerate iff $\eta_x \neq 0$ for all x.*

Example D.58. We first specialize the above discussion to ξ and η given by (D.28). The bilinear form on $\Bbbk P$ is given by

$$\langle \mathtt{H}_x, \mathtt{H}_y \rangle := \begin{cases} 1 & \text{if } x \vee y = \top, \\ 0 & \text{otherwise,} \end{cases} \qquad \text{or} \qquad \langle \mathtt{Q}_x, \mathtt{Q}_y \rangle = \begin{cases} \mu(x, \top) & \text{if } x = y, \\ 0 & \text{otherwise,} \end{cases}$$

and its determinant is $\prod_{x \in P} \mu(x, \top)$. It is nondegenerate iff $\mu(x, \top) \neq 0$ for all $x \in P$.

In view of Theorem D.51, the bilinear form in the special case (D.29) is nondegenerate iff the module M is faithful.

Since $\Bbbk P$ is an algebra, it is a (left) module over itself. We view $(\Bbbk P)^*$ as a $\Bbbk P$-module with the dual action (D.2). Let $\mathtt{M}$ be the basis of $(\Bbbk P)^*$ dual to $\mathtt{H}$, and $\mathtt{P}$ be the basis dual to $\mathtt{Q}$. The action on these bases is given by

$$\mathtt{H}_y \cdot \mathtt{M}_w = \sum_{x:\, x \vee y = w} \mathtt{M}_x \quad \text{and} \quad \mathtt{H}_y \cdot \mathtt{P}_x = \begin{cases} \mathtt{P}_x & \text{if } x \geq y, \\ 0 & \text{otherwise.} \end{cases} \tag{D.34}$$

This is straightforward to check. Observe that the actions on the $\mathtt{Q}$- and $\mathtt{P}$-bases are identical. This also follows from the fact that $\mathtt{Q}$ is a basis of primitive idempotents.

Theorem D.59. *For any ξ and η related by* (D.27)*, the linear map $\Bbbk P \to (\Bbbk P)^*$ given by*

$$\mathtt{H}_x \mapsto \sum_y \xi_{x \vee y} \mathtt{M}_y \qquad \textit{or equivalently} \qquad \mathtt{Q}_x \mapsto \eta_x \mathtt{P}_x$$

is a morphism of $\Bbbk P$-modules. In particular, if $\eta_x \neq 0$ for all x, then this map is an isomorphism.

PROOF. By Lemma D.11 or directly from (D.25) and (D.34), it follows that $\mathtt{Q}_x \mapsto \eta_x \mathtt{P}_x$ is a morphism of $\Bbbk P$-modules. The form of the map from the $\mathtt{H}$-basis to the $\mathtt{M}$-basis is a simple calculation:

$$\mathtt{H}_x = \sum_{y:\, y \geq x} \mathtt{Q}_y \mapsto \sum_{y:\, y \geq x} \eta_y \mathtt{P}_y = \sum_{w} \Big(\sum_{y \geq x \vee w} \eta_y \Big) \mathtt{M}_w = \sum_{w} \xi_{x \vee w} \mathtt{M}_w.$$

Alternatively: The map is induced by the bilinear form (D.33). The fact that it is morphism of $\Bbbk P$-modules is equivalent to associativity of the bilinear form. □

D.9.8. Join-preserving maps. Suppose $\lambda : P \to Q$ is a join-preserving map between finite lattices, that is, it preserves finite joins. Then its linearization

$$\Bbbk P \to \Bbbk Q, \qquad \mathtt{H}_x \mapsto \mathtt{H}_{\lambda(x)} \tag{D.35}$$

is an algebra homomorphism.

Lemma D.60. *On the* $\mathtt{Q}$*-basis of primitive idempotents, the map* (D.35) *is given by*

$$\mathtt{Q}_z \mapsto \sum_{w:\, \rho(w)=z} \mathtt{Q}_w, \tag{D.36}$$

where $\rho : Q \to P$ *is the right adjoint of* λ.

This particular form of the map is expected since a primitive idempotent must map to a sum of primitive idempotents under an algebra homomorphism.

PROOF. We use the formula in the $\mathtt{Q}$-basis to deduce the formula in the $\mathtt{H}$-basis:

$$\mathtt{H}_x = \sum_{z: x \leq z} \mathtt{Q}_z \mapsto \sum_{w:\, x \leq \rho(w)} \mathtt{Q}_w = \sum_{w:\, \lambda(x) \leq w} \mathtt{Q}_w = \mathtt{H}_{\lambda(x)}.$$

The second step used (B.2). □

The map $\lambda : P \to Q$ turns Q into a P-set. The action is given by

$$x \cdot y := \lambda(x) \vee y \tag{D.37}$$

for $x \in P$ and $y \in Q$. Note that

$$\rho(y) = \max\{x \in P \mid x \cdot y = y\}. \tag{D.38}$$

In words, $\rho(y)$ is the maximum element of P which stabilizes y. By linearizing (D.37), $\Bbbk Q$ is a module over the algebra $\Bbbk P$.

Lemma D.61. *The set* $\{\mathtt{Q}_y \mid \rho(y) = \bot\}$ *is a basis for the primitive part* $\mathcal{P}(\Bbbk Q)$*. More generally,* $\{\mathtt{Q}_y \mid \rho(y) \leq x\}$ *is a basis for* $\mathcal{P}_x(\Bbbk Q)$.

PROOF. Formula (D.36) shows that for $z \in P$,

$$\mathtt{Q}_z \cdot \Bbbk Q = \Big(\sum_{w:\, \rho(w)=z} \mathtt{Q}_w \Big) \cdot \Bbbk Q = \bigoplus_{w:\, \rho(w)=z} \mathtt{Q}_w \cdot \Bbbk Q.$$

The second step used that $\Bbbk Q$ is a split-semisimple commutative algebra. Clearly, the set $\{\mathtt{Q}_w \mid \rho(w) = z\}$ is a basis for the rhs. Now apply Lemma D.54 to deduce the result. □

Lemma D.62. *For* $x \in P$, $\xi_x(\Bbbk Q)$ *is the number of elements* $y \in Q$ *such that* $y \geq \lambda(x)$ *or equivalently* $\rho(y) \geq x$*, and* $\eta_x(\Bbbk Q)$ *is the number of elements* $y \in Q$ *such that* $\rho(y) = x$.

PROOF. The claim about $\xi_x(\Bbbk Q)$ is clear from (D.29) and (B.2). The second claim then follows from (D.27). Alternatively, use (D.36) as in the proof above. □

We record some special cases.

- $P = Q$ and $\lambda = \mathrm{id}$. In this case, $\xi_x(\Bbbk P)$ is the number of elements in P which are greater than x, and $\eta_x(\Bbbk P) \equiv 1$.
- Q is a singleton, λ is the unique map from P to Q, and ρ sends the unique element of Q to the top element of P. Then $\xi_x(\Bbbk Q) \equiv 1$ and $\eta_x(\Bbbk Q)$ is 1 if x is the top element of P and zero otherwise.
- P is a singleton, ρ is the unique map from Q to P, and λ sends the unique element of P to the bottom element of Q. Then both $\xi_\bot(\Bbbk Q)$ and $\eta_\bot(\Bbbk Q)$ equal the number of elements in Q.

D.9.9. Naturality of linear functionals. It is of interest to consider commutative diagrams of the form

$$\begin{array}{ccc} \Bbbk P & \longrightarrow & \Bbbk Q \\ & {\scriptstyle f_P}\searrow \quad \swarrow{\scriptstyle f_Q} & \\ & \Bbbk, & \end{array} \tag{D.39}$$

with the horizontal map obtained by linearizing a join-preserving map $\lambda : P \to Q$. We will discuss below some interesting instances of the above diagram.

Recall from (D.32) that any linear functional f_P induces a bilinear form on $\Bbbk P$. The above diagram says that the bilinear form on $\Bbbk Q$ induced by f_Q pulls back under λ to the bilinear form on $\Bbbk P$ induced by f_P.

Lemma D.63. *Suppose $\lambda : P \to Q$ is a join-preserving map between finite lattices satisfying $\lambda(x) = \top$ iff $x = \top$. Then*

$$\mu_P(z, \top) = \sum_{w:\, \rho(w)=z} \mu_Q(w, \top), \tag{D.40}$$

where $\rho : Q \to P$ is the right adjoint of λ.

PROOF. This is a special case of the Rota formula (C.14). □

Note that the hypothesis on λ is stronger than the tightness hypothesis (Definition B.13).

Lemma D.64. *For a join-preserving map $\lambda : P \to Q$ satisfying $\lambda(x) = \top$ iff $x = \top$, diagram* (D.39) *commutes, where the map f_P (respectively f_Q) sends Q_x to $\mu_P(x, \top)$ (respectively $\mu_Q(x, \top)$).*

PROOF. This follows from (D.36) and Lemma D.63. □

Lemma D.65. *Let $\lambda : P \to Q$ be a join-preserving map, and M be a $\Bbbk Q$-module (and hence a $\Bbbk P$-module). Then diagram* (D.39) *commutes, where*

$$f_P(\mathsf{Q}_x) = \eta_x(M) \textit{ for } x \in P \textit{ and } M \textit{ viewed as a } \Bbbk P\textit{-module},$$

$$f_Q(\mathsf{Q}_x) = \eta_x(M) \textit{ for } x \in Q \textit{ and } M \textit{ viewed as a } \Bbbk Q\textit{-module}.$$

PROOF. Let $\varphi : A \to B$ be an algebra homomorphism, and let M be a B-module with character χ_M. Then the character of M viewed as an A-module is $\chi_M \varphi$. Now specialize this to φ equal to the linearization of λ, and recall from Example D.50 that f_P and f_Q as given above are the characters of M as a $\Bbbk P$-module and as a $\Bbbk Q$-module, respectively. □

Notes

Finite-dimensional algebras. The study of idempotent and nilpotent elements and Peirce decompositions in algebras goes back to Peirce [**335**]. The Wedderburn structure theorem was proved by Wedderburn [**426**, Theorems 17 and 22]. Special cases were obtained earlier by Molien [**315**] and Cartan [**110**]. It was later generalized by Artin to a more general class of rings [**21**]. This result is treated in many books usually under the name Artin-Wedderburn theorem, see for instance, [**49**, Chapter 3], [**162**, Section 2.4], [**224**, Section 2.2], [**242**, Chapter 4], [**259**, Section II.2], [**269**, Chapters 1 and 2] or [**343**, Chapter 3].

The idempotent lifting problem is discussed in many places, see for instance, [**162**, Section 3.2], [**224**, Section 10.3], [**241**, Section III.8, Propositions 3 and 4], [**251**, Chapter II.4], [**259**, Section II.5], [**269**, Section 21], [**270**, Sections 3.6 and 3.7] or [**280**, Proposition 4.4]. An early reference on this problem is [**149**, Section II.6, Satz 1]. Theorem D.31 is part of the Wedderburn-Malcev theorem which appeared in the combined work of Wedderburn [**426**, Theorems 24 and 28] and Malcev [**297**, Theorem 2, page 42]. It is also called the Wedderburn principal theorem. It is treated for instance in [**162**, Section 6.2] or [**343**, Section 11.6]. An early reference is the book by Albert [**14**, Chapter III, Theorem 23]. Elementary algebras and quivers are discussed in [**32**] and [**25**]. In the literature, elementary algebras are also called basic algebras or split-basic algebras. Quivers go back to work of Gabriel [**193**]. Characters of algebras over modules are briefly treated in [**269**, pages 111 and 112]. References for the radical and socle of a module are [**17**, Sections 9 and 32], [**25**, Sections V.1 and V.2], [**162**, Sections 3.1 and 9.1] or [**270**, Section 3.3]. There are alternative characterizations of the radical of a ring. The radical is the intersection of all maximal left ideals. An element belongs to the radical iff it annihilates all simple left A-modules. (This is a consequence of the Nakayama Lemma. For finite dimensional algebras, this lemma is a simple consequence of nilpotency: see the proof of Theorem D.35.) The same properties hold for right ideals and modules. The notion of socle of a ring appeared in work of Dieudonné [**154**]. An early reference for Loewy series is the book by Artin, Nesbitt, and Thrall [**22**, Section IX.4]. For information on Frobenius algebras, see [**135**, Chapter 61], [**225**, Chapter 4] or [**261**, Chapter 2]. References for the Jordan-Hölder theorem are [**17**, Section 11], [**162**, Section 1.5], [**224**, Section 3.2] or [**343**, Section 2.6]. Theorem D.20 is equivalent to [**237**, Proposition in Section 4.2].

Lattices. Theorem D.47 is a special case of a result of Solomon [**387**, Theorem 1]; see also [**207**, Section 1] and [**401**, Theorem 3.9.2]. A specialization of Lemma D.57 is given in the second part of [**8**, Lemma 2.5.4]. The proof is the same. Dowling and Wilson [**160**, Lemma 1, page 506] show that the first bilinear form in Example D.58 is nondegenerate if $\mu(x, \top) \neq 0$ for all $x \in P$. Our argument coincides with the one given in the proof of [**418**, Theorem 25.5].

APPENDIX E

Bands

E.1. Bands

We briefly review bands and left regular bands.

E.1.1. Bands. A *band* or *idempotent monoid* is a monoid in which every element is idempotent: $x^2 = x$.

Let Σ be a band. For $x, y \in \Sigma$, set

$$x \leq y \quad \text{if} \quad xy = yx = y. \tag{E.1}$$

This defines a partial order on Σ. The necessary checks are done below.

- $x \leq x$ because $x^2 = x$.
- If $x \leq y$ and $y \leq z$, then $x \leq z$. This is because $xz = x(yz) = (xy)z = yz = z$, and similarly, $zx = z$.
- If $x \leq y$ and $y \leq x$, then $x = y$. This is because $xy = yx = x = y$.

The unit element of Σ is the unique minimum element of this partial order. We call an element of Σ a *face* and a maximal element a *chamber*.

Exercise E.1. Check that for any $x, y \in \Sigma$, $x \leq y \iff xyx = y$.

E.1.2. Left regular bands. A *left regular band* or *LRB* is a monoid in which

$$xyx = xy$$

for all elements x and y. By letting y be the unit element, we obtain $x^2 = x$. Thus, a left regular band is, in particular, a band.

It follows from Exercise E.1 that for a left regular band, the partial order (E.1) can be expressed more simply as

$$x \leq y \quad \text{if} \quad xy = y. \tag{E.2}$$

Example E.2. Any join-semilattice P is a (commutative) semigroup via

$$xy := x \vee y.$$

If P has a minimum element $\perp$ (in particular, if P is a lattice), we obtain a monoid. Note that

$$xyx = x \vee y \vee x = x \vee y = xy.$$

Thus P is a LRB. The partial order (E.2) coincides with that of P since

$$x \leq y \iff x \vee y = y.$$

For example, for the Boolean poset, the product is given by union, with the empty set as the unit element, and the partial order is inclusion.

E.1.3. Support map. Let Σ be a band. Define a relation $\rightharpoonup$ on Σ by

$$x \rightharpoonup y \text{ if } y = yxy.$$

This is reflexive and transitive, but not necessarily antisymmetric. We therefore obtain a poset Π by identifying x and y if $x \rightharpoonup y$ and $y \rightharpoonup x$. We denote the quotient map by

$$\mathrm{s} : \Sigma \to \Pi$$

and call it the *support map*. It is order-preserving. For any $x, y \in \Sigma$,

$$\mathrm{s}(x) \leq \mathrm{s}(y) \iff y = yxy, \tag{E.3}$$

$$\mathrm{s}(x) = \mathrm{s}(y) \iff y = yxy \text{ and } x = xyx, \tag{E.4}$$

$$\mathrm{s}(xy) = \mathrm{s}(yx) = \mathrm{s}(xyx) = \mathrm{s}(yxy). \tag{E.5}$$

Further, the poset Π is a join-semilattice, $\mathrm{s}(1)$ is the minimum element, and

$$\mathrm{s}(xy) = \mathrm{s}(x) \vee \mathrm{s}(y). \tag{E.6}$$

We refer to Π as the *support semilattice* of Σ. If Π is finite, it is a lattice.

If x and x' have the same support, and $x \leq y$, then y and $x'yx'$ have the same support. If y and y' have the same support, and $x \leq y$, then $xy'x$ has the same support as y and y'.

For any band, the support map is a morphism of monoids, with the support semilattice viewed as a monoid as in Example E.2. This is the content of (E.6).

If P is a join-semilattice viewed as a semigroup as in Example E.2, then the support semilattice of P is P itself, with the support map being the identity.

E.1.4. Unital augmentation. To any semigroup, one can associate a monoid by adjoining a unit element. The resulting monoid is the *unital augmentation* of that semigroup. An idempotent monoid Σ is always the unital augmentation of an idempotent semigroup. This is because if x and y are idempotent and $xy = 1$, then $x = y = 1$. So $\Sigma \setminus \{1\}$ is a subsemigroup and Σ is its unital augmentation.

E.1.5. Bands and partial orders. Let Σ be a band, endowed with the partial order (E.1). Let P be a poset and let $s, t : P \to \Sigma$ be maps such that

$$t \text{ is order-preserving,} \tag{E.7a}$$

$$t(a) \leq s(a) \text{ for all } a \in P. \tag{E.7b}$$

Define a new relation on P as follows:

$$a \preceq a' \text{ if } a \leq a' \quad \text{and} \quad t(a')s(a) = s(a') = s(a)t(a'). \tag{E.8}$$

Lemma E.3. *The relation $\preceq$ is a partial order on P. Moreover, both s and t are order-preserving for the partial order $\preceq$.*

Proof. Reflexivity for $\preceq$ follows from (E.7b) and reflexivity for $\leq$. Antisymmetry for $\preceq$ follows from that for $\leq$. To check transitivity, suppose $a \preceq a' \preceq a''$. By (E.7a), we have $t(a')t(a'') = t(a'')t(a') = t(a'')$. Hence, by (E.8),

$$s(a'') = t(a'')s(a') = t(a'')t(a')s(a) = t(a'')s(a)$$

and

$$s(a'') = s(a')t(a'') = s(a)t(a')t(a'') = s(a)t(a'').$$

Together with transitivity for $\leq$, this yields $a \preceq a''$. This proves the first assertion.

Suppose $a \preceq a'$. Then $a \le a'$ and hence $t(a) \le t(a')$, by (E.7a). In addition, by (E.7b), $t(a')s(a') = s(a')$. Hence, by (E.8),

$$s(a)s(a') = s(a)t(a')s(a') = s(a')s(a') = s(a')$$

and

$$s(a')s(a) = t(a')s(a)s(a) = t(a')s(a) = s(a').$$

This shows that $s(a) \le s(a')$, proving the second assertion. □

E.2. Distance functions

We introduce the notion of an abstract distance function on a band. This can be approached in two equivalent ways, either as a distance function on chambers, or as a distance function on faces. We also consider two weaker notions, namely, that of a left distance function and a right distance function.

For the free LRB, we give an explicit example of a distance function and relate it to the problem of sorting lists. We also classify all left distance functions on the free LRB.

E.2.1. Distance function on chambers. Let Σ be a finite band. A *distance function on chambers* is a function υ on pairs of chambers of Σ which satisfies the following conditions. For any chamber c,

$$\upsilon_{c,c} = 1. \tag{E.9a}$$

For x and x' with the same support, and $x \le c, d$,

$$\upsilon_{c,x'cx'} = \upsilon_{d,x'dx'}. \tag{E.9b}$$

For any c and $x' \le c'$,

$$\upsilon_{c,c'} = \upsilon_{c,x'cx'}\,\upsilon_{x'cx',c'}. \tag{E.9c}$$

For any c' and $x \le c$,

$$\upsilon_{c,c'} = \upsilon_{c,xc'x}\,\upsilon_{xc'x,c'}. \tag{E.9d}$$

For a LRB, repetitions can be deleted, for example, $x'cx'$ can be replaced by $x'c$; so in this case, the above conditions assume a slightly simpler form.

In certain situations, one may also want to consider the following addtional axiom. For x and x' with the same support, and $x \le c, d$,

$$\upsilon_{c,d} = \upsilon_{x'cx',x'dx'}. \tag{E.10}$$

A *left distance function on chambers* is a function υ on pairs of chambers of Σ which satisfies (E.9a), (E.9b) and (E.9c). Similarly, a *right distance function on chambers* is a function which satisfies (E.9a), (E.9b) and (E.9d).

Any function υ on pairs of chambers has a transpose υ^t defined by $(\upsilon^t)_{c,d} := \upsilon_{d,c}$. Observe that the transpose of a left distance function is a right distance function and vice-versa.

E.2.2. Distance function on faces. A *distance function on faces* is a function v on pairs of faces with the same support which satisfies the following conditions. For any face x,

$$v_{x,x} = 1. \tag{E.11a}$$

For x and x' with the same support, and $x \leq y$,

$$v_{x,x'} = v_{y,x'yx'}. \tag{E.11b}$$

For y and y' with the same support, and $x' \leq y'$,

$$v_{y,y'} = v_{y,x'yx'}\, v_{x'yx',y'}. \tag{E.11c}$$

For y and y' with the same support, and $x \leq y$,

$$v_{y,y'} = v_{y,xy'x}\, v_{xy'x,y'}. \tag{E.11d}$$

A *left distance function on faces* is a function v on pairs of faces with the same support which satisfies (E.11a), (E.11b) and (E.11c). Similarly, a *right distance function on faces* is a function which satisfies (E.11a), (E.11b) and (E.11d).

A distance function on faces clearly restricts to a distance function on chambers. Conversely, a distance function on chambers extends canonically to a distance function on faces via

$$v_{x,y} := v_{c,ycy}, \tag{E.12}$$

where c is any chamber greater than x. So the two notions are equivalent.

E.2.3. Free LRB. Let Σ be the free LRB on a finite generating set S. We refer to elements of S as letters. Elements of Σ are words with no repetition of letters. Product xy of x and y is obtained by concatenating x and y, and removing those letters from y which have appeared in x. We have $x \leq y$ if x is an initial segment of y. Chambers are words in which all letters appear exactly once, which is the same as linear orders on S. The poset Σ is a meet-semilattice. The meet $x \wedge y$ of two words x and y is the largest initial segment common to x and y. Two words x and y have the same support iff the same set of letters appear in x and in y.

Example E.4. To any pair of letters (a, b) assign a scalar $w(a, b)$. Now for any pair of chambers (c, d), define

$$v_{c,d} := \prod w(a, b), \tag{E.13}$$

where the product is over all pairs (a, b) such that the letter a appears before b in the word c but it appears after b in the word d.

We think of $v_{c,d}$ as the cost of sorting the list c according to the order specified by the list d. The sorting can be done by adjacent transpositions with the cost of changing ab to ba being $w(a, b)$.

Formula (E.13) defines a distance function on the free LRB. The axioms are verified below.

- (E.9a) is clear.
- Suppose x and y have the same support. So the same set of letters appear in x and in y. Let z be any word written using all the remaining letters. (E.9b) is equivalent to showing that $v_{xz,yz}$ does not depend on the particular choice of z. This is clear since to sort xz, we only need to sort x.

- Consider (E.9c). Suppose $y \leq d$. Then $v_{c,d} = v_{c,yc}v_{yc,d}$ says that the sorting can be done by first sorting the letters in y and bringing them to the beginning of the list and then sorting the rest. This is known as *selection sorting* [**260**, Section 5.2.3].
- Consider (E.9d). Suppose $x \leq c$. Then $v_{c,d} = v_{c,xd}v_{xd,d}$ says that the sorting can be done by first sorting z where $c = xz$ without touching x and then sorting the rest. This is known as *insertion sorting* [**260**, Section 5.2.1].

The additional axiom (E.10) also holds in this case.

We now proceed to describe all left distance functions on the free LRB. Consider pairs of the form (xya, xay), where x and y are words with y nonempty, and a is a letter, and there are no letter repetitions. We call (xya, xay) a basic pair.

Proposition E.5. *Any function on the set of basic pairs extends uniquely to a left distance function on Σ.*

PROOF. Let v denote the given function. We extend it to pairs of chambers (c, d) by a backward induction on the length of $x := c \wedge d$. Let c and d be any two chambers. If $x = d$, then $c = d$, and we define $v_{c,d} = 1$; so suppose not. Let a be the letter after x in d. Write $c = xyaz$ and $d = xaz'$ for unique choices of y, z and z', and define

$$v_{c,d} := v_{xya,xay}v_{xayz,xaz'}. \tag{A}$$

In the rhs, the first term is defined since (xya, xay) is a basic pair, and the second term is defined by the induction hypothesis. We now check that v is a left distance function.

- (E.9a) is clear.
- Suppose x and y have the same support. So the same set of letters appear in x and in y. Let z be any word written using all the remaining letters. (E.9b) is equivalent to showing that $v_{xz,yz}$ does not depend on the particular choice of z. This can be established by a backward induction on the length of $x \wedge y$.
- The above along with (A) implies that whenever $x = c \wedge d$ and the letter a follows x in d,

$$v_{c,d} = v_{c,xac}v_{xac,d}. \tag{B}$$

  (E.9c) says that $v_{c,d} = v_{c,yc}v_{yc,d}$ whenever $y \leq d$. This is clear if $y \leq c$. So suppose not. Write $y = xay'$ where $x := c \wedge d$. Then

$$v_{c,d} = v_{c,xac}v_{xac,d} = v_{c,xac}v_{xac,yc}v_{yc,d} = v_{c,yc}v_{yc,d}.$$

  The first and third equality follows from (B), while the second equality follows from the induction hypothesis, see illustration below.

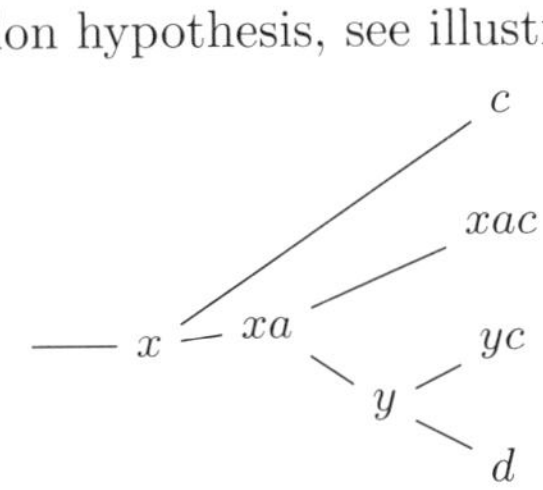

This shows that v is a left distance function. Since the construction only made use of its properties, uniqueness follows. $\square$

E.2.4. Karnofsky-Rhodes expansions. Fix a lattice Π and a set of generators S (with product being join). The Karnofsky-Rhodes expansion Σ is the freest LRB generated by S whose support lattice is Π. Explicitly: Call a word reduced if all its initial segments represent distinct elements of Π. Elements of Σ consist of all reduced words. The product is concatenation followed by reducing the word (in the obvious sense) from left to right.

We obtain the free LRB by taking Π to be the Boolean lattice, with S being its set of atoms. More generally, we can take Π to be the lattice of flats of a matroid, with S being its ground set. We then obtain the first matroid semigroup of Brown [**103**, Section 6].

The result of Proposition E.5 (with the same proof) generalizes to Karnofsky-Rhodes expansions: A pair is basic if it has the form (xz, xaz) where xz and xa are reduced words, xz and xaz have the same support, and z does not start with a. Further, z is optimal in the sense that no initial segment of z has these properties.

Notes

The partial order on a band (E.1) can be traced to Rees [**355**]; see also Clifford [**119**]. The origin of the axiom $xyx = xy$ can be traced to Schützenberger [**379**] and Klein-Barmen [**256**]. The construction of the support map associated to a band is due to McLean [**310**, Theorem 1]. It is discussed further by Clifford [**120**, Theorem 3] and Kimura [**255**, Theorem 1]. Example E.2 is discussed in detail by Birkhoff [**70**, Section I.5]. A unital LRB is called a *graphic monoid* by Lawvere in his work on topos theory [**278, 279**]. The work of Brown on Markov chains [**103**] brought in new examples and ideas to the study of LRBs. See for instance [**8**, Chapter 2]. Margolis, Saliola and Steinberg have undertaken an in-depth study of LRBs in a series of papers including [**302, 303, 304**].

General references for LRBs and other topics in semigroup theory are [**211, 340, 341**]. A modern reference is the recent book of Steinberg [**405**].

References

Bibliography

1. P. Abramenko, *Walls in Coxeter complexes*, Geom. Dedicata **49** (1994), no. 1, 71–84. 134
2. P. Abramenko and K. S. Brown, *Buildings*, Graduate Texts in Mathematics, vol. 248, Springer, New York, 2008, Theory and applications. 44, 74, 134, 175, 516
3. K. Adiprasito, J. Huh, and E. Katz, *Hodge theory for combinatorial geometries*, 2015. Available at arXiv:1511.02888. 45
4. M. Aguiar, *Internal categories and quantum groups*, ProQuest LLC, Ann Arbor, MI, 1997, Thesis (Ph.D.)–Cornell University. 544
5. M. Aguiar, N. Bergeron, and K. Nyman, *The peak algebra and the descent algebras of types B and D*, Trans. Amer. Math. Soc. **356** (2004), no. 7, 2781–2824. 510
6. M. Aguiar and W. Ferrer Santos, *Galois connections for incidence Hopf algebras of partially ordered sets*, Adv. Math. **151** (2000), no. 1, 71–100. 544
7. M. Aguiar and A. Lauve, *The characteristic polynomial of the Adams operators on graded connected Hopf algebras*, Algebra Number Theory **9** (2015), no. 3, 547–583. 365
8. M. Aguiar and S. Mahajan, *Coxeter groups and Hopf algebras*, Fields Monograph Series, vol. 23, AMS, Providence, RI, 2006. 44, 45, 99, 118, 134, 174, 209, 273, 508, 509, 570, 576
9. ______, *Monoidal functors, species and Hopf algebras*, CRM Monograph Series, vol. 29, American Mathematical Society, Providence, RI, 2010, With forewords by Kenneth Brown and Stephen Chase and André Joyal. 99, 118, 174, 209, 238, 273, 274, 364, 446, 478, 508, 509
10. ______, *Hopf monoids in the category of species*, Hopf algebras and tensor categories, Contemp. Math., vol. 585, Amer. Math. Soc., Providence, RI, 2013, pp. 17–124. 174, 273, 274, 365, 445
11. M. Aguiar, K. Nyman, and R. Orellana, *New results on the peak algebra*, J. Algebraic Combin. **23** (2006), no. 2, 149–188. 510
12. M. Aguiar and T. K. Petersen, *The Steinberg torus of a Weyl group as a module over the Coxeter complex*, J. Algebraic Combin. **42** (2015), no. 4, 1135–1175. 510
13. M. Aigner, *Combinatorial theory*, Springer-Verlag, Berlin, 1997, Reprint of the 1979 original. 45, 544
14. A. A. Albert, *Structure of algebras*, Revised printing of the 1939 original. American Mathematical Society Colloquium Publications, Vol. XXIV, American Mathematical Society, Providence, R.I., 1961. 570
15. V. Alexeev, *Moduli of weighted hyperplane arrangements*, Advanced Courses in Mathematics. CRM Barcelona, Birkhäuser/Springer, Basel, 2015, Edited by Gilberto Bini, Martí Lahoz, Emanuele Macrì and Paolo Stellari. xiii
16. J. Almeida, S. Margolis, B. Steinberg, and M. Volkov, *Representation theory of finite semigroups, semigroup radicals and formal language theory*, Trans. Amer. Math. Soc. **361** (2009), no. 3, 1429–1461. 273
17. F. W. Anderson and K. R. Fuller, *Rings and categories of modules*, second ed., Graduate Texts in Mathematics, vol. 13, Springer-Verlag, New York, 1992. 570
18. F. Ardila, *Computing the Tutte polynomial of a hyperplane arrangement*, Pacific J. Math. **230** (2007), no. 1, 1–26. 45
19. ______, *Algebraic and geometric methods in enumerative combinatorics*, Handbook of enumerative combinatorics, Discrete Math. Appl. (Boca Raton), CRC Press, Boca Raton, FL, 2015, pp. 3–172. 44, 45
20. D. Armstrong, *Hyperplane arrangements and diagonal harmonics*, J. Comb. **4** (2013), no. 2, 157–190. 175

21. E. Artin, *Zur Theorie der hyperkomplexen Zahlen*, Abh. Math. Sem. Univ. Hamburg **5** (1927), no. 1, 251–260. 570
22. E. Artin, C. J. Nesbitt, and R. M. Thrall, *Rings with Minimum Condition*, University of Michigan Publications in Mathematics, no. 1, University of Michigan Press, Ann Arbor, Mich., 1944. 570
23. W. A. Arvola, *Complexified real arrangements of hyperplanes*, Manuscripta Math. **71** (1991), no. 3, 295–306. 99
24. S. Assaf, P. Diaconis, and K. Soundararajan, *A rule of thumb for riffle shuffling*, Ann. Appl. Probab. **21** (2011), no. 3, 843–875. 365
25. I. Assem, D. Simson, and A. Skowroński, *Elements of the representation theory of associative algebras. Vol. 1*, London Mathematical Society Student Texts, vol. 65, Cambridge University Press, Cambridge, 2006, Techniques of representation theory. 570
26. C. A. Athanasiadis, *Characteristic polynomials of subspace arrangements and finite fields*, Adv. Math. **122** (1996), no. 2, 193–233. 45, 364
27. ———, *On free deformations of the braid arrangement*, European J. Combin. **19** (1998), no. 1, 7–18. 175
28. ———, *Extended Linial hyperplane arrangements for root systems and a conjecture of Postnikov and Stanley*, J. Algebraic Combin. **10** (1999), no. 3, 207–225. 175
29. C. A. Athanasiadis and P. Diaconis, *Functions of random walks on hyperplane arrangements*, Adv. in Appl. Math. **45** (2010), no. 3, 410–437. 45, 273, 364
30. M. D. Atkinson, *A new proof of a theorem of Solomon*, Bull. London Math. Soc. **18** (1986), no. 4, 351–354. 508
31. ———, *Solomon's descent algebra revisited*, Bull. London Math. Soc. **24** (1992), no. 6, 545–551. 174, 509, 510
32. M. Auslander, I. Reiten, and S. O. Smalø, *Representation theory of Artin algebras*, Cambridge Studies in Advanced Mathematics, vol. 36, Cambridge University Press, Cambridge, 1997, Corrected reprint of the 1995 original. 562, 570
33. S. P. Avann, *Application of the join-irreducible excess function to semi-modular lattices*, Math. Ann. **142** (1960/1961), 345–354. 524
34. A. Ayyer, A. Schilling, B. Steinberg, and N. M. Thiéry, *Markov chains, $\mathcal{R}$-trivial monoids and representation theory*, Internat. J. Algebra Comput. **25** (2015), no. 1-2, 169–231. 364
35. S. Backman and S. Hopkins, *Fourientations and the Tutte Polynomial*, 2015. Available at arXiv:1503.05885. 175
36. K. Baclawski, *Whitney numbers of geometric lattices*, Advances in Math. **16** (1975), 125–138. 444
37. K. Baclawski and A. Björner, *Fixed points in partially ordered sets*, Adv. in Math. **31** (1979), no. 3, 263–287. 445
38. K. A. Baker, P. C. Fishburn, and F. S. Roberts, *A new characterization of partial orders of dimension two*, Ann. New York Acad. Sci. **175** (1970), 23–24. 174
39. ———, *Partial orders of dimension* 2, Networks **2** (1972), 11–28. 174
40. H. Barcelo, *On the action of the symmetric group on the free Lie algebra and the partition lattice*, J. Combin. Theory Ser. A **55** (1990), no. 1, 93–129. 445, 446
41. H. Barcelo and N. Bergeron, *The Orlik-Solomon algebra on the partition lattice and the free Lie algebra*, J. Combin. Theory Ser. A **55** (1990), no. 1, 80–92. 446
42. H. Barcelo and E. Ihrig, *Lattices of parabolic subgroups in connection with hyperplane arrangements*, J. Algebraic Combin. **9** (1999), no. 1, 5–24. 175
43. H. Barcelo and S. Sundaram, *On some submodules of the action of the symmetric group on the free Lie algebra*, J. Algebra **154** (1993), no. 1, 12–26. 445
44. M. Barr, *Harrison homology, Hochschild homology and triples*, J. Algebra **8** (1968), 314–323. 364
45. T. Bauer, *Über die Struktur der Solomon-Algebren: ein Zugang über einen Differential und Integralkalkül*, Bayreuth. Math. Schr. (2001), no. 63, 1–102. 510
46. P. Baumann and C. Hohlweg, *A Solomon descent theory for the wreath products $G \wr S_n$*, Trans. Amer. Math. Soc. **360** (2008), no. 3, 1475–1538. 510
47. D. Bayer and P. Diaconis, *Trailing the dovetail shuffle to its lair*, Ann. Appl. Probab. **2** (1992), no. 2, 294–313. 364, 365, 366, 509
48. M. Bayer and B. Sturmfels, *Lawrence polytopes*, Canad. J. Math. **42** (1990), no. 1, 62–79. 46

49. J. A. Beachy, *Introductory lectures on rings and modules*, London Mathematical Society Student Texts, vol. 47, Cambridge University Press, Cambridge, 1999. 570
50. D. J. Benson, *Representations and cohomology. I*, second ed., Cambridge Studies in Advanced Mathematics, vol. 30, Cambridge University Press, Cambridge, 1998, Basic representation theory of finite groups and associative algebras. 365
51. ______, *Representations and cohomology. II*, second ed., Cambridge Studies in Advanced Mathematics, vol. 31, Cambridge University Press, Cambridge, 1998, Cohomology of groups and modules. 365
52. C. Berg, N. Bergeron, S. Bhargava, and F. Saliola, *Primitive orthogonal idempotents for R-trivial monoids*, J. Algebra **348** (2011), 446–461. 332
53. F. Bergeron and N. Bergeron, *A decomposition of the descent algebra of the hyperoctahedral group. I*, J. Algebra **148** (1992), no. 1, 86–97. 510
54. ______, *Orthogonal idempotents in the descent algebra of B_n and applications*, J. Pure Appl. Algebra **79** (1992), no. 2, 109–129. 366, 510
55. F. Bergeron, N. Bergeron, R. B. Howlett, and D. E. Taylor, *A decomposition of the descent algebra of a finite Coxeter group*, J. Algebraic Combin. **1** (1992), no. 1, 23–44. 508
56. N. Bergeron, *A hyperoctahedral analogue of the free Lie algebra*, J. Combin. Theory Ser. A **58** (1991), no. 2, 256–278. 446
57. ______, *A decomposition of the descent algebra of the hyperoctahedral group. II*, J. Algebra **148** (1992), no. 1, 98–122. 510
58. N. Bergeron, F. Hivert, and J.-Y. Thibon, *The peak algebra and the Hecke-Clifford algebras at $q = 0$*, J. Combin. Theory Ser. A **107** (2004), no. 1, 1–19. 509, 510
59. N. Bergeron and C. Hohlweg, *Coloured peak algebras and Hopf algebras*, J. Algebraic Combin. **24** (2006), no. 3, 299–330. 510
60. N. Bergeron and S. van Willigenburg, *A multiplication rule for the descent algebra of type D*, J. Algebra **206** (1998), no. 2, 699–705. 510
61. T. P. Bidigare, P. Hanlon, and D. N. Rockmore, *A combinatorial description of the spectrum for the Tsetlin library and its generalization to hyperplane arrangements*, Duke Math. J. **99** (1999), no. 1, 135–174. 99, 273, 364, 365
62. T. P. Bidigare, *Hyperplane arrangement face algebras and their associated Markov chains*, ProQuest LLC, Ann Arbor, MI, 1997, Thesis (Ph.D.)–University of Michigan. 273, 508
63. L. J. Billera, K. S. Brown, and P. Diaconis, *Random walks and plane arrangements in three dimensions*, Amer. Math. Monthly **106** (1999), no. 6, 502–524. 99, 364
64. L. J. Billera, R. Ehrenborg, and M. Readdy, *The c-2d-index of oriented matroids*, J. Combin. Theory Ser. A **80** (1997), no. 1, 79–105. 46
65. L. J. Billera, J. Tatch Moore, C. Dufort Moraites, Y. Wang, and K. Williams, *Maximal unbalanced families*, 2012. Available at arXiv:1209.2309. 174
66. G. Birkhoff, *On the combination of subalgebras*, Mathematical Proceedings of the Cambridge Philosophical Society **29** (1933), no. 4, 441–464. 524
67. ______, *Abstract Linear Dependence and Lattices*, Amer. J. Math. **57** (1935), no. 4, 800–804. 524
68. ______, *On the structure of abstract algebras*, Mathematical Proceedings of the Cambridge Philosophical Society **31** (1935), no. 4, 433–454. 174
69. ______, *Lattice Theory*, American Mathematical Society, New York, 1940. 524
70. ______, *Lattice theory*, third ed., American Mathematical Society Colloquium Publications, vol. 25, American Mathematical Society, Providence, R.I., 1979. 174, 519, 524, 576
71. G. D. Birkhoff, *A determinant formula for the number of ways of coloring a map*, Ann. of Math. (2) **14** (1912/13), no. 1-4, 42–46. 175
72. M. Bishop, *On the quiver presentation of the descent algebra of the hyperoctahedral group*, Algebr. Represent. Theory **18** (2015), no. 2, 397–420. 510
73. M. Bishop and G. Pfeiffer, *On the quiver presentation of the descent algebra of the symmetric group*, J. Algebra **383** (2013), 212–231. 510
74. A. Björner, *Shellable and Cohen-Macaulay partially ordered sets*, Trans. Amer. Math. Soc. **260** (1980), no. 1, 159–183. 409
75. A. Björner, *On complements in lattices of finite length*, Discrete Math. **36** (1981), no. 3, 325–326. 519
76. A. Björner, *On the homology of geometric lattices*, Algebra Universalis **14** (1982), no. 1, 107–128. 444, 445

77. ———, *Posets, regular CW complexes and Bruhat order*, European J. Combin. **5** (1984), no. 1, 7–16. 516
78. ———, *The homology and shellability of matroids and geometric lattices*, Matroid applications, Encyclopedia Math. Appl., vol. 40, Cambridge Univ. Press, Cambridge, 1992, pp. 226–283. 444, 445
79. ———, *Positive sum systems*, Combinatorial methods in topology and algebra, Springer INdAM Ser., vol. 12, Springer, Cham, 2015, pp. 157–171. 174, 444
80. A. Björner and F. Brenti, *Combinatorics of Coxeter groups*, Graduate Texts in Mathematics, vol. 231, Springer, New York, 2005. 45, 134, 175, 485, 508
81. A. Björner, P. H. Edelman, and G. M. Ziegler, *Hyperplane arrangements with a lattice of regions*, Discrete Comput. Geom. **5** (1990), no. 3, 263–288. 45
82. A. Björner, M. Las Vergnas, B. Sturmfels, N. White, and G. M. Ziegler, *Oriented matroids*, Encyclopedia of Mathematics and its Applications, vol. 46, Cambridge University Press, Cambridge, 1993. 19, 44, 45, 46, 405, 444, 516
83. A. Björner and M. L. Wachs, *Permutation statistics and linear extensions of posets*, J. Combin. Theory Ser. A **58** (1991), no. 1, 85–114. 174
84. ———, *Geometrically constructed bases for homology of partition lattices of type A, B and D*, Electron. J. Combin. **11** (2004/06), no. 2, Research Paper 3, 26. 175, 410, 445, 446
85. A. Björner and V. Welker, *Complexes of directed graphs*, SIAM J. Discrete Math. **12** (1999), no. 4, 413–424 (electronic). 174
86. A. Björner and G. M. Ziegler, *Introduction to greedoids*, Matroid applications, Encyclopedia Math. Appl., vol. 40, Cambridge Univ. Press, Cambridge, 1992, pp. 284–357. 524
87. R. G. Bland, *Complementary orthogonal subspaces of R^n and Orientability of Matroids*, ProQuest LLC, Ann Arbor, MI, 1974, Thesis (Ph.D.)–Cornell University. 44
88. D. Blessenohl and H. Laue, *Algebraic combinatorics related to the free Lie algebra*, Séminaire Lotharingien de Combinatoire (Thurnau, 1992), Prépubl. Inst. Rech. Math. Av., vol. 1993/33, Univ. Louis Pasteur, Strasbourg, 1993, pp. 1–21. 509
89. D. Blessenohl, C. Hohlweg, and M. Schocker, *A symmetry of the descent algebra of a finite Coxeter group*, Adv. Math. **193** (2005), no. 2, 416–437. 510
90. D. Blessenohl and H. Laue, *On the descending Loewy series of Solomon's descent algebra*, J. Algebra **180** (1996), no. 3, 698–724. 509, 510
91. ———, *The module structure of Solomon's descent algebra*, J. Aust. Math. Soc. **72** (2002), no. 3, 317–333. 510
92. D. Blessenohl and M. Schocker, *Noncommutative character theory of the symmetric group*, Imperial College Press, London, 2005. 510
93. T. S. Blyth, *Lattices and ordered algebraic structures*, Universitext, Springer-Verlag London, Ltd., London, 2005. 524
94. C. Bonnafé and G. Pfeiffer, *Around Solomon's descent algebras*, Algebr. Represent. Theory **11** (2008), no. 6, 577–602. 508
95. C. Bonnafé and C. Hohlweg, *Generalized descent algebra and construction of irreducible characters of hyperoctahedral groups*, Ann. Inst. Fourier (Grenoble) **56** (2006), no. 1, 131–181, With an appendix by Pierre Baumann and Hohlweg. 510
96. G. Boole, *The mathematical analysis of logic, Being an essay towards a calculus of deductive reasoning*, Cambridge: Macmillan, Barclay and Macmillan, 1847. 524
97. A. V. Borovik and A. Borovik, *Mirrors and reflections*, Universitext, Springer, New York, 2010, The geometry of finite reflection groups. 134
98. A. V. Borovik, I. M. Gelfand, and N. White, *Coxeter matroids*, Progress in Mathematics, vol. 216, Birkhäuser Boston, Inc., Boston, MA, 2003. 134
99. N. Bourbaki, *Éléments de mathématique*, Masson, Paris, 1981, Groupes et algèbres de Lie. Chapitres 4, 5 et 6. [Lie groups and Lie algebras. Chapters 4, 5 and 6]. 134
100. A. Brandt, *The free Lie ring and Lie representations of the full linear group*, Trans. Amer. Math. Soc. **56** (1944), 528–536. 445
101. G. E. Bredon, *Introduction to compact transformation groups*, Academic Press, New York-London, 1972, Pure and Applied Mathematics, Vol. 46. 544
102. E. Brieskorn, *Sur les groupes de tresses [d'après V. I. Arnol'd]*, Séminaire Bourbaki, 24ème année (1971/1972), Exp. No. 401, Springer, Berlin, 1973, pp. 21–44. Lecture Notes in Math., Vol. 317. 134

103. K. S. Brown, *Semigroups, rings, and Markov chains*, J. Theoret. Probab. **13** (2000), no. 3, 871–938. 44, 209, 273, 331, 363, 508, 576

104. ———, *Semigroup and ring theoretical methods in probability*, Representations of finite dimensional algebras and related topics in Lie theory and geometry, Fields Inst. Commun., vol. 40, Amer. Math. Soc., Providence, RI, 2004, pp. 3–26. 274

105. K. S. Brown and P. Diaconis, *Random walks and hyperplane arrangements*, Ann. Probab. **26** (1998), no. 4, 1813–1854. 19, 273, 363, 364, 395

106. T. H. Brylawski and A. N. Varchenko, *The determinant formula for a matroid bilinear form*, Adv. Math. **129** (1997), no. 1, 1–24. 238

107. T. H. Brylawski, *A decomposition for combinatorial geometries*, Trans. Amer. Math. Soc. **171** (1972), 235–282. 364

108. R. C. Buck, *Partition of space*, Amer. Math. Monthly **50** (1943), 541–544. 45

109. A. R. Calderbank, P. Hanlon, and S. Sundaram, *Representations of the symmetric group in deformations of the free Lie algebra*, Trans. Amer. Math. Soc. **341** (1994), no. 1, 315–333. 509

110. E. Cartan, *Les groupes bilinéaires et les systèmes de nombres complexes*, Ann. Fac. Sci. Toulouse Sci. Math. Sci. Phys. **12** (1898), B1–B99. 570

111. P. Cartier, *Les arrangements d'hyperplans: un chapitre de géométrie combinatoire*, Bourbaki Seminar, Vol. 1980/81, Lecture Notes in Math., vol. 901, Springer, Berlin-New York, 1981, pp. 1–22. 44

112. ———, *A primer of Hopf algebras*, Frontiers in number theory, physics, and geometry. II, Springer, Berlin, 2007, pp. 537–615. 365

113. P. Cellini, *A general commutative descent algebra*, J. Algebra **175** (1995), no. 3, 990–1014. 510

114. ———, *Cyclic Eulerian elements*, European J. Combin. **19** (1998), no. 5, 545–552. 510

115. F. Chapoton, *A bijection between shrubs and series-parallel posets*, 20th Annual International Conference on Formal Power Series and Algebraic Combinatorics (FPSAC 2008), Discrete Math. Theor. Comput. Sci. Proc., AJ, Assoc. Discrete Math. Theor. Comput. Sci., Nancy, 2008, pp. 227–234. 174

116. A. L. C. Cheung, *Adjoints of a geometry*, Canad. Math. Bull. **17** (1974), no. 3, 363–365; correction, ibid. 17 (1974), no. 4, 623. 174

117. C. Chevalley, *Sur certains groupes simples*, Tôhoku Math. J. (2) **7** (1955), 14–66. 508

118. C. Cibils, *Cohomology of incidence algebras and simplicial complexes*, J. Pure Appl. Algebra **56** (1989), no. 3, 221–232. 478

119. A. H. Clifford, *Semigroups admitting relative inverses*, Ann. of Math. (2) **42** (1941), 1037–1049. 576

120. ———, *Bands of semigroups*, Proc. Amer. Math. Soc. **5** (1954), 499–504. 576

121. D. Cohen, G. Denham, M. Falk, H. Schenck, A. Suciu, H. Terao, and S. Yuzvinsky, *Complex arrangements: Algebra, Geometry, Topology*, 2009, Preliminary draft. xiii, 44

122. P. M. Cohn, *Sur le critère de Friedrichs pour les commutateurs dans une algèbre associative libre*, C. R. Acad. Sci. Paris **239** (1954), 743–745. 296

123. M. A. Conger and J. Howald, *A better way to deal the cards*, Amer. Math. Monthly **117** (2010), no. 8, 686–700. 365

124. M. Content, F. Lemay, and P. Leroux, *Catégories de Möbius et fonctorialités: un cadre général pour l'inversion de Möbius*, J. Combin. Theory Ser. A **28** (1980), no. 2, 169–190. 544

125. G. E. Cooke and R. L. Finney, *Homology of cell complexes*, Princeton University Press, Princeton, N.J., 1967, Based on lectures by Norman E. Steenrod. 516

126. H. S. M. Coxeter, *The product of the generators of a finite group generated by reflections*, Duke Math. J. **18** (1951), 765–782. 134

127. H. H. Crapo, *A higher invariant for matroids*, J. Combinatorial Theory **2** (1967), 406–417. 238

128. ———, *Möbius inversion in lattices*, Arch. Math. (Basel) **19** (1968), 595–607 (1969). 45

129. ———, *Orthogonal representations of combinatorial geometries*, Atti del Convegno di Geometria Combinatoria e sue Applicazioni (Univ. Perugia, Perugia, 1970), Inst. Mat., Univ. Perugia, Perugia, 1971, pp. 175–186. 174

130. S. A. Csar, *Root and weight semigroup rings for signed posets*, ProQuest LLC, Ann Arbor, MI, 2014, Thesis (Ph.D.)–University of Minnesota. 175

131. M. Cuntz, *Crystallographic arrangements: Weyl groupoids and simplicial arrangements*, Bull. Lond. Math. Soc. **43** (2011), no. 4, 734–744. 45

132. ———, *Simplicial arrangements with up to 27 lines*, Discrete Comput. Geom. **48** (2012), no. 3, 682–701. 45

133. M. Cuntz and I. Heckenberger, *Finite Weyl groupoids*, J. Reine Angew. Math. **702** (2015), 77–108. 45

134. C. W. Curtis and G. I. Lehrer, *Homology representations of finite groups of Lie type*, Papers in algebra, analysis and statistics (Hobart, 1981), Contemp. Math., vol. 9, Amer. Math. Soc., Providence, R.I., 1981, pp. 1–28. 544

135. C. W. Curtis and I. Reiner, *Representation theory of finite groups and associative algebras*, AMS Chelsea Publishing, Providence, RI, 2006, Reprint of the 1962 original. 570

136. B. A. Davey and H. A. Priestley, *Introduction to lattices and order*, second ed., Cambridge University Press, New York, 2002. 524

137. M. W. Davis, *The geometry and topology of Coxeter groups*, London Mathematical Society Monographs Series, vol. 32, Princeton University Press, Princeton, NJ, 2008. 134

138. C. De Concini and C. Procesi, *Topics in hyperplane arrangements, polytopes and box-splines*, Universitext, Springer, New York, 2011. xiii

139. R. A. Dean and G. Keller, *Natural partial orders*, Canad. J. Math. **20** (1968), 535–554. 174

140. R. Dedekind, *Über zerlegungen von zahlen durch ihre grössten gemeinsamen teiler*, 1897. 524

141. ———, *Über die von drei Moduln erzeugte Dualgruppe*, Math. Ann. **53** (1900), no. 3, 371–403. 524

142. G. Denham, *Hanlon and Stanley's conjecture and the Milnor fibre of a braid arrangement*, J. Algebraic Combin. **11** (2000), no. 3, 227–240. 238

143. ———, *Eigenvectors for a random walk on a hyperplane arrangement*, Adv. in Appl. Math. **48** (2012), no. 2, 312–324. 395

144. G. Denham and P. Hanlon, *On the Smith normal form of the Varchenko bilinear form of a hyperplane arrangement*, Pacific J. Math. (1997), no. Special Issue, 123–146, Olga Taussky-Todd: in memoriam. 238

145. ———, *Some algebraic properties of the Schechtman-Varchenko bilinear forms*, New perspectives in algebraic combinatorics (Berkeley, CA, 1996–97), Math. Sci. Res. Inst. Publ., vol. 38, Cambridge Univ. Press, Cambridge, 1999, pp. 149–176. 238

146. T. Denton, F. Hivert, A. Schilling, and N. M. Thiéry, *On the representation theory of finite $\mathcal{J}$-trivial monoids*, Sém. Lothar. Combin. **64** (2010/11), Art. B64d, 44. 395

147. P. Deshpande, *On a generalization of Zaslavsky's theorem for hyperplane arrangements*, Ann. Comb. **18** (2014), no. 1, 35–55. 46

148. ———, *Arrangements of spheres and projective spaces*, Rocky Mountain J. Math. **46** (2016), no. 5, 1447–1487. 99

149. M. Deuring, *Algebren*, Zweite, korrigierte auflage. Ergebnisse der Mathematik und ihrer Grenzgebiete, Band 41, Springer-Verlag, Berlin-New York, 1968, Die 1. Auflage erschien im Jahre 1935 als Band 4, Heft 1 dieser Reihe. 570

150. M. Develin, M. Macauley, and V. Reiner, *Toric partial orders*, Trans. Amer. Math. Soc. **368** (2016), no. 4, 2263–2287. 175

151. P. Diaconis, *Mathematical developments from the analysis of riffle shuffling*, Groups, combinatorics & geometry (Durham, 2001), World Sci. Publ., River Edge, NJ, 2003, pp. 73–97. 365

152. P. Diaconis, C. Y. A. Pang, and A. Ram, *Hopf algebras and Markov chains: two examples and a theory*, J. Algebraic Combin. **39** (2014), no. 3, 527–585. 365

153. L. E. Dickson, *History of the theory of numbers. Vol. II: Diophantine analysis*, Chelsea Publishing Co., New York, 1966. 509

154. J. Dieudonné, *Sur le socle d'un anneau et les anneaux simples infinis*, Bull. Soc. Math. France **70** (1942), 46–75. 570

155. K. Dilks, T. K. Petersen, and J. R. Stembridge, *Affine descents and the Steinberg torus*, Adv. in Appl. Math. **42** (2009), no. 4, 423–444. 510

156. R. P. Dilworth, *Lattices with unique irreducible decompositions*, Ann. of Math. (2) **41** (1940), 771–777. 524

157. A. Dimca, *Hyperplane arrangements*, Universitext, Springer, Cham, 2017, An introduction. xiii

158. X. Dong, *The bounded complex of a uniform affine oriented matroid is a ball*, J. Combin. Theory Ser. A **115** (2008), no. 4, 651–661. 444

159. P. Doubilet, G.-C. Rota, and R. Stanley, *On the foundations of combinatorial theory. VI. The idea of generating function*, Proceedings of the Sixth Berkeley Symposium on Mathematical Statistics and Probability (Univ. California, Berkeley, Calif., 1970/1971), Vol. II: Probability theory, Univ. California Press, Berkeley, Calif., 1972, pp. 267–318. 544

160. T. A. Dowling and R. M. Wilson, *Whitney number inequalities for geometric lattices*, Proc. Amer. Math. Soc. **47** (1975), 504–512. 570

161. A. W. M. Dress and R. Scharlau, *Gated sets in metric spaces*, Aequationes Math. **34** (1987), no. 1, 112–120. 516

162. Y. A. Drozd and V. V. Kirichenko, *Finite-dimensional algebras*, Springer-Verlag, Berlin, 1994, Translated from the 1980 Russian original and with an appendix by Vlastimil Dlab. 570

163. G. Duchamp, A. Klyachko, D. Krob, and J.-Y. Thibon, *Noncommutative symmetric functions. III. Deformations of Cauchy and convolution algebras*, Discrete Math. Theor. Comput. Sci. **1** (1997), no. 1, 159–216, Lie computations (Marseille, 1994). 238, 509

164. A. Dür, *Möbius functions, incidence algebras and power series representations*, Lecture Notes in Mathematics, vol. 1202, Springer-Verlag, Berlin, 1986. 544

165. B. Dushnik and E. W. Miller, *Partially ordered sets*, Amer. J. Math. **63** (1941), 600–610. 174

166. E. B. Dynkin, *Calculation of the coefficients in the Campbell-Hausdorff formula*, Doklady Akad. Nauk SSSR (N.S.) **57** (1947), 323–326. 445

167. R. Eager and J. Lawrence, *The odd-even invariant for graphs*, European J. Combin. **50** (2015), 87–96. 238

168. V. Eberhard, *Eine Classification der allgemeinen Ebenensysteme*, J. Reine Angew. Math. **106** (1890), 89–120. 45

169. P. H. Edelman, *Meet-distributive lattices and the anti-exchange closure*, Algebra Universalis **10** (1980), no. 3, 290–299. 524

170. ______, *A partial order on the regions of* $\mathbf{R}^n$ *dissected by hyperplanes*, Trans. Amer. Math. Soc. **283** (1984), no. 2, 617–631. 45

171. ______, *Abstract convexity and meet-distributive lattices*, Combinatorics and ordered sets (Arcata, Calif., 1985), Contemp. Math., vol. 57, Amer. Math. Soc., Providence, RI, 1986, pp. 127–150. 519, 528

172. P. H. Edelman and R. E. Jamison, *The theory of convex geometries*, Geom. Dedicata **19** (1985), no. 3, 247–270. 74, 524

173. P. H. Edelman and P. Klingsberg, *The subposet lattice and the order polynomial*, European J. Combin. **3** (1982), no. 4, 341–346. 174

174. P. H. Edelman, V. Reiner, and V. Welker, *Convex, acyclic, and free sets of an oriented matroid*, Discrete Comput. Geom. **27** (2002), no. 1, 99–116, Geometric combinatorics (San Francisco, CA/Davis, CA, 2000). 174

175. R. Ehrenborg, M. Readdy, and M. Slone, *Affine and toric hyperplane arrangements*, Discrete Comput. Geom. **41** (2009), no. 4, 481–512. 46, 175

176. P. Eklund and W. Gähler, *Generalized Cauchy spaces*, Math. Nachr. **147** (1990), 219–233. 174

177. F. El Zein, A. I. Suciu, M. Tosun, A. M. Uludağ, and S. Yuzvinsky (eds.), *Arrangements, local systems and singularities*, Progress in Mathematics, vol. 283, Birkhäuser Verlag, Basel, 2010, Lectures notes from the CIMPA Summer School held at Galatasaray University, Istanbul, June 11–22, 2007. xiii

178. K. Erdmann and M. Schocker, *Modular Lie powers and the Solomon descent algebra*, Math. Z. **253** (2006), no. 2, 295–313. 510

179. D. C. Ernst, *Cell complexes for arrangements with group actions*, Master's thesis, Northern Arizona University, 2000. Available at arXiv:0905.4434. 99, 134

180. G. Etienne, *Linear extensions of finite posets and a conjecture of G. Kreweras on permutations*, Discrete Math. **52** (1984), no. 1, 107–111. 508

181. C. J. Everett, *Closure operators and Galois theory in lattices*, Trans. Amer. Math. Soc. **55** (1944), 514–525. 524

182. B. L. Feĭgin and B. L. Tsygan, *Additive K-theory*, K-theory, arithmetic and geometry (Moscow, 1984–1986), Lecture Notes in Math., vol. 1289, Springer, Berlin, 1987, pp. 67–209. 364
183. D. Finkelstein, *On relations between commutators*, Comm. Pure Appl. Math. **8** (1955), 245–250. 296
184. S. Foldes and R. Woodroofe, *Antichain cutsets of strongly connected posets*, Order **30** (2013), no. 2, 351–361. 524
185. J. Folkman, *The homology groups of a lattice*, J. Math. Mech. **15** (1966), 631–636. 444
186. B. Fresse, *On the homotopy of simplicial algebras over an operad*, Trans. Amer. Math. Soc. **352** (2000), no. 9, 4113–4141. 509
187. ———, *Koszul duality of operads and homology of partition posets*, Homotopy theory: relations with algebraic geometry, group cohomology, and algebraic K-theory, Contemp. Math., vol. 346, Amer. Math. Soc., Providence, RI, 2004, pp. 115–215. 445
188. K. O. Friedrichs, *Mathematical aspects of the quantum theory of fields. V. Fields modified by linear homogeneous forces*, Comm. Pure Appl. Math. **6** (1953), 1–72. 296
189. J. Fulman, *Semisimple orbits of Lie algebras and card-shuffling measures on Coxeter groups*, J. Algebra **224** (2000), no. 1, 151–165. 364
190. ———, *Applications of the Brauer complex: card shuffling, permutation statistics, and dynamical systems*, J. Algebra **243** (2001), no. 1, 96–122. 366
191. ———, *Descent algebras, hyperplane arrangements, and shuffling cards*, Proc. Amer. Math. Soc. **129** (2001), no. 4, 965–973. 364
192. P. Gabriel and M. Zisman, *Calculus of fractions and homotopy theory*, Ergebnisse der Mathematik und ihrer Grenzgebiete, Band 35, Springer-Verlag New York, Inc., New York, 1967. 544
193. P. Gabriel, *Unzerlegbare Darstellungen. I*, Manuscripta Math. **6** (1972), 71–103; correction, ibid. 6 (1972), 309. 570
194. P. Gabriel, *Des catégories abéliennes*, Bull. Soc. Math. France **90** (1962), 323–448. 544
195. O. Ganyushkin, V. Mazorchuk, and B. Steinberg, *On the irreducible representations of a finite semigroup*, Proc. Amer. Math. Soc. **137** (2009), no. 11, 3585–3592. 273
196. A. M. Garsia, *Combinatorics of the free Lie algebra and the symmetric group*, Analysis, et cetera, Academic Press, Boston, MA, 1990, pp. 309–382. 296, 364, 445, 509
197. A. M. Garsia and J. Remmel, *Shuffles of permutations and the Kronecker product*, Graphs Combin. **1** (1985), no. 3, 217–263. 509
198. A. M. Garsia and C. Reutenauer, *A decomposition of Solomon's descent algebra*, Adv. Math. **77** (1989), no. 2, 189–262. 508, 509, 510
199. I. M. Gelfand, D. Krob, A. Lascoux, B. Leclerc, V. S. Retakh, and J.-Y. Thibon, *Noncommutative symmetric functions*, Adv. Math. **112** (1995), no. 2, 218–348. 509
200. R. Gente, *The Varchenko matrix for cones*, Ph.D. thesis, Philipps-Universität Marburg, 2013. 238
201. M. Gerstenhaber and S. D. Schack, *A Hodge-type decomposition for commutative algebra cohomology*, J. Pure Appl. Algebra **48** (1987), no. 3, 229–247. 364
202. ———, *The shuffle bialgebra and the cohomology of commutative algebras*, J. Pure Appl. Algebra **70** (1991), no. 3, 263–272. 365
203. I. M. Gessel, *Multipartite P-partitions and inner products of skew Schur functions*, Combinatorics and algebra (Boulder, Colo., 1983), Amer. Math. Soc., Providence, RI, 1984, pp. 289–317. 508
204. E. Gottlieb and M. L. Wachs, *Cohomology of Dowling lattices and Lie (super)algebras*, Adv. in Appl. Math. **24** (2000), no. 4, 301–336. 446
205. G. Grätzer, *General lattice theory*, Birkhäuser Verlag, Basel, 2003, With appendices by B. A. Davey, R. Freese, B. Ganter, M. Greferath, P. Jipsen, H. A. Priestley, H. Rose, E. T. Schmidt, S. E. Schmidt, F. Wehrung and R. Wille, Reprint of the 1998 second edition. 519, 524
206. J. A. Green, *On the structure of semigroups*, Ann. of Math. (2) **54** (1951), 163–172. 45, 134
207. C. Greene, *On the Möbius algebra of a partially ordered set*, Advances in Math. **10** (1973), 177–187. 570
208. ———, *Acyclic orientations (notes)*, Higher Combinatorics (M. Aigner, ed.), D. Reidel, Dordrecht, 1977, pp. 65–68. 175

209. ______, *The Möbius function of a partially ordered set*, Ordered sets (Banff, Alta., 1981), NATO Adv. Study Inst. Ser. C: Math. Phys. Sci., vol. 83, Reidel, Dordrecht, 1982, pp. 555–581. 544
210. C. Greene and T. Zaslavsky, *On the interpretation of Whitney numbers through arrangements of hyperplanes, zonotopes, non-Radon partitions, and orientations of graphs*, Trans. Amer. Math. Soc. **280** (1983), no. 1, 97–126. 173, 175, 238, 444
211. P. A. Grillet, *Semigroups. an introduction to the structure theory*, Monographs and Textbooks in Pure and Applied Mathematics, vol. 193, Marcel Dekker Inc., New York, 1995. 576
212. L. C. Grove and C. T. Benson, *Finite reflection groups*, second ed., Graduate Texts in Mathematics, vol. 99, Springer-Verlag, New York, 1985. 134
213. B. Grünbaum, *Two-coloring the faces of arrangements*, Period. Math. Hungar. **11** (1980), no. 3, 181–185. 45
214. B. Grünbaum, *Arrangements of hyperplanes*, Proceedings of the Second Louisiana Conference on Combinatorics, Graph Theory and Computing (Louisiana State Univ., Baton Rouge, La., 1971), Louisiana State Univ., Baton Rouge, La., 1971, pp. 41–106. 44, 45
215. ______, *Convex polytopes*, second ed., Graduate Texts in Mathematics, vol. 221, Springer-Verlag, New York, 2003, Prepared and with a preface by Volker Kaibel, Victor Klee and Günter M. Ziegler. 44, 520
216. ______, *A catalogue of simplicial arrangements in the real projective plane*, Ars Math. Contemp. **2** (2009), no. 1, 1–25. 45, 329
217. J. Haigh, *On the Möbius algebra and the Grothendieck ring of a finite category*, J. London Math. Soc. (2) **21** (1980), no. 1, 81–92. 544
218. R. M. Hain, *On the indecomposable elements of the bar construction*, Proc. Amer. Math. Soc. **98** (1986), no. 2, 312–316. 364
219. P. Hall, *The eulerian functions of a group*, Quarterly J. Math. **7** (1936), no. 1, 134–151. 544
220. P. Hanlon, *The fixed-point partition lattices*, Pacific J. Math. **96** (1981), no. 2, 319–341. 445
221. ______, *The characters of the wreath product group acting on the homology groups of the Dowling lattices*, J. Algebra **91** (1984), no. 2, 430–463. 446
222. ______, *The action of S_n on the components of the Hodge decomposition of Hochschild homology*, Michigan Math. J. **37** (1990), no. 1, 105–124. 364
223. P. Hanlon and R. P. Stanley, *A q-deformation of a trivial symmetric group action*, Trans. Amer. Math. Soc. **350** (1998), no. 11, 4445–4459. 238, 509
224. M. Hazewinkel, N. Gubareni, and V. V. Kirichenko, *Algebras, rings and modules. Vol. 1*, Mathematics and its Applications, vol. 575, Kluwer Academic Publishers, Dordrecht, 2004. 570
225. ______, *Algebras, rings and modules. Vol. 2*, Mathematics and Its Applications (Springer), vol. 586, Springer, Dordrecht, 2007. 570
226. I. Heckenberger and V. Welker, *Geometric combinatorics of Weyl groupoids*, J. Algebraic Combin. **34** (2011), no. 1, 115–139. 45
227. J. Hedlíková, *Ternary spaces, media, and Chebyshev sets*, Czechoslovak Math. J. **33(108)** (1983), no. 3, 373–389. 516
228. J. Helmstetter, *Série de Hausdorff d'une algèbre de Lie et projections canoniques dans l'algèbre enveloppante*, J. Algebra **120** (1989), no. 1, 170–199. 365
229. C. Hohlweg, *Generalized descent algebras*, Canad. Math. Bull. **50** (2007), no. 4, 535–546. 510
230. S. Hopkins and D. Perkinson, *Bigraphical arrangements*, Trans. Amer. Math. Soc. **368** (2016), no. 1, 709–725. 175
231. R. A. Horn and C. R. Johnson, *Topics in matrix analysis*, Cambridge University Press, Cambridge, 1994, Corrected reprint of the 1991 original. 217
232. ______, *Matrix analysis*, second ed., Cambridge University Press, Cambridge, 2013. 338
233. S. K. Hsiao, *A semigroup approach to wreath-product extensions of Solomon's descent algebras*, Electron. J. Combin. **16** (2009), no. 1, Research Paper 21, 9. 174, 510
234. A. Hubery, *The cluster complex of an hereditary Artin algebra*, Algebr. Represent. Theory **14** (2011), no. 6, 1163–1185. 524
235. J. Huh, *Milnor numbers of projective hypersurfaces and the chromatic polynomial of graphs*, J. Amer. Math. Soc. **25** (2012), no. 3, 907–927. 45

236. J. Huh and E. Katz, *Log-concavity of characteristic polynomials and the Bergman fan of matroids*, Math. Ann. **354** (2012), no. 3, 1103–1116. 45
237. J. E. Humphreys, *Introduction to Lie algebras and representation theory*, Graduate Texts in Mathematics, vol. 9, Springer-Verlag, New York-Berlin, 1978, Second printing, revised. 570
238. ———, *Reflection groups and Coxeter groups*, Cambridge Studies in Advanced Mathematics, vol. 29, Cambridge University Press, Cambridge, 1990. 134, 209, 485, 508
239. M. C. Iovanov and G. D. Koffi, *On incidence algebras and their representations*, 2017. Available at arXiv:1702.03356. 544
240. J. R. Isbell, *Median algebra*, Trans. Amer. Math. Soc. **260** (1980), no. 2, 319–362. 516
241. N. Jacobson, *Structure of rings*, American Mathematical Society Colloquium Publications, Vol. 37. Revised edition, American Mathematical Society, Providence, R.I., 1964. 570
242. ———, *Basic algebra. II*, second ed., W. H. Freeman and Company, New York, 1989. 570
243. M. Jambu, *Arrangements d'hyperplans. I. Les groupes de réflexions*, Ann. Sci. Math. Québec **12** (1988), no. 1, 73–99. 44
244. R. E. Jamison, *Copoints in antimatroids*, Proceedings of the Eleventh Southeastern Conference on Combinatorics, Graph Theory and Computing (Florida Atlantic Univ., Boca Raton, Fla., 1980), Vol. II, vol. 29, 1980, pp. 535–544. 524
245. ———, *A perspective on abstract convexity: Classifying alignments by varieties*, Convexity and related combinatorial geometry. Proceedings of the Second University of Oklahoma Conference held in Norman, Okla., March 13–15, 1980 (D. C. Kay and M. Breen, eds.), Lecture Notes in Pure and Appl. Math., vol. 76, Marcel Dekker, Inc., New York, 1982, pp. 113–150. 524
246. A. Jöllenbeck and C. Reutenauer, *Eine Symmetrieeigenschaft von Solomons Algebra und der höheren Lie-Charaktere*, Abh. Math. Sem. Univ. Hamburg **71** (2001), 105–111. 510
247. A. Joyal, *Foncteurs analytiques et espèces de structures*, Combinatoire énumérative (Montreal, Que., 1985/Quebec, Que., 1985), Lecture Notes in Math., vol. 1234, Springer, Berlin, 1986, pp. 126–159. 445
248. H. Kamiya, A. Takemura, and H. Terao, *Ranking patterns of unfolding models of codimension one*, Adv. in Appl. Math. **47** (2011), no. 2, 379–400. 174
249. ———, *Arrangements stable under the Coxeter groups*, Configuration spaces, CRM Series, vol. 14, Ed. Norm., Pisa, 2012, pp. 327–354. 174
250. R. Kane, *Reflection groups and invariant theory*, CMS Books in Mathematics/Ouvrages de Mathématiques de la SMC, vol. 5, Springer-Verlag, New York, 2001. 134
251. I. Kaplansky, *Fields and rings*, second ed., The University of Chicago Press, Chicago, Ill.-London, 1972, Chicago Lectures in Mathematics. 570
252. Y. Kashina, S. Montgomery, and S.-H. Ng, *On the trace of the antipode and higher indicators*, Israel J. Math. **188** (2012), 57–89. 365
253. Y. Kashina, Y. Sommerhäuser, and Y. Zhu, *On higher Frobenius-Schur indicators*, Mem. Amer. Math. Soc. **181** (2006), no. 855, viii+65. 365
254. Y. Kemper and J. Lawrence, *The odd-even invariant and hamiltonian circuits in tope graphs*, 2016. Available at arXiv:1610.05039. 238
255. N. Kimura, *The structure of idempotent semigroups. I*, Pacific J. Math. **8** (1958), 257–275. 576
256. F. Klein-Barmen, *Über eine weitere Verallgemeinerung des Verbandsbegriffes*, Math. Z. **46** (1940), 472–480. 576
257. C. J. Klivans and E. Swartz, *Projection volumes of hyperplane arrangements*, Discrete Comput. Geom. **46** (2011), no. 3, 417–426. 45
258. A. A. Klyachko, *Lie elements in the tensor algebra*, Siberian Math. J. **15** (1974), 914–920. 297, 445
259. A. W. Knapp, *Advanced algebra*, Cornerstones, Birkhäuser Boston, Inc., Boston, MA, 2007, Along with a companion volume *Basic algebra*. 570
260. D. E. Knuth, *The art of computer programming. Vol. 3: Sorting and searching*, 2nd ed., Addison-Wesley, Reading, MA, 1998. 575
261. J. Kock, *Frobenius algebras and 2D topological quantum field theories*, London Mathematical Society Student Texts, vol. 59, Cambridge University Press, Cambridge, 2004. 570
262. C. Krattenthaler, *Advanced determinant calculus*, Sém. Lothar. Combin. **42** (1999), Art. B42q, 67 pp. (electronic), The Andrews Festschrift (Maratea, 1998). 238

263. ______, *Advanced determinant calculus: a complement*, Linear Algebra Appl. **411** (2005), 68–166. 238
264. A. Kriegl, *A characterization of reduced incidence algebras*, Discrete Math. **34** (1981), no. 2, 141–144. 544
265. D. Krob, B. Leclerc, and J.-Y. Thibon, *Noncommutative symmetric functions. II. Transformations of alphabets*, Internat. J. Algebra Comput. **7** (1997), no. 2, 181–264. 509, 510
266. D. Krob and J.-Y. Thibon, *Higher order peak algebras*, Ann. Comb. **9** (2005), no. 4, 411–430. 510
267. J. P. S. Kung, *Critical problems*, Matroid theory (Seattle, WA, 1995), Contemp. Math., vol. 197, Amer. Math. Soc., Providence, RI, 1996, pp. 1–127. 45
268. T. Y. Lam, *Lectures on modules and rings*, Graduate Texts in Mathematics, vol. 189, Springer-Verlag, New York, 1999. 268
269. ______, *A first course in noncommutative rings*, second ed., Graduate Texts in Mathematics, vol. 131, Springer-Verlag, New York, 2001. 570
270. J. Lambek, *Lectures on rings and modules*, With an appendix by Ian G. Connell, Blaisdell Publishing Co. Ginn and Co., Waltham, Mass.-Toronto, Ont.-London, 1966. 570
271. M. Las Vergnas, *Matroïdes orientables*, C. R. Acad. Sci. Paris Sér. A-B **280** (1975), Ai, A61–A64. 45
272. ______, *Acyclic and totally cyclic orientations of combinatorial geometries*, Discrete Math. **20** (1977/78), no. 1, 51–61. 175, 238
273. ______, *Sur les activités des orientations d'une géométrie combinatoire*, Cahiers Centre Études Rech. Opér. **20** (1978), no. 3-4, 293–300, Colloque Mathématiques Discrètes: Codes et Hypergraphes (Brussels, 1978). 238
274. ______, *Convexity in oriented matroids*, J. Combin. Theory Ser. B **29** (1980), no. 2, 231–243. 45
275. E. L. Lawler, *Sequencing jobs to minimize total weighted completion time subject to precedence constraints*, Ann. Discrete Math. **2** (1978), 75–90, Algorithmic aspects of combinatorics (Conf., Vancouver Island, B.C., 1976). 174
276. J. Lawrence, *The odd-even invariant of an oriented matroid*, European J. Combin. **27** (2006), no. 6, 806–813. 238
277. M. V. Lawson, *Semigroups and ordered categories. I. The reduced case*, J. Algebra **141** (1991), no. 2, 422–462. 118
278. F. W. Lawvere, *More on graphic toposes*, Cahiers Topologie Géom. Différentielle Catég. **32** (1991), no. 1, 5–10, International Category Theory Meeting (Bangor, 1989 and Cambridge, 1990). 576
279. ______, *Linearization of graphic toposes via Coxeter groups*, J. Pure Appl. Algebra **168** (2002), no. 2-3, 425–436, Category theory 1999 (Coimbra). 576
280. L. Le Bruyn, *Noncommutative geometry and Cayley-smooth orders*, Pure and Applied Mathematics (Boca Raton), vol. 290, Chapman & Hall/CRC, Boca Raton, FL, 2008. 570
281. G. I. Lehrer, *On hyperoctahedral hyperplane complements*, The Arcata Conference on Representations of Finite Groups (Arcata, Calif., 1986), Proc. Sympos. Pure Math., vol. 47, Amer. Math. Soc., Providence, RI, 1987, pp. 219–234. 446
282. G. I. Lehrer and L. Solomon, *On the action of the symmetric group on the cohomology of the complement of its reflecting hyperplanes*, J. Algebra **104** (1986), no. 2, 410–424. 446
283. T. Leinster, *Higher operads, higher categories*, London Mathematical Society Lecture Note Series, vol. 298, Cambridge University Press, Cambridge, 2004. 478
284. ______, *The Euler characteristic of a category*, Doc. Math. **13** (2008), 21–49. 544
285. ______, *Notions of Möbius inversion*, Bull. Belg. Math. Soc. Simon Stevin **19** (2012), no. 5, 911–935. 544
286. D. A. Levin, Y. Peres, and E. L. Wilmer, *Markov chains and mixing times*, American Mathematical Society, Providence, RI, 2009, With a chapter by James G. Propp and David B. Wilson. 365
287. J.-L. Loday, *Opérations sur l'homologie cyclique des algèbres commutatives*, Invent. Math. **96** (1989), no. 1, 205–230. 364, 509
288. ______, *Cyclic homology*, second ed., Grundlehren der Mathematischen Wissenschaften [Fundamental Principles of Mathematical Sciences], vol. 301, Springer-Verlag, Berlin, 1998, Appendix E by María O. Ronco, Chapter 13 by the author in collaboration with Teimuraz Pirashvili. 365, 509

289. ———, *Dialgebras*, Dialgebras and related operads, Lecture Notes in Math., vol. 1763, Springer, Berlin, 2001, pp. 7–66. 191, 193
290. J.-L. Loday and B. Vallette, *Algebraic operads*, Grundlehren der Mathematischen Wissenschaften [Fundamental Principles of Mathematical Sciences], vol. 346, Springer, Heidelberg, 2012. 174, 446, 478
291. M. Lothaire, *Combinatorics on words*, Cambridge Mathematical Library, Cambridge University Press, Cambridge, 1997, With a foreword by Roger Lyndon and a preface by Dominique Perrin, Corrected reprint of the 1983 original, with a new preface by Perrin. 510
292. A. T. Lundell and S. Weingram, *The topology of CW complexes*, Van Nostrand Reinhold Company, New York, 1969. 516
293. R. C. Lyndon, *A theorem of Friedrichs*, Michigan Math. J. **3** (1955), 27–29. 296
294. S. Mac Lane, *A lattice formulation for transcendence degrees and p-bases*, Duke Math. J. **4** (1938), no. 3, 455–468. 524
295. W. Magnus, *On the exponential solution of differential equations for a linear operator*, Comm. Pure Appl. Math. **7** (1954), 649–673. 296
296. S. A. Mahajan, *Shuffles, shellings and projections*, ProQuest LLC, Ann Arbor, MI, 2002, Thesis (Ph.D.)–Cornell University. 45, 99, 175, 209, 365, 366
297. A. Malcev, *On the representation of an algebra as a direct sum of the radical and a semi-simple subalgebra*, C. R. (Doklady) Acad. Sci. URSS (N.S.) **36** (1942), 42–45. 570
298. C. Malvenuto, *Produits et coproduits des fonctions quasi-symétriques et de l'algèbre des descents*, Ph.D. thesis, Laboratoire de combinatoire et d'informatique mathématique (LACIM), Univ. du Québec à Montréal, 1994. 509
299. C. Malvenuto and C. Reutenauer, *Duality between quasi-symmetric functions and the Solomon descent algebra*, J. Algebra **177** (1995), no. 3, 967–982. 509
300. A. Mandel, *Topology of oriented matroids*, ProQuest LLC, Ann Arbor, MI, 1982, Thesis (Ph.D.)–University of Waterloo (Canada). 45
301. R. Mantaci and C. Reutenauer, *A generalization of Solomon's algebra for hyperoctahedral groups and other wreath products*, Comm. Algebra **23** (1995), no. 1, 27–56. 510
302. S. Margolis, F. Saliola, and B. Steinberg, *Semigroups embeddable in hyperplane face monoids*, Semigroup Forum **89** (2014), no. 1, 236–248. 576
303. ———, *Combinatorial topology and the global dimension of algebras arising in combinatorics*, J. Eur. Math. Soc. (JEMS) **17** (2015), no. 12, 3037–3080. 332, 395, 444, 576
304. ———, *Cell complexes, poset topology and the representation theory of algebras arising in algebraic combinatorics and discrete geometry*, 2015. Available at arXiv:1508.05446. 46, 332, 395, 444, 478, 524, 576
305. S. Margolis and B. Steinberg, *The quiver of an algebra associated to the Mantaci-Reutenauer descent algebra and the homology of regular semigroups*, Algebr. Represent. Theory **14** (2011), no. 1, 131–159. 395, 510
306. ———, *Quivers of monoids with basic algebras*, Compos. Math. **148** (2012), no. 5, 1516–1560. 395
307. M. Markl, S. Shnider, and J. Stasheff, *Operads in algebra, topology and physics*, Mathematical Surveys and Monographs, vol. 96, American Mathematical Society, Providence, RI, 2002. 174, 446, 478
308. A. Mathas and R. C. Orellana, *Cyclotomic Solomon algebras*, Adv. Math. **219** (2008), no. 2, 450–487. 510
309. T. McConville, *Biclosed sets in real hyperplane arrangements*, 2014. Available at arXiv:1411.1305. 45
310. D. McLean, *Idempotent semigroups*, Amer. Math. Monthly **61** (1954), 110–113. 576
311. P. McMullen and E. Schulte, *Abstract regular polytopes*, Encyclopedia of Mathematics and its Applications, vol. 92, Cambridge University Press, Cambridge, 2002. 44, 524
312. J. W. Milnor and J. C. Moore, *On the structure of Hopf algebras*, Ann. of Math. (2) **81** (1965), 211–264. 364
313. B. Mitchell, *Rings with several objects*, Advances in Math. **8** (1972), 1–161. 544
314. A. F. Möbius, *Über eine besondere Art von Umkehrung der Reihen*, J. Reine Angew. Math. **9** (1832), 105–123. 544
315. T. Molien, *Über Systeme höherer complexer Zahlen*, Math. Ann. **41** (1892), no. 1, 83–156. 570

316. S. Montgomery, *Hopf algebras and their actions on rings*, CBMS Regional Conference Series in Mathematics, vol. 82, Published for the Conference Board of the Mathematical Sciences, Washington, DC, 1993. 364
317. E. H. Moore, *Introduction to a form of general analysis*, New Haven, Yale University Press, 1910. 524
318. P. Moszkowski, *Puissance d'un triplet de formes*, C. R. Acad. Sci. Paris Sér. I Math. **299** (1984), no. 4, 93–95. 508
319. M. Moynihan, *The colored Eulerian descent algebra*, J. Algebraic Combin. **42** (2015), no. 3, 671–694. 510
320. S.-H. Ng and P. Schauenburg, *Central invariants and higher indicators for semisimple quasi-Hopf algebras*, Trans. Amer. Math. Soc. **360** (2008), no. 4, 1839–1860. 365
321. I. Novik, A. Postnikov, and B. Sturmfels, *Syzygies of oriented matroids*, Duke Math. J. **111** (2002), no. 2, 287–317. 175
322. K. L. Nyman, *The peak algebra of the symmetric group*, J. Algebraic Combin. **17** (2003), no. 3, 309–322. 510
323. O. Ore, *Galois connexions*, Trans. Amer. Math. Soc. **55** (1944), 493–513. 524
324. P. Orlik, *Introduction to arrangements*, CBMS Regional Conference Series in Mathematics, vol. 72, American Mathematical Society, Providence, R.I., 1989. 44
325. ______, *Lectures on arrangements: combinatorics and topology*, Algebraic combinatorics, Universitext, Springer, Berlin, 2007, pp. 1–79. 44
326. P. Orlik and L. Solomon, *Combinatorics and topology of complements of hyperplanes*, Invent. Math. **56** (1980), no. 2, 167–189. 134
327. ______, *Unitary reflection groups and cohomology*, Invent. Math. **59** (1980), no. 1, 77–94. 134
328. P. Orlik and H. Terao, *Arrangements of hyperplanes*, Grundlehren der Mathematischen Wissenschaften, vol. 300, Springer-Verlag, Berlin, 1992. 44, 45, 46, 99, 134, 175, 364, 544
329. ______, *Arrangements and hypergeometric integrals*, second ed., MSJ Memoirs, vol. 9, Mathematical Society of Japan, Tokyo, 2007. xiii
330. A. Pang, *Hopf algebras and Markov chains*, Ph.D. thesis, Stanford University, 2014. 365
331. F. Patras, *Construction géométrique des idempotents eulériens. Filtration des groupes de polytopes et des groupes d'homologie de Hochschild*, Bull. Soc. Math. France **119** (1991), no. 2, 173–198. 364
332. ______, *La décomposition en poids des algèbres de Hopf*, Ann. Inst. Fourier (Grenoble) **43** (1993), no. 4, 1067–1087. 365
333. ______, *Lambda-rings*, Handbook of algebra, Vol. 3, North-Holland, Amsterdam, 2003, pp. 961–986. 365
334. F. Patras and C. Reutenauer, *Higher Lie idempotents*, J. Algebra **222** (1999), no. 1, 51–64. 509
335. B. Peirce, *Linear Associative Algebra*, Amer. J. Math. **4** (1881), no. 1-4, 97–229. 570
336. C. S. Peirce, *On the Algebra of Logic*, Amer. J. Math. **3** (1880), no. 1, 15–57. 524
337. T. K. Petersen, *Cyclic descents and P-partitions*, J. Algebraic Combin. **22** (2005), no. 3, 343–375. 510
338. ______, *Enriched P-partitions and peak algebras*, Adv. Math. **209** (2007), no. 2, 561–610. 510
339. ______, *Eulerian numbers*, Birkhäuser Advanced Texts: Basler Lehrbücher. [Birkhäuser Advanced Texts: Basel Textbooks], Birkhäuser/Springer, New York, 2015, With a foreword by Richard Stanley. 134, 175, 510
340. M. Petrich, *A construction and a classification of bands*, Math. Nachr. **48** (1971), 263–274. 576
341. ______, *Lectures in semigroups*, John Wiley & Sons, London-New York-Sydney, 1977. 576
342. G. Pfeiffer, *A quiver presentation for Solomon's descent algebra*, Adv. Math. **220** (2009), no. 5, 1428–1465. 508
343. R. S. Pierce, *Associative algebras*, Graduate Texts in Mathematics, vol. 88, Springer-Verlag, New York-Berlin, 1982, Studies in the History of Modern Science, 9. 570
344. S. Poirier, *Cycle type and descent set in wreath products*, Proceedings of the 7th Conference on Formal Power Series and Algebraic Combinatorics (Noisy-le-Grand, 1995), vol. 180, 1998, pp. 315–343. 510

345. P. Polo, *On Cohen-Macaulay posets, Koszul algebras and certain modules associated to Schubert varieties*, Bull. London Math. Soc. **27** (1995), no. 5, 425–434. 478
346. A. Postnikov, V. Reiner, and L. Williams, *Faces of generalized permutohedra*, Doc. Math. **13** (2008), 207–273. 174, 175
347. A. Postnikov and R. P. Stanley, *Deformations of Coxeter hyperplane arrangements*, J. Combin. Theory Ser. A **91** (2000), no. 1-2, 544–597, In memory of Gian-Carlo Rota. 175
348. G. B. Preston, *Inverse semi-groups*, J. London Math. Soc. **29** (1954), 396–403. 134
349. E. M. Rains, *The homology of real subspace arrangements*, J. Topol. **3** (2010), no. 4, 786–818. 478
350. H. Randriamaro, *Spectral properties of descent algebra elements*, J. Algebraic Combin. **39** (2014), no. 1, 127-139. 508
351. ———, *The Varchenko determinants of the classical Coxeter groups*, 2016. Available at arXiv:1411.1647. 238
352. N. Reading, *Lattice and order properties of the poset of regions in a hyperplane arrangement*, Algebra Universalis **50** (2003), no. 2, 179–205. 45
353. ———, *Noncrossing partitions and the shard intersection order*, J. Algebraic Combin. **33** (2011), no. 4, 483–530. 45
354. R. Ree, *Lie elements and an algebra associated with shuffles*, Ann. of Math. (2) **68** (1958), 210–220. 296, 445
355. D. Rees, *On semi-groups*, Proc. Cambridge Philos. Soc. **36** (1940), 387–400. 576
356. V. Reiner, *Quotients of Coxeter complexes and P-partitions*, ProQuest LLC, Ann Arbor, MI, 1990, Thesis (Ph.D.)–Massachusetts Institute of Technology. 174, 175
357. ———, *Signed posets*, J. Combin. Theory Ser. A **62** (1993), no. 2, 324–360. 175
358. V. Reiner, F. Saliola, and V. Welker, *Spectra of symmetrized shuffling operators*, Mem. Amer. Math. Soc. **228** (2014), no. 1072, vi+109. 99, 363
359. C. Reutenauer, *Theorem of Poincaré-Birkhoff-Witt, logarithm and symmetric group representations of degrees equal to Stirling numbers*, Combinatoire énumérative (Montreal, Que., 1985/Quebec, Que., 1985), Lecture Notes in Math., vol. 1234, Springer, Berlin, 1986, pp. 267–284. 364, 365
360. ———, *Free Lie algebras*, London Mathematical Society Monographs. New Series, vol. 7, The Clarendon Press Oxford University Press, New York, 1993, Oxford Science Publications. 174, 296, 364, 365, 446, 509, 510
361. J. Rhodes and B. Steinberg, *The q-theory of finite semigroups*, Springer Monographs in Mathematics, Springer, New York, 2009. 134
362. A. Robinson and S. Whitehouse, *The tree representation of Σ_{n+1}*, J. Pure Appl. Algebra **111** (1996), no. 1-3, 245–253. 445
363. O. Rodrigues, *Note sur les inversions, ou dérangements produits dans les permutations*, J. de Math. **4** (1839), 236–240. 509
364. G.-C. Rota, *On the foundations of combinatorial theory. I. Theory of Möbius functions*, Z. Wahrscheinlichkeitstheorie und Verw. Gebiete **2** (1964), 340–368 (1964). 544
365. F. V. Saliola, *The face semigroup algebra of a hyperplane arrangement*, ProQuest LLC, Ann Arbor, MI, 2006, Thesis (Ph.D.)–Cornell University. 331, 332, 508
366. ———, *The quiver of the semigroup algebra of a left regular band*, Internat. J. Algebra Comput. **17** (2007), no. 8, 1593–1610. 331, 332, 395
367. ———, *On the quiver of the descent algebra*, J. Algebra **320** (2008), no. 11, 3866–3894. 134, 331, 332, 394, 444, 508, 510
368. ———, *The face semigroup algebra of a hyperplane arrangement*, Canad. J. Math. **61** (2009), no. 4, 904–929. 331, 332, 394, 444, 478
369. ———, *The Loewy length of the descent algebra of type D*, Algebr. Represent. Theory **13** (2010), no. 2, 243–254. 508
370. ———, *Eigenvectors for a random walk on a left-regular band*, Adv. in Appl. Math. **48** (2012), no. 2, 306–311. 363, 395
371. R. Scharlau, *Metrical shellings of simplicial complexes*, European J. Combin. **6** (1985), no. 3, 265–269. 516
372. V. V. Schechtman and A. N. Varchenko, *Arrangements of hyperplanes and Lie algebra homology*, Invent. Math. **106** (1991), no. 1, 139–194. 238

373. ______, *Quantum groups and homology of local systems*, Algebraic geometry and analytic geometry (Tokyo, 1990), ICM-90 Satell. Conf. Proc., Springer, Tokyo, 1991, pp. 182–197. 238
374. M. Schocker, *The descent algebra of the symmetric group*, Representations of finite dimensional algebras and related topics in Lie theory and geometry, Fields Inst. Commun., vol. 40, Amer. Math. Soc., Providence, RI, 2004, pp. 145–161. 509, 510
375. ______, *The peak algebra of the symmetric group revisited*, Adv. Math. **192** (2005), no. 2, 259–309. 509, 510
376. ______, *The module structure of the Solomon-Tits algebra of the symmetric group*, J. Algebra **301** (2006), no. 2, 554–586. 332, 365, 394, 445, 509, 510
377. ______, *Radical of weakly ordered semigroup algebras*, J. Algebraic Combin. **28** (2008), no. 1, 231–234, With a foreword by Nantel Bergeron. 273
378. E. Schröder, *Vorlesungen über die algebra der logik*, Leipzig, 1890-1895. 524
379. M. P. Schützenberger, *Sur certains treillis gauches*, C. R. Acad. Sci. Paris **224** (1947), 776–778. 576
380. E. D. Schwab, *Möbius categories as reduced standard division categories of combinatorial inverse monoids*, Semigroup Forum **69** (2004), no. 1, 30–40. 544
381. ______, *The Möbius category of some combinatorial inverse semigroups*, Semigroup Forum **69** (2004), no. 1, 41–50. 544
382. G. C. Shephard and J. A. Todd, *Finite unitary reflection groups*, Canadian J. Math. **6** (1954), 274–304. 134
383. G. J. Simmons and J. E. Wetzel, *A two-coloring inequality for Euclidean two-arrangements*, Proc. Amer. Math. Soc. **77** (1979), no. 1, 124–127. 45
384. N. J. A. Sloane, *The on-line encyclopedia of integer sequences (OEIS)*, Published electronically at `http://oeis.org`. 156
385. L. Solomon, *Invariants of finite reflection groups*, Nagoya Math. J. **22** (1963), 57–64. 134
386. ______, *The orders of the finite Chevalley groups*, J. Algebra **3** (1966), 376–393. 508
387. ______, *The Burnside algebra of a finite group*, J. Combinatorial Theory **2** (1967), 603–615. 570
388. ______, *On the Poincaré-Birkhoff-Witt theorem*, J. Combinatorial Theory **4** (1968), 363–375. 364
389. ______, *A Mackey formula in the group ring of a Coxeter group*, J. Algebra **41** (1976), no. 2, 255–264. 508
390. W. Specht, *Die linearen Beziehungen zwischen höheren Kommutatoren*, Math. Z. **51** (1948), 367–376. 445
391. E. Spiegel and C. J. O'Donnell, *Incidence algebras*, Monographs and Textbooks in Pure and Applied Mathematics, vol. 206, Marcel Dekker, Inc., New York, 1997. 544
392. R. P. Stanley, *Acyclic orientations of graphs*, Discrete Math. **5** (1973), 171–178. 175
393. ______, *Combinatorial reciprocity theorems*, Advances in Math. **14** (1974), 194–253. 174
394. ______, *Enumeration of posets generated by disjoint unions and ordinal sums*, Proc. Amer. Math. Soc. **45** (1974), 295–299. 174
395. ______, *Some aspects of groups acting on finite posets*, J. Combin. Theory Ser. A **32** (1982), no. 2, 132–161. 445, 544
396. ______, *A survey of Eulerian posets*, Polytopes: abstract, convex and computational (Scarborough, ON, 1993), NATO Adv. Sci. Inst. Ser. C Math. Phys. Sci., vol. 440, Kluwer Acad. Publ., Dordrecht, 1994, pp. 301–333. 544
397. ______, *A symmetric function generalization of the chromatic polynomial of a graph*, Adv. Math. **111** (1995), no. 1, 166–194. 175
398. ______, *Hyperplane arrangements, interval orders, and trees*, Proc. Nat. Acad. Sci. U.S.A. **93** (1996), no. 6, 2620–2625. 175
399. ______, *Enumerative combinatorics. Vol. 2*, Cambridge Studies in Advanced Mathematics, vol. 62, Cambridge University Press, Cambridge, 1999, With a foreword by Gian-Carlo Rota and appendix 1 by Sergey Fomin. 174, 235, 509
400. ______, *An introduction to hyperplane arrangements*, Geometric combinatorics, IAS/Park City Math. Ser., vol. 13, Amer. Math. Soc., Providence, RI, 2007, pp. 389–496. 44, 45, 134, 175, 364, 444

401. ———, *Enumerative combinatorics. Volume 1*, second ed., Cambridge Studies in Advanced Mathematics, vol. 49, Cambridge University Press, Cambridge, 2012. 44, 46, 174, 519, 524, 544, 570
402. B. Steinberg, *Möbius functions and semigroup representation theory*, J. Combin. Theory Ser. A **113** (2006), no. 5, 866–881. 273
403. ———, *Möbius functions and semigroup representation theory. II. Character formulas and multiplicities*, Adv. Math. **217** (2008), no. 4, 1521–1557. 273
404. ———, *A simple proof of Brown's diagonalizability theorem*, 2010. Available at arXiv:1010.0716. 364
405. ———, *Representation theory of finite monoids*, Universitext, Springer, Cham, 2016. 45, 273, 364, 395, 576
406. J. Steiner, *Einige Gesetze über die Theilung der Ebene und des Raumes*, J. Reine Angew. Math. **1** (1826), 349–364. 45
407. J. R. Stembridge, *Coxeter cones and their h-vectors*, Adv. Math. **217** (2008), no. 5, 1935–1961. 175
408. M. Stern, *Semimodular lattices*, Encyclopedia of Mathematics and its Applications, vol. 73, Cambridge University Press, Cambridge, 1999, Theory and applications. 524
409. S. Sundaram, *The homology representations of the symmetric group on Cohen-Macaulay subposets of the partition lattice*, Adv. Math. **104** (1994), no. 2, 225–296. 445
410. M. E. Sweedler, *Hopf algebras*, Mathematics Lecture Note Series, W. A. Benjamin, Inc., New York, 1969. 364, 394
411. M. Takeuchi, *Free Hopf algebras generated by coalgebras*, J. Math. Soc. Japan **23** (1971), 561–582. 364
412. H. Terao, *Generalized exponents of a free arrangement of hyperplanes and Shepherd-Todd-Brieskorn formula*, Invent. Math. **63** (1981), no. 1, 159–179. 134
413. H. Terao and S. Yuzvinsky (eds.), *Arrangements of hyperplanes—Sapporo 2009*, Advanced Studies in Pure Mathematics, vol. 62, Mathematical Society of Japan, Tokyo, 2012. xiii
414. J.-Y. Thibon, *Lectures on noncommutative symmetric functions*, Interaction of combinatorics and representation theory, MSJ Mem., vol. 11, Math. Soc. Japan, Tokyo, 2001, pp. 39–94. 509
415. J. Tits, *Buildings of spherical type and finite BN-pairs*, Springer-Verlag, Berlin, 1974, Lecture Notes in Mathematics, Vol. 386. 44, 74, 134, 516
416. ———, *Two properties of Coxeter complexes*, J. Algebra **41** (1976), no. 2, 265–268, Appendix to "A Mackey formula in the group ring of a Coxeter group" (J. Algebra **41** (1976), no. 2, 255–264) by Louis Solomon. 44, 508
417. V. V. Vagner, *Generalized groups*, Doklady Akad. Nauk SSSR (N.S.) **84** (1952), 1119–1122. 134
418. J. H. van Lint and R. M. Wilson, *A course in combinatorics*, second ed., Cambridge University Press, Cambridge, 2001. 570
419. S. van Willigenburg, *Noncommutative irreducible characters of the symmetric group and noncommutative Schur functions*, J. Comb. **4** (2013), no. 4, 403–418. 510
420. A. N. Varchenko, *Bilinear form of real configuration of hyperplanes*, Adv. Math. **97** (1993), no. 1, 110–144. 99, 238
421. ———, *Multidimensional hypergeometric functions and representation theory of Lie algebras and quantum groups*, Adv. Ser. Math. Phys., vol. 21, World Sci. Publ., River Edge, NJ, 1995. 238
422. M. L. Wachs, *On the (co)homology of the partition lattice and the free Lie algebra*, Discrete Math. **193** (1998), no. 1-3, 287–319, Selected papers in honor of Adriano Garsia (Taormina, 1994). 445, 446
423. ———, *Poset topology: tools and applications*, Geometric combinatorics, IAS/Park City Math. Ser., vol. 13, Amer. Math. Soc., Providence, RI, 2007, pp. 497–615. 444, 445
424. M. Ward, *The algebra of lattice functions*, Duke Math. J. **5** (1939), no. 2, 357–371. 544
425. ———, *The closure operators of a lattice*, Ann. of Math. (2) **43** (1942), 191–196. 524
426. J. H. M. Wedderburn, *On Hypercomplex Numbers*, Proc. London Math. Soc. **S2-6** (1908), no. 1, 77–118. 570
427. L. Weisner, *Abstract theory of inversion of finite series*, Trans. Amer. Math. Soc. **38** (1935), no. 3, 474–484. 544

428. V. Welker, *Colored partitions and a generalization of the braid arrangement*, Electron. J. Combin. **4** (1997), no. 1, Research Paper 4, approx. 12 pp. 175

429. F. Wever, *Operatoren in Lieschen Ringen*, J. Reine Angew. Math. **187** (1949), 44–55. 445

430. G. W. Whitehead, *Elements of homotopy theory*, Graduate Texts in Mathematics, vol. 61, Springer-Verlag, New York-Berlin, 1978. 364

431. J. H. C. Whitehead, *Combinatorial homotopy. I*, Bull. Amer. Math. Soc. **55** (1949), 213–245. 516

432. H. Whitney, *The coloring of graphs*, Ann. of Math. (2) **33** (1932), no. 4, 688–718. 175, 364

433. ______, *On the Abstract Properties of Linear Dependence*, Amer. J. Math. **57** (1935), no. 3, 509–533. 524

434. R. O. Winder, *Partitions of N-space by hyperplanes*, SIAM J. Appl. Math. **14** (1966), 811–818. 46

435. E. Witt, *Spiegelungsgruppen und Aufzählung halbeinfacher Liescher Ringe*, Abh. Math. Sem. Hansischen Univ. **14** (1941), 289–322. 209

436. D. Woodcock, *Cohen-Macaulay complexes and Koszul rings*, J. London Math. Soc. (2) **57** (1998), no. 2, 398–410. 478

437. P. Young, U. S. R. Murty, and J. Edmonds, *Equicardinal matroids and matroid-designs*, Proc. Second Chapel Hill Conf. on Combinatorial Mathematics and its Applications (Univ. North Carolina, Chapel Hill, N.C., 1970), Univ. North Carolina, Chapel Hill, N.C., 1970, pp. 498–542. 544

438. D. Zagier, *Realizability of a model in infinite statistics*, Comm. Math. Phys. **147** (1992), no. 1, 199–210. 238

439. T. Zaslavsky, *Counting the faces of cut-up spaces*, Bull. Amer. Math. Soc. **81** (1975), no. 5, 916–918. 46

440. ______, *Facing up to arrangements: face-count formulas for partitions of space by hyperplanes*, Mem. Amer. Math. Soc. **1** (1975), no. issue 1, 154, vii+102. 44, 45, 238, 364, 444

441. ______, *A combinatorial analysis of topological dissections*, Advances in Math. **25** (1977), no. 3, 267–285. 46, 238

442. ______, *Arrangements of hyperplanes; matroids and graphs*, Proceedings of the Tenth Southeastern Conference on Combinatorics, Graph Theory and Computing (Florida Atlantic Univ., Boca Raton, Fla., 1979), Congress. Numer., XXIII–XXIV, Utilitas Math., Winnipeg, Man., 1979, pp. 895–911. 175

443. ______, *The geometry of root systems and signed graphs*, Amer. Math. Monthly **88** (1981), no. 2, 88–105. 175

444. ______, *Signed graph coloring*, Discrete Math. **39** (1982), no. 2, 215–228. 175, 364

445. ______, *Signed graphs*, Discrete Appl. Math. **4** (1982), no. 1, 47–74. 175

446. ______, *A mathematical bibliography of signed and gain graphs and allied areas*, Electron. J. Combin. (1998), Dynamic Survey 8, 8th ed. (2012), vi + 340 pp. (electronic). 175

447. G. M. Ziegler, *The face lattice of hyperplane arrangements*, Proceedings of the Oberwolfach Meeting "Kombinatorik" (1986), vol. 73, 1989, pp. 233–238. 44

448. ______, *Lectures on polytopes*, Graduate Texts in Mathematics, vol. 152, Springer-Verlag, New York, 1995. 19, 44, 520

Notation Index

Abbreviations

iff	if and only if
lhs	left hand side
rhs	right hand side
wrt	with respect to

Number systems

$\mathbb{N}$	set of nonnegative integers $\{0, 1, 2, \dots\}$
$\mathbb{Z}$	set of integer numbers
$\mathbb{Q}$	set of rational numbers
$\mathbb{R}$	set of real numbers
$\Bbbk$	field or commutative ring
$\mathbb{A}$	abelian group

Posets

P, Q	posets
$x \leq y$	x is smaller than y
$x < y$	x is strictly smaller than y
$x \lessdot y$	x is covered by y, or, y covers x
$[x, y]$	interval consisting of all elements which lie between x and y
$\mathrm{rk}(x)$	rank of the element x in a graded poset
$\mathrm{rk}(P)$	rank of the poset P
$\bot$, $\top$, $\wedge$, $\vee$	minimum element, maximum element, meet, join
(λ, ρ)	Galois connection or adjunction between posets
Σ	band
Π	support lattice of a band

Homology.

$\Delta(P)$	order complex of the poset P
$\mathcal{H}_k(P)$, $\mathcal{H}^k(P)$	order homology, cohomology groups of the poset P
$\mathcal{WH}_k(P)$, $\mathcal{WH}^k(P)$	Whitney homology, cohomology groups of the poset P

Incidence algebras.

$\mathrm{I}(P)$	incidence algebra of the poset P
$\mathrm{M}(P)$	incidence module of the poset P
$\mathrm{I}(\varphi, \psi)$	incidence bimodule
ζ	zeta function

μ	Möbius function
$\mathrm{I}_{\sim}(P)$	reduced incidence algebra of the poset P for the order-compatible relation $\sim$
$\mathrm{I}(\mathsf{C})$	incidence algebra of the category C
$\partial(\alpha)$	coboundary of a cochain α
$\mathrm{I}(P;\gamma)$	deformation of incidence algebra of P by cocycle γ
$\Gamma_{x,z}(v)$	fiber of order-preserving map
γ_*	transfer of cocycle γ

Algebras and modules

A	algebra over a field $\Bbbk$
M	(left or right) A-module
$\mathrm{End}_{\Bbbk}(M)$	algebra of endomorphisms of the module M
Ψ_M	representation of A associated to the module M
$\Psi_M(w)$	linear operator of the action of $w \in A$ on the module M
wM, Mw	image of the linear operator $\Psi_M(w)$
$\mathrm{ann}(M)$	annihilator of M
χ_M	character of the module M
M^*	linear dual of the module M
$\Bbbk^n$	n-dimensional $\Bbbk$-algebra with coordinatewise addition and multiplication
$\Bbbk[w]$	subalgebra of A generated by $w \in A$
$\Bbbk[x]$	algebra of polynomials in the variable x
w_d, w_n	diagonalizable, nilpotent part of w in its Jordan-Chevalley decomposition
I	ideal of A
N	nilpotent ideal of A
$\mathrm{rad}(A)$	radical of A
e, f	idempotents in A
$e \cong f$	isomorphic idempotents in A
$e_1, \dots, e_n$	family of mutually orthogonal idempotents in A
$A^\times$	group of invertible elements of A
$\bar{A}$	split-semisimple quotient of an elementary algebra A
χ_i	multiplicative characters of an elementary algebra
$\eta_i(M)$	generic multiplicity of χ_i in M
A^G	subalgebra of A invariant under action of a finite group G
Q	quiver
$\Bbbk Q$	path algebra of the quiver Q
$\mathrm{rad}(M)$	radical of the module M
$\mathrm{soc}(M)$	socle of the module M
$\Bbbk P$	linearization of a lattice P over the field $\Bbbk$
$\mathtt{H}_x$	element of the $\mathtt{H}$-basis of $\Bbbk P$
$\mathtt{Q}_x$	element of the $\mathtt{Q}$-basis of $\Bbbk P$
$\mathcal{P}(M)$	primitive part of the module M over $\Bbbk P$
$\mathcal{D}(M)$	decomposable part of the module M over $\Bbbk P$

Cell complexes

X	cell complex
$\chi(X)$	reduced Euler characteristic of X
(X, A)	relative pair of cell complexes
F, G, H, K	faces
C, D, E	chambers

$\mathrm{dist}(C, D)$	minimum length of a gallery connecting C and D
C -- E -- D	minimal gallery from C to D passing through E
$[C\!:\!D]$	gallery interval
Σ	set of faces
Γ	set of chambers
Σ_F	star of the face F
Γ_F	top-star of the face F

Arrangements

dim	dimension
$\mathcal{A}$	hyperplane arrangement
$\mathcal{A} \times \mathcal{A}'$	cartesian product of arrangements $\mathcal{A}$ and $\mathcal{A}'$
$\mathcal{A}^{\mathrm{X}}$	arrangement under the flat X of $\mathcal{A}$
$\mathcal{A}_{\mathrm{X}}$	arrangement over the flat X of $\mathcal{A}$
$\mathcal{A}^F$	arrangement under the support of F
$\mathcal{A}_F$	arrangement over the support of F
$\mathcal{A}^{\mathrm{X}}_{\mathrm{Y}}$, $\mathcal{A}^{\mathrm{X}}_F$, $\mathcal{A}^G_F$	arrangements between flats
$\widehat{\mathcal{A}}$	adjoint of the arrangement $\mathcal{A}$

Faces, flats, cones and lunes.

O	central face
P, Q	vertices
F, G, H, K	faces
C, D, E	chambers
$\overline{F}$	face opposite to F
FG	Tits product of F and G, or Tits projection of G on F
$[F\!:\!G]$	gallery interval with F and G of the same support
X, Y, Z	flats
$\bot$	minimum flat
$\top$	maximum flat
Cl(X)	closure of the flat X
V, W	cones
$\overline{\mathrm{V}}$	cone opposite to V
Cl(V)	closure of the cone V
V^o	interior of the cone V
V^b	boundary of the cone V
W_F	restriction of the top-cone W to the face F
${}_F\mathrm{V}$	extension of the top-cone V from the face F
(H, G)	nested face
(H, D)	top-nested face
L, M, N	lunes
$\mathrm{L} \circ \mathrm{M}$	composite of the lunes L and M
$\mathrm{rk}(F)$	rank of the face F
rk(X)	rank of the flat X
sk(L)	slack of the lune L

Projective objects.

$\{F, \overline{F}\}$	projective face
$\{C, \overline{C}\}$	projective chamber
$\{\mathrm{V}, \overline{\mathrm{V}}\}$	projective cone
$\{\mathrm{L}, \overline{\mathrm{L}}\}$	projective lune

Hyperplanes and half-spaces.

H	hyperplane
h	half-space
H^+, H^-	the two half-spaces bounded by the hyperplane H
$\overline{h}$	half-space opposite to h
$g(C, D)$	set of hyperplanes which separate chambers C and D
$r(C, D)$	set of half-spaces which contain C but do not contain D
wt(h)	weight assigned to the half-space h
$h(D)$	largest face of D which is contained in the half-space h

Charts and dicharts.

g, h	charts
$\rho(g)$	center of g
cG	set of connected charts
$(H_1, \dots, H_r)$	ordered coordinate chart
G	set of charts
$\overrightarrow{G}$	set of dicharts

Sets.

Σ	set of faces, Tits monoid
Γ	set of chambers, two-sided ideal of Σ
Π	set of flats, Birkhoff monoid
J	set of bi-faces, Janus monoid
Ω	set of cones
$\widehat{\Omega}$	set of top-cones
$\widehat{\Omega}_F$	set of top-cones contained in the top-star of F
${}_F\widehat{\Omega}$	set of top-cones whose closure contains F
Q	set of nested faces
$\widehat{Q}$	set of top-nested faces
P	set of nested flats
Λ	set of lunes
$\widehat{\Lambda}$	set of top-lunes
$\Gamma \times \Gamma$	set of pairs of chambers
$\Sigma \times \Sigma$	set of pairs of faces
$\Pi \times \Pi$	set of pairs of flats

The above sets are all associated to an arrangement $\mathcal{A}$. If we wish to show this dependence explicitly, we write $\Sigma[\mathcal{A}]$, $\Gamma[\mathcal{A}]$, $\Pi[\mathcal{A}]$ and so on.

Action of monoids.

${}^h\Sigma$	set of h-faces
${}^h\Pi$	set of h-flats
$F \cdot x$, $x \cdot F$	left, right action of the face F on the element x
$X \cdot x$	action of the flat X on the element x
$\Sigma_{x,y}$	set of all faces F such that $F \cdot x = y$
$\ell(F, y)$	set of all elements x such that $F \cdot x = y$
${}_{x,y}\Sigma$	set of all faces F such that $x \cdot F = y$
$\Pi_{x,y}$	set of all flats X such that $X \cdot x = y$
h_x	star of x in the right Σ-set h
h_F	star of the face F in the left Σ-set h

Maps.

s	support map
$\mathrm{s}(F)$	support of the face F
$\mathrm{s}(H, D)$	support of the top-nested face (H, D)
$\mathrm{s}(H, G)$	support of the nested face (H, G)
c(V)	support or case of the cone V
c(L)	case of the lune L
b	base map
b(V)	base of the cone V
b(L)	base of the lune L
bc	base-case map
Des	descent map
$\mathrm{Des}(C, D)$	descent of D wrt C
v	distance function
$v_{C,D}$	distance from C to D for a distance function v
$(v_{C,D})$	Varchenko matrix indexed by chambers
$(q_{C,D})$	q-Varchenko matrix
(v_{ℓ_1,ℓ_2})	Varchenko matrix indexed by linear extensions of a poset

Enumeration.

$\mu(\mathcal{A})$	Möbius number of $\mathcal{A}$
$c(\mathcal{A})$	number of chambers in $\mathcal{A}$
$f(\mathcal{A})$	number of faces in $\mathcal{A}$
$\chi(\mathcal{A}, t)$	characteristic polynomial of $\mathcal{A}$
$\mathrm{wy}(\mathcal{A}, k)$	Whitney numbers of the first kind of $\mathcal{A}$
β_{X}	Crapo invariant for the lattice of flats of $\mathcal{A}_{\mathrm{X}}$

Reflection arrangements

$\boldsymbol{\alpha} = (\mathcal{A}, C)$	reflection arrangement $\mathcal{A}$ with reference chamber C
W	Coxeter group
S	generating set of the Coxeter group W
(W, S)	Coxeter system
W_F, W_T	parabolic subgroup of W
$\widehat{W}_{\mathrm{X}}$	subgroup of W which leaves X invariant
W_{X}	subgroup of $\widehat{W}_{\mathrm{X}}$ which fixes X pointwise
W_{L}	subgroup of W which leaves the top-lune L invariant

Group elements, face-types and flat-types.

u, v, w, σ	elements of W
T, U, V	face-types
λ, μ	flat-types
$d(C, D)$	W-valued gallery distance between chambers C and D
$l(w)$	length of the element $w \in W$
$\mathrm{Des}(\sigma)$	descent of the element $\sigma \in W$

Sets and maps.

$\mathrm{W}\Sigma$	Coxeter-Tits monoid
$\mathrm{W}\Pi$	Coxeter-Birkhoff monoid
WJ	Coxeter-Janus monoid
Σ^W	set of face-types
Π^W	set of flat-types

Sh_T	set of T-shuffles for a face-type T
t	type map

Enumeration.

d_T	number of faces of type T
$a^{TUT'}$	structure constants of the invariant Tits algebra
$d_T(q)$, $a^{TUT'}(q)$	q-analogues of d_T and $a^{TUT'}$
$d_S(q)$	Poincaré polynomial of a Coxeter group W

Braid arrangement and related examples

$\mathbb{R}^I$	vector space of functions from I to $\mathbb{R}$
S_n	symmetric group on n letters
$F = (I_1, \ldots, I_k)$	set composition
$\alpha = (a_1, \ldots, a_k)$	integer composition
$\lambda = (l_1, \ldots, l_k)$	integer partition
$F \vDash I$	F is a composition of the set I
$\mathrm{X} \vdash I$	X is a partition of the set I
$\alpha \vDash n$	α is a composition of n
$\lambda \vdash n$	λ is a partition of n
Par_n	set of integer partitions of n
$\mathrm{Inv}(C, D)$	inversion set of (C, D)
$\mathrm{Inv}(F, G)$	inversion set of (F, G)
$\deg(G)$	number of blocks of a set composition G
$\deg!(G)$	factorial of the number of blocks of a set composition G
$s(m, k)$	Stirling numbers of the first kind
$(n)_q!$	q-factorial
$\binom{n}{i}_q$	q-binomial coefficient

Arrangement of type B.

$\mathrm{S}_n^{\pm}$	signed symmetric group on n letters
$z(F)$	zero block of the type B set composition F
$z(\mathrm{X})$	zero block of the type B set partition X
$\alpha = (a_0, a_1, \ldots, a_k)$	type B composition
$\lambda = (l_0, l_1, \ldots, l_k)$	type B partition
$s^{\pm}(m, k)$	type B Stirling numbers
$(2k)!!$, $(2k+1)!!$	double factorials

Graphic arrangements.

g, h	graphs
$\mathcal{A}(g)$	graphic arrangement of g
$c(g)$	number of connected components of g
k_I	complete graph on a set I
d_I	discrete graph on a set I
g_{X}	restriction of a graph g to a flat X
g^{X}	contraction of a flat X from a graph g
$\mathcal{O}$	orientation of a graph
$\chi(g, t)$	chromatic polynomial of g

Birkhoff algebra and Tits algebra

Π	Birkhoff algebra
Σ	Tits algebra
J	Janus algebra
J_q	q-Janus algebra for the scalar q
J_v	v-Janus algebra for the distance function v
Σ_0	diagonal 0-Janus algebra
$\mathrm{rad}(\Sigma)$	radical of the Tits algebra
Π^W	invariant Birkhoff algebra
Σ^W	invariant Tits algebra
$\mathrm{rad}(\Sigma^W)$	radical of the invariant Tits algebra
$\widehat{\mathsf{Q}}$	linear space indexed by top-nested faces
W	group algebra of the Coxeter group W
$\mathsf{W}\Sigma$	Coxeter-Tits algebra
$\mathsf{W}\Pi$	Coxeter-Birkhoff algebra
WJ	Coxeter-Janus algebra
G	algebra of charts
$\vec{\mathsf{G}}$	algebra of dicharts

The above vector spaces are all associated to an arrangement $\mathcal{A}$. If we wish to show this dependence explicitly, we write $\Pi[\mathcal{A}]$, $\Sigma[\mathcal{A}]$, $\mathsf{J}[\mathcal{A}]$ and so on.

Modules.

Γ	left module of chambers
h, k	(left or right) modules over the Tits algebra
$\mathcal{P}(\mathsf{h})$	primitive part of the left module h over the Tits algebra
$\mathcal{D}(\mathsf{h})$	decomposable part of the right module h over the Tits algebra
$\mathcal{P}_k(\mathsf{h})$	k-th term of the primitive series of the left module h
$\mathcal{D}_k(\mathsf{h})$	k-th term of the decomposable series of the right module h

Elements and maps.

$\mathtt{H}_{\mathrm{X}}$	element of the $\mathtt{H}$-basis of the Birkhoff algebra
$\mathtt{Q}_{\mathrm{X}}$	element of the $\mathtt{Q}$-basis of the Birkhoff algebra
$\mathtt{H}_F$	element of the $\mathtt{H}$-basis of the Tits algebra
$\mathtt{Q}_F$	element of the $\mathtt{Q}$-basis of the Tits algebra
$\mathtt{H}_{\{F,\overline{F}\}}$	element of the $\mathtt{H}$-basis of the projective Tits algebra
$\mathtt{Q}_{\{F,\overline{F}\}}$	element of the $\mathtt{Q}$-basis of the projective Tits algebra
$\mathtt{u}$	homogeneous section
$\mathtt{E}$	Eulerian family
$\mathtt{E}_{\mathrm{X}}$	Eulerian idempotents
$\mathtt{E}_\perp$	first Eulerian idempotent
$\mathtt{E}_k$	idempotent obtained by summing certain $\mathtt{E}_{\mathrm{X}}$
P	special Zie family
$\mathtt{u}_H$, $\mathtt{E}_H$, P_H	induced $\mathtt{u}$, $\mathtt{E}$ and P on $\mathcal{A}_H$
Δ_{X}, μ_{X}	maps relating the Birkhoff algebras of $\mathcal{A}_{\mathrm{X}}$ and $\mathcal{A}$
$\beta_{G,F}$	isomorphism between the Tits algebras of $\mathcal{A}_F$ and $\mathcal{A}_G$
Δ_F, μ_F	maps relating the Tits algebras of $\mathcal{A}_F$ and $\mathcal{A}$
Tak	Takeuchi element of the Tits algebra
$\mathbf{Tak}$	two-sided Takeuchi element of the Janus algebra

$\mathtt{Ful}_t$	Fulman element of parameter t
$\mathtt{Ads}_n$	Adams element of parameter n
$\mathtt{Ads}_n^{\pm}$	Type B Adams element of parameter n

Lie and Zie elements

$\mathsf{Lie}[\mathcal{A}]$	space of Lie elements of $\mathcal{A}$
$\mathsf{Zie}[\mathcal{A}]$	space of Zie elements of $\mathcal{A}$
$\mathsf{Lie}[\mathcal{A}]^W$	space of invariant Lie elements of $\mathcal{A}$
$\mathsf{Zie}[\mathcal{A}]^W$	space of invariant Zie elements of $\mathcal{A}$
$\mathsf{Lie}[I]$, $\mathsf{Lie}[n]$	Classical (type A) Lie elements
$\mathsf{Lie}[\mathbf{I}]$, $\mathsf{Lie}[\mathbf{n}]$	Type B Lie elements
θ_{h}	Dynkin element associated to the generic half-space h
d_{h}	symmetrized Dynkin element associated to h
$\mathtt{BW}_C$	an element of the Björner-Wachs basis
$\mathtt{B}_g$	an element of the Björner basis
$\mathtt{L}_g$	an element of the Lyndon basis
$d_{I,q}$	q-Dynkin element
$\mathsf{E}^{\mathrm{o}}[\mathcal{A}]$	orientation space of $\mathcal{A}$
$\mathsf{E}^{-}[\mathcal{A}]$	signature space of $\mathcal{A}$

Incidence algebras

$\mathrm{I}_{\mathrm{face}}$	face-incidence algebra
$\mathrm{I}_{\mathrm{flat}}$	flat-incidence algebra
$\mathrm{I}_{\mathrm{lune}}$	lune-incidence algebra
$\mathrm{M}_{\mathrm{lune}}$	lune-incidence module
$\mathrm{I}_{\mathrm{lunetype}}$	invariant lune-incidence algebra
$\mathrm{M}_{\mathrm{lunetype}}$	invariant lune-incidence module
$\mathrm{I}_{\mathrm{Lie}}$	Lie-incidence algebra
$\mathrm{M}_{\mathrm{Lie}}$	Lie-incidence module
$\boldsymbol{\zeta}$	noncommutative zeta function
$\boldsymbol{\mu}$	noncommutative Möbius function
bc	base-case map
Zet	space of all additive functions on lunes
Zet_1	space of all noncommutative zeta functions
Mob	space of all Weisner functions on lunes
Mob_1	space of all noncommutative Möbius functions

Subject Index

Selected Published Titles in This Series

226 **Marcelo Aguiar and Swapneel Mahajan,** Topics in Hyperplane Arrangements, 2017
222 **Alexandru Buium,** Foundations of Arithmetic Differential Geometry, 2017
221 **Dennis Gaitsgory and Nick Rozenblyum,** A Study in Derived Algebraic Geometry, 2017
219 **Richard Evan Schwartz,** The Projective Heat Map, 2017
218 **Tushar Das, David Simmons, and Mariusz Urbański,** Geometry and Dynamics in Gromov Hyperbolic Metric Spaces, 2017
217 **Benoit Fresse,** Homotopy of Operads and Grothendieck–Teichmüller Groups, 2017
216 **Frederick W. Gehring, Gaven J. Martin, and Bruce P. Palka,** An Introduction to the Theory of Higher-Dimensional Quasiconformal Mappings, 2017
215 **Robert Bieri and Ralph Strebel,** On Groups of PL-homeomorphisms of the Real Line, 2016
214 **Jared Speck,** Shock Formation in Small-Data Solutions to 3D Quasilinear Wave Equations, 2016
213 **Harold G. Diamond and Wen-Bin Zhang (Cheung Man Ping),** Beurling Generalized Numbers, 2016
212 **Pandelis Dodos and Vassilis Kanellopoulos,** Ramsey Theory for Product Spaces, 2016
211 **Charlotte Hardouin, Jacques Sauloy, and Michael F. Singer,** Galois Theories of Linear Difference Equations: An Introduction, 2016
210 **Jason P. Bell, Dragos Ghioca, and Thomas J. Tucker,** The Dynamical Mordell–Lang Conjecture, 2016
209 **Steve Y. Oudot,** Persistence Theory: From Quiver Representations to Data Analysis, 2015
208 **Peter S. Ozsváth, András I. Stipsicz, and Zoltán Szabó,** Grid Homology for Knots and Links, 2015
207 **Vladimir I. Bogachev, Nicolai V. Krylov, Michael Röckner, and Stanislav V. Shaposhnikov,** Fokker–Planck–Kolmogorov Equations, 2015
206 **Bennett Chow, Sun-Chin Chu, David Glickenstein, Christine Guenther, James Isenberg, Tom Ivey, Dan Knopf, Peng Lu, Feng Luo, and Lei Ni,** The Ricci Flow: Techniques and Applications: Part IV: Long-Time Solutions and Related Topics, 2015
205 **Pavel Etingof, Shlomo Gelaki, Dmitri Nikshych, and Victor Ostrik,** Tensor Categories, 2015
204 **Victor M. Buchstaber and Taras E. Panov,** Toric Topology, 2015
203 **Donald Yau and Mark W. Johnson,** A Foundation for PROPs, Algebras, and Modules, 2015
202 **Shiri Artstein-Avidan, Apostolos Giannopoulos, and Vitali D. Milman,** Asymptotic Geometric Analysis, Part I, 2015
201 **Christopher L. Douglas, John Francis, André G. Henriques, and Michael A. Hill, Editors,** Topological Modular Forms, 2014
200 **Nikolai Nadirashvili, Vladimir Tkachev, and Serge Vlăduţ,** Nonlinear Elliptic Equations and Nonassociative Algebras, 2014
199 **Dmitry S. Kaliuzhnyi-Verbovetskyi and Victor Vinnikov,** Foundations of Free Noncommutative Function Theory, 2014
198 **Jörg Jahnel,** Brauer Groups, Tamagawa Measures, and Rational Points on Algebraic Varieties, 2014
197 **Richard Evan Schwartz,** The Octagonal PETs, 2014
196 **Silouanos Brazitikos, Apostolos Giannopoulos, Petros Valettas, and Beatrice-Helen Vritsiou,** Geometry of Isotropic Convex Bodies, 2014
195 **Ching-Li Chai, Brian Conrad, and Frans Oort,** Complex Multiplication and Lifting Problems, 2014